W9-AFJ-188

Introduction to
LINEAR ALGEBRA

DONALD J. WRIGHT
University of Cincinnati

Boston Burr Ridge, IL Dubuque, IA Madison, WI New York San Francisco St. Louis
Bangkok Bogotá Caracas Lisbon London Madrid
Mexico City Milan New Delhi Seoul Singapore Sydney Taipei Toronto

WCB/McGraw-Hill

*A Division of The **McGraw-Hill** Companies*

INTRODUCTION TO LINEAR ALGEBRA

This book is printed on acid-free paper.

1 2 3 4 5 6 7 8 9 0 QPF/QPF 9 3 2 1 0 9 8

ISBN 0–07–072098–3

Vice president and editorial director: *Kevin T. Kane*
Publisher: *JP Lenney*
Sponsoring editor: *Maggie Rogers*
Editorial assistant: *Amy Upgren*
Marketing manager: *Mary K. Kittell*
Project manager: *Marilyn M. Sulzer*
Senior production supervisor: *Mary E. Haas*
Designer: *Lorna Lo*
Supplement coordinator: *Stacy A. Patch*
Compositor: *Interactive Composition Corporation*
Typeface: *10.5/12 Times*
Printer: *Quebecor Printing Book Group/Fairfield, PA*

Library of Congress Cataloging-in-Publication Data

Wright, Donald J.
Introduction to linear algebra / Donald J. Wright. — 1st ed.
p. cm.
Includes index.
ISBN 0–07–072098–3
1. Algebra, Linear. I. Title.
QA184.W75 1999
512'.5—dc21 98–22676
CIP

www.mhhe.com

CONTENTS

PREFACE

As a standard course in the second-year mathematics curriculum, Linear Algebra delivers material that is important to quite a variety of disciplines. But it also serves as a bridge between the intuitive and computational treatment usually found in freshman calculus and the more formal atmosphere of upper-division courses. For most students, the transition between these two modes of operation is something of a coming-of-age experience. They enter Linear Algebra unaccustomed to working with abstract concepts and inexperienced in communicating their thoughts in a clear and precise manner, yet their success in subsequent courses depends to a large extent on mastery of these skills. The challenge then is to develop their mathematical sophistication while exploring the landscape of linear algebra.

Abstraction is probably the most formidable feature on the horizon. In practice, it evolves from concrete experience. To appreciate the beauty and formalism of an axiomatically defined structure like a vector space, it is very helpful (maybe even necessary) to have experienced a concrete prototype in which the main ideas arise as natural responses to compelling real-life problems. Consequently, this book devotes its first six chapters to a thorough discussion of the basic facts of linear algebra in the context of Euclidean n-space. It begins with the task of describing the algebraic structure of the set of solutions of a system of linear equations, a problem that is both ubiquitous and compelling. Its solution involves a modest step into the realm of abstraction (adapting to the environment of n-space), but the ideas that arise (subspace, spanning set, linear independence, etc.) are firmly grounded in concrete three-dimensional experience, with ample support from geometry to help visualize the results. Formulating and exploring these concepts in higher dimensions is a substantive but manageable project for a second-year student. It provides plenty of opportunities to develop mathematical sophistication without resorting to a purely abstract structure that is sufficiently far removed from everyday experience as to seem foreign, threatening, or even irrelevant. With the foundation having been firmly established in n-space, the last chapter introduces the notion of an abstract vector space. At this point, the ideas surrounding linearity have become familiar. What remains is to point out their occurrence in other contexts and observe how the abstract structure serves as a unifying force that highlights the common features of different settings.

Chapter 1 discusses linear systems and the matrix algebra that is used to describe and manipulate them. Students find the computational aspects of this material straightforward, but when a proper response requires something more than a numerical calculation, they have difficulty expressing their insights in a comprehensible manner. Consequently, this chapter places a fair amount of emphasis on using notation to express oneself and on presenting one's thoughts in an organized sequence of steps that lead logically from the hypotheses to the conclusion.

The second chapter introduces Euclidean n-space as the context in which to study the solutions of a linear system. The fundamental notion of subspace arises from the properties of the solution set of a homogeneous linear system, and notions such as spanning set and linear independence are seen as natural attempts to describe those solutions in an efficient manner. A full discussion of the subspace structure unfolds, including coordinate systems and direct sum decompositions. The presentation draws heavily on the geometric interpretation of the ideas in 2- and 3-space. The level of sophistication is equal to that of customary treatments in abstract vector spaces, but this approach has the advantage that the practicality of the ideas is apparent from the start and the setting is not intimidating.

Orthogonality is the subject of Chapter 3. This topic occurs rather early in the text for several reasons. First, in exploring the advantages that acrue to working with a "rectangular" coordinate system, the big ideas developed in the previous chapter receive solid reinforcement and start to become part of the thinking process. Second, early treatment of orthogonal projection makes available an important geometric operation that is very useful in illustrating later concepts (such as linear transformations and eigenvalues). Finally, to demonstrate the use of the ideas of Chapter 2 with an important application, I wanted to include as early as possible the least-squares linear fit problem, and orthogonality plays a central role in that discussion.

Chapter 4 deals with linear transformations. The treatment places special emphasis on understanding geometrically how these functions rearrange the subspaces of 2- and 3-space, making considerable use of rotations, projections, and reflections to provide insight into concepts such as one-to-one, onto, and invertible. Matrix representations are discussed in detail, and in preparation for the later appearance of eigenvalues, a fair amount of the attention is given to utilizing coordinate systems with respect to which the action of the function is particularly easy to describe. The last section on isometries and similarities is a nice application of many previously developed ideas.

A fairly concise and calculational treatment of the determinant occupies Chapter 5. The concept is defined inductively by Laplace expansion, the object being to quickly produce the properties of the determinant for future use in finding eigenvalues. For those with the time and inclination, the last section explains the determinant's use in calculating the "volume" of an n-dimensional parallelepiped and in determining the orientation of a basis.

In Chapter 6, virtually all the concepts developed earlier come together in the analysis of linear operators on n-space. The notion of eigenvector arises from an attempt to find a coordinate system whose axes are invariant under the action of a given function. As in the earlier chapters, the presentation has a strong geometric flavor, and the treatment is therefore restricted to the case of real eigenvalues. It culminates in a discussion of symmetric matrices and the role orthogonal diagonalization plays in the analysis of quadratic forms.

The final chapter focuses on the process of capturing the essence of a particular concept via a set of axioms, which can then be used to extend the idea to a purely abstract setting. Guided by previous experience in n-space, it develops the abstract notions of vector space, norm, and inner product. Along the way, the student sees interesting manifestations of linearity in various contexts and discovers that at least algebraically all n-dimensional vector spaces are only superficially different from Euclidean n-space. The latter discussion emphasizes the use of coordinate isomorphisms to perform tasks in the abstract setting via familiar routines in n-space.

Throughout this book, the emphasis is on understanding and using concepts. The goal is to learn to think in terms of linear algebra notions such as linear combination and linear independence, and that sort of familiarity comes from using the ideas in a substantive way. That is not to say that the book is especially theoretical in tone. The exercise sets contain an ample number of routine problems designed to develop competence in standard calculational techniques, but they also include a rich assortment of problems whose solutions are one or two steps removed from a routine calculation. These latter exercises tend to be concrete (so they seem doable), but they force the student to engage the ideas to make the necessary connections. It is in wrestling with these exercises that the desired maturation starts to take place. Once students adjust to using the definitions and theorems as tools to solve such problems, they are well on their way to constructing proofs of general results.

The text contains a variety of examples and exercises that illustrate the usefulness of linear algebra, including a discussion of the adjacency matrix for a graph, analysis of predator-prey models, least-squares approximation, eigenvalue techniques for solving linear difference equations, and the classification of critical points of a function of two variables. These topics are not developed to any great depth, but sufficient exposure is provided to give the reader an appreciation of the many ways in which this subject connects with other disciplines.

A number of the exercises, particularly those involving applications, require the use of technology to carry out the computations. Most of the routine problems, however, are designed so as to yield readily to paper and pencil calculation. It is important to recognize that algorithms such as Gauss-Jordan elimination are not just means to produce answers. They are valuable ways of thinking about problems, and I believe one has to perform them a fair number of times by hand before a real appreciation of their conceptual and practical usefulness sinks in. On the other hand, once students demonstrate that they've digested a particular algorithm, I'm content to have them carry out the procedure with the push of a button. Ultimately, the decision as to when and how much to incorporate computational technology into the course depends on the taste and judgment of the instructor. The problem sets provide some opportunities to exercise both modes of operation.

In each exercise set some of the problems are marked with bold numerals or letters, indicating that the answer or, in some cases, a full solution can be found in the back of the book. A complete solutions manual is available to adopting instructors.

ACKNOWLEDGEMENTS

I am indebted to a number of my colleagues for patiently class testing early versions of this text. I would especially like to thank Joe Fisher, David Herron, David Minda, and Steve Pelikan for many helpful suggestions and discussions. I also appreciate the thoughtful comments of the following reviewers, many of whose recommendations are reflected in the final version of the book.

Robert Beezer, *University of Puget Sound*
Keith Chavey, *University of Wisconsin–River Falls*
Francis Conlan, *Santa Clara University*
Daniel Drucker, *Wayne State University*
Michael Ecker, *Pennsylvania State University*
Murray Eisenberg, *University of Massachusetts*
Sheldon Eisenberg, *University of Hartford*
John Ellison, *Grove City College*
S. B. Khleif, *Tennessee Tech University*
Patricia Johnson, *Ohio State University*
Christopher McCord, *University of Cincinnati*
David Meredith, *San Francisco State University*
Tom Metzger, *University of Pittsburgh*
David Weinberg, *Texas Tech University*

Finally, it has been a pleasure working with the editors and production staff at McGraw-Hill. I am particularly grateful to Jack Shira for his confidence in the early versions of this project and to Marilyn Sulzer for her careful attention to detail during production.

Donald J. Wright

1

Linear Systems and Matrices

1.1 LINEAR EQUATIONS

A line in the Cartesian plane has equation

$$ax + by = c, \tag{1.1}$$

where a, b, and c are real numbers and at least one of a and b is not zero. That is to say, the line consists of the points in the plane whose Cartesian coordinates (x,y) satisfy (1.1). The analogous equation in three variables is

$$ax + by + cz = d, \tag{1.2}$$

and the points in space with Cartesian coordinates satisfying (1.2) constitute a plane (as long as one of a, b, and c is not zero). Equations (1.1) and (1.2) are the general linear equations in two and three variables, respectively. This notion of a linear equation can be extended to accommodate a larger number of variables, and in doing so it is helpful to establish some notation. The set of real numbers is denoted by $\mathcal{R}$. It is customary to use subscripts to distinguish notationally between four or more variables, so n distinct variables are indicated by writing $x_1, \ldots, x_n$. This convention also serves to order the variables, that is, x_1 is considered the first, x_2 the second, and so forth.

DEFINITION. Given a positive integer n, the general *linear equation* in $x_1, \ldots, x_n$ is

$$a_1x_1 + a_2x_2 + \cdots + a_nx_n = b,$$

where $a_1, \ldots, a_n, b \in \mathcal{R}$ and at least one of $a_1, \ldots, a_n$ is not zero.

Interest in linear equations stems from the fact that there are many practical problems in which these linear relationships are the appropriate means of expressing the interaction between the variables.

■ **EXAMPLE 1.1.** *Allocation of Resources.* A small business has a monthly advertising budget of B dollars, and the cost per ad in the local media is

1. Television . . . . a dollars
2. Radio . . . . b dollars
3. Newspaper . . . . c dollars
4. Neighborhood Newsletter . . . . d dollars.

Let x_1, x_2, x_3, and x_4 be the numbers of monthly ads placed in these media, respectively. Assuming the entire budget is used each month, the equation governing the allocation of monthly ads is

$$ax_1 + bx_2 + cx_3 + dx_4 = B. \ \square$$

■ **EXAMPLE 1.2.** *Networks.* Figure 1.1 is a schematic drawing of a hypothetical distribution network for some commodity, with arrows indicating the direction of flow through each branch of the network.

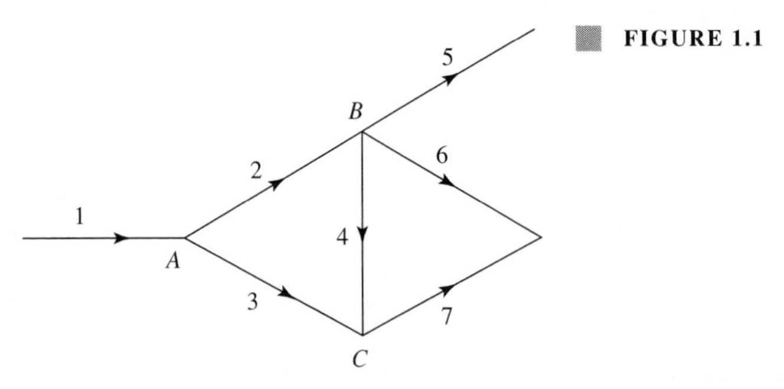

FIGURE 1.1

Suppose $x_1, \ldots, x_7$ are the amounts of the commodity distributed through branches $1, \ldots, 7$, respectively. Since the amount arriving at a given node is the same as the amount leaving that point, $x_1, \ldots, x_7$ are related as follows:

$$A: \quad x_2 + x_3 = x_1$$

$$B: \quad x_4 + x_5 + x_6 = x_2$$

$$C: \quad x_7 = x_3 + x_4.$$

Also, the amount arriving at A equals the total amount reaching the final destinations, so

$$x_1 = x_5 + x_6 + x_7.$$

The possible distribution scenarios are thus described by the linear equations

$$\begin{aligned} x_1 - x_2 - x_3 &= 0 \\ x_2 - x_4 - x_5 - x_6 &= 0 \\ x_3 + x_4 - x_7 &= 0 \\ x_1 - x_5 - x_6 - x_7 &= 0. \ \square \end{aligned}$$

■ **EXAMPLE 1.3.** *Interpolation of Data.* Consider the problem of fitting a given type of curve to experimentally obtained data points. For example, it seems likely that there is a parabola with a vertical axis that passes through (1,1), (2,3), and (3,−1). Such a curve is described by

$$y = a + bx + cx^2,$$

where $a, b, c \in \mathcal{R}$. That the curve is to pass through the given points is expressed by the following linear equations in the variables a, b, and c:

$$\begin{aligned}(1,1)&: \quad 1 = a + b + c \\ (2,3)&: \quad 3 = a + 2b + 4c \\ (3,-1)&: \quad -1 = a + 3b + 9c. \ \square\end{aligned}$$

■ **EXAMPLE 1.4.** *Population Dynamics.* A given population is counted periodically and the kth census records n_k individuals, $k = 0, 1, \ldots$. A simple growth model assumes that the change in the population during the kth time period is proportional to the population size at the beginning of the period, that is, that $n_{k+1} - n_k = dn_k$, $d \in \mathcal{R}$. Then $n_{k+1} = (1 + d)n_k$, $k = 0, 1, \ldots$, so

$$n_k = (1 + d)n_{k-1} = (1 + d)^2 n_{k-2} = \cdots = (1 + d)^k n_0.$$

This model predicts exponential growth or decay, depending on whether d is positive or negative, respectively. A more interesting situation arises when two species interact, the classic case being the predator-prey relationship. Suppose f_k and r_k are census results for populations of foxes and rabbits, respectively. In the absence of rabbits, the fox population declines, say $f_{k+1} = \alpha f_k$, $0 < \alpha < 1$, and without foxes, the rabbit population expands, say $r_{k+1} = \delta r_k$, $\delta > 1$. If you assume that the presence of rabbits benefits the fox population in proportion to the number of rabbits and that foxes have a similar but negative influence on the rabbit count, then you get the following interaction model:

$$\begin{aligned}f_{k+1} &= \alpha f_k + \beta r_k \\ r_{k+1} &= -\gamma f_k + \delta r_k,\end{aligned}$$

where $\beta > 0$ and $\gamma > 0$. Note that the two equations are linear in r_k and f_k. $\square$

A *solution* to

$$a_1x_1 + a_2x_2 + \cdots + a_nx_n = b \tag{1.3}$$

is an ordered set of n real numbers, which, when substituted for their associated variables, satisfy the equation. An ordered set of n real numbers is called an *n-tuple,* written $(x_1, \ldots, x_n)$. As long as one of $a_1, \ldots, a_n$ is not zero, (1.3) has a solution. For example, if $a_1 \neq 0$, then (1.3) can be solved for x_1 to obtain

$$x_1 = (1/a_1)\{b - a_2x_2 - \cdots - a_nx_n\}.$$

Any choice of values for $x_2, \ldots, x_n$ generates a corresponding value for x_1, and the resulting n-tuple is a solution. If $a_i \neq 0$ for some $i \in \{1, \ldots, n\}$, then a similar expression can be obtained for x_i. In that event, the variables other than x_i can be assigned arbitrary values, and they in turn determine x_i. There are, in fact, infinitely many solutions (provided $n \geq 2$). Each description of the solutions involves assigning values independently to $n - 1$ of the variables, so the general solution is said to depend on $n - 1$ parameters.

■ **EXAMPLE 1.5.** Solving

$$-x + 2y - 3z = 6$$

for x yields $x = -6 + 2y - 3z$. If y and z are assigned arbitrary values s and t, respectively, then the set of solutions is

$$\{(-6 + 2s - 3t, s, t) \ : \ s, t \in \mathcal{R}\}.$$

Alternatively, setting $x = s$ and $z = t$ and solving for y produces

$$\{(s, 3 + s/2 + 3t/2, t) \;:\; s,t \in \mathcal{R}\}.$$

In either event, the solutions are described in terms of two parameters, s and t. A third description, namely

$$\{(s, t, -2 - s/3 + 2t/3) \;:\; s,t \in \mathcal{R}\},$$

results from setting $x = s$ and $y = t$ and solving for z. □

Exercises 1.1

1. Which of the following equations are linear in the variables x, y, and z?
   - **a.** $2x - 3y = \ln(2)$
   - b. $5z - \frac{1}{2}x = 3y + 2$
   - c. $\pi x + \sqrt{2}y + \dfrac{z}{e} = 0$
   - d. $2xy + 3z = -1$
   - **e.** $2\cos(x) + 3\sin(y) = \pi$
   - f. $x\cos(2) + y\sin(3) = \pi$
   - **g.** $x + \dfrac{2}{y} = 3$
   - h. $z - 4 = \dfrac{x + 3y}{2}$
   - **i.** $\sqrt{x} - \sqrt{y} = 1$
   - j. $3y - 5 = 2\dfrac{x}{z}$
   - **k.** $z = 4$
   - l. $y - 1 = 3z^2 - 2x$

2. Interpreting each as a linear equation in x, y, and z, describe the solution set.
   - **a.** $3x + 2y - z = 0$
   - b. $2 - x + 3z = 4y$
   - **c.** $x - \frac{1}{2}y = z + 2$
   - d. $2y - z = 1$
   - **e.** $x + 2z = 0$
   - f. $2x = 3y$
   - **g.** $3y = 4$
   - h. $4 + 3x = -1$

3. A cash box containing p pennies, n nickels, d dimes, and q quarters has a total of \$37.65 in coins. Find an equation relating the values of p, n, d, and q. Is the equation linear?

4. A parabola with a horizontal axis is to be fitted to the data points in Example 1.3. State the general equation of such a curve and find the conditions the coefficients must satisfy if the curve is to pass through the given points.

5. The general gas law states that the pressure P, volume V, and temperature T, of a sample consisting of n moles of an ideal gas are related by

$$PV = n(8.317)T. \qquad (*)$$

   - **a.** Is $(*)$ a linear equation in P, V, and T?
   - b. When T is held constant, $(*)$ is known as Boyle's law. Is Boyle's law a linear equation in P and V?
   - c. When P is held constant, $(*)$ is known as Charles's law. Is Charles's law a linear equation in V and T?

6. Masses m_1, m_2, and m_3 are located on a balance beam at distances x_1, x_2, and x_3, respectively, from the fulcrum (as illustrated in the figure).

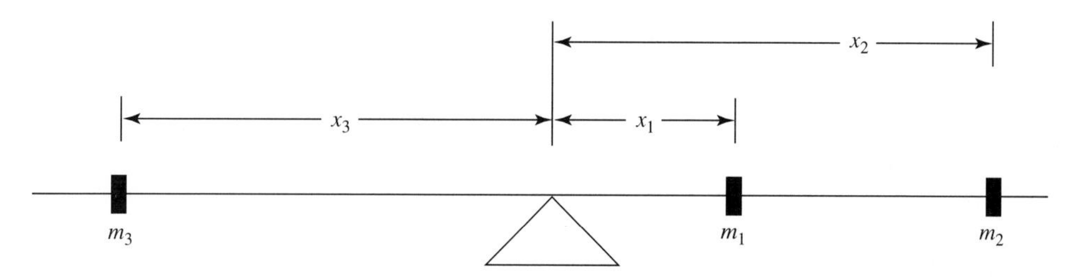

What condition must be satisfied for this system of masses to be in equilibrium (i.e., for the beam to balance)? Is this condition a linear equation in the variables

a. $m_1, m_2,$ and m_3? **b.** $x_1, x_2,$ and x_3? c. $m_1, m_2, m_3, x_1, x_2,$ and x_3?

7. Consider a mechanical system that consists of a spring connecting a fixed support to a piston that slides back and forth in a stationary cylinder. Figure (a) shows the system in equilibrium, with 0 indicating the rest position of the piston. When displaced and released, the piston exhibits an oscillatory motion with position x, velocity v, and acceleration a. Hooke's law states that the spring exerts a restoring force on the piston opposite in direction to the displacement with magnitude proportional to the size of the displacement. The constant of proportionality, called the spring constant, is denoted by $k > 0$. The cylinder produces a damping effect by exerting a force on the piston directed opposite to the direction of motion, with magnitude proportional to the magnitude of the velocity. Assume $d > 0$ is the constant of proportionality. By Newton's second law of motion, the sum of the forces acting on the piston is equal to the product of its mass m and its acceleration.

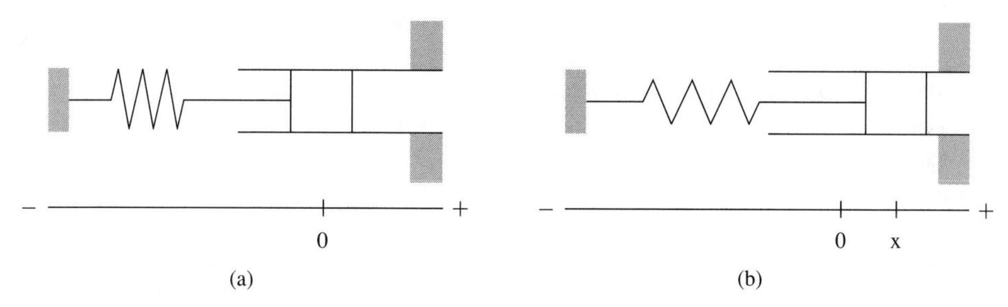

a. Find the equation governing the motion of the piston.

b. Is the answer in part (a) a linear equation in x, v, and a?

1.2 LINEAR SYSTEMS

Given positive integers m and n, a collection of m linear equations in the same n variables is called a *linear system*. A *solution* for such a system is an n-tuple of real numbers that satisfies all the equations. Each equation generally has infinitely many solutions, but there may not be any n-tuples that satisfy all m equations, so the first issue regarding linear systems is the question of existence of solutions. The second, given that there are solutions, is the problem of identifying and describing them. When the system involves only two or three variables, these issues have simple geometric interpretations.

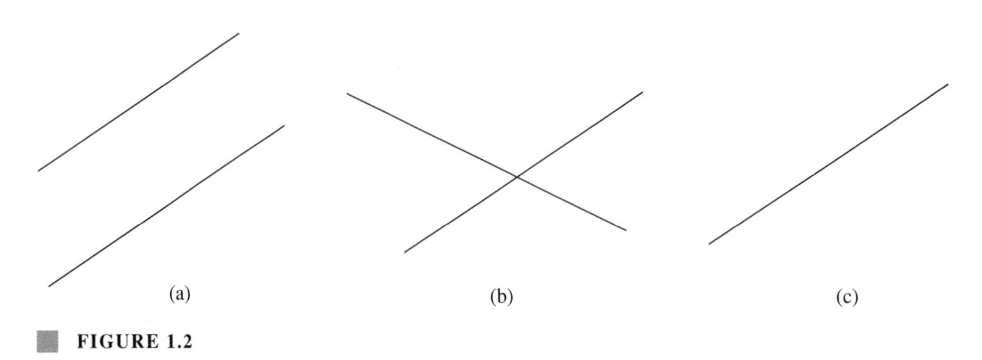

FIGURE 1.2

When $n = 2$, each equation represents a line in the plane, and a solution is a point that is common to all of the lines. For a system of two equations $(m = 2)$, there are three possible outcomes, as illustrated in Figure 1.2. The two lines may be distinct but parallel (Figure 1.2(a)), in which case the system has no solutions. Alternatively, the lines intersect and, therefore, either meet in a single point (Figure 1.2(b)) or are identical (Figure 1.2(c)). Of the latter two possibilities, the first corresponds to a system with a unique solution, the second to a system with infinitely many solutions that constitute a line.

When the system has three or more equations, it is still the case that the corresponding lines either have no points in common, have a single point in common, or are all the same. In general then, a system of m equations in two unknowns has either no solution, a unique solution, or infinitely many solutions that form a line.

When $n = 3$, each equation represents a plane in space. Suppose first that there are two equations. If they represent the same plane, then the solutions of the system are just the points of that plane. Alternatively, they represent distinct planes that either intersect in a line or are parallel. The first of the latter options corresponds to a system with infinitely many solutions (that constitute a line), the second corresponds to a system with no solutions. Note that having a unique solution is not one of the possibilities.

Now suppose the system has three equations. The case when two or more of them represent the same plane is covered by the previous discussion, so assume the planes are distinct. If they are all parallel, then the system has no solution. Otherwise, at least two of them meet. In that event, the remaining plane either is or is not parallel to the line of intersection of the two. The first of these possibilities can occur in three essentially different ways, as illustrated in Figure 1.3. Two of them (Figures 1.3(a) and 1.3(b)) are situations in which there are no solutions, and the third (Figure 1.3(c)) corresponds to a system whose solutions form a line. Finally, when two of the planes meet and the line of intersection is not parallel to the third plane, that line intersects the third plane in a single point. In this case, the system has a unique solution.

Regardless of the number of equations in the system, there are always only four possible outcomes: the planes may have nothing in common, a point in common, a line in common, or a plane in common.

The foregoing discussion indicates that when $n = 2$ or $n = 3$, the solution set comes in one of a few very special forms (a point, a line, a plane, or the empty set). For

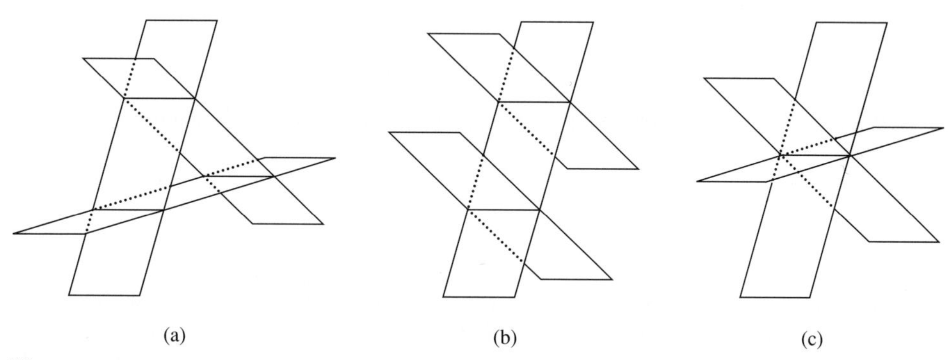

FIGURE 1.3

larger values of n, the equations in the system no longer represent familiar geometric objects, but the solution set still has a certain form that can be characterized algebraically. The concepts that have been developed to describe the "algebraic structure" of such a set make up the core of the subject known as *linear algebra*. The present study of that subject begins therefore by examining algebraic techniques for solving linear systems.

A general linear system of m equations in n unknowns looks like

$$\begin{aligned} a_{11}x_1 + a_{12}x_2 + \cdots + a_{1n}x_n &= b_1 \\ a_{21}x_1 + a_{22}x_2 + \cdots + a_{2n}x_n &= b_2 \\ &\vdots \\ a_{m1}x_1 + a_{m2}x_2 + \cdots + a_{mn}x_n &= b_m. \end{aligned}$$

The real numbers, a_{ij}, $1 \le i \le m$, $1 \le j \le n$, are called the *coefficients* of the system. Note that the first index in a_{ij} indicates the equation in which the coefficient occurs, whereas the second identifies the variable that it multiplies. The numbers $b_1, \ldots, b_n$ are called the *constant terms*. A system that has a solution is said to be *consistent*, whereas one that does not is called *inconsistent*.

The strategy for finding solutions is to replace the given system by another system that has the same solutions but whose form is sufficiently simple that the solutions are apparent. The simpler system is obtained by repeating one basic process. Solve one equation for one of the variables, producing an expression for it in terms of the others. Then substitute that expression in the other equations. The effect is to eliminate the one variable from the other equations. The new system consists of a single equation in n variables and some replacement equations involving only $n - 1$ variables.

EXAMPLE 1.6. Consider

$$\begin{aligned} 2x_1 + 3x_2 - x_3 &= 3 \\ x_1 - x_2 + 2x_3 &= -1. \end{aligned}$$

The second equation gives $x_1 = -1 + x_2 - 2x_3$, and substituting in the first equation yields

$$2(-1 + x_2 - 2x_3) + 3x_2 - x_3 = 3,$$

or

$$5x_2 - 5x_3 = 5.$$

The last line is further simplified by dividing both sides by 5, whereupon the original system has been replaced by

$$\begin{aligned} x_1 - x_2 + 2x_3 &= -1 \\ x_2 - x_3 &= 1. \end{aligned}$$

In the latter system, each choice of x_3 in the second equation determines a value of x_2, and the values of x_2 and x_3, together with the first equation, determine x_1. There are infinitely many solutions depending on one parameter. Setting $x_3 = t$ gives $x_2 = 1 + t$ and $x_1 = -1 + x_2 - 2x_3 = -1 + (1 + t) - 2t = -t$. The solutions are the 3-tuples $(-t, 1 + t, t), t \in \mathcal{R}$. In this example, the two equations represent planes in space, and the solution set is the line of intersection of those planes. It is described parametrically by

$$x_1 = -t, \qquad x_2 = 1 + t, \qquad x_3 = t, \qquad t \in \mathcal{R}. \ \square$$

■ **EXAMPLE 1.7.** Recall the system generated in Example 1.3, that is,

$$\begin{aligned} a + b + c &= 1 \\ a + 2b + 4c &= 3 \\ a + 3b + 9c &= -1. \end{aligned}$$

From the first equation you have $a = 1 - b - c$, which, when substituted in the other equations, produces

$$\begin{aligned} b + 3c &= 2 \\ 2b + 8c &= -2. \end{aligned}$$

Multiplying through the last line by 1/2, the new system becomes

$$\begin{aligned} a + b + c &= 1 \\ b + 3c &= 2 \\ b + 4c &= -1. \end{aligned}$$

Note that the variable a no longer occurs in the last two equations. Now repeat the process, using the second equation of the new system to eliminate b from the third equation. The result is

$$\begin{aligned} a + b + c &= 1 \\ b + 3c &= 2 \\ c &= -3. \end{aligned}$$

Since $c = -3$, the second equation gives $b = 11$, and the first equation then yields $a = -7$. The system has a unique solution, $(-7,11,-3)$. Referring again to Example 1.3, this conclusion confirms that there is a unique parabola with a vertical axis that passes through $(1,1)$, $(2,3)$, and $(3,-1)$. Its equation is $y = -7 + 11x - 3x^2$. $\square$

■ **EXAMPLE 1.8.** Suppose you attempt to fit a line to the points in Example 1.3. Since the three points are noncollinear, you cannot succeed, and the impossibility of the task should become algebraically apparent while trying to solve the resulting system of equations. The proposed line has equation $y = mx + b$, and $(1,1)$, $(2,3)$, and $(3,-1)$ lie on the line if and only if the variables m and b satisfy the linear system

$$\begin{aligned} (1,1)&: \quad m + b = 1 \\ (2,3)&: \quad 2m + b = 3 \\ (3,-1)&: \quad 3m + b = -1. \end{aligned}$$

According to the first equation, $m = 1 - b$, and substituting in the second and third equations produces $-b = 1$ and $-2b = -4$, respectively. Of course b cannot be both -1 and 2, so the system is inconsistent. □

The main feature of the elimination process is: solving one equation for one of the variables and substituting the result in the other equations. To analyze this process in general, consider a pair of equations in n variables, say,

$$\begin{aligned} a_1x_1 + a_2x_2 + \cdots + a_nx_n &= b \\ c_1x_1 + c_2x_2 + \cdots + c_nx_n &= d. \end{aligned}$$

Suppose you solve the first equation for x_1 (assuming $a_1 \neq 0$), obtaining

$$x_1 = -(a_2/a_1)x_2 - \cdots - (a_n/a_1)x_n + (b/a_1).$$

Substituting this expression in the second equation yields

$$c_1[-(a_2/a_1)x_2 - \cdots - (a_n/a_1)x_n + (b/a_1)] + c_2x_2 + \cdots + c_nx_n = d,$$

which, on collecting like terms, becomes

$$[-(c_1/a_1)a_2 + c_2]x_2 + \cdots + [-(c_1/a_1)a_n + c_n]x_n = -(c_1/a_1)b + d.$$

Now, observe that the last equation can also be obtained by the following algebraic procedure: multiply the first equation by $-(c_1/a_1)$ and then add the left and right sides of the resulting equation to the left and right sides, respectively, of the second equation. The multiplication changes the coefficient of x_1 in the first equation to the negative of the coefficient of x_1 in the second equation. When the addition is performed, x_1 cancels out of the second equation.

■ **EXAMPLE 1.9.** Consider

$$\begin{aligned} 2x_1 - 3x_2 + 4x_3 + x_4 &= 4 \\ 3x_1 + 2x_2 - x_3 + 2x_4 &= 5. \end{aligned}$$

In the notation above, $-(c_1/a_1) = -3/2$. Multiplying the first equation by $-3/2$ and adding the result to the second produces

$$(9/2 + 2)x_2 + (-6 - 1)x_3 + (-3/2 + 2)x_4 = -6 + 5,$$

or

$$(13/2)x_2 - 7x_3 + (1/2)x_4 = -1.$$

You might check that solving the first equation for x_1 and substituting the result in the second equation produces the same outcome. □

The point of the previous observation is that the process of eliminating a variable from one of the equations can be viewed as a special case of a general algebraic operation, namely, adding a constant multiple of one equation to another.

Two other kinds of steps were used in the preceding examples to convert the original system to a simpler one. The first has to do with performing the elimination process in a systematic manner. You might try, as a standard procedure, to solve the first equation for x_1 and then eliminate that variable from the remaining equations. However, it is possible that x_1 does not occur in the first equation, or that some other equation has a simpler coefficient of x_1. The latter situation occurred in Example 1.6, where the second equation was solved for x_1 because the coefficient there was 1. With an eye toward implementing the standard procedure, you could begin by rearranging the equations so that the one with the most convenient coefficient of x_1 appears as the first equation. Thus, the first type of algebraic manipulation is

I. interchanging two equations.

The second is to simplify one of the equations by removing a factor that is common to each term. This is accomplished algebraically by

II. multiplying each term in one equation by a nonzero constant.

The primary step in the process is

III. adding a constant multiple of one equation to another.

These procedures are known as *elementary operations* of types I, II, and III. By performing a sequence of them in a systematic manner, the original system can be replaced by a system whose solutions are readily obtained. Of course, it is crucial that the new system have the same solutions as the original. That issue is addressed after the following example.

■ **EXAMPLE 1.10.** To solve

$$\begin{aligned} -2x_1 + 4x_3 - x_4 + 5x_5 &= -2 \\ 2x_1 + x_2 - x_3 - 4x_5 &= -1 \\ x_1 + x_2 + x_3 + x_4 &= 1, \end{aligned}$$

begin by interchanging the first and third equations, thereby obtaining a first equation in which the coefficient of x_1 is 1. At the same time it is helpful to insert 0's for the coefficients of the missing variables, producing

$$\begin{aligned} x_1 + x_2 + x_3 + x_4 + 0x_5 &= 1 \\ 2x_1 + x_2 - x_3 + 0x_4 - 4x_5 &= -1 \\ -2x_1 + 0x_2 + 4x_3 - x_4 + 5x_5 &= -2. \end{aligned}$$

Now eliminate x_1 from the last two equations by performing two type III operations. Multiply the first equation by -2 and add the result to the second. Then multiply the first equation by 2 and add the result to the third. The new system is

$$\begin{aligned} x_1 + x_2 + x_3 + x_4 + 0x_5 &= 1 \\ - x_2 - 3x_3 - 2x_4 - 4x_5 &= -3 \\ 2x_2 + 6x_3 + x_4 + 5x_5 &= 0. \end{aligned}$$

To eliminate x_2 from the third equation, multiply the second equation by 2 and add the result to the third, generating

$$\begin{aligned} x_1 + x_2 + \ x_3 + \ x_4 + 0x_5 &= \ \ 1 \\ - x_2 - 3x_3 - 2x_4 - 4x_5 &= -3 \\ - 3x_4 - 3x_5 &= -6. \end{aligned}$$

The last system can be cleaned up a bit by multiplying its second and third equations by -1 and $-1/3$, respectively, to get

$$\begin{aligned} x_1 + x_2 + \ x_3 + \ x_4 + 0x_5 &= 1 \\ x_2 + 3x_3 + 2x_4 + 4x_5 &= 3 \\ x_4 + \ x_5 &= 2. \end{aligned}$$

The solutions can now be described. First, x_5 is assigned an arbitrary value, t, and the third equation yields $x_4 = 2 - t$. Substituting these results in the second equation and assigning an arbitrary value, s, to x_3 produces

$$x_2 = 3 - 3s - 2(2 - t) - 4t = -1 - 3s - 2t.$$

Finally, the first equation gives

$$x_1 = 1 - (-1 - 3s - 2t) - s - (2 - t) = 2s + 3t.$$

The solutions, described in terms of the parameters s and t, are

$$(2s + 3t, -1 - 3s - 2t, s, 2 - t, t), s,t \in \mathcal{R}. \ \square$$

The concern as to whether the original and simplified systems have the same solutions remains to be dealt with. Since the process involves a sequence of elementary operations, it suffices to show that a single such operation generates a new system with the same solutions as the previous one.

Certainly interchanging the order in which the equations appear has no effect on the solutions, so the conclusion holds for a type I operation.

A type II operation changes one of the equations, say the ith one, by multiplying each of its terms by a nonzero constant, t. Any n-tuple that satisfies the original equation satisfies the new one. Moreover, this step can be reversed by a type II operation. Indeed, multiplying the ith equation of the new system by $1/t$ converts it back to the original. Thus, each solution of the new system is a solution of the previous one.

Now, consider a type III operation. If you add t times the ith equation to the jth equation, then, except for its jth equation, the new system is identical to the original. Any n-tuple that statisfies the original system certainly satisfies its ith and jth equations and, consequently, satisfies the jth equation of the new system. Conversely, the original system can be recovered from the new one by a type III operation, namely, add $-t$ times the ith equation to the jth equation.

DEFINITION. Two linear systems of m equations in n unknowns are said to be *equivalent* if one can be obtained from the other by a finite sequence of elementary operations.

The foregoing observations can be summarized as follows.

THEOREM 1.1. Equivalent linear systems have the same solutions.

The operations used to solve the system in Example 1.10 were selected to conform to a standard procedure for eliminating variables. There is much more to be said about this process, but first it is helpful to introduce a notationally simpler format, called a matrix, for managing the information in the system. That's a project for the next section.

Exercises 1.2

1. Describe the solutions of the following systems and if possible interpret your conclusions geometrically.

a. $\begin{aligned} 2x - y &= 3 \\ x + 2y &= -1 \end{aligned}$

b. $\begin{aligned} x + 3y &= \tfrac{1}{2} \\ 6y - 1 &= -2x \end{aligned}$

c. $\begin{aligned} 3 - y &= \tfrac{1}{2}x \\ 2y + x &= 4 \end{aligned}$

d. $\begin{aligned} x + 2y - 3z &= 1 \\ 3x - y + z &= 6 \\ -x + 5y - 7z &= 4 \end{aligned}$

e. $\begin{aligned} 2x - 3y - z &= 4 \\ x - 2y - 3z &= 1 \end{aligned}$

f. $\begin{aligned} x - 2y &= 0 \\ y - 2z &= 1 \\ z - 2x &= -1 \end{aligned}$

g. $\begin{aligned} 6x - 3y + 2z &= -1 \\ z + 3x &= \tfrac{3}{2}y - \tfrac{1}{2} \\ \tfrac{1}{3} - \tfrac{2}{3}z &= 2x - y \end{aligned}$

h. $\begin{aligned} x - y &= 1 \\ y - z &= 1 \\ z - w &= 1 \\ w - x &= 1 \end{aligned}$

i. $\begin{aligned} x + y + z - w &= 0 \\ y + z + w - x &= 0 \\ z + w + x - y &= 0 \end{aligned}$

2. Consider a system of three equations in two unknowns and assume the three lines described by these equations are distinct. Sketch the different ways in which these lines can be related to one another and, in each case, comment on the nature of the solution set.

3. Suppose $b, c \in \mathcal{R}$ and consider the two planes described by the equations

$$\begin{aligned} x - 2y + cz &= -2 \\ -2x + by - z &= 4. \end{aligned}$$

Find conditions on b and c such that
a. the two equations describe the same plane.
b. the two planes intersect in a line.
c. Are there choices of b and c for which the system is inconsistent?

4. Every circle in the plane has an equation of the form

$$x^2 + y^2 + ax + by + c = 0,$$

and any three noncollinear points in the plane determine a unique circle. Find the equation of the circle passing through (1,0), (0,1), and (−1,−3).

5. Three noncollinear points in space determine a unique plane. Find the equation of the plane that passes through (1,1,2), (−1,0,1), and (0,1,−1).

6. Suppose x_1, x_2, x_3, and x_4 are real numbers with the property that each is the average of the other three. Find a linear system satisfied by the four numbers and solve the system.

7. Let p be the plane with equation $ax + by + cz = d$. What conditions on $a, b, c,$ and/or d correspond to
a. p being parallel to the x axis? The y axis? The z axis?
b. p being parallel to the xy plane? The yz plane? The xz plane?

c. Explain geometrically why the linear system

$$\begin{aligned} a_1x + b_1y + c_1z &= d_1 \\ b_2y + c_2z &= d_2 \\ c_3z &= d_3 \end{aligned}$$

has a unique solution when $a_1 \neq 0$, $b_2 \neq 0$, and $c_3 \neq 0$.

8. The kth annual census of a certain mammal population records y_k immature females and a_k adult females, $k = 0, 1, 2, \ldots$. Individuals of this species require a year to develop to sexual maturity, and thereafter a female produces, on average, t female offspring per year. Suppose s and r are the annual survival rates (expressed as decimals) for immatures and adults, respectively. Find a pair of linear equations relating y_{k+1} and a_{k+1} to y_k and a_k.

1.3 MATRICES

The operations used in simplifying a linear system are determined at each stage by the coefficients in the current system. The goal of the process is to systematically generate as many zero coefficients as possible, thereby producing an equivalent system in which the interaction between the variables is relatively simple. Since the coefficients generate the operations, you don't really have to write the variables at each stage. All that is needed is a scheme for keeping track of the association between the coefficients and their variables. The standard way of doing this is to delete from

$$\begin{aligned} a_{11}x_1 + a_{12}x_2 + \cdots + a_{1n}x_n &= b_1 \\ a_{21}x_1 + a_{22}x_2 + \cdots + a_{2n}x_n &= b_2 \\ &\vdots \\ a_{m1}x_1 + a_{m2}x_2 + \cdots + a_{mn}x_n &= b_m \end{aligned} \tag{1.4}$$

the variables and the "+" and "=" signs, leaving the coefficients and constant terms in their same relative positions as the rectangular array

$$\begin{matrix} a_{11} & a_{12} & \cdots & a_{1n} & b_1 \\ a_{21} & a_{22} & \cdots & a_{2n} & b_2 \\ & & \vdots & & \\ a_{m1} & a_{m2} & \cdots & a_{mn} & b_m. \end{matrix}$$

In this format, the rows of the array correspond to the equations, and each column is associated with a particular variable (except, of course, the last column, which contains the constant terms). Rectangular arrays are very useful means of organizing information, and there is a fair amount of notation and terminology associated with them.

DEFINITION. Given positive integers, m and n, an $m \times n$ (pronounced m by n) *matrix* is a collection of mn real numbers arranged in a rectangular array of m rows and n columns. A matrix is typically denoted by a capital letter such as A. The number in the ith row, jth column of A, called the *ij-entry* of A, is denoted by a_{ij}. In displaying the entries in A, the array is enclosed in square brackets, that is,

$$A = \begin{bmatrix} a_{11} & \cdots & a_{1n} \\ \vdots & & \vdots \\ a_{m1} & \cdots & a_{mn} \end{bmatrix}.$$

The entries with first index i make up the ith *row* of A, denoted by $\text{Row}_i(A)$, and the entries with second index j constitute the jth *column* of A, denoted by $\text{Col}_j(A)$. More precisely, $\text{Row}_i(A)$ and $\text{Col}_j(A)$ are the $1 \times n$ and $m \times 1$ matrices

$$\text{Row}_i(A) = [a_{i1} \quad \cdots \quad a_{in}] \quad \text{and} \quad \text{Col}_j(A) = \begin{bmatrix} a_{1j} \\ \vdots \\ a_{mj} \end{bmatrix},$$

$1 \le i \le m$, $1 \le j \le n$. The symbol $\mathcal{M}_{m\times n}$ stands for the set of all $m \times n$ matrices.

■ **EXAMPLE 1.11.** If $A = \begin{bmatrix} 2 & 0 & -3 & 12 & 7 \\ -1 & 4 & 0 & -2 & 1 \\ 0 & 25 & 8 & \pi & -5 \\ 3 & \sqrt{2} & -1 & 9 & 0 \end{bmatrix}$, then $A \in \mathcal{M}_{4\times 5}$, $a_{41} = 3$,

$a_{34} = \pi$, $\text{Col}_3(A) = \begin{bmatrix} -3 \\ 0 \\ 8 \\ -1 \end{bmatrix}$, and $\text{Row}_2(A) = [-1 \quad 4 \quad 0 \quad -2 \quad 1]$. □

Associated with system (1.4) are two matrices of interest. The first, called the *matrix of coefficients,* is

$$A = \begin{bmatrix} a_{11} & \cdots & a_{1n} \\ \vdots & & \vdots \\ a_{m1} & \cdots & a_{mn} \end{bmatrix} \in \mathcal{M}_{m\times n}.$$

The second is an $m \times (n+1)$ matrix, which includes both the coefficients and the constant terms, namely,

$$\left[\begin{array}{ccc:c} a_{11} & \cdots & a_{1n} & b_1 \\ \vdots & & \vdots & \vdots \\ a_{m1} & \cdots & a_{mn} & b_m \end{array}\right].$$

It is called the *augmented matrix* for the system. The dashed vertical line helps distinguish between the coefficients and the constants. By setting

$$B = \begin{bmatrix} b_1 \\ \vdots \\ b_m \end{bmatrix},$$

the augmented matrix can be abbreviated as $[A \mid B]$.

Now, the idea is to algebraically simplify the augmented matrix for a linear system in a manner that corresponds to performing elementary operations on the system, and then interpret the result as the augmented matrix of an equivalent system. Elementary operations act on equations, so their matrix counterparts act on rows. In the matrix setting, they are called *elementary row operations,* and, as before, they come in three types:

I. interchange two rows,
II. multiply each entry in one row by a nonzero constant,
III. add a constant multiple of one row to another row.

In adding two rows, the entry in the jth position of the sum is the sum of the entries in the jth positions of the two rows. When you perform an elementary operation on a linear system and also perform the corresponding elementary row operation on the augmented matrix for that system, the new matrix is the augmented matrix of the new system.

EXAMPLE 1.12. The augmented matrix for

$$\begin{aligned} x_1 + 2x_2 - 2x_3 &= 1 \\ 2x_1 + 2x_2 - 2x_3 + x_4 &= 9 \\ -x_1 - x_2 + x_3 + x_4 &= 0 \end{aligned}$$

is

$$\left[\begin{array}{rrrr|r} 1 & 2 & -2 & 0 & 1 \\ 2 & 2 & -2 & 1 & 9 \\ -1 & -1 & 1 & 1 & 0 \end{array}\right].$$

Eliminating x_1 from equations two and three corresponds to performing row operations on the augmented matrix that produce 0's in rows 2 and 3 of column 1. First add -2 times row 1 to row 2. Then add the first row to row 3 (as a type III operation, this amounts to adding 1 times the first row to row 3). The outcome is

$$\left[\begin{array}{rrrr|r} 1 & 2 & -2 & 0 & 1 \\ 0 & -2 & 2 & 1 & 7 \\ 0 & 1 & -1 & 1 & 1 \end{array}\right].$$

At this point it is convenient to interchange rows 2 and 3 to get

$$\left[\begin{array}{rrrr|r} 1 & 2 & -2 & 0 & 1 \\ 0 & 1 & -1 & 1 & 1 \\ 0 & -2 & 2 & 1 & 7 \end{array}\right].$$

Then, add 2 times row 2 to row 3, producing

$$\left[\begin{array}{rrrr|r} 1 & 2 & -2 & 0 & 1 \\ 0 & 1 & -1 & 1 & 1 \\ 0 & 0 & 0 & 3 & 9 \end{array}\right],$$

and, finally, multiply the last row by $1/3$. The result is the augmented matrix for

$$\begin{aligned} x_1 + 2x_2 - 2x_3 + 0x_4 &= 1 \\ x_2 - x_3 + x_4 &= 1 \\ x_4 &= 3. \end{aligned}$$

The solutions are described as follows: $x_4 = 3$, x_3 can be assigned an arbitrary value, t, the value of x_2 is then determined by the second equation to be

$$x_2 = 1 + t - 3 = t - 2,$$

and substituting these results in the first equation yields

$$x_1 = 1 - 2(t - 2) + 2t = -5.$$

That is, the solutions are $(-5, t - 2, t, 3), t \in \mathcal{R}$. □

It helps to have a shorthand for describing the elementary row operations being performed on a matrix, so hereafter

I. r_{ij} stands for interchanging rows i and j.
II. $r_i(t)$ refers to multiplying row i by t.
III. $r_{ij}(t)$ denotes adding t times row i to row j.

Note that the order of the subscripts in $r_{ij}(t)$ is crucial in distinguishing which of two possible operations is being described. With this notation, the elementary row operations used in Example 1.12 are (in the order of their occurrence)

$$r_{12}(-2),\quad r_{13}(1),\quad r_{23},\quad r_{23}(2),\quad \text{and } r_3(1/3).$$

DEFINITION. Suppose A,B $\in \mathcal{M}_{m\times n}$. If B can be obtained from A by a finite sequence of elementary row operations, then B is said to be *row equivalent* to A. This relationship is expressed by writing A $\equiv_r$ B. The process of converting A to B by a sequence of elementary row operations is indicated by A $\to$ B.

Let A $\in \mathcal{M}_{m\times n}$ and suppose B is obtained by performing a single elementary row operation on A. Then A can be recovered by performing a row operation on B. In fact, the operations that reverse the effect of r_{ij}, $r_i(t)$, and $r_{ij}(t)$ are r_{ij}, $r_i(1/t)$, and $r_{ij}(-t)$, respectively. Each is of the same type as the one it reverses. More generally, if A $\equiv_r$ B, then the sequence of elementary row operations that converts A to B can be reversed to convert B to A, that is, B $\equiv_r$ A.

Theorem 1.1 can be reformulated as follows.

THEOREM 1.2. Linear systems whose augmented matrices are row equivalent have the same solutions.

The next example illustrates the shorthand used to record the elementary row operations performed while simplifying a matrix. The operations are listed to the right of the resulting matrix in the order in which they are performed.

■ **EXAMPLE 1.13.** The augmented matrix of

$$\begin{aligned} x_1 - 2x_2 + 3x_3 + x_4 &= 5 \\ 2x_1 - x_3 &= 4 \\ -x_1 + x_2 + 2x_4 &= -1 \end{aligned}$$

is

$$\left[\begin{array}{cccc|c} 1 & -2 & 3 & 1 & 5 \\ 2 & 0 & -1 & 0 & 4 \\ -1 & 1 & 0 & 2 & -1 \end{array}\right].$$

Observe that

$$\left[\begin{array}{cccc|c} 1 & -2 & 3 & 1 & 5 \\ 2 & 0 & -1 & 0 & 4 \\ -1 & 1 & 0 & 2 & -1 \end{array}\right] \longrightarrow \left[\begin{array}{cccc|c} 1 & -2 & 3 & 1 & 5 \\ 0 & 4 & -7 & -2 & -6 \\ 0 & -1 & 3 & 3 & 4 \end{array}\right] \begin{array}{l} r_{12}(-2) \\ r_{13}(1) \end{array}$$

$$\longrightarrow \left[\begin{array}{cccc|c} 1 & -2 & 3 & 1 & 5 \\ 0 & -1 & 3 & 3 & 4 \\ 0 & 4 & -7 & -2 & -6 \end{array}\right] r_{23}$$

$$\longrightarrow \left[\begin{array}{cccc|c} 1 & -2 & 3 & 1 & 5 \\ 0 & 1 & -3 & -3 & -4 \\ 0 & 0 & 5 & 10 & 10 \end{array}\right] \begin{array}{l} r_{23}(4) \\ r_2(-1) \end{array}$$

$$\longrightarrow \left[\begin{array}{cccc|c} 1 & -2 & 3 & 1 & 5 \\ 0 & 1 & -3 & -3 & -4 \\ 0 & 0 & 1 & 2 & 2 \end{array}\right] r_3(1/5).$$

The final result is the augmented matrix of an equivalent system, whose solutions are described by $x_4 = t$, $x_3 = 2 - 2t$, $x_2 = -4 + 3(2 - 2t) + 3t = 2 - 3t$, and $x_1 = 5 + 2(2 - 3t) - 3(2 - 2t) - t = 3 - t$. Thus, the original system has solutions $(3 - t, 2 - 3t, 2 - 2t, t)$, $t \in \mathcal{R}$. □

Now that the machinery for solving a linear system is in place, it is time to consider the form of the simplified matrix. In the examples so far, the operations were selected so that the final matrix was in so-called *row echelon form.* The features that characterize row echelon form are

1. Any zero rows (i.e., rows in which all entries are zero) are located at the bottom of the matrix.
2. The first nonzero entry (reading from left to right) in any nonzero row is a 1. Such an entry is called a *leading one.*
3. In any two adjacent nonzero rows, the leading one in the lower row is located to the right of the leading one in the upper row.

EXAMPLE 1.14.

$$\begin{bmatrix} 1 & 2 & 3 & 4 & 5 \\ 0 & 0 & 1 & 2 & 3 \\ 0 & 0 & 0 & 0 & 1 \end{bmatrix}, \quad \begin{bmatrix} 1 & 2 & 3 \\ 0 & 1 & 2 \\ 0 & 0 & 1 \\ 0 & 0 & 0 \end{bmatrix}, \quad \text{and} \quad \begin{bmatrix} 0 & 1 & 0 & -2 & 4 & 7 \\ 0 & 0 & 0 & 1 & -1 & 0 \\ 0 & 0 & 0 & 0 & 1 & 3 \\ 0 & 0 & 0 & 0 & 0 & 0 \end{bmatrix} \quad \text{are in}$$

row echelon form, but

$$\begin{bmatrix} 1 & 1 & 2 & 2 \\ 0 & 2 & 2 & 1 \\ 0 & 0 & 1 & 2 \end{bmatrix}, \quad \begin{bmatrix} 1 & 0 & 0 & 0 \\ 0 & 0 & 1 & 0 \\ 0 & 1 & 0 & 0 \\ 0 & 0 & 0 & 1 \end{bmatrix}, \quad \text{and} \quad \begin{bmatrix} 1 & -1 \\ 0 & 1 \\ 0 & 1 \\ 0 & 0 \end{bmatrix}$$

are not. □

It is a simple matter to describe the solutions of a linear system when its augmented matrix is in row echelon form. Beginning with the last nonzero row, variables associated with entries to the right of the leading one are independently assigned arbitrary values. The variable associated with the leading one is then expressed in terms of those arbitrary values and the constant term in the last column. Moving to the next-to-last nonzero row, variables associated with positions lying between the leading one of that row and the leading one of the row below it are also assigned arbitrary values. The variable associated with the leading one is then expressed in terms of all the previously assigned values and the constant term at the end of the row. Continuing in this manner leads to a complete description of the solutions. Each step uses the information obtained previously, so this process is called *back substitution.*

EXAMPLE 1.15. Note that

$$\left[\begin{array}{ccccccc|c} 1 & -2 & 1 & 0 & -1 & 3 & 0 & 1 \\ 0 & 0 & 1 & 2 & 0 & 2 & 1 & 2 \\ 0 & 0 & 0 & 1 & 2 & 4 & 0 & 3 \\ 0 & 0 & 0 & 0 & 0 & 1 & -1 & 4 \end{array}\right]$$

is in row echelon form. As an augmented matrix, it represents a system of four equations in seven unknowns. The solutions of the system are described as follows. Starting with the bottom row, the lack of a leading one in column 7 indicates that x_7 can be assigned an arbitrary value, t. Then row 4 gives $x_6 = 4 + t$. Moving to row 3, the entry 2 is associated with a variable, x_5, that can be assigned an arbitrary value, s. Then row 3 yields

$$x_4 = 3 - 2s - 4(4 + t) = -13 - 2s - 4t.$$

Using this information in row 2 produces

$$x_3 = 2 - 2(-13 - 2s - 4t) - 2(4 + t) - t = 20 + 4s + 5t.$$

Finally, assigning x_2 the arbitrary value r, row 1 gives

$$x_1 = 1 + 2r - (20 + 4s + 5t) + s - 3(4 + t) = -31 + 2r - 3s - 8t.$$

The solutions, described in terms of parameters r, s, and t, are

$$(-31 + 2r - 3s - 8t, r, 20 + 4s + 5t, -13 - 2s - 4t, s, 4 + t, t). \square$$

One desirable feature of row echelon form is that any matrix can be systematically converted to a matrix of that form by a sequence of elementary row operations. The process of doing so is called *Gaussian elimination,* in honor of the mathematician, Carl Friederich Gauss (1777–1855). The steps in the Gaussian elimination algorithm are as follows:

1. Locate the first column (counting from left to right) that contains a nonzero entry.
2. If necessary, interchange rows so that a nonzero entry occurs in the top position of that column.
3. Multiply the first row, if necessary, by an appropriate constant so the first nonzero entry in the resulting row is 1. (This is the first leading one in the row echelon matrix.)
4. Perform type III operations $r_{1j}(t)$, $j \geq 2$, to produce zeros below the leading one. (At this point, row 1 and the columns up to and including the first nonzero column have the desired form.)
5. Now, ignore the first row and the columns up to and including the column that holds the first leading one and repeat the previous steps on the smaller matrix that remains.

When the original matrix has a nonzero entry in its first column (which is usually the case), the output from step 4 looks like

$$\left[\begin{array}{c|c} 1 & ? \quad ? \quad \cdots \quad ? \\ \hline 0 & \text{remaining} \\ \vdots & \text{smaller} \\ 0 & \text{matrix} \end{array}\right].$$

Otherwise, there are one or more columns of zeros preceding the first nonzero column. As long as the remaining smaller matrix has nonzero entries, steps 1 through 4 can be applied to it, as illustrated in the next example.

EXAMPLE 1.16.

$$\begin{bmatrix} 2 & 0 & 2 & 7 & -2 \\ 1 & -1 & 2 & 0 & -1 \\ -1 & 2 & -3 & 4 & 1 \\ 2 & -1 & 3 & 3 & -2 \end{bmatrix} \longrightarrow \begin{bmatrix} 1 & -1 & 2 & 0 & -1 \\ 2 & 0 & 2 & 7 & -2 \\ -1 & 2 & -3 & 4 & 1 \\ 2 & -1 & 3 & 3 & -2 \end{bmatrix} \quad r_{12}$$

$$\longrightarrow \left[\begin{array}{c|cccc} 1 & -1 & 2 & 0 & -1 \\ \hline 0 & 2 & -2 & 7 & 0 \\ 0 & 1 & -1 & 4 & 0 \\ 0 & 1 & -1 & 3 & 0 \end{array}\right] \quad \begin{array}{l} r_{12}(-2) \\ r_{13}(1) \\ r_{14}(-2) \end{array}$$

$$\longrightarrow \left[\begin{array}{c|c|ccc} 1 & -1 & 2 & 0 & -1 \\ \hline 0 & 1 & -1 & 4 & 0 \\ \hline 0 & 0 & 0 & -1 & 0 \\ 0 & 0 & 0 & -1 & 0 \end{array}\right] \quad \begin{array}{l} r_{23} \\ r_{23}(-2) \\ r_{24}(-1) \end{array}$$

$$\longrightarrow \left[\begin{array}{cc|cc|c} 1 & -1 & 2 & 0 & -1 \\ 0 & 1 & -1 & 4 & 0 \\ \hline 0 & 0 & 0 & 1 & 0 \\ \hline 0 & 0 & 0 & 0 & 0 \end{array}\right] \quad \begin{array}{l} r_3(-1) \\ r_{34}(1) \end{array}. \quad \square$$

Assume $A \in \mathcal{M}_{m \times n}$, $B \in \mathcal{M}_{m \times 1}$, and $[A \mid B]$ is the augmented matrix of a linear system. Suppose Gaussian elimination, applied to $[A \mid B]$, produces $[C \mid D]$, where $C \in \mathcal{M}_{m \times n}$ and $D \in \mathcal{M}_{m \times 1}$. The locations of the leading ones in $[C \mid D]$ provide qualitative information about the solution set. For example, when D contains a leading one the system is inconsistent. Indeed, in this case the last nonzero row of $[C \mid D]$ is $[0 \ \cdots \ 0 \mid 1]$, and its associated equation, $0x_1 + \cdots + 0x_n = 1$, has no solutions. Suppose D does not contain a leading one. Any column of C that doesn't have a leading one corresponds to a variable that can be assigned an arbitrary value. If there are such columns, then the system has infinitely many solutions, and the number of parameters involved in their description is the number of such columns. Alternatively, every column of C contains a leading one, and in that event $[C \mid D]$ looks like

$$\left[\begin{array}{cccccc|c} 1 & ? & ? & \cdots & & ? & ? \\ 0 & 1 & ? & \cdots & & ? & ? \\ 0 & 0 & 1 & \cdots & & ? & ? \\ \vdots & & & \ddots & & \vdots & \vdots \\ 0 & 0 & 0 & 0 & & 1 & ? \\ \hline \multicolumn{7}{c}{\mathbf{0}} \end{array}\right],$$

where the large zero at the bottom indicates any zero rows that might occur. When $[C \mid D]$ has this form, the last nonzero row identifies x_n, the next-to-last nonzero row then determines a unique value for x_{n-1}, and so on, that is, the system has a unique solution.

THEOREM 1.3. Assume $A \in \mathcal{M}_{m\times n}$, $B \in \mathcal{M}_{m\times 1}$, and $[A \mid B]$ is the augmented matrix of a linear system. Suppose $[A \mid B] \rightarrow [C \mid D]$, where $C \in \mathcal{M}_{m\times n}$, $D \in \mathcal{M}_{m\times 1}$, and $[C \mid D]$ is in row echelon form. The system is consistent if and only if D does not contain a leading one. If the system is consistent, then it

(1) has a unique solution when each column of C contains a leading one, and
(2) has infinitely many solutions, depending on q parameters, when C has exactly q columns that do not contain a leading one.

A linear system in which the constant terms are all zeros, that is,

$$\begin{aligned} a_{11}x_1 + a_{12}x_2 + \cdots + a_{1n}x_n &= 0 \\ a_{21}x_1 + a_{22}x_2 + \cdots + a_{2n}x_n &= 0 \\ &\vdots \\ a_{m1}x_1 + a_{m2}x_2 + \cdots + a_{mn}x_n &= 0, \end{aligned}$$

is said to be *homogeneous*. Every homogeneous system is consistent, because $x_1 = 0$, $x_2 = 0, \ldots, x_n = 0$ is a solution, regardless of the values of the coefficients. This solution is referred to as the *zero* or *trivial solution*. There are only two possible alternatives for the nature of the solution set of a homogeneous system. There is either a unique solution, namely $(0, \ldots, 0)$, or there are infinitely many solutions.

The entries in the last column of the augmented matrix for a homogeneous system are all zeros. No change occurs in that column as row operations are performed, so there is no need to carry it along through Gaussian elimination. For homogeneous systems, then, Theorem 1.3 can be simplified as follows.

COROLLARY. Suppose $A \in \mathcal{M}_{m\times n}$ is the coefficient matrix of a homogeneous linear system and C is a matrix in row echelon form that is row equivalent to A. The system has the zero solution as its only solution if and only if each column of C contains a leading one. Otherwise, the system has infinitely many solutions, and the number of parameters required to describe them is the number of columns of C that do not contain a leading one.

Having simplified an augmented matrix to a row equivalent matrix in row echelon form, the next step in describing the solutions is to assign arbitrary values to the variables not associated with leading ones. The other variables are then expressed in terms of those arbitrary values by back substitution. Back substitution is needed whenever the row echelon matrix has a nonzero entry in a position above a leading one. By further simplifying the matrix so as to eliminate such entries, back substitution can be avoided. The next example illustrates the latter simplification.

■ **EXAMPLE 1.17.** Suppose Gaussian elimination has produced the row echelon augmented matrix

$$\left[\begin{array}{rrrrr|r} 1 & 3 & -4 & 1 & -1 & 1 \\ 0 & 1 & 1 & 2 & 0 & 2 \\ 0 & 0 & 0 & 1 & -2 & 2 \\ 0 & 0 & 0 & 0 & 1 & 1 \end{array}\right].$$

The entries above the leading ones can be converted to 0's by performing type III operations. Starting with the bottom row and working upwards,

$$\left[\begin{array}{ccccc|c} 1 & 3 & -4 & 1 & -1 & 1 \\ 0 & 1 & 1 & 2 & 0 & 2 \\ 0 & 0 & 0 & 1 & -2 & 2 \\ 0 & 0 & 0 & 0 & 1 & 1 \end{array}\right] \longrightarrow \left[\begin{array}{ccccc|c} 1 & 3 & -4 & 1 & 0 & 2 \\ 0 & 1 & 1 & 2 & 0 & 2 \\ 0 & 0 & 0 & 1 & 0 & 4 \\ 0 & 0 & 0 & 0 & 1 & 1 \end{array}\right] \quad \begin{array}{l} r_{43}(2) \\ r_{41}(1) \end{array}$$

$$\longrightarrow \left[\begin{array}{ccccc|c} 1 & 3 & -4 & 0 & 0 & -2 \\ 0 & 1 & 1 & 0 & 0 & -6 \\ 0 & 0 & 0 & 1 & 0 & 4 \\ 0 & 0 & 0 & 0 & 1 & 1 \end{array}\right] \quad \begin{array}{l} r_{32}(-2) \\ r_{31}(-1) \end{array}$$

$$\longrightarrow \left[\begin{array}{ccccc|c} 1 & 0 & -7 & 0 & 0 & 16 \\ 0 & 1 & 1 & 0 & 0 & -6 \\ 0 & 0 & 0 & 1 & 0 & 4 \\ 0 & 0 & 0 & 0 & 1 & 1 \end{array}\right] \quad r_{21}(-3).$$

In the original matrix, the entries to the left of the leading ones are zeros, so type III operations that utilize a particular leading one to produce 0's in the positions above it leave the columns to its left unchanged. In this case, x_3 can be given arbitrary value t, whereupon the solutions are described by $x_5 = 1$, $x_4 = 4$, $x_3 = t$, $x_2 = -6 - t$, and $x_1 = 16 + 7t$. No back substitution is needed. In each equation, the variable associated with the leading one is directly expressible in terms of the constant term and the parameter t. □

The last matrix in the previous example is said to be in *reduced row echelon form.* In addition to having the three features that characterize row echelon form, a matrix in reduced row echelon form also satisfies condition:

4. In each column containing a leading one, the entries above the leading one are zeros.

The algorithm for transforming a given matrix into a row equivalent matrix in reduced row echelon form is referred to as *Gauss-Jordan elimination.*

EXAMPLE 1.18.

$$\begin{bmatrix} 1 & 2 & 0 & -3 & 0 \\ 0 & 0 & 1 & 7 & 0 \\ 0 & 0 & 0 & 0 & 1 \end{bmatrix}, \quad \begin{bmatrix} 1 & 0 & -1 & 0 \\ 0 & 1 & 3 & 0 \\ 0 & 0 & 0 & 1 \\ 0 & 0 & 0 & 0 \end{bmatrix}, \quad \text{and} \quad \begin{bmatrix} 0 & 1 & 2 & 0 \\ 0 & 0 & 0 & 1 \end{bmatrix} \quad \text{are in}$$

reduced row echelon form, but

$$\begin{bmatrix} 1 & 0 & 2 & -1 \\ 0 & 1 & 1 & 0 \\ 0 & 0 & 0 & 1 \end{bmatrix} \quad \text{and} \quad \begin{bmatrix} 1 & 0 & 1 & 0 \\ 0 & 1 & 0 & 1 \\ 0 & 0 & 1 & 0 \\ 0 & 0 & 0 & 1 \end{bmatrix}$$

are not. □

Reduced row echelon form has the following uniqueness property.

THEOREM 1.4. For each $A \in \mathcal{M}_{m\times n}$, there is a unique $B \in \mathcal{M}_{m\times n}$ such that $A \equiv_r B$ and B is in reduced row echelon form.

Proof. Omitted. ■

There are many different sequences of row operations that can convert a given A to a matrix in reduced row echelon form. The point of Theorem 1.4 is that they all produce the same outcome, and hereafter that resulting matrix will be denoted by rref(A). The locations of its leading ones are uniquely determined by A and are called *pivot positions*. The columns in which they reside are known as *pivot columns*. Thanks to its algorithmic nature, Gauss-Jordon elimination is a process that lends itself to machine calculation, and there are numerous calculators and software packages that return rref(A) at the push of a button.

For linear systems, matrices serve as bookkeeping devices; they organize information in a concise format. They play an organizational role in other contexts as well, as the next example illustrates.

EXAMPLE 1.19. Figure 1.4 is a schematic drawing of a transportation system. Each line segment represents a highway joining two cities, the latter being indicated by dots labeled $1, \dots, 7$. The information on the various connections is displayed in the 7×7 matrix, A, where

$$a_{ij} = \begin{cases} 1, & \text{if a highway joins city } i \text{ to city } j \\ 0, & \text{otherwise.} \end{cases}$$

That is,

$$A = \begin{bmatrix} 0 & 1 & 0 & 0 & 0 & 1 & 0 \\ 1 & 0 & 1 & 0 & 0 & 1 & 1 \\ 0 & 1 & 0 & 1 & 0 & 0 & 1 \\ 0 & 0 & 1 & 0 & 1 & 0 & 1 \\ 0 & 0 & 0 & 1 & 0 & 1 & 0 \\ 1 & 1 & 0 & 0 & 1 & 0 & 1 \\ 0 & 1 & 1 & 1 & 0 & 1 & 0 \end{bmatrix}.$$

A mathematical structure, called a *graph,* is used to describe this type of situation. It consists of a finite number of points, called *vertices,* and connections, called *edges,* between various pairs of vertices. Two vertices joined by an edge are said to be *adjacent,* and each graph has an associated *adjacency matrix,* like A. Its *ij*-entry is 1 when vertex *i* is adjacent to vertex *j* and 0 otherwise. Graphs and their adjacency matrices are also used in the social sciences to model relationships between individuals in a group. □

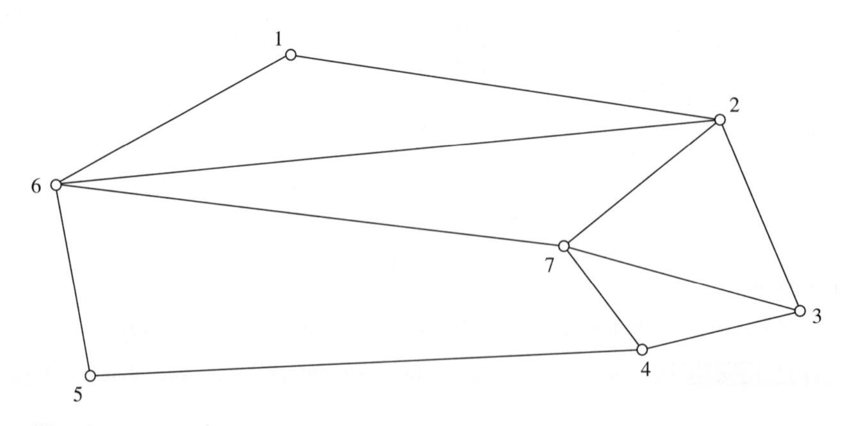

FIGURE 1.4

Exercises 1.3

1. Find the augmented matrix for the system, use Gaussian elimination to get an equivalent matrix in row echelon form, and describe the solutions.

a. $\begin{aligned} 3x - y &= 3 \\ x + 2y &= 3 \end{aligned}$
b. $\begin{aligned} 3y &= 1 + 2x \\ x &= y - 2 \end{aligned}$
c. $\begin{aligned} 6 + x - 2y &= 0 \\ 4y &= 2x - 5 \end{aligned}$

d. $\begin{aligned} x - y + 2z &= -6 \\ 2x + y + z &= -3 \\ 2x + 2y + z &= -1 \end{aligned}$
e. $\begin{aligned} 2x + 4y - z &= 8 \\ -4x + y + 5z &= -7 \\ 2x + y - 2z &= 5 \end{aligned}$
f. $\begin{aligned} 2x + y + 4z + w &= 0 \\ y - 2z &= 1 \\ x - y + 5z &= -2 \end{aligned}$

g. $\begin{aligned} 2x_1 - 2x_2 + 8x_3 + x_4 + 6x_5 &= -3 \\ -2x_1 - x_2 - 5x_3 - 5x_5 &= 2 \\ x_1 - 2x_2 + 5x_3 + x_4 + 4x_5 &= 0 \\ x_1 + x_2 + 2x_3 + 3x_5 &= 1 \end{aligned}$
h. $\begin{aligned} x + 2y &= 3 \\ 4x + 5y &= 6 \\ 6x + 5y &= 4 \\ 3x + 2y &= 1 \end{aligned}$

2. Find rref(A) when A is

a. $\begin{bmatrix} 1 & 2 \\ 3 & 4 \end{bmatrix}$.
b. $\begin{bmatrix} -8 & 4 & 2 \\ 4 & -2 & 0 \end{bmatrix}$.
c. $\begin{bmatrix} 0 & 0 \\ 0 & 7 \end{bmatrix}$.

d. $\begin{bmatrix} 1 & -2 & -1 \\ -1 & 2 & 2 \\ 3 & -6 & -4 \end{bmatrix}$.
e. $\begin{bmatrix} 0 & 1 & 2 \\ 1 & 0 & 1 \\ 2 & 1 & 0 \end{bmatrix}$.
f. $\begin{bmatrix} 1 & 2 & 3 & 4 & 5 \\ 6 & 7 & 8 & 9 & 10 \\ 11 & 12 & 13 & 14 & 15 \end{bmatrix}$.

g. $\begin{bmatrix} 1 & 2 & 3 & 4 & 5 \\ 0 & 1 & 2 & 3 & 4 \\ 0 & 0 & 1 & 2 & 3 \\ 0 & 0 & 0 & 1 & 2 \end{bmatrix}$.
h. $\begin{bmatrix} 2 & -3 \\ -2 & 1 \\ 0 & 4 \\ 1 & 5 \end{bmatrix}$.
i. $\begin{bmatrix} 2 & -3 & -1 \\ -3 & 1 & -2 \\ -1 & 4 & 3 \\ 4 & -2 & 2 \end{bmatrix}$.

j. $\begin{bmatrix} 1 & 2 & 3 & 4 & 5 \\ 2 & 1 & 2 & 3 & 4 \\ 3 & 2 & 1 & 2 & 3 \\ 4 & 3 & 2 & 1 & 2 \\ 5 & 4 & 3 & 2 & 1 \end{bmatrix}$.
k. $\begin{bmatrix} 1 & 2 & 3 & 4 & 5 \\ 2 & 3 & 4 & 5 & 6 \\ 3 & 4 & 5 & 6 & 7 \\ 4 & 5 & 6 & 7 & 8 \\ 5 & 6 & 7 & 8 & 9 \end{bmatrix}$.
l. $\begin{bmatrix} 5 & 4 & 3 & 2 & 1 \\ 4 & 4 & 3 & 2 & 2 \\ 3 & 3 & 3 & 3 & 3 \\ 2 & 2 & 3 & 4 & 4 \\ 1 & 2 & 3 & 4 & 5 \end{bmatrix}$.

m. $\begin{bmatrix} 1 & 1 & 1 & 1 & 1 & \cdots & 1 \\ 1 & 2 & 2 & 2 & 2 & \cdots & 2 \\ 2 & 2 & 3 & 3 & 3 & \cdots & 3 \\ 3 & 3 & 3 & 4 & 4 & \cdots & 4 \\ 4 & 4 & 4 & 4 & 5 & \cdots & 5 \\ \vdots & & & & & \ddots & \\ n-1 & n-1 & n-1 & n-1 & \cdots & n-1 & n \end{bmatrix}$.

3. Find the 2×2 matrix A with the property that the sequence of row operations $r_{12}(-1)$, $r_{21}(2), r_2(2)$, and $r_{12}(1)$ transforms A into $\begin{bmatrix} 1 & 2 \\ 3 & 4 \end{bmatrix}$.

4. Assuming $A = \begin{bmatrix} 2 & 1 & 0 & 3 \\ -1 & 0 & 1 & 2 \\ 0 & 2 & 1 & -1 \end{bmatrix}$, list a sequence of row operations that

a. transforms A into rref(A). b. transforms rref(A) into A.

5. Assuming $A \in \mathcal{M}_{3\times3}$, find all the possibilities for rref(A) (insert ? for the entries that aren't necessarily 0 or 1).

6. Let $a,b,c \in \mathcal{R}$. Show that the system is consistent if and only if (a,b,c) lies on a plane and find the equation of that plane. Does the number of solutions change with different choices of (a,b,c) on that plane?

   a. $\begin{aligned} x - y + 2z &= a \\ -x + 2y - z &= b \\ -x + 4y + z &= c \end{aligned}$

   b. $\begin{aligned} x - 2y &= a \\ 2x + y &= b \\ -x + 3y &= c \end{aligned}$

   c. $\begin{aligned} 2x - 2y + z + w &= a \\ -x + y - w &= b \\ x - y + z &= c \end{aligned}$

7. Suppose $A \in \mathcal{M}_{m\times n}$ and $m < n$.
   a. What can you say about the solution set for the homogeneous system with coefficient matrix A? Explain!
   b. Suppose $B \in \mathcal{M}_{m\times 1}$ and $B \neq 0$. Does the same conclusion hold for the system with augmented matrix $[A \mid B]$? Explain!

8. Suppose $A \in \mathcal{M}_{n\times n}$. Show that if $AX = B$ has a unique solution for one $B \in \mathcal{M}_{n\times 1}$, then it has a unique solution for every $B \in \mathcal{M}_{n\times 1}$.

9. Suppose $x_1,x_2,x_3 \in \mathcal{R}$ are distinct, $y_1,y_2,y_3 \in \mathcal{R}$ and $f(x) = a + bx + cx^2$ is to be fitted to the data points (x_1, y_1), (x_2, y_2), and (x_3, y_3).
   a. Find the augmented matrix for the linear system that (a,b,c) must satisfy.
   b. Does this system have a unique solution?

10. Consider a communication system consisting of six stations labeled $1, \ldots, 6$. That station i can contact station j is indicated by writing $i \to j$, and $i \leftrightarrow j$ means that each of stations i and j can contact the other. No station contacts itself except through an intermediary.
    a. Sketch a diagram, consisting of six dots and some directed line segments, representing a system with the following structure:

    $$1 \to 2,\ 1 \to 6,\ 2 \to 3,\ 2 \leftrightarrow 4,\ 3 \to 4,\ 4 \leftrightarrow 5,\ 5 \to 6, \text{ and } 6 \to 2.$$

    b. Find the 6×6 matrix, A, defined by $a_{ij} = \begin{cases} 1, & \text{if } i \to j \\ 0, & \text{otherwise} \end{cases}$.

1.4
MATRIX ALGEBRA

Gaussian elimination involves two algebraic operations on the rows of a matrix: multiplying a row by a constant and adding two rows. They were designed so that their application to the augmented matrix for a linear system would correspond to performing elementary operations on the equations in the system, thus generating a streamlined procedure for obtaining solutions. The next task is to extend these operations to matrices and to investigate their algebraic properties. In so doing it will be helpful to have the following alternative notation for the entries in a matrix: given $A \in \mathcal{M}_{m\times n}$, the ij-entry of A will be denoted by

$$[A]_{ij},$$

that is, $[A]_{ij}$ is synonymous with a_{ij}. The utility of this notation will become apparent shortly.

Two matrices are equal if they have the same size and the same entries. Thus, for $A,B \in \mathcal{M}_{m\times n}$, $A = B$ when $[A]_{ij} = [B]_{ij}$, $1 \leq i \leq m$, $1 \leq j \leq n$.

DEFINITION. For A,B $\in \mathcal{M}_{m\times n}$ and $t \in \mathcal{R}$, A + B and tA denote the $m \times n$ matrices with entries

$$[A+B]_{ij} = [A]_{ij} + [B]_{ij}$$

and

$$[tA]_{ij} = t[A]_{ij},$$

$1 \le i \le m, 1 \le j \le n.$

This definition establishes two operations, called *addition* and *scalar multiplication,* on the set $\mathcal{M}_{m\times n}$. The matrix A + B is called the *sum* of A and B, and tA is said to be obtained by *scalar multiplying* A by the scalar t.

■ **EXAMPLE 1.20.** If $A = \begin{bmatrix} 1 & 0 & -2 \\ 2 & -1 & 3 \end{bmatrix}$, $B = \begin{bmatrix} -2 & 1 & 0 \\ 0 & -3 & 4 \end{bmatrix}$, and $C = \begin{bmatrix} 2 & 5 \\ -3 & 1 \end{bmatrix}$, then

$$A + B = \begin{bmatrix} -1 & 1 & -2 \\ 2 & -4 & 7 \end{bmatrix}, \qquad 3C = \begin{bmatrix} 6 & 15 \\ -9 & 3 \end{bmatrix},$$

and

$$2A + (-1)B = \begin{bmatrix} 2 & 0 & -4 \\ 4 & -2 & 6 \end{bmatrix} + \begin{bmatrix} 2 & -1 & 0 \\ 0 & 3 & -4 \end{bmatrix} = \begin{bmatrix} 4 & -1 & -4 \\ 4 & 1 & 2 \end{bmatrix}.$$

The sum of A and C is not defined because A and C are not of the same size. □

■ **EXAMPLE 1.21.** Assume A $\in \mathcal{M}_{m\times n}$, $t \in \mathcal{R}$, and $i,j \in \{1,\ldots,m\}$, with $i \neq j$. Then $\mathrm{Row}_i(A)$, $\mathrm{Row}_j(A) \in \mathcal{M}_{1\times n}$. If B and C are obtained by performing $r_i(t)$ and $r_{ij}(t)$, respectively, on A, then $\mathrm{Row}_i(B) = t\mathrm{Row}_i(A)$ and $\mathrm{Row}_j(C) = t\mathrm{Row}_i(A) + \mathrm{Row}_j(A)$. □

The definition of matrix addition involves some ambiguity in the use of the plus sign. In $[A]_{ij} + [B]_{ij}$, the plus sign denotes addition of two real numbers, but in A + B it stands for addition of matrices. Although these are different operations, the meaning of + is clear from the context, so it is customary not to complicate matters notationally by introducing extra symbols. The algebraic rules governing matrix addition and scalar multiplication are listed in the next theorem.

THEOREM 1.5. If A,B,C $\in \mathcal{M}_{m\times n}$ and $s,t \in \mathcal{R}$, then

(1) A + B = B + A, (commutativity)
(2) A + (B + C) = (A + B) + C, (associativity)
(3) there is a unique matrix in $\mathcal{M}_{m\times n}$, denoted by 0, such that A + 0 = A for each A $\in \mathcal{M}_{m\times n}$, (additive identity)
(4) for each A $\in \mathcal{M}_{m\times n}$ there is a unique matrix in $\mathcal{M}_{m\times n}$, denoted by −A, such that A + (−A) = 0, (additive inverse)
(5) $(s + t)A = sA + tA$,
(6) $s(A + B) = sA + sB$, and
(7) $s(tA) = (st)A = t(sA)$.

Proof. Each property states that certain matrices are the same, so each is justified by showing that those matrices have the same entries.

(1) For $1 \le i \le m, 1 \le j \le n$,

$$\begin{aligned} [A+B]_{ij} &= [A]_{ij} + [B]_{ij} && \text{(defn. of matrix addition)} \\ &= [B]_{ij} + [A]_{ij} && \text{(algebra in } \mathcal{R}) \\ &= [B+A]_{ij}. && \text{(defn. of matrix addition)} \end{aligned}$$

(2) See Exercise 3.

(3) Let 0 be the $m \times n$ matrix whose entries are all zeros. Since

$$\begin{aligned}[\mathrm{A}+0]_{ij} &= [\mathrm{A}]_{ij} + [0]_{ij} && \text{(defn. of matrix addition)} \\ &= [\mathrm{A}]_{ij} + 0 && \text{(defn. of the matrix 0)} \\ &= [\mathrm{A}]_{ij}, && \text{(algebra in } \mathcal{R}\text{)}\end{aligned}$$

$1 \leq i \leq m, 1 \leq j \leq n$, 0 has the desired property. If $\mathrm{X} \in \mathcal{M}_{m\times n}$ satisfies $\mathrm{A} + \mathrm{X} = \mathrm{A}$, then $[\mathrm{A}]_{ij} + [\mathrm{X}]_{ij} = [\mathrm{A} + \mathrm{X}]_{ij} = [\mathrm{A}]_{ij}$, $1 \leq i \leq m$, $1 \leq j \leq n$, and therefore $[\mathrm{X}]_{ij} = 0$. Thus, $\mathrm{X} = 0$.

(4) Let $-\mathrm{A}$ be the $m \times n$ matrix with entries $[-\mathrm{A}]_{ij} = -[\mathrm{A}]_{ij}$. Then

$$\begin{aligned}[\mathrm{A}+(-\mathrm{A})]_{ij} &= [\mathrm{A}]_{ij} + [-\mathrm{A}]_{ij} && \text{(defn. of matrix addition)} \\ &= [\mathrm{A}]_{ij} + (-[A]_{ij}) && \text{(defn. of } -\mathrm{A}\text{)} \\ &= 0 && \text{(algebra in } \mathcal{R}\text{)} \\ &= [0]_{ij}, && \text{(defn. of the matrix 0)}\end{aligned}$$

$1 \leq i \leq m, 1 \leq j \leq n$. Justification for the uniqueness claim is left as Exercise 4.

(5) See Exercise 3.

(6) For $1 \leq i \leq m, 1 \leq j \leq n$,

$$\begin{aligned}[s(\mathrm{A}+\mathrm{B})]_{ij} &= s[\mathrm{A}+\mathrm{B}]_{ij} && \text{(defn. of scalar multiplication)} \\ &= s([\mathrm{A}]_{ij} + [\mathrm{B}]_{ij}) && \text{(defn. of matrix addition)} \\ &= s[\mathrm{A}]_{ij} + s[\mathrm{B}]_{ij} && \text{(algebra in } \mathcal{R}\text{)} \\ &= [s\mathrm{A}]_{ij} + [s\mathrm{B}]_{ij} && \text{(defn. of scalar multiplication)} \\ &= [s\mathrm{A} + s\mathrm{B}]_{ij}. && \text{(defn. of matrix addition)}\end{aligned}$$

(7) See Exercise 3. ■

The matrix 0 in property (3) is called the *additive identity* in $\mathcal{M}_{m\times n}$, and $-\mathrm{A}$ in property (4) is referred to as the *additive inverse* of A.

THEOREM 1.6. If $\mathrm{A} \in \mathcal{M}_{m\times n}$ and $t \in \mathcal{R}$, then

(1) $1\mathrm{A} = \mathrm{A}$,

(2) $(-1)\mathrm{A} = -\mathrm{A}$,

(3) $t\mathrm{A} = 0$ if and only if either $t = 0$ or $\mathrm{A} = 0$.

Proof. Properties (1) and (2) are established by the following steps. For $1 \leq i \leq m$, $1 \leq j \leq n$,

$$[1\mathrm{A}]_{ij} = 1[\mathrm{A}]_{ij} = [\mathrm{A}]_{ij},$$

and

$$[(-1)\mathrm{A}]_{ij} = (-1)[\mathrm{A}]_{ij} = -[\mathrm{A}]_{ij} = [-\mathrm{A}]_{ij}.$$

You should supply the reason for each step. As for property (3), $t\mathrm{A} = 0$ if and only if

$$t[\mathrm{A}]_{ij} = [t\mathrm{A}]_{ij} = [0]_{ij} = 0,$$

$1 \leq i \leq m, 1 \leq j \leq n$. The latter condition holds precisely when either $t = 0$ or $[\mathrm{A}]_{ij} = 0$ for each i and j, that is, when either $t = 0$ or $\mathrm{A} = 0$. ■

If you interpret the symbols A, B, and C in Theorems 1.5 and 1.6 as real numbers instead of matrices, then the statements in the theorems are standard properties of addition and multiplication in the real number system. By now these latter operations are so familiar to you that you routinely perform algebraic manipulations involving real numbers and variables representing real numbers without much attention to the rules that justify the individual steps. Thanks to Theorems 1.5 and 1.6, these same instincts serve you well in adding and scalar multiplying matrices. The next example gives a routine algebraic manipulation involving matrices, and to emphasize the role played by Theorems 1.5 and 1.6, each step is listed together with the property that justifies it.

■ **EXAMPLE 1.22.** If $\mathrm{A,B} \in \mathcal{M}_{m\times n}$, then

$$\begin{aligned}
2\mathrm{A} + 3[2\mathrm{B} + (-\mathrm{A})] &= 2\mathrm{A} + [3(2\mathrm{B}) + 3(-\mathrm{A})] && \text{(Theorem 1.5(6))}\\
&= 2\mathrm{A} + [3(2\mathrm{B}) + 3((-1)\mathrm{A})] && \text{(Theorem 1.6(2))}\\
&= 2\mathrm{A} + [(3\cdot 2)\mathrm{B} + (3(-1))\mathrm{A}] && \text{(Theorem 1.5(7))}\\
&= 2\mathrm{A} + [6\mathrm{B} + (-3)\mathrm{A}] && \text{(algebra in } \mathcal{R})\\
&= 2\mathrm{A} + [(-3)\mathrm{A} + 6\mathrm{B}] && \text{(Theorem 1.5(1))}\\
&= [2\mathrm{A} + (-3)\mathrm{A}] + 6\mathrm{B} && \text{(Theorem 1.5(2))}\\
&= [2 + (-3)]\mathrm{A} + 6\mathrm{B} && \text{(Theorem 1.5(5))}\\
&= (-1)\mathrm{A} + 6\mathrm{B} && \text{(algebra in } \mathcal{R})\\
&= -\mathrm{A} + 6\mathrm{B}. && \text{(Theorem 1.6(2))} \quad \square
\end{aligned}$$

The operation of addition (be it matrix addition or addition of real numbers) is a way of combining two objects to produce a third. You can use the operation to combine three or more objects, but formally it is accomplished by grouping them so the outcome is obtained by successively combining two at a time. For example, if A, B, and C, are three matrices of the same size, then you can add them via the grouping $\mathrm{A} + (\mathrm{B} + \mathrm{C})$. The other grouping that maintains the same order of the matrices is $(\mathrm{A} + \mathrm{B}) + \mathrm{C}$. The point of Theorem 1.5(2) is that both produce the same result, so you might as well write $\mathrm{A} + \mathrm{B} + \mathrm{C}$ for the sum. In fact, thanks to Theorem 1.5(1), the order in which the matrices occur is also unimportant. Similar remarks apply to adding more than three matrices. For example, $\mathrm{A} + \mathrm{B} + \mathrm{C} + \mathrm{D}$ will indicate the result of adding A, B, C, and D, even though formally the addition is performed two matrices at a time according to a grouping like $(\mathrm{A} + (\mathrm{B} + (\mathrm{C} + \mathrm{D})))$.

For $\mathrm{A} \in \mathcal{M}_{m\times n}$, $2\mathrm{A} = (1+1)\mathrm{A} = 1\mathrm{A} + 1\mathrm{A} = \mathrm{A} + \mathrm{A}$. That is, scalar multiplying A by 2 produces the same result as adding A to itself. Similarly,

$$3\mathrm{A} = (1+2)\mathrm{A} = 1\mathrm{A} + 2\mathrm{A} = \mathrm{A} + \mathrm{A} + \mathrm{A}.$$

More generally, if t is a positive integer, then $t\mathrm{A}$ is the result of adding t copies of A.

The notion of additive inverse (Theorem 1.5(4)), in conjunction with addition, produces an operation on matrices called *subtraction*.

DEFINITION. If $\mathrm{A,B} \in \mathcal{M}_{m\times n}$, then $\mathrm{A} - \mathrm{B}$ stands for $\mathrm{A} + (-\mathrm{B})$.

According to Theorem 1.6(2), $-\mathrm{B} = (-1)\mathrm{B}$, so $\mathrm{A} - \mathrm{B} = \mathrm{A} + (-1)\mathrm{B}$. For $1 \le i \le m$, $1 \le j \le n$,

$$[\mathrm{A} - \mathrm{B}]_{ij} = [\mathrm{A}]_{ij} - [\mathrm{B}]_{ij}.$$

EXAMPLE 1.23. If $A, B \in \mathcal{M}_{m\times n}$, then

$$
\begin{aligned}
2A - (2B - A) &= 2A + (-1)[2B + (-1)A] && \text{(defn. of matrix subtraction)}\\
&= 2A + [(-1)(2B) + (-1)((-1)A)] && \text{(Theorem 1.5(6))}\\
&= 2A + [(-1)(2B) + ((-1)(-1))A] && \text{(Theorem 1.5(7))}\\
&= 2A + [(-1)(2B) + 1A] && \text{(algebra in } \mathcal{R})\\
&= 2A + [1A + (-1)(2B)] && \text{(Theorem 1.5(1))}\\
&= (2A + 1A) + (-1)(2B) && \text{(Theorem 1.5(2))}\\
&= (2+1)A + (-1)(2B) && \text{(Theorem 1.5(5))}\\
&= 3A + (-1)(2B) && \text{(algebra in } \mathcal{R})\\
&= 3A - 2B. && \text{(defn. of matrix subtraction)} \quad \square
\end{aligned}
$$

The next project is to introduce a method of multiplying matrices. As with addition, multiplication is not defined for just any two matrices; there are size considerations involved. At first glance, the operation doesn't seem very natural, but its usefulness will soon be apparent. The first step is to understand how to multiply a row times a column. If

$$A = [a_1 \quad \cdots \quad a_n] \in \mathcal{M}_{1\times n} \qquad \text{and} \qquad B = \begin{bmatrix} b_1 \\ \vdots \\ b_n \end{bmatrix} \in \mathcal{M}_{n\times 1},$$

then AB is the 1×1 matrix defined by

$$AB = [a_1b_1 + a_2b_2 + \cdots + a_nb_n].$$

Note that A and B are required to have the same number of entries so the entries in one can pair off with those in the other.

EXAMPLE 1.24. If $A = [1 \quad 2 \quad 3 \quad 4]$, $B = \begin{bmatrix} 2 \\ -2 \\ 1 \\ -1 \end{bmatrix}$, $C = [2 \quad 0 \quad 3]$, and $D = \begin{bmatrix} -2 \\ 1 \\ 3 \end{bmatrix}$,

then

$$AB = [1 \quad 2 \quad 3 \quad 4]\begin{bmatrix} 2 \\ -2 \\ 1 \\ -1 \end{bmatrix} = [1\cdot 2 + 2\cdot(-2) + 3\cdot 1 + 4\cdot(-1)] = [-3],$$

and

$$CD = [2 \quad 0 \quad 3]\begin{bmatrix} -2 \\ 1 \\ 3 \end{bmatrix} = [2\cdot(-2) + 0\cdot 1 + 3\cdot 3] = [5].$$

The sizes of A and D are not appropriate for multiplying A times D. □

In general, AB will be defined only when the number of columns in A is the same as the number of rows in B, that is, when the sizes of A and B are ___ $\times\, n$ and $n \times$ ___, respectively. Here, n is a positive integer, and the blanks indicate positive integers that need not be the same. Under these circumstances, each row of A is a $1 \times n$ matrix and each column of B is an $n \times 1$ matrix, so a row of A can multiply a column of B as indicated above.

DEFINITION. If $A \in \mathcal{M}_{m\times n}$ and $B \in \mathcal{M}_{n\times p}$, then AB denotes the $m \times p$ matrix with entries

$$[AB]_{ij} = \text{Row}_i(A)\,\text{Col}_j(B),$$

$1 \le i \le m, 1 \le j \le p$.

There is a minor abuse of notation in the last statement. The expression on the left side of the equality sign is a real number, but that on the right is a 1×1 matrix. What is being ignored here is the distinction between a real number, x, and the 1×1 matrix $[x]$. In terms of the entries in A and B,

$$[AB]_{ij} = [a_{i1} \quad \cdots \quad a_{in}]\begin{bmatrix} b_{1j} \\ \vdots \\ b_{nj} \end{bmatrix} = a_{i1}b_{1j} + a_{i2}b_{2j} + \cdots + a_{in}b_{nj} = \sum_{k=1}^{n} a_{ik}b_{kj},$$

or alternatively,

$$[AB]_{ij} = \sum_{k=1}^{n} [A]_{ik}[B]_{kj}.$$

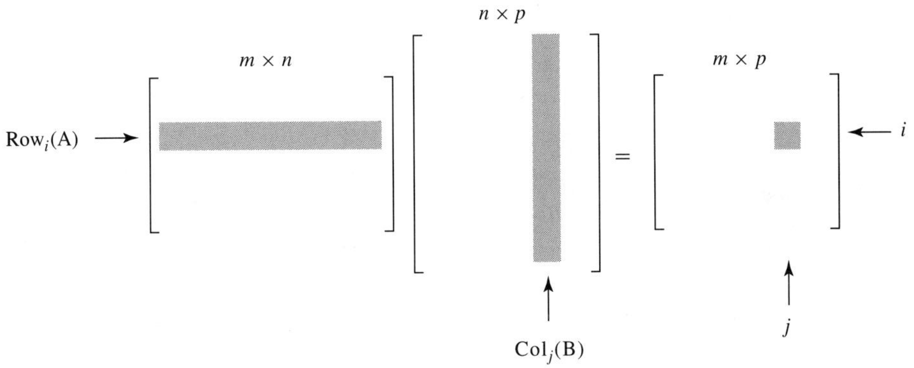

$n \times p$

$m \times n$

$m \times p$

$\text{Row}_i(A) \longrightarrow$

$=$

$\longleftarrow i$

$\text{Col}_j(B)$

j

■ **EXAMPLE 1.25.** Let $A = \begin{bmatrix} 1 & 2 & 3 \\ 3 & 2 & 1 \end{bmatrix}$ and $B = \begin{bmatrix} 1 & -1 & 2 & -2 \\ 3 & -1 & 3 & 1 \\ 0 & -2 & 4 & -1 \end{bmatrix}$. Since $A \in \mathcal{M}_{2\times 3}$ and $B \in \mathcal{M}_{3\times 4}$, the sizes are appropriate for calculating AB. The result is the 2×4 matrix with entries

$$[AB]_{11} = \text{Row}_1(A)\,\text{Col}_1(B) = 1\cdot 1 + 2\cdot 3 + 3\cdot 0 = 7$$

$$[AB]_{12} = \text{Row}_1(A)\,\text{Col}_2(B) = 1\cdot(-1) + 2\cdot(-1) + 3\cdot(-2) = -9$$

$$[AB]_{13} = \text{Row}_1(A)\,\text{Col}_3(B) = 1\cdot 2 + 2\cdot 3 + 3\cdot 4 = 20$$

$$[AB]_{14} = \text{Row}_1(A)\,\text{Col}_4(B) = 1\cdot(-2) + 2\cdot 1 + 3\cdot(-1) = -3$$

$$[AB]_{21} = \text{Row}_2(A)\,\text{Col}_1(B) = 3\cdot 1 + 2\cdot 3 + 1\cdot 0 = 9$$

$$[AB]_{22} = \text{Row}_2(A)\,\text{Col}_2(B) = 3\cdot(-1) + 2\cdot(-1) + 1\cdot(-2) = -7$$

$$[AB]_{23} = \text{Row}_2(A)\,\text{Col}_3(B) = 3\cdot 2 + 2\cdot 3 + 1\cdot 4 = 16$$

$$[AB]_{24} = \text{Row}_2(A)\,\text{Col}_4(B) = 3\cdot(-2) + 2\cdot 1 + 1\cdot(-1) = -5,$$

that is,

$$\mathrm{AB} = \begin{bmatrix} 7 & -9 & 20 & -3 \\ 9 & -7 & 16 & -5 \end{bmatrix}. \square$$

EXAMPLE 1.26. If $\mathrm{A} = \begin{bmatrix} 1 & 0 & -1 \\ 2 & 1 & 3 \end{bmatrix}$ and $\mathrm{B} = \begin{bmatrix} -2 & 1 \\ 1 & -4 \\ -1 & 3 \end{bmatrix}$, then AB and BA are both meaningful, and

$$\mathrm{AB} = \begin{bmatrix} 1 & 0 & -1 \\ 2 & 1 & 3 \end{bmatrix} \begin{bmatrix} -2 & 1 \\ 1 & -4 \\ -1 & 3 \end{bmatrix} = \begin{bmatrix} -1 & -2 \\ -6 & 7 \end{bmatrix},$$

whereas

$$\mathrm{BA} = \begin{bmatrix} -2 & 1 \\ 1 & -4 \\ -1 & 3 \end{bmatrix} \begin{bmatrix} 1 & 0 & -1 \\ 2 & 1 & 3 \end{bmatrix} = \begin{bmatrix} 0 & 1 & 5 \\ -7 & -4 & -13 \\ 5 & 3 & 10 \end{bmatrix}. \square$$

Matrix multiplication provides a convenient means of describing a linear system

$$\begin{aligned} a_{11}x_1 + a_{12}x_2 + \cdots + a_{1n}x_n &= b_1 \\ a_{21}x_1 + a_{22}x_2 + \cdots + a_{2n}x_n &= b_2 \\ &\vdots \\ a_{m1}x_1 + a_{m2}x_2 + \cdots + a_{mn}x_n &= b_m. \end{aligned}$$

Let A be the coefficient matrix and set

$$\mathrm{X} = \begin{bmatrix} x_1 \\ \vdots \\ x_n \end{bmatrix} \quad \text{and} \quad \mathrm{B} = \begin{bmatrix} b_1 \\ \vdots \\ b_m \end{bmatrix}.$$

Note that $\mathrm{A} \in \mathcal{M}_{m\times n}$, $\mathrm{X} \in \mathcal{M}_{n\times 1}$, and $\mathrm{B} \in \mathcal{M}_{m\times 1}$. Since

$$\mathrm{AX} = \begin{bmatrix} a_{11} & \cdots & a_{1n} \\ \vdots & & \vdots \\ a_{m1} & \cdots & a_{mn} \end{bmatrix} \begin{bmatrix} x_1 \\ \vdots \\ x_n \end{bmatrix} = \begin{bmatrix} a_{11}x_1 + \cdots + a_{1n}x_n \\ \vdots \\ a_{m1}x_1 + \cdots + a_{mn}x_n \end{bmatrix} \in \mathcal{M}_{m\times 1},$$

the system can be expressed by the matrix equation $\mathrm{AX} = \mathrm{B}$.

REMARK. Because of this compact means of writing linear systems, n-tuples in the future will usually be written as $n \times 1$ matrices.

In multiplying two real numbers, the order of the factors is unimportant, that is, if $a, b \in \mathcal{R}$, then $ab = ba$. This fact is known as the commutative law of multiplication. With multiplication of matrices, however, the situation is different. If $\mathrm{A} \in \mathcal{M}_{m\times n}$ and $\mathrm{B} \in \mathcal{M}_{p\times q}$, then AB makes sense only if $p = n$, and BA is meaningful only if $q = m$. When both conditions hold, AB and BA are $m \times m$ and $n \times n$ matrices, respectively, so dimensional considerations alone rule out the possibility of equality between them unless $m = n$. Even then it is usually the case that $\mathrm{AB} \neq \mathrm{BA}$. The next example illustrates this point in the simplest of settings, that of 2×2 matrices.

■ **EXAMPLE 1.27.** If $A = \begin{bmatrix} 1 & 2 \\ 3 & 4 \end{bmatrix}$ and $B = \begin{bmatrix} 1 & -1 \\ -2 & 3 \end{bmatrix}$, then

$$AB = \begin{bmatrix} 1 & 2 \\ 3 & 4 \end{bmatrix}\begin{bmatrix} 1 & -1 \\ -2 & 3 \end{bmatrix} = \begin{bmatrix} -3 & 5 \\ -5 & 9 \end{bmatrix}$$

and

$$BA = \begin{bmatrix} 1 & -1 \\ -2 & 3 \end{bmatrix}\begin{bmatrix} 1 & 2 \\ 3 & 4 \end{bmatrix} = \begin{bmatrix} -2 & -2 \\ 7 & 8 \end{bmatrix}.$$

Not only are AB and BA different matrices, there is not a single position in which they agree. □

There are, however, plenty of examples of matrices $A, B \in \mathcal{M}_{n\times n}$ such that $AB = BA$. (You might try to find some 2×2's with this property.) When $AB = BA$, A and B are said to *commute*.

Because matrix multiplication is not commutative, the rules governing the use of this operation are not identical to those governing multiplication of real numbers. Consequently, you will have to develop an appreciation for the sometimes subtle role commutativity plays in your algebraic habits. The basic properties of matrix multiplication are recorded in the next theorem.

THEOREM 1.7. Suppose A, B, and C are matrices and $t \in \mathcal{R}$. In each statement assume the sizes of A, B, and C are appropriate for the operations indicated there. Then

(1) $A(BC) = (AB)C$,
(2) $A(B + C) = AB + AC$,
(3) $(B + C)A = BA + CA$,
(4) $(tA)B = t(AB) = A(tB)$,
(5) there are unique matrices $I_m \in \mathcal{M}_{m\times m}$ and $I_n \in \mathcal{M}_{n\times n}$ such that $I_m A = A = A I_n$ for every $A \in \mathcal{M}_{m\times n}$.

Proof. Property (1) is technically more complicated than the others, and its proof is saved till last.

(2) Here it is appropriate to assume $A \in \mathcal{M}_{m\times n}$ and $B, C \in \mathcal{M}_{n\times p}$, in which case $B + C \in \mathcal{M}_{n\times p}$, $A(B + C) \in \mathcal{M}_{m\times p}$ and AB, AC and $AB + AC \in \mathcal{M}_{m\times p}$. Note that equality in statement (2) is possible from a dimensional standpoint. If $1 \le i \le m$ and $1 \le j \le p$, then

$$\begin{aligned}
[A(B+C)]_{ij} &= \sum_{k=1}^{n} [A]_{ik}[B+C]_{kj} && \text{(defn. of matrix mult.)} \\
&= \sum_{k=1}^{n} [A]_{ik}([B]_{kj} + [C]_{kj}) && \text{(defn. of matrix add.)} \\
&= \sum_{k=1}^{n} ([A]_{ik}[B]_{kj} + [A]_{ik}[C]_{kj}) && \text{(algebra in } \mathcal{R}) \\
&= \sum_{k=1}^{n} [A]_{ik}[B]_{kj} + \sum_{k=1}^{n} [A]_{ik}[C]_{kj} && \text{(algebra in } \mathcal{R}) \\
&= [AB]_{ij} + [AC]_{ij} && \text{(defn. of matrix mult.)} \\
&= [AB + AC]_{ij}, && \text{(defn. of matrix add.)}
\end{aligned}$$

so the left and right sides of statement (2) agree entry by entry.

(3) See Exercise 8.

(4) Assume $A \in \mathcal{M}_{m\times n}$ and $B \in \mathcal{M}_{n\times p}$. The three products in statement (4) are then $m \times p$ matrices. For $1 \le i \le m$ and $1 \le j \le p$,

$$\begin{aligned}
[(tA)B]_{ij} &= \sum_{k=1}^{n} [tA]_{ik}[B]_{kj} && \text{(defn. of matrix mult.)}\\
&= \sum_{k=1}^{n} (t[A]_{ik})[B]_{kj} && \text{(defn. of scalar mult.)}\\
&= \sum_{k=1}^{n} t([A]_{ik}[B]_{kj}) && \text{(algebra in } \mathcal{R})\\
&= t\sum_{k=1}^{n} [A]_{ik}[B]_{kj} && \text{(algebra in } \mathcal{R})\\
&= t[AB]_{ij} && \text{(defn. of matrix mult.)}\\
&= [t(AB)]_{ij}. && \text{(defn. of scalar mult.)}
\end{aligned}$$

This establishes part of statement (4). The rest is left as Exercise 9.

(5) Let I_m be the $m \times m$ matrix with entries $[I_m]_{ij} = \begin{cases} 1, & \text{if } i = j \\ 0, & \text{if } i \neq j \end{cases}$, that is,

$$I_m = \begin{bmatrix} 1 & 0 & \cdots & 0 & 0 \\ 0 & 1 & & & 0 \\ \vdots & & \ddots & & \vdots \\ 0 & & & 1 & 0 \\ 0 & 0 & \cdots & 0 & 1 \end{bmatrix}.$$

Then,

$$\begin{aligned}
[I_m A]_{ij} &= \sum_{k=1}^{n} [I_m]_{ik}[A]_{kj} && \text{(defn. of matrix mult.)}\\
&= [I_m]_{ii}[A]_{ij} && ([I_m]_{ik} = 0 \text{ if } k \neq i)\\
&= [A]_{ij}. && ([I_m]_{ii} = 1)
\end{aligned}$$

The matrix I_n is defined similarly, and the proof that $AI_n = A$ involves only minor modifications of the steps above. To establish that I_m and I_n are the only matrices satisfying $I_m A = A = AI_n$ for all $A \in \mathcal{M}_{m\times n}$, see Exercise 22.

(1) Here the dimensional assumptions are $A \in \mathcal{M}_{m\times n}$, $B \in \mathcal{M}_{n\times p}$, and $C \in \mathcal{M}_{p\times q}$. Then $BC \in \mathcal{M}_{n\times q}$, $A(BC) \in \mathcal{M}_{m\times q}$, $AB \in \mathcal{M}_{m\times p}$, and $(AB)C \in \mathcal{M}_{m\times q}$. For $1 \le i \le m$ and $1 \le j \le q$,

$$\begin{aligned}
[A(BC)]_{ij} &= \sum_{k=1}^{n} [A]_{ik}[BC]_{kj} && \text{(defn. of matrix mult.)}\\
&= \sum_{k=1}^{n} [A]_{ik}\left(\sum_{l=1}^{p} [B]_{kl}[C]_{lj}\right) && \text{(defn. of matrix mult.)}\\
&= \sum_{k=1}^{n}\sum_{l=1}^{p} [A]_{ik}[B]_{kl}[C]_{lj} && \text{(algebra in } \mathcal{R})\\
&= \sum_{l=1}^{p}\sum_{k=1}^{n} [A]_{ik}[B]_{kl}[C]_{lj} && \text{(algebra in } \mathcal{R})
\end{aligned}$$

$$= \sum_{l=1}^{p} \left(\sum_{k=1}^{n} [A]_{ik}[B]_{kl} \right) [C]_{lj} \qquad \text{(algebra in } \mathcal{R}\text{)}$$

$$= \sum_{l=1}^{p} [AB]_{il}[C]_{lj} \qquad \text{(defn. of matrix mult.)}$$

$$= [(AB)C]_{ij}. \qquad \text{(defn. of matrix mult.)} \blacksquare$$

The matrix I_n in property (5) is called the $n \times n$ *identity* matrix. It is often denoted simply by I when its size is clear from context.

A matrix having the same number of rows as columns is said to be *square,* and members of $\mathcal{M}_{n\times n}$ are referred to as square matrices of *size n*. If $A, B \in \mathcal{M}_{n\times n}$ and $t \in \mathcal{R}$, then $A + B$, AB, and tA are all defined, and each is an $n \times n$ matrix. Thus, when performing algebraic manipulations involving elements of $\mathcal{M}_{n\times n}$, you need not be concerned at each step about the sizes of the resulting matrices. In particular, it makes sense to multiply A by itself, and the result is denoted by A^2. Similarly, $A^3 = AAA$, and A^p (p a positive integer) stands for the product of p copies of A.

The next example illustrates the use of some of the rules governing matrix multiplication and at the same time points out a commonly encountered situation in which the noncommutativity of this operation produces an unexpected outcome.

■ **EXAMPLE 1.28.** Suppose A and B are $n \times n$ matrices and consider the task of expanding $(A + B)^2$. The formal steps are

$$\begin{aligned} (A+B)^2 &= (A+B)(A+B) && \text{(defn. of the square of a matrix)} \\ &= A(A+B) + B(A+B) && \text{(Theorem 1.7(3))} \\ &= AA + AB + BA + BB && \text{(Theorem 1.7(2))} \\ &= A^2 + AB + BA + B^2. && \text{(defn. of the square of a matrix)} \end{aligned}$$

If A and B commute, that is, if $AB = BA$, then the last expression simplifies to $A^2 + 2AB + B^2$. On the other hand, if A and B are known to satisfy

$$(A+B)^2 = A^2 + 2AB + B^2, \tag{1.5}$$

then the above expansion of the left side produces

$$A^2 + AB + BA + B^2 = A^2 + 2AB + B^2,$$

from which it follows that $BA = AB$. Thus, (1.5) holds if and only if A and B commute. □

■ **EXAMPLE 1.29.** The linear system

$$\begin{aligned} a_{11}x_1 + a_{12}x_2 + \cdots + a_{1n}x_n &= b_1 \\ a_{21}x_1 + a_{22}x_2 + \cdots + a_{2n}x_n &= b_2 \\ &\vdots \\ a_{m1}x_1 + a_{m2}x_2 + \cdots + a_{mn}x_n &= b_m \end{aligned}$$

can be written concisely as $AX = B$, where A is the coefficient matrix,

$$X = \begin{bmatrix} x_1 \\ \vdots \\ x_n \end{bmatrix} \quad \text{and} \quad B = \begin{bmatrix} b_1 \\ \vdots \\ b_m \end{bmatrix}.$$

Suppose each of $x_1, \ldots, x_n$ depends linearly on variables $y_1, \ldots, y_p$, that is, there are real numbers c_{ij}, $1 \le i \le n$, $1 \le j \le p$, such that

$$\begin{aligned} c_{11}y_1 + c_{12}y_2 + \cdots + c_{1p}y_p &= x_1 \\ c_{21}y_1 + c_{22}y_2 + \cdots + c_{2p}y_p &= x_2 \\ &\vdots \\ c_{n1}y_1 + c_{n2}y_2 + \cdots + c_{np}y_p &= x_n. \end{aligned} \tag{1.6}$$

If you substitute the latter expressions in the original system and simplify the resulting equations by collecting coefficients of $y_1, \ldots, y_p$, the outcome is a system of m linear equations in $y_1, \ldots, y_p$, with the same constant terms, $b_1, \ldots, b_m$. The new system is said to be obtained from the original by a *linear change of variables*. This straightforward but tedious calculation can be performed efficiently as follows. Let C be the coefficient matrix for system (1.6) and set

$$\mathrm{Y} = \begin{bmatrix} y_1 \\ \vdots \\ y_p \end{bmatrix}.$$

Then (1.6) can be abbreviated as X = CY. Substituting the latter expression in AX = B yields A(CY) = B, which can be rewritten as (AC)Y = B. Thus, AC is the coefficient matrix for the new system. □

Given $\mathrm{A} \in \mathcal{M}_{m\times n}$ and $\mathrm{B} \in \mathcal{M}_{n\times p}$, consider for a moment how the columns and rows of AB are generated. For $1 \le j \le p$,

$$\mathrm{Col}_j(\mathrm{AB}) = \begin{bmatrix} \mathrm{Row}_1(\mathrm{A})\mathrm{Col}_j(\mathrm{B}) \\ \mathrm{Row}_2(\mathrm{A})\mathrm{Col}_j(\mathrm{B}) \\ \vdots \\ \mathrm{Row}_m(\mathrm{A})\mathrm{Col}_j(\mathrm{B}) \end{bmatrix} = \mathrm{ACol}_j(\mathrm{B}),$$

and for $1 \le i \le m$,

$$\mathrm{Row}_i(\mathrm{AB}) = [\mathrm{Row}_i(\mathrm{A})\mathrm{Col}_1(\mathrm{B}) \quad \cdots \quad \mathrm{Row}_i(\mathrm{A})\mathrm{Col}_p(\mathrm{B})] = \mathrm{Row}_i(\mathrm{A})\mathrm{B}.$$

These observations are recorded in the next theorem for future reference.

THEOREM 1.8. If $\mathrm{A} \in \mathcal{M}_{m\times n}$ and $\mathrm{B} \in \mathcal{M}_{n\times p}$, then for $1 \le j \le p$ and $1 \le i \le m$,

(1) $\mathrm{Col}_j(\mathrm{AB}) = \mathrm{ACol}_j(\mathrm{B})$, and
(2) $\mathrm{Row}_i(\mathrm{AB}) = \mathrm{Row}_i(\mathrm{A})\mathrm{B}$.

As a consequence of Theorem 1.8, AB can be computed in either of the following ways:

$$\mathrm{AB} = \mathrm{A}[\mathrm{Col}_1(\mathrm{B}) \quad \cdots \quad \mathrm{Col}_p(\mathrm{B})] = [\mathrm{ACol}_1(\mathrm{B}) \quad \cdots \quad \mathrm{ACol}_p(\mathrm{B})],$$

or

$$\mathrm{AB} = \begin{bmatrix} \mathrm{Row}_1(\mathrm{A}) \\ \vdots \\ \mathrm{Row}_m(\mathrm{A}) \end{bmatrix} \mathrm{B} = \begin{bmatrix} \mathrm{Row}_1(\mathrm{A})\mathrm{B} \\ \vdots \\ \mathrm{Row}_m(\mathrm{A})\mathrm{B} \end{bmatrix}.$$

Exercises 1.4

1. If $A=\begin{bmatrix} 1 & 0 & -1 \\ -1 & 2 & 1 \end{bmatrix}$, $B=\begin{bmatrix} 1 & -1 & 0 \\ -3 & 1 & 2 \end{bmatrix}$, $C=\begin{bmatrix} 0 & 1 & -3 \\ 1 & 2 & 0 \end{bmatrix}$, and $D=\begin{bmatrix} 2 & -1 & 3 \\ -2 & 1 & 1 \end{bmatrix}$, find

   a. $A + 2B$. b. $2(2B + C) + D$. **c.** $3A - D$.
   d. $2A - [B - (-1)C]$. **e.** $3A - B - 2C + D$. f. $-[(D - 2A) - B]$.
   g. $-A + [B - 3(B - 2A)] - [4(A - B) + 3B]$.

2. Assuming $A=\begin{bmatrix} 1 & 1 & 0 \\ 1 & 0 & -1 \\ 0 & -1 & 1 \\ -1 & 1 & 0 \end{bmatrix}$, $B=\begin{bmatrix} 2 & 0 & -1 \\ 1 & 2 & 1 \\ -1 & 0 & 2 \\ 0 & 1 & 0 \end{bmatrix}$, and $C=\begin{bmatrix} 2 & 1 & 2 \\ 1 & 0 & -1 \\ 0 & 2 & 0 \\ -1 & 0 & 1 \end{bmatrix}$, find

   a. $A + B + C$. b. $A - (B - C)$. **c.** $2(3B - A) - [C - (A - 2B)]$.

3. Write formal proofs of statements (2), (5), and (7) of Theorem 1.5, listing each step together with its justification.

4. Show that the additive inverse of an $m \times n$ matrix is unique, that is, show that if $A \in \mathcal{M}_{m\times n}$ and X is an element of $\mathcal{M}_{m\times n}$ with the property that $A + X = 0$, then $X = -A$. (This completes the proof of statement (4) of Theorem 1.5.)

5. If $A=\begin{bmatrix} 1 & -1 \\ 0 & 2 \\ 3 & -2 \end{bmatrix}$, $B=\begin{bmatrix} 0 & 1 & 0 \\ 1 & 0 & 0 \\ 0 & 0 & -1 \end{bmatrix}$, $C=[-1 \quad 1 \quad 1]$, $D=\begin{bmatrix} 1 & 2 & 0 \\ 0 & 1 & -1 \\ -2 & -1 & -1 \\ 0 & 1 & 1 \end{bmatrix}$,

   and $E=\begin{bmatrix} -1 & 0 & 2 & 4 \\ 2 & -1 & 3 & 2 \end{bmatrix}$, find

   a. AE. b. CA. **c.** EDA. d. DBA. **e.** DI_3. f. I_4D.

6. If $A=\begin{bmatrix} 1 \\ -1 \\ 2 \end{bmatrix}$, $B=\begin{bmatrix} 1 & 1 & 2 \\ -2 & 1 & 0 \end{bmatrix}$, $C=[2 \quad 3]$, $D=\begin{bmatrix} 0 & 0 & 1 \\ 0 & -1 & 0 \\ 1 & 0 & 0 \end{bmatrix}$,

   and $E=\begin{bmatrix} 1 & -2 \\ -1 & 0 \\ 1 & 2 \end{bmatrix}$, find

   a. BDE. b. D^2. **c.** AC. d. CBA. **e.** $EB + 2D$.

7. Find matrices $A, B \in \mathcal{M}_{2\times 2}$ such that $A \neq 0$ and $B \neq 0$, but $AB = 0$.

8. State the dimensional conditions on matrices A, B, and C under which $(B + C)A = BA + CA$ makes sense and then prove the assertion.

9. Show that if $A \in \mathcal{M}_{m\times n}$, $B \in \mathcal{M}_{n\times p}$, and $t \in \mathcal{R}$, then $t(AB) = A(tB)$.

10. Assuming $A \in \mathcal{M}_{m\times n}$ and $B \in \mathcal{M}_{n\times p}$, show that
    a. if A has a row of zeros, then AB has a row of zeros.
    b. if B has a column of zeros, then AB has a column of zeros.

11. Suppose A and B are $n \times n$ matrices.
    a. Is $(AB)^2 = A^2B^2$? Explain!
    b. Is $(A + B)(A - B) = A^2 - B^2$? Explain!

12. Find all $B \in \mathcal{M}_{2\times 2}$ such that $AB = BA$ when

    a. $A = \begin{bmatrix} 1 & 2 \\ 3 & 4 \end{bmatrix}$. b. $A = \begin{bmatrix} 1 & 1 \\ 1 & 1 \end{bmatrix}$. c. $A = \begin{bmatrix} 0 & 0 \\ 1 & 1 \end{bmatrix}$.

13. Show that if $A, B, C \in \mathcal{M}_{n\times n}$ and B and C both commute with A, then
    a. tB commutes with A for each $t \in \mathcal{R}$.
    b. $B + C$ commutes with A.
    c. BC commutes with A.

14. Assume $A \in \mathcal{M}_{m\times n}$, $B \in \mathcal{M}_{m\times 1}$, and $X_1, X_2, X \in \mathcal{M}_{n\times 1}$.
    a. Show that if X_1 and X_2 are solutions of the homogeneous system $AX = 0$, then so is $X_1 + X_2$.
    b. Show that if X_1 is a solution of $AX = 0$, then so is tX_1 for each $t \in \mathcal{R}$.
    c. Do similar conclusions hold for the system $AX = B$ when $B \neq 0$? Explain!

15. Assume $A \in \mathcal{M}_{m\times n}$, $X \in \mathcal{M}_{n\times 1}$, and $B_1, \ldots, B_p \in \mathcal{M}_{m\times 1}$. To solve $AX = B_1$, $AX = B_2, \ldots, AX = B_p$, you row reduce $[A \mid B_1], [A \mid B_2], \ldots, [A \mid B_p]$. The steps used to row reduce A are the same in each case, so all p systems can be solved simultaneously by row reducing $[A \mid B_1 \mid B_2 \mid \cdots \mid B_p]$ and interpreting the results appropriately. Assuming
    $A = \begin{bmatrix} 1 & 2 & 1 \\ 2 & 3 & 3 \\ 0 & 1 & -1 \end{bmatrix}$, solve $AX = B$ when

    a. $B = \begin{bmatrix} 1 \\ 1 \\ 1 \end{bmatrix}$. b. $B = \begin{bmatrix} -1 \\ 2 \\ -3 \end{bmatrix}$. **c.** $B = \begin{bmatrix} 0 \\ 1 \\ -1 \end{bmatrix}$. d. $B = \begin{bmatrix} 2 \\ 3 \\ 1 \end{bmatrix}$.

16. Suppose $a, b \in \mathcal{R}$,

$$\begin{aligned} x_1 - 2x_2 + 3x_3 &= a \\ -2x_1 + x_2 - x_3 &= b \end{aligned} \quad \text{and} \quad \begin{aligned} 2y_1 - 3y_2 &= x_1 \\ -y_1 + 2y_2 &= x_2 . \\ y_1 + y_2 &= x_3 \end{aligned}$$

    Find the coefficient matrix for the linear system that expresses a and b in terms of y_1 and y_2.

17. Recall the predator-prey model in Example 1.4. Set $X_k = \begin{bmatrix} f_k \\ r_k \end{bmatrix}$ and observe that $X_{k+1} = AX_k$, where $A = \begin{bmatrix} \alpha & \beta \\ -\gamma & \delta \end{bmatrix}$.

    a. Show that $X_k = A^kX_0$, $k = 1, 2, \ldots$.
    b. Let $\alpha = 0.4$, $\beta = 0.5$, $\gamma = 0.1$, $\delta = 1.2$, and compute A^5, A^{10}, and A^{20}. Calculate X_5, X_{10}, and X_{20} when the initial data is $f_0 = 100$ and $r_0 = 1000$.

18. Recall the population model in Problem 8, Exercises 1.2. Set $X_k = \begin{bmatrix} y_k \\ a_k \end{bmatrix}$.

    a. Show that $X_k = A^kX_0$, where $A = \begin{bmatrix} 0 & t \\ s & r \end{bmatrix}$.

b. Compute A^5, A^{10}, and A^{20} when $r = s = 0.5$ and $t = 1$. Make several choices for the initial data X_0 and examine the corresponding long-term distribution of immature and adult females. What conclusion is suggested by your calculations? How does it depend on the initial data?

19. Assume A is an $n \times n$ adjacency matrix for some transportation system (as in Example 1.19).
   a. What is the meaning of $[A^2]_{ij}$ in the context of the system? (Hint: interpret each of the terms in $\sum_{k=1}^{n} a_{ik}a_{kj}$).
   b. Compute A^2 for the special case in Example 1.19 and check the result against your conclusion in part (a).
   c. How do you interpret $[A^3]_{ij}$?

20. Recall the communication system in Problem 10, Exercises 1.3. A *path* connects station i to station j if a message can be sent from the former to the latter by relaying it through one or more other stations. The *length* of a path is the number of intermediary links. For example, a path from station i through station k to station j has length 2.
   a. Using the matrix A from Problem 10, Exercises 1.3, calculate A^2, A^3, $A + A^2$, and $A + A^2 + A^3$. Interpret the ij-entry of each in the context of the communication system.
   b. What feature of A, A^2, A^3, ... indicates that no station can contact station 1, regardless of how long a path is used?
   c. What is the smallest integer m with the property that for each $i, j \geq 2$, station i is connected to station j by a path of length at most m?

21. Suppose A and B are $n \times n$ matrices such that the sum of the entries in each of their rows is 1, that is, $\sum_{j=1}^{n} a_{ij} = 1 = \sum_{j=1}^{n} b_{ij}$, $1 \leq i \leq n$. Show that AB also has this property.

22. Suppose $X \in \mathcal{M}_{m\times m}$ has the property that $XA = A$ for each $A \in \mathcal{M}_{m\times n}$. Show that $X = I_m$.

1.5

SPECIAL MATRICES

DEFINITION. A matrix obtained by performing a single elementary row operation on I_n is called an $n \times n$ *elementary matrix*. Operations $r_{ij}, r_i(t)$, and $r_{ij}(t)$ produce elementary matrices denoted by E_{ij}, $E_i(t)$, and $E_{ij}(t)$, respectively. This notation does not indicate the size of the matrix, but that information can usually be inferred from context.

EXAMPLE 1.30. In the 4×4 case,

$$E_{13} = \begin{bmatrix} 0 & 0 & 1 & 0 \\ 0 & 1 & 0 & 0 \\ 1 & 0 & 0 & 0 \\ 0 & 0 & 0 & 1 \end{bmatrix}, \qquad E_2(-3) = \begin{bmatrix} 1 & 0 & 0 & 0 \\ 0 & -3 & 0 & 0 \\ 0 & 0 & 1 & 0 \\ 0 & 0 & 0 & 1 \end{bmatrix},$$

$$E_{42}(5) = \begin{bmatrix} 1 & 0 & 0 & 0 \\ 0 & 1 & 0 & 5 \\ 0 & 0 & 1 & 0 \\ 0 & 0 & 0 & 1 \end{bmatrix}, \qquad E_{23}(-1) = \begin{bmatrix} 1 & 0 & 0 & 0 \\ 0 & 1 & 0 & 0 \\ 0 & -1 & 1 & 0 \\ 0 & 0 & 0 & 1 \end{bmatrix},$$

and

$$\mathrm{E}_{14} = \begin{bmatrix} 0 & 0 & 0 & 1 \\ 0 & 1 & 0 & 0 \\ 0 & 0 & 1 & 0 \\ 1 & 0 & 0 & 0 \end{bmatrix}. \square$$

Consider the effect of multiplying $\mathrm{A} \in \mathcal{M}_{m\times n}$ on the left by an $m \times m$ elementary matrix E. According to Theorem 1.8(2), the rows of the product are

$$\mathrm{Row}_k(\mathrm{EA}) = \mathrm{Row}_k(\mathrm{E})\mathrm{A},$$

$1 \le k \le m$. Note that E differs from I_m in only one or two rows. If k is an index for which $\mathrm{Row}_k(\mathrm{E}) = \mathrm{Row}_k(\mathrm{I}_m)$, then by Theorem 1.8(2),

$$\mathrm{Row}_k(\mathrm{EA}) = \mathrm{Row}_k(\mathrm{E})\mathrm{A} = \mathrm{Row}_k(\mathrm{I}_m)\mathrm{A} = \mathrm{Row}_k(\mathrm{I}_m\mathrm{A}) = \mathrm{Row}_k(\mathrm{A}),$$

that is, these rows of EA are identical to their counterparts in A. The remaining rows of EA depend on the type of E. For example, if $\mathrm{E} = \mathrm{E}_{ij}$, then $\mathrm{Row}_k(\mathrm{E}_{ij}\mathrm{A}) = \mathrm{Row}_k(\mathrm{A})$ for $k \neq i,j$,

$$\mathrm{Row}_i(\mathrm{E}_{ij}\mathrm{A}) = \mathrm{Row}_i(\mathrm{E}_{ij})\mathrm{A} = \mathrm{Row}_j(\mathrm{I}_m)\mathrm{A} = \mathrm{Row}_j(\mathrm{I}_m\mathrm{A}) = \mathrm{Row}_j(\mathrm{A}),$$

and

$$\mathrm{Row}_j(\mathrm{E}_{ij}\mathrm{A}) = \mathrm{Row}_j(\mathrm{E}_{ij})\mathrm{A} = \mathrm{Row}_i(\mathrm{I}_m)\mathrm{A} = \mathrm{Row}_i(\mathrm{I}_m\mathrm{A}) = \mathrm{Row}_i(\mathrm{A}).$$

Observe that $\mathrm{E}_{ij}\mathrm{A}$ is the matrix that results from interchanging rows i and j of A. Thus, multiplying A on the left by E_{ij} has the same effect as performing on A the elementary row operation that turns I_m into E_{ij}. A similar conclusion holds for elementary matrices of types II or III.

THEOREM 1.9. Suppose r is an elementary row operation and let E be the corresponding $m \times m$ elementary matrix. If $\mathrm{A} \in \mathcal{M}_{m\times n}$, then EA is the matrix obtained by performing r on A.

Proof. There are two cases remaining to be checked. Suppose $r = r_{ij}(t)$, that is, $\mathrm{E} = \mathrm{E}_{ij}(t)$. For $k \neq j$, $\mathrm{Row}_k(\mathrm{E}_{ij}(t)) = \mathrm{Row}_k(\mathrm{I}_m)$, so $\mathrm{Row}_k(\mathrm{E}_{ij}(t)\mathrm{A}) = \mathrm{Row}_k(\mathrm{A})$. Moreover,

$$\begin{aligned}\mathrm{Row}_j(\mathrm{E}_{ij}(t)\mathrm{A}) &= \mathrm{Row}_j(\mathrm{E}_{ij}(t))\mathrm{A} = [t\mathrm{Row}_i(\mathrm{I}_m) + \mathrm{Row}_j(\mathrm{I}_m)]\mathrm{A} \\ &= t[\mathrm{Row}_i(\mathrm{I}_m)\mathrm{A}] + \mathrm{Row}_j(\mathrm{I}_m)\mathrm{A} \\ &= t[\mathrm{Row}_i(\mathrm{I}_m\mathrm{A})] + \mathrm{Row}_j(\mathrm{I}_m\mathrm{A}) \\ &= t\mathrm{Row}_i(\mathrm{A}) + \mathrm{Row}_j(\mathrm{A}).\end{aligned}$$

Thus, $\mathrm{E}_{ij}(t)\mathrm{A}$ is the $m \times n$ matrix obtained by performing $r_{ij}(t)$ on A. The case $\mathrm{E} = \mathrm{E}_i(t)$ is left as an exercise. ■

■ **EXAMPLE 1.31.** If $\mathrm{A} = \begin{bmatrix} a & b \\ c & d \\ e & f \end{bmatrix}$, then

$$\mathrm{E}_{13}\mathrm{A} = \begin{bmatrix} 0 & 0 & 1 \\ 0 & 1 & 0 \\ 1 & 0 & 0 \end{bmatrix} \begin{bmatrix} a & b \\ c & d \\ e & f \end{bmatrix} = \begin{bmatrix} e & f \\ c & d \\ a & b \end{bmatrix},$$

which is the matrix obtained by interchanging rows 1 and 3 of A. Also,

$$E_{12}(-1)A = \begin{bmatrix} 1 & 0 & 0 \\ -1 & 1 & 0 \\ 0 & 0 & 1 \end{bmatrix} \begin{bmatrix} a & b \\ c & d \\ e & f \end{bmatrix} = \begin{bmatrix} a & b \\ c-a & d-b \\ e & f \end{bmatrix},$$

the result of applying $r_{12}(-1)$ to A. □

Suppose A,B $\in \mathcal{M}_{m\times n}$. If A $\equiv_r$ B, then there is a sequence of elementary row operations, $r_1, \ldots, r_q$, which transforms A into B. Let $E_1, \ldots, E_q$ be the corresponding $m \times m$ elementary matrices. According to Theorem 1.9, performing $r_1, \ldots, r_q$ in succession has the same effect as multiplying on the left successively by the $E_1, \ldots, E_q$, that is,

$$B = E_q \cdots E_2E_1A.$$

Conversely, if $B = E_q \cdots E_2E_1A$, then B is obtained from A by performing a sequence of elementary row operations. These observations establish the following characterization of row equivalence.

THEOREM 1.10. If A,B $\in \mathcal{M}_{m\times n}$, then A $\equiv_r$ B if and only if there is a sequence of elementary matrices, $E_1, \ldots, E_q$, such that $B = E_q \cdots E_2E_1A$.

■ **EXAMPLE 1.32.** Since

$$A = \begin{bmatrix} 1 & 2 & 3 \\ 4 & 5 & 6 \end{bmatrix} \longrightarrow \begin{bmatrix} 1 & 2 & 3 \\ 0 & -3 & -6 \end{bmatrix} \quad r_{12}(-4)$$

$$\longrightarrow \begin{bmatrix} 1 & 2 & 3 \\ 0 & 1 & 2 \end{bmatrix} \quad r_2(-1/3)$$

$$\longrightarrow \begin{bmatrix} 1 & 0 & -1 \\ 0 & 1 & 2 \end{bmatrix} \quad r_{21}(-2),$$

A is row equivalent to $B = \begin{bmatrix} 1 & 0 & -1 \\ 0 & 1 & 2 \end{bmatrix}$, and $B = E_{21}(-2)E_2(-1/3)E_{12}(-4)A$. That is,

$$\begin{bmatrix} 1 & 0 & -1 \\ 0 & 1 & 2 \end{bmatrix} = \begin{bmatrix} 1 & -2 \\ 0 & 1 \end{bmatrix} \begin{bmatrix} 1 & 0 \\ 0 & -1/3 \end{bmatrix} \begin{bmatrix} 1 & 0 \\ -4 & 1 \end{bmatrix} \begin{bmatrix} 1 & 2 & 3 \\ 4 & 5 & 6 \end{bmatrix}. \ \square$$

A square matrix of size n looks like

$$A = \begin{bmatrix} a_{11} & a_{12} & \cdots & a_{1n} \\ a_{21} & a_{22} & & \vdots \\ \vdots & & \ddots & \vdots \\ a_{n1} & a_{n2} & \cdots & a_{nn} \end{bmatrix}.$$

The numbers a_{ii}, $1 \le i \le n$, are the *diagonal* entries; they constitute the so-called *main diagonal* of A. When all the entries off the main diagonal are zeros, A is said to be *diagonal*. For example, the general 3×3 diagonal matrix is

$$\begin{bmatrix} a & 0 & 0 \\ 0 & b & 0 \\ 0 & 0 & c \end{bmatrix},$$

$a,b,c \in \mathcal{R}$. Observe that A is diagonal if and only if $a_{ij} = 0$ whenever $i \neq j$.

A square matrix whose entries below the main diagonal are zeros is said to be *upper triangular*. Formally, $A \in \mathcal{M}_{n\times n}$ is upper triangular if and only if $a_{ij} = 0$ for $1 \le j < i \le n$. *Lower triangular* matrices are defined in a similar manner. A diagonal matrix is both upper and lower triangular.

■ **EXAMPLE 1.33.** $\begin{bmatrix} 1 & 0 & 0 \\ 0 & 2 & 0 \\ 0 & 0 & 3 \end{bmatrix}$ and $\begin{bmatrix} 3 & 0 & 0 & 0 \\ 0 & -2 & 0 & 0 \\ 0 & 0 & 0 & 0 \\ 0 & 0 & 0 & 1 \end{bmatrix}$ are diagonal, whereas $\begin{bmatrix} 1 & 2 \\ 0 & 3 \end{bmatrix}$ and $\begin{bmatrix} 2 & 0 & 0 \\ -1 & 1 & 0 \\ 3 & -2 & 3 \end{bmatrix}$ are upper and lower triangular, respectively. □

THEOREM 1.11. Assume $A, B \in \mathcal{M}_{n\times n}$ and $t \in \mathcal{R}$.

(1) If A and B are diagonal, then so are $A + B$, tA, and AB.
(2) If A and B are upper triangular, then so are $A + B$, tA, and AB.

Proof. Suppose A and B are diagonal. For $i \neq j$, $[A]_{ij} = 0 = [B]_{ij}$, so

$$[A + B]_{ij} = [A]_{ij} + [B]_{ij} = 0,$$

and

$$[tA]_{ij} = t[A]_{ij} = 0.$$

The same steps, with $i \neq j$ replaced by $1 \le j < i \le n$, show that the sum of two upper triangular matrices and a scalar multiple of an upper triangular matrix are upper triangular. Justification for the conclusions regarding AB is left to the exercises. ■

Exercises 1.5

1. Which of the following are elementary matrices? Describe the ones that are in terms of the notation E_{ij}, $E_i(t)$, and $E_{ij}(t)$.

a. $\begin{bmatrix} 0 & 1 \\ 1 & 0 \end{bmatrix}$ b. $\begin{bmatrix} 1 & 4 \\ 0 & 1 \end{bmatrix}$ **c.** $\begin{bmatrix} 0 & 0 \\ 0 & 1 \end{bmatrix}$ d. $\begin{bmatrix} 1 & 1 \\ 1 & 1 \end{bmatrix}$ **e.** $\begin{bmatrix} -2 & 0 \\ 0 & 1 \end{bmatrix}$

f. $\begin{bmatrix} 0 & 0 & 1 \\ 0 & 1 & 0 \\ 1 & 0 & 0 \end{bmatrix}$ **g.** $\begin{bmatrix} 1 & 0 & -3 \\ 0 & 1 & 0 \\ 0 & 0 & 1 \end{bmatrix}$ h. $\begin{bmatrix} 1 & 0 & 0 \\ 0 & 1 & 0 \\ 0 & 0 & 1 \end{bmatrix}$ **i.** $\begin{bmatrix} 1 & 0 & 0 \\ 4 & 1 & 0 \\ 0 & 0 & 1 \end{bmatrix}$

j. $\begin{bmatrix} 0 & 0 & 0 & 1 \\ 0 & 0 & 1 & 0 \\ 0 & 1 & 0 & 0 \\ 1 & 0 & 0 & 0 \end{bmatrix}$ **k.** $\begin{bmatrix} 1 & 0 & 0 & 0 \\ 0 & 0 & 1 & 0 \\ 0 & 1 & 0 & 0 \\ 0 & 0 & 0 & 1 \end{bmatrix}$ l. $\begin{bmatrix} 1 & 0 & 0 & 0 \\ 0 & 1 & 0 & 0 \\ 0 & 0 & 1 & 0 \\ 5 & 0 & 0 & 1 \end{bmatrix}$

m. $\begin{bmatrix} 1 & 0 & 0 & 0 \\ 0 & 2 & 0 & 0 \\ 0 & 0 & 3 & 0 \\ 0 & 0 & 0 & 4 \end{bmatrix}$

2. Assuming $A = \begin{bmatrix} 1 & 2 \\ 3 & 4 \\ 5 & 6 \end{bmatrix}$, find an elementary matrix E such that EA is

a. $\begin{bmatrix} 5 & 6 \\ 3 & 4 \\ 1 & 2 \end{bmatrix}$. b. $\begin{bmatrix} -2 & -2 \\ 3 & 4 \\ 5 & 6 \end{bmatrix}$. **c.** $\begin{bmatrix} 1 & 2 \\ -6 & -8 \\ 5 & 6 \end{bmatrix}$. d. $\begin{bmatrix} 1 & 2 \\ 1 & 0 \\ 5 & 6 \end{bmatrix}$.

3. Assuming $A = \begin{bmatrix} 1 & 1 & 1 \\ 2 & 2 & 2 \\ 1 & 1 & 1 \\ 2 & 2 & 2 \end{bmatrix}$, find all elementary matrices E such that EA is

a. $\begin{bmatrix} 1 & 1 & 1 \\ 0 & 0 & 0 \\ 1 & 1 & 1 \\ 2 & 2 & 2 \end{bmatrix}$. b. $\begin{bmatrix} 1 & 1 & 1 \\ 1 & 1 & 1 \\ 1 & 1 & 1 \\ 2 & 2 & 2 \end{bmatrix}$. **c.** $\begin{bmatrix} 2 & 2 & 2 \\ 1 & 1 & 1 \\ 1 & 1 & 1 \\ 2 & 2 & 2 \end{bmatrix}$. d. $\begin{bmatrix} 1 & 1 & 1 \\ 2 & 2 & 2 \\ -1 & -1 & -1 \\ 2 & 2 & 2 \end{bmatrix}$.

4. Suppose $A = \begin{bmatrix} 1 & 3 & 5 \\ 2 & 4 & 6 \end{bmatrix}$ and $B = \text{rref}(A)$.
   a. Find elementary matrices E_1, E_2, E_3, such that $B = E_3E_2E_1A$.
   b. Find elementary matrices E'_1, E'_2, E'_3, such that $A = E'_3E'_2E'_1B$.

5. Write a formal proof of the case $E = E_i(t)$ in Theorem 1.9.

6. Fill in the blanks.
   a. $E_{ij}^2 =$ _____
   b. $E_i(1/t)E_i(t) =$ _____
   c. $E_i(t)^2 = E_i($_____$)$
   d. $E_i(s)E_i(t) = E_i($_____$)$
   e. $E_{ij}(-3)E_{ij}(2) = E_{ij}($_____$)$
   f. $E_{ij}(t)^2 = E_{ij}($_____$)$
   g. $E_i(-1)E_{ij}(1)E_{ji}(-1)E_{ij}(1) =$ _____

7. Can $\begin{bmatrix} 1 & 1 \\ 1 & 1 \end{bmatrix}$ be written as a product of elementary matrices? (Hint: refer to Theorem 1.10.)

8. Let $A = \begin{bmatrix} a & 0 & 0 \\ 0 & b & 0 \\ 0 & 0 & c \end{bmatrix}$, $a,b,c \in \mathcal{R}$ and suppose $B \in \mathcal{M}_{3\times n}$. Describe the rows of AB in terms of the rows of B.

9. Assume $A,B \in \mathcal{M}_{m\times n}$ and $C \in \mathcal{M}_{n\times p}$. Show that if $A \equiv_r B$, then $AC \equiv_r BC$.

10. Which of the following matrices are diagonal? Upper triangular? Lower triangular?

a. $\begin{bmatrix} 0 & 0 \\ 0 & 0 \end{bmatrix}$ b. $\begin{bmatrix} 0 & 0 & 0 \\ 1 & 2 & 0 \\ 3 & 4 & 0 \end{bmatrix}$ **c.** $\begin{bmatrix} 1 & 2 & 4 & 0 \\ 0 & 2 & 1 & 0 \\ 0 & 0 & 3 & 0 \end{bmatrix}$ d. $\begin{bmatrix} 0 & 0 & 1 \\ 0 & 2 & 3 \\ 4 & 5 & 6 \end{bmatrix}$

e. I_n f. $\begin{bmatrix} 0 & 1 & 2 & 3 \\ 0 & 0 & 1 & 2 \\ 0 & 0 & 0 & 1 \\ 0 & 0 & 0 & 0 \end{bmatrix}$ **g.** $\begin{bmatrix} 0 & 0 & 0 & 0 \\ 0 & 2 & 0 & 0 \\ 0 & 0 & 2 & 0 \\ 0 & 0 & 0 & 0 \end{bmatrix}$ h. $\begin{bmatrix} 1 & 0 & 0 & 1 \\ 0 & 1 & 0 & 0 \\ 0 & 0 & 1 & 0 \\ 1 & 0 & 0 & 1 \end{bmatrix}$

11. **a.** Show that if A and B are $n \times n$ diagonal matrices, then AB is diagonal. What are the diagonal entries of AB?
    b. Show that if A and B are $n \times n$ upper triangular matrices, then AB is upper triangular. What are the diagonal entries of AB?

12. Find all upper triangular 3×3 matrices that commute with $\begin{bmatrix} 1 & -1 & 1 \\ 0 & 1 & -1 \\ 0 & 0 & 1 \end{bmatrix}$.

13. The sum of the diagonal entries of $A \in \mathcal{M}_{n \times n}$ is called the *trace* of A, that is,

$$\text{trace}(A) = \sum_{i=1}^{n} a_{ii}.$$

Assuming $A,B \in \mathcal{M}_{n \times n}$ and $t \in \mathcal{R}$, show that
a. $\text{trace}(tA) = t[\text{trace}(A)]$.
b. $\text{trace}(A + B) = \text{trace}(A) + \text{trace}(B)$.
c. $\text{trace}(AB) = \text{trace}(BA)$.

1.6 INVERTIBILITY

One of the primary features of the real number system is the fact that every nonzero real number has a unique multiplicative inverse, that is, if $a \in \mathcal{R}$ and $a \neq 0$, then there is a unique $c \in \mathcal{R}$ such that $ca = 1$. This c, sometimes called the reciprocal of a, is denoted by a^{-1}. The practical significance of this property is that

$$ax = b, \quad a \neq 0, \tag{1.7}$$

has a unique solution. You obtain it by multiplying both sides of (1.7) by a^{-1}, the formal steps being

$$x = 1x = (a^{-1}a)x = a^{-1}(ax) = a^{-1}b.$$

Note that (1.7) is the general linear equation in one unknown.

Matrix multiplication lends itself to expressing a linear system by an equation of the same form as (1.7), namely,

$$AX = B,$$

where $A \in \mathcal{M}_{m \times n}$, $X \in \mathcal{M}_{n \times 1}$, and $B \in \mathcal{M}_{m \times 1}$. If A had a multiplicative inverse, then the system could be solved by the same algebraic steps as used for (1.7). The next project is to explore the issues involved here.

How does one formulate a notion of multiplicative inverse for a matrix A? In $\mathcal{R}$, the idea is based on the equation

$$ca = 1.$$

Of course, 1 is the multiplicative identity in $\mathcal{R}$, and it has a matrix analogue, the identity matrix. In $\mathcal{R}$, if $ca = 1$, then $ac = 1$, because multiplication of real numbers is commutative. This commutativity, however, cannot be taken for granted in the matrix context, so perhaps one should consider the conditions

$$\mathrm{CA} = \mathrm{I}_n \quad \text{and} \quad \mathrm{AC} = \mathrm{I}_n.$$

Observe that for these equations to make sense dimensionally, A and C must be $n \times n$ matrices. Thus, the notion of multiplicative inverse for A is reserved for the case when A is square.

DEFINITION. $\mathrm{A} \in \mathcal{M}_{n\times n}$ is said to be *invertible* or *nonsingular* if there is a $\mathrm{C} \in \mathcal{M}_{n\times n}$ such that

$$\mathrm{CA} = \mathrm{I}_n = \mathrm{AC},$$

and a C with this property is called a *multiplicative inverse* (or simply an *inverse*) of A. If there is no such C, then A is said to be *singular*.

Since matrix multiplication is a fairly complicated operation, a C satisfying $\mathrm{CA} = \mathrm{I}_n = \mathrm{AC}$ is not likely to be found by inspection. The next example illustrates a direct approach to obtaining such a matrix.

■ **EXAMPLE 1.34.** Let $\mathrm{A} = \begin{bmatrix} 1 & 2 \\ 2 & 3 \end{bmatrix}$ and set $\mathrm{C} = \begin{bmatrix} a & b \\ c & d \end{bmatrix}$. To be an inverse of A, C must satisfy

$$\begin{bmatrix} a & b \\ c & d \end{bmatrix}\begin{bmatrix} 1 & 2 \\ 2 & 3 \end{bmatrix} = \begin{bmatrix} 1 & 0 \\ 0 & 1 \end{bmatrix} = \begin{bmatrix} 1 & 2 \\ 2 & 3 \end{bmatrix}\begin{bmatrix} a & b \\ c & d \end{bmatrix}.$$

Comparing entries across the left equality sign produces

$$\begin{aligned} a + 2b &= 1 \\ 2a + 3b &= 0 \\ c + 2d &= 0 \\ 2c + 3d &= 1, \end{aligned}$$

and row reducing the augmented matrix of this system yields

$$\left[\begin{array}{cccc|c} 1 & 2 & 0 & 0 & 1 \\ 2 & 3 & 0 & 0 & 0 \\ 0 & 0 & 1 & 2 & 0 \\ 0 & 0 & 2 & 3 & 1 \end{array}\right] \longrightarrow \left[\begin{array}{cccc|c} 1 & 2 & 0 & 0 & 1 \\ 0 & -1 & 0 & 0 & -2 \\ 0 & 0 & 1 & 2 & 0 \\ 0 & 0 & 0 & -1 & 1 \end{array}\right] \begin{array}{l} r_{12}(-2) \\ r_{34}(-2) \end{array}$$

$$\longrightarrow \left[\begin{array}{cccc|c} 1 & 0 & 0 & 0 & -3 \\ 0 & -1 & 0 & 0 & -2 \\ 0 & 0 & 1 & 0 & 2 \\ 0 & 0 & 0 & -1 & 1 \end{array}\right] \begin{array}{l} r_{21}(2) \\ r_{43}(2) \end{array}$$

$$\longrightarrow \left[\begin{array}{cccc|c} 1 & 0 & 0 & 0 & -3 \\ 0 & 1 & 0 & 0 & 2 \\ 0 & 0 & 1 & 0 & 2 \\ 0 & 0 & 0 & 1 & -1 \end{array}\right] \begin{array}{l} r_2(-1) \\ r_4(-1) \end{array}.$$

The system has the solution $a=-3$, $b=2$, $c=2$, and $d=-1$, so the left half of $CA=I_2=AC$ is satisfied by

$$C=\begin{bmatrix} -3 & 2 \\ 2 & -1 \end{bmatrix}.$$

The right half can then be checked directly. Indeed,

$$AC=\begin{bmatrix} 1 & 2 \\ 2 & 3 \end{bmatrix}\begin{bmatrix} -3 & 2 \\ 2 & -1 \end{bmatrix}=\begin{bmatrix} 1 & 0 \\ 0 & 1 \end{bmatrix},$$

so C satisfies both conditions required of an inverse of A. Note that the system that produced C has only one solution, so A has a unique inverse. □

■ **EXAMPLE 1.35.** If $A=\begin{bmatrix} 1 & 0 \\ 2 & 0 \end{bmatrix}$ and $C=\begin{bmatrix} a & b \\ c & d \end{bmatrix}$, then

$$CA=\begin{bmatrix} a & b \\ c & d \end{bmatrix}\begin{bmatrix} 1 & 0 \\ 2 & 0 \end{bmatrix}=\begin{bmatrix} a+2b & 0 \\ c+2d & 0 \end{bmatrix}.$$

There is no choice of a, b, c, and d such that $CA=I_2$, so A is singular. □

As you might suspect, given the outcome of Example 1.34, a nonsingular matrix can have only one inverse. This, in fact, is a direct consequence of the definition of invertibility. If $A \in \mathcal{M}_{n\times n}$ is nonsingular and C and D are both inverses of A, then $CA=I_n=AC$ and $DA=I_n=AD$, so

$$C=CI_n=C(AD)=(CA)D=I_nD=D.$$

The unique inverse of A is denoted by A^{-1}.

Suppose $A \in \mathcal{M}_{n\times n}$ is nonsingular and consider the linear system

$$AX=B, \tag{1.8}$$

where $B \in \mathcal{M}_{n\times 1}$. Substituting $X=A^{-1}B$ yields

$$A(A^{-1}B)=(AA^{-1})B=I_nB=B,$$

so $A^{-1}B$ is a solution of (1.8). Moreover, if X_0 is a solution of (1.8), then multiplying both sides on the left by A^{-1} produces

$$X_0=I_nX_0=(A^{-1}A)X_0=A^{-1}(AX_0)=A^{-1}B,$$

so $A^{-1}B$ is the only solution. Thus, nonsingularity of A implies (1.8) has a unique solution, regardless of the choice of B. That solution can be obtained by row reducing $[A \mid B]$. Since each column of rref(A) contains a leading one (Theorem 1.3), rref(A), being square, must be I_n. Thus, the unique solution, X_0, is generated by the row reduction

$$[A \mid B] \longrightarrow [I_n \mid X_0]. \tag{1.9}$$

These observations lead to a procedure for calculating the inverse of A. Assuming $A \in \mathcal{M}_{n\times n}$ is nonsingular, consider the matrix equation $AA^{-1}=I_n$. Comparing columns of both sides you get

$$A\,\mathrm{Col}_j(A^{-1})=\mathrm{Col}_j(AA^{-1})=\mathrm{Col}_j(I_n),$$

$1 \le j \le n$ (Theorem 1.8), that is, $X_0 = \mathrm{Col}_j(A^{-1})$ is the unique solution of $AX = \mathrm{Col}_j(I_n)$. By (1.9), with $B = \mathrm{Col}_j(I_n)$ and $X_0 = \mathrm{Col}_j(A^{-1})$,

$$[A \mid \mathrm{Col}_j(I_n)] \longrightarrow [I_n \mid \mathrm{Col}_j(A^{-1})].$$

There are n such calculations to be made (one for each value of j), each producing a different column of A^{-1}. The same row operations are used each time to reduce A to I_n, so the n calculations can be performed simultaneously by

$$[A \mid \mathrm{Col}_1(I_n) \mid \cdots \mid \mathrm{Col}_n(I_n)] \longrightarrow [I_n \mid \mathrm{Col}_1(A^{-1}) \mid \cdots \mid \mathrm{Col}_n(A^{-1})].$$

Thus,

$$[A \mid I_n] \longrightarrow [I_n \mid A^{-1}]. \tag{1.10}$$

In other words, any sequence of row operations that transforms A into I_n also transforms I_n into A^{-1}.

The discussion above shows that (1.10) generates A^{-1} as long as A is known to be nonsingular, but it does not directly address the question of whether A is nonsingular. It is shown later (Theorem 1.14) that A is nonsingular if and only if $A \equiv_r I_n$. Thus, row reduction of $[A \mid I_n]$ either reveals that A is singular or establishes that A is nonsingular, and in the latter case it produces A^{-1}.

■ **EXAMPLE 1.36.** Consider $A = \begin{bmatrix} 1 & 2 & 3 \\ 0 & 1 & 1 \\ 2 & 0 & 1 \end{bmatrix}$. Since

$$\left[\begin{array}{ccc|ccc} 1 & 2 & 3 & 1 & 0 & 0 \\ 0 & 1 & 1 & 0 & 1 & 0 \\ 2 & 0 & 1 & 0 & 0 & 1 \end{array}\right] \longrightarrow \left[\begin{array}{ccc|ccc} 1 & 2 & 3 & 1 & 0 & 0 \\ 0 & 1 & 1 & 0 & 1 & 0 \\ 0 & -4 & -5 & -2 & 0 & 1 \end{array}\right] \; r_{13}(-2)$$

$$\longrightarrow \left[\begin{array}{ccc|ccc} 1 & 2 & 3 & 1 & 0 & 0 \\ 0 & 1 & 1 & 0 & 1 & 0 \\ 0 & 0 & -1 & -2 & 4 & 1 \end{array}\right] \; r_{23}(4)$$

$$\longrightarrow \left[\begin{array}{ccc|ccc} 1 & 2 & 0 & -5 & 12 & 3 \\ 0 & 1 & 0 & -2 & 5 & 1 \\ 0 & 0 & -1 & -2 & 4 & 1 \end{array}\right] \; \begin{array}{l} r_{32}(1) \\ r_{31}(3) \end{array}$$

$$\longrightarrow \left[\begin{array}{ccc|ccc} 1 & 0 & 0 & -1 & 2 & 1 \\ 0 & 1 & 0 & -2 & 5 & 1 \\ 0 & 0 & 1 & 2 & -4 & -1 \end{array}\right] \; \begin{array}{l} r_{21}(-2) \\ r_{3}(-1) \end{array},$$

A is nonsingular, and

$$A^{-1} = \begin{bmatrix} -1 & 2 & 1 \\ -2 & 5 & 1 \\ 2 & -4 & -1 \end{bmatrix}.$$

You can check the conclusion by multiplying the last matrix by A. □

■ **EXAMPLE 1.37.** Since $\left[\begin{array}{cc|cc} 2 & 4 & 1 & 0 \\ 3 & 6 & 0 & 1 \end{array}\right] \longrightarrow \left[\begin{array}{cc|cc} 1 & 2 & 1/2 & 0 \\ 0 & 0 & -3/2 & 1 \end{array}\right]$, $\begin{bmatrix} 2 & 4 \\ 3 & 6 \end{bmatrix}$ is singular. □

■ **EXAMPLE 1.38.** For 2×2 matrices there is a simple test for invertibility and a formula for the inverse. Suppose

$$\mathrm{A} = \begin{bmatrix} a & b \\ c & d \end{bmatrix}$$

is nonsingular. Then A is row equivalent to I_2, and consequently at least one of the entries in its first column is nonzero. Suppose $a \neq 0$. Then

$$\left[\begin{array}{cc|cc} a & b & 1 & 0 \\ c & d & 0 & 1 \end{array}\right] \longrightarrow \left[\begin{array}{cc|cc} 1 & b/a & 1/a & 0 \\ 0 & d-(cb/a) & -c/a & 1 \end{array}\right] \begin{array}{l} r_1(1/a) \\ r_{12}(-c) \end{array}$$

$$\longrightarrow \left[\begin{array}{cc|cc} 1 & b/a & 1/a & 0 \\ 0 & ad-bc & -c & a \end{array}\right] r_2(a) \qquad \left(\begin{array}{l}\text{Note that } ad-bc \neq 0, \\ \text{for otherwise } \mathrm{A} \not\equiv_r \mathrm{I}_2.\end{array}\right)$$

$$\longrightarrow \left[\begin{array}{cc|cc} 1 & b/a & 1/a & 0 \\ 0 & 1 & -c/(ad-bc) & a/(ad-bc) \end{array}\right] r_2(1/(ad-bc))$$

$$\longrightarrow \left[\begin{array}{cc|cc} 1 & 0 & d/(ad-bc) & -b/(ad-bc) \\ 0 & 1 & -c/(ad-bc) & a/(ad-bc) \end{array}\right] r_{21}(-b/a).$$

If $a = 0$, then $c \neq 0$, and in that case similar calculations give $ad - bc \neq 0$ and

$$\left[\begin{array}{cc|cc} a & b & 1 & 0 \\ c & d & 0 & 1 \end{array}\right] \longrightarrow \left[\begin{array}{cc|cc} 1 & 0 & d/(ad-bc) & -b/(ad-bc) \\ 0 & 1 & -c/(ad-bc) & a/(ad-bc) \end{array}\right].$$

Thus, nonsingularity of A implies $ad - bc \neq 0$ and

$$\mathrm{A}^{-1} = \frac{1}{(ad-bc)} \begin{bmatrix} d & -b \\ -c & a \end{bmatrix}. \tag{1.11}$$

Conversely, if $ad - bc \neq 0$, then direct calculation shows that

$$\frac{1}{(ad-bc)} \begin{bmatrix} d & -b \\ -c & a \end{bmatrix} \begin{bmatrix} a & b \\ c & d \end{bmatrix} = \begin{bmatrix} 1 & 0 \\ 0 & 1 \end{bmatrix} = \begin{bmatrix} a & b \\ c & d \end{bmatrix} \left(\frac{1}{(ad-bc)} \begin{bmatrix} d & -b \\ -c & a \end{bmatrix} \right),$$

so A is nonsingular with inverse (1.11). In particular,

$$\begin{bmatrix} 1 & 2 \\ 3 & 4 \end{bmatrix}^{-1} = -\frac{1}{2} \begin{bmatrix} 4 & -2 \\ -3 & 1 \end{bmatrix}. \ \square$$

There are some instances in which the inverse of a matrix can be found without resorting to calculation. Consider the elementary matrix $\mathrm{E}_i(t)$, obtained by multiplying $\mathrm{Row}_i(\mathrm{I}_n)$ by $t \neq 0$. You can change $\mathrm{E}_i(t)$ back to I_n by multiplying its ith row by $1/t$, which by Theorem 1.9 can be accomplished by multiplying $\mathrm{E}_i(t)$ on the left by $\mathrm{E}_i(1/t)$. That is,

$$\mathrm{E}_i(1/t)\mathrm{E}_i(t) = \mathrm{I}_n.$$

Similarly, $\mathrm{E}_i(t)\mathrm{E}_i(1/t) = \mathrm{I}_n$. Thus, $\mathrm{E}_i(t)$ is nonsingular and

$$\mathrm{E}_i(t)^{-1} = \mathrm{E}_i(1/t).$$

Similar considerations show that elementary matrices of types I and III are nonsingular.

THEOREM 1.12. Elementary matrices are nonsingular. Moreover,

(1) $E_{ij}^{-1} = E_{ij}$,
(2) $E_i(t)^{-1} = E_i(1/t)$, and
(3) $E_{ij}(t)^{-1} = E_{ij}(-t)$.

Proof.

(1) E_{ij} is the result of interchanging rows i and j of I_n. Multiplying E_{ij} on the left by E_{ij} has the effect of interchanging rows i and j of E_{ij}, so $E_{ij}E_{ij} = I_n$. In this case, reversing the order of the factors produces the same product. Thus, E_{ij} is nonsingular and $E_{ij}^{-1} = E_{ij}$.
(2) This conclusion was justified in the discussion preceding the theorem.
(3) The details in this case are left to the reader. ■

■ **EXAMPLE 1.39.**

$$\begin{bmatrix} 0 & 1 & 0 \\ 1 & 0 & 0 \\ 0 & 0 & 1 \end{bmatrix}^{-1} = \begin{bmatrix} 0 & 1 & 0 \\ 1 & 0 & 0 \\ 0 & 0 & 1 \end{bmatrix}, \quad \begin{bmatrix} 1 & 0 & 0 & 0 \\ 0 & 1 & 0 & 0 \\ 0 & 0 & 4 & 0 \\ 0 & 0 & 0 & 1 \end{bmatrix}^{-1} = \begin{bmatrix} 1 & 0 & 0 & 0 \\ 0 & 1 & 0 & 0 \\ 0 & 0 & 1/4 & 0 \\ 0 & 0 & 0 & 1 \end{bmatrix},$$

$$\begin{bmatrix} 1 & 0 & 3 \\ 0 & 1 & 0 \\ 0 & 0 & 1 \end{bmatrix}^{-1} = \begin{bmatrix} 1 & 0 & -3 \\ 0 & 1 & 0 \\ 0 & 0 & 1 \end{bmatrix}, \quad \begin{bmatrix} 0 & 1 \\ 1 & 0 \end{bmatrix}^{-1} = \begin{bmatrix} 0 & 1 \\ 1 & 0 \end{bmatrix},$$

and

$$\begin{bmatrix} 1 & 0 \\ -1/2 & 1 \end{bmatrix}^{-1} = \begin{bmatrix} 1 & 0 \\ 1/2 & 1 \end{bmatrix}. \square$$

The next theorem records some useful facts about nonsingular matrices.

THEOREM 1.13. If $A, B \in \mathcal{M}_{n\times n}$ are nonsingular, $t \in \mathcal{R}$, and $t \neq 0$, then

(1) A^{-1} is nonsingular and $(A^{-1})^{-1} = A$,
(2) tA is nonsingular and $(tA)^{-1} = (1/t)A^{-1}$, and
(3) AB is nonsingular and $(AB)^{-1} = B^{-1}A^{-1}$.

Proof.

(1) A^{-1} is nonsingular if and only if there is a $C \in \mathcal{M}_{n\times n}$ such that $CA^{-1} = I_n = A^{-1}C$. Since $C = A$ satisfies the latter equations, A^{-1} is nonsingular with inverse A.
(2) See Exercise 3.
(3) Since

$$(B^{-1}A^{-1})(AB) = B^{-1}(A^{-1}A)B = B^{-1}I_nB = B^{-1}B = I_n$$

and

$$(AB)(B^{-1}A^{-1}) = A(BB^{-1})A^{-1} = AI_nA^{-1} = AA^{-1} = I_n,$$

AB is nonsingular and its inverse is $B^{-1}A^{-1}$. ■

Statement (3) of Theorem 1.13 can be extended as follows. If $A_1, A_2, \ldots, A_q \in \mathcal{M}_{n\times n}$ are nonsingular, then so is $A_1A_2\cdots A_q$, and

$$(A_1A_2\cdots A_q)^{-1} = A_q^{-1}\cdots A_2^{-1}A_1^{-1}.$$

You can check this conclusion by simplifying the products

$$(A_q^{-1}\cdots A_2^{-1}A_1^{-1})(A_1A_2\cdots A_q) \quad \text{and} \quad (A_1A_2\cdots A_q)(A_q^{-1}\cdots A_2^{-1}A_1^{-1}).$$

Multiplying the factors in pairs from the inside out and utilizing the fact that $A_j^{-1}A_j = I_n = A_jA_j^{-1}$, $1 \leq j \leq q$, you see that both products reduce to I_n.

THEOREM 1.14. Assuming $A \in \mathcal{M}_{n\times n}$, the following are equivalent:

(1) A is nonsingular,
(2) there is a $C \in \mathcal{M}_{n\times n}$ such that $CA = I_n$,
(3) the homogeneous system $AX = 0$ has a unique solution (namely, $X = 0$),
(4) $A \equiv_r I_n$,
(5) A is a product of elementary matrices.

Proof.
(1) ⇒ (2) If A is nonsingular, then $A^{-1}A = I_n$, so $C = A^{-1}$.
(2) ⇒ (3) Assume $C \in \mathcal{M}_{n\times n}$ and $CA = I_n$. If $X \in \mathcal{M}_{n\times 1}$ and $AX = 0$, then

$$X = I_nX = (CA)X = C(AX) = C0 = 0.$$

(3) ⇒ (4) Suppose the only solution to $AX = 0$ is $X = 0$. Let $D = \text{rref}(A)$. By the Corollary to Theorem 1.3, each column of D contains a leading one, so D, being square, is the identity matrix. Thus, $A \equiv_r D = I_n$.
(4) ⇒ (5) Suppose $A \equiv_r I_n$. By Theorem 1.10, there are elementary matrices, $E_1, \ldots, E_q$, such that

$$I_n = E_q \cdots E_2E_1A.$$

Each elementary matrix is nonsingular, so the latter equation can be solved for A by multiplying on the left successively by $E_q^{-1}, \ldots, E_2^{-1}, E_1^{-1}$ to get

$$A = E_1^{-1}E_2^{-1} \cdots E_q^{-1}.$$

The conclusion then follows from the fact that the inverse of an elementary matrix is elementary (Theorem 1.12).
(5) ⇒ (1) Elementary matrices are nonsingular and a product of nonsingular matrices is nonsingular. ■

REMARK. According to Theorem 1.14, $CA = I_n$ by itself implies A is invertible. Then, multiplying on the right by A^{-1} gives $C = A^{-1}$. Thus, one can establish the nonsingularity of A and identify its inverse by verifying just one of the two conditions for invertibility.

THEOREM 1.15. $A \in \mathcal{M}_{n\times n}$ is nonsingular if and only if for each $B \in \mathcal{M}_{n\times 1}$ the linear system $AX = B$ has a unique solution. When A is nonsingular, the unique solution is $X_0 = A^{-1}B$.

Proof. It has already been pointed out that if A is nonsingular, then for each $B \in \mathcal{M}_{n\times 1}$, $AX = B$ has the unique solution $X_0 = A^{-1}B$. On the other hand, if $AX = B$ has a unique solution for every $B \in \mathcal{M}_{n\times 1}$, then in particular $AX = 0$ has a unique solution, so by Theorem 1.14(3), A is nonsingular. ■

THEOREM 1.16. Given $A,B \in \mathcal{M}_{m\times n}$, $A \equiv_r B$ if and only if there is a nonsingular $m \times m$ matrix P such that $B = PA$. If $A \equiv_r B$, then

$$[A \mid I_m] \longrightarrow [B \mid P].$$

Proof. Suppose $A \equiv_r B$. By Theorem 1.10, there are elementary matrices $E_1, \ldots, E_q$ such that $B = E_q \cdots E_1A$. Thus, $B = PA$, where $P = E_q \cdots E_1$, and by Theorem 1.14, P is nonsingular. Conversely, if $B = PA$ and P is nonsingular, then P is a product of elementary matrices, so $A \equiv_r B$. Observe that If $B = PA$, $P = E_q \cdots E_1$, then

$$P = E_q \cdots E_1 I_m,$$

so P is obtained by performing on I_m the same sequence of elementary row operations that transforms A into B, that is, $[A \mid I_m] \longrightarrow [B \mid P]$. ■

■ **EXAMPLE 1.40.** Since

$$A = \begin{bmatrix} 1 & 0 & 0 & 1 \\ 1 & 1 & 0 & 0 \\ 0 & 1 & 1 & 0 \\ 0 & 0 & 1 & 1 \end{bmatrix} \longrightarrow \begin{bmatrix} 1 & 0 & 0 & 1 \\ 0 & 1 & 0 & -1 \\ 0 & 1 & 1 & 0 \\ 0 & 0 & 1 & 1 \end{bmatrix} \; r_{12}(-1)$$

$$\longrightarrow \begin{bmatrix} 1 & 0 & 0 & 1 \\ 0 & 1 & 0 & -1 \\ 0 & 0 & 1 & 1 \\ 0 & 0 & 0 & 0 \end{bmatrix} \; \begin{matrix} r_{23}(-1) \\ r_{34}(-1) \end{matrix},$$

A is row equivalent to

$$B = \begin{bmatrix} 1 & 0 & 0 & 1 \\ 0 & 1 & 0 & -1 \\ 0 & 0 & 1 & 1 \\ 0 & 0 & 0 & 0 \end{bmatrix},$$

and $B = E_{34}(-1)E_{23}(-1)E_{12}(-1)A$. Thus,

$$P = E_{34}(-1)E_{23}(-1)E_{12}(-1)$$

$$= \begin{bmatrix} 1 & 0 & 0 & 0 \\ 0 & 1 & 0 & 0 \\ 0 & 0 & 1 & 0 \\ 0 & 0 & -1 & 1 \end{bmatrix} \begin{bmatrix} 1 & 0 & 0 & 0 \\ 0 & 1 & 0 & 0 \\ 0 & -1 & 1 & 0 \\ 0 & 0 & 0 & 1 \end{bmatrix} \begin{bmatrix} 1 & 0 & 0 & 0 \\ -1 & 1 & 0 & 0 \\ 0 & 0 & 1 & 0 \\ 0 & 0 & 0 & 1 \end{bmatrix}$$

$$= \begin{bmatrix} 1 & 0 & 0 & 0 \\ -1 & 1 & 0 & 0 \\ 1 & -1 & 1 & 0 \\ -1 & 1 & -1 & 1 \end{bmatrix}$$

is a nonsingular matrix such that $B = PA$. Observe also that

$$[A \mid I_4] = \left[\begin{array}{cccc|cccc} 1 & 0 & 0 & 1 & 1 & 0 & 0 & 0 \\ 1 & 1 & 0 & 0 & 0 & 1 & 0 & 0 \\ 0 & 1 & 1 & 0 & 0 & 0 & 1 & 0 \\ 0 & 0 & 1 & 1 & 0 & 0 & 0 & 1 \end{array}\right]$$

$$\longrightarrow \left[\begin{array}{cccc|cccc} 1 & 0 & 0 & 1 & 1 & 0 & 0 & 0 \\ 0 & 1 & 0 & -1 & -1 & 1 & 0 & 0 \\ 0 & 0 & 1 & 1 & 1 & -1 & 1 & 0 \\ 0 & 0 & 0 & 0 & -1 & 1 & -1 & 1 \end{array}\right] = [B \mid P]. \; \square$$

Exercises 1.6

1. Determine if the matrix is nonsingular and, if so, find its inverse.

a. $\begin{bmatrix} 0 & -3 \\ 2 & 1 \end{bmatrix}$ b. $\begin{bmatrix} 12 & 6 \\ 16 & 8 \end{bmatrix}$ **c.** $\begin{bmatrix} 5 & 4 \\ 3 & 2 \end{bmatrix}$

d. $\begin{bmatrix} 1 & 2 & 2 \\ 3 & 3 & 4 \\ 4 & 5 & 5 \end{bmatrix}$ e. $\begin{bmatrix} 1 & 2 & 3 \\ 2 & 0 & 2 \\ 3 & 2 & 1 \end{bmatrix}$ **f.** $\begin{bmatrix} 1 & 2 & 3 \\ 2 & 3 & 4 \\ 4 & 5 & 6 \end{bmatrix}$

g. $\begin{bmatrix} 1 & 1 & -1 & 0 \\ 0 & -2 & -2 & -1 \\ 2 & 2 & -1 & 3 \\ -1 & 0 & 2 & 1 \end{bmatrix}$ **h.** $\begin{bmatrix} 1 & 0 & 1 & 1 \\ 1 & 1 & 0 & 1 \\ 1 & 1 & 1 & 0 \\ 0 & 1 & 1 & 1 \end{bmatrix}$ i. $\begin{bmatrix} 3 & 2 & 1 \\ 0 & 2 & 1 \\ 0 & 0 & 1 \end{bmatrix}$

j. $\begin{bmatrix} 1 & 0 & 0 \\ 0 & 2 & 0 \\ 0 & 0 & 3 \end{bmatrix}$ k. $\begin{bmatrix} 0 & 0 & 1 \\ 0 & 2 & 0 \\ 3 & 0 & 0 \end{bmatrix}$ **l.** $\begin{bmatrix} 0 & 0 & 1 \\ 0 & 1 & 1 \\ 1 & 1 & 1 \end{bmatrix}$

2. Show that I_n is nonsingular and find its inverse.

3. Show that if $A \in \mathcal{M}_{n\times n}$ is nonsingular and $t \neq 0$, then tA is nonsingular and

$$(tA)^{-1} = (1/t)A^{-1}.$$

4. Assuming A, B, and C are nonsingular $n \times n$ matrices, simplify

$$\left[2C^{-1}[A(2B)]^{-1}A\right]^{-1}.$$

5. Assuming $A,B,C \in \mathcal{M}_{n\times n}$ are nonsingular, solve $[(2C)^{-1}B]^{-1} + B^{-1}A = I_n$ for C.

6. a. State conditions under which $\begin{bmatrix} a & 0 & 0 \\ 0 & b & 0 \\ 0 & 0 & c \end{bmatrix}$ is nonsingular, and find its inverse.

b. Generalize part (a) to obtain a theorem about invertibility of $n \times n$ diagonal matrices.

7. Suppose A and B are nonsingular $n \times n$ matrices.
   a. Show that $A + B$ need not be invertible.
   b. Assuming $A + B$ is nonsingular, show that $(A^{-1} + B^{-1})^{-1} = A(A + B)^{-1}B$.

8. Let A be the nonsingular 3×3 matrix with inverse $\begin{bmatrix} 7 & -4 & 3 \\ 2 & 1 & -5 \\ 6 & -3 & 2 \end{bmatrix}$. Solve $AX = B$ when

a. $B = \begin{bmatrix} 1 \\ 1 \\ 1 \end{bmatrix}$. b. $B = \begin{bmatrix} 3 \\ 2 \\ 1 \end{bmatrix}$. **c.** $B = \begin{bmatrix} -1 \\ 0 \\ 1 \end{bmatrix}$.

9. Suppose $A \in \mathcal{M}_{n\times n}$ is upper triangular.
   a. Under what conditions will A be nonsingular?
   b. Show that if A is nonsingular, then A^{-1} is also upper triangular.

c. What are the diagonal entries of A^{-1}?

10. Let $A = \begin{bmatrix} 0 & 0 & a \\ 0 & b & d \\ c & e & f \end{bmatrix}$.

a. Make up several specific matrices of this form and find their inverses.
b. What are necessary and sufficient conditions for A to be invertible?
c. Does A^{-1} have any special form?

11. Assume $A \in \mathcal{M}_{m \times m}$, $B, C \in \mathcal{M}_{m \times n}$, and $AB = AC$.
a. Show that if A is invertible, then $B = C$.
b. Find 2×2 matrices A, B, and C such that $AB = AC$ but $B \neq C$.

12. Assume $i \neq j$ and $A = E_{ji}(2)^{-1} E_{ij} E_j(-2) \in \mathcal{M}_{n \times n}$. Express A^{-1} as a product of elementary matrices.

13. Suppose $A, B \in \mathcal{M}_{n \times n}$, A is nonsingular, and $A \equiv_r B$. Show that B is nonsingular.

14. Show that if A and B are nonsingular $n \times n$ matrices, then $A \equiv_r B$.

15. Assuming $A \in \mathcal{M}_{n \times n}$, show that each condition implies A is singular.
a. A has a row of zeros.
b. A has a column of zeros.
c. A has two identical rows.
d. A has two identical columns.

16. Suppose A and B are nonsingular $n \times n$ matrices and assume A commutes with B. Show that A^{-1} commutes with B^{-1}.

17. Assume $A, B \in \mathcal{M}_{n \times n}$ are nonsingular. Show that if $(AB)^2 = A^2B^2$, then A commutes with B.

18. Find a nonsingular P such that $B = PA$ when

a. $A = \begin{bmatrix} 1 & 2 & 3 \\ 4 & 5 & 6 \end{bmatrix}$, $B = \begin{bmatrix} 1 & 0 & -1 \\ 0 & 1 & 2 \end{bmatrix}$. b. $A = \begin{bmatrix} 1 & 2 \\ -2 & -1 \\ 1 & 3 \end{bmatrix}$, $B = \begin{bmatrix} 1 & 0 \\ 0 & 1 \\ 0 & 0 \end{bmatrix}$.

c. $A = \begin{bmatrix} 0 & 0 & 1 \\ 0 & 1 & 1 \\ 1 & 1 & 1 \end{bmatrix}$, $B = \begin{bmatrix} 1 & 0 & 0 \\ 0 & 1 & 0 \\ 0 & 0 & 1 \end{bmatrix}$.

d. $A = \begin{bmatrix} -1 & 3 & 0 \\ 0 & 2 & 1 \\ 1 & -1 & 0 \\ 3 & 0 & -2 \end{bmatrix}$, $B = \begin{bmatrix} 1 & 0 & 0 \\ 0 & 1 & 0 \\ 0 & 0 & 1 \\ 0 & 0 & 0 \end{bmatrix}$.

e. $A = \begin{bmatrix} 1 & -1 & -3 & -2 & 7 & -1 \\ 1 & 2 & 0 & 1 & -3 & -2 \\ 2 & 0 & -4 & 0 & 6 & 2 \\ 2 & -1 & -5 & -3 & 11 & -4 \\ 1 & 1 & -1 & 1 & 0 & 1 \end{bmatrix}$, $B = \begin{bmatrix} 1 & 0 & -2 & 0 & 0 & 0 \\ 0 & 1 & 1 & 0 & 0 & 0 \\ 0 & 0 & 0 & 1 & 0 & 0 \\ 0 & 0 & 0 & 0 & 1 & 0 \\ 0 & 0 & 0 & 0 & 0 & 1 \end{bmatrix}$.

19. Find a nonsingular P such that B = PA when

a. $A = \begin{bmatrix} 1 & 2 \\ 2 & 3 \end{bmatrix}$, $B = \begin{bmatrix} -2 & 1 \\ 1 & 3 \end{bmatrix}$. b. $A = \begin{bmatrix} 1 & 1 & 3 \\ -2 & 1 & 0 \\ -1 & 1 & 1 \end{bmatrix}$, $B = \begin{bmatrix} 1 & -2 & -3 \\ 0 & 1 & 2 \\ -1 & 3 & 5 \end{bmatrix}$.

20. Suppose $A, B \in \mathcal{M}_{n \times n}$. Show that
  a. B singular implies AB is singular.
  b. A singular implies AB is singular.

1.7 THE TRANSPOSE

The custom of writing the linear system

$$\begin{gathered} a_{11}x_1 + a_{12}x_2 + \cdots + a_{1n}x_n = b_1 \\ \vdots \\ a_{m1}x_1 + a_{m2}x_2 + \cdots + a_{mn}x_n = b_m \end{gathered}$$

as

$$AX = B,$$

where

$$A = \begin{bmatrix} a_{11} & \cdots & a_{1n} \\ \vdots & & \vdots \\ a_{m1} & \cdots & a_{mn} \end{bmatrix}, \quad X = \begin{bmatrix} x_1 \\ \vdots \\ x_n \end{bmatrix}, \quad \text{and} \quad B = \begin{bmatrix} b_1 \\ \vdots \\ b_m \end{bmatrix},$$

had a pervasive influence on the development of the material in the previous sections. A great deal of attention was given to row features of matrices (elementary row operations, row echelon form, etc.) because the rows of $[A \,\vdots\, B]$ carry the information about the equations in the system.

Here is an alternative approach. If

$$X' = [x_1 \quad \cdots \quad x_n], \quad A' = \begin{bmatrix} a_{11} & \cdots & a_{m1} \\ \vdots & & \vdots \\ a_{1n} & \cdots & a_{mn} \end{bmatrix}, \quad \text{and} \quad B' = [b_1 \quad \cdots \quad b_m],$$

then

$$X'A' = [x_1 \quad \cdots \quad x_n] \begin{bmatrix} a_{11} & \cdots & a_{m1} \\ \vdots & & \vdots \\ a_{1n} & \cdots & a_{mn} \end{bmatrix} = \left[\sum_{k=1}^{n} a_{1k}x_k \quad \cdots \quad \sum_{k=1}^{n} a_{mk}x_k \right],$$

so the same system can be described by the matrix equation

$$X'A' = B'.$$

The coefficients in the various equations are now organized as columns in the coefficient matrix A′. To align them with the constant terms in their associated equations, the augmented matrix becomes

$$\begin{bmatrix} A' \\ \hline B' \end{bmatrix},$$

and column operations are then required to algebraically simplify it in a manner that preserves the connection with the equations. Notions like column echelon form and reduced column echelon form become natural objectives in the simplification process.

Although rows play the central role in the customary description of the system, each row notion has its corresponding column notion. The latter ideas can be developed along the same lines as for rows, but there is an alternative approach that utilizes what has already been accomplished.

Assume $A \in \mathcal{M}_{m\times n}$ and consider the following associated matrix: the entries in row k of the new matrix are the entries from the column k of A, in the same order. The new matrix is $n \times m$, and its ij-entry is the ji-entry of A. Operating on its rows corresponds to operating on columns of A.

DEFINITION. If $A \in \mathcal{M}_{m\times n}$, then the *transpose* of A, denoted by A^T, is the $n \times m$ matrix with entries

$$[A^T]_{ij} = [A]_{ji},$$

$1 \le i \le n$, $1 \le j \le m$.

■ **EXAMPLE 1.41.** If $A = \begin{bmatrix} 1 & 2 & 3 \\ 4 & 5 & 6 \end{bmatrix}$, $B = \begin{bmatrix} 1 & 0 & 2 \\ 0 & 3 & 0 \\ 4 & 0 & 5 \\ 0 & 6 & 0 \end{bmatrix}$, and $C = [1 \quad 2 \quad 3 \quad 4]$, then

$$A^T = \begin{bmatrix} 1 & 4 \\ 2 & 5 \\ 3 & 6 \end{bmatrix}, \quad B^T = \begin{bmatrix} 1 & 0 & 4 & 0 \\ 0 & 3 & 0 & 6 \\ 2 & 0 & 5 & 0 \end{bmatrix}, \quad \text{and} \quad C^T = \begin{bmatrix} 1 \\ 2 \\ 3 \\ 4 \end{bmatrix}. \square$$

The transpose of a row is a column, and vice versa. For $A \in \mathcal{M}_{m\times n}$,

$$\text{Row}_k(A^T) = \big[[A^T]_{k1} \quad \cdots \quad [A^T]_{km}\big] = \big[[A]_{1k} \quad \cdots \quad [A]_{mk}\big] = \big(\text{Col}_k(A)\big)^T,$$

and similarly,

$$\text{Col}_k(A^T) = \big(\text{Row}_k(A)\big)^T.$$

The next theorem lists the rules governing the interaction between this transposition process and the algebraic operations on matrices.

THEOREM 1.17. Suppose A and B are matrices and $t \in \mathcal{R}$. In each statement, assume the sizes of A and B are appropriate for the operations in which they are involved.

(1) $(A^T)^T = A$.
(2) $(A + B)^T = A^T + B^T$.
(3) $(tA)^T = tA^T$.
(4) $(AB)^T = B^T A^T$.

Proof.
(1) If $A \in \mathcal{M}_{m\times n}$, then $A^T \in \mathcal{M}_{n\times m}$, and $(A^T)^T \in \mathcal{M}_{m\times n}$. For $1 \le i \le m$ and $1 \le j \le n$,

$$[(A^T)^T]_{ij} = [A^T]_{ji} = [A]_{ij}.$$

(2) See Exercise 2.
(3) See Exercise 2.

(4) Assume $A \in \mathcal{M}_{m\times n}$, and $B \in \mathcal{M}_{n\times p}$. Then $AB \in \mathcal{M}_{m\times p}$, $(AB)^T \in \mathcal{M}_{p\times m}$, $A^T \in \mathcal{M}_{n\times m}$, $B^T \in \mathcal{M}_{p\times n}$, and $B^T A^T \in \mathcal{M}_{p\times m}$. For $1 \leq i \leq p$ and $1 \leq j \leq m$,

$$\begin{aligned}
[(AB)^T]_{ij} &= [AB]_{ji} && \text{(defn. of transpose)} \\
&= \sum_{k=1}^{n} [A]_{jk}[B]_{ki} && \text{(defn. of matrix mult.)} \\
&= \sum_{k=1}^{n} [A^T]_{kj}[B^T]_{ik} && \text{(defn. of transpose)} \\
&= \sum_{k=1}^{n} [B^T]_{ik}[A^T]_{kj} && \text{(algebra in } \mathcal{R}) \\
&= [B^T A^T]_{ij}. && \text{(defn. of matrix mult.)} \ \blacksquare
\end{aligned}$$

There are three types of elementary column operations:

I. c_{ij} denotes the interchange of columns i and j.
II. $c_i(t)$ stands for scalar multiplying column i by t.
III. $c_{ij}(t)$ refers to adding t times column i to column j.

Two $m \times n$ matrices A and B are said to be *column equivalent,* written $A \equiv_c B$, if one can be obtained from the other by a finite sequence of elementary column operations. The formal descriptions of column echelon form and reduced column echelon form are left for the reader to formulate, but the following example illustrates the process of column reducing a matrix to reduced column echelon form.

EXAMPLE 1.42.

$$\begin{bmatrix} 1 & 2 & -1 \\ 0 & 1 & -2 \\ 1 & -1 & 5 \\ 2 & 2 & 1 \end{bmatrix} \longrightarrow \begin{bmatrix} 1 & 0 & 0 \\ 0 & 1 & -2 \\ 1 & -3 & 6 \\ 2 & -2 & 3 \end{bmatrix} \begin{matrix} c_{12}(-2) \\ c_{13}(1) \end{matrix}$$

$$\longrightarrow \begin{bmatrix} 1 & 0 & 0 \\ 0 & 1 & 0 \\ 1 & -3 & 0 \\ 2 & -2 & 1 \end{bmatrix} \begin{matrix} c_{23}(2) \\ c_3(-1) \end{matrix}$$

$$\longrightarrow \begin{bmatrix} 1 & 0 & 0 \\ 0 & 1 & 0 \\ 1 & -3 & 0 \\ 0 & 0 & 1 \end{bmatrix} \begin{matrix} c_{32}(2) \\ c_{31}(-2) \end{matrix}. \ \square$$

The matrices that result from performing c_{ij}, $c_i(t)$, and $c_{ij}(t)$ on an identity matrix will be denoted by F_{ij}, $F_i(t)$, and $F_{ij}(t)$, respectively. They are called *elementary matrices,* and they have the following properties:

a. $F_{ij} = E_{ij}$. b. $F_i(t) = E_i(t)$. c. $F_{ij}(t) = E_{ji}(t)$.
d. $F_{ij}^T = F_{ij}$. e. $F_i(t)^T = F_i(t)$. f. $F_{ij}(t)^T = F_{ji}(t)$.

Proofs of these statements are omitted, but you can easily convince yourself informally that they are correct. For example, if $i < j$, then $F_{ij}(t)$ and $E_{ji}(t)$ both look like

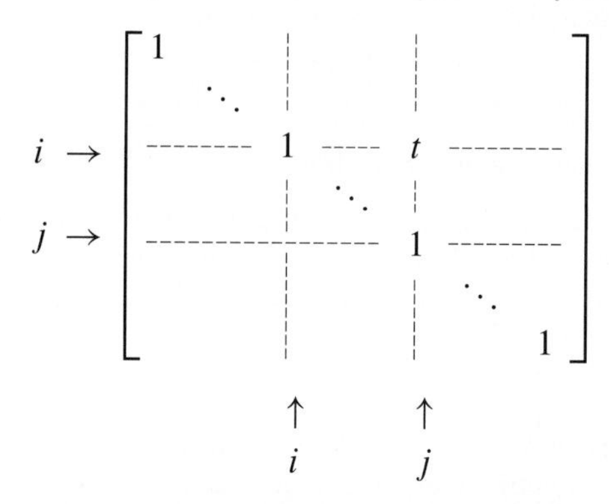

A column operation can be performed on A by multiplying by the associated elementary matrix. This time, however, you must multiply on the right.

THEOREM 1.18. If $A \in \mathcal{M}_{m\times n}$, c is an elementary column operation, and F is the elementary matrix obtained by performing c on I_n, then AF is the matrix that results from performing c on A.

Proof. Consider $F = F_{ij}$. By Theorem 1.17 and properties (a) through (f) above,

$$(AF_{ij})^T = F_{ij}^T A^T = F_{ij}A^T = E_{ij}A^T,$$

so $AF_{ij} = ((AF_{ij})^T)^T = (E_{ij}A^T)^T$. The conclusion then follows from Theorem 1.9 and the fact that the rows of A^T are the transposed columns of A. The conclusions in the other two cases are based on the following observations:

$$(AF_i(t))^T = F_i(t)^T A^T = F_i(t)A^T = E_i(t)A^T,$$

and

$$(AF_{ij}(t))^T = F_{ij}(t)^T A^T = F_{ji}(t)A^T = E_{ij}(t)A^T. \ \blacksquare$$

It is easy to see that F_{ij}, $F_i(t)$, and $F_{ij}(t)$ are nonsingular. In fact $F_{ij}^{-1} = F_{ij}$, $F_i(t)^{-1} = F_i(1/t)$, and $F_{ij}(t)^{-1} = F_{ij}(-t)$.

THEOREM 1.19. If $A \in \mathcal{M}_{m\times n}$ is nonsingular, then A^T is also nonsingular, and

$$(A^T)^{-1} = (A^{-1})^T.$$

Proof. Observe that $(A^{-1})^T A^T = (AA^{-1})^T = I_n^T = I_n$. Thus, by Theorem 1.14, A^T is nonsingular with inverse $(A^{-1})^T$. $\blacksquare$

The next two results are column analogues of Theorems 1.14 and 1.16, and their proofs are omitted.

THEOREM 1.20. Assuming $A \in \mathcal{M}_{n\times n}$, the following are equivalent:

(1) A is nonsingular.
(2) There is a $C \in \mathcal{M}_{n\times n}$ such that $AC = I_n$.
(3) If $Y \in \mathcal{M}_{1\times n}$ and $YA = 0$, then $Y = 0$.
(4) $A \equiv_c I_n$.
(5) A is a product of elementary matrices.

THEOREM 1.21. Suppose A,B $\in \mathcal{M}_{m\times n}$. Then A $\equiv_c$ B if and only if there is a nonsingular $n \times n$ matrix Q such that B = AQ. If A $\equiv_c$ B, then

$$\begin{bmatrix} A \\ \hline I_n \end{bmatrix} \longrightarrow \begin{bmatrix} B \\ \hline Q \end{bmatrix},$$

where the arrow indicates column reduction.

If A $\in \mathcal{M}_{n\times n}$, then $A^T \in \mathcal{M}_{n\times n}$. You can think of A^T as being obtained by reflecting A about its diagonal. The ij- and ji-entries in A are located symmetrically with respect to the diagonal, and passage from A to A^T amounts to interchanging these entries for each choice of i and j.

DEFINITION. Suppose A $\in \mathcal{M}_{n\times n}$. If $A^T = A$, then A is said to be *symmetric*. In terms of its entries, A is symmetric if and only if $a_{ji} = a_{ij}$, $1 \le i, j \le n$. A square matrix satisfying $A^T = -A$ is called *skew-symmetric* and is characterized by the conditions $a_{ji} = -a_{ij}$, $1 \le i, j \le n$.

Symmetry puts no restriction on the diagonal entries of A, but if A is skew-symmetric, then $a_{ii} = -a_{ii}$, $1 \le i \le n$, so the diagonal entries must be zeros. General symmetric and skew-symmetric 3×3 matrices look like

$$\begin{bmatrix} d & a & b \\ a & e & c \\ b & c & f \end{bmatrix} \quad \text{and} \quad \begin{bmatrix} 0 & a & b \\ -a & 0 & c \\ -b & -c & 0 \end{bmatrix},$$

respectively.

THEOREM 1.22. If A,B $\in \mathcal{M}_{n\times n}$ are symmetric and $t \in \mathcal{R}$, then A + B and tA are symmetric.

Proof. Since A and B are symmetric, $A^T = A$ and $B^T = B$. By Theorem 1.17,

$$(A + B)^T = A^T + B^T = A + B$$

and

$$(tA)^T = tA^T = tA. \blacksquare$$

Exercises 1.7

1. If $A = \begin{bmatrix} 1 & -1 & 2 \\ -2 & 3 & -3 \end{bmatrix}$, $B = \begin{bmatrix} 3 & 0 \\ 1 & 2 \\ 0 & 4 \end{bmatrix}$, $C = \begin{bmatrix} 1 & 1 & 1 \\ 0 & 1 & 1 \\ 0 & 0 & 1 \end{bmatrix}$, $D = \begin{bmatrix} 1 & 2 \\ 2 & 3 \end{bmatrix}$,

   and $X = \begin{bmatrix} x \\ y \end{bmatrix}$, find

   a. A^T. b. $A + B^T$. **c.** CC^T. d. X^TDX. **e.** B^TA^T.

2. Show that if A,B $\in \mathcal{M}_{m\times n}$ and $t \in \mathcal{R}$, then
   a. $(A + B)^T = A^T + B^T$. **b.** $(tA)^T = tA^T$.

3. Assuming A,B,C $\in \mathcal{M}_{n\times n}$, express $(ABC)^T$ in terms of A^T, B^T, and C^T.

4. Suppose A,B $\in \mathcal{M}_{n\times n}$ are symmetric. What additional restrictions on A and B are required for AB to be symmetric?

5. Show that if A $\in \mathcal{M}_{n\times n}$ is nonsingular and symmetric, then A^{-1} is symmetric.

6. Show that for A $\in \mathcal{M}_{n\times n}$,
   a. AA^T is symmetric. b. $A + A^T$ is symmetric.

7. Assuming A $\in \mathcal{M}_{n\times n}$, what can you conclude about $A^2, A^3, A^4, \ldots$ when
   a. A is symmetric? b. A is skew-symmetric?

8. List the four properties that characterize column reduced echelon form.

9. Find a matrix in column reduced echelon form that is column equivalent to

a. $\begin{bmatrix} 1 & 0 & 2 \\ -1 & 2 & -1 \\ -3 & 4 & -4 \\ 2 & -4 & -1 \end{bmatrix}$. b. $\begin{bmatrix} 1 & -1 & 0 & 1 \\ 2 & 0 & 2 & 2 \\ 3 & 1 & 2 & 1 \\ 4 & 2 & 6 & 5 \end{bmatrix}$. **c.** $\begin{bmatrix} 1 & 3 \\ 2 & 2 \\ 3 & 1 \end{bmatrix}$.

d. $\begin{bmatrix} 1 & 2 & -1 & 3 \\ 2 & 0 & 1 & -1 \end{bmatrix}$. **e.** $\begin{bmatrix} 2 & 1 & 1 \\ -1 & 1 & -2 \\ 2 & 3 & -1 \end{bmatrix}$. f. $\begin{bmatrix} -1 & 1 & 1 & -2 \\ 1 & -1 & -1 & 2 \\ 1 & 1 & 0 & 1 \\ -1 & 3 & 2 & -3 \end{bmatrix}$.

10. Find an elementary matrix F such that $\begin{bmatrix} 1 & 2 \\ 3 & 4 \\ 5 & 6 \end{bmatrix}$ F is

a. $\begin{bmatrix} 2 & 1 \\ 4 & 3 \\ 6 & 5 \end{bmatrix}$. b. $\begin{bmatrix} -1 & 2 \\ -1 & 4 \\ -1 & 6 \end{bmatrix}$. c. $\begin{bmatrix} 2 & 2 \\ 6 & 4 \\ 10 & 6 \end{bmatrix}$. d. $\begin{bmatrix} 1 & 0 \\ 3 & -2 \\ 5 & -4 \end{bmatrix}$.

11. Suppose $A = \begin{bmatrix} 1 & 2 & 0 & -1 \\ 0 & 0 & 1 & 3 \\ 0 & 0 & 0 & 0 \end{bmatrix}$ and $B = \begin{bmatrix} 1 & 0 & 0 & 0 \\ 0 & 1 & 0 & 0 \\ 0 & 0 & 0 & 0 \end{bmatrix}$.

    a. Find elementary matrices $F_1, \ldots, F_q$, such that $AF_1 \cdots F_q = B$.
    b. Find elementary matrices $F'_1, \ldots, F'_q$, such that $BF'_1 \cdots F'_q = A$.

12. For each A in Problem 9, find a nonsingular Q such that AQ is in column reduced echelon form.

13. Let U_i be the $n \times 1$ matrix with a one in the ith position and zeros elsewhere. Describe $U_iU_j^T, i \neq j$, and show that $E_{ji}(t) = I_n + tU_iU_j^T$.

14. Suppose A $\in \mathcal{M}_{n\times n}$ is nonsingular and B is the $n \times n$ matrix obtained by interchanging rows i and j of A ($i \neq j$). Show that B is nonsingular and that B^{-1} is obtained by interchanging columns i and j of A^{-1}.

15. Assuming A,B $\in \mathcal{M}_{m\times n}$, show that $A \equiv_r B$ if and only if $A^T \equiv_c B^T$.

1.8 PARTITIONED MATRICES

Sometimes it is useful to subdivide a matrix into smaller rectangular blocks of entries, called *submatrices,* by inserting horizontal and/or vertical separating lines. If

$$\mathrm{A} = \begin{bmatrix} 1 & 0 & -3 & 2 & 0 \\ 0 & 1 & 4 & 1 & 3 \\ 0 & -1 & 0 & 2 & 0 \\ -1 & 0 & -2 & 3 & 1 \end{bmatrix},$$

then

$$\left[\begin{array}{cc|ccc} 1 & 0 & -3 & 2 & 0 \\ 0 & 1 & 4 & 1 & 3 \\ \hline 0 & -1 & 0 & 2 & 0 \\ -1 & 0 & -2 & 3 & 1 \end{array}\right] \quad \text{and} \quad \left[\begin{array}{c|ccc|c} 1 & 0 & -3 & 2 & 0 \\ 0 & 1 & 4 & 1 & 3 \\ 0 & -1 & 0 & 2 & 0 \\ \hline -1 & 0 & -2 & 3 & 1 \end{array}\right]$$

are two such *partitionings* of A. The first, with submatrices

$$\mathrm{A}_{11} = \begin{bmatrix} 1 & 0 \\ 0 & 1 \end{bmatrix}, \quad \mathrm{A}_{12} = \begin{bmatrix} -3 & 2 & 0 \\ 4 & 1 & 3 \end{bmatrix}, \quad \mathrm{A}_{21} = \begin{bmatrix} 0 & -1 \\ -1 & 0 \end{bmatrix},$$

and

$$\mathrm{A}_{22} = \begin{bmatrix} 0 & 2 & 0 \\ -2 & 3 & 1 \end{bmatrix},$$

is written

$$\mathrm{A} = \begin{bmatrix} \mathrm{A}_{11} & \mathrm{A}_{12} \\ \mathrm{A}_{21} & \mathrm{A}_{22} \end{bmatrix}.$$

The second is described by

$$\mathrm{A} = \begin{bmatrix} \mathrm{A}_{11} & \mathrm{A}_{12} & \mathrm{A}_{13} \\ \mathrm{A}_{21} & \mathrm{A}_{22} & \mathrm{A}_{23} \end{bmatrix},$$

where

$$\mathrm{A}_{11} = \begin{bmatrix} 1 \\ 0 \\ 0 \end{bmatrix}, \qquad \mathrm{A}_{12} = \begin{bmatrix} 0 & -3 & 2 \\ 1 & 4 & 1 \\ -1 & 0 & 2 \end{bmatrix}, \qquad \mathrm{A}_{13} = \begin{bmatrix} 0 \\ 3 \\ 0 \end{bmatrix},$$

$$\mathrm{A}_{21} = [-1], \qquad \mathrm{A}_{22} = [0 \quad -2 \quad 3], \qquad \text{and} \qquad \mathrm{A}_{23} = [1].$$

In a general partitioning of an $m \times n$ matrix A, there are positive integers p and q and submatrices A_{ij}, $1 \le i \le p$, $1 \le j \le q$, such that

$$\mathrm{A} = \begin{bmatrix} \mathrm{A}_{11} & \cdots & \mathrm{A}_{1q} \\ \vdots & & \vdots \\ \mathrm{A}_{p1} & \cdots & \mathrm{A}_{pq} \end{bmatrix}.$$

Moreover, there are positive integers $m_1, \ldots, m_p$ and $n_1, \ldots, n_q$ such that A_{ij} has m_i rows and n_j columns, $1 \le j \le q$, $1 \le i \le p$,

$$m_1 + \cdots + m_p = m \quad \text{and} \quad n_1 + \cdots + n_q = n.$$

The sizes of the submatrices are indicated schematically by

$$\left[\begin{array}{c:c:c} m_1 \times n_1 & \cdots & m_1 \times n_q \\ \hdashline \vdots & & \vdots \\ \hdashline m_p \times n_1 & \cdots & m_p \times n_q \end{array}\right].$$

Suppose A and B are $m \times n$ matrices with identical partitionings, say

$$\mathrm{A} = \begin{bmatrix} \mathrm{A}_{11} & \cdots & \mathrm{A}_{1q} \\ \vdots & & \vdots \\ \mathrm{A}_{p1} & \cdots & \mathrm{A}_{pq} \end{bmatrix} \quad \text{and} \quad \mathrm{B} = \begin{bmatrix} \mathrm{B}_{11} & \cdots & \mathrm{B}_{1q} \\ \vdots & & \vdots \\ \mathrm{B}_{p1} & \cdots & \mathrm{B}_{pq} \end{bmatrix},$$

where A_{ij} and B_{ij} are $m_i \times n_j$ submatrices, $1 \le i \le p,\ 1 \le j \le q$. If $\mathrm{C} = \mathrm{A} + \mathrm{B}$, and C is partitioned in the same manner as A and B, then

$$\mathrm{C}_{ij} = \mathrm{A}_{ij} + \mathrm{B}_{ij}.$$

Similarly, if $t \in \mathcal{R}$ and $t\mathrm{A}$ is partitioned in the same manner as A, then the submatrix in the "i, j position" of $t\mathrm{A}$ is $t\mathrm{A}_{ij}$, $1 \le i \le p,\ 1 \le j \le q$.

The interaction between partitioning and matrix multiplication is a little more complicated. Assume $\mathrm{A} \in \mathcal{M}_{m\times n}$ and $\mathrm{B} \in \mathcal{M}_{n\times p}$. To begin, suppose A is partitioned by means of a single vertical subdivision. In that case, there is a positive integer q satisfying $1 \le q < n$ such that

$$\mathrm{A} = [\mathrm{A}_1 \mid \mathrm{A}_2],$$

where $\mathrm{A}_1 \in \mathcal{M}_{m\times q}$ and $\mathrm{A}_2 \in \mathcal{M}_{m\times(n-q)}$. This presents A as a "$1 \times 2$" matrix with "entries" A_1 and A_2. Now suppose B is partitioned with one horizontal subdivision, such as

$$\mathrm{B} = \left[\begin{array}{c} \mathrm{B}_1 \\ \hdashline \mathrm{B}_2 \end{array}\right].$$

The idea is to try to compute the product, AB, by the calculation

$$\mathrm{AB} = [\mathrm{A}_1 \mid \mathrm{A}_2] \left[\begin{array}{c} \mathrm{B}_1 \\ \hdashline \mathrm{B}_2 \end{array}\right] = \mathrm{A}_1\mathrm{B}_1 + \mathrm{A}_2\mathrm{B}_2.$$

The latter two products make sense only if $\mathrm{B}_1 \in \mathcal{M}_{q\times p}$ and $\mathrm{B}_2 \in \mathcal{M}_{(n-q)\times p}$, in which case $\mathrm{A}_1\mathrm{B}_1$ and $\mathrm{A}_2\mathrm{B}_2$ are both $m \times p$ matrices. Under these circumstances, the calculation is at least dimensionally feasible. To check that it actually works requires a comparison of the entries of AB and $\mathrm{A}_1\mathrm{B}_1 + \mathrm{A}_2\mathrm{B}_2$.

$$\mathrm{Row}_i(\mathrm{A}) \rightarrow \underset{\mathrm{A}_1 \qquad\qquad\qquad \mathrm{A}_2}{\left[\begin{array}{ccc:ccc} a_{i1} & \cdots & a_{iq} & a_{i,q+1} & \cdots & a_{in} \end{array}\right]} \underset{\uparrow \atop \mathrm{Col}_j(\mathrm{B})}{\left[\begin{array}{c} b_{1j} \\ \vdots \\ b_{qj} \\ \hdashline b_{q+1,j} \\ \vdots \\ b_{nj} \end{array}\right]} \begin{array}{c} \mathrm{B}_1 \\ \\ \\ \mathrm{B}_2 \end{array}$$

For $1 \leq i \leq m,\ 1 \leq j \leq p$,

$$\begin{aligned}\text{Row}_i(\text{A})\text{Col}_j(\text{B}) &= \sum_{k=1}^{n} a_{ik}b_{kj} = \sum_{k=1}^{q} a_{ik}b_{kj} + \sum_{k=q+1}^{n} a_{ik}b_{kj} \\ &= \text{Row}_i(\text{A}_1)\text{Col}_j(\text{B}_1) + \text{Row}_i(\text{A}_2)\text{Col}_j(\text{B}_2),\end{aligned}$$

and consequently,

$$[\text{AB}]_{ij} = [\text{A}_1\text{B}_1]_{ij} + [\text{A}_2\text{B}_2]_{ij} = [\text{A}_1\text{B}_1 + \text{A}_2\text{B}_2]_{ij}.$$

■ **EXAMPLE 1.43.** Suppose

$$\text{A} = \left[\begin{array}{rrr:r} 1 & 1 & -1 & 1 \\ 2 & -1 & 1 & -1 \end{array}\right] = [\text{A}_1 \mid \text{A}_2]$$

and

$$\text{B} = \left[\begin{array}{rrr} 1 & 0 & 1 \\ 3 & -1 & 2 \\ 0 & 1 & 1 \\ \hdashline -1 & 0 & 2 \end{array}\right] = \left[\begin{array}{c} \text{B}_1 \\ \hdashline \text{B}_2 \end{array}\right].$$

Then

$$\text{AB} = \begin{bmatrix} 1 & 1 & -1 & 1 \\ 2 & -1 & 1 & -1 \end{bmatrix}\begin{bmatrix} 1 & 0 & 1 \\ 3 & -1 & 2 \\ 0 & 1 & 1 \\ -1 & 0 & 2 \end{bmatrix} = \begin{bmatrix} 3 & -2 & 4 \\ 0 & 2 & -1 \end{bmatrix},$$

and

$$\begin{aligned}\text{A}_1\text{B}_1 + \text{A}_2\text{B}_2 &= \begin{bmatrix} 1 & 1 & -1 \\ 2 & -1 & 1 \end{bmatrix}\begin{bmatrix} 1 & 0 & 1 \\ 3 & -1 & 2 \\ 0 & 1 & 1 \end{bmatrix} + \begin{bmatrix} 1 \\ -1 \end{bmatrix}[-1 \quad 0 \quad 2] \\ &= \begin{bmatrix} 4 & -2 & 2 \\ -1 & 2 & 1 \end{bmatrix} + \begin{bmatrix} -1 & 0 & 2 \\ 1 & 0 & -2 \end{bmatrix} \\ &= \begin{bmatrix} 3 & -2 & 4 \\ 0 & 2 & -1 \end{bmatrix}. \square\end{aligned}$$

These ideas extend easily to the case when A has more than one vertical subdivision. The only requirement is that B have correspondingly placed horizontal subdivisions. For example, suppose

$$\text{A} = [\underset{m\times n_1}{\text{A}_1} \mid \underset{m\times n_2}{\text{A}_2} \mid \cdots \mid \underset{m\times n_q}{\text{A}_q}],$$

where A_k is an $m \times n_k$ matrix, $k = 1, \ldots, q$, and $n_1 + n_2 + \cdots + n_q = n$. If

$$\text{B} = \left[\begin{array}{c} \text{B}_1 \\ \hdashline \text{B}_2 \\ \hdashline \vdots \\ \hdashline \text{B}_q \end{array}\right] \begin{array}{l} n_1\times p \\ n_2\times p \\ \\ n_q\times p \end{array},$$

where B_k is an $n_k \times p$ matrix, then $\text{A}_k\text{B}_k \in \mathcal{M}_{m\times p}$ for each k, and

$$\text{AB} = \sum_{k=1}^{q} \text{A}_k\text{B}_k.$$

As long as the vertical partitioning of A agrees with the horizontal partitioning of B, AB can be computed in this manner. That is, the submatrices can be treated as if they were entries in a "row" and "column," respectively. If A also has horizontal subdivisions and B has vertical subdivisions, then there are more "rows" and "columns," but the sizes of the submatrices are still appropriate for calculating AB as if they were entries. The general situation is described as follows.

Suppose $A \in \mathcal{M}_{m\times n}$, $B \in \mathcal{M}_{n\times p}$,

$$A = \begin{bmatrix} A_{11} & \cdots & A_{1q} \\ \vdots & & \vdots \\ A_{r1} & \cdots & A_{rq} \end{bmatrix}, \quad \text{and} \quad B = \begin{bmatrix} B_{11} & \cdots & B_{1s} \\ \vdots & & \vdots \\ B_{q1} & \cdots & B_{qs} \end{bmatrix}.$$

Assume A_{ik} is an $m_i \times n_k$ matrix, $1 \le i \le r$, $1 \le k \le q$, B_{kj} is an $n_k \times p_j$ matrix, $1 \le k \le q$, $1 \le j \le s$, and

$$m_1 + \cdots + m_r = m, \quad n_1 + \cdots + n_q = n, \quad \text{and} \quad p_1 + \cdots + p_s = p.$$

The sizes of the "entries" in the ith "row" of the partitioned matrix A and in the jth "column" of the partitioned matrix B are given schematically by

$$\begin{bmatrix} m_i\times n_1 & m_i\times n_2 & \cdots & m_i\times n_q \end{bmatrix} \begin{bmatrix} n_1\times p_j \\ n_2\times p_j \\ \vdots \\ n_q\times p_j \end{bmatrix} = \begin{bmatrix} & & \\ & m_i\times p_j & \\ & & \end{bmatrix}.$$

Let $C = AB$ and suppose

$$C = \begin{bmatrix} C_{11} & \cdots & C_{1s} \\ \vdots & & \vdots \\ C_{r1} & \cdots & C_{rs} \end{bmatrix},$$

where C_{ij} is an $m_i \times p_j$ matrix, $1 \le i \le r$, $1 \le j \le s$. Then

$$C_{ij} = \sum_{k=1}^{q} A_{ik}B_{kj}.$$

■ **EXAMPLE 1.44.** Let

$$A = \left[\begin{array}{ccc|c|cc} 1 & 0 & 0 & 0 & 0 & -1 \\ 0 & 1 & 0 & 0 & 1 & 0 \\ 0 & 0 & 1 & 0 & 0 & -1 \\ \hline 0 & 0 & 0 & 1 & 0 & 1 \\ 0 & 0 & 0 & 1 & 1 & 0 \end{array}\right] = \begin{bmatrix} A_{11} & A_{12} & A_{13} \\ A_{21} & A_{22} & A_{23} \end{bmatrix}$$

and

$$B = \left[\begin{array}{c|ccc} 1 & 1 & 0 & 2 \\ 1 & 0 & 3 & 0 \\ 1 & 4 & 0 & 5 \\ \hline 1 & 0 & 0 & 0 \\ \hline 0 & 1 & 2 & 3 \\ 0 & 3 & 2 & 1 \end{array}\right] = \begin{bmatrix} B_{11} & B_{12} \\ B_{21} & B_{22} \\ B_{31} & B_{32} \end{bmatrix}.$$

The vertical partitioning of A matches the horizontal partitioning of B. As partitioned matrices with matrix entries, A is 2×3, B is 3×2, and $C = AB$ is 2×2. If

$$C = \begin{bmatrix} C_{11} & C_{12} \\ C_{21} & C_{22} \end{bmatrix},$$

then

$$C_{11} = A_{11}B_{11} + A_{12}B_{21} + A_{13}B_{31}$$
$$= \begin{bmatrix} 1 & 0 & 0 \\ 0 & 1 & 0 \\ 0 & 0 & 1 \end{bmatrix}\begin{bmatrix} 1 \\ 1 \\ 1 \end{bmatrix} + \begin{bmatrix} 0 \\ 0 \\ 0 \end{bmatrix}[1] + \begin{bmatrix} 0 & -1 \\ 1 & 0 \\ 0 & -1 \end{bmatrix}\begin{bmatrix} 0 \\ 0 \end{bmatrix} = \begin{bmatrix} 1 \\ 1 \\ 1 \end{bmatrix} + \begin{bmatrix} 0 \\ 0 \\ 0 \end{bmatrix} + \begin{bmatrix} 0 \\ 0 \\ 0 \end{bmatrix}$$
$$= \begin{bmatrix} 1 \\ 1 \\ 1 \end{bmatrix},$$

$$C_{12} = A_{11}B_{12} + A_{12}B_{22} + A_{13}B_{32}$$
$$= \begin{bmatrix} 1 & 0 & 0 \\ 0 & 1 & 0 \\ 0 & 0 & 1 \end{bmatrix}\begin{bmatrix} 1 & 0 & 2 \\ 0 & 3 & 0 \\ 4 & 0 & 5 \end{bmatrix} + \begin{bmatrix} 0 \\ 0 \\ 0 \end{bmatrix}[0 \quad 0 \quad 0] + \begin{bmatrix} 0 & -1 \\ 1 & 0 \\ 0 & -1 \end{bmatrix}\begin{bmatrix} 1 & 2 & 3 \\ 3 & 2 & 1 \end{bmatrix}$$
$$= \begin{bmatrix} 1 & 0 & 2 \\ 0 & 3 & 0 \\ 4 & 0 & 5 \end{bmatrix} + \begin{bmatrix} 0 & 0 & 0 \\ 0 & 0 & 0 \\ 0 & 0 & 0 \end{bmatrix} + \begin{bmatrix} -3 & -2 & -1 \\ 1 & 2 & 3 \\ -3 & -2 & -1 \end{bmatrix} = \begin{bmatrix} -2 & -2 & 1 \\ 1 & 5 & 3 \\ 1 & -2 & 4 \end{bmatrix},$$

$$C_{21} = A_{21}B_{11} + A_{22}B_{21} + A_{23}B_{31}$$
$$= \begin{bmatrix} 0 & 0 & 0 \\ 0 & 0 & 0 \end{bmatrix}\begin{bmatrix} 1 \\ 1 \\ 1 \end{bmatrix} + \begin{bmatrix} 1 \\ 1 \end{bmatrix}[1] + \begin{bmatrix} 0 & 1 \\ 1 & 0 \end{bmatrix}\begin{bmatrix} 0 \\ 0 \end{bmatrix} = \begin{bmatrix} 0 \\ 0 \end{bmatrix} + \begin{bmatrix} 1 \\ 1 \end{bmatrix} + \begin{bmatrix} 0 \\ 0 \end{bmatrix} = \begin{bmatrix} 1 \\ 1 \end{bmatrix},$$

$$C_{22} = A_{21}B_{12} + A_{22}B_{22} + A_{23}B_{32}$$
$$= \begin{bmatrix} 0 & 0 & 0 \\ 0 & 0 & 0 \end{bmatrix}\begin{bmatrix} 1 & 0 & 2 \\ 0 & 3 & 0 \\ 4 & 0 & 5 \end{bmatrix} + \begin{bmatrix} 1 \\ 1 \end{bmatrix}[0 \quad 0 \quad 0] + \begin{bmatrix} 0 & 1 \\ 1 & 0 \end{bmatrix}\begin{bmatrix} 1 & 2 & 3 \\ 3 & 2 & 1 \end{bmatrix}$$
$$= \begin{bmatrix} 0 & 0 & 0 \\ 0 & 0 & 0 \end{bmatrix} + \begin{bmatrix} 0 & 0 & 0 \\ 0 & 0 & 0 \end{bmatrix} + \begin{bmatrix} 3 & 2 & 1 \\ 1 & 2 & 3 \end{bmatrix} = \begin{bmatrix} 3 & 2 & 1 \\ 1 & 2 & 3 \end{bmatrix},$$

so

$$AB = \left[\begin{array}{c|ccc} 1 & -2 & -2 & 1 \\ 1 & 1 & 5 & 3 \\ 1 & 1 & -2 & 4 \\ \hline 1 & 3 & 2 & 1 \\ 1 & 1 & 2 & 3 \end{array}\right]. \quad \square$$

■ **EXAMPLE 1.45.** Suppose $A,B \in \mathcal{M}_{m\times m}$ and $C,D \in \mathcal{M}_{n\times n}$, and consider the $(m+n) \times (m+n)$ "diagonal" block matrices

$$\begin{bmatrix} A & 0 \\ 0 & C \end{bmatrix} \quad \text{and} \quad \begin{bmatrix} B & 0 \\ 0 & D \end{bmatrix}.$$

The zeros in the upper right and lower left corners represent zero matrices of sizes $m \times n$ and $n \times m$, respectively. Observe that

$$\begin{bmatrix} A & 0 \\ 0 & C \end{bmatrix}\begin{bmatrix} B & 0 \\ 0 & D \end{bmatrix} = \begin{bmatrix} AB+00 & A0+0D \\ 0B+C0 & 00+CD \end{bmatrix} = \begin{bmatrix} AB & 0 \\ 0 & CD \end{bmatrix}.$$

In particular,

$$\left[\begin{array}{cc|ccc} 0 & 1 & 0 & 0 & 0 \\ 1 & 0 & 0 & 0 & 0 \\ \hline 0 & 0 & 1 & 0 & 0 \\ 0 & 0 & 0 & 2 & 0 \\ 0 & 0 & 0 & 0 & 3 \end{array}\right]\left[\begin{array}{cc|ccc} 1 & 2 & 0 & 0 & 0 \\ 4 & 3 & 0 & 0 & 0 \\ \hline 0 & 0 & 1 & 0 & 1 \\ 0 & 0 & 0 & 1 & 0 \\ 0 & 0 & 1 & 0 & 1 \end{array}\right] = \left[\begin{array}{cc|ccc} 4 & 3 & 0 & 0 & 0 \\ 1 & 2 & 0 & 0 & 0 \\ \hline 0 & 0 & 1 & 0 & 1 \\ 0 & 0 & 0 & 2 & 0 \\ 0 & 0 & 3 & 0 & 3 \end{array}\right]. \ \square$$

■ **EXAMPLE 1.46.** Suppose $A \in \mathcal{M}_{m\times m}$ and $B \in \mathcal{M}_{n\times n}$ are nonsingular. By Example 1.45,

$$\begin{bmatrix} A & 0 \\ 0 & B \end{bmatrix}\begin{bmatrix} A^{-1} & 0 \\ 0 & B^{-1} \end{bmatrix} = \begin{bmatrix} AA^{-1} & 0 \\ 0 & BB^{-1} \end{bmatrix} = \begin{bmatrix} I_m & 0 \\ 0 & I_n \end{bmatrix} = I_{m+n},$$

so $\begin{bmatrix} A & 0 \\ 0 & B \end{bmatrix}$ is nonsingular, and

$$\begin{bmatrix} A & 0 \\ 0 & B \end{bmatrix}^{-1} = \begin{bmatrix} A^{-1} & 0 \\ 0 & B^{-1} \end{bmatrix}.$$

In particular,

$$\left[\begin{array}{cc|cc} 1 & 2 & 0 & 0 \\ 3 & 4 & 0 & 0 \\ \hline 0 & 0 & 2 & 1 \\ 0 & 0 & 3 & 0 \end{array}\right]^{-1} = \left[\begin{array}{cc|cc} -2 & 1 & 0 & 0 \\ 3/2 & -1/2 & 0 & 0 \\ \hline 0 & 0 & 0 & 1/3 \\ 0 & 0 & 1 & -2/3 \end{array}\right],$$

where the inverses of the 2×2 blocks were obtained via Example 1.36. □

■ **EXAMPLE 1.47.** Partitioning can be used to prove inductively that the product of upper triangular matrices is upper triangular. The primary task in such a proof is to show that if the statement is true for $n \times n$ matrices, then it must also be true for $(n+1) \times (n+1)$ matrices. To that end, suppose $A, B \in \mathcal{M}_{(n+1)\times(n+1)}$ are upper triangular and consider the partitionings

$$A = \left[\begin{array}{c|c} a_{11} & A_{12} \\ \hline 0 & A_{22} \end{array}\right] \quad \text{and} \quad B = \left[\begin{array}{c|c} b_{11} & B_{12} \\ \hline 0 & B_{22} \end{array}\right].$$

Note that A_{22} and B_{22} are $n \times n$ upper triangular matrices, $A_{12}, B_{12} \in \mathcal{M}_{1\times n}$, and $0 \in \mathcal{M}_{n\times 1}$. Then

$$AB = \left[\begin{array}{c|c} a_{11}b_{11} & a_{11}B_{12} + A_{12}B_{22} \\ \hline 0 & A_{22}B_{22} \end{array}\right].$$

Assuming that the product of any two $n \times n$ upper triangular matrices is upper triangular, the lower right-hand block of the latter matrix must be upper triangular, and consequently AB is upper triangular. This takes care of the inductive step. What remains is to show that the statement holds for the first reasonable value of n. When $n = 2$,

$$\mathrm{A} = \begin{bmatrix} a_{11} & a_{12} \\ 0 & a_{22} \end{bmatrix} \quad \text{and} \quad \mathrm{B} = \begin{bmatrix} b_{11} & b_{12} \\ 0 & b_{22} \end{bmatrix},$$

and

$$\mathrm{AB} = \begin{bmatrix} a_{11}b_{11} & a_{11}b_{12} + a_{12}b_{22} \\ 0 & a_{22}b_{22} \end{bmatrix}.$$

Thus, the truth of the statement when $n = 2$ implies the truth for $n = 3$, which, in turn, yields the truth for $n = 4$ and so forth. □

Exercises 1.8

1. Calculate AB directly and by using the given partitionings of A and B. Indicate the induced partitioning of the product matrix.

a. $\mathrm{A} = \left[\begin{array}{c|cc} 0 & 0 & 1 \\ 1 & 2 & 0 \end{array}\right], \quad \mathrm{B} = \left[\begin{array}{cccc} 1 & 1 & 2 & 2 \\ \hline 0 & 0 & -1 & -1 \\ 3 & 3 & 4 & 4 \end{array}\right]$

b. $\mathrm{A} = \left[\begin{array}{cc|c|c} -3 & -1 & -2 & 1 \\ 1 & 0 & 1 & -1 \\ -2 & 1 & 0 & 2 \end{array}\right], \quad \mathrm{B} = \left[\begin{array}{cc} 0 & 1 \\ -1 & 2 \\ \hline 3 & -1 \\ \hline -1 & 1 \end{array}\right]$

c. $\mathrm{A} = \left[\begin{array}{cc|c} 1 & 0 & -1 \\ 0 & 1 & 1 \\ \hline 0 & 0 & 2 \end{array}\right], \quad \mathrm{B} = \left[\begin{array}{cc|cc} 1 & 2 & 0 & 1 \\ -2 & 1 & 3 & 2 \\ \hline 1 & 1 & -1 & 0 \end{array}\right]$

d. $\mathrm{A} = \left[\begin{array}{ccc|c} 1 & -1 & 0 & -1 \\ 2 & 0 & 1 & 1 \\ \hline 0 & 0 & 1 & 3 \\ 0 & 0 & 2 & -2 \\ 1 & 2 & 1 & 1 \end{array}\right], \quad \mathrm{B} = \left[\begin{array}{cc|c} 0 & 2 & -1 \\ 1 & 3 & 1 \\ -2 & 0 & -1 \\ \hline 1 & 1 & 2 \end{array}\right]$

e. $\mathrm{A} = \left[\begin{array}{cc|c|ccc} -1 & 2 & 0 & 0 & 0 & 0 \\ 2 & -3 & 0 & 0 & 0 & 0 \\ \hline 0 & 0 & 3 & 0 & 0 & 0 \\ \hline 0 & 0 & 0 & 1 & -1 & 1 \\ 0 & 0 & 0 & 0 & -1 & 1 \\ 0 & 0 & 0 & 1 & 1 & 0 \end{array}\right], \quad \mathrm{B} = \left[\begin{array}{cc|c|ccc} 1 & 1 & 0 & 0 & 0 & 0 \\ 0 & 1 & 0 & 0 & 0 & 0 \\ \hline 0 & 0 & 1/3 & 0 & 0 & 0 \\ \hline 0 & 0 & 0 & 1 & 0 & 0 \\ 0 & 0 & 0 & 1 & 1 & 0 \\ 0 & 0 & 0 & 1 & 1 & 1 \end{array}\right]$

f. $\mathrm{A} = \left[\begin{array}{cc|ccc} -1 & 1 & 0 & 0 & 0 \\ 1 & 1 & 0 & 0 & 0 \\ \hline -2 & 1 & 3 & -2 & 1 \\ 1 & 0 & 1 & -1 & 1 \\ 2 & -1 & 0 & 1 & -1 \\ 0 & 1 & 2 & 0 & -3 \end{array}\right], \quad \mathrm{B} = \left[\begin{array}{ccc|cc} -3 & 2 & -1 & 0 & -2 \\ 1 & -3 & 2 & 1 & 0 \\ \hline -1 & 1 & 0 & 0 & 0 \\ 1 & -2 & 1 & 0 & 0 \\ 2 & 1 & -1 & 0 & 0 \end{array}\right]$

2. Use partitioning to find the inverse of

a. $\begin{bmatrix} 3 & 2 & 0 \\ 2 & 1 & 0 \\ 0 & 0 & 4 \end{bmatrix}$. b. $\begin{bmatrix} 3 & 0 & 0 & 0 \\ 0 & -2 & 0 & 0 \\ 0 & 0 & 0 & 1 \\ 0 & 0 & 1 & 0 \end{bmatrix}$.

c. $\begin{bmatrix} 0 & 0 & 1 & 0 & 0 \\ 0 & 1 & 1 & 0 & 0 \\ 1 & 1 & 1 & 0 & 0 \\ 0 & 0 & 0 & 4 & 3 \\ 0 & 0 & 0 & 2 & 1 \end{bmatrix}$. d. $\begin{bmatrix} 1 & 1 & 1 & 0 & 0 & 0 \\ 0 & 2 & 2 & 0 & 0 & 0 \\ 0 & 0 & 3 & 0 & 0 & 0 \\ 0 & 0 & 0 & 1 & 0 & 0 \\ 0 & 0 & 0 & 2 & 2 & 0 \\ 0 & 0 & 0 & 3 & 3 & 3 \end{bmatrix}$.

3. Suppose $A \in \mathcal{M}_{m\times m}$ and $B \in \mathcal{M}_{n\times n}$ are nonsingular, and let $C \in \mathcal{M}_{m\times n}$. Show that $\begin{bmatrix} A & C \\ 0 & B \end{bmatrix}$ has an inverse of the form $\begin{bmatrix} A^{-1} & X \\ 0 & B^{-1} \end{bmatrix}$ and find an expression for X in terms of A^{-1}, B^{-1}, and C.

4. Assuming $A \in \mathcal{M}_{m\times m}$ and $B \in \mathcal{M}_{n\times n}$ are nonsingular, show that $\begin{bmatrix} 0 & A \\ B & 0 \end{bmatrix}$ is nonsingular and find an expression for the inverse.

5. Assuming $A \in \mathcal{M}_{m\times p}$, $B \in \mathcal{M}_{n\times p}$, and $t \in \mathcal{R}$, find a partitioned matrix E such that

a. $E\begin{bmatrix} A \\ \hline B \end{bmatrix} = \begin{bmatrix} B \\ \hline A \end{bmatrix}$. **b.** $E\begin{bmatrix} A \\ \hline B \end{bmatrix} = \begin{bmatrix} tA \\ \hline B \end{bmatrix}$.

If, in addition, $m = n$, find a partitioned matrix E such that

c. $E\begin{bmatrix} A \\ \hline B \end{bmatrix} = \begin{bmatrix} A \\ \hline tA + B \end{bmatrix}$.

6. Suppose $A \in \mathcal{M}_{m\times m}$ is nonsingular, $B \in \mathcal{M}_{m\times n}$, $C \in \mathcal{M}_{n\times m}$, and $D \in \mathcal{M}_{n\times n}$. Find a nonsingular $(m+n) \times (m+n)$ matrix P such that

$$P\left[\begin{array}{c|c} A & B \\ \hline C & D \end{array}\right] = \left[\begin{array}{c|c} A & B \\ \hline 0 & M \end{array}\right],$$

and identify M.

7. Assuming $A \in \mathcal{M}_{m\times n}$, $B \in \mathcal{M}_{m\times p}$, and $C \in \mathcal{M}_{m\times q}$, find an $(n+p+q) \times (n+p+q)$ matrix Q such that

$$[A \mid B \mid C]Q = [A \mid C \mid B].$$

8. Give an inductive argument for the fact that the inverse of a nonsingular upper triangular matrix is upper triangular. (You may find the result of Problem 3 helpful.)

2

Euclidean n-Space

2.1 INTRODUCTION

This chapter explores some ideas that arise in analyzing the set of solutions for a linear system. A solution for a system in n unknowns is an ordered n-tuple, so the natural context in which to carry out this endeavor is the collection of all ordered n-tuples, called Euclidean n-space.

A particular linear system can have no solutions, a unique solution, or infinitely many solutions. The first two possibilities present situations in which the solution set is very easy to describe; it is either the empty set or a set consisting of a single element. In the third case, the description involves one or more parameters, and consequently the set possesses a certain "structure," which has yet to be investigated. Recall, for example, that when a system involving two or three variables has infinitely many solutions, those solutions constitute a line or a plane. That is, when $n = 2$ or $n = 3$, the structure of the set is expressible in terms of familiar notions in Euclidean geometry. For larger values of n, it will have to be characterized algebraically.

The algebraic notions encountered in the general case have simple geometric interpretations when $n = 2$ or $n = 3$, and the intuition associated with those geometric considerations often suggests appropriate lines of development in the general context. Consequently, the discussion begins with a presentation of the geometric analogue of Euclidean 2-space, namely, the collection of two-dimensional geometric vectors. The objective initially is to gain a clear understanding of the connections and the distinctions between the geometric objects and the algebraic vehicles used to describe them.

2.2 GEOMETRIC VECTORS

Motivation for the vector concept comes from physical notions like force and velocity, where a full description of the item in question requires that you specify both a magnitude and a direction. This section presents two mathematical models that embody these features. The first is a geometric formulation of "vector," the second an algebraic one. For convenience (e.g., drawing illustrations) the details are carried out in the two-dimensional case, but the development for three-dimensional vectors is entirely similar.

The geometric context for the two-dimensional case is a plane, and a natural geometric vehicle for expressing magnitude and direction is an arrow. You describe an arrow by specifying the points corresponding to its base and tip, say P and Q, respectively. Notationally, the base is distinguished from the tip by listing them in order, and it is customary to list the base first. Thus, $\overrightarrow{PQ}$ indicates the arrow based at P with tip Q. The length of the line segment joining P to Q determines a magnitude, and the position of Q relative to P fixes a direction. In spite of the geometric appeal of the terms arrow, base, and tip, it is more common to refer to $\overrightarrow{PQ}$ as the directed line segment with initial point P and terminal point Q, so the latter terminology is used hereafter.

Although $\overrightarrow{PQ}$ is associated with a particular magnitude and direction, it is not the only directed line segment with that property (Figure 2.1). In fact, each point of the plane is the initial point of a directed line segment that determines that same magnitude and direction. You can think of all these directed line segments as representing a single vector, each acting at a different point.

DEFINITION. A two-dimensional *geometric vector* is a set of directed line segments, one based at each point of the plane, all having the same direction and magnitude. The set of geometric vectors will be denoted by $\mathcal{G}^2$. If $\mathbf{x} \in \mathcal{G}^2$ and P is a point of the plane, then there is a Q in the plane such that $\overrightarrow{PQ} \in \mathbf{x}$, and $\overrightarrow{PQ}$ is called the *representative of* $\mathbf{x}$ *acting at P.*

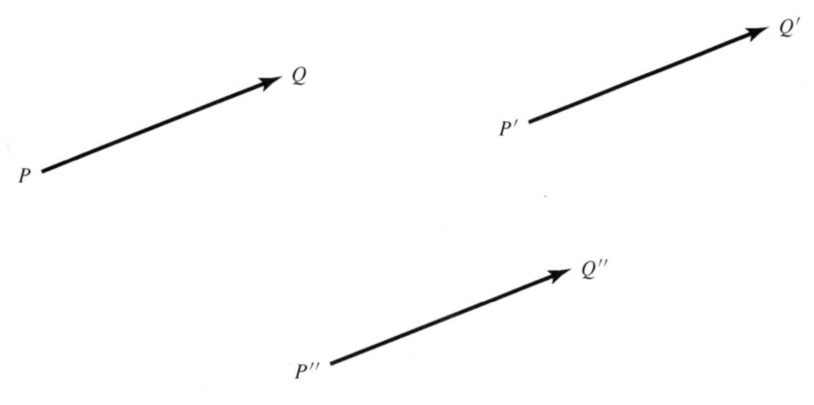

FIGURE 2.1

Associated with any two forces is a third force, called the resultant, which produces the same effect as the combination of the two. Moreover, there is a simple geometric way to represent it. Suppose the two forces are modeled by geometric vectors $\mathbf{x}$ and $\mathbf{y}$, and let $\overrightarrow{PQ_1}$ and $\overrightarrow{PQ_2}$ be the representatives of $\mathbf{x}$ and $\mathbf{y}$, respectively, acting at P. If Q is the point diagonally opposite P on the parallelogram determined by P, Q_1, and Q_2 (Figure 2.2), then $\overrightarrow{PQ}$ represents a geometric vector that models the resultant force. Here lies the motivation for a notion of addition of geometric vectors that serves a useful purpose. However, this approach to adding vectors presents some complications. For example, if $\overrightarrow{PQ_1}$ and $\overrightarrow{PQ_2}$ appear as in Figure 2.3, then what parallelogram do they determine and how do you locate Q? This difficulty is overcome as follows. Given $\overrightarrow{PQ_1} \in \mathbf{x}$, choose Q so that $\overrightarrow{Q_1Q}$ is the representative of $\mathbf{y}$ acting at Q_1. Then $\overrightarrow{PQ}$ represents the resultant force (Figure 2.4). In this way, reference to a parallelogram is avoided, and yet the result is the same as that of the parallelogram procedure when P, Q_1, and Q_2 are noncollinear.

There is another concern that must be faced before one can be satisfied with this notion of addition. If $\mathbf{x}$ and $\mathbf{y}$ are represented by directed line segments of the same length but opposite directions (Figure 2.5), then the point Q such that $\overrightarrow{Q_1Q} \in \mathbf{y}$ is P. That is, $\overrightarrow{PQ}$ is a point rather than a directed line segment. A point has zero length, but it doesn't indicate a direction. If a point is not viewed as determining a vector, then it will not always be possible to add two geometric vectors as indicated above and obtain another geometric vector as a result. The solution to this problem lies in creating a "vector of convenience" called the zero vector.

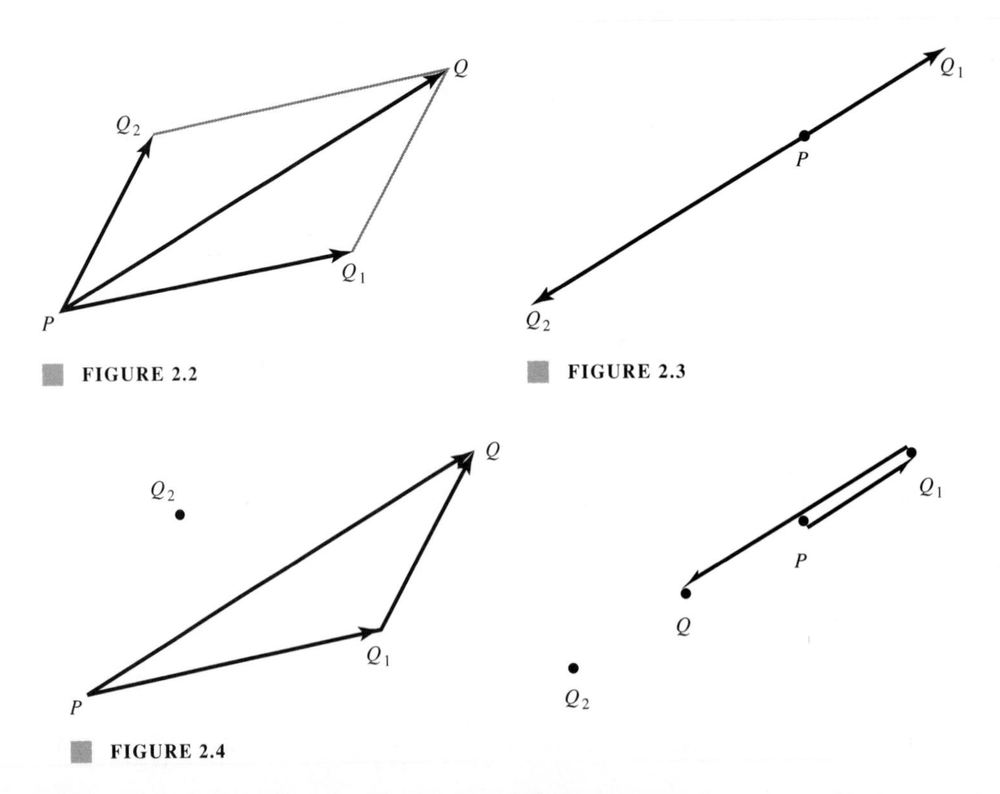

FIGURE 2.2

FIGURE 2.3

FIGURE 2.4

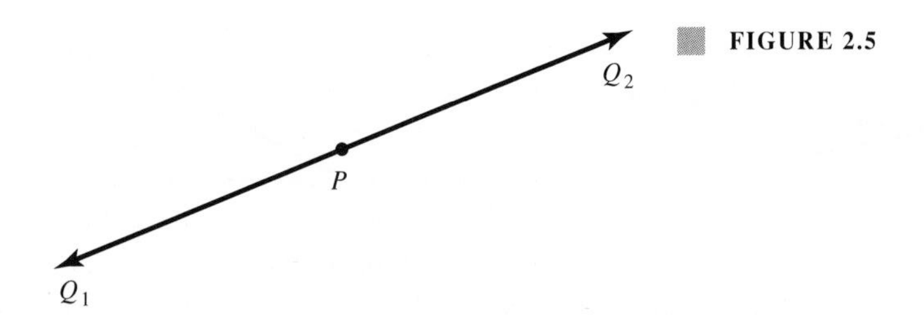

FIGURE 2.5

DEFINITION. The *zero geometric vector,* denoted by $\mathbf{0}$, is the set of all points in the plane. The representative of $\mathbf{0}$ acting at P is simply P, and $\overrightarrow{PP}$ will be interpreted as a degenerate directed line segment representing $\mathbf{0}$. Hereafter, $\mathbf{0}$ is also considered to be an element of $\mathcal{G}^2$.

Addition of geometric vectors can now be formally defined.

DEFINITION. Suppose $\mathbf{x},\mathbf{y} \in \mathcal{G}^2$ and let P be any point of the plane. Choose points Q and R such that $\overrightarrow{PQ} \in \mathbf{x}$ and $\overrightarrow{QR} \in \mathbf{y}$. The *sum* of $\mathbf{x}$ and $\mathbf{y}$, denoted by $\mathbf{x}+\mathbf{y}$, is the geometric vector represented by $\overrightarrow{PR}$.

There is one remaining concern regarding addition of geometric vectors. It is conceivable that if you select different starting points, P and P', and proceed to determine $\mathbf{x}+\mathbf{y}$ from each, you might get different results. If Q and Q' are chosen so that $\overrightarrow{PQ}, \overrightarrow{P'Q'} \in \mathbf{x}$ and R and R' are chosen so that $\overrightarrow{QR}, \overrightarrow{Q'R'} \in \mathbf{y}$, then according to the definition, $\overrightarrow{PR}$ and $\overrightarrow{P'R'}$ are both directed line segments representing $\mathbf{x}+\mathbf{y}$. For this addition to make sense, $\overrightarrow{PR}$ and $\overrightarrow{P'R'}$ must have the same direction and magnitude, and indeed they do (Figure 2.6).

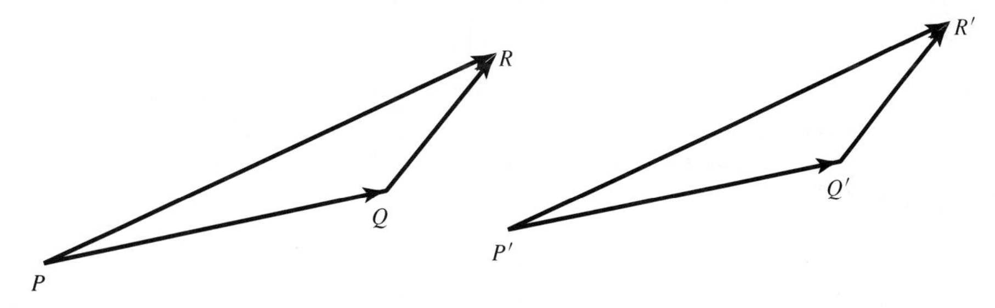

FIGURE 2.6

The next theorem lists the rules governing addition of geometric vectors. Each conclusion contains a statement that two vectors are equal, and to prove such a statement it suffices to show that there is one directed line segment that represents both vectors.

THEOREM 2.1. If $\mathbf{x},\mathbf{y},\mathbf{z} \in \mathcal{G}^2$, then

(1) $\mathbf{x}+\mathbf{y}=\mathbf{y}+\mathbf{x}$,
(2) $\mathbf{x}+(\mathbf{y}+\mathbf{z})=(\mathbf{x}+\mathbf{y})+\mathbf{z}$, and
(3) $\mathbf{x}+\mathbf{0}=\mathbf{x}$.
(4) For each $\mathbf{x} \in \mathcal{G}^2$ there is an element of $\mathcal{G}^2$, denoted by $-\mathbf{x}$, such that $\mathbf{x}+(-\mathbf{x})=\mathbf{0}$.

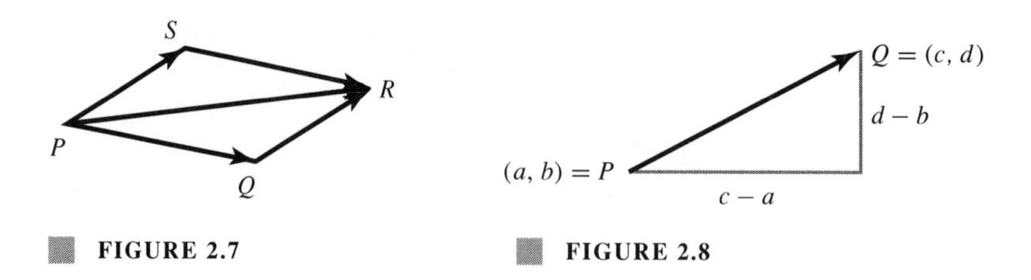

FIGURE 2.7

FIGURE 2.8

Proof.

(1) Let $\overrightarrow{PQ} \in \mathbf{x}$ and $\overrightarrow{QR} \in \mathbf{y}$. Then $\overrightarrow{PR} \in \mathbf{x}+\mathbf{y}$. Assume P, Q, and R are noncollinear and choose S so that P, Q, R, and S are the vertices of a parallelogram (Figure 2.7). Observe that $\overrightarrow{PS} \in \mathbf{y}$ and $\overrightarrow{SR} \in \mathbf{x}$, so $\overrightarrow{PR} \in \mathbf{y}+\mathbf{x}$. Since $\mathbf{x}+\mathbf{y}$ and $\mathbf{y}+\mathbf{x}$ are both represented by $\overrightarrow{PR}$, $\mathbf{x}+\mathbf{y} = \mathbf{y}+\mathbf{x}$. The case when P, Q, and R are collinear is left for the reader.

(2) See Exercise 1.

(3) Assume $\overrightarrow{PQ} \in \mathbf{x}$. Since $\overrightarrow{QQ} \in \mathbf{0}$, $\overrightarrow{PQ}$ also represents $\mathbf{x}+\mathbf{0}$, so $\mathbf{x} = \mathbf{x}+\mathbf{0}$.

(4) Suppose $\overrightarrow{PQ} \in \mathbf{x}$ and let $-\mathbf{x}$ denote the geometric vector represented by $\overrightarrow{QP}$. Then the representative of $\mathbf{x}+(-\mathbf{x})$ acting at P is $\overrightarrow{PP}$, so $\mathbf{x}+(-\mathbf{x}) = \mathbf{0}$. ■

The next project is to produce a formulation of the vector concept in which vector manipulations are reduced to calculations with real numbers. To that end, let $\mathbf{x} \in \mathcal{G}^2$ and suppose $\overrightarrow{PQ} \in \mathbf{x}$. In the usual rectangular coordinate system, P and Q are identified with ordered pairs of real numbers, say $P = (a,b)$ and $Q = (c,d)$. Then $c - a$ and $d - b$ are the horizontal and vertical displacements, respectively, from P to Q (Figure 2.8). These numbers determine both the position of Q relative to P and the length of the line segment that joins P to Q. Thus,

$$\mathrm{X} = \begin{bmatrix} x_1 \\ x_2 \end{bmatrix} = \begin{bmatrix} c-a \\ d-b \end{bmatrix}$$

is an ordered pair associated with $\mathbf{x}$ whose entries determine the magnitude and direction of $\mathbf{x}$. Actually X is associated with $\overrightarrow{PQ}$, but if $\overrightarrow{P'Q'}$ is any other representative of $\mathbf{x}$, with $P' = (a',b')$ and $Q' = (c',d')$, then $c' - a' = c - a$ and $d' - b' = d - b$ (Figure 2.9), so X does characterize $\mathbf{x}$.

Let

$$\mathcal{R}^2 = \left\{ \begin{bmatrix} x_1 \\ x_2 \end{bmatrix} : x_1, x_2 \in \mathcal{R} \right\}.$$

Note that $\mathcal{R}^2$ is just the collection of 2×1 matrices. To each $\mathbf{x} \in \mathcal{G}^2$ there is associated an element of $\mathcal{R}^2$ as indicated above. Conversely, each ordered pair $[x_1 \;\; x_2]^T$ determines exactly one element of $\mathcal{G}^2$ in this manner. Indeed, if $P = (a,b)$ and $Q = (a + x_1, b + x_2)$, then the element of $\mathcal{G}^2$ associated with $[x_1 \;\; x_2]^T$ is the geometric vector represented by $\overrightarrow{PQ}$. Thus, the association

$$\mathbf{x} \longleftrightarrow \begin{bmatrix} x_1 \\ x_2 \end{bmatrix}$$

establishes a one-to-one pairing of the elements of $\mathcal{G}^2$ with those of $\mathcal{R}^2$.

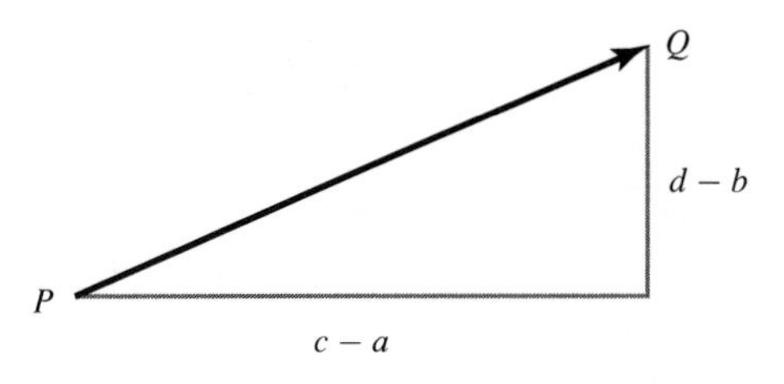

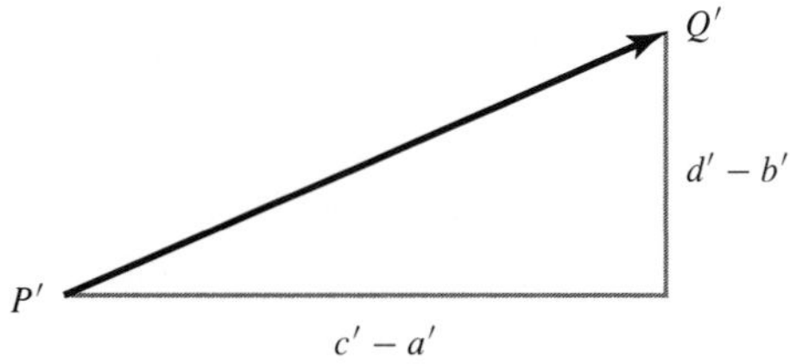

FIGURE 2.9

■ **EXAMPLE 2.1.** Let $P = (3,2)$ and $Q = (-1,4)$. If $\mathbf{x}$ is the geometric vector determined by $\overrightarrow{PQ}$, then $x_1 = -1 - 3 = -4$, $x_2 = 4 - 2 = 2$, and $\mathbf{x} \longleftrightarrow \begin{bmatrix} -4 \\ 2 \end{bmatrix}$. □

■ **EXAMPLE 2.2.** Suppose $\mathbf{x} \in \mathcal{G}^2$, $\mathbf{x} \longleftrightarrow [2 \quad -1]^T$, and $P = (-3,6)$. If $Q = (-3+2, 6+(-1)) = (-1,5)$, then $\overrightarrow{PQ}$ represents $\mathbf{x}$. □

■ **EXAMPLE 2.3.** Let $P = (a,b)$. The zero vector is represented by $\overrightarrow{PP}$. If $[x_1 \quad x_2]^T$ is the element of $\mathcal{R}^2$ associated with $\mathbf{0}$, then $x_1 = a - a = 0$, and $x_2 = b - b = 0$, that is,

$$\mathbf{0} \longleftrightarrow \begin{bmatrix} 0 \\ 0 \end{bmatrix}. \ \square$$

■ **EXAMPLE 2.4.** Suppose $\mathbf{x} \in \mathcal{G}^2$ and $\overrightarrow{PQ} \in \mathbf{x}$. If $P = (a,b)$ and $Q = (c,d)$, then $\mathbf{x} \longleftrightarrow [x_1 \quad x_2]^T$, where $x_1 = c - a$ and $x_2 = d - b$. Since $-\mathbf{x}$ is determined by $\overrightarrow{QP}$,

$$-\mathbf{x} \longleftrightarrow \begin{bmatrix} a - c \\ b - d \end{bmatrix} = \begin{bmatrix} -x_1 \\ -x_2 \end{bmatrix}.$$

In particular, if $\mathbf{x} \longleftrightarrow \begin{bmatrix} -4 \\ 7 \end{bmatrix}$, then $-\mathbf{x} \longleftrightarrow \begin{bmatrix} 4 \\ -7 \end{bmatrix}$. □

Consider now the connection between addition for geometric vectors and addition for 2×1 matrices. Suppose $\mathbf{x},\mathbf{y} \in \mathcal{G}^2$, $\mathbf{x} \longleftrightarrow [x_1 \quad x_2]^T$ and $\mathbf{y} \longleftrightarrow [y_1 \quad y_2]^T$. If $\overrightarrow{PQ} \in \mathbf{x}$ and $\overrightarrow{QR} \in \mathbf{y}$, then $\overrightarrow{PR}$ represents $\mathbf{x} + \mathbf{y}$. Figure 2.10 suggests that the element of $\mathcal{R}^2$ associated with $\mathbf{x} + \mathbf{y}$ is $[x_1 + y_1 \quad x_2 + y_2]^T$. Actually, it depicts only the case when x_1, x_2, y_1, and y_2 are all positive, but similar figures would convince you that this conclusion holds in the other cases as well. Thus, under the pairing of the elements of $\mathcal{G}^2$ with those of $\mathcal{R}^2$, the sum of two vectors in $\mathcal{G}^2$ corresponds to the matrix sum of the associated elements of $\mathcal{R}^2$.

THEOREM 2.2. If $\mathbf{x},\mathbf{y} \in \mathcal{G}^2$, $\mathbf{x} \longleftrightarrow \begin{bmatrix} x_1 \\ x_2 \end{bmatrix}$ and $\mathbf{y} \longleftrightarrow \begin{bmatrix} y_1 \\ y_2 \end{bmatrix}$, then $\mathbf{x}+\mathbf{y} \longleftrightarrow \begin{bmatrix} x_1 + y_1 \\ x_2 + y_2 \end{bmatrix}$.

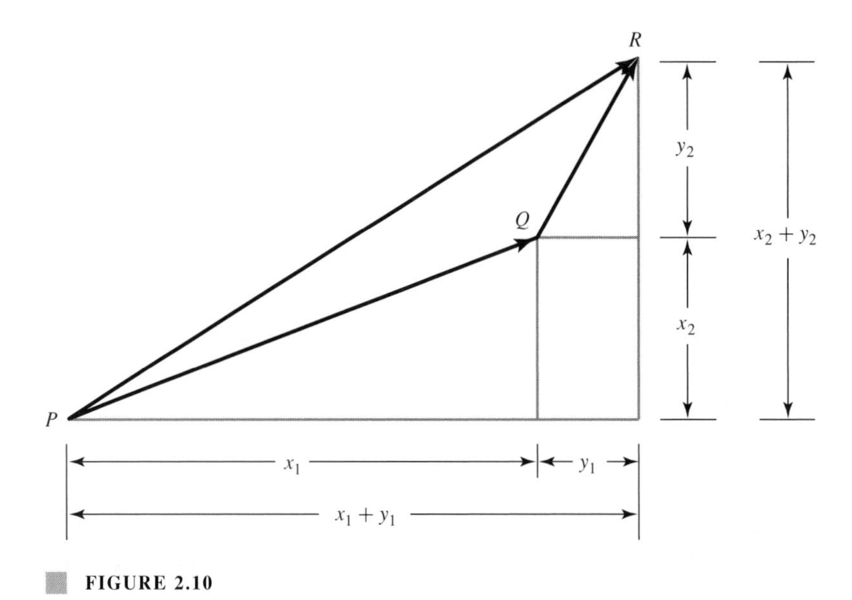

FIGURE 2.10

EXAMPLE 2.5. If $\mathbf{x} \longleftrightarrow \begin{bmatrix} 2 \\ -3 \end{bmatrix}$ and $\mathbf{y} \longleftrightarrow \begin{bmatrix} 0 \\ 5 \end{bmatrix}$, then $\mathbf{x}+\mathbf{y} \longleftrightarrow \begin{bmatrix} 2+0 \\ -3+5 \end{bmatrix} = \begin{bmatrix} 2 \\ 2 \end{bmatrix}$. □

The directed line segments representing a given $\mathbf{x} \in \mathcal{G}^2$ all have the same length, and that common value is the *magnitude* of $\mathbf{x}$, denoted by $\|\mathbf{x}\|$. Suppose

$$\mathbf{x} \longleftrightarrow \begin{bmatrix} x_1 \\ x_2 \end{bmatrix},$$

and $\overrightarrow{PQ}$ is a representative of $\mathbf{x}$. If neither x_1 nor x_2 is 0, then $\overrightarrow{PQ}$ is the hypotenuse of a right triangle with horizontal and vertical sides of lengths $|x_1|$ and $|x_2|$, respectively, and according to the Pythagorean theorem,

$$\|\mathbf{x}\| = \sqrt{|x_1|^2 + |x_2|^2} = \sqrt{x_1^2 + x_2^2}.$$

When x_1 or x_2 is zero, the triangle degenerates, but the last formula still gives the correct value for $\|\mathbf{x}\|$. For example, If $x_1 = 0$, then

$$\|\mathbf{x}\| = |x_2| = \sqrt{x_2^2} = \sqrt{0^2 + x_2^2}.$$

EXAMPLE 2.6. If $\mathbf{x}$ is the geometric vector represented by $\overrightarrow{PQ}$, where $P = (0,-2)$ and $Q = (2,4)$, then $\mathbf{x} \longleftrightarrow \begin{bmatrix} 2 \\ 6 \end{bmatrix}$, and $\|\mathbf{x}\| = \sqrt{4+36} = 2\sqrt{10}$. □

EXAMPLE 2.7. Suppose $\mathbf{x} \in \mathcal{G}^2$ and $\mathbf{x} \longleftrightarrow \begin{bmatrix} x_1 \\ x_2 \end{bmatrix}$. If $\|\mathbf{x}\| = \sqrt{x_1^2 + x_2^2} = 0$, then $x_1 = 0 = x_2$, so $\mathbf{0}$ is the only geometric vector with length 0. □

Two nonzero geometric vectors are *parallel* if the directed line segments representing them are parallel. Suppose $\mathbf{x}$ and $\mathbf{y}$ are nonzero elements of $\mathcal{G}^2$, with representatives $\overrightarrow{PQ}$ and $\overrightarrow{PR}$, respectively. Then $\mathbf{x}$ is parallel to $\mathbf{y}$ if and only if P, Q, and R are collinear. Assuming P, Q, and R do lie on a line, the direction of $\mathbf{x}$ is either the same as

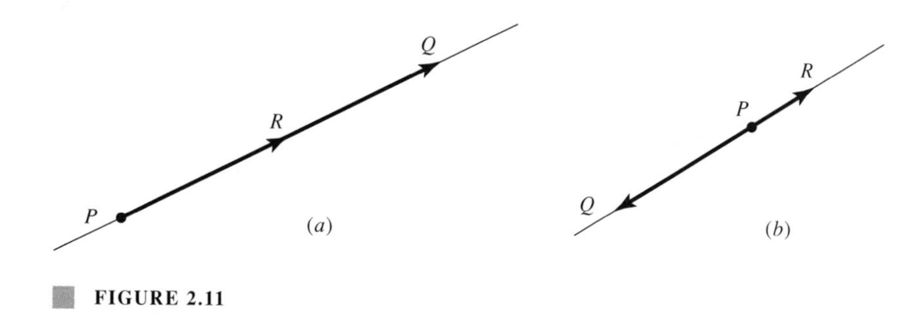

FIGURE 2.11

or the opposite of that of $\mathbf{y}$, depending on the relative positions of the points. When P lies between Q and R, $\mathbf{x}$ and $\mathbf{y}$ have opposite directions (Figure 2.11(b)); otherwise they have the same direction (Figure 2.11(a)).

If $\mathbf{y}$ has the same direction as $\mathbf{x}$, then you can think of $\mathbf{y}$ as being obtained by stretching or shrinking $\mathbf{x}$, and the "scaling factor" relating their lengths is the positive real number, t, satisfying $\|\mathbf{y}\| = t\|\mathbf{x}\|$. Under these circumstances it is suggestive to write $\mathbf{y} = t\mathbf{x}$ and refer to $\mathbf{y}$ as a "scalar multiple" of $\mathbf{x}$. Similarly, negative "scalars" are utilized to describe the situation when $\mathbf{x}$ and $\mathbf{y}$ have opposite directions.

DEFINITION. If $\mathbf{x} \in \mathcal{G}^2$ and $t \in \mathcal{R}$, then $t\mathbf{x}$ denotes the vector in $\mathcal{G}^2$ defined as follows. If $t = 0$ or $\mathbf{x} = \mathbf{0}$, then $t\mathbf{x} = \mathbf{0}$. Otherwise, $t\mathbf{x}$ is the unique vector satisfying $\|t\mathbf{x}\| = |t|\|\mathbf{x}\|$ which has either the same or opposite direction as $\mathbf{x}$, depending on whether t is positive or negative, respectively.

The preceding definition establishes an operation on geometric vectors called *scalar multiplication*. The interaction between this operation and the correspondence between $\mathcal{G}^2$ and $\mathcal{R}^2$ is recorded in the next theorem.

THEOREM 2.3. If $\mathbf{x} \in \mathcal{G}^2$, $t \in \mathcal{R}$, and $\mathbf{x} \longleftrightarrow \begin{bmatrix} x_1 \\ x_2 \end{bmatrix}$, then $t\mathbf{x} \longleftrightarrow \begin{bmatrix} tx_1 \\ tx_2 \end{bmatrix}$.

Proof. If $\mathbf{x} = \mathbf{0}$ or $t = 0$, then $t\mathbf{x} = \mathbf{0}$ and $[tx_1 \quad tx_2]^T = [0 \quad 0]^T$. In this case, the conclusion follows from the fact that $\mathbf{0} \longleftrightarrow [0 \quad 0]^T$. Assume then that $\mathbf{x}$ and t are nonzero. Set $\mathbf{y} = t\mathbf{x}$ and suppose $\mathbf{y} \longleftrightarrow [y_1 \quad y_2]^T$. The object is to show that $y_1 = tx_1$ and $y_2 = tx_2$.

Suppose $t > 0$. Then $\mathbf{x}$ and $\mathbf{y}$ have the same direction, and $\|\mathbf{y}\| = t\|\mathbf{x}\|$. Let $\overrightarrow{PQ}$ and $\overrightarrow{PR}$ be representatives of $\mathbf{x}$ and $\mathbf{y}$. Since P does not lie between Q and R, y_1 and x_1 have the same sign, as do y_2 and x_2. It may happen that x_1 or x_2 is zero (not both, since $\mathbf{x} \neq \mathbf{0}$), whereupon the corresponding y_1 or y_2 is also zero. Figure 2.12 illustrates the situation when $x_1 \neq 0$ and $x_2 \neq 0$. In that case, it follows from properties of similar triangles that

$$\frac{|y_1|}{|x_1|} = \frac{|y_2|}{|x_2|} = \frac{\|\mathbf{y}\|}{\|\mathbf{x}\|} = t.$$

Since y_1 and x_1 have the same sign, $|y_1|/|x_1| = y_1/x_1$, and similarly $|y_2|/|x_2| = y_2/x_2$. Thus, $y_1/x_1 = t = y_2/x_2$, that is, $y_1 = tx_1$ and $y_2 = tx_2$.

If $x_1 = 0$, then $y_1 = 0$, and setting $t = |y_2|/|x_2|$ yields $y_2/x_2 = |y_2|/|x_2| = t$, so $y_2 = tx_2$. This time $y_1 = tx_1$ because y_1 and x_1 are both 0. A similar argument handles the case when $x_2 = 0$.

The argument when $t < 0$ is similar, and the details are omitted. ■

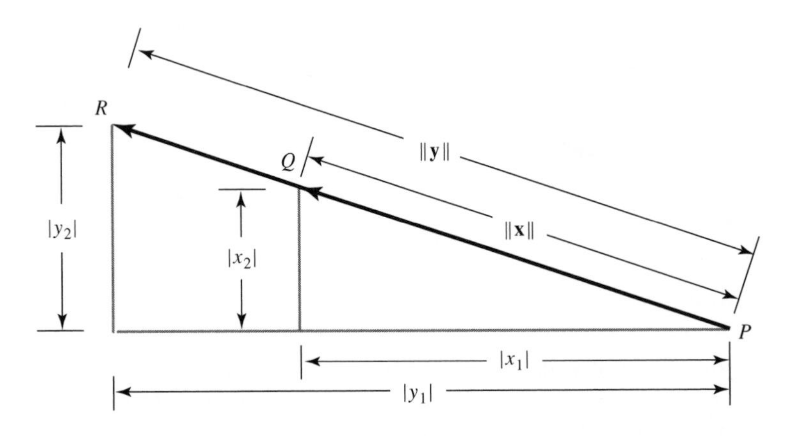

FIGURE 2.12

The rules governing scalar multiplication are listed in the next theorem.

THEOREM 2.4. If $\mathbf{x},\mathbf{y} \in \mathcal{G}^2$ and $s,t \in \mathcal{R}$, then

(1) $t(\mathbf{x} + \mathbf{y}) = t\mathbf{x} + t\mathbf{y}$,
(2) $(s + t)\mathbf{x} = s\mathbf{x} + t\mathbf{x}$,
(3) $s(t\mathbf{x}) = (st)\mathbf{x} = t(s\mathbf{x})$, and
(4) $(-1)\mathbf{x} = -\mathbf{x}$.

Proof. Assume $\mathbf{x} \leftrightarrow \begin{bmatrix} x_1 \\ x_2 \end{bmatrix}$ and $\mathbf{y} \leftrightarrow \begin{bmatrix} y_1 \\ y_2 \end{bmatrix}$. According to Theorem 2.2,

$$\mathbf{x} + \mathbf{y} \leftrightarrow \begin{bmatrix} x_1 + y_1 \\ x_2 + y_2 \end{bmatrix},$$

so by Theorem 2.3,

$$t(\mathbf{x} + \mathbf{y}) \leftrightarrow \begin{bmatrix} t(x_1 + y_1) \\ t(x_2 + y_2) \end{bmatrix} = \begin{bmatrix} tx_1 + ty_1 \\ tx_2 + ty_2 \end{bmatrix}.$$

Also

$$t\mathbf{x} \leftrightarrow \begin{bmatrix} tx_1 \\ tx_2 \end{bmatrix} \quad \text{and} \quad t\mathbf{y} \leftrightarrow \begin{bmatrix} ty_1 \\ ty_2 \end{bmatrix},$$

so

$$t\mathbf{x} + t\mathbf{y} \leftrightarrow \begin{bmatrix} tx_1 + ty_1 \\ tx_2 + ty_2 \end{bmatrix}.$$

Thus, $t(\mathbf{x} + \mathbf{y})$ and $t\mathbf{x} + t\mathbf{y}$ are associated with the same element of $\mathcal{R}^2$ and are therefore the same geometric vector. This establishes conclusion (1). Proofs for the remaining conclusions are omitted. ■

Given $\mathbf{x},\mathbf{y} \in \mathcal{G}^2$, the *difference* $\mathbf{x} - \mathbf{y}$ is defined as

$$\mathbf{x} - \mathbf{y} = \mathbf{x} + (-\mathbf{y}),$$

where $-\mathbf{y}$ is the element of $\mathcal{G}^2$ satisfying $\mathbf{y} + (-\mathbf{y}) = \mathbf{0}$. If

$$\mathbf{x} \leftrightarrow \begin{bmatrix} x_1 \\ x_2 \end{bmatrix} \quad \text{and} \quad \mathbf{y} \leftrightarrow \begin{bmatrix} y_1 \\ y_2 \end{bmatrix},$$

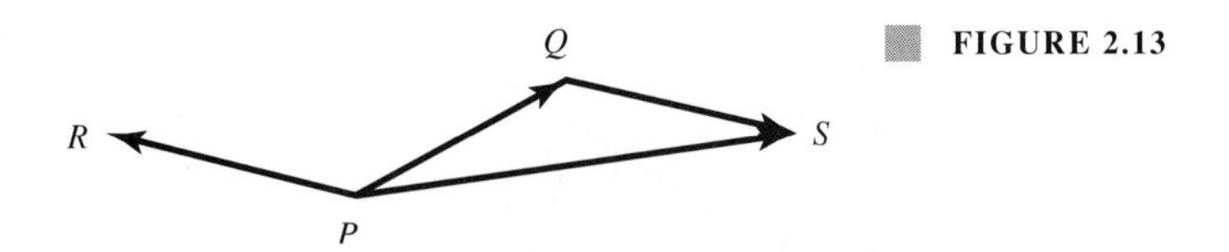

FIGURE 2.13

then $-\mathbf{y} \longleftrightarrow \begin{bmatrix} -y_1 \\ -y_2 \end{bmatrix}$ (Example 2.4), and therefore

$$\mathbf{x} - \mathbf{y} \longleftrightarrow \begin{bmatrix} x_1 - y_1 \\ x_2 - y_2 \end{bmatrix}.$$

Suppose $\overrightarrow{PQ} \in \mathbf{x}$ and $\overrightarrow{PR} \in \mathbf{y}$. Then the representative of $\mathbf{x} - \mathbf{y}$ acting at P is the directed line segment $\overrightarrow{PS}$ illustrated in Figure 2.13. Note that $\overrightarrow{PS}$ and $\overrightarrow{RQ}$ have the same direction and magnitude, so $\overrightarrow{RQ}$ also represents $\mathbf{x} - \mathbf{y}$.

EXAMPLE 2.8. Assume P, Q, R, and S are vertices of a parallelogram (Figure 2.14) and let $\mathbf{x}$ and $\mathbf{y}$ be the geometric vectors represented by $\overrightarrow{PQ}$ and $\overrightarrow{PR}$, respectively.

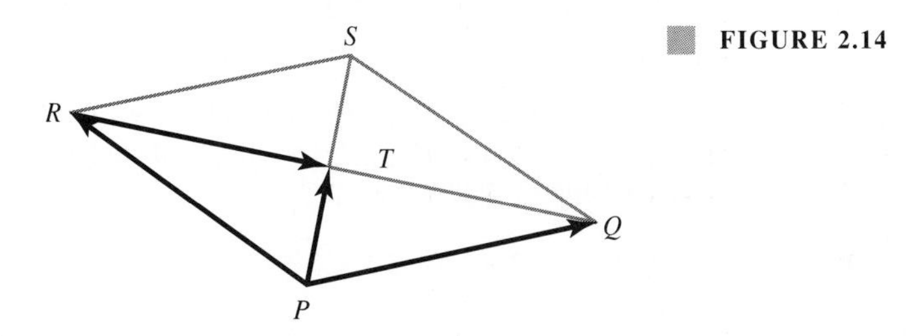

FIGURE 2.14

Note that $\overrightarrow{RQ} \in \mathbf{x} - \mathbf{y}$. If T is the midpoint of diagonal $\overrightarrow{RQ}$, then $\overrightarrow{RT} \in \frac{1}{2}(\mathbf{x} - \mathbf{y})$. Thus,

$$\overrightarrow{PT} \in \mathbf{y} + \tfrac{1}{2}(\mathbf{x} - \mathbf{y}) = \tfrac{1}{2}(\mathbf{x} + \mathbf{y}),$$

so T is the midpoint of $\overrightarrow{PS}$. The diagonals of a parallelogram are therefore bisectors of one another. □

Had the discussion in this section been carried out in space instead of the plane, the outcome would have been a set, $\mathcal{G}^3$, of three-dimensional geometric vectors and an associated set of ordered triples,

$$\mathcal{R}^3 = \left\{ \begin{bmatrix} x_1 \\ x_2 \\ x_3 \end{bmatrix} : x_1, x_2, x_3 \in \mathcal{R} \right\}.$$

Each $\mathbf{x} \in \mathcal{G}^3$ is represented by infinitely many directed line segments, one based at each point in space. If $\overrightarrow{PQ} \in \mathbf{x}$, $P = (a,b,c)$ and $Q = (d,e,f)$, then

$$\mathbf{x} \longleftrightarrow \begin{bmatrix} d - a \\ e - b \\ f - c \end{bmatrix}$$

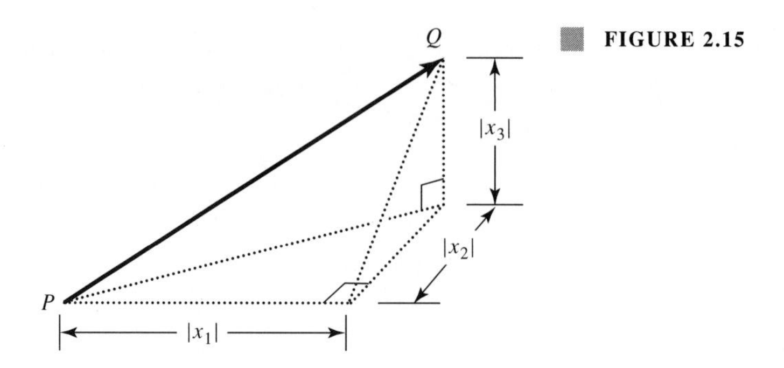

FIGURE 2.15

pairs the elements of $\mathcal{G}^3$ with those of $\mathcal{R}^3$. In $\mathcal{G}^3$ the operations of addition and scalar multiplication are defined in a manner analogous to that in $\mathcal{G}^2$, and the three-dimensional analogues of Theorems 2.1 through 2.4 continue to hold. Also, if $\mathbf{x} \in \mathcal{G}^3$ and $\mathbf{x} \longleftrightarrow [x_1 \quad x_2 \quad x_3]^T$, then $\|\mathbf{x}\| = \sqrt{x_1^2 + x_2^2 + x_3^2}$ (Figure 2.15).

Exercises 2.2

1. Assume **x**, **y**, and **z** are nonzero elements of $\mathcal{G}^2$, with $\overrightarrow{PQ} \in \mathbf{x}$, $\overrightarrow{QR} \in \mathbf{y}$, and $\overrightarrow{RS} \in \mathbf{z}$. Using directed line segment representatives of the different vectors occurring in $\mathbf{x} + (\mathbf{y} + \mathbf{z}) = (\mathbf{x} + \mathbf{y}) + \mathbf{z}$, sketch a figure that supports this statement.

2. Let $P = (2,-1,3)$, $Q = (-1,2,0)$, $R = (3,-2,4)$, and $S = (3,2,1)$.
   a. Are P, Q, and R collinear?
   b. How about Q, R, and S?

3. If $\mathbf{x,y,z} \in \mathcal{G}^3$, $\mathbf{x} \longleftrightarrow [1 \quad 0 \quad -1]^T$, $\mathbf{y} \longleftrightarrow [3 \quad 1 \quad 1]^T$, and $\mathbf{z} \longleftrightarrow [2 \quad 1 \quad 0]^T$, find
   a. $\|2\mathbf{z} - \mathbf{y}\|$. b. $\|\mathbf{x} - \mathbf{y} + 2\mathbf{z}\|$. **c.** $\|2(\mathbf{y} - \mathbf{x}) - 3\mathbf{z}\|$.

4. Assume P, Q, R, and S are noncoplanar points in space and consider the parallelepiped illustrated in the accompanying figure. Let **x**, **y**, and **z** be the geometric vectors represented

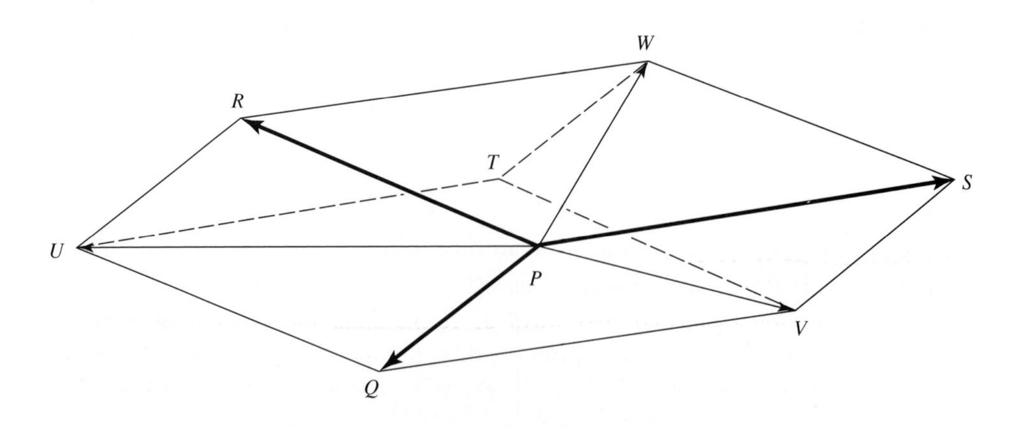

by $\overrightarrow{PQ}$, $\overrightarrow{PR}$, and $\overrightarrow{PS}$, respectively. Find the relationship between the geometric vector represented by $\overrightarrow{PT}$ and the sum of the geometric vectors represented by $\overrightarrow{PU}$, $\overrightarrow{PV}$, and $\overrightarrow{PW}$.

5. Assume P and Q are distinct points in space and let **m** and **b** be the vectors in $\mathcal{G}^3$ represented by $\overrightarrow{PQ}$ and $\overrightarrow{OP}$, respectively. The line ℓ passing through P and Q is described by $\mathbf{x}(t) = t\mathbf{m} + \mathbf{b}$, $t \in \mathcal{R}$, in the sense that the tip of the directed line segment based at 0 that represents $\mathbf{x}(t)$ traces out the line as t varies from $-\infty$ to ∞. The corresponding equation in $\mathcal{R}^3$ is $\mathrm{X}(t) = t\mathrm{M} + \mathrm{B}$, $t \in \mathcal{R}$, where $\mathbf{m} \leftrightarrow \mathrm{M}$ and $\mathbf{b} \leftrightarrow \mathrm{B}$. The vectors M and B (or **m** and **b**) are called *direction* and *position* vectors, respectively, for ℓ. Find the equation in $\mathcal{R}^3$ for the line passing through P and Q when
   a. $P = (1,2,3)$, $Q = (-2,1,-3)$. b. $P = (0,0,0)$, $Q = (1,-2,4)$.
   c. $P = (2,-1,1)$, $Q = (1,2,1)$. d. $P = (1,-1,3)$, $Q = (-1,3,-5)$.
   e. $P = (0,1,1)$, $Q = (1,-2,1)$. f. $P = (-1,1,-2)$, $Q = (0,2,-1)$.

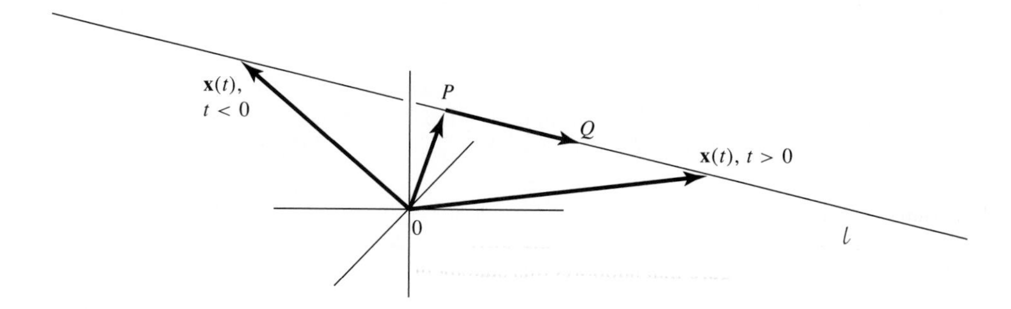

6. Which of the lines in Problem 5 are parallel?

7. Assume X and Y are elements of $\mathcal{R}^2$ such that 0, X, and Y are noncollinear.
   a. Draw a figure supporting the statement that for each $\mathrm{Z} \in \mathcal{R}^2$, there are scalars s and t such that $\mathrm{Z} = s\mathrm{X} + t\mathrm{Y}$.
   b. Assuming instead that X and Y are in $\mathcal{R}^3$, describe geometrically the set $\{s\mathrm{X} + t\mathrm{Y} : s,t \in \mathcal{R}\}$.

8. Suppose $\mathrm{M},\mathrm{B} \in \mathcal{R}^3$ and ℓ is the line with equation $\mathrm{X}(t) = t\mathrm{M} + \mathrm{B}$, $t \in \mathcal{R}$. Show that ℓ passes through 0 if and only if B is a scalar multiple of M.

9. Suppose $\mathrm{M},\mathrm{B},\mathrm{B}' \in \mathcal{R}^3$ and ℓ is the line with equation $\mathrm{X}(t) = t\mathrm{M} + \mathrm{B}$, $t \in \mathcal{R}$. Show that $\mathrm{Y}(s) = s\mathrm{M} + \mathrm{B}'$, $s \in \mathcal{R}$, describes ℓ if and only if $\mathrm{B} - \mathrm{B}'$ is a scalar multiple of M.

2.3 EUCLIDEAN n-SPACE

The previous section discussed two ways of formulating the concept of a two-dimensional vector. The first, resulting in $\mathcal{G}^2$ and its accompanying notions of addition and scalar multiplication, was motivated by physical and geometric ideas. The second was algebraic, utilizing elements of $\mathcal{R}^2$ as vectors, with addition and scalar multiplication being that of 2×1 matrices. Thanks to the correspondence $\mathbf{x} \leftrightarrow \mathrm{X} = [x_1 \quad x_2]^T$, $\mathcal{G}^2$ and $\mathcal{R}^2$ are just different expressions of a single fundamental concept. Theorems 2.2 and 2.3 state that you can algebraically manipulate elements of one of these sets using the operations in that setting or you can perform the same manipulation on the corresponding elements in the other set using the operations there, and the results will correspond to one another. This means the two approaches are algebraically indistinguishable. Each has its advantages. The first allows you to visualize vectors and engage your geometric and physical intuition; the second facilitates calculation and lends itself to extending vector notions to higher dimensional settings.

DEFINITION. Given a positive integer n, the term *Euclidean n-space* refers to the set

$$\mathcal{R}^n = \left\{ \begin{bmatrix} x_1 \\ \vdots \\ x_n \end{bmatrix} : x_1, \ldots, x_n \in \mathcal{R} \right\}.$$

An $\mathrm{X} = [x_1 \quad \cdots \quad x_n]^T \in \mathcal{R}^n$ is called an *n-dimensional vector,* and x_j is its jth *coordinate*, $1 \leq j \leq n$. Addition and scalar multiplication for elements of $\mathcal{R}^n$ are defined as follows. If $t \in \mathcal{R}$, $\mathrm{X} = [x_1 \quad \cdots \quad x_n]^T$, and $\mathrm{Y} = [y_1 \quad \cdots \quad y_n]^T$, then

$$\mathrm{X} + \mathrm{Y} = \begin{bmatrix} x_1 + x_1 \\ \vdots \\ x_n + y_n \end{bmatrix} \quad \text{and} \quad t\mathrm{X} = \begin{bmatrix} tx_1 \\ \vdots \\ tx_n \end{bmatrix}.$$

Elements of $\mathcal{R}^n$ are simply $n \times 1$ matrices. Since the operations of addition and scalar multiplication for these vectors are identical to those for $n \times 1$ matrices, the rules governing their use have already been established. They are listed below for future reference.

THEOREM 2.5. If X,Y,Z $\in \mathcal{R}^n$ and $s,t \in \mathcal{R}$, then

(1) $\mathrm{X} + \mathrm{Y} = \mathrm{Y} + \mathrm{X}$.
(2) $\mathrm{X} + (\mathrm{Y} + \mathrm{Z}) = (\mathrm{X} + \mathrm{Y}) + \mathrm{Z}$.
(3) there is a unique element of $\mathcal{R}^n$, denoted by 0, such that $\mathrm{X} + 0 = \mathrm{X}$ for each $\mathrm{X} \in \mathcal{R}^n$.
(4) for each $\mathrm{X} \in \mathcal{R}^n$, there is a unique element of $\mathcal{R}^n$, denoted by $-\mathrm{X}$, such that $\mathrm{X} + (-\mathrm{X}) = 0$.
(5) $(s + t)\mathrm{X} = s\mathrm{X} + t\mathrm{X}$.
(6) $s(\mathrm{X} + \mathrm{Y}) = s\mathrm{X} + s\mathrm{Y}$.
(7) $s(t\mathrm{X}) = (st)\mathrm{X}$.
(8) $1\mathrm{X} = \mathrm{X}$.
(9) $(-1)\mathrm{X} = -\mathrm{X}$.
(10) $t\mathrm{X} = 0$ if and only if either $t = 0$ or $\mathrm{X} = 0$.

The vector 0 in property (3) is the $n \times 1$ matrix with zero entries, that is, $0 = [0 \quad \cdots \quad 0]^T$. Notice that 0 is used here in two different ways. On the left it stands for an n-dimensional vector; on the right it denotes the real number occuring in each coordinate of that vector. This ambiguity is acceptable because the reader can determine the correct interpretation of the symbol by its use in the sentence.

The vector $-X$ in property (4) is called the additive inverse of X. If

$$X = \begin{bmatrix} x_1 \\ \vdots \\ x_n \end{bmatrix}, \qquad \text{then} \qquad -X = \begin{bmatrix} -x_1 \\ \vdots \\ -x_n \end{bmatrix}.$$

■ **EXAMPLE 2.9.** In $\mathcal{R}^4$, if $X = \begin{bmatrix} 1 \\ 2 \\ 3 \\ 4 \end{bmatrix}$, $Y = \begin{bmatrix} 4 \\ 3 \\ 2 \\ 1 \end{bmatrix}$, and $Z = \begin{bmatrix} 0 \\ -1 \\ -2 \\ -3 \end{bmatrix}$, then

$$3X + ((-1)Y + 2Z) = \begin{bmatrix} 3 \\ 6 \\ 9 \\ 12 \end{bmatrix} + \left(\begin{bmatrix} -4 \\ -3 \\ -2 \\ -1 \end{bmatrix} + \begin{bmatrix} 0 \\ -2 \\ -4 \\ -6 \end{bmatrix} \right)$$

$$= \begin{bmatrix} 3 \\ 6 \\ 9 \\ 12 \end{bmatrix} + \begin{bmatrix} -4 \\ -5 \\ -6 \\ -7 \end{bmatrix} = \begin{bmatrix} -1 \\ 1 \\ 3 \\ 5 \end{bmatrix}. \square$$

It follows from properties (1) and (2) of Theorem 2.5 that the result of summing three or more elements of $\mathcal{R}^n$ is independent of both the ordering and the grouping of the vectors, so the sum of $X_1, \ldots, X_k \in \mathcal{R}^n$ will be written simply as $X_1 + X_2 + \cdots + X_k$.

There is no inherent notion of direction for elements of $\mathcal{R}^n$ when $n \geq 4$, but there is an algebraically natural way to extend the concept of magnitude to vectors in $\mathcal{R}^n$.

DEFINITION. If $X = [x_1 \quad \cdots \quad x_n]^T \in \mathcal{R}^n$, then the *length* or *magnitude* of X, denoted by $\|X\|$, is

$$\|X\| = \sqrt{x_1^2 + \cdots + x_n^2}.$$

■ **EXAMPLE 2.10.** In $\mathcal{R}^5$, the length of $X = [1 \quad 0 \quad -1 \quad 2 \quad -3]^T$ is

$$\|X\| = \sqrt{1^2 + 0^2 + (-1)^2 + 2^2 + (-3)^2} = \sqrt{15}. \square$$

The next result verifies that this notion of length has properties that are to be expected of a concept bearing that name.

THEOREM 2.6. If $X \in \mathcal{R}^n$ and $t \in \mathcal{R}$, then

(1) $\|X\| \geq 0$,
(2) $\|X\| = 0$ if and only if $X = 0$, and
(3) $\|tX\| = |t|\|X\|$.

Proof. See Exercise 4.

DEFINITION. Nonzero vectors $X, Y \in \mathcal{R}^n$ are said to be *parallel* if one is a scalar multiple of the other. When $Y = tX$, X and Y have the *same* or *opposite directions*, depending on whether $t > 0$ or $t < 0$, respectively.

■ **EXAMPLE 2.11.** Let $X = \begin{bmatrix} 1 \\ 0 \\ -1 \\ 2 \\ -3 \end{bmatrix}$, $Y = \begin{bmatrix} -2 \\ 0 \\ 2 \\ -4 \\ 6 \end{bmatrix}$, and $Z = \begin{bmatrix} 3 \\ 0 \\ -3 \\ 5 \\ -9 \end{bmatrix}$. Observe that X and Y have opposite directions, because $Y = -2X$. If there were a $t \in \mathcal{R}$ such that $Z = tX$, then comparing first coordinates of Z and tX reveals that t would have to be 3. But the fourth coordinates of Z and 3X do not agree, so Z is not parallel to X. Is Z parallel to Y? □

Subtraction for n-dimensional vectors is the same as subtraction for $n \times 1$ matrices. If

$$X = \begin{bmatrix} x_1 \\ \vdots \\ x_n \end{bmatrix} \quad \text{and} \quad Y = \begin{bmatrix} y_1 \\ \vdots \\ y_n \end{bmatrix}, \quad \text{then} \quad X - Y = \begin{bmatrix} x_1 - y_1 \\ \vdots \\ x_n - y_n \end{bmatrix}.$$

■ **EXAMPLE 2.12.** If $X = \begin{bmatrix} -1 \\ 2 \\ -3 \end{bmatrix}$, $Y = \begin{bmatrix} 0 \\ 1 \\ -1 \end{bmatrix}$, and $Z = \begin{bmatrix} 2 \\ 1 \\ 2 \end{bmatrix}$, then

$$X - (2Y - 3Z) = \begin{bmatrix} -1 \\ 2 \\ -3 \end{bmatrix} - \left(\begin{bmatrix} 0 \\ 2 \\ -2 \end{bmatrix} - \begin{bmatrix} 6 \\ 3 \\ 6 \end{bmatrix} \right) = \begin{bmatrix} -1 \\ 2 \\ -3 \end{bmatrix} - \begin{bmatrix} -6 \\ -1 \\ -8 \end{bmatrix} = \begin{bmatrix} 5 \\ 3 \\ 5 \end{bmatrix}. \square$$

Exercises 2.3

1. Assuming $X = \begin{bmatrix} 1 \\ -2 \\ 0 \\ 3 \end{bmatrix}$, $Y = \begin{bmatrix} -3 \\ 2 \\ 1 \\ -1 \end{bmatrix}$, and $Z = \begin{bmatrix} 0 \\ 1 \\ -2 \\ 1 \end{bmatrix}$, find

   a. $(-2)X + 3Y$,
   b. $Y - 2(X - Z)$,
   c. $\|Z + X\|$,
   d. a vector of length $3\sqrt{14}$ that has the same direction as X, and
   e. a vector of length 1 whose direction is opposite that of $Y - Z$.

2. If $X = \begin{bmatrix} -2 \\ 1 \\ 4 \\ -3 \\ -4 \end{bmatrix}$, $Y = \begin{bmatrix} 1 \\ 0 \\ -1 \\ 2 \\ 1 \end{bmatrix}$, and $Z = \begin{bmatrix} 0 \\ -1 \\ 0 \\ 1 \\ 1 \end{bmatrix}$, find

   a. $Y - 3Z + X$.
   b. a vector of length 10 that is parallel to $Y - Z$.

c. all W $\in \mathcal{R}^5$ such that X + 2Y + W is parallel to Z.
d. Which, if any, of the W's in part (c) have length 3?

3. Suppose $X = \begin{bmatrix} 1 \\ -1 \\ 1 \\ 1 \end{bmatrix}$, $Y = \begin{bmatrix} 1 \\ 1 \\ 1 \\ -1 \end{bmatrix}$, $Z = \begin{bmatrix} -1 \\ -1 \\ 1 \\ 1 \end{bmatrix}$, and $W = \begin{bmatrix} 1 \\ -1 \\ -1 \\ 1 \end{bmatrix}$. Can you find
   a. scalars r, s, and t such that $rX + sY + tZ = W$?
   b. nonzero scalars r and s such that $rX + sY$ is parallel to Z?
   c. scalars r and s such that $\|rX + sY\| = 6$?

4. Prove Theorem 2.6.

5. Show that if $X \in \mathcal{R}^n$ and $X \neq 0$, then $U = \left(\frac{1}{\|X\|}\right)X$ is a vector of length 1 that has the same direction as X.

6. Assuming $X \in \mathcal{R}^n$, show that $X^T X = \|X\|^2$.

7. Show that $\|X - Y\|^2 + \|X + Y\|^2 = 2(\|X\|^2 + \|Y\|^2)$, $X, Y \in \mathcal{R}^n$. Interpret this conclusion in $\mathcal{R}^2$ as a statement about the diagonals of a parallelogram.

8. The distance between X and Y in $\mathcal{R}^n$ is defined to be $d(X,Y) = \|X - Y\|$.
   a. Show that $d(X,Y) = d(Y,X)$.
   b. Assuming $t \in \mathcal{R}$, what is the relationship between $d(tX,tY)$ and $d(X,Y)$?
   c. Assuming X and Y are distinct, show that $\frac{1}{2}(X + Y)$ is equidistant from X and Y.
   d. Sketch a figure in $\mathcal{R}^2$ illustrating X, Y, and $\frac{1}{2}(X + Y)$, and explain the geometric significance of the last point.

9. In $\mathcal{R}^2$, a point mass m located at $X = [x_1 \quad x_2]^T$ has a *moment* about the vertical axis given by mx_1. This quantity measures the tendency of the mass to rotate about the axis. The moment about the vertical axis of a system consisting of a finite number of point masses is

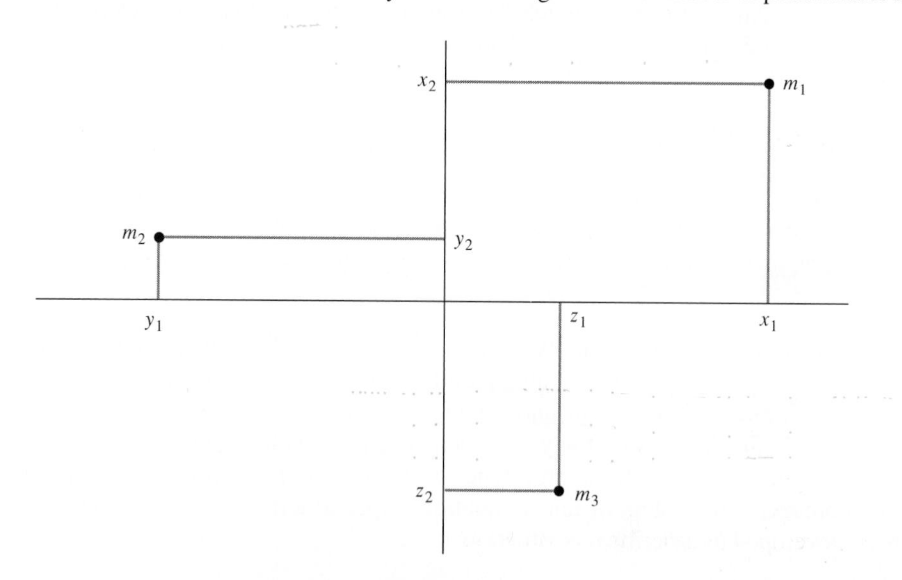

defined to be the sum of the moments of the individual masses. Moments about the horizontal axis are defined similarly. Given a system of masses m_1, m_2, and m_3 located at X, Y, and Z, respectively, show that a single mass of size $m = m_1 + m_2 + m_3$ located at

$$C = \frac{1}{m}(m_1\mathrm{X} + m_2\mathrm{Y} + m_3\mathrm{Z})$$

has the same moments as the system. (C is called the *center of mass* or *centroid* of the system.)

10. In $\mathcal{R}^3$, the sphere centered at 0 of radius 1 is $\mathcal{S} = \{\mathrm{X} \in \mathcal{R}^3 : \|\mathrm{X}\| = 1\}$. Assuming

$$\mathrm{X}_0 = \begin{bmatrix} a \\ b \\ c \end{bmatrix} \in \mathcal{S},$$

show that the plane tangent to $\mathcal{S}$ at X_0 has equation $ax + by + cz = 1$.

2.4

SUBSPACES

For $\mathrm{A} \in \mathcal{M}_{m\times n}$,

$$\mathcal{V} = \{\mathrm{X} \in \mathcal{R}^n : \mathrm{AX} = 0\}$$

consists of the solutions of the linear homogeneous system with coefficient matrix A. Every linear homogeneous system is consistent (0 is a solution), so $\mathcal{V}$ has at least one element. If $\mathrm{X,Y} \in \mathcal{V}$, then $\mathrm{AX} = 0$ and $\mathrm{AY} = 0$, so

$$\mathrm{A(X + Y) = AX + AY} = 0 + 0 = 0,$$

and therefore X + Y is an element of $\mathcal{V}$. Moreover, if $\mathrm{X} \in \mathcal{V}$ and $t \in \mathcal{R}$, then

$$\mathrm{A}(t\mathrm{X}) = t(\mathrm{AX}) = t0 = 0,$$

so $t\mathrm{X} \in \mathcal{V}$ as well. These observations indicate that the solution set for a linear homogeneous system has a certain "algebraic structure." The sum of any two elements of the set is another element of the set, and any scalar multiple of an element of the set is an element of the set.

DEFINITION. Suppose $\mathcal{V}$ is a nonempty subset of $\mathcal{R}^n$.

1. If $\mathrm{X + Y} \in \mathcal{V}$ whenever $\mathrm{X} \in \mathcal{V}$ and $\mathrm{Y} \in \mathcal{V}$, then $\mathcal{V}$ is *closed under addition.*
2. If $t\mathrm{X} \in \mathcal{V}$ whenever $\mathrm{X} \in \mathcal{V}$ and $t \in \mathcal{R}$, then $\mathcal{V}$ is *closed under scalar multiplication.*

A *subspace* of $\mathcal{R}^n$ is a subset that is closed under both operations.

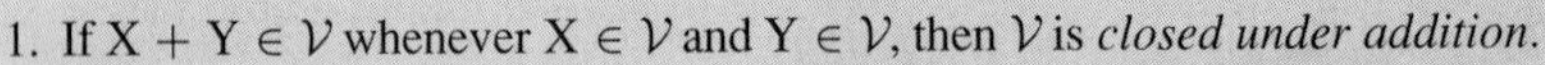

A subspace, $\mathcal{V}$, of $\mathcal{R}^n$ is simply a subset of $\mathcal{R}^n$ that is "algebraically self-contained." That is to say, any algebraic manipulation of elements of $\mathcal{V}$ using addition and/or scalar multiplication results in another element of $\mathcal{V}$. In particular, since $\mathrm{X - Y = X + (-1)Y}$, the difference of any two vectors in $\mathcal{V}$ is an element of $\mathcal{V}$.

As pointed out above, the prototype for the notion of subspace is the solution set of a homogeneous system of linear equations. Special terminology and notation have been developed to describe this situation.

DEFINITION. Given $A \in \mathcal{M}_{m\times n}$, the set of solutions of the homogeneous linear system with coefficient matrix A is called the *null space* of A, denoted by $\mathcal{N}(A)$. That is,

$$\mathcal{N}(A) = \{X \in \mathcal{R}^n \ : \ AX = 0\}.$$

THEOREM 2.7. If $A \in \mathcal{M}_{m\times n}$, then $\mathcal{N}(A)$ is a subspace of $\mathcal{R}^n$.

■ **EXAMPLE 2.13.** Consider the plane $\mathcal{V} = \{X \in \mathcal{R}^3 \ : \ ax_1 + bx_2 + cx_3 = d\}$ and suppose $X, Y \in \mathcal{V}$. Then $ax_1 + bx_2 + cx_3 = d$ and $ay_1 + by_2 + cy_3 = d$. The coordinates of X + Y satisfy the equation of the plane precisely when

$$\begin{aligned} d &= a(x_1 + y_1) + b(x_2 + y_2) + c(x_3 + y_3) \\ &= (ax_1 + bx_2 + cx_3) + (ay_1 + by_2 + cy_3) \\ &= d + d, \end{aligned}$$

that is, when $d = 0$. Moreover, $tX \in \mathcal{V}$ if and only if

$$d = a(tx_1) + b(tx_2) + c(tx_3) = t(ax_1 + bx_2 + cx_3) = td,$$

and d must be 0 if this equation is to hold for all $t \in \mathcal{R}$. Thus, $\mathcal{V}$ is closed under addition and scalar multiplication if and only if $d = 0$, a condition that characterizes planes passing through the origin. In other words, the planes in $\mathcal{R}^3$, which are subspaces of $\mathcal{R}^3$, are those

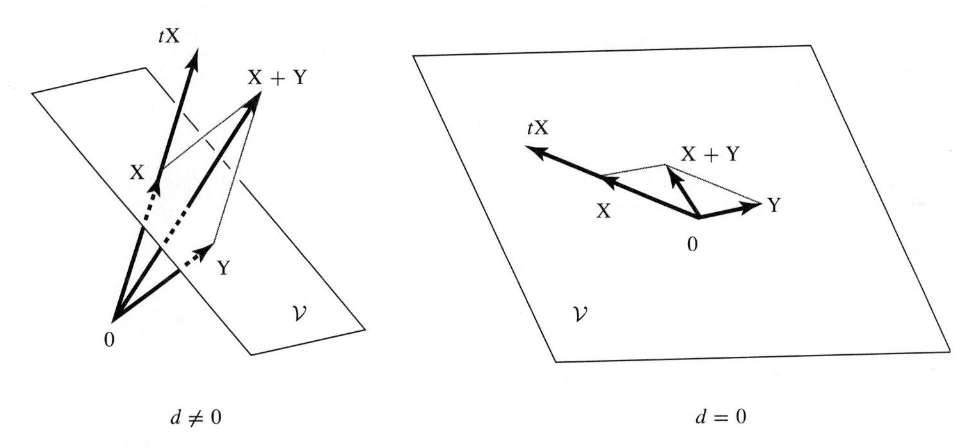

FIGURE 2.16

that contain 0. Figure 2.16 gives a geometric interpretation of this conclusion. Note that if $A = [a \quad b \quad c]$, then

$$ax_1 + bx_2 + cx_3 = [a \quad b \quad c]\begin{bmatrix} x_1 \\ x_2 \\ x_3 \end{bmatrix} = AX,$$

so when $d = 0$, $\mathcal{V} = \mathcal{N}(A)$. □

■ **EXAMPLE 2.14.** Let $\mathcal{V} = \{X \in \mathcal{R}^2 \ : \ x_1x_2 \geq 0\}$. This set consists of the first and third quadrants together with the two coordinate axes (Figure 2.17). If $X \in \mathcal{V}$ (i.e., $x_1x_2 \geq 0$) and $t \in \mathcal{R}$, then $tX = \begin{bmatrix} tx_1 \\ tx_2 \end{bmatrix}$, and $(tx_1)(tx_2) = t^2(x_1x_2) \geq 0$, so $tX \in \mathcal{V}$. Thus, $\mathcal{V}$ is closed under

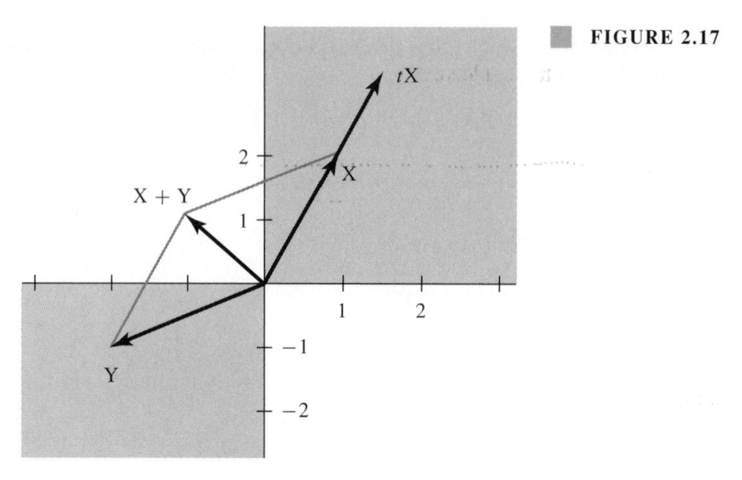

FIGURE 2.17

scalar multiplication. However, $X = \begin{bmatrix} 1 \\ 2 \end{bmatrix} \in \mathcal{V}$, $Y = \begin{bmatrix} -2 \\ -1 \end{bmatrix} \in \mathcal{V}$, and $X + Y = \begin{bmatrix} -1 \\ 1 \end{bmatrix} \notin \mathcal{V}$, so $\mathcal{V}$ is not closed under addition. Thus, $\mathcal{V}$ is not a subspace of $\mathcal{R}^2$. □

■ **EXAMPLE 2.15.** Let $\mathcal{V}$ be a line in $\mathcal{R}^3$ passing through 0 with direction vector M. Then $M \neq 0$, and $\mathcal{V} = \{tM \ : \ t \in \mathcal{R}\}$. If $X, Y \in \mathcal{V}$, then $X = sM$ and $Y = tM$, $s,t \in \mathcal{R}$, so

$$X + Y = sM + tM = (s + t)M \in \mathcal{V}.$$

Moreover, for $r \in \mathcal{R}$,

$$rX = r(sM) = (rs)M \in \mathcal{V}.$$

Thus, $\mathcal{V}$, being closed under addition and scalar multiplication, is a subspace of $\mathcal{R}^3$. These conclusions also follow readily from geometric considerations. Figure 2.18 illustrates the

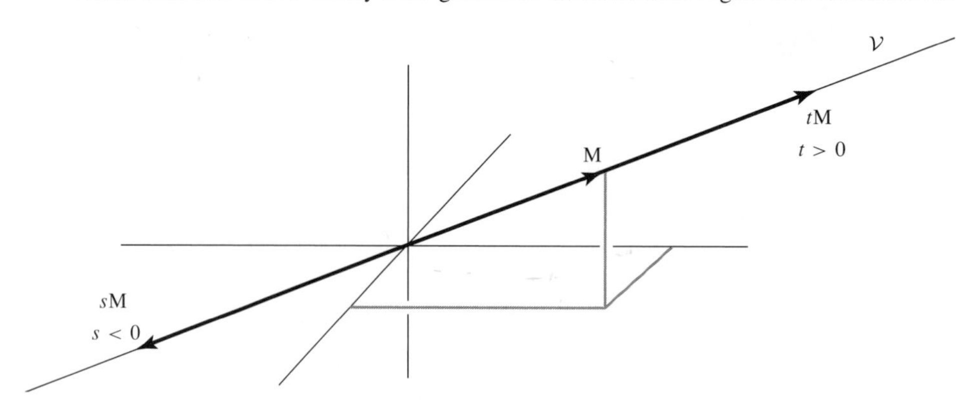

FIGURE 2.18

fact that $\mathcal{V}$ is closed under scalar multiplication, and the reader is invited to supply a figure supporting the claim that $\mathcal{V}$ is closed under addition. □

■ **EXAMPLE 2.16.** There are two "extreme" subsets of $\mathcal{R}^n$ that obviously satisfy the subspace criteria, namely $\{0\}$ and $\mathcal{R}^n$ itself. They are the "smallest" and "largest" subspaces of

$\mathcal{R}^n$, respectively, and are often referred to as the *trivial subspaces*. Each can be viewed as the null space of a matrix. Indeed,

$$\{0\} = \{X \in \mathcal{R}^n : I_n X = 0\} = \mathcal{N}(I_n),$$

and $\mathcal{R}^n$ is the null space of the $m \times n$ zero matrix. □

Examples 2.13 and 2.15 show that any line or plane in $\mathcal{R}^3$ that passes through 0 is a subspace of $\mathcal{R}^3$. These are, in fact, the only nontrivial subspaces of $\mathcal{R}^3$.

THEOREM 2.8. If $\mathcal{V}$ is a subspace of $\mathcal{R}^3$, then $\mathcal{V}$ is either $\{0\}$, a line through 0, a plane through 0, or $\mathcal{R}^3$ itself.

Proof. Suppose $\mathcal{V}$ is a subspace of $\mathcal{R}^3$ other than $\{0\}$. Then $\mathcal{V}$ contains a nonzero vector X, and since $\mathcal{V}$ is closed under scalar multiplication,

$$\{tX : t \in \mathcal{R}\} \subseteq \mathcal{V}.$$

It may be that $\mathcal{V} = \{tX : t \in \mathcal{R}\}$, in which case $\mathcal{V}$ is the line through 0 with direction vector X. Alternatively, $\mathcal{V}$ contains a vector Y that is not a scalar multiple of X. In that event,

$$\{sX : s \in \mathcal{R}\} \subseteq \mathcal{V} \qquad \text{and} \qquad \{tY : t \in \mathcal{R}\} \subseteq \mathcal{V},$$

and, since $\mathcal{V}$ is closed under addition,

$$\{sX + tY : s,t \in \mathcal{R}\} \subseteq \mathcal{V}.$$

Note that the vectors $V = sX + tY$, $s,t \in \mathcal{R}$, constitute the plane p passing through 0, X, and Y (Figure 2.19). If $\mathcal{V} = p$, then, of course, $\mathcal{V}$ is a plane through 0. Otherwise, $\mathcal{V}$ contains

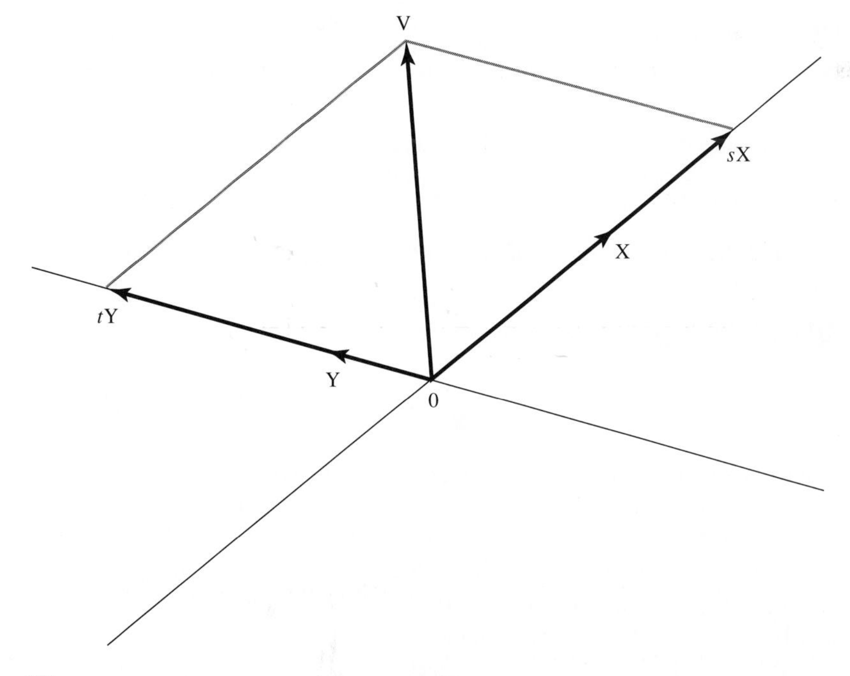

FIGURE 2.19

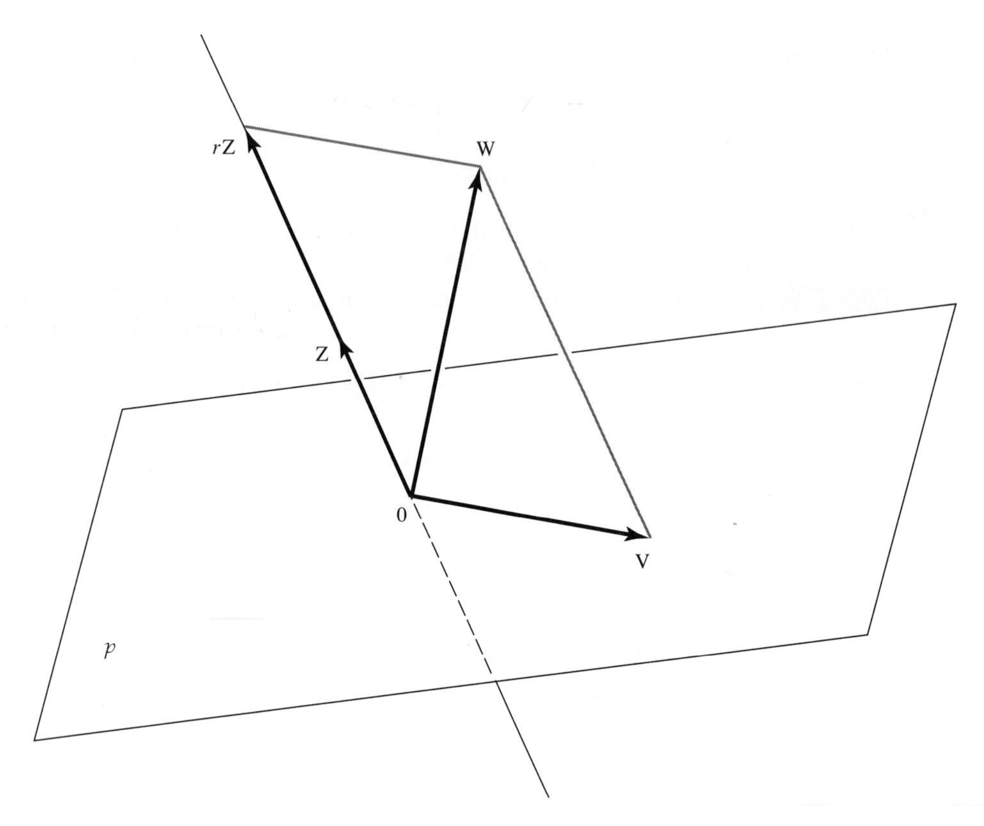

FIGURE 2.20

a vector Z that does not lie in $\mathcal{p}$. In that case, each $W \in \mathcal{R}^3$ can be written as $W = V + rZ$ for some $r \in \mathcal{R}$ and $V \in \mathcal{p}$ (Figure 2.20), and since $\mathcal{V}$ is closed under addition and scalar multiplication, $W \in \mathcal{V}$. Thus, $\mathcal{V} = \mathcal{R}^3$. ■

Observe that each subspace of $\mathcal{R}^3$ contains the zero vector. Subspaces in general have this feature.

THEOREM 2.9. If $\mathcal{V}$ is a subspace of $\mathcal{R}^n$, then $0 \in \mathcal{V}$.

Proof. Let $X \in \mathcal{V}$. Since $\mathcal{V}$ is closed under scalar multiplication, $tX \in \mathcal{V}$ for each $t \in \mathcal{R}$. In particular, when $t = 0$, $0 = 0X \in \mathcal{V}$. ■

Consider the nonhomogeneous system $AX = B$, where $A \in \mathcal{M}_{m\times n}$ and $B \in \mathcal{R}^m$. Assume it is consistent and set $\mathcal{V} = \{X \in \mathcal{R}^n : AX = B\}$. If $X,Y \in \mathcal{V}$, then $AX = B$, $AY = B$, and $A(X + Y) = AX + AY = B + B = 2B \neq B$, so $X + Y \notin \mathcal{V}$. Moreover, if $X \in \mathcal{V}$ and $t \in \mathcal{R}$, then $A(tX) = t(AX) = tB$, so $tX = \mathcal{V}$ only when $t = 1$. Thus, $\mathcal{V}$ is not a subspace of $\mathcal{R}^n$. It is, however, closely related to the null space of A. Suppose $X_0 \in \mathcal{V}$. If $Y \in \mathcal{N}(A)$ and $X = X_0 + Y$, then

$$AX = A(X_0 + Y) = AX_0 + AY = B + 0 = B,$$

so $X \in \mathcal{V}$. Conversely, if $X \in \mathcal{V}$, then

$$A(X - X_0) = AX - AX_0 = B - B = 0,$$

so $Y = X - X_0 \in \mathcal{N}(A)$ and $X = X_0 + Y$. These observations are recorded in the next theorem.

THEOREM 2.10. If $A \in \mathcal{M}_{m\times n}$, $B \in \mathcal{R}^m$, and X_0 is a solution of $AX = B$, then

$$\{X : AX = B\} = \{X_0 + Y : Y \in \mathcal{N}(A)\}.$$

■ **EXAMPLE 2.17.** Let $\mathcal{V}$ be the set of solutions of

$$\begin{aligned} x + 2y &= -1 \\ x + \quad z &= \ \ 3. \end{aligned}$$

These equations represent planes in $\mathcal{R}^3$, and $\mathcal{V}$ is their line of intersection. The matrix description for the system is $AX = B$, where

$$A = \begin{bmatrix} 1 & 2 & 0 \\ 1 & 0 & 1 \end{bmatrix}, \quad X = \begin{bmatrix} x \\ y \\ z \end{bmatrix}, \quad \text{and} \quad B = \begin{bmatrix} -1 \\ 3 \end{bmatrix}.$$

Since

$$[A \mid B] = \left[\begin{array}{ccc|c} 1 & 2 & 0 & -1 \\ 1 & 0 & 1 & 3 \end{array}\right] \longrightarrow \left[\begin{array}{ccc|c} 1 & 0 & 1 & 3 \\ 0 & 1 & -1/2 & -2 \end{array}\right],$$

$z = t$, $y = -2 + t/2$, and $x = 3 - t$, or equivalently,

$$X = \begin{bmatrix} 3 - t \\ -2 + t/2 \\ t \end{bmatrix} = \begin{bmatrix} 3 \\ -2 \\ 0 \end{bmatrix} + t \begin{bmatrix} -1 \\ 1/2 \\ 1 \end{bmatrix}.$$

Thus, $\mathcal{V} = \{X_0 + tM : t \in \mathcal{R}\}$, where $X_0 = \begin{bmatrix} 3 \\ -2 \\ 0 \end{bmatrix}$ and $M = \begin{bmatrix} -1 \\ 1/2 \\ 1 \end{bmatrix}$. Since

$$A = \begin{bmatrix} 1 & 2 & 0 \\ 1 & 0 & 1 \end{bmatrix} \longrightarrow \begin{bmatrix} 1 & 0 & 1 \\ 0 & 1 & -1/2 \end{bmatrix},$$

the solutions of $AX = 0$ are $z = t$, $y = t/2$, and $x = -t$, that is,

$$\mathcal{N}(A) = \left\{ \begin{bmatrix} -t \\ t/2 \\ t \end{bmatrix} : t \in \mathcal{R} \right\} = \{tM : t \in \mathcal{R}\}.$$

Thus, each X in $\mathcal{V}$ is the sum of X_0 and an element of $\mathcal{N}(A)$. Geometrically, $\mathcal{N}(A)$ is the line through 0 with direction vector M, and $\mathcal{V}$ is its translation by X_0 (Figure 2.21). □

The solutions of a linear homogeneous system constitute the subspace $\mathcal{N}(A)$, A being the coefficient matrix. When the nonhomogeneous system $AX = B$ is consistent, its solution set is related to $\mathcal{N}(A)$ in the manner indicated in Theorem 2.10. Thus, the problem of understanding the "form" of such a solution set can be addressed by exploring the "algebraic structure" of a subspace. The next three sections are devoted to the latter project.

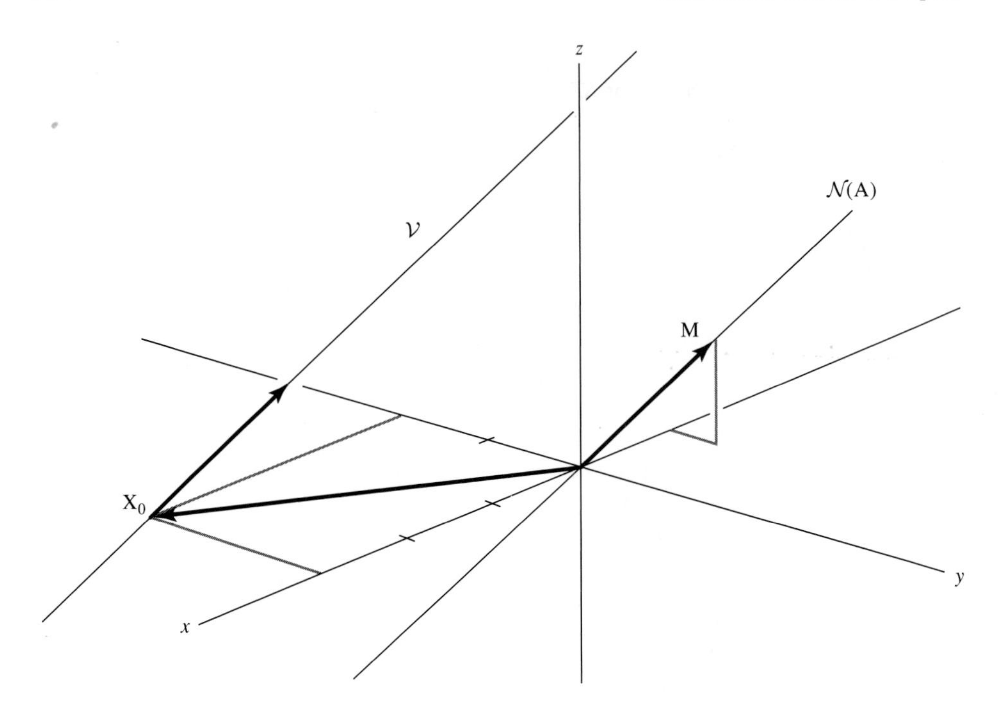

FIGURE 2.21

Exercises 2.4

1. Which of the following sets are closed under addition? Which are closed under scalar multiplication? Which are subspaces?
   a. $\{X \in \mathcal{R}^2 : x_1 + 2x_2 = 0\}$
   b. $\{X \in \mathcal{R}^2 : x_1 + 2x_2 \geq 3\}$
   c. $\{X \in \mathcal{R}^2 : |x_1| = |x_2|\}$
   d. $\{X \in \mathcal{R}^2 : x_1 \neq x_2\}$
   e. $\{X \in \mathcal{R}^3 : x_1 + 2x_2 = 0\}$
   f. $\{X \in \mathcal{R}^3 : \|X\| = 1\}$
   g. $\{X \in \mathcal{R}^3 : x_3 = 0\}$
   h. $\{X \in \mathcal{R}^3 : x_1 \geq 0, x_2 \geq 0, x_3 \geq 0\}$
   i. $\{X \in \mathcal{R}^3 : x_3^2 = x_1^2 + x_2^2\}$
   j. $\{X \in \mathcal{R}^3 : x_1x_2x_3 = 0\}$
   k. $\{X \in \mathcal{R}^3 : x_1 = x_2 = -x_3\}$

2. Suppose $X_0 \in \mathcal{R}^n$, $d \in \mathcal{R}$, and $\mathcal{V} = \{X \in \mathcal{R}^n : X^T X_0 = d\}$.
   a. Show that $\mathcal{V}$ is a subspace of $\mathcal{R}^n$ if and only if $d = 0$.
   b. What is the geometric interpretation of $\mathcal{V}$ when $n = 2$? When $n = 3$?

3. Make a sketch that illustrates geometrically why a line in $\mathcal{R}^2$ that does not pass through 0 is not a suspace of $\mathcal{R}^2$.

4. Assuming X,Y $\in \mathcal{R}^n$, show that $\mathcal{V} = \{sX + tY : s,t \in \mathcal{R}\}$ is a subspace of $\mathcal{R}^n$.

5. If $\mathcal{U}$ and $\mathcal{V}$ are subspaces of $\mathcal{R}^n$, show that
   a. $\mathcal{U} \cap \mathcal{V}$ is a subspace of $\mathcal{R}^n$.
   b. $\mathcal{U} \cup \mathcal{V}$ is closed under scalar multiplication.
   c. Find an example in $\mathcal{R}^2$ which shows that $\mathcal{U} \cup \mathcal{V}$ need not be a subspace.

6. Suppose $\mathcal{U}$ and $\mathcal{V}$ are subspaces of $\mathcal{R}^n$ and let $\mathcal{W} = \{U + V : U \in \mathcal{U} \text{ and } V \in \mathcal{V}\}$.
   a. Show that $\mathcal{W}$ is a subspace of $\mathcal{R}^n$.
   b. Describe $\mathcal{W}$ when $\mathcal{U},\mathcal{V} \subseteq \mathcal{R}^3$ are two different lines through 0.
   c. Describe $\mathcal{W}$ when $\mathcal{U},\mathcal{V} \subseteq \mathcal{R}^3$, $\mathcal{U}$ is a plane through 0, $\mathcal{V}$ is a line through 0, and $\mathcal{V}$ does not lie in $\mathcal{U}$.

7. Let $M = \begin{bmatrix} 2 \\ 1 \\ -3 \end{bmatrix}$ and $\mathcal{V} = \{tM : t \in \mathcal{R}\}$. Find $A \in \mathcal{M}_{2\times 3}$ such that $\mathcal{V} = \mathcal{N}(A)$.

8. What is $\mathcal{N}(A)$ when $A \in \mathcal{M}_{n\times n}$ is nonsingular?

9. Suppose $A \in \mathcal{M}_{m\times n}$ and $\mathcal{W} = \{B \in \mathcal{R}^m : AX = B \text{ is consistent}\}$. Show that $\mathcal{W}$ is a subspace of $\mathcal{R}^m$.

10. Suppose $A \in \mathcal{M}_{m\times n}$ and $P \in \mathcal{M}_{m\times m}$.
   a. Show that $\mathcal{N}(A) \subseteq \mathcal{N}(PA)$.
   b. Show that $\mathcal{N}(PA) = \mathcal{N}(A)$ when P is nonsingular.

11. Solve $AX = B$ and verify that each solution is of the form $X_0 + Y$, where X_0 is a particular solution of the system and $Y \in \mathcal{N}(A)$.

   a. $A = \begin{bmatrix} 1 & -1 & 2 \\ -1 & 2 & -1 \end{bmatrix}$, $B = \begin{bmatrix} 7 \\ -9 \end{bmatrix}$ b. $A = \begin{bmatrix} 1 & 1 & -2 & 1 \\ -1 & 1 & 0 & -3 \\ 0 & -1 & 1 & 1 \end{bmatrix}$, $B = \begin{bmatrix} -2 \\ 4 \\ -1 \end{bmatrix}$.

2.5

SPANNING SETS

The term "algebraic structure," as applied to a subset of $\mathcal{R}^n$, refers to the properties of the subset associated with the operations of addition and scalar multiplication. The subsets of interest here are the subspaces of $\mathcal{R}^n$, which are distinguished by being closed under the two operations, and the present section begins the study of their structure by exploring the most immediate consequences of these closure properties.

Suppose $\mathcal{V}$ is a subspace of $\mathcal{R}^n$ and $X \in \mathcal{V}$. Since $\mathcal{V}$ is closed under scalar multiplication, $tX \in \mathcal{V}$ for each $t \in \mathcal{R}$. The collection of all scalar multiples of X is called the span of X, denoted by $\mathcal{S}pan\{X\}$, that is,

$$\mathcal{S}pan\{X\} = \{tX : t \in \mathcal{R}\} \subseteq \mathcal{V}.$$

If $X,Y \in \mathcal{V}$, then $\mathcal{V}$ contains both $\mathcal{S}pan\{X\}$ and $\mathcal{S}pan\{Y\}$. Moreover, being closed under addition, $\mathcal{V}$ contains any vector expressible as the sum of an element of $\mathcal{S}pan\{X\}$ and an element of $\mathcal{S}pan\{Y\}$, that is,

$$\{sX + tY : s,t \in \mathcal{R}\} \subseteq \mathcal{V}.$$

The vectors $sX + tY$ are called linear combinations of X and Y, and the set of all such vectors is referred to as the span of $\{X,Y\}$, denoted by $\mathcal{S}pan\{X,Y\}$. These ideas extend readily to sets consisting of a finite number of vectors.

DEFINITION. If $X_1, \ldots, X_p \in \mathcal{R}^n$ and $t_1, \ldots, t_p \in \mathcal{R}$, then

$$t_1X_1 + \cdots + t_pX_p = \sum_1^p t_kX_k$$

is a *linear combination* of $X_1, \ldots, X_p$, and $\mathcal{S}pan\{X_1, \ldots, X_p\}$ is the set of all such linear combinations. If $\mathcal{V} = \mathcal{S}pan\{X_1, \ldots, X_p\}$, then $X_1, \ldots, X_p$ are said to *span* $\mathcal{V}$, or alternatively, $\{X_1, \ldots, X_p\}$ is called a *spanning set* for $\mathcal{V}$.

■ **EXAMPLE 2.18.** A nonzero $X \in \mathcal{R}^3$ spans the subspace represented geometrically by the line through 0 with direction vector X (Example 2.15). If $X, Y \in \mathcal{R}^3$ are nonzero and not parallel, then $\mathcal{S}pan\{X,Y\}$ is the plane determined by 0, X, and Y (Figure 2.19). □

The solution set for a linear homogeneous system is a subspace. Its description is obtained by row reducing the coefficient matrix, and as the next two examples illustrate, that description implicitly involves a spanning set for the subspace.

■ **EXAMPLE 2.19.** Since $A = \begin{bmatrix} -1 & 1 & -2 & -5 & -2 \\ -2 & 2 & 1 & 0 & -4 \\ 1 & -1 & 1 & 3 & 2 \end{bmatrix} \longrightarrow \begin{bmatrix} 1 & -1 & 0 & 1 & 2 \\ 0 & 0 & 1 & 2 & 0 \\ 0 & 0 & 0 & 0 & 0 \end{bmatrix}$,

the solutions of $AX = 0$ are

$$X = \begin{bmatrix} x_1 \\ x_2 \\ x_3 \\ x_4 \\ x_5 \end{bmatrix} = \begin{bmatrix} r - s - 2t \\ r \\ -2s \\ s \\ t \end{bmatrix} = r\begin{bmatrix} 1 \\ 1 \\ 0 \\ 0 \\ 0 \end{bmatrix} + s\begin{bmatrix} -1 \\ 0 \\ -2 \\ 1 \\ 0 \end{bmatrix} + t\begin{bmatrix} -2 \\ 0 \\ 0 \\ 0 \\ 1 \end{bmatrix},$$

$r,s,t \in \mathcal{R}$. Thus, $\mathcal{N}(A) = \mathcal{S}pan\{X_1,X_2,X_3\}$, where

$$X_1 = \begin{bmatrix} 1 \\ 1 \\ 0 \\ 0 \\ 0 \end{bmatrix}, \quad X_2 = \begin{bmatrix} -1 \\ 0 \\ -2 \\ 1 \\ 0 \end{bmatrix}, \quad \text{and} \quad X_3 = \begin{bmatrix} -2 \\ 0 \\ 0 \\ 0 \\ 1 \end{bmatrix}. \ \square$$

■ **EXAMPLE 2.20.** Let $\mathcal{V} = \{X \in \mathcal{R}^3 : x_1 - 2x_2 + 3x_3 = 0\}$. Then $\mathcal{V}$, being a plane containing 0, is a subspace of $\mathcal{R}^3$. Its elements are the solutions of the homogeneous "system" with coefficient matrix $A = [1 \ \ -2 \ \ 3]$, and you will note that A is in reduced row echelon form. The solutions are

$$\begin{bmatrix} x_1 \\ x_2 \\ x_3 \end{bmatrix} = \begin{bmatrix} 2s - 3t \\ s \\ t \end{bmatrix} = s\begin{bmatrix} 2 \\ 1 \\ 0 \end{bmatrix} + t\begin{bmatrix} -3 \\ 0 \\ 1 \end{bmatrix},$$

$s,t \in \mathcal{R}$, so

$$\mathcal{V} = \mathcal{S}pan\left\{\begin{bmatrix} 2 \\ 1 \\ 0 \end{bmatrix}, \begin{bmatrix} -3 \\ 0 \\ 1 \end{bmatrix}\right\}. \ \square$$

The span of $\{X_1, \ldots, X_p\} \subseteq \mathcal{R}^n$ is simply the collection of all vectors that can be generated from $X_1, \ldots, X_p$ using addition and scalar multiplication. It should come as no surprise that this set is closed under the two operations.

THEOREM 2.11. If $X_1, \ldots, X_p \in \mathcal{R}^n$, then $\mathcal{S}pan\{X_1, \ldots, X_p\}$ is a subspace of $\mathcal{R}^n$.

Proof. Suppose $X, Y \in \mathcal{S}pan\{X_1, \ldots, X_p\}$, that is, $X = \sum_1^p s_k X_k$ and $Y = \sum_1^p t_k X_k$, $s_k, t_k \in \mathcal{R}$. Then

$$X + Y = \sum_1^p s_k X_k + \sum_1^p t_k X_k = \sum_1^p (s_k + t_k) X_k.$$

The last equation displays $X + Y$ as a linear combination of $X_1, \ldots, X_p$, so $X + Y \in \mathcal{S}pan\{X_1, \ldots, X_p\}$. For each $r \in \mathcal{R}$,

$$rX = r\left(\sum_1^p s_k X_k\right) = \sum_1^p (r s_k) X_k,$$

so rX is also a linear combination of $X_1, \ldots, X_p$. ■

Given $X_1, \ldots, X_p \in \mathcal{R}^n$, each X_j can be expressed as a linear combination of $X_1, \ldots, X_p$ by writing

$$X_j = \sum_1^p t_k X_k,$$

where $t_j = 1$ and $t_k = 0$ for $k \neq j$. Thus, $X_j \in \mathcal{S}pan\{X_1, \ldots, X_p\}$. In fact, $\mathcal{S}pan\{X_1, \ldots, X_p\}$ is the "smallest" subspace of $\mathcal{R}^n$ containing $X_1, \ldots, X_p$. The precise meaning of this statement is the subject of the next theorem.

THEOREM 2.12. If $\mathcal{V}$ is a subspace of $\mathcal{R}^n$ containing $X_1, \ldots, X_p$, then

$$\mathcal{S}pan\{X_1, \ldots, X_p\} \subseteq \mathcal{V}.$$

Proof. Let $t_1, \ldots, t_p \in \mathcal{R}$. Since $X_1, \ldots, X_p \in \mathcal{V}$ and $\mathcal{V}$ is closed under scalar multiplication, $t_1 X_1, \ldots, t_p X_p \in \mathcal{V}$. Moreover, $\mathcal{V}$ is closed under addition, so $t_1 X_1 + \cdots + t_p X_p \in \mathcal{V}$. Thus, each element of $\mathcal{S}pan\{X_1, \ldots, X_p\}$ is in $\mathcal{V}$. ■

■ **EXAMPLE 2.21.** Let $\mathcal{V} = \mathcal{S}pan\{X, Y\}$, where $X = \begin{bmatrix} 1 \\ 2 \\ 1 \end{bmatrix}$ and $Y = \begin{bmatrix} -1 \\ 1 \\ -2 \end{bmatrix}$, and consider the problem of deciding whether a given $B \in \mathcal{R}^3$ is an element of $\mathcal{V}$. For example, $B = \begin{bmatrix} 3 \\ 2 \\ 1 \end{bmatrix} \in \mathcal{V}$ if and only if there are scalars s and t such that

$$\begin{bmatrix} 3 \\ 2 \\ 1 \end{bmatrix} = s\begin{bmatrix} 1 \\ 2 \\ 1 \end{bmatrix} + t\begin{bmatrix} -1 \\ 1 \\ -2 \end{bmatrix} = \begin{bmatrix} s - t \\ 2s + t \\ s - 2t \end{bmatrix} = \begin{bmatrix} 1 & -1 \\ 2 & 1 \\ 1 & -2 \end{bmatrix}\begin{bmatrix} s \\ t \end{bmatrix}.$$

You solve for s and t by row reducing the augmented matrix of the system. Since

$$\left[\begin{array}{cc|c} 1 & -1 & 3 \\ 2 & 1 & 2 \\ 1 & -2 & 1 \end{array}\right] \longrightarrow \left[\begin{array}{cc|c} 1 & 0 & 5 \\ 0 & 1 & 2 \\ 0 & 0 & 10 \end{array}\right],$$

the system is inconsistent, and therefore $B \notin \mathcal{V}$. However, if $B = \begin{bmatrix} -4 \\ 1 \\ -7 \end{bmatrix}$ then

$$\left[\begin{array}{cc|c} 1 & -1 & -4 \\ 2 & 1 & 1 \\ 1 & -2 & -7 \end{array}\right] \longrightarrow \left[\begin{array}{cc|c} 1 & 0 & -1 \\ 0 & 1 & 3 \\ 0 & 0 & 0 \end{array}\right].$$

This time the system is consistent, the unique solution is $s = -1, t = 3$, and

$$B = (-1)X + 3Y \in \mathcal{S}pan\{X, Y\}.$$

Observe that 0, X, and Y are noncollinear, so $\mathcal{V}$ is the plane containing these three points (Example 2.18). Moreover, $B = [x \quad y \quad z]^T \in \mathcal{V}$ if and only if

$$\begin{bmatrix} 1 & -1 \\ 2 & 1 \\ 1 & -2 \end{bmatrix} \begin{bmatrix} s \\ t \end{bmatrix} = \begin{bmatrix} x \\ y \\ z \end{bmatrix}$$

is consistent. Since

$$\left[\begin{array}{cc|c} 1 & -1 & x \\ 2 & 1 & y \\ 1 & -2 & z \end{array}\right] \longrightarrow \left[\begin{array}{cc|c} 1 & 0 & 2x - z \\ 0 & 1 & x - z \\ 0 & 0 & -5x + y + 3z \end{array}\right],$$

the system has a solution precisely when $-5x + y + 3z = 0$. Note that the coordinates of $\begin{bmatrix} -4 \\ 1 \\ -7 \end{bmatrix}$ satisfy this equation, but those of $\begin{bmatrix} 3 \\ 2 \\ 1 \end{bmatrix}$ do not. □

In general, to decide if $B \in \mathcal{S}pan\{X_1, \ldots, X_p\}$, where $X_1, \ldots, X_p, B \in \mathcal{R}^n$, set $A = [X_1 \quad \cdots \quad X_p]$ and $T = \begin{bmatrix} t_1 \\ \vdots \\ t_p \end{bmatrix}$ and observe that

$$AT = \begin{bmatrix} a_{11} & \cdots & a_{1p} \\ \vdots & & \vdots \\ a_{n1} & \cdots & a_{np} \end{bmatrix} \begin{bmatrix} t_1 \\ \vdots \\ t_p \end{bmatrix} = \begin{bmatrix} a_{11}t_1 + \cdots + a_{1p}t_p \\ \vdots \\ a_{n1}t_1 + \cdots + a_{np}t_p \end{bmatrix}$$

$$= t_1 \begin{bmatrix} a_{11} \\ \vdots \\ a_{n1} \end{bmatrix} + \cdots + t_p \begin{bmatrix} a_{1p} \\ \vdots \\ a_{np} \end{bmatrix}$$

$$= t_1 X_1 + \cdots + t_p X_p,$$

that is,

$$\boxed{AT = \sum_{1}^{p} t_k X_k}. \tag{2.1}$$

Thus, $B \in \mathcal{Span}\{X_1, \ldots, X_p\}$ if and only if there is a choice of T such that $B = \sum_1^p t_k X_k = AT$. Row reducing $[A \mid B]$ decides the issue. In fact, you can test several vectors, $B_1, \ldots, B_q$, simultaneously for membership in $\mathcal{Span}\{X_1, \ldots, X_p\}$ by row reducing $[A \mid B_1 \quad \cdots \quad B_q]$.

■ **EXAMPLE 2.22.** Let

$$X_1 = \begin{bmatrix} 1 \\ -1 \\ 0 \\ 2 \end{bmatrix}, \quad X_2 = \begin{bmatrix} 1 \\ 0 \\ 2 \\ 1 \end{bmatrix}, \quad X_3 = \begin{bmatrix} -1 \\ 1 \\ 0 \\ 0 \end{bmatrix}, \quad B_1 = \begin{bmatrix} 1 \\ 0 \\ 3 \\ 0 \end{bmatrix}, \quad B_2 = \begin{bmatrix} -2 \\ -1 \\ -6 \\ 1 \end{bmatrix},$$

and set $A = [X_1 \quad X_2 \quad X_3]$. Then

$$[A \mid B_1 \quad B_2] = \left[\begin{array}{rrr|rr} 1 & 1 & -1 & 1 & -2 \\ -1 & 0 & 1 & 0 & -1 \\ 0 & 2 & 0 & 3 & -6 \\ 2 & 1 & 0 & 0 & 1 \end{array}\right] \longrightarrow \left[\begin{array}{rrr|rr} 1 & 0 & 0 & 1/2 & 2 \\ 0 & 1 & 0 & 1 & -3 \\ 0 & 0 & 1 & -1/2 & 1 \\ 0 & 0 & 0 & 1 & 0 \end{array}\right].$$

Referring to the last matrix, the nonzero entry in the 4,4 position implies that $B_1 \notin \mathcal{Span}\{X_1, X_2, X_3\}$. On the other hand, the last column indicates that $B_2 \in \mathcal{Span}\{X_1, X_2, X_3\}$ and that $B_2 = 2X_1 - 3X_2 + X_3$. □

Equation (2.1) is a very useful way of viewing linear combinations, and hereafter you will see it used repeatedly in these pages. Expressed slightly differently, it states that if $A \in \mathcal{M}_{m\times n}$ and $X = [x_1 \quad \cdots \quad x_n]^T$, then

$$AX = \sum_1^n x_k \mathrm{Col}_k(A).$$

In particular, this offers a new perspective on the question of consistency for nonhomogeneous linear systems. Indeed, for $B \in \mathcal{R}^m$, it follows that $AX = B$ is consistent if and only if B is a linear combination of the columns of A.

DEFINITION. Given $A \in \mathcal{M}_{m\times n}$, $\mathcal{Span}\{\mathrm{Col}_1(A), \ldots, \mathrm{Col}_n(A)\}$ is called the *column space* of A, denoted by $\mathcal{C}(A)$.

THEOREM 2.13. If $A \in \mathcal{M}_{m\times n}$, then $\mathcal{C}(A)$ is a subspace of $\mathcal{R}^m$.

Proof. This conclusion is a direct consequence of Theorem 2.11. ■

THEOREM 2.14. If $A \in \mathcal{M}_{m\times n}$ and $B \in \mathcal{R}^m$, then $AX = B$ is consistent if and only if $B \in \mathcal{C}(A)$.

Exercises 2.5

1. Describe geometrically (in words or with a sketch)

a. $\mathcal{Span}\left\{\begin{bmatrix} 2 \\ -1 \end{bmatrix}\right\}$, **b.** $\mathcal{Span}\left\{\begin{bmatrix} 2 \\ -1 \end{bmatrix}, \begin{bmatrix} 1 \\ 3 \end{bmatrix}\right\}$ c. $\mathcal{Span}\left\{\begin{bmatrix} 2 \\ -1 \end{bmatrix}, \begin{bmatrix} -4 \\ 2 \end{bmatrix}\right\}$.

2. Find all values of $y \in \mathcal{R}$ such that $\begin{bmatrix} 5 \\ y \\ -4 \end{bmatrix} \in \mathcal{S}pan\left\{\begin{bmatrix} 1 \\ 1 \\ 0 \end{bmatrix}, \begin{bmatrix} 0 \\ 1 \\ -1 \end{bmatrix}\right\}$.

3. **a.** If $E_1 = \begin{bmatrix} 1 \\ 0 \\ 0 \end{bmatrix}$, $E_2 = \begin{bmatrix} 0 \\ 1 \\ 0 \end{bmatrix}$, and $E_3 = \begin{bmatrix} 0 \\ 0 \\ 1 \end{bmatrix}$, show that $\mathcal{R}^3 = \mathcal{S}pan\{E_1, E_2, E_3\}$.

   b. For $k = 1, \ldots, n$, let E_k be the vector in $\mathcal{R}^n$ with kth coordinate 1 and 0's for its other coordinates. Show that $\mathcal{S}pan\{E_1, \ldots, E_n\} = \mathcal{R}^n$.

4. Find a spanning set for the null space of each matrix.

   a. $\begin{bmatrix} 1 & -2 & 3 \\ -4 & 5 & -6 \\ 7 & -8 & 9 \end{bmatrix}$ b. $\begin{bmatrix} 1 & 3 & -1 & 2 & 1 \\ -2 & -6 & 2 & 0 & 2 \\ 0 & 0 & 1 & -1 & 7 \\ 2 & 6 & -2 & 2 & 0 \end{bmatrix}$ **c.** $\begin{bmatrix} 1 & 1 & 2 \\ 1 & 2 & 1 \\ 2 & 3 & 3 \\ 0 & 1 & -1 \end{bmatrix}$

   d. $\begin{bmatrix} 1 & 1 & 1 & -1 \\ 2 & 3 & 0 & -2 \\ 1 & 2 & -1 & 1 \end{bmatrix}$ **e.** $\begin{bmatrix} 1 & -3 & 0 & -1 & 0 & -1 \\ -1 & 3 & 1 & 3 & 0 & 3 \\ 2 & -6 & 1 & 0 & -1 & -1 \\ -1 & 3 & 2 & 5 & 0 & 5 \end{bmatrix}$ f. $\begin{bmatrix} 1 & 2 \\ 3 & 4 \\ 5 & 6 \end{bmatrix}$

5. Let $\mathcal{V}$ be the subspace spanned by the X's. For each B, determine whether $B \in \mathcal{V}$, and if so, find all the ways in which B can be expressed as a linear combination of the X's.

   a. $X_1 = \begin{bmatrix} 1 \\ 0 \\ 1 \\ 0 \end{bmatrix}$, $X_2 = \begin{bmatrix} -1 \\ 2 \\ 0 \\ 1 \end{bmatrix}$, $X_3 = \begin{bmatrix} 0 \\ 1 \\ 2 \\ -1 \end{bmatrix}$;

   $B_1 = \begin{bmatrix} 3 \\ -1 \\ 4 \\ 2 \end{bmatrix}$, $B_2 = \begin{bmatrix} 1 \\ 1 \\ 1 \\ 1 \end{bmatrix}$, $B_3 = \begin{bmatrix} -3 \\ 1 \\ -7 \\ 5 \end{bmatrix}$

   b. $X_1 = \begin{bmatrix} 1 \\ -2 \\ 3 \end{bmatrix}$, $X_2 = \begin{bmatrix} 0 \\ 1 \\ -1 \end{bmatrix}$, $X_3 = \begin{bmatrix} 1 \\ 0 \\ 1 \end{bmatrix}$;

   $B_1 = \begin{bmatrix} 2 \\ -8 \\ 10 \end{bmatrix}$, $B_2 = \begin{bmatrix} 2 \\ 2 \\ 0 \end{bmatrix}$, $B_3 = \begin{bmatrix} 1 \\ 1 \\ 1 \end{bmatrix}$

   c. $X_1 = \begin{bmatrix} 1 \\ 1 \\ -1 \end{bmatrix}$, $X_2 = \begin{bmatrix} 2 \\ 0 \\ 1 \end{bmatrix}$; $B_1 = \begin{bmatrix} 1 \\ -3 \\ 5 \end{bmatrix}$, $B_2 = \begin{bmatrix} 1 \\ 3 \\ 3 \end{bmatrix}$, $B_3 = \begin{bmatrix} 7 \\ 1 \\ 2 \end{bmatrix}$

   d. $X_1 = \begin{bmatrix} 1 \\ 0 \\ -1 \end{bmatrix}$, $X_2 = \begin{bmatrix} 2 \\ 2 \\ 1 \end{bmatrix}$, $X_3 = \begin{bmatrix} 1 \\ -2 \\ -4 \end{bmatrix}$, $X_4 = \begin{bmatrix} 0 \\ 2 \\ 3 \end{bmatrix}$;

   $B_1 = \begin{bmatrix} 3 \\ 2 \\ 3 \end{bmatrix}$, $B_2 = \begin{bmatrix} 7 \\ 6 \\ 2 \end{bmatrix}$, $B_3 = \begin{bmatrix} 0 \\ 2 \\ 3 \end{bmatrix}$

6. Let $\mathcal{V} = \mathcal{S}pan\{X_1, X_2\}$ and $\mathcal{W} = \mathcal{S}pan\{Y_1, Y_2\}$, where

$$X_1 = \begin{bmatrix} 1 \\ 0 \\ 1 \end{bmatrix}, \quad X_2 = \begin{bmatrix} 2 \\ 1 \\ 1 \end{bmatrix}, \quad Y_1 = \begin{bmatrix} 1 \\ 1 \\ 0 \end{bmatrix}, \quad Y_2 = \begin{bmatrix} 0 \\ 1 \\ -1 \end{bmatrix}.$$

Show that $\mathcal{V} = \mathcal{W}$. (Hint: Show that $X_1, X_2 \in \mathcal{W}$ and $Y_1, Y_2 \in \mathcal{V}$ and explain why these conclusions imply $\mathcal{V} \subseteq \mathcal{W}$ and $\mathcal{W} \subseteq \mathcal{V}$, respectively.)

7. Show that $\begin{bmatrix} a \\ b \\ c \end{bmatrix} \in \mathcal{S}pan\left\{ \begin{bmatrix} 1 \\ 1 \\ 0 \end{bmatrix}, \begin{bmatrix} 0 \\ 1 \\ 1 \end{bmatrix}, \begin{bmatrix} 1 \\ 0 \\ 1 \end{bmatrix} \right\}$ for all $a,b,c \in \mathcal{R}$.

8. What conditions on x, y, and z distinguish the vectors $\begin{bmatrix} x \\ y \\ z \end{bmatrix}$ in $\mathcal{C}(A)$?

a. $A = \begin{bmatrix} 1 & 2 & 5 \\ -1 & 1 & 4 \\ 1 & 1 & 2 \end{bmatrix}$ b. $A = \begin{bmatrix} 1 & 2 \\ 2 & 4 \\ -3 & -6 \end{bmatrix}$ **c.** $A = \begin{bmatrix} 1 & 0 & -1 & -1 \\ -1 & 1 & 0 & -1 \\ 0 & -1 & 1 & -1 \end{bmatrix}$

9. Find a spanning set for $\mathcal{U} \cap \mathcal{V}$ when
   a. $\mathcal{U} = \{X \in \mathcal{R}^3 : x_1 - 2x_2 + x_3 = 0\}$ and $\mathcal{V} = \{X \in \mathcal{R}^3 : 2x_1 - 2x_2 + 3x_3 = 0\}$.
   b. $\mathcal{U} = \mathcal{N}\left(\begin{bmatrix} 1 & 1 & 2 & 0 & 0 \\ -2 & 1 & 2 & 3 & -3 \end{bmatrix}\right)$ and $\mathcal{V} = \mathcal{N}\left(\begin{bmatrix} 0 & 1 & 2 & 1 & 3 \\ -1 & 0 & 0 & 1 & 2 \\ 2 & 1 & 2 & -1 & 0 \end{bmatrix}\right)$.

10. Let A be a nonzero 3×3 matrix. $\mathcal{C}(A)$ is a subspace of $\mathcal{R}^3$, and since $A \neq 0$, $\mathcal{C}(A) \neq \{0\}$. By Theorem 2.8, $\mathcal{C}(A)$ is a line through 0, a plane through 0, or $\mathcal{R}^3$ itself. How can you tell, based on the number of leading ones in rref(A), which of these options occurs?

11. Assuming $A \in \mathcal{M}_{m \times n}$ and $Q \in \mathcal{M}_{n \times n}$, show that
   a. $\text{Col}_j(AQ) \in \mathcal{C}(A)$, $j = 1, \ldots, n$.
   b. $\mathcal{C}(AQ) \subseteq \mathcal{C}(A)$.
   c. $\mathcal{C}(AQ) = \mathcal{C}(A)$ when Q is nonsingular.

2.6

LINEAR INDEPENDENCE

Suppose $\mathcal{V}$ is a plane in $\mathcal{R}^3$ that passes through 0 and let X and Y be nonzero, nonparallel vectors in $\mathcal{V}$. Then $\mathcal{V} = \mathcal{S}pan\{X, Y\}$. In fact, for each $Z \in \mathcal{V}$, there is exactly one parallelogram in $\mathcal{V}$ with diagonal $\overrightarrow{0Z}$ and sides parallel to $\overrightarrow{0X}$ and $\overrightarrow{0Y}$, so there is exactly one choice of scalars, r and s, such that $Z = rX + sY$ (Figure 2.22). Given $Z \in \mathcal{V}$, each linear combination of X and Y can be viewed as a linear combination of X, Y, and Z by writing

$$rX + sY = rX + sY + 0Z,$$

so $\{X,Y,Z\}$ is also a spanning set for $\mathcal{V}$. However, the vectors in $\mathcal{V}$ do not have unique representations as linear combinations of X, Y, and Z. Indeed, if $V \in \mathcal{V}$ and $t \in \mathcal{R}$, then

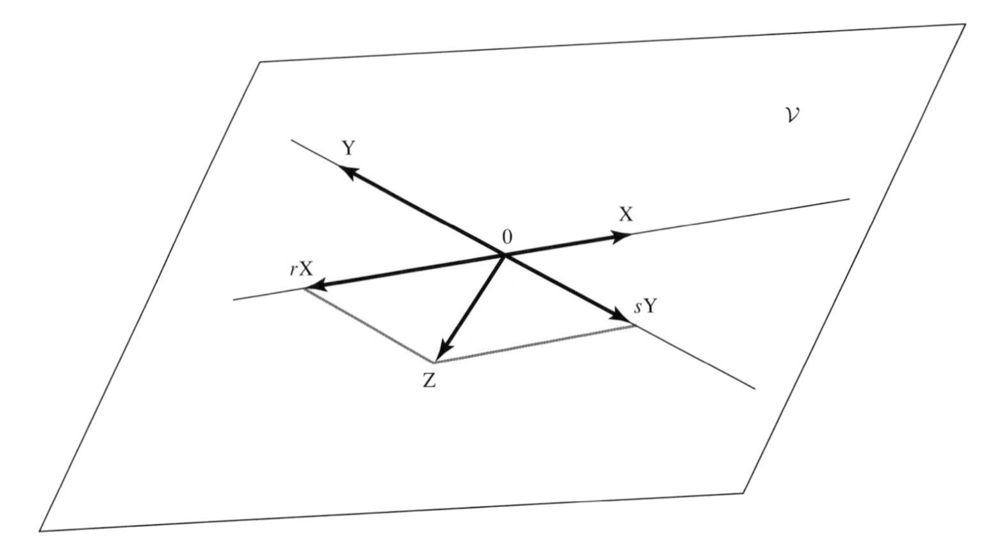

FIGURE 2.22

$V - tZ \in \mathcal{V}$, so there are scalars $r(t)$ and $s(t)$ such that

$$V - tZ = r(t)X + s(t)Y,$$

and therefore

$$V = r(t)X + s(t)Y + tZ.$$

Different values of t produce different expressions for V.

Similar reasoning shows that if $p \geq 2$ and $\{X_1, \ldots, X_p\}$ is a subset of $\mathcal{V}$ containing two nonzero nonparallel vectors, then $\mathcal{V} = \mathcal{S}pan\{X_1, \ldots, X_p\}$. However, only the spanning sets consisting of two vectors have the additional feature that each element of $\mathcal{V}$ can be uniquely written as a linear combination of the spanning vectors.

In general, a spanning set for a subspace is a vehicle for describing all members of the subspace in terms of a few select vectors. Maximal efficiency in this regard is achieved by using spanning sets with a minimal number of elements. For the plane discussed above, that optimal number is two. What is emerging here is a notion of "minimal spanning set." The next theorem introduces some key considerations involved in its formulation.

THEOREM 2.15. If $p \geq 2$, $\{X_1, \ldots, X_p\} \subseteq \mathcal{R}^n$, and $\mathcal{V} = \mathcal{S}pan\{X_1, \ldots, X_p\}$, then the following statements are equivalent:

(1) One of $X_1, \ldots, X_p$ is a linear combination of the others.
(2) A proper subset of $\{X_1, \ldots, X_p\}$ spans $\mathcal{V}$.
(3) There are scalars $t_1, \ldots, t_p$, at least one of which is not zero, such that $t_1X_1 + \cdots + t_pX_p = 0$.

Proof.

(1) $\Rightarrow$ (2). Assume one of $X_1, \ldots, X_p$ is a linear combination of the others. For the sake of argument, let it be X_p (otherwise $X_1, \ldots, X_p$ could be relabeled so that the vector

in question has index p). Then

$$X_p = t_1X_1 + \cdots + t_{p-1}X_{p-1},$$

where $t_1, \ldots, t_{p-1} \in \mathcal{R}$. For each $V \in \mathcal{V}$ there are scalars $s_1, \ldots, s_p$ such that $V = \sum_1^p s_kX_k$, and therefore

$$\begin{aligned} V &= s_1X_1 + \cdots + s_{p-1}X_{p-1} + s_pX_p \\ &= s_1X_1 + \cdots + s_{p-1}X_{p-1} + s_p(t_1X_1 + \cdots + t_{p-1}X_{p-1}) \\ &= (s_1 + s_pt_1)X_1 + \cdots + (s_{p-1} + s_pt_{p-1})X_{p-1}. \end{aligned}$$

The last expression displays V as a linear combination of $X_1, \ldots, X_{p-1}$, so $V \in \mathcal{S}pan\{X_1, \ldots, X_{p-1}\}$. This establishes that $\mathcal{V} \subseteq \mathcal{S}pan\{X_1, \ldots, X_{p-1}\}$. The reverse containment follows from Theorem 2.12, and therefore $\mathcal{V} = \mathcal{S}pan\{X_1, \ldots, X_{p-1}\}$.

(2) $\Rightarrow$ (3). Suppose a proper subset of $\{X_1, \ldots, X_p\}$ spans $\mathcal{V}$. It is then possible to remove one of $X_1, \ldots, X_p$ and have the remaining $p - 1$ vectors span $\mathcal{V}$. Assume the vectors are so labeled that X_p can be removed, that is, that $\mathcal{V} = \mathcal{S}pan\{X_1, \ldots, X_{p-1}\}$. Since $X_p \in \mathcal{V}$, there are scalars $t_1, \ldots, t_{p-1}$ such that

$$X_p = t_1X_1 + \cdots + t_{p-1}X_{p-1},$$

and therefore

$$t_1X_1 + \cdots + t_{p-1}X_{p-1} - X_p = 0.$$

The last equation expresses 0 as a linear combination of $X_1, \ldots, X_p$ with at least one of the scalars (the coefficient of X_p) being nonzero.

(3) $\Rightarrow$ (1). Assume there are scalars $t_1, \ldots, t_p$, at least one of which is not 0, such that

$$t_1X_1 + \cdots + t_pX_p = 0.$$

By relabeling, if necessary, you may assume $t_p \neq 0$. Then solving for X_p yields

$$X_p = (-t_1/t_p)X_1 + \cdots + (-t_{p-1}/t_p)X_{p-1},$$

which displays X_p as a linear combination of $X_1, \ldots, X_{p-1}$. ■

REMARK. The foregoing argument contains a very useful piece of information. It shows that you can remove from $\{X_1, \ldots, X_p\}$ any vector that is a linear combination of the others, and what remains is still a spanning set for $\mathcal{V}$.

The three statements in Theorem 2.15 are alternative formulations of a single fundamental property of $\{X_1, \ldots, X_p\}$. The condition $p \geq 2$ is needed for statements (1) and (2), but statement (3) also makes sense when $p = 1$. The last of the three is usually preferred for the formal definition of the property.

DEFINITION. Given $X_1, \ldots, X_p \in \mathcal{R}^n$, if there are scalars $t_1, \ldots, t_p$, at least one of which is not zero, such that

$$t_1X_1 + \cdots + t_pX_p = 0,$$

then $X_1, \ldots, X_p$ are said to be *linearly dependent*. Otherwise, $X_1, \ldots, X_p$ are *linearly independent*. One also speaks of the set $\{X_1, \ldots, X_p\}$ being linearly dependent or independent.

The zero vector can always be written as a linear combination of $X_1, \dots, X_p$ by setting $t_1 = \cdots = t_p = 0$. At issue here is whether that is the only way. If so, $\{X_1, \dots, X_p\}$ is linearly independent; if not, it is linearly dependent.

■ **EXAMPLE 2.23.** A set consisting of a single element, X, in $\mathcal{R}^n$ is linearly dependent if and only if there is a nonzero $t \in \mathcal{R}$ such that $tX = 0$. When $X = 0$, $tX = 0$ for any $t \in \mathcal{R}$, but when $X \neq 0$, $tX = 0$ only for $t = 0$. Thus, $\{X\}$ is linearly dependent if and only if $X = 0$. □

■ **EXAMPLE 2.24.** For nonzero vectors $X, Y \in \mathcal{R}^n$, statement (1) of Theorem 2.15 amounts to saying that one of X and Y is a scalar multiple of the other. Thus, $\{X,Y\}$ is linearly dependent precisely when X is parallel to Y. In $\mathcal{R}^2$ or $\mathcal{R}^3$, the latter condition is equivalent to 0, X, and Y being collinear. □

■ **EXAMPLE 2.25.** Suppose $X_1, \dots, X_p \in \mathcal{R}^n$ and $X_j = 0$ for some $j \in \{1, \dots, p\}$. If t_j is any nonzero scalar and $t_k = 0$ when $k \neq j$, then

$$\sum_1^p t_k X_k = 0X_1 + \cdots + 0X_{j-1} + t_j 0 + 0X_{j+1} + \cdots + 0X_p = 0.$$

One of the scalars is not zero, so $X_1, \dots, X_p$ are linearly dependent. Thus, any finite subset of $\mathcal{R}^n$ containing 0 is linearly dependent. □

You cannot usually tell by inspection whether a set consisting of three or more nonzero vectors is linearly independent, so it would be helpful to have an algorithm for deciding that issue. To that end, suppose $X_1, \dots, X_p \in \mathcal{R}^n$ and assume $t_1, \dots, t_p$ are scalars such that

$$0 = t_1X_1 + \cdots + t_pX_p.$$

Setting $A = [X_1 \quad \cdots \quad X_p]$ and $T = \begin{bmatrix} t_1 \\ \vdots \\ t_p \end{bmatrix}$, the last equation can be rewritten as

$$AT = 0$$

(recall equation (2.1)). Thus, linear dependence or independence of $X_1, \dots, X_p$ depends on whether or not the homogeneous system $AT = 0$ has solutions other than $T = 0$. Row reducing A tells the tale.

■ **EXAMPLE 2.26.** Suppose $X_1 = \begin{bmatrix} 1 \\ -1 \\ 0 \\ 1 \end{bmatrix}$, $X_2 = \begin{bmatrix} 2 \\ 0 \\ 1 \\ 1 \end{bmatrix}$, and $X_3 = \begin{bmatrix} -1 \\ 1 \\ 0 \\ 2 \end{bmatrix}$. Since

$$A = [X_1 \quad X_2 \quad X_3] = \begin{bmatrix} 1 & 2 & -1 \\ -1 & 0 & 1 \\ 0 & 1 & 0 \\ 1 & 1 & 2 \end{bmatrix} \longrightarrow \begin{bmatrix} 1 & 0 & 0 \\ 0 & 1 & 0 \\ 0 & 0 & 1 \\ 0 & 0 & 0 \end{bmatrix},$$

the only solution of $AT = 0$ is $T = 0$, and therefore the only scalars t_1, t_2, and t_3 such that $t_1X_1 + t_2X_2 + t_3X_3 = 0$ are $t_1 = t_2 = t_3 = 0$. Thus, $\{X_1, X_2, X_3\}$ is linearly independent. □

■ **EXAMPLE 2.27.** Let $\mathcal{V} = \mathcal{S}pan\{X_1,X_2,X_3\}$, where

$$X_1 = \begin{bmatrix} 1 \\ 2 \\ -1 \end{bmatrix}, \quad X_2 = \begin{bmatrix} 0 \\ 1 \\ -1 \end{bmatrix}, \quad \text{and} \quad X_3 = \begin{bmatrix} 2 \\ 1 \\ 1 \end{bmatrix}.$$

Observe that

$$A = [X_1 \quad X_2 \quad X_3] = \begin{bmatrix} 1 & 0 & 2 \\ 2 & 1 & 1 \\ -1 & -1 & 1 \end{bmatrix} \longrightarrow \begin{bmatrix} 1 & 0 & 2 \\ 0 & 1 & -3 \\ 0 & 0 & 0 \end{bmatrix} \begin{matrix} r_{12}(-2) \\ r_{13}(1) \\ r_{23}(1) \end{matrix},$$

so the solutions of $AT = 0$ are $t_1 = -2s$, $t_2 = 3s$, $t_3 = s$, $s \in \mathcal{R}$. Thus,

$$(-2s)\begin{bmatrix} 1 \\ 2 \\ -1 \end{bmatrix} + (3s)\begin{bmatrix} 0 \\ 1 \\ -1 \end{bmatrix} + s\begin{bmatrix} 2 \\ 1 \\ 1 \end{bmatrix} = \begin{bmatrix} 0 \\ 0 \\ 0 \end{bmatrix}.$$

Since 0 can be written as a linear combination of X_1, X_2, and X_3 in a nontrivial manner, $\{X_1,X_2,X_3\}$ is linearly dependent.

As long as $s \neq 0$, the last equation can be solved for any one of X_1, X_2, or X_3, expressing it as a linear combination of the other two. Thus, any one of the three can be discarded and the remaining vectors still span $\mathcal{V}$ (Remark following Theorem 2.15). That is,

$$\mathcal{S}pan\{X_1,X_2,X_3\} = \mathcal{S}pan\{X_1,X_2\} = \mathcal{S}pan\{X_1,X_3\} = \mathcal{S}pan\{X_2,X_3\}.$$

Remove X_3 and note that $\{X_1,X_2\}$ is linearly independent. In fact, the row reduction

$$[X_1 \quad X_2] = \begin{bmatrix} 1 & 0 \\ 2 & 1 \\ -1 & -1 \end{bmatrix} \longrightarrow \begin{bmatrix} 1 & 0 \\ 0 & 1 \\ 0 & 0 \end{bmatrix} \begin{matrix} r_{12}(-2) \\ r_{13}(1) \\ r_{23}(1) \end{matrix}$$

utilizes the same sequence of row operations that reduced $[X_1 \quad X_2 \quad X_3]$. This is because column three of $[X_1 \quad X_2 \quad X_3]$ is not a pivot column and, therefore, does not influence the choice of operations. In effect, $\text{rref}([X_1 \quad X_2])$ is obtained by striking the third column of $\text{rref}([X_1 \quad X_2 \quad X_3])$. You can check that $\{X_1,X_3\}$ and $\{X_2,X_3\}$ are also linearly independent, but you will see that the Gauss-Jordan algorithm uses different row operations in those cases. □

The system $AT = 0$ has a unique solution if and only if each column of rref(A) contains a leading one (Corollary to Theorem 1.3). This fact, together with equation (2.1), yields the following test for linear independence.

THEOREM 2.16. Let $A = [X_1 \quad \cdots \quad X_p]$, where $X_1, \ldots, X_p \in \mathcal{R}^n$. Then $\{X_1, \ldots, X_p\}$ is linearly independent if and only if each column of rref(A) contains a leading one.

Let $\mathcal{V} = \mathcal{S}pan\{X_1, \ldots, X_p\} \subseteq \mathcal{R}^n$. If no proper subset of $\{X_1, \ldots, X_p\}$ spans $\mathcal{V}$, then $\{X_1, \ldots, X_p\}$ is called a *minimal* spanning set for $\mathcal{V}$. By Theorem 2.15, $\{X_1, \ldots, X_p\}$ is minimal if and only if it is linearly independent. When $\{X_1, \ldots, X_p\}$ is linearly dependent, you can remove any vector that is a linear combination of the others and obtain a proper subset that spans $\mathcal{V}$ (Remark after Theorem 2.15). Thus, you might seek a minimal spanning set by repeating this process until the remaining set is linearly independent. The next example illustrates a means of choosing vectors that can be removed.

■ **EXAMPLE 2.28.** Let $X_1 = \begin{bmatrix} 1 \\ 1 \\ 0 \\ -1 \end{bmatrix}$, $X_2 = \begin{bmatrix} 0 \\ 1 \\ 1 \\ 0 \end{bmatrix}$, $X_3 = \begin{bmatrix} -1 \\ 1 \\ 2 \\ 1 \end{bmatrix}$, $X_4 = \begin{bmatrix} 0 \\ -1 \\ 1 \\ 1 \end{bmatrix}$,

$X_5 = \begin{bmatrix} 1 \\ 7 \\ 2 \\ -3 \end{bmatrix}$, and set $T = \begin{bmatrix} t_1 \\ \vdots \\ t_5 \end{bmatrix}$. Observe that

$$A = [X_1 \quad \cdots \quad X_5] = \begin{bmatrix} 1 & 0 & -1 & 0 & 1 \\ 1 & 1 & 1 & -1 & 7 \\ 0 & 1 & 2 & 1 & 2 \\ -1 & 0 & 1 & 1 & -3 \end{bmatrix} \longrightarrow \begin{bmatrix} 1 & 0 & -1 & 0 & 1 \\ 0 & 1 & 2 & 0 & 4 \\ 0 & 0 & 0 & 1 & -2 \\ 0 & 0 & 0 & 0 & 0 \end{bmatrix},$$

so the solutions of $AT = 0$ are

$$t_1 = s - t, \qquad t_2 = -2s - 4t, \qquad t_3 = s, \qquad t_4 = 2t, \qquad t_5 = t,$$

and

$$(s - t)X_1 - (2s + 4t)X_2 + sX_3 + (2t)X_4 + tX_5 = 0.$$

Setting $s = -1$ and $t = 0$, you get $X_3 = -X_1 + 2X_2 + 0X_4$, whereas $s = 0$ and $t = -1$ gives $X_5 = X_1 + 4X_2 - 2X_4$. Since X_3 and X_5 are linear combinations of X_1, X_2, and X_4,

$$\mathcal{Span}\{X_1,X_2,X_4\} = \mathcal{Span}\{X_1,X_2,X_3,X_4,X_5\}.$$

Note that columns 1, 2, and 4 are the pivot columns of A. They determine a sequence of operations that row reduces A, and those same operations yield

$$[X_1 \quad X_2 \quad X_4] \longrightarrow \begin{bmatrix} 1 & 0 & 0 \\ 0 & 1 & 0 \\ 0 & 0 & 1 \\ 0 & 0 & 0 \end{bmatrix}.$$

Thus, $\{X_1,X_2,X_4\}$ is linearly independent and, therefore, is a minimal spanning set for $\mathcal{V} = \mathcal{Span}\{X_1,X_2,X_3,X_4,X_5\}$. □

THEOREM 2.17. If $\{X_1, \ldots, X_p\} \subseteq \mathcal{R}^n$ and at least one of $X_1, \ldots, X_p$ is not 0, then the pivot columns of $A = [X_1 \quad \cdots \quad X_p]$ form a linearly independent spanning set for $\mathcal{V} = \mathcal{Span}\{X_1, \ldots, X_p\}$.

Proof. If $X_1, \ldots, X_p$ are linearly independent, then each column of A is a pivot column (Theorem 2.16), and there is nothing more to prove. Suppose then that $\{X_1, \ldots, X_p\}$ is linearly dependent. Let $B = \text{rref}(A)$ and set $T = \begin{bmatrix} t_1 \\ \vdots \\ t_p \end{bmatrix}$. The solutions of $AT = 0$ identify the scalars $t_1, \ldots, t_p$ such that

$$t_1X_1 + \cdots + t_pX_p = 0.$$

Let $j_1, \ldots, j_q$ be the indices of the pivot columns, $1 \leq j_1 < \cdots < j_q \leq p$, that is, $\text{Col}_k(B)$ contains a leading one for $k = j_1, \ldots, j_q$ and the remaining $p - q$ columns of B do not. In describing the solutions of $AT = 0$, the coordinates of T other than $t_{j_1}, \ldots, t_{j_q}$ are assigned arbitrary values, and $t_{j_1}, \ldots, t_{j_q}$ are then determined by those values and the entries in B. If

$i \neq j_1, \ldots, j_q$, then setting $t_i = -1$ and $t_k = 0$ for $k \neq i, j_1, \ldots, j_q$ produces values of $t_{j_1}, \ldots, t_{j_q}$ such that

$$0 = \mathrm{AT} = \sum_{1}^{p} t_k \mathrm{X}_k = \sum_{k=1}^{q} t_{j_k} \mathrm{X}_{j_k} - \mathrm{X}_i.$$

Thus, X_i is a linear combination of $\mathrm{X}_{j_1}, \ldots, \mathrm{X}_{j_q}$ and, therefore, can be deleted from $\{\mathrm{X}_1, \ldots, \mathrm{X}_p\}$ without diminishing the span. These remarks apply to each index other than $j_1, \ldots, j_q$, so

$$\mathcal{V} = \mathcal{S}pan\{\mathrm{X}_1, \ldots, \mathrm{X}_p\} = \mathcal{S}pan\{\mathrm{X}_{j_1}, \ldots, \mathrm{X}_{j_q}\}.$$

It remains to show that $\mathrm{X}_{j_1}, \ldots, \mathrm{X}_{j_q}$ are linearly independent. Keep in mind that $\mathrm{X}_{j_1}, \ldots, \mathrm{X}_{j_q}$ are the columns of A that produce the leading ones during Gauss-Jordan elimination, and as such they determine the row operations in the algorithm. Those same row operations transform $[\mathrm{X}_{j_1} \ \cdots \ \mathrm{X}_{j_q}]$ into the $n \times q$ matrix obtained by deleting from B the columns of index other than $j_1, \ldots, j_q$. That matrix is

$$\left[\begin{array}{c} I_q \\ \hdashline 0 \end{array}\right],$$

where $0 \in \mathcal{M}_{(n-q)\times q}$, and each of its columns contains a leading one. Thus, $\{\mathrm{X}_{j_1}, \ldots, \mathrm{X}_{j_q}\}$ is linearly independent. ■

REMARK. Referring to Theorem 2.17, $\mathcal{S}pan\{\mathrm{X}_1, \ldots, \mathrm{X}_p\} = \mathcal{C}(\mathrm{A})$, so the pivot columns of A form a linearly independent spanning set for $\mathcal{C}(\mathrm{A})$.

Besides its minimal size, a linearly independent spanning set has the feature that each vector in its span has a unique representation as a linear combination of the spanning vectors.

THEOREM 2.18. If $\{\mathrm{X}_1, \ldots, \mathrm{X}_p\} \subseteq \mathcal{R}^n$ is linearly independent, then for each $\mathrm{V} \in \mathcal{S}pan\{\mathrm{X}_1, \ldots, \mathrm{X}_p\}$ there are unique scalars, $t_1, \ldots, t_p$, such that

$$\mathrm{V} = t_1\mathrm{X}_1 + \cdots + t_p\mathrm{X}_p.$$

Proof. Suppose $\mathrm{V} \in \mathcal{V}$. If there are scalars $t_1, \ldots, t_p$ and $s_1, \ldots, s_p$ such that

$$t_1\mathrm{X}_1 + \cdots + t_p\mathrm{X}_p = \mathrm{V} = s_1\mathrm{X}_1 + \cdots + s_p\mathrm{X}_p,$$

then

$$0 = (t_1 - s_1)\mathrm{X}_1 + \cdots + (t_p - s_p)\mathrm{X}_p.$$

Since $\mathrm{X}_1, \ldots, \mathrm{X}_p$ are linearly independent, the coefficients in the last equation must all be zeros, that is, $t_1 = s_1, \ldots, t_p = s_p$. ■

According to Theorem 2.16, the columns of A are linearly independent if and only if each column of rref(A) contains a leading one. When A is square, that occurs precisely when rref(A) $= \mathrm{I}_n$. This observation together with Theorem 1.14 establishes another characterization of nonsingular matrices.

THEOREM 2.19. $\mathrm{A} \in \mathcal{M}_{n\times n}$ is nonsingular if and only if its columns are linearly independent.

Exercises 2.6

1. Check $\mathcal{X}$ for linear independence. If $\mathcal{X}$ is linearly dependent, find a linearly independent subset, $\mathcal{Y}$, having the same span as $\mathcal{X}$ and express each of the other vectors in $\mathcal{X}$ as a linear combination of those in $\mathcal{Y}$.

a. $\mathcal{X}=\left\{\begin{bmatrix}1\\1\end{bmatrix},\begin{bmatrix}-1\\1\end{bmatrix}\right\}$

b. $\mathcal{X}=\left\{\begin{bmatrix}1\\-1\\0\\2\end{bmatrix},\begin{bmatrix}1\\0\\0\\1\end{bmatrix},\begin{bmatrix}1\\2\\0\\-1\end{bmatrix}\right\}$

c. $\mathcal{X}=\left\{\begin{bmatrix}1\\-1\\0\end{bmatrix},\begin{bmatrix}0\\1\\-1\end{bmatrix},\begin{bmatrix}2\\1\\3\end{bmatrix},\begin{bmatrix}4\\6\\2\end{bmatrix}\right\}$

d. $\mathcal{X}=\left\{\begin{bmatrix}1\\2\\1\end{bmatrix},\begin{bmatrix}0\\-1\\1\end{bmatrix},\begin{bmatrix}2\\7\\-1\end{bmatrix},\begin{bmatrix}3\\5\\4\end{bmatrix}\right\}$

e. $\mathcal{X}=\left\{\begin{bmatrix}1\\-2\end{bmatrix},\begin{bmatrix}-1\\2\end{bmatrix},\begin{bmatrix}2\\-2\end{bmatrix}\right\}$

f. $\mathcal{X}=\left\{\begin{bmatrix}1\\3\\-1\end{bmatrix},\begin{bmatrix}0\\-2\\1\end{bmatrix}\right\}$

g. $\mathcal{X}=\left\{\begin{bmatrix}-1\\2\\0\\1\end{bmatrix},\begin{bmatrix}1\\0\\2\\-1\end{bmatrix},\begin{bmatrix}2\\1\\-1\\0\end{bmatrix}\right\}$

h. $\mathcal{X}=\left\{\begin{bmatrix}1\\1\\1\\2\end{bmatrix},\begin{bmatrix}1\\0\\2\\1\end{bmatrix},\begin{bmatrix}1\\4\\-2\\5\end{bmatrix}\right\}$

i. $\mathcal{X}=\left\{\begin{bmatrix}1\\0\\2\\2\end{bmatrix},\begin{bmatrix}1\\2\\-1\\1\end{bmatrix},\begin{bmatrix}1\\-2\\5\\3\end{bmatrix},\begin{bmatrix}-1\\2\\4\\5\end{bmatrix}\right\}$

j. $\mathcal{X}=\left\{\begin{bmatrix}1\\1\\1\\1\end{bmatrix},\begin{bmatrix}1\\1\\1\\2\end{bmatrix},\begin{bmatrix}1\\1\\2\\2\end{bmatrix},\begin{bmatrix}1\\2\\2\\2\end{bmatrix}\right\}$

k. $\mathcal{X}=\left\{\begin{bmatrix}1\\1\\0\\-1\\1\end{bmatrix},\begin{bmatrix}-1\\0\\1\\1\\1\end{bmatrix},\begin{bmatrix}1\\-1\\1\\0\\1\end{bmatrix},\begin{bmatrix}1\\1\\-1\\1\\0\end{bmatrix}\right\}$

l. $\mathcal{X}=\left\{\begin{bmatrix}1\\1\\-1\\0\\1\\2\end{bmatrix},\begin{bmatrix}0\\1\\-1\\1\\1\\-1\end{bmatrix},\begin{bmatrix}3\\1\\-1\\-2\\1\\8\end{bmatrix},\begin{bmatrix}0\\-1\\0\\0\\1\\0\end{bmatrix},\begin{bmatrix}-1\\-2\\0\\1\\2\\-3\end{bmatrix},\begin{bmatrix}1\\6\\-3\\1\\0\\0\end{bmatrix}\right\}.$

2. Assuming X, Y, and Z are nonzero vectors in $\mathcal{R}^3$, explain geometrically why {X,Y,Z} is linearly dependent if and only if 0, X, Y, and Z are coplanar.

3. Let X,Y,Z $\in \mathcal{R}^3$ be the vectors illustrated in Figure 2.22 and assume Z $= r$X $+ s$Y is the unique expression for Z as a linear combination of X and Y.
   a. Find rref([X Y ⋮ Z]). Explain!
   b. Assuming V $\in$ $\mathcal{S}pan${X,Y} and V $= a$X $+ b$Y, find rref([X Y Z ⋮ V]). Explain!
   c. Find functions $r(t)$ and $s(t)$ such that V $= r(t)$X $+ s(t)$Y $+ t$Z, $t \in \mathcal{R}$.

4. Suppose $\mathcal{V}$ is a subspace of $\mathcal{R}^n$ with spanning set $\{X_1, \ldots, X_p\}$ and assume each $V \in \mathcal{V}$ can be written as a linear combination of $X_1, \ldots, X_p$ in only one way. Explain why $\{X_1, \ldots, X_p\}$ is linearly independent.

5. Let X, Y, and Z be vectors in $\mathcal{R}^3$ such that {X,Y}, {X,Z}, and {Y,Z} are linearly independent. Is {X,Y,Z} linearly independent? Explain!

6. **a.** Suppose X and Y are linearly independent vectors in $\mathcal{R}^2$ and s and t are nonzero scalars. Explain geometrically (in words or with a sketch) why sX and tY are linearly independent.
   b. Assuming $X_1, \ldots, X_p \in \mathcal{R}^n$ are linearly independent and $s_1, \ldots, s_p$ are nonzero scalars, show that $s_1X_1, \ldots, s_pX_p$ are linearly independent.

7. Assume X and Y are linearly independent vectors in $\mathcal{R}^2$.
   a. Explain geometrically why $U = X + Y$ and $V = X - Y$ must also be linearly independent.
   b. Supply an algebraic argument for the conclusion in part (a).
   c. What conditions on $a,b,c,d \in \mathcal{R}$ imply $U = aX + bY$ and $V = cX + dY$ are linearly independent?

8. Assuming $\{X_1, \ldots, X_p\} \subseteq \mathcal{R}^n$ is linearly independent and $Y_k = X_1 + \cdots + X_k$, $k = 1, \ldots, p$, show that $\{Y_1, \ldots, Y_p\}$ is linearly independent.

9. Show that if $\{X_1, \ldots, X_p\} \subseteq \mathcal{R}^n$ and $p > n$, then $\{X_1, \ldots, X_p\}$ is linearly dependent.

10. Suppose $\mathcal{X} = \{X_1, \ldots, X_p\} \subseteq \mathcal{R}^n$ and $\mathcal{Y} \subseteq \mathcal{X}$. Show that
   a. if $\mathcal{Y}$ is linearly dependent, then so is $\mathcal{X}$.
   b. if $\mathcal{X}$ is linearly independent, then so is $\mathcal{Y}$.

11. Assume $P \in \mathcal{M}_{n\times n}$, $A \in \mathcal{M}_{n\times p}$, and $1 \leq j_1 < j_2 < \cdots < j_q \leq p$.
   a. Show that if the columns of PA are linearly independent, then so are the columns of A.
   b. Show that if columns $j_1, j_2, \ldots, j_q$ of PA are linearly independent, then so are columns $j_1, j_2, \ldots, j_q$ of A.
   c. Suppose the leading ones in rref(A) occur in columns $j_1, j_2, \ldots, j_q$. Use part (b) to show that $\mathrm{Col}_{j_1}(A), \ldots, \mathrm{Col}_{j_q}(A)$ are linearly independent.

2.7

BASIS AND DIMENSION

Until now the word dimension has been used rather loosely. Planes were said to be two-dimensional, space three-dimensional, and so forth, without ever establishing a strict meaning for the word. The current section rectifies this situation by presenting the concept of dimension used to describe subspaces of $\mathcal{R}^n$.

Suppose $\mathcal{B} = \{X_1, \ldots, X_p\}$ is a minimal spanning set for $\mathcal{V} \subseteq \mathcal{R}^n$. Then $\mathcal{B}$ is linearly independent, and each vector in $\mathcal{V}$ is uniquely expressible as a linear combination of $X_1, \ldots, X_p$. These two features make minimal spanning sets especially useful.

DEFINITION. If $\mathcal{V}$ is a subspace of $\mathcal{R}^n$ and $\mathcal{B} = \{X_1, \ldots, X_p\}$ is a linearly independent set that spans $\mathcal{V}$, then $\mathcal{B}$ is called a *basis* for $\mathcal{V}$.

■ **EXAMPLE 2.29.** Bases for subspaces of $\mathcal{R}^3$ are easily identified through geometric considerations. There are four cases to consider:

1. $\mathcal{V} = \{0\}$. This subspace does not have a basis. Its only spanning set is $\{0\}$, and $\{0\}$ is not linearly independent (Example 2.23).
2. $\mathcal{V}$ is a line through 0. If $X \in \mathcal{V}$ and $X \neq 0$, then $\mathcal{S}pan\{X\} = \mathcal{V}$ and $\{X\}$ is linearly independent, so $\{X\}$ is a basis for $\mathcal{V}$. In fact, every basis for $\mathcal{V}$ consists of a single nonzero vector. Indeed, if a spanning set for $\mathcal{V}$ contains two or more vectors, then those vectors lie on a line through 0, so one must be a scalar multiple of another. Such a set, being linearly dependent, is not a basis for $\mathcal{V}$.
3. $\mathcal{V}$ is a plane through 0. If X and Y are nonzero, nonparallel vectors in $\mathcal{V}$, then $\{X,Y\}$ is linearly independent and spans $\mathcal{V}$. Any spanning set for $\mathcal{V}$ contains two such vectors. If it contains a third vector, then the third is a linear combination of the other two, rendering the set linearly dependent. Thus, a basis for $\mathcal{V}$ consists of two vectors X and Y such that 0, X, and Y are noncollinear.
4. $\mathcal{V} = \mathcal{R}^3$. Suppose $\{X,Y,Z\} \subseteq \mathcal{R}^3$ and 0, X, Y, and Z are noncoplanar. Under these conditions none of X, Y, or Z is a linear combination of the others, so $\{X,Y,Z\}$ is linearly independent. Moreover, each vector in $\mathcal{R}^3$ is a linear combination of X, Y, and Z (recall Figure 2.20), so $\{X,Y,Z\}$ spans $\mathcal{R}^3$. A subset of $\mathcal{R}^3$ must contain three such vectors if it is to span $\mathcal{R}^3$ and it cannot contain a fourth without the fourth being a linear combination of the other three. Thus, bases for $\mathcal{R}^3$ consist of vectors X, Y, and Z, where 0, X, Y, and Z are noncoplanar. □

■ **EXAMPLE 2.30.** For $1 \leq j \leq n$, E_j denotes the vector in $\mathcal{R}^n$ whose jth coordinate is 1 and whose other coordinates are zeros, that is,

$$E_1 = \begin{bmatrix} 1 \\ 0 \\ 0 \\ \vdots \\ 0 \end{bmatrix}, \quad E_2 = \begin{bmatrix} 0 \\ 1 \\ 0 \\ \vdots \\ 0 \end{bmatrix}, \quad \ldots, \quad E_n = \begin{bmatrix} 0 \\ \vdots \\ 0 \\ 0 \\ 1 \end{bmatrix}.$$

If $t_1, \ldots, t_n \in \mathcal{R}$ and

$$0 = t_1E_1 + \cdots + t_nE_n = t_1 \begin{bmatrix} 1 \\ 0 \\ 0 \\ \vdots \\ 0 \end{bmatrix} + t_2 \begin{bmatrix} 0 \\ 1 \\ 0 \\ \vdots \\ 0 \end{bmatrix} + \cdots + t_n \begin{bmatrix} 0 \\ 0 \\ \vdots \\ 0 \\ 1 \end{bmatrix} = \begin{bmatrix} t_1 \\ t_2 \\ \cdot \\ \cdot \\ \cdot \\ t_n \end{bmatrix},$$

then $t_1 = \cdots = t_n = 0$, so $\mathcal{E} = \{E_1, \ldots, E_n\}$ is linearly independent. Moreover, each $X \in \mathcal{R}^n$ can be expressed as a linear combination of $E_1, \ldots, E_n$ by writing $X = x_1E_1 + \cdots + x_nE_n$, so $\mathcal{R}^n = \mathcal{S}pan\{E_1, \ldots, E_n\}$. Thus, $\mathcal{E}$ is a basis for $\mathcal{R}^n$. It is called the *usual* or *standard basis* for $\mathcal{R}^n$. □

Each subspace of $\mathcal{R}^3$ other than $\{0\}$ has infinitely many bases, and any two of these bases have the same number of elements. Indeed, a basis for a line through 0 contains a single vector; planes through 0 have bases consisting of two vectors; and all bases for $\mathcal{R}^3$ consist of three vectors. These observations suggest the following question: given a subspace $\mathcal{V}$ of $\mathcal{R}^n$ and bases $\mathcal{B}$ and $\mathcal{B}'$ for $\mathcal{V}$, do $\mathcal{B}$ and $\mathcal{B}'$ have the same number of elements? The answer is yes. When $n \geq 4$ this conclusion is based on an algebraic argument, the key ingredient of which is the content of the next theorem.

THEOREM 2.20. Assume $\mathcal{V}$ is a subspace of $\mathcal{R}^n$, $\mathcal{B} = \{X_1, \ldots, X_p\}$ is a basis for $\mathcal{V}$, and $Y_1, \ldots, Y_q \in \mathcal{V}$. If $q > p$, then $Y_1, \ldots, Y_q$ are linearly dependent.

Proof. The object is to show that there are scalars $t_1, \ldots, t_q$ that are not all zero such that

$$0 = t_1 Y_1 + \cdots + t_q Y_q. \tag{2.2}$$

Since $\mathcal{B}$ spans $\mathcal{V}$, each vector in $\mathcal{V}$ is a linear combination of $X_1, \ldots, X_p$. In particular, there are scalars s_{ij}, $1 \le i \le p$, $1 \le j \le q$, such that

$$\begin{aligned} Y_1 &= s_{11}X_1 + \cdots + s_{p1}X_p \\ Y_2 &= s_{12}X_1 + \cdots + s_{p2}X_p \\ &\vdots \\ Y_q &= s_{1q}X_1 + \cdots + s_{pq}X_p, \end{aligned}$$

and substituting these expressions in (2.2) produces

$$\begin{aligned} 0 &= t_1(s_{11}X_1 + \cdots + s_{p1}X_p) + \cdots + t_q(s_{1q}X_1 + \cdots + s_{pq}X_p) \\ &= (s_{11}t_1 + \cdots + s_{1q}t_q)X_1 + \cdots + (s_{p1}t_1 + \cdots + s_{pq}t_q)X_p. \end{aligned}$$

By assumption, $X_1, \ldots, X_p$ are linearly independent, so the last equation holds only if

$$\begin{aligned} s_{11}t_1 + \cdots + s_{1q}t_q &= 0 \\ &\vdots \\ s_{p1}t_1 + \cdots + s_{pq}t_q &= 0, \end{aligned}$$

that is, only if $[t_1 \ \cdots \ t_q]^T$ is a solution of the linear homogeneous system with coefficient matrix

$$A = \begin{bmatrix} s_{11} & \cdots & s_{1q} \\ \vdots & & \vdots \\ s_{p1} & \cdots & s_{pq} \end{bmatrix}.$$

When $q > p$, A has more columns than rows, and consequently rref(A) cannot have a leading one in each column. Thus, there are nontrivial choices of $t_1, \ldots, t_q$ such that $t_1 Y_1 + \cdots + t_q Y_q = 0$ (Corollary to Theorem 1.3). ■

COROLLARY. If $\{X_1, \ldots, X_p\}$ and $\{Y_1, \ldots, Y_q\}$ are bases for a subspace of $\mathcal{R}^n$, then $p = q$.

Proof. Since $Y_1, \ldots, Y_q$ are linearly independent vectors in a subspace with basis $\mathcal{B} = \{X_1, \ldots, X_p\}$, $q \le p$ (Theorem 2.20). Interchanging the roles of the X's and Y's, the same reasoning yields $p \le q$. Thus, $p = q$. ■

The number of elements in a basis for $\mathcal{V}$ conveys a fair amount of information about the subspace. In a sense it measures the "size" of $\mathcal{V}$.

DEFINITION. A subspace $\mathcal{V}$ of $\mathcal{R}^n$ having a basis consisting of p vectors is said to be *p-dimensional,* written dim $\mathcal{V} = p$. Although $\{0\}$ has no basis, it is considered to have dimension 0, so dim $\mathcal{V} = 0$ simply means $\mathcal{V} = \{0\}$.

■ **EXAMPLE 2.31.** Lines and planes in $\mathcal{R}^3$ that contain 0 are subspaces of dimensions 1 and 2, respectively. $\mathcal{R}^3$ itself has dimension 3. □

■ **EXAMPLE 2.32.** Since $\mathcal{E} = \{E_1, \ldots, E_n\}$ is a basis for $\mathcal{R}^n$ (Example 2.30), dim $\mathcal{R}^n = n$. □

■ **EXAMPLE 2.33.** Let $\mathcal{V} = Span\{X_1, X_2, X_3\}$, where

$$X_1 = \begin{bmatrix} -1 \\ 2 \\ 1 \end{bmatrix}, \quad X_2 = \begin{bmatrix} 2 \\ -1 \\ -3 \end{bmatrix}, \quad \text{and} \quad X_3 = \begin{bmatrix} 1 \\ 4 \\ -3 \end{bmatrix}.$$

Since

$$\begin{bmatrix} -1 & 2 & 1 \\ 2 & -1 & 4 \\ 1 & -3 & -3 \end{bmatrix} \longrightarrow \begin{bmatrix} 1 & 0 & 3 \\ 0 & 1 & 2 \\ 0 & 0 & 0 \end{bmatrix},$$

$\mathcal{B} = \{X_1, X_2\}$ is a linearly independent spanning set for $\mathcal{V}$ (Theorem 2.17). Thus, $\mathcal{B}$ is a basis for $\mathcal{V}$ and dim $\mathcal{V} = 2$. □

In effect, Theorem 2.17 shows how to extract a basis from a spanning set. In the opposite direction, it may be necessary at times to enlarge a linearly independent set to achieve a basis. The next result indicates how to expand without losing independence.

THEOREM 2.21. If $\{X_1, \ldots, X_p\} \subseteq \mathcal{R}^n$ is linearly independent and $X \in \mathcal{R}^n$, then $\{X_1, \ldots, X_p, X\}$ is linearly independent if and only if $X \notin Span\{X_1, \ldots, X_p\}$.

Proof. If $\{X_1, \ldots, X_p, X\}$ is linearly independent, then none of its vectors is a linear combination of the others (Theorem 2.15), and in particular, $X \notin Span\{X_1, \ldots, X_p\}$. For the converse, assume $X \notin Span\{X_1, \ldots, X_p\}$ and suppose $t_1, \ldots, t_p, t$ are scalars such that

$$t_1X_1 + \cdots + t_pX_p + tX = 0. \tag{2.3}$$

If $t \neq 0$, then X can be expressed as a linear combination of $X_1, \ldots, X_p$, contrary to the assumption that $X \notin Span\{X_1, \ldots, X_p\}$. Thus, $t = 0$ and (2.3) reduces to

$$t_1X_1 + \cdots + t_pX_p = 0.$$

It then follows that $t_1, \ldots, t_p$ are all zeros (because $X_1, \ldots, X_p$ are linearly independent), so (2.3) holds only when $t_1, \ldots, t_p, t$ are all zeros. ■

■ **EXAMPLE 2.34.** Observe that $\mathcal{B} = \left\{ \begin{bmatrix} 1 \\ -1 \\ 2 \end{bmatrix}, \begin{bmatrix} -1 \\ 2 \\ -1 \end{bmatrix} \right\}$, being linearly independent, spans a two-dimensional subspace $\mathcal{V}$ of $\mathcal{R}^3$. Moreover, $X = [x \quad y \quad z]^T \in \mathcal{V}$ if and only if

$$\begin{bmatrix} 1 & -1 \\ -1 & 2 \\ 2 & -1 \end{bmatrix} \begin{bmatrix} s \\ t \end{bmatrix} = \begin{bmatrix} x \\ y \\ z \end{bmatrix}$$

is consistent. Since

$$\left[\begin{array}{cc|c} 1 & -1 & x \\ -1 & 2 & y \\ 2 & -1 & z \end{array}\right] \longrightarrow \left[\begin{array}{cc|c} 1 & 0 & 2x + y \\ 0 & 1 & x + y \\ 0 & 0 & -3x - y + z \end{array}\right],$$

$\mathcal{V}$ is the plane with equation $-3x - y + z = 0$. Any $X \notin \mathcal{V}$ can be adjoined to $\mathcal{B}$ to obtain a larger linearly independent set (Theorem 2.21). Moreover, the resulting three vectors, being noncoplanar, constitute a basis for $\mathcal{R}^3$. In particular, $X = [1 \quad 1 \quad 1]^T \notin \mathcal{V}$, so $\mathcal{B} \cup \{X\}$ is a basis for $\mathcal{R}^3$ containing $\mathcal{B}$. □

Suppose $\mathcal{V}$ is a subspace of $\mathcal{R}^n$. To be a basis for $\mathcal{V}$, $\mathcal{B} = \{X_1, \ldots, X_p\}$ must be linearly independent and span $\mathcal{V}$. The next theorem shows that if you know $\mathcal{V}$ has dimension p, that is, that the number of vectors in $\mathcal{B}$ is dim $\mathcal{V}$, then $\mathcal{B}$ cannot have one of the two basis properties without having the other.

THEOREM 2.22. If $\mathcal{V}$ is a subspace of $\mathcal{R}^n$ other than $\{0\}$, $\{X_1, \ldots, X_p\} \subseteq \mathcal{V}$, and $\dim \mathcal{V} = p$, then the following statements are equivalent:

(1) $\{X_1, \ldots, X_p\}$ is linearly independent.
(2) $Span\{X_1, \ldots, X_p\} = \mathcal{V}$.

Proof. (1) $\Rightarrow$ (2). Assume $\{X_1, \ldots, X_p\}$ is linearly independent. Of course, $Span\{X_1, \ldots, X_p\} \subseteq \mathcal{V}$. If there is an $X \in \mathcal{V}$ such that $X \notin Span\{X_1, \ldots, X_p\}$, then $\{X_1, \ldots, X_p, X\}$ is a linearly independent subset of $\mathcal{V}$ containing more than p vectors (Theorem 2.21), contrary to Theorem 2.20. Thus, $Span\{X_1, \ldots, X_p\} = \mathcal{V}$.

(2) $\Rightarrow$ (1). Assume $Span\{X_1, \ldots, X_p\} = \mathcal{V}$. If $X_1, \ldots, X_p$ are linearly dependent, then some linearly independent subset of $\{X_1, \ldots, X_p\}$ spans $\mathcal{V}$, and $\mathcal{V}$ therefore has a basis consisting of fewer than p vectors, contrary to $\dim \mathcal{V} = p$. Thus, $\{X_1, \ldots, X_p\}$ is linearly independent. ■

As a consequence of Theorem 2.22, any set of n linearly independent vectors in $\mathcal{R}^n$ is a basis for $\mathcal{R}^n$, and any spanning set for $\mathcal{R}^n$ consisting of n vectors is a basis for $\mathcal{R}^n$.

■ **EXAMPLE 2.35.** Let $A = \begin{bmatrix} -1 & 0 & -2 & -1 \\ -1 & 1 & -3 & 0 \\ 2 & 0 & 4 & 1 \\ 1 & 1 & 1 & 0 \end{bmatrix}$ and consider $\mathcal{V} = \mathcal{C}(A) \subseteq \mathcal{R}^4$. According to Theorem 2.17, the pivot columns of A form a linearly independent spanning set for $\mathcal{V}$. Since

$$\begin{bmatrix} -1 & 0 & -2 & -1 \\ -1 & 1 & -3 & 0 \\ 2 & 0 & 4 & 1 \\ 1 & 1 & 1 & 0 \end{bmatrix} \longrightarrow \begin{bmatrix} 1 & 0 & 2 & 0 \\ 0 & 1 & -1 & 0 \\ 0 & 0 & 0 & 1 \\ 0 & 0 & 0 & 0 \end{bmatrix},$$

$\mathcal{B} = \{Col_1(A), Col_2(A), Col_4(A)\}$ is a basis for $\mathcal{V}$, and $\dim \mathcal{V} = 3$. Note also that

$$\begin{bmatrix} 0 & -2 & -1 \\ 1 & -3 & 0 \\ 0 & 4 & 1 \\ 1 & 1 & 0 \end{bmatrix} \longrightarrow \begin{bmatrix} 1 & 0 & 0 \\ 0 & 1 & 0 \\ 0 & 0 & 1 \\ 0 & 0 & 0 \end{bmatrix},$$

so $\mathcal{A} = \{Col_2(A), Col_3(A), Col_4(A)\}$ is linearly independent. Since $\mathcal{A}$ consists of three vectors and $\dim \mathcal{V} = 3$, $\mathcal{A}$ is also be a spanning set for $\mathcal{V}$ (Theorem 2.22). Thus, $\mathcal{A}$ is a second basis for $\mathcal{V}$. □

It seems reasonable to expect that each subspace of $\mathcal{R}^n$ other than $\{0\}$ has a basis and hence has dimension. This is certainly true when $n = 3$, but it has not yet been established in general. In fact, the definition of dimension was carefully phrased so as not to presume it is so. The proof in the general case involves generating a basis for the given subspace.

THEOREM 2.23. If $\mathcal{V}$ is a subspace of $\mathcal{R}^n$ other than $\{0\}$, then $\mathcal{V}$ has a basis, and $1 \le \dim \mathcal{V} \le n$.

Proof. Suppose $\mathcal{V}$ is a subspace of $\mathcal{R}^n$ and $\mathcal{V} \neq \{0\}$. Let X_1 be a nonzero vector in $\mathcal{V}$. Then $\{X_1\}$ is linearly independent and $\mathcal{S}pan\{X_1\} \subseteq \mathcal{V}$. If $\mathcal{S}pan\{X_1\} = \mathcal{V}$, then $\{X_1\}$ is a basis for $\mathcal{V}$ and $\dim \mathcal{V} = 1$. Alternatively, $\mathcal{S}pan\{X_1\}$ is a proper subset of $\mathcal{V}$, and in that event there is an $X_2 \in \mathcal{V}$ such that $X_2 \notin \mathcal{S}pan\{X_1\}$. By Theorem 2.21, $\{X_1,X_2\}$ is linearly independent, and of course $\mathcal{S}pan\{X_1,X_2\} \subseteq \mathcal{V}$. If $\mathcal{S}pan\{X_1,X_2\} = \mathcal{V}$, then $\{X_1,X_2\}$ is a basis for $\mathcal{V}$ and $\dim \mathcal{V} = 2$. Otherwise, there is an $X_3 \in \mathcal{V}$ such that $X_3 \notin \mathcal{S}pan\{X_1,X_2\}$. Under these circumstances, $\{X_1,X_2,X_3\}$ is linearly independent (Theorem 2.21), and $\mathcal{S}pan\{X_1,X_2,X_3\} \subseteq \mathcal{V}$. Continuing in this manner eventually produces a linearly independent spanning set for $\mathcal{V}$. Such a set will be achieved after at most n steps, for otherwise the process would generate a linearly independent subset of $\mathcal{R}^n$ consisting of more than n vectors, contrary to Theorem 2.20. Thus, $\mathcal{V}$ has a basis consisting of p vectors, for some $p \in \{1, \ldots, n\}$. ■

For a subspace $\mathcal{V}$ of $\mathcal{R}^n$, the term algebraic structure refers collectively to the properties of $\mathcal{V}$ associated with addition and scalar multiplication. The source of that structure is the assumption that $\mathcal{V}$ is closed under the two operations, and those closure conditions produce a positive integer, $\dim \mathcal{V}$, which is the key feature of the subspace. It is the minimal number of vectors needed to generate the entire subspace via addition and scalar multiplication. It is shown later (Chapter 7) that any two subspaces of $\mathcal{R}^n$ having the same dimension are in a sense algebraically indistinguishable.

The impetus for studying subspaces was the fact that the solution set for a homogeneous system of linear equations has such a structure. In fact, since the solution set for a nonhomogeneous linear system is simply related to that of the associated homogeneous system (Theorem 2.10), the task of describing the "form" of such sets is now essentially accomplished.

Exercises 2.7

1. Is the given set a basis for its associated Euclidean space? Explain!

a. $\left\{\begin{bmatrix}1\\-2\end{bmatrix}, \begin{bmatrix}-3\\4\end{bmatrix}\right\}$

b. $\left\{\begin{bmatrix}4\\5\end{bmatrix}\right\}$

c. $\left\{\begin{bmatrix}2/3\\-1/2\end{bmatrix}, \begin{bmatrix}-4\\3\end{bmatrix}\right\}$

d. $\left\{\begin{bmatrix}1\\-1\\0\end{bmatrix}, \begin{bmatrix}1\\0\\-1\end{bmatrix}, \begin{bmatrix}0\\1\\-1\end{bmatrix}\right\}$

e. $\left\{\begin{bmatrix}\pi\\1\\\sin(2)\end{bmatrix}, \begin{bmatrix}\sqrt{3}\\\ln(2)\\1\end{bmatrix}\right\}$

f. $\left\{\begin{bmatrix}1\\1\\2\end{bmatrix}, \begin{bmatrix}1\\2\\2\end{bmatrix}, \begin{bmatrix}2\\1\\1\end{bmatrix}, \begin{bmatrix}2\\2\\1\end{bmatrix}\right\}$

g. $\left\{\begin{bmatrix}1\\1\\-1\\0\end{bmatrix}, \begin{bmatrix}0\\-1\\1\\1\end{bmatrix}, \begin{bmatrix}-1\\0\\1\\1\end{bmatrix}, \begin{bmatrix}1\\1\\0\\-1\end{bmatrix}\right\}$

h. $\left\{\begin{bmatrix}0\\1\\0\\1\end{bmatrix}, \begin{bmatrix}1\\0\\1\\0\end{bmatrix}, \begin{bmatrix}0\\1\\1\\0\end{bmatrix}, \begin{bmatrix}1\\0\\0\\1\end{bmatrix}\right\}$

i. $\left\{\begin{bmatrix}1\\1\\2\\2\end{bmatrix}, \begin{bmatrix}1\\2\\2\\1\end{bmatrix}, \begin{bmatrix}2\\1\\1\\2\end{bmatrix}\right\}$.

2. Find dim $\mathcal{V}$. Check that $\mathcal{B} \subseteq \mathcal{V}$ and determine whether $\mathcal{B}$ is a basis for $\mathcal{V}$.

a. $\mathcal{V} = \mathcal{N}\left(\begin{bmatrix} 1 & -1 & 2 & -3 \\ -2 & 1 & 0 & 4 \end{bmatrix}\right), \quad \mathcal{B} = \left\{ \begin{bmatrix} 2 \\ 4 \\ 1 \\ 0 \end{bmatrix}, \begin{bmatrix} 1 \\ -2 \\ 0 \\ 1 \end{bmatrix} \right\}$

b. $\mathcal{V} = \mathcal{N}\left(\begin{bmatrix} 1 & 2 & 0 \\ 1 & 0 & -3 \end{bmatrix}\right), \quad \mathcal{B} = \left\{ \begin{bmatrix} 1 \\ -1/2 \\ 1/3 \end{bmatrix}, \begin{bmatrix} -3 \\ 3/2 \\ -1 \end{bmatrix} \right\}$

c. $\mathcal{V} = \{X \in \mathcal{R}^4 \,:\, x_4 = 0\}, \quad \mathcal{B} = \left\{ \begin{bmatrix} 1 \\ 0 \\ 0 \\ 0 \end{bmatrix}, \begin{bmatrix} 1 \\ 1 \\ 0 \\ 0 \end{bmatrix}, \begin{bmatrix} 1 \\ 1 \\ 1 \\ 0 \end{bmatrix} \right\}$

d. $\mathcal{V} = \{X \in \mathcal{R}^4 \,:\, x_1 = x_4\}, \quad \mathcal{B} = \left\{ \begin{bmatrix} 1 \\ 1 \\ 0 \\ 1 \end{bmatrix}, \begin{bmatrix} 2 \\ 0 \\ 1 \\ 2 \end{bmatrix} \right\}$

e. $\mathcal{V} = \{X \in \mathcal{R}^4 \,:\, 2x_1 - x_2 + 3x_3 - x_4 = 0\}, \quad \mathcal{B} = \left\{ \begin{bmatrix} -1 \\ 0 \\ 1 \\ 1 \end{bmatrix}, \begin{bmatrix} 0 \\ 1 \\ 0 \\ -1 \end{bmatrix}, \begin{bmatrix} 2 \\ 1 \\ -1 \\ 0 \end{bmatrix} \right\}$

f. $\mathcal{V} = \mathcal{C}\left(\begin{bmatrix} 2 & -6 & 1 & 2 \\ -1 & 3 & 2 & -6 \\ 1 & -3 & 1 & 0 \end{bmatrix}\right), \quad \mathcal{B} = \left\{ \begin{bmatrix} -6 \\ 3 \\ -3 \end{bmatrix}, \begin{bmatrix} 2 \\ -6 \\ 0 \end{bmatrix} \right\}$

g. $\mathcal{V} = \mathcal{C}\left(\begin{bmatrix} 1 & -1 & 1 \\ 1 & -2 & -1 \\ -1 & 1 & -1 \end{bmatrix}\right), \quad \mathcal{B} = \left\{ \begin{bmatrix} -1 \\ -3 \\ 1 \end{bmatrix}, \begin{bmatrix} 1 \\ -3 \\ -1 \end{bmatrix} \right\}$

3. Assuming $A \in \mathcal{M}_{n \times n}$ is nonsingular, find dim $\mathcal{N}(A)$ and dim $\mathcal{C}(A)$.

4. Suppose $\{X,Y\} \subseteq \mathcal{R}^n$ is linearly independent and let $U = X - Y$, $V = 2X + Y$, and $W = X + 3Y$. Is $\{U,V,W\}$ linearly independent? Explain!

5. Let $\mathcal{V} = \mathcal{S}pan\left\{ \begin{bmatrix} 2 \\ 1 \\ -3 \\ 4 \end{bmatrix}, \begin{bmatrix} -4 \\ 2 \\ 5 \\ -3 \end{bmatrix}, \begin{bmatrix} 2 \\ 5 \\ -4 \\ 9 \end{bmatrix} \right\}$ and $\mathcal{W} = \mathcal{S}pan\left\{ \begin{bmatrix} -6 \\ 5 \\ 7 \\ -2 \end{bmatrix}, \begin{bmatrix} 2 \\ -7 \\ -1 \\ -6 \end{bmatrix} \right\}$.

Show that $\mathcal{V} = \mathcal{W}$.

6. Let $\mathcal{V}$ be a three-dimensional subspace of $\mathcal{R}^n$ with basis $\{X,Y,Z\}$. Assume U and V are linearly independent vectors in $\mathcal{V}$. Show that at least one of $\{U,V,X\}$, $\{U,V,Y\}$, and $\{U,V,Z\}$ is a basis for $\mathcal{V}$.

7. Let $\mathcal{V}$ be a subspace of $\mathcal{R}^n$ other than $\{0\}$ and suppose $X_1, \ldots, X_p \in \mathcal{V}$.
   a. Show that if $\mathcal{V} = \mathcal{S}pan\{X_1, \ldots, X_p\}$, then dim $\mathcal{V} \le p$.
   b. Show that if $\{X_1, \ldots, X_p\}$ is linearly independent, then dim $\mathcal{V} \ge p$.

8. Let $\mathcal{U}$ and $\mathcal{V}$ be subspaces of $\mathcal{R}^n$ such that $\mathcal{U} \subseteq \mathcal{V}$ and suppose $\mathcal{B} = \{X_1, \ldots, X_p\}$ is a basis for $\mathcal{U}$.
   a. Show that there is a basis for $\mathcal{V}$ that contains $\mathcal{B}$ and conclude that $\dim \mathcal{U} \leq \dim \mathcal{V}$.
   b. Show that $\dim \mathcal{U} = \dim \mathcal{V}$ if and only if $\mathcal{U} = \mathcal{V}$.

9. Suppose $\mathcal{U}$ and $\mathcal{V}$ are subspaces of $\mathcal{R}^n$, $\mathcal{U} \subseteq \mathcal{V}$, and $\mathcal{B} = \{X_1, \ldots, X_p\}$ is a basis for $\mathcal{V}$. Does $\mathcal{B}$ have a subset that forms a basis for $\mathcal{U}$? Explain!

10. Assume $\mathcal{V}$ is a subspace of $\mathcal{R}^n$ and $\mathcal{B} = \{X_1, \ldots, X_p\}$ is a linearly independent subset of $\mathcal{V}$.
   a. What should it mean to say that $\mathcal{B}$ is a *maximal* linearly independent subset of $\mathcal{V}$?
   b. Why must a maximal linearly independent subset of $\mathcal{V}$ be a basis for $\mathcal{V}$?

2.8

RANK AND NULLITY

Associated with $A \in \mathcal{M}_{m\times n}$ are two subspaces of special interest, $\mathcal{N}(A)$ and $\mathcal{C}(A)$. The first is the solution set of the homogeneous system with coefficient matrix A; the second consists of those $B \in \mathcal{R}^m$ for which $AX = B$ is consistent. Their dimensions have acquired special names.

DEFINITION. If $A \in \mathcal{M}_{m\times n}$, then $\dim \mathcal{C}(A)$ is called the *rank* of A, written rank(A), and $\dim \mathcal{N}(A)$ is called the *nullity* of A, written nul(A).

THEOREM 2.24. For $A \in \mathcal{M}_{m\times n}$, rank(A) is the number of leading ones in rref(A).

Proof. If $A \neq 0$, then the pivot columns of A form a basis for $\mathcal{C}(A)$ (Theorem 2.17). When $A = 0$, $\mathcal{C}(A) = \{0\}$ and $\dim \mathcal{C}(A) = 0$. In either event, rank(A) is the number of leading ones in rref(A). ■

You have seen how Gauss-Jordan elimination generates a spanning set for $\mathcal{N}(A)$. As it turns out, that spanning set is linearly independent. The next example illustrates the line of reasoning used to justify that statement.

■ **EXAMPLE 2.36.** Let A be any 4×7 matrix that row reduces to

$$B = \begin{bmatrix} 1 & 2 & 0 & 0 & -1 & 0 & 5 \\ 0 & 0 & 1 & 0 & 1 & 0 & -4 \\ 0 & 0 & 0 & 1 & -1 & 0 & -3 \\ 0 & 0 & 0 & 0 & 0 & 1 & 2 \end{bmatrix}.$$

The solutions of $BX = 0$ are

$$X = \begin{bmatrix} x_1 \\ x_2 \\ x_3 \\ x_4 \\ x_5 \\ x_6 \\ x_7 \end{bmatrix} = \begin{bmatrix} -2r + s - 5t \\ r \\ -s + 4t \\ s + 3t \\ s \\ -2t \\ t \end{bmatrix} = r\begin{bmatrix} -2 \\ 1 \\ 0 \\ 0 \\ 0 \\ 0 \\ 0 \end{bmatrix} + s\begin{bmatrix} 1 \\ 0 \\ -1 \\ 1 \\ 1 \\ 0 \\ 0 \end{bmatrix} + t\begin{bmatrix} -5 \\ 0 \\ 4 \\ 3 \\ 0 \\ -2 \\ 1 \end{bmatrix},$$

$r,s,t, \in \mathcal{R}$, and consequently $\mathcal{N}(A) = \mathcal{S}pan\{X_1,X_2,X_3\}$, where

$$X_1 = \begin{bmatrix} -2 \\ 1 \\ 0 \\ 0 \\ 0 \\ 0 \\ 0 \end{bmatrix}, \quad X_2 = \begin{bmatrix} 1 \\ 0 \\ -1 \\ 1 \\ 1 \\ 0 \\ 0 \end{bmatrix}, \quad \text{and} \quad X_3 = \begin{bmatrix} -5 \\ 0 \\ 4 \\ 3 \\ 0 \\ -2 \\ 1 \end{bmatrix}.$$

Since the second, fifth, and seventh coordinates of $rX_1 + sX_2 + tX_3$ are r, s, and t, respectively,

$$rX_1 + sX_2 + tX_3 = 0$$

only when $r = s = t = 0$. Thus, $\{X_1,X_2,X_3\}$ is a basis for $\mathcal{N}(A)$. Note that dim $\mathcal{N}(A)$ is the number of columns of B that do not contain a leading one. □

THEOREM 2.25. If $A \in \mathcal{M}_{m\times n}$, then dim $\mathcal{N}(A) = n - q$, where q is the number of pivot columns of A.

Proof. If $A = 0$, then $\mathcal{N}(A) = \mathcal{R}^n$ and dim $\mathcal{N}(A) = n$. Suppose $A \neq 0$ and the leading ones in rref(A) occur in columns $j_1, \ldots, j_q$, $1 \le j_1 < j_2 < \cdots < j_q \le n$. If $q = n$, then $j_1 = 1$, $j_2 = 2, \ldots, j_n = n$, that is, each column of A is a pivot column. In this case, $\mathcal{N}(A) = \{0\}$ and dim $\mathcal{N}(A) = 0$. Assume hereafter that $1 \le q < n$. To describe the vectors in $\mathcal{N}(A)$ you assign arbitrary values to the coordinates having indices other than $j_1, \ldots, j_q$, and coordinates $j_1, \ldots, j_q$ are then determined by those values and the entries in rref(A). There are $n - q$ coordinates that become parameters. If they have indices $k_1, \ldots, k_{n-q}$, where $1 \le k_1 < k_2 < \cdots < k_{n-q} \le n$, then a typical X in $\mathcal{N}(A)$ looks like

$$X = \begin{bmatrix} \vdots \\ t_1 \\ \vdots \\ t_2 \\ \vdots \\ t_{n-q} \\ \vdots \end{bmatrix} \begin{matrix} \\ \leftarrow k_1 \ \text{position} \\ \\ \leftarrow k_2 \ \text{position.} \\ \\ \leftarrow k_{n-q} \ \text{position} \\ \\ \end{matrix}$$

Here $t_1, \ldots, t_{n-q}$ are arbitrary parameters and the other coordinates of X are linear combinations of them. This expression for X can be rewritten as

$$X = t_1X_1 + \cdots + t_{n-q}X_{n-q},$$

where $X_1, \ldots, X_{n-q} \in \mathcal{N}(A)$. Thus, $\mathcal{N}(A) = \mathcal{S}pan\{X_1, \ldots, X_{n-q}\}$. For $j = 1, \ldots, n - q$, the k_j coordinate of X_j is 1, and the k_j coordinates of the other vectors in $\{X_1, \ldots, X_{n-q}\}$ are 0's, so the k_j coordinate of $t_1X_1 + \cdots + t_{n-q}X_{n-q}$ is t_j. Thus, $t_1X_1 + \cdots + t_{n-q}X_{n-q} = 0$ implies $t_1 = \cdots = t_{n-q} = 0$, that is, $\{X_1, \ldots, X_{n-q}\}$ is linearly independent. It follows that $\{X_1, \ldots, X_{n-q}\}$ is a basis for $\mathcal{N}(A)$, and therefore dim $\mathcal{N}(A) = n - q$. ■

■ **EXAMPLE 2.37.** Since

$$A = \begin{bmatrix} 1 & 0 & -1 & -1 & 1 \\ 1 & 1 & 0 & -1 & 0 \\ -1 & 2 & 3 & 2 & -1 \end{bmatrix} \longrightarrow \begin{bmatrix} 1 & 0 & -1 & 0 & 3 \\ 0 & 1 & 1 & 0 & -1 \\ 0 & 0 & 0 & 1 & 2 \end{bmatrix} = B,$$

and B has three leading ones, $\dim \mathcal{N}(A) = 5 - 3 = 2$. □

Theorems 2.24 and 2.25 establish an important complementary relationship between the rank and nullity of a matrix.

THEOREM 2.26. **(Rank-Nullity Theorem).** For $A \in \mathcal{M}_{m\times n}$,

$$\dim \mathcal{C}(A) + \dim \mathcal{N}(A) = n,$$

that is, rank(A) + nul(A) = the number of columns of A.

If A and B are row equivalent $m \times n$ matrices, then $AX = 0$ and $BX = 0$ have the same solutions, that is, $\mathcal{N}(A) = \mathcal{N}(B)$. This conclusion can be rephrased as follows.

THEOREM 2.27. If $A \in \mathcal{M}_{m\times n}$ and $P \in \mathcal{M}_{m\times m}$ is nonsingular, then $\mathcal{N}(PA) = \mathcal{N}(A)$.

Proof. If $B = PA$ and P is nonsingular, then $B \equiv_r A$ (Theorem 1.16). ■

By Theorem 1.21, $A \equiv_c B$ if and only if there is a nonsingular Q such that $B = AQ$. This connection leads to the conclusion that column equivalent matrices have the same column space.

THEOREM 2.28. If $A \in \mathcal{M}_{m\times n}$ and $Q \in \mathcal{M}_{n\times n}$ is nonsingular, then $\mathcal{C}(AQ) = \mathcal{C}(A)$.

Proof. Set $B = AQ$. According to Theorem 1.8, $\text{Col}_j(B) = A\text{Col}_j(Q)$, $1 \le j \le n$. If $\text{Col}_j(Q) = [t_1 \ \cdots \ t_n]^T$, then

$$\text{Col}_j(B) = [\text{Col}_1(A) \ \cdots \ \text{Col}_n(A)] \begin{bmatrix} t_1 \\ \vdots \\ t_n \end{bmatrix} = \sum_1^n t_k \text{Col}_k(A) \in \mathcal{C}(A).$$

This conclusion holds for each $j = 1, \ldots, n$, so by Theorem 2.12,

$$\mathcal{C}(B) = \mathcal{S}pan\{\text{Col}_1(B), \ldots, \text{Col}_n(B)\} \subseteq \mathcal{C}(A).$$

On the other hand, since Q is nonsingular, $A = BQ^{-1}$, and the same argument, with A and B interchanged and Q^{-1} replacing Q, leads to the conclusion that $\mathcal{C}(A) \subseteq \mathcal{C}(B)$. Thus, $\mathcal{C}(B) = \mathcal{C}(A)$. ■

THEOREM 2.29. Suppose $A \in \mathcal{M}_{m\times n}$, $P \in \mathcal{M}_{m\times m}$, and $Q \in \mathcal{M}_{n\times n}$. If P and Q are nonsingular, then A, PA, AQ, and PAQ all have the same rank.

Proof. It follows directly from Theorem 2.28 that rank(A) = rank(AQ). Since $\mathcal{N}(PA) = \mathcal{N}(A)$, the rank-nullity theorem gives $\text{rank}(PA) = n - \dim \mathcal{N}(PA) = n - \dim \mathcal{N}(A) = \text{rank}(A)$. Finally, rank(PAQ) = rank((PA)Q) = rank(PA) = rank(A). ■

■ **EXAMPLE 2.38.** Observe that

$$A = \begin{bmatrix} -1 & 2 & 1 & 0 \\ 1 & -2 & 1 & -2 \\ -1 & 2 & 2 & -1 \end{bmatrix} \longrightarrow \begin{bmatrix} 1 & -2 & 0 & -1 \\ 0 & 0 & 1 & -1 \\ 0 & 0 & 0 & 0 \end{bmatrix} = B,$$

so rank(A) = 2. The row reduction $[A \mid I_3] \to [B \mid P]$ produces a nonsingular P such that

B = PA (Theorem 1.16), and in this instance the computations yield

$$\left[\begin{array}{rrrr|rrr} -1 & 2 & 1 & 0 & 1 & 0 & 0 \\ 1 & -2 & 1 & -2 & 0 & 1 & 0 \\ -1 & 2 & 2 & -1 & 0 & 0 & 1 \end{array}\right] \longrightarrow \left[\begin{array}{rrrr|rrr} 1 & -2 & 0 & -1 & -1/2 & 1/2 & 0 \\ 0 & 0 & 1 & -1 & 1/2 & 1/2 & 0 \\ 0 & 0 & 0 & 0 & -3/2 & -1/2 & 1 \end{array}\right].$$

Thus,

$$\mathrm{P} = \frac{1}{2}\begin{bmatrix} -1 & 1 & 0 \\ 1 & 1 & 0 \\ -3 & -1 & 2 \end{bmatrix}.$$

Column reducing B to a column equivalent matrix, C, in reduced column echelon form produces

$$\mathrm{C} = \begin{bmatrix} 1 & 0 & 0 & 0 \\ 0 & 1 & 0 & 0 \\ 0 & 0 & 0 & 0 \end{bmatrix}.$$

According to Theorem 1.21, there is a nonsingular Q such that C = BQ and

$$\left[\begin{array}{c} \mathrm{B} \\ \hdashline \mathrm{I}_4 \end{array}\right] \longrightarrow \left[\begin{array}{c} \mathrm{C} \\ \hdashline \mathrm{Q} \end{array}\right],$$

where the arrow indicates column reduction. In this case,

$$\left[\begin{array}{c} \mathrm{B} \\ \hdashline \mathrm{I}_4 \end{array}\right] = \left[\begin{array}{rrrr} 1 & -2 & 0 & -1 \\ 0 & 0 & 1 & -1 \\ 0 & 0 & 0 & 0 \\ \hdashline 1 & 0 & 0 & 0 \\ 0 & 1 & 0 & 0 \\ 0 & 0 & 1 & 0 \\ 0 & 0 & 0 & 1 \end{array}\right] \longrightarrow \left[\begin{array}{rrrr} 1 & 0 & 0 & 0 \\ 0 & 1 & 0 & 0 \\ 0 & 0 & 0 & 0 \\ \hdashline 1 & 0 & 2 & 1 \\ 0 & 0 & 1 & 0 \\ 0 & 1 & 0 & 1 \\ 0 & 0 & 0 & 1 \end{array}\right],$$

so

$$\mathrm{Q} = \begin{bmatrix} 1 & 0 & 2 & 0 \\ 0 & 0 & 1 & 0 \\ 0 & 1 & 0 & 1 \\ 0 & 0 & 0 & 1 \end{bmatrix}.$$

Thus,

$$\mathrm{PAQ} = \begin{bmatrix} 1 & 0 & 0 & 0 \\ 0 & 1 & 0 & 0 \\ 0 & 0 & 0 & 0 \end{bmatrix}. \ \square$$

Suppose A is a nonzero $m \times n$ matrix of rank r and B = rref(A). There is a nonsingular P such that B = PA. Since rank(A) = r, B has a leading one in each of its first r rows, and its remaining $m - r$ rows consist of zeros. Let C be the matrix in reduced column echelon form that is column equivalent to B. Then columns 1 through r of C contain leading ones, and the remaining $n - r$ columns consist of zeros. There is a nonsingular Q such that C = BQ, so

$$\mathrm{C} = \mathrm{PAQ} = \left[\begin{array}{c|c} \mathrm{I}_r & 0 \\ \hline 0 & 0 \end{array}\right]. \qquad (2.4)$$

It may happen that some or all of the zero blocks do not occur, depending on the relative sizes of r, m, and n. For example, if $r = m < n$, then

$$\mathrm{PAQ} = \left[\, \mathrm{I}_r \;\vdots\; 0 \,\right],$$

whereas $r = n < m$, implies

$$\mathrm{PAQ} = \left[\begin{array}{c} \mathrm{I}_r \\ \hline 0 \end{array}\right].$$

Strictly speaking, (2.4) represents the case when $r < m$ and $r < n$, but it will be used to indicate the general form of PAQ, with the understanding that the reader can make the appropriate adjustments in the other cases.

THEOREM 2.30. Suppose $\mathrm{A} \in \mathcal{M}_{m\times n}$ and r is a positive integer such that $r \leq m$ and $r \leq n$. Then $\mathrm{rank(A)} = r$ if and only if there are nonsingular matrices P and Q such that

$$\mathrm{PAQ} = \left[\begin{array}{c|c} \mathrm{I}_r & 0 \\ \hline 0 & 0 \end{array}\right]. \tag{2.5}$$

Proof. If $\mathrm{rank(A)} = r$, then the existence of nonsingular matrices P and Q satisfying (2.5) was established in the discussion preceding the theorem. On the other hand, if there are such matrices, then solving (2.5) for A gives

$$\mathrm{A} = \mathrm{P}^{-1} \left[\begin{array}{c|c} \mathrm{I}_r & 0 \\ \hline 0 & 0 \end{array}\right] \mathrm{Q}^{-1},$$

and by Theorem 2.29,

$$\mathrm{rank(A)} = \mathrm{rank}\left(\mathrm{P}^{-1} \left[\begin{array}{c|c} \mathrm{I}_r & 0 \\ \hline 0 & 0 \end{array}\right] \mathrm{Q}^{-1}\right) = \mathrm{rank}\left(\left[\begin{array}{c|c} \mathrm{I}_r & 0 \\ \hline 0 & 0 \end{array}\right]\right) = r. \quad \blacksquare$$

THEOREM 2.31. If $\mathrm{A} \in \mathcal{M}_{m\times n}$, then $\mathrm{rank(A)} = \mathrm{rank}(\mathrm{A}^T)$.

Proof. If $\mathrm{A} = 0$, then $\mathrm{A}^T = 0$, so $\mathrm{rank(A)} = 0 = \mathrm{rank}(\mathrm{A}^T)$. Suppose $\mathrm{A} \neq 0$ and let $r = \mathrm{rank(A)}$. By Theorem 2.30, there are nonsingular matrices P and Q such that

$$\mathrm{PAQ} = \left[\begin{array}{c|c} \mathrm{I}_r & 0 \\ \hline 0 & 0 \end{array}\right].$$

Recall that P^T and Q^T are also nonsingular (Theorem 1.19). Since

$$\mathrm{Q}^T\mathrm{A}^T\mathrm{P}^T = (\mathrm{PAQ})^T = \left[\begin{array}{c|c} \mathrm{I}_r & 0 \\ \hline 0 & 0 \end{array}\right]^T = \left[\begin{array}{c|c} \mathrm{I}_r & 0 \\ \hline 0 & 0 \end{array}\right],$$

another application of Theorem 2.30 gives $\mathrm{rank}(\mathrm{A}^T) = r$. $\blacksquare$

Exercises 2.8

1. Find the rank and nullity of each matrix.

a. $\begin{bmatrix} 1 & 1 \\ -1 & 0 \\ 1 & 2 \end{bmatrix}$ **b.** $\begin{bmatrix} 1 & -1 & 0 & 2 \\ 1 & 0 & 1 & -1 \\ 2 & -1 & 1 & 1 \end{bmatrix}$ c. $\begin{bmatrix} 1 & 2 & 1 & 2 \\ 1 & 1 & 1 & 1 \end{bmatrix}$

d. $\begin{bmatrix} 1 & -1 & 0 & 1 & -1 \\ -2 & 2 & 1 & 0 & 3 \\ 1 & -1 & 1 & 3 & 0 \\ 0 & 0 & 1 & 2 & 2 \end{bmatrix}$ **e.** $\begin{bmatrix} 1 & -1 & -3 \\ 2 & 3 & 4 \\ 1 & 1 & 1 \\ 4 & 2 & 0 \end{bmatrix}$ f. $\begin{bmatrix} 1 & 0 & 1 & 1 \\ 1 & 1 & -2 & 0 \\ 2 & 1 & -1 & 2 \\ 1 & 1 & -2 & 1 \end{bmatrix}$

g. $\begin{bmatrix} 1 & 1 & 2 \\ 2 & 3 & 3 \\ 4 & 4 & 5 \\ 5 & 6 & 6 \end{bmatrix}$ h. $\begin{bmatrix} 1 & 2 & 3 \\ 4 & 5 & 6 \\ 7 & 8 & 9 \end{bmatrix}$ **i.** $\begin{bmatrix} 0 & 0 & 0 \\ 0 & 0 & -1 \\ 0 & 1 & 1 \\ 0 & 2 & 3 \end{bmatrix}$

2. If $A \in \mathcal{M}_{4\times 3}$, then $\mathcal{N}(A)$ is a subspace of $\mathcal{R}^3$. Find rank(A) when $\mathcal{N}(A)$ is a plane through 0.

3. Suppose $A \in \mathcal{M}_{3\times 5}$, in which case $\mathcal{C}(A)$ is a subspace of $\mathcal{R}^3$. What is the dimension of $\mathcal{N}(A)$ when $\mathcal{C}(A)$ is a line through 0?

4. Can you find a 5×14 matrix whose nullity is 8?

5. Suppose A is a 15×23 matrix with the property that for each $B \in \mathcal{R}^{15}$ the linear system $AX = B$ is consistent. How many parameters are involved in the description of the solutions of the homogeneous system $AX = 0$?

6. Assume $A \in \mathcal{M}_{3\times 5}$ and rref(A) has two leading ones. What's the nullity of A^T?

7. Assuming at least one of $a_1, \ldots, a_n \in \mathcal{R}$ is not zero, find the dimension of $\mathcal{V} = \{X \in \mathcal{R}^n : a_1x_1 + \cdots + a_nx_n = 0\}$.

8. For each matrix in Problem 1, find nonsingular matrices P and Q such that

$$PAQ = \left[\begin{array}{c|c} I_r & 0 \\ \hline 0 & 0 \end{array}\right].$$

9. Suppose A and B are $m \times n$ matrices of equal rank. Show that there are nonsingular matrices P and Q such that $PAQ = B$.

10. Assuming $A \in \mathcal{M}_{m\times n}$ and $B \in \mathcal{M}_{n\times p}$, show that
   a. $\text{rank}(AB) \le \text{rank}(A)$. b. $\text{rank}(AB) \le \text{rank}(B)$.

11. Find matrices A and B of rank 2 such that
   a. $\text{rank}(AB) = 0$. b. $\text{rank}(AB) = 1$. c. $\text{rank}(AB) = 2$.

12. Suppose $A \in \mathcal{M}_{m\times n}$, $B \in \mathcal{M}_{n\times p}$, $C \in \mathcal{M}_{p\times m}$, and $ABC = I_m$.
   a. Show that A and C have rank m.
   b. What can you say about the sizes of n and p relative to m?

13. Assuming $A, B \in \mathcal{M}_{m\times n}$, show that $\text{rank}(A + B) \le \text{rank}(A) + \text{rank}(B)$. Hint: Show that $\mathcal{C}(A + B) \subseteq \mathcal{C}([A \mid B])$.

14. Find matrices A and B of ranks 1 and 2, respectively, such that
   a. $\text{rank}(A + B) = 1$. b. $\text{rank}(A + B) = 2$. c. $\text{rank}(A + B) = 3$.

15. If $A \in \mathcal{M}_{m\times n}$ and $B \in \mathcal{M}_{p\times q}$, show that $\text{rank}\left(\left[\begin{array}{c|c} A & 0 \\ \hline 0 & B \end{array}\right]\right) = \text{rank}(A) + \text{rank}(B)$.

16. Given $A \in \mathcal{M}_{m\times n}$ and $B \in \mathcal{R}^m$, show that $AX = B$ is consistent if and only if $\text{rank}([A \mid B]) = \text{rank}(A)$.

2.9

COORDINATES

A coordinate system for the plane is a scheme for describing the location of each point. The usual rectangular coordinate system employs a central reference point, denoted by 0, and a pair of perpendicular axes through that point, one horizontal, the other vertical. A fixed distance is designated as the standard unit of length, and points are then located by specifying their directed distances from the reference axes. If P is the point whose directed distances from the vertical and horizontal axes are x_1 and x_2, respectively, then the location of P is described by $\mathrm{X} = [x_1 \quad x_2]^T$, and x_1 and x_2 are called the *usual* or *standard* coordinates of P.

This procedure for describing points in the plane has the effect of identifying them with the elements of $\mathcal{R}^2$. If $\mathcal{E} = \{\mathrm{E}_1,\mathrm{E}_2\}$ is the standard basis for $\mathcal{R}^2$, that is,

$$\mathrm{E}_1 = \begin{bmatrix} 1 \\ 0 \end{bmatrix} \quad \text{and} \quad \mathrm{E}_2 = \begin{bmatrix} 0 \\ 1 \end{bmatrix},$$

then

$$\mathrm{X} = \begin{bmatrix} x_1 \\ x_2 \end{bmatrix} = x_1\mathrm{E}_1 + x_2\mathrm{E}_2,$$

so the standard rectangular coordinates of P are just the scalars used to express X as a linear combination of the vectors in $\mathcal{E}$.

There are other ways to identify points in the plane with ordered pairs of real numbers. Let $\mathcal{B} = \{\mathrm{X}_1,\mathrm{X}_2\}$ be a basis for $\mathcal{R}^2$. Since $\{\mathrm{X}_1,\mathrm{X}_2\}$ is linearly independent, $\mathcal{S}pan\{\mathrm{X}_1\}$ and $\mathcal{S}pan\{\mathrm{X}_2\}$ are distinct lines that intersect at 0 (Figure 2.23). Think of

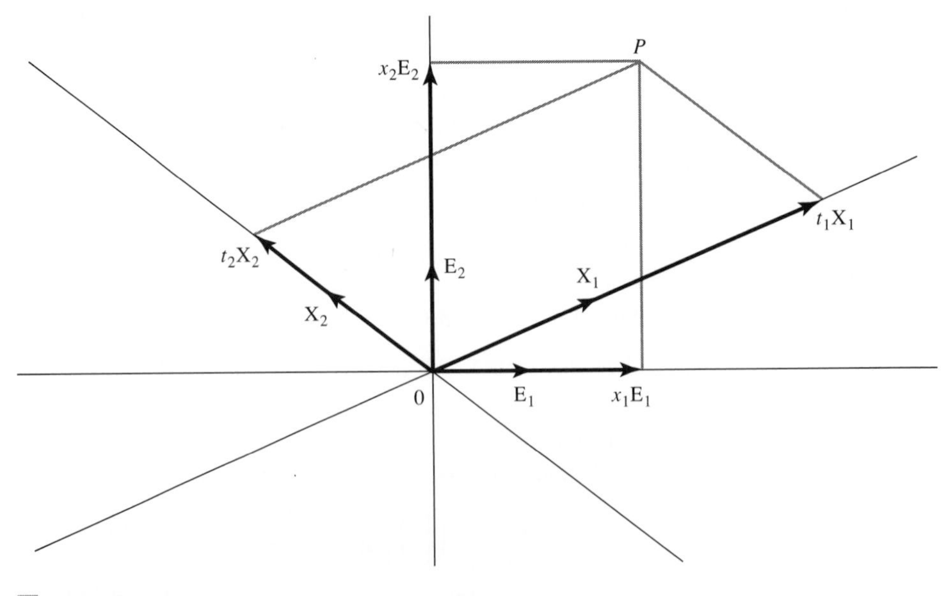

FIGURE 2.23

these lines as axes of an "oblique coordinate system." If P is a point of the plane, X is the corresponding element of $\mathcal{R}^2$, and t_1 and t_2 are the unique scalars such that $X = t_1X_1 + t_2X_2$, then P is determined by the ordered pair $[t_1 \quad t_2]^T$. This time, t_1 and t_2 are not directed distances from P to the coordinate axes, but instead indicate how to scale X_1 and X_2 so the line segment from 0 to P can be interpreted as the diagonal of a parallelogram with sides parallel to $\overrightarrow{0X_1}$ and $\overrightarrow{0X_2}$.

A coordinate system for $\mathcal{R}^n$ or, for that matter, any subspace of $\mathcal{R}^n$ can be established in a similar manner. Given a basis for the subspace, each basis vector is thought of as determining a coordinate axis, and the coordinates of a particular element of the subspace are the scalars required to express that vector as a linear combination of the basis vectors.

DEFINITION. Suppose $\mathcal{V} \subseteq \mathcal{R}^n$ is a subspace of dimension p, $1 \le p \le n$. If $\mathcal{B} = \{X_1, \ldots, X_p\}$ is a basis for $\mathcal{V}$ and $X \in \mathcal{V}$, then the unique scalars $t_1, \ldots, t_p$ such that $X = t_1X_1 + \cdots + t_pX_p$ are called *the coordinates of* X *relative to* $\mathcal{B}$. The p-tuple

$$[X]_{\mathcal{B}} = \begin{bmatrix} t_1 \\ \vdots \\ t_p \end{bmatrix}$$

is referred to as the *coordinate vector for* X *relative to* $\mathcal{B}$. Note that $X \in \mathcal{R}^n$, but $[X]_{\mathcal{B}} \in \mathcal{R}^p$.

■ **EXAMPLE 2.39.** If $\mathcal{B} = \left\{ \begin{bmatrix} 1 \\ 1 \\ 0 \end{bmatrix}, \begin{bmatrix} 1 \\ 2 \\ 1 \end{bmatrix}, \begin{bmatrix} 0 \\ 1 \\ 2 \end{bmatrix} \right\}$ and $X = \begin{bmatrix} 1 \\ 2 \\ 3 \end{bmatrix}$, then $\mathcal{B}$ is a basis for $\mathcal{R}^3$, and the coordinates of X relative to $\mathcal{B}$ are the unique scalars r, s, and t satisfying

$$\begin{bmatrix} 1 \\ 2 \\ 3 \end{bmatrix} = r\begin{bmatrix} 1 \\ 1 \\ 0 \end{bmatrix} + s\begin{bmatrix} 1 \\ 2 \\ 1 \end{bmatrix} + t\begin{bmatrix} 0 \\ 1 \\ 2 \end{bmatrix} = \begin{bmatrix} 1 & 1 & 0 \\ 1 & 2 & 1 \\ 0 & 1 & 2 \end{bmatrix}\begin{bmatrix} r \\ s \\ t \end{bmatrix}.$$

Since

$$\left[\begin{array}{ccc|c} 1 & 1 & 0 & 1 \\ 1 & 2 & 1 & 2 \\ 0 & 1 & 2 & 3 \end{array}\right] \longrightarrow \left[\begin{array}{ccc|c} 1 & 0 & 0 & 2 \\ 0 & 1 & 0 & -1 \\ 0 & 0 & 1 & 2 \end{array}\right],$$

$r = 2$, $s = -1$, and $t = 2$. Thus,

$$X = \begin{bmatrix} 1 \\ 2 \\ 3 \end{bmatrix} = 2\begin{bmatrix} 1 \\ 1 \\ 0 \end{bmatrix} - \begin{bmatrix} 1 \\ 2 \\ 1 \end{bmatrix} + 2\begin{bmatrix} 0 \\ 1 \\ 2 \end{bmatrix} \quad \text{and} \quad [X]_{\mathcal{B}} = \begin{bmatrix} 2 \\ -1 \\ 2 \end{bmatrix}. \square$$

■ **EXAMPLE 2.40.** Let $\mathcal{V}$ be the plane in $\mathcal{R}^3$ with equation $3x + y - 2z = 0$ and set $\mathcal{B} = \{X_1, X_2\}$, where

$$X_1 = \begin{bmatrix} 1 \\ -1 \\ 1 \end{bmatrix} \quad \text{and} \quad X_2 = \begin{bmatrix} -1 \\ 5 \\ 1 \end{bmatrix}.$$

Observe that $\mathcal{B} \subseteq \mathcal{V}$ (both points satisfy the equation of the plane) and $\mathcal{B}$ is linearly independent (X_1 is not a scalar multiple of X_2), so $\mathcal{B}$ is a basis for $\mathcal{V}$. Note also that $X = [1 \quad 1 \quad 2]^T \in \mathcal{V}$. The unique scalars s and t satisfying $X = sX_1 + tX_2$ are the solutions of

$$\begin{bmatrix} 1 & -1 \\ -1 & 5 \\ 1 & 1 \end{bmatrix} \begin{bmatrix} s \\ t \end{bmatrix} = \begin{bmatrix} 1 \\ 1 \\ 2 \end{bmatrix},$$

and since

$$\left[\begin{array}{cc:c} 1 & -1 & 1 \\ -1 & 5 & 1 \\ 1 & 1 & 2 \end{array}\right] \longrightarrow \left[\begin{array}{cc:c} 1 & 0 & 3/2 \\ 0 & 1 & 1/2 \\ 0 & 0 & 0 \end{array}\right],$$

$s = 3/2$ and $t = 1/2$. Thus, $[X]_{\mathcal{B}} = \begin{bmatrix} 3/2 \\ 1/2 \end{bmatrix}$. □

■ **EXAMPLE 2.41.** Suppose $\mathcal{V}$ is a subspace of $\mathcal{R}^n$ and $\mathcal{B} = \{X_1, \ldots, X_p\}$ is a basis for $\mathcal{V}$. The unique representation of the zero vector as a linear combination of $X_1, \ldots, X_p$ is $0 = 0X_1 + \cdots + 0X_p$, so

$$[0]_{\mathcal{B}} = \begin{bmatrix} 0 \\ \vdots \\ 0 \end{bmatrix}.$$

Also, for $1 \leq j \leq p$,

$$X_j = \sum_{1}^{p} t_k X_k, \qquad \text{where} \qquad t_k = \begin{cases} 1, & \text{if } k = j \\ 0, & \text{if } k \neq j \end{cases},$$

so

$$[X_j]_{\mathcal{B}} = E_j. \ \square$$

Given a subspace $\mathcal{V}$ of $\mathcal{R}^n$ with basis $\mathcal{B} = \{X_1, \ldots, X_p\}$, the $\mathcal{B}$ coordinates of $X \in \mathcal{V}$ are found by solving the linear system $AT = X$, where

$$A = [X_1 \quad \cdots \quad X_p] \qquad \text{and} \qquad T = \begin{bmatrix} t_1 \\ \vdots \\ t_p \end{bmatrix},$$

that is, by row reducing $[A \mid X]$. Note that A is an $n \times p$ matrix with linearly independent columns. If $p = n$, then $\mathcal{V} = \mathcal{R}^n$. In that event, $\text{rref}(A) = I_n$, and

$$[A \mid X] \longrightarrow [I_n \mid [X]_{\mathcal{B}}].$$

If $p < n$, then

$$[A \mid X] \longrightarrow \left[\begin{array}{c:c} I_p & [X]_{\mathcal{B}} \\ \hdashline 0 & 0 \end{array}\right],$$

where the 0's represent blocks of zeros with $n - p$ rows.

■ **EXAMPLE 2.42.** Let $\mathcal{B} = \left\{ \begin{bmatrix} 1 \\ 1 \\ 0 \\ 0 \end{bmatrix}, \begin{bmatrix} 1 \\ 0 \\ 0 \\ 1 \end{bmatrix}, \begin{bmatrix} 0 \\ 0 \\ 1 \\ -1 \end{bmatrix}, \begin{bmatrix} 0 \\ 0 \\ -1 \\ 2 \end{bmatrix} \right\}$ and set $X = \begin{bmatrix} 1 \\ 2 \\ 3 \\ 4 \end{bmatrix}$. The row reduction

$$\left[\begin{array}{cccc|c} 1 & 1 & 0 & 0 & 1 \\ 1 & 0 & 0 & 0 & 2 \\ 0 & 0 & 1 & -1 & 3 \\ 0 & 1 & -1 & 2 & 4 \end{array}\right] \longrightarrow \left[\begin{array}{cccc|c} 1 & 0 & 0 & 0 & 2 \\ 0 & 1 & 0 & 0 & -1 \\ 0 & 0 & 1 & 0 & 11 \\ 0 & 0 & 0 & 1 & 8 \end{array}\right]$$

confirms that the elements of $\mathcal{B}$ are linearly independent (and therefore form a basis for $\mathcal{R}^4$) and simultaneously shows that

$$[\mathrm{X}]_{\mathcal{B}} = \begin{bmatrix} 2 \\ -1 \\ 11 \\ 8 \end{bmatrix}, \quad \text{that is,} \quad \begin{bmatrix} 1 \\ 2 \\ 3 \\ 4 \end{bmatrix} = 2\begin{bmatrix} 1 \\ 1 \\ 0 \\ 0 \end{bmatrix} - \begin{bmatrix} 1 \\ 0 \\ 0 \\ 1 \end{bmatrix} + 11\begin{bmatrix} 0 \\ 0 \\ 1 \\ -1 \end{bmatrix} + 8\begin{bmatrix} 0 \\ 0 \\ -1 \\ 2 \end{bmatrix}. \ \square$$

■ **EXAMPLE 2.43.** Let $\mathcal{V} = \mathcal{S}pan\{\mathrm{X}_1,\mathrm{X}_2\}$, where

$$\mathrm{X}_1 = \begin{bmatrix} 1 \\ 1 \\ 0 \\ -1 \end{bmatrix} \quad \text{and} \quad \mathrm{X}_2 = \begin{bmatrix} 2 \\ 1 \\ 1 \\ -1 \end{bmatrix}.$$

Since X_1 and X_2 are linearly independent, $\mathcal{V}$ is a two-dimensional subspace of $\mathcal{R}^4$ with basis $\mathcal{B} = \{\mathrm{X}_1,\mathrm{X}_2\}$. If $\mathrm{X} = [1 \quad -1 \quad 2 \quad 1]^T$, then

$$[\mathrm{X}_1 \quad \mathrm{X}_2 \mid \mathrm{X}] = \left[\begin{array}{cc|c} 1 & 2 & 1 \\ 1 & 1 & -1 \\ 0 & 1 & 2 \\ -1 & -1 & 1 \end{array}\right] \longrightarrow \left[\begin{array}{cc|c} 1 & 0 & -3 \\ 0 & 1 & 2 \\ 0 & 0 & 0 \\ 0 & 0 & 0 \end{array}\right],$$

so $\mathrm{X} \in \mathcal{V}$, and

$$[\mathrm{X}]_{\mathcal{B}} = \begin{bmatrix} -3 \\ 2 \end{bmatrix}. \ \square$$

The standard basis for $\mathcal{R}^n$ produces the *standard coordinate system* for $\mathcal{R}^n$. If $\mathrm{X} = x_1\mathrm{E}_1 + \cdots + x_n\mathrm{E}_n$, then

$$[\mathrm{X}]_{\mathcal{E}} = \begin{bmatrix} x_1 \\ \vdots \\ x_n \end{bmatrix} = \mathrm{X},$$

so there is actually no distinction between X and its $\mathcal{E}$ coordinates.

THEOREM 2.32. Assume $\mathcal{V}$ is a subspace of $\mathcal{R}^n$ with basis $\mathcal{B} = \{\mathrm{X}_1, \ldots, \mathrm{X}_p\}$. If $\mathrm{X},\mathrm{Y} \in \mathcal{V}$ and $t \in \mathcal{R}$, then
(1) $[\mathrm{X}+\mathrm{Y}]_{\mathcal{B}} = [\mathrm{X}]_{\mathcal{B}} + [\mathrm{Y}]_{\mathcal{B}}$, and
(2) $[t\mathrm{X}]_{\mathcal{B}} = t[\mathrm{X}]_{\mathcal{B}}$.

Proof. (1) If $[\mathrm{X}]_{\mathcal{B}} = [t_1 \quad \cdots \quad t_p]^T$ and $[\mathrm{Y}]_{\mathcal{B}} = [s_1 \quad \cdots \quad s_p]^T$, then $\mathrm{X} = t_1\mathrm{X}_1 + \cdots + t_p\mathrm{X}_p, \mathrm{Y} = s_1\mathrm{X}_1 + \cdots + s_p\mathrm{X}_p$, and

$$\mathrm{X}+\mathrm{Y} = (t_1+s_1)\mathrm{X}_1 + \cdots + (t_p+s_p)\mathrm{X}_p.$$

Thus,

$$[X+Y]_{\mathcal{B}} = \begin{bmatrix} t_1+s_1 \\ \vdots \\ t_p+s_p \end{bmatrix} = \begin{bmatrix} t_1 \\ \vdots \\ t_p \end{bmatrix} + \begin{bmatrix} s_1 \\ \vdots \\ s_p \end{bmatrix} = [X]_{\mathcal{B}} + [Y]_{\mathcal{B}}.$$

(2) See Exercise 5. ■

Let $\mathcal{V}$ be a subspace of $\mathcal{R}^n$ with basis $\mathcal{B}$. If X,Y $\in \mathcal{V}$ and $s,t \in \mathcal{R}$, then by Theorem 2.32,

$$[sX+tY]_{\mathcal{B}} = [sX]_{\mathcal{B}} + [tY]_{\mathcal{B}} = s[X]_{\mathcal{B}} + t[Y]_{\mathcal{B}}.$$

This relationship extends to general linear combinations. For $X_1, \ldots, X_q \in \mathcal{V}$ and $t_1, \ldots, t_q \in \mathcal{R}$,

$$\boxed{[t_1X_1 + \cdots + t_qX_q]_{\mathcal{B}} = t_1[X_1]_{\mathcal{B}} + \cdots + t_q[X_q]_{\mathcal{B}}}. \tag{2.6}$$

Equation (2.6) is the fundamental algebraic rule governing calculations with coordinates.

Suppose $\mathcal{V}$ is a p-dimensional subspace of $\mathcal{R}^n$ with bases $\mathcal{A} = \{X_1, \ldots, X_p\}$ and $\mathcal{B} = \{Y_1, \ldots, Y_p\}$. For each X $\in \mathcal{V}$, $[X]_{\mathcal{A}}$ and $[X]_{\mathcal{B}}$ are two descriptions of the same vector. If

$$[X]_{\mathcal{A}} = \begin{bmatrix} t_1 \\ \vdots \\ t_p \end{bmatrix},$$

then $X = t_1X_1 + \cdots + t_pX_p$, and by (2.6),

$$\begin{aligned} [X]_{\mathcal{B}} &= [t_1X_1 + \cdots + t_pX_p]_{\mathcal{B}} = t_1[X_1]_{\mathcal{B}} + \cdots + t_p[X_p]_{\mathcal{B}} \\ &= \begin{bmatrix} [X_1]_{\mathcal{B}} & \cdots & [X_p]_{\mathcal{B}} \end{bmatrix} \begin{bmatrix} t_1 \\ \vdots \\ t_p \end{bmatrix} = \begin{bmatrix} [X_1]_{\mathcal{B}} & \cdots & [X_p]_{\mathcal{B}} \end{bmatrix} [X]_{\mathcal{A}}. \end{aligned}$$

Since $\dim \mathcal{V} = p$, $[X_j]_{\mathcal{B}} \in \mathcal{R}^p$ for each $j = 1, \ldots, p$, and consequently

$$\begin{bmatrix} [X_1]_{\mathcal{B}} & \cdots & [X_p]_{\mathcal{B}} \end{bmatrix}$$

is a $p \times p$ matrix. It depends only on the bases $\mathcal{A}$ and $\mathcal{B}$ (its columns are the $\mathcal{B}$ coordinates of the $\mathcal{A}$ vectors), and it provides a mechanism for converting $[X]_{\mathcal{A}}$ to $[X]_{\mathcal{B}}$ for any X $\in \mathcal{V}$.

THEOREM 2.33. Suppose $\mathcal{V}$ is a p-dimensional subspace of $\mathcal{R}^n$ with bases $\mathcal{A} = \{X_1, \ldots, X_p\}$ and $\mathcal{B} = \{Y_1, \ldots, Y_p\}$. There is a unique nonsingular $p \times p$ matrix, $P_{\mathcal{BA}}$, with the property that for each X $\in \mathcal{V}$,

$$[X]_{\mathcal{B}} = P_{\mathcal{BA}}[X]_{\mathcal{A}}. \tag{2.7}$$

In fact,

$$P_{\mathcal{BA}} = \begin{bmatrix} [X_1]_{\mathcal{B}} & \cdots & [X_p]_{\mathcal{B}} \end{bmatrix}. \tag{2.8}$$

Proof. The discussion above shows that matrix (2.8) has property (2.7). It remains to establish that $\mathrm{P}_{\mathcal{BA}}$ is unique and nonsingular. For uniqueness, suppose Q is a $p \times p$ matrix satisfying $[\mathrm{X}]_{\mathcal{B}} = \mathrm{Q}[\mathrm{X}]_{\mathcal{A}}$, $\mathrm{X} \in \mathcal{V}$. Since $[\mathrm{X}_j]_{\mathcal{A}} = \mathrm{E}_j$ (Example 2.41),

$$\mathrm{Col}_j(\mathrm{P}_{\mathcal{BA}}) = [\mathrm{X}_j]_{\mathcal{B}} = \mathrm{Q}[\mathrm{X}_j]_{\mathcal{A}} = \mathrm{QE}_j = \mathrm{Col}_j(\mathrm{Q}).$$

This conclusion holds for each $j = 1, \ldots, p$, so $\mathrm{P}_{\mathcal{BA}} = \mathrm{Q}$. For nonsingularity, it suffices to show that the columns of $\mathrm{P}_{\mathcal{BA}}$ are linearly independent. Suppose $t_1, \ldots, t_p$ are scalars such that $t_1\mathrm{Col}_1(\mathrm{P}_{\mathcal{BA}}) + \cdots + t_p\mathrm{Col}_p(\mathrm{P}_{\mathcal{BA}}) = 0$. By (2.8) and (2.6),

$$\begin{aligned} 0 &= t_1\mathrm{Col}_1(\mathrm{P}_{\mathcal{BA}}) + \cdots + t_p\mathrm{Col}_p(\mathrm{P}_{\mathcal{BA}}) \\ &= t_1[\mathrm{X}_1]_{\mathcal{B}} + \cdots + t_p[\mathrm{X}_p]_{\mathcal{B}} \\ &= [t_1\mathrm{X}_1 + \cdots + t_p\mathrm{X}_p]_{\mathcal{B}}, \end{aligned}$$

which in turn implies $t_1\mathrm{X}_1 + \cdots + t_p\mathrm{X}_p = 0$. Since $\{\mathrm{X}_1, \ldots, \mathrm{X}_p\}$ is linearly independent, $t_1 = \cdots = t_p = 0$. ■

■ **EXAMPLE 2.44.** Let $\mathcal{V} = \mathcal{R}^3$,

$$\mathcal{A} = \left\{ \begin{bmatrix} 1 \\ 0 \\ 1 \end{bmatrix}, \begin{bmatrix} 0 \\ 1 \\ 1 \end{bmatrix}, \begin{bmatrix} 1 \\ 1 \\ 0 \end{bmatrix} \right\}, \text{ and } \mathcal{B} = \left\{ \begin{bmatrix} 1 \\ 0 \\ 0 \end{bmatrix}, \begin{bmatrix} 1 \\ 1 \\ 0 \end{bmatrix}, \begin{bmatrix} 1 \\ 1 \\ 1 \end{bmatrix} \right\}.$$

In the notation of Theorem 2.33,

$$\mathrm{X}_1 = \begin{bmatrix} 1 \\ 0 \\ 1 \end{bmatrix}, \mathrm{X}_2 = \begin{bmatrix} 0 \\ 1 \\ 1 \end{bmatrix}, \mathrm{X}_3 = \begin{bmatrix} 1 \\ 1 \\ 0 \end{bmatrix} \text{ and } \mathrm{Y}_1 = \begin{bmatrix} 1 \\ 0 \\ 0 \end{bmatrix}, \mathrm{Y}_2 = \begin{bmatrix} 1 \\ 1 \\ 0 \end{bmatrix}, \mathrm{Y}_3 = \begin{bmatrix} 1 \\ 1 \\ 1 \end{bmatrix}.$$

The columns of $\mathrm{P}_{\mathcal{BA}}$, namely $[\mathrm{X}_1]_{\mathcal{B}}$, $[\mathrm{X}_2]_{\mathcal{B}}$, and $[\mathrm{X}_3]_{\mathcal{B}}$, are obtained from the row reductions

$$\begin{aligned} [\mathrm{Y}_1 \quad \mathrm{Y}_2 \quad \mathrm{Y}_3 \mid \mathrm{X}_1] &\longrightarrow [\mathrm{I}_3 \mid [\mathrm{X}_1]_{\mathcal{B}}] \\ [\mathrm{Y}_1 \quad \mathrm{Y}_2 \quad \mathrm{Y}_3 \mid \mathrm{X}_2] &\longrightarrow [\mathrm{I}_3 \mid [\mathrm{X}_2]_{\mathcal{B}}] \\ [\mathrm{Y}_1 \quad \mathrm{Y}_2 \quad \mathrm{Y}_3 \mid \mathrm{X}_3] &\longrightarrow [\mathrm{I}_3 \mid [\mathrm{X}_3]_{\mathcal{B}}]. \end{aligned}$$

These three calculations can be performed simultaneously via

$$[\mathrm{Y}_1 \quad \mathrm{Y}_2 \quad \mathrm{Y}_3 \mid \mathrm{X}_1 \quad \mathrm{X}_2 \quad \mathrm{X}_3] \longrightarrow [\mathrm{I}_3 \mid [\mathrm{X}_1]_{\mathcal{B}} \quad [\mathrm{X}_2]_{\mathcal{B}} \quad [\mathrm{X}_3]_{\mathcal{B}}] = [\mathrm{I}_3 \mid \mathrm{P}_{\mathcal{BA}}],$$

which in this case produces

$$\left[\begin{array}{ccc|ccc} 1 & 1 & 1 & 1 & 0 & 1 \\ 0 & 1 & 1 & 0 & 1 & 1 \\ 0 & 0 & 1 & 1 & 1 & 0 \end{array}\right] \longrightarrow \left[\begin{array}{ccc|ccc} 1 & 0 & 0 & 1 & -1 & 0 \\ 0 & 1 & 0 & -1 & 0 & 1 \\ 0 & 0 & 1 & 1 & 1 & 0 \end{array}\right].$$

Thus,

$$\mathrm{P}_{\mathcal{BA}} = \begin{bmatrix} 1 & -1 & 0 \\ -1 & 0 & 1 \\ 1 & 1 & 0 \end{bmatrix}.$$

This outcome is easy to check. The first column of $\mathrm{P}_{\mathcal{BA}}$ should be $[\mathrm{X}_1]_{\mathcal{B}}$, and indeed,

$$1\mathrm{Y}_1 + (-1)\mathrm{Y}_2 + 1\mathrm{Y}_3 = \begin{bmatrix} 1 \\ 0 \\ 0 \end{bmatrix} - \begin{bmatrix} 1 \\ 1 \\ 0 \end{bmatrix} + \begin{bmatrix} 1 \\ 1 \\ 1 \end{bmatrix} = \begin{bmatrix} 1 \\ 0 \\ 1 \end{bmatrix} = \mathrm{X}_1.$$

Columns 2 and 3 of $P_{\mathcal{BA}}$ should be $[X_2]_{\mathcal{B}}$ and $[X_3]_{\mathcal{B}}$, respectively, and in fact,

$$(-1)Y_1 + 0Y_2 + 1Y_3 = (-1)\begin{bmatrix}1\\0\\0\end{bmatrix} + 0\begin{bmatrix}1\\1\\0\end{bmatrix} + \begin{bmatrix}1\\1\\1\end{bmatrix} = \begin{bmatrix}0\\1\\1\end{bmatrix} = X_2,$$

and

$$0Y_1 + 1Y_2 + 0Y_3 = 0\begin{bmatrix}1\\0\\0\end{bmatrix} + \begin{bmatrix}1\\1\\0\end{bmatrix} + 0\begin{bmatrix}1\\1\\1\end{bmatrix} = \begin{bmatrix}1\\1\\0\end{bmatrix} = X_3.$$

To illustrate the use of (2.7), consider $X = X_1 + 2X_2 + 3X_3$. Then

$$[X]_{\mathcal{A}} = \begin{bmatrix}1\\2\\3\end{bmatrix} \quad \text{and} \quad X = \begin{bmatrix}1\\0\\1\end{bmatrix} + 2\begin{bmatrix}0\\1\\1\end{bmatrix} + 3\begin{bmatrix}1\\1\\0\end{bmatrix} = \begin{bmatrix}4\\5\\3\end{bmatrix}.$$

According to (2.7),

$$[X]_{\mathcal{B}} = P_{\mathcal{BA}}[X]_{\mathcal{A}} = \begin{bmatrix}1 & -1 & 0\\-1 & 0 & 1\\1 & 1 & 0\end{bmatrix}\begin{bmatrix}1\\2\\3\end{bmatrix} = \begin{bmatrix}-1\\2\\3\end{bmatrix},$$

and indeed,

$$(-1)Y_1 + 2Y_2 + 3Y_3 = (-1)\begin{bmatrix}1\\0\\0\end{bmatrix} + 2\begin{bmatrix}1\\1\\0\end{bmatrix} + 3\begin{bmatrix}1\\1\\1\end{bmatrix} = \begin{bmatrix}4\\5\\3\end{bmatrix} = X. \;\square$$

■ **EXAMPLE 2.45.** Let $\mathcal{V}$ be the plane in $\mathcal{R}^3$ with equation $3x + y - 2z = 0$, and observe that

$$\mathcal{B} = \{Y_1, Y_2\} = \left\{\begin{bmatrix}1\\-1\\1\end{bmatrix}, \begin{bmatrix}-1\\5\\1\end{bmatrix}\right\} \quad \text{and} \quad \mathcal{A} = \{X_1, X_2\} = \left\{\begin{bmatrix}1\\1\\2\end{bmatrix}, \begin{bmatrix}0\\2\\1\end{bmatrix}\right\}$$

are bases for $\mathcal{V}$. According to (2.8), $P_{\mathcal{BA}}$ is the 2×2 matrix whose columns are the $\mathcal{B}$ coordinates of the $\mathcal{A}$ vectors. Since

$$[Y_1 \;\; Y_2 \mid X_1 \;\; X_2] = \left[\begin{array}{rr|rr}1 & -1 & 1 & 0\\-1 & 5 & 1 & 2\\1 & 1 & 2 & 1\end{array}\right] \longrightarrow \left[\begin{array}{rr|rr}1 & 0 & 3/2 & 1/2\\0 & 1 & 1/2 & 1/2\\0 & 0 & 0 & 0\end{array}\right],$$

$$[X_1]_{\mathcal{B}} = \begin{bmatrix}3/2\\1/2\end{bmatrix}, \; [X_2]_{\mathcal{B}} = \begin{bmatrix}1/2\\1/2\end{bmatrix}, \quad \text{and} \quad P_{\mathcal{BA}} = \begin{bmatrix}3/2 & 1/2\\1/2 & 1/2\end{bmatrix} = \frac{1}{2}\begin{bmatrix}3 & 1\\1 & 1\end{bmatrix}. \;\square$$

For a p-dimensional subspace $\mathcal{V}$ of $\mathcal{R}^n$ with bases $\mathcal{A} = \{X_1, \ldots, X_p\}$ and $\mathcal{B} = \{Y_1, \ldots, Y_p\}$, row reduction of $[Y_1 \;\cdots\; Y_p \mid X_1 \;\cdots\; X_p]$ produces $P_{\mathcal{BA}}$. If $p = n$, then $\mathcal{V} = \mathcal{R}^n$, and

$$[Y_1 \;\cdots\; Y_n \mid X_1 \;\cdots\; X_n] \longrightarrow [I_n \mid P_{\mathcal{BA}}].$$

When $p < n$,

$$[Y_1 \ \cdots \ Y_p \mid X_1 \ \cdots \ X_p] \longrightarrow \left[\begin{array}{c|c} I_p & P_{\mathcal{BA}} \\ \hline 0 & 0 \end{array}\right].$$

Here the 0's represent $(n-p) \times p$ blocks of zeros.

$P_{\mathcal{BA}}$ is called the *$\mathcal{A}$ to $\mathcal{B}$ change of coordinates matrix*. It converts the $\mathcal{A}$ description of X to the $\mathcal{B}$ description of X via $[X]_{\mathcal{B}} = P_{\mathcal{BA}}[X]_{\mathcal{A}}$. Note that the order of the subscripts in $P_{\mathcal{BA}}$ corresponds to the change that takes place in the representation of X as you read from right to left. Interchanging roles of $\mathcal{A}$ and $\mathcal{B}$ produces another matrix, $P_{\mathcal{AB}}$, such that $[X]_{\mathcal{A}} = P_{\mathcal{AB}}[X]_{\mathcal{B}}$, $X \in \mathcal{V}$. It is calculated by row reducing $[X_1 \ \cdots \ X_p \mid Y_1 \ \cdots \ Y_p]$.

THEOREM 2.34. If $\mathcal{V}$ is a p-dimensional subspace of $\mathcal{R}^n$ with bases $\mathcal{A}$ and $\mathcal{B}$, then

$$P_{\mathcal{AB}} = P_{\mathcal{BA}}^{-1}.$$

Proof. For each $X \in \mathcal{V}$, $[X]_{\mathcal{B}} = P_{\mathcal{BA}}[X]_{\mathcal{A}}$, so $[X]_{\mathcal{A}} = P_{\mathcal{BA}}^{-1}[X]_{\mathcal{B}}$. Since $P_{\mathcal{AB}}$ is the unique $M \in \mathcal{M}_{p\times p}$ such that $[X]_{\mathcal{A}} = M[X]_{\mathcal{B}}$ for all $X \in \mathcal{V}$, $P_{\mathcal{AB}} = P_{\mathcal{BA}}^{-1}$. ■

■ **EXAMPLE 2.46.** For $\mathcal{V} = \mathcal{R}^2$, consider $\mathcal{A} = \{X_1, X_2\} = \left\{\begin{bmatrix} 1 \\ 1 \end{bmatrix}, \begin{bmatrix} 1 \\ -1 \end{bmatrix}\right\}$ and $\mathcal{B} = \{Y_1, Y_2\} = \left\{\begin{bmatrix} 1 \\ -2 \end{bmatrix}, \begin{bmatrix} -1 \\ 3 \end{bmatrix}\right\}$. Since

$$[Y_1 \ \ Y_2 \mid X_1 \ \ X_2] = \left[\begin{array}{cc|cc} 1 & -1 & 1 & 1 \\ -2 & 3 & 1 & -1 \end{array}\right] \longrightarrow \left[\begin{array}{cc|cc} 1 & 0 & 4 & 2 \\ 0 & 1 & 3 & 1 \end{array}\right],$$

$$P_{\mathcal{BA}} = \begin{bmatrix} 4 & 2 \\ 3 & 1 \end{bmatrix}.$$

Moreover,

$$[X_1 \ \ X_2 \mid Y_1 \ \ Y_2] = \left[\begin{array}{cc|cc} 1 & 1 & 1 & -1 \\ 1 & -1 & -2 & 3 \end{array}\right] \longrightarrow \left[\begin{array}{cc|cc} 1 & 0 & -1/2 & 1 \\ 0 & 1 & 3/2 & -2 \end{array}\right],$$

so

$$P_{\mathcal{AB}} = -\frac{1}{2}\begin{bmatrix} 1 & -2 \\ -3 & 4 \end{bmatrix}.$$

Note that $P_{\mathcal{AB}}$ is the inverse of $P_{\mathcal{BA}}$. □

■ **EXAMPLE 2.47.** Suppose $\mathcal{A} = \{X_1, \ldots, X_n\}$ is a basis for $\mathcal{R}^n$ and $\mathcal{E} = \{E_1, \ldots, E_n\}$. Since $[X]_{\mathcal{E}} = X$, $X \in \mathcal{R}^n$,

$$P_{\mathcal{EA}} = \big[[X_1]_{\mathcal{E}} \ \ \cdots \ \ [X_n]_{\mathcal{E}}\big] = [X_1 \ \ \cdots \ \ X_n].$$

In particular, if $\mathcal{A}$ and $\mathcal{B}$ are the bases for $\mathcal{R}^2$ in the previous example, then

$$P_{\mathcal{EA}} = \begin{bmatrix} 1 & 1 \\ 1 & -1 \end{bmatrix} \quad \text{and} \quad P_{\mathcal{EB}} = \begin{bmatrix} 1 & -1 \\ -2 & 3 \end{bmatrix}. \ \square$$

Exercises 2.9

1. Assuming $\mathcal{B} = \left\{ \begin{bmatrix} 1 \\ -1 \\ 2 \end{bmatrix}, \begin{bmatrix} -1 \\ 2 \\ 1 \end{bmatrix}, \begin{bmatrix} -1 \\ 2 \\ 3 \end{bmatrix} \right\}$, find $X \in \mathcal{R}^3$ such that

   a. $[X]_{\mathcal{B}} = \begin{bmatrix} 1 \\ 2 \\ 3 \end{bmatrix}$. b. $[X]_{\mathcal{B}} = \begin{bmatrix} -1 \\ 1 \\ 0 \end{bmatrix}$. **c.** $[X]_{\mathcal{B}} = \begin{bmatrix} -3 \\ 2 \\ -1 \end{bmatrix}$.

   Find $[X]_{\mathcal{B}}$ when

   d. $X = \begin{bmatrix} 1 \\ 2 \\ 3 \end{bmatrix}$. e. $X = \begin{bmatrix} 3 \\ 2 \\ 1 \end{bmatrix}$. **f.** $X = \begin{bmatrix} 1 \\ 1 \\ 1 \end{bmatrix}$.

2. Let $\mathcal{V}$ be the subspace of $\mathcal{R}^4$ spanned by $\mathcal{B} = \left\{ \begin{bmatrix} 1 \\ -1 \\ 2 \\ 3 \end{bmatrix}, \begin{bmatrix} 1 \\ 1 \\ 2 \\ 1 \end{bmatrix} \right\}$. Find $X \in \mathcal{V}$ when

   a. $[X]_{\mathcal{B}} = \begin{bmatrix} -1 \\ 2 \end{bmatrix}$. b. $[X]_{\mathcal{B}} = \begin{bmatrix} 0 \\ 1 \end{bmatrix}$. c. $[X]_{\mathcal{B}} = \begin{bmatrix} 2 \\ -3 \end{bmatrix}$.

   Find $[X]_{\mathcal{B}}$ when

   d. $X = \begin{bmatrix} 2 \\ 3 \\ 4 \\ 1 \end{bmatrix}$. e. $X = \begin{bmatrix} 3 \\ 1 \\ 8 \\ 5 \end{bmatrix}$. f. $X = \begin{bmatrix} 2 \\ 4 \\ 4 \\ 0 \end{bmatrix}$.

3. Let $\mathcal{V}$ be the line through zero in $\mathcal{R}^3$ spanned by $\mathcal{B} = \left\{ \begin{bmatrix} -1 \\ 2 \\ -3 \end{bmatrix} \right\}$. Show that

   $X = \begin{bmatrix} 3 \\ -6 \\ 9 \end{bmatrix} \in \mathcal{V}$ and find $[X]_{\mathcal{B}}$.

4. Assume $\mathcal{V}$ is a three-dimensional subspace of $\mathcal{R}^n$ with basis $\mathcal{B} = \{X_1, X_2, X_3\}$ and let X be the element of $\mathcal{V}$ such that $[X]_{\mathcal{B}} = [3 \quad 1 \quad 2]^T$. Find $[X]_{\mathcal{A}}$ when

   a. $\mathcal{A} = \{X_3, X_1, X_2\}$. b. $\mathcal{A} = \{-3X_2, 2X_3, \frac{1}{2}X_1\}$.

5. Prove part (2) of Theorem 2.32.

6. If $\mathcal{V} = \mathcal{R}^2$, $\mathcal{A} = \left\{ \begin{bmatrix} 2 \\ 4 \end{bmatrix}, \begin{bmatrix} -1 \\ 2 \end{bmatrix} \right\}$, and $\mathcal{B} = \left\{ \begin{bmatrix} -1 \\ -1 \end{bmatrix}, \begin{bmatrix} 1 \\ -2 \end{bmatrix} \right\}$, find

   a. $P_{\mathcal{AB}}$. b. $P_{\mathcal{BA}}$. **c.** $P_{\mathcal{AE}}$. d. $P_{\mathcal{BE}}$.

7. If $\mathcal{V} = \mathcal{R}^3$, $\mathcal{A} = \left\{ \begin{bmatrix} 1 \\ 0 \\ 1 \end{bmatrix}, \begin{bmatrix} -1 \\ 1 \\ 0 \end{bmatrix}, \begin{bmatrix} -1 \\ -1 \\ 1 \end{bmatrix} \right\}$, $\mathcal{B} = \left\{ \begin{bmatrix} 1 \\ 1 \\ 1 \end{bmatrix}, \begin{bmatrix} 0 \\ 1 \\ 1 \end{bmatrix}, \begin{bmatrix} 1 \\ 1 \\ 0 \end{bmatrix} \right\}$, and

   $\mathcal{C} = \left\{ \begin{bmatrix} 1 \\ 1 \\ 2 \end{bmatrix}, \begin{bmatrix} 1 \\ 2 \\ 1 \end{bmatrix}, \begin{bmatrix} 2 \\ 1 \\ 1 \end{bmatrix} \right\}$, find

   a. $P_{\mathcal{BA}}$. b. $P_{\mathcal{CB}}$. **c.** $P_{\mathcal{AC}}$. d. $P_{\mathcal{EC}}$. **e.** $P_{\mathcal{EA}}$. f. $P_{\mathcal{BE}}$.

8. Assuming $\mathcal{V} = \mathcal{R}^4$, $\mathcal{A} = \left\{ \begin{bmatrix} 1 \\ 0 \\ 1 \\ 0 \end{bmatrix}, \begin{bmatrix} 0 \\ 1 \\ -1 \\ 0 \end{bmatrix}, \begin{bmatrix} -1 \\ 1 \\ 0 \\ 1 \end{bmatrix}, \begin{bmatrix} 0 \\ 0 \\ 0 \\ 1 \end{bmatrix} \right\}$, and

   $\mathcal{B} = \left\{ \begin{bmatrix} 1 \\ 0 \\ 0 \\ 0 \end{bmatrix}, \begin{bmatrix} 1 \\ 1 \\ 0 \\ 0 \end{bmatrix}, \begin{bmatrix} 1 \\ 1 \\ 1 \\ 0 \end{bmatrix}, \begin{bmatrix} 1 \\ 1 \\ 1 \\ 1 \end{bmatrix} \right\}$, find $P_{\mathcal{AB}}$ and $P_{\mathcal{BA}}$.

9. Let $\mathcal{V}$ be a two-dimensional subspace of $\mathcal{R}^n$ with basis $\mathcal{A} = \{X_1, X_2\}$. Set $Y_1 = 3X_1 - 2X_2$, $Y_2 = -X_1 + 4X_2$, and $\mathcal{B} = \{Y_1, Y_2\}$.
   a. Show that $\mathcal{B}$ is a basis for $\mathcal{V}$.
   b. Find $P_{\mathcal{BA}}$ and $P_{\mathcal{AB}}$.

10. Let $\mathcal{B} = \left\{ \begin{bmatrix} 1 \\ 2 \\ 0 \\ 1 \end{bmatrix}, \begin{bmatrix} -1 \\ 1 \\ 1 \\ 0 \end{bmatrix}, \begin{bmatrix} 0 \\ 2 \\ -1 \\ 3 \end{bmatrix} \right\}$ and $P_{\mathcal{BA}} = \begin{bmatrix} 1 & 1 & 2 \\ 2 & -1 & 2 \\ 2 & 1 & 1 \end{bmatrix}$, where $\mathcal{A}$ is another basis

    for $\mathcal{V} = \mathcal{S}pan\{\mathcal{B}\}$. Find the $X \in \mathcal{V}$ such that $[X]_{\mathcal{A}} = [1 \quad 2 \quad 3]^T$.

11. If $\mathcal{A} = \left\{ \begin{bmatrix} -1 \\ 1 \end{bmatrix}, \begin{bmatrix} 2 \\ -1 \end{bmatrix} \right\}$, find the basis $\mathcal{B}$ for $\mathcal{R}^2$ such that $P_{\mathcal{BA}} = \begin{bmatrix} 1 & 2 \\ 3 & 4 \end{bmatrix}$.

12. Let $A = \begin{bmatrix} 1 & 1 & 1 & 3 \\ 2 & 1 & 0 & 5 \end{bmatrix}$.

    a. Show that $\mathcal{A} = \left\{ \begin{bmatrix} 1 \\ 2 \end{bmatrix}, \begin{bmatrix} 1 \\ 1 \end{bmatrix} \right\}$ and $\mathcal{B} = \left\{ \begin{bmatrix} 1 \\ 0 \end{bmatrix}, \begin{bmatrix} 3 \\ 5 \end{bmatrix} \right\}$ are bases for $\mathcal{C}(A)$ and find $P_{\mathcal{AB}}$ and $P_{\mathcal{BA}}$.

    b. Show that $\mathcal{A} = \left\{ \begin{bmatrix} -1 \\ -3 \\ 1 \\ 1 \end{bmatrix}, \begin{bmatrix} -3 \\ 1 \\ -1 \\ 1 \end{bmatrix} \right\}$ and $\mathcal{B} = \left\{ \begin{bmatrix} 1 \\ -2 \\ 1 \\ 0 \end{bmatrix}, \begin{bmatrix} -2 \\ -1 \\ 0 \\ 1 \end{bmatrix} \right\}$ are bases for $\mathcal{N}(A)$,

    and find $P_{\mathcal{BA}}$ and $P_{\mathcal{AB}}$.

13. Suppose $\mathcal{B} = \{X_1, X_2, X_3\}$ is a linearly independent subset of $\mathcal{R}^n$ and $\mathcal{V}$ is the span of $\mathcal{B}$. Find $P_{\mathcal{BA}}$ and $P_{\mathcal{AB}}$ when $\mathcal{A} = \{X_3, X_1, X_2\}$.

14. If $\mathcal{V}$ is a subspace of $\mathcal{R}^n$ with bases $\mathcal{A}$, $\mathcal{B}$, and $\mathcal{C}$, show that $P_{\mathcal{CA}} = P_{\mathcal{CB}}P_{\mathcal{BA}}$.

15. Suppose $\mathcal{B}$ is a basis for $\mathcal{R}^n$ and $X_1, \ldots, X_q \in \mathcal{R}^n$. Show that $X_1, \ldots, X_q$ are linearly independent if and only if $[X_1]_{\mathcal{B}}, \ldots, [X_q]_{\mathcal{B}}$ are linearly independent.

16. Let $\mathcal{A} = \{X_1, \ldots, X_n\}$ and $\mathcal{B} = \{Y_1, \ldots, Y_n\}$ be bases for $\mathcal{R}^n$ and set $A = [X_1 \ \cdots \ X_n]$ and $B = [Y_1 \ \cdots \ Y_n]$. Show that $P_{\mathcal{BA}} = B^{-1}A$.

2.10
DIRECT SUM DECOMPOSITION

If $X, Y \in \mathcal{R}^n$, $\mathcal{U} = \mathcal{S}pan\{X\}$, $\mathcal{V} = \mathcal{S}pan\{Y\}$, and $\mathcal{W} = \mathcal{S}pan\{X, Y\}$, then

$$\mathcal{W} = \{sX + tY \ : \ s, t \in \mathcal{R}\} = \{U + V \ : \ U \in \mathcal{U}, V \in \mathcal{V}\}.$$

In other words, $\mathcal{W}$ is the set of vectors in $\mathcal{R}^n$ that can be expressed as the sum of a vector in $\mathcal{U}$ and a vector in $\mathcal{V}$. It is natural to think of $\mathcal{W}$ as the sum of $\mathcal{U}$ and $\mathcal{V}$.

DEFINITION. Given $\mathcal{U}, \mathcal{V} \subseteq \mathcal{R}^n$, $\mathcal{U} + \mathcal{V} = \{U + V \ : \ U \in \mathcal{U} \text{ and } V \in \mathcal{V}\}$ is called the *sum* of $\mathcal{U}$ and $\mathcal{V}$.

THEOREM 2.35. If $\mathcal{U}$ and $\mathcal{V}$ are subspaces of $\mathcal{R}^n$, then so is $\mathcal{U} + \mathcal{V}$. Moreover, $\mathcal{U} \subseteq \mathcal{U} + \mathcal{V}$ and $\mathcal{V} \subseteq \mathcal{U} + \mathcal{V}$.

Proof. Set $\mathcal{W} = \mathcal{U} + \mathcal{V}$ and assume $X, Y \in \mathcal{W}$. Then $X = U_1 + V_1$ and $Y = U_2 + V_2$, where $U_1, U_2 \in \mathcal{U}$ and $V_1, V_2 \in \mathcal{V}$, and consequently,

$$X + Y = (U_1 + V_1) + (U_2 + V_2) = (U_1 + U_2) + (V_1 + V_2).$$

Since $U_1 + U_2 \in \mathcal{U}$ and $V_1 + V_2 \in \mathcal{V}$, $X + Y \in \mathcal{W}$. For $t \in \mathcal{R}$,

$$tX = t(U_1 + V_1) = tU_1 + tV_1.$$

Since $tU_1 \in \mathcal{U}$ and $tV_1 \in \mathcal{V}$, $tX \in \mathcal{W}$. Thus, $\mathcal{W}$ is a subspace. For each $U \in \mathcal{U}$, $U = U + 0 \in \mathcal{U} + \mathcal{V}$ (since $0 \in \mathcal{V}$), so $\mathcal{U} \subseteq \mathcal{U} + \mathcal{V}$. Similarly, $\mathcal{V} \subseteq \mathcal{U} + \mathcal{V}$. ■

■ **EXAMPLE 2.48.** Let $\mathcal{U}$, $\mathcal{V}$, and $\mathcal{Z}$ be noncoplanar lines in $\mathcal{R}^3$ passing through 0. If $\mathcal{W}$ is the plane containing $\mathcal{U}$ and $\mathcal{V}$, then each vector in $\mathcal{W}$ can be expressed as the sum of a vector in $\mathcal{U}$ and a vector in $\mathcal{V}$, so $\mathcal{W} = \mathcal{U} + \mathcal{V}$ (Figure 2.24(a)). Moreover, each element of $\mathcal{R}^3$ is the sum of a vector in $\mathcal{W}$ and a vector in $\mathcal{Z}$, so $\mathcal{R}^3 = \mathcal{W} + \mathcal{Z}$ (Figure 2.24(b)). □

■ **EXAMPLE 2.49.** Suppose $\mathcal{U} = \{X \in \mathcal{R}^3 \ : \ x_2 = 0\}$ and $\mathcal{V} = \{X \in \mathcal{R}^3 \ : \ x_1 = 0\}$. Note that $E_1 \in \mathcal{U}$, $E_2 \in \mathcal{V}$, and $\mathcal{U} \cap \mathcal{V} = \mathcal{S}pan\{E_3\}$ (Figure 2.25). For each $X \in \mathcal{R}^3$,

$$X = x_1E_1 + x_2E_2 + x_3E_3. \tag{2.9}$$

Since $U = x_1E_1 \in \mathcal{U}$ and $V = x_2E_2 + x_3E_3 \in \mathcal{V}$,

$$X = U + V \in \mathcal{U} + \mathcal{V}.$$

Thus, $\mathcal{U} + \mathcal{V} = \mathcal{R}^3$. Equation (2.9) provides a second way of writing X as a vector in $\mathcal{U}$ plus a vector in $\mathcal{V}$, namely, $X = (x_1E_1 + x_3E_3) + x_2E_2$ (Figure 2.26). In fact, there are infinitely many ways to accomplish this. For each $Z \in \mathcal{U} \cap \mathcal{V}$, $U + Z \in \mathcal{U}$, $V - Z \in \mathcal{V}$, and $(U + Z) + (V - Z) = X$ (Figure 2.27). □

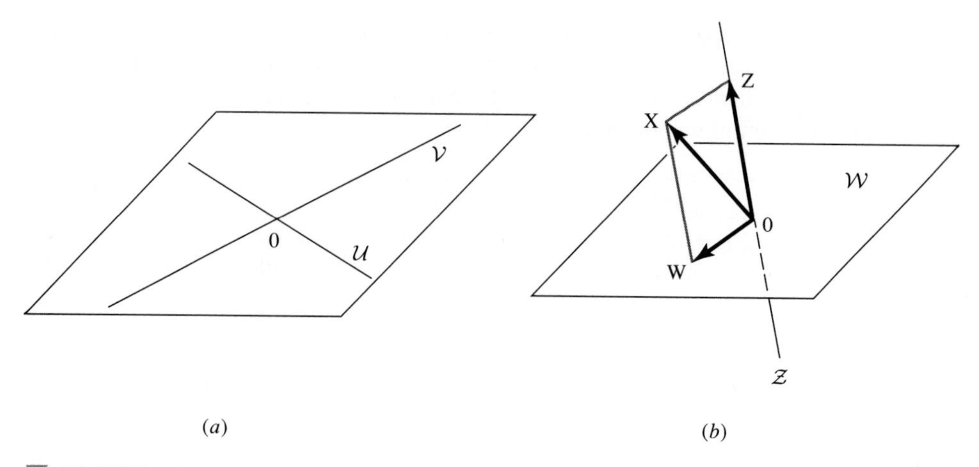

FIGURE 2.24

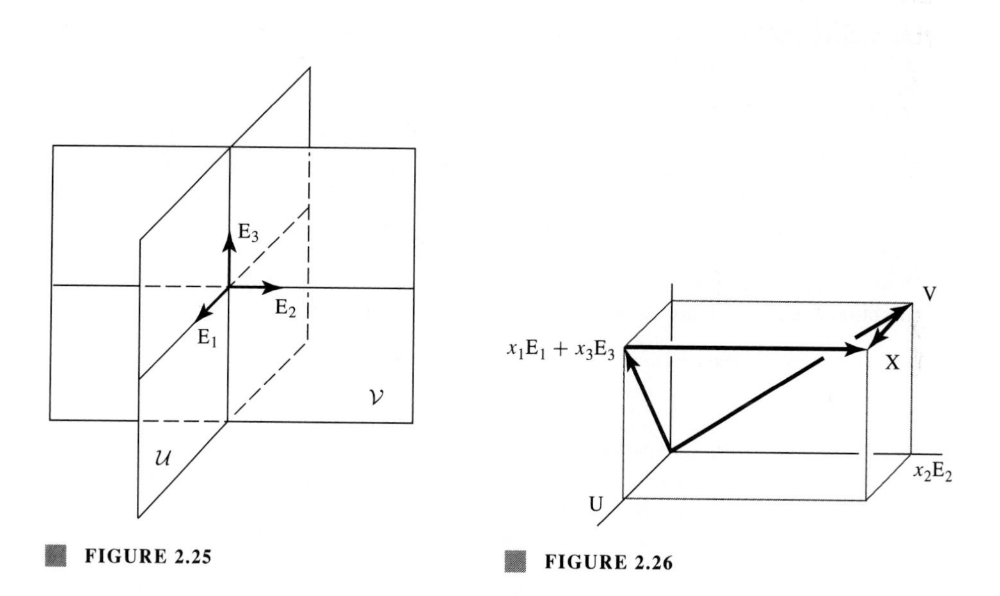

FIGURE 2.25

FIGURE 2.26

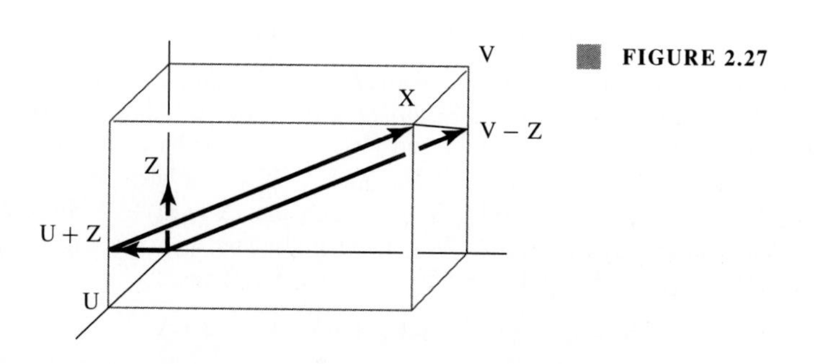

FIGURE 2.27

Expressing a subspace as a sum of two smaller subspaces amounts to algebraically decomposing the larger one into two "component" parts. This can be a very useful way of viewing the original subspace, especially when each of its vectors has a unique representation as a sum of vectors in the smaller spaces.

DEFINITION. Suppose $\mathcal{U}$ and $\mathcal{V}$ are subspaces of $\mathcal{R}^n$. If each vector in $\mathcal{U} + \mathcal{V}$ is uniquely expressible as the sum of a vector in $\mathcal{U}$ and a vector in $\mathcal{V}$, then $\mathcal{U}$ and $\mathcal{V}$ are said to be *independent*. The sum of independent subspaces $\mathcal{U}$ and $\mathcal{V}$ is called the *direct sum* of $\mathcal{U}$ and $\mathcal{V}$, written $\mathcal{U} \oplus \mathcal{V}$.

In Example 2.48, each vector in $\mathcal{W}$ is uniquely expressible as the sum of a vector in $\mathcal{U}$ and a vector in $\mathcal{V}$, so $\mathcal{W} = \mathcal{U} \oplus \mathcal{V}$. Similarly, $\mathcal{W}$ and $\mathcal{Z}$ are independent, so $\mathcal{R}^3 = \mathcal{W} \oplus \mathcal{Z}$. In Example 2.49, $\mathcal{R}^3 = \mathcal{U} + \mathcal{V}$, but $\mathcal{U}$ and $\mathcal{V}$ are not independent. The different decompositions of X in the latter example suggest that $\mathcal{U} \cap \mathcal{V}$ plays a role in deciding whether the sum is a direct sum.

THEOREM 2.36. Assuming $\mathcal{U}$ and $\mathcal{V}$ are subspaces of $\mathcal{R}^n$, $\mathcal{U}$ and $\mathcal{V}$ are independent if and only if $\mathcal{U} \cap \mathcal{V} = \{0\}$.

Proof. First assume $\mathcal{U} \cap \mathcal{V} = \{0\}$. If $\mathrm{W} \in \mathcal{U} + \mathcal{V}$ and $\mathrm{U}_1 + \mathrm{V}_1 = \mathrm{W} = \mathrm{U}_2 + \mathrm{V}_2$, where $\mathrm{U}_1, \mathrm{U}_2 \in \mathcal{U}$ and $\mathrm{V}_1, \mathrm{V}_2 \in \mathcal{V}$, then $\mathrm{U}_1 - \mathrm{U}_2 = \mathrm{V}_2 - \mathrm{V}_1$. Note that $\mathrm{U}_1 - \mathrm{U}_2 \in \mathcal{U}$ and $\mathrm{V}_2 - \mathrm{V}_1 \in \mathcal{V}$, so $\mathrm{U}_1 - \mathrm{U}_2 = \mathrm{V}_2 - \mathrm{V}_1 \in \mathcal{U} \cap \mathcal{V}$. Since $\mathcal{U} \cap \mathcal{V} = \{0\}$, $\mathrm{U}_1 - \mathrm{U}_2 = 0$ and $\mathrm{V}_2 - \mathrm{V}_1 = 0$, that is, $\mathrm{U}_1 = \mathrm{U}_2$ and $\mathrm{V}_2 = \mathrm{V}_1$. Thus, the decomposition of W is unique. Conversely, suppose $\mathcal{W} = \mathcal{U} \oplus \mathcal{V}$ and let $\mathrm{W} = \mathrm{U} + \mathrm{V} \in \mathcal{W}$, $\mathrm{U} \in \mathcal{U}$, $\mathrm{V} \in \mathcal{V}$. If $\mathcal{U} \cap \mathcal{V}$ contains a nonzero vector Z, then $\mathrm{U} + \mathrm{Z} \in \mathcal{U}$, $\mathrm{V} - \mathrm{Z} \in \mathcal{V}$, and $\mathrm{W} = \mathrm{U} + \mathrm{V} = (\mathrm{U} + \mathrm{Z}) + (\mathrm{V} - \mathrm{Z})$ displays W as the sum of a vector in $\mathcal{U}$ and a vector in $\mathcal{V}$ in two different ways, contrary to the assumption that $\mathcal{U}$ and $\mathcal{V}$ are independent. Thus, $\mathcal{U} \cap \mathcal{V} = \{0\}$. ■

COROLLARY. Subspaces $\mathcal{U}$ and $\mathcal{V}$ of $\mathcal{R}^n$ are independent if and only if

$$0 = \mathrm{U} + \mathrm{V}, \quad \mathrm{U} \in \mathcal{U}, \quad \mathrm{V} \in \mathcal{V} \qquad \text{implies} \qquad \mathrm{U} = 0 \quad \text{and} \quad \mathrm{V} = 0. \tag{2.10}$$

Proof. If $\mathcal{U}$ and $\mathcal{V}$ are independent, then the only way to decompose 0 into the sum of a vector in $\mathcal{U}$ and a vector in $\mathcal{V}$ is $0 = 0 + 0$. On the other hand, if (2.10) holds, then for each $\mathrm{X} \in \mathcal{U} \cap \mathcal{V}$, $\mathrm{X} \in \mathcal{U}$, $-\mathrm{X} \in \mathcal{V}$, and $0 = \mathrm{X} + (-\mathrm{X})$, so $\mathrm{X} = 0$. Thus, $\mathcal{U} \cap \mathcal{V} = \{0\}$. By Theorem 2.36, $\mathcal{U}$ and $\mathcal{V}$ are independent. ■

The point of the corollary is that uniqueness of decomposition for 0 implies uniqueness of decomposition for every element of $\mathcal{U} + \mathcal{V}$.

THEOREM 2.37. If $\mathcal{U}$ and $\mathcal{V}$ are subspaces of $\mathcal{R}^n$ such that $\mathcal{U} \cap \mathcal{V} = \{0\}$, then

$$\dim \mathcal{U} \oplus \mathcal{V} = \dim \mathcal{U} + \dim \mathcal{V}.$$

Proof. The conclusion follows trivially when $\mathcal{U}$ or $\mathcal{V}$ is $\{0\}$. For example, if $\mathcal{V} = \{0\}$, then $\mathcal{U} \oplus \mathcal{V} = \mathcal{U}$, and $\dim \mathcal{U} \oplus \mathcal{V} = \dim \mathcal{U} = \dim \mathcal{U} + 0 = \dim \mathcal{U} + \dim \mathcal{V}$. Assume then that $\mathcal{U} \neq \{0\}$ and $\mathcal{V} \neq \{0\}$. Let $\{\mathrm{U}_1, \ldots, \mathrm{U}_p\}$ and $\{\mathrm{V}_1, \ldots, \mathrm{V}_q\}$ be bases for $\mathcal{U}$ and $\mathcal{V}$, respectively. The idea is to show that $\{\mathrm{U}_1, \ldots, \mathrm{U}_p, \mathrm{V}_1, \ldots, \mathrm{V}_q\}$ is a basis for $\mathcal{U} \oplus \mathcal{V}$. For each $\mathrm{X} \in \mathcal{U} \oplus \mathcal{V}$ there is a $\mathrm{U} \in \mathcal{U}$ and a $\mathrm{V} \in \mathcal{V}$ such that $\mathrm{X} = \mathrm{U} + \mathrm{V}$. Since $\mathcal{U} = \mathit{Span}\{\mathrm{U}_1, \ldots, \mathrm{U}_p\}$ and $\mathcal{V} = \mathit{Span}\{\mathrm{V}_1, \ldots, \mathrm{V}_q\}$, there are scalars $s_1, \ldots, s_p$ and $t_1, \ldots, t_q$ such that $\mathrm{U} = s_1\mathrm{U}_1 + \cdots + s_p\mathrm{U}_p$ and $\mathrm{V} = t_1\mathrm{V}_1 + \cdots + t_q\mathrm{V}_q$. Thus,

$$\mathrm{X} = s_1\mathrm{U}_1 + \cdots + s_p\mathrm{U}_p + t_1\mathrm{V}_1 + \cdots + t_q\mathrm{V}_q.$$

This shows that $\mathcal{U} \oplus \mathcal{V} = \mathcal{S}pan\{U_1, \ldots, U_p, V_1, \ldots, V_q\}$. If $s_1, \ldots, s_p, t_1, \ldots, t_q \in \mathcal{R}$ and

$$0 = s_1U_1 + \cdots + s_pU_p + t_1V_1 + \cdots + t_qV_q,$$

then $U = s_1U_1 + \cdots + s_pU_p \in \mathcal{U}$, $V = t_1V_1 + \cdots + t_qV_q \in \mathcal{V}$, and $0 = U + V$. By the Corollary to Theorem 2.36,

$$s_1U_1 + \cdots + s_pU_p = 0 = t_1V_1 + \cdots + t_qV_q,$$

and the linear independence of $\{U_1, \ldots, U_p\}$ and $\{V_1, \ldots, V_q\}$ then implies that $s_1, \ldots, s_p$ and $t_1, \ldots, t_q$ are zeros. Thus, $\{U_1, \ldots, U_p, V_1, \ldots, V_q\}$ is linearly independent. ■

Much of the foregoing discussion applies equally well to three or more subspaces of $\mathcal{R}^n$. Notions of sum, independence, and direct sum in this context are defined analogously.

THEOREM 2.38. If $\mathcal{V}_1, \ldots, \mathcal{V}_q$ are subspaces of $\mathcal{R}^n$, then $\mathcal{V}_1 + \cdots + \mathcal{V}_q$ is a subspace of $\mathcal{R}^n$. Moreover, if $\mathcal{V}_1, \ldots, \mathcal{V}_q$ are independent, then

$$\dim \mathcal{V}_1 \oplus \mathcal{V}_2 \oplus \cdots \oplus \mathcal{V}_q = \dim \mathcal{V}_1 + \cdots + \dim \mathcal{V}_q.$$

Proof. Omitted. ■

Exercises 2.10

1. Identify $\mathcal{U} + \mathcal{V}$ when $\mathcal{U}$ and $\mathcal{V}$ are subspaces of $\mathcal{R}^n$ and $\mathcal{U} \subseteq \mathcal{V}$.

2. Let $\mathcal{U}$ and $\mathcal{V}$ be planes in $\mathcal{R}^3$ given by $x + y - z = 0$ and $2x + y - 3z = 0$.
   a. Show that $\mathcal{R}^3 = \mathcal{U} + \mathcal{V}$.
   b. Are $\mathcal{U}$ and $\mathcal{V}$ independent? Explain!

3. Assuming $X, Y, Z \in \mathcal{R}^n$, show that
   a. $\mathcal{S}pan\{X,Y,Z\} = \mathcal{S}pan\{X\} + \mathcal{S}pan\{Y\} + \mathcal{S}pan\{Z\}$.
   b. $\mathcal{S}pan\{X,Y,Z\} = \mathcal{S}pan\{X\} \oplus \mathcal{S}pan\{Y\} \oplus \mathcal{S}pan\{Z\}$ if and only if X, Y, and Z are linearly independent.

4. a. Find two-dimensional subspaces, $\mathcal{V}_1$, $\mathcal{V}_2$, $\mathcal{V}_3$, and $\mathcal{V}_4$, of $\mathcal{R}^8$ such that $\mathcal{R}^8 = \mathcal{V}_1 \oplus \mathcal{V}_2 \oplus \mathcal{V}_3 \oplus \mathcal{V}_4$.
   b. Can you decompose $\mathcal{R}^8$ into the direct sum of three three-dimensional subspaces? Explain!

5. Suppose $\mathcal{U}$ is a subspace of $\mathcal{R}^n$ of dimension p, $0 < p < n$. Show that there is a subspace $\mathcal{V}$ of $\mathcal{R}^n$ such that $\mathcal{R}^n = \mathcal{U} \oplus \mathcal{V}$.

6. Show that subspaces $\mathcal{U}$, $\mathcal{V}$, and $\mathcal{W}$ of $\mathcal{R}^n$ are independent if and only if $0 = U + V + W$, $U \in \mathcal{U}$, $V \in \mathcal{V}$, $W \in \mathcal{W}$ implies $U = V = W = 0$.

7. Let $\mathcal{U} = \{X \in \mathcal{R}^3 : x_1 = 0\}$, $\mathcal{V} = \{X \in \mathcal{R}^3 : x_2 = 0\}$, and $\mathcal{W} = \{X \in \mathcal{R}^3 : x_3 = 0\}$.
   a. Show that $\mathcal{R}^3 = \mathcal{U} + \mathcal{V} + \mathcal{W}$.
   b. Observe that $\mathcal{U} \cap \mathcal{V} \cap \mathcal{W} = \{0\}$. Are $\mathcal{U}$, $\mathcal{V}$, and $\mathcal{W}$ independent?

8. Show that if $\mathcal{U}$, $\mathcal{V}$, and $\mathcal{W}$ are independent subspaces of $\mathcal{R}^n$, then
   a. $\mathcal{U} \cap \mathcal{V} = \mathcal{U} \cap \mathcal{W} = \mathcal{V} \cap \mathcal{W} = \{0\}$.
   b. $\mathcal{U} \cap (\mathcal{V} \oplus \mathcal{W}) = \mathcal{V} \cap (\mathcal{U} \oplus \mathcal{W}) = \mathcal{W} \cap (\mathcal{U} \oplus \mathcal{V}) = \{0\}$.

9. Suppose $\mathcal{U}$, $\mathcal{V}$, and $\mathcal{W}$ are subspaces of $\mathcal{R}^n$ such that $\mathcal{U} \cap \mathcal{V} = \{0\} = \mathcal{W} \cap (\mathcal{U} \oplus \mathcal{V})$. Show that $\mathcal{U}$, $\mathcal{V}$, and $\mathcal{W}$ are independent.

10. Assume $\mathcal{U}$ and $\mathcal{V}$ are nontrivial subspaces of $\mathcal{R}^n$ such that $\mathcal{U} \cap \mathcal{V} = \{0\}$. Let $\mathcal{A} = \{\mathrm{U}_1, \ldots, \mathrm{U}_p\}$ and $\mathcal{B} = \{\mathrm{V}_1, \ldots, \mathrm{V}_q\}$ be bases for $\mathcal{U}$ and $\mathcal{V}$, respectively, and set $\mathcal{C} = \{\mathrm{U}_1, \ldots, \mathrm{U}_p, \mathrm{V}_1, \ldots, \mathrm{V}_q\}$. The proof of Theorem 2.37 showed that $\mathcal{C}$ is a basis for $\mathcal{U} \oplus \mathcal{V}$.
   a. Show that if $\mathrm{X} \in \mathcal{U} \oplus \mathcal{V}$ and $\mathrm{X} = \mathrm{U} + \mathrm{V}$, where $\mathrm{U} \in \mathcal{U}$ and $\mathrm{V} \in \mathcal{V}$, then

$$[\mathrm{X}]_{\mathcal{C}} = \left[\begin{array}{c} [\mathrm{U}]_{\mathcal{A}} \\ \hdashline [\mathrm{V}]_{\mathcal{B}} \end{array}\right].$$

   b. Show that if $\mathcal{A}' = \{\mathrm{U}'_1, \ldots, \mathrm{U}'_p\}$ and $\mathcal{B}' = \{\mathrm{V}'_1, \ldots, \mathrm{V}'_q\}$ are bases for $\mathcal{U}$ and $\mathcal{V}$, respectively, and $\mathcal{C}' = \{\mathrm{U}'_1, \ldots, \mathrm{U}'_p, \mathrm{V}'_1, \ldots, \mathrm{V}'_q\}$, then

$$\mathrm{P}_{\mathcal{C}'\mathcal{C}} = \left[\begin{array}{c:c} \mathrm{P}_{\mathcal{A}'\mathcal{A}} & 0 \\ \hdashline 0 & \mathrm{P}_{\mathcal{B}'\mathcal{B}} \end{array}\right].$$

3

Orthogonality

3.1 INTRODUCTION

There is an inherent notion of perpendicularity for nonzero vectors X,Y in $\mathcal{R}^2$. If **x** and **y** are the corresponding geometric vectors, with directed line segment representatives $\overrightarrow{PQ}$ and $\overrightarrow{PR}$, then X is considered to be perpendicular to Y when $\overrightarrow{PQ}$ is perpendicular to $\overrightarrow{PR}$ (Figure 3.1). There is also an algebraic way of expressing this relationship. It involves an operation called the dot product, an idea that extends easily to $\mathcal{R}^n$. This chapter explores the resulting generalized notion of perpendicularity.

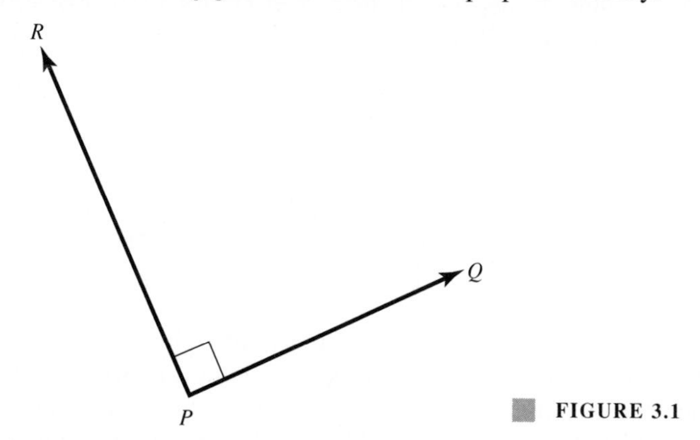

FIGURE 3.1

3.2 THE SCALAR PRODUCT

Suppose X and Y are nonzero vectors in $\mathcal{R}^2$, with corresponding geometric vectors **x** and **y**. Let $\overrightarrow{PQ} \in \mathbf{x}$ and $\overrightarrow{PR} \in \mathbf{y}$. If X and Y are linearly independent, then P, Q, and R are vertices of a triangle, and the interior angle of that triangle at P is considered to be

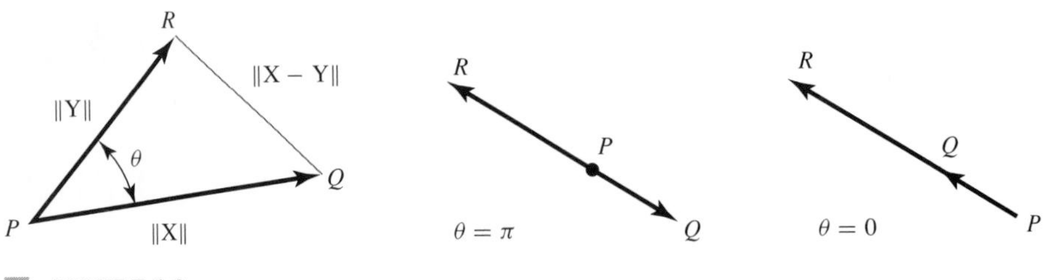

FIGURE 3.2

the angle between X and Y. If X and Y are linearly dependent, then X is a scalar multiple of Y. In that case the angle between X and Y is either 0 or π, depending on whether the scalar is positive or negative, respectively. In either event, θ will denote the angle between X and Y (Figure 3.2). Note that $\theta \in [0,\pi]$.

When X and Y are linearly independent, θ is related to the lengths of the sides of $\triangle PQR$ by the law of cosines, that is,

$$\|\mathrm{X} - \mathrm{Y}\|^2 = \|\mathrm{X}\|^2 + \|\mathrm{Y}\|^2 - 2\|\mathrm{X}\|\,\|\mathrm{Y}\|\cos(\theta).$$

Expressing the latter equation in standard coordinates yields

$$(x_1 - y_1)^2 + (x_2 - y_2)^2 = x_1^2 + x_2^2 + y_1^2 + y_2^2 - 2\|\mathrm{X}\|\,\|\mathrm{Y}\|\cos(\theta),$$

which simplifies to

$$\cos(\theta) = \frac{x_1 y_1 + x_2 y_2}{\|\mathrm{X}\|\,\|\mathrm{Y}\|}. \tag{3.1}$$

Thus, the angle between X and Y is the unique θ in $[0,\pi]$ satisfying (3.1).

If X and Y are linearly dependent, then $\mathrm{X} = t\mathrm{Y}$, for some $t \in \mathcal{R}$, so $x_1 = ty_1$, $x_2 = ty_2$, and $\|\mathrm{X}\| = |t|\,\|\mathrm{Y}\|$. Then

$$x_1 y_1 + x_2 y_2 = (ty_1)y_1 + (ty_2)y_2 = t\|\mathrm{Y}\|^2,$$

and the right side of (3.1) becomes

$$\frac{x_1 y_1 + x_2 y_2}{\|\mathrm{X}\|\,\|\mathrm{Y}\|} = \frac{t\|\mathrm{Y}\|^2}{|t|\,\|\mathrm{Y}\|\,\|\mathrm{Y}\|} = \begin{cases} 1, & \text{if } t > 0 \\ -1, & \text{if } t < 0 \end{cases} = \begin{cases} \cos(0), & \text{if } t > 0 \\ \cos(\pi), & \text{if } t < 0 \end{cases}.$$

Again, the angle between X and Y is the $\theta \in [0,\pi]$ satisfying (3.1).

The geometry of $\mathcal{R}^2$ played a role in the derivation of (3.1), but the result is a strictly algebraic equation that determines θ. The numerator on the right side of (3.1) is called the dot product of X and Y, denoted by $\mathrm{X} \cdot \mathrm{Y}$. It is an interesting quantity precisely because of its connection with θ.

The dot product of two vectors is an idea that can be extended to $\mathcal{R}^n$. After doing so and after developing some of the properties of this generalized dot product, it will be possible to use an n-dimensional analogue of (3.1) to extend the concept of "angle between two vectors" to Euclidean n-space.

DEFINITION. If X,Y $\in \mathcal{R}^n$, then the *scalar product* or *dot product* of X and Y is

$$\mathrm{X} \cdot \mathrm{Y} = x_1 y_1 + \cdots + x_n y_n.$$

This "product" assigns to each pair of vectors, X,Y $\in \mathcal{R}^n$, the real number $x_1 y_1 + \cdots + x_n y_n$, hence the name scalar product.

■ **EXAMPLE 3.1.** $\begin{bmatrix} 2 \\ -1 \\ 3 \\ -2 \end{bmatrix} \cdot \begin{bmatrix} 1 \\ 0 \\ 1 \\ 2 \end{bmatrix} = 2 \cdot 1 + (-1) \cdot 0 + 3 \cdot 1 + (-2) \cdot 2 = 1.$ □

If X,Y $\in \mathcal{R}^n$, then

$$\mathrm{X}^T \mathrm{Y} = [x_1 \quad \cdots \quad x_n] \begin{bmatrix} y_1 \\ \vdots \\ y_n \end{bmatrix} = [x_1 y_1 + \cdots + x_n y_n] = [\mathrm{X} \cdot \mathrm{Y}].$$

There is no real harm in confusing the 1×1 matrix $[\mathrm{X} \cdot \mathrm{Y}]$ with its entry $\mathrm{X} \cdot \mathrm{Y}$, so it is customary (though slightly inaccurate) to write $\mathrm{X}^T \mathrm{Y} = \mathrm{X} \cdot \mathrm{Y}$. This observation allows some features of the scalar product to be interpreted as special cases of the rules governing matrix multiplication.

THEOREM 3.1. If X,Y,Z $\in \mathcal{R}^n$ and $t \in \mathcal{R}$, then

(1) $\mathrm{X} \cdot \mathrm{Y} = \mathrm{Y} \cdot \mathrm{X}$.
(2) $\mathrm{X} \cdot (\mathrm{Y} + \mathrm{Z}) = \mathrm{X} \cdot \mathrm{Y} + \mathrm{X} \cdot \mathrm{Z}$.
(3) $\mathrm{X} \cdot (t\mathrm{Y}) = t(\mathrm{X} \cdot \mathrm{Y}) = (t\mathrm{X}) \cdot \mathrm{Y}$.
(4) $\mathrm{X} \cdot \mathrm{X} \geq 0$, and $\|\mathrm{X}\| = \sqrt{\mathrm{X} \cdot \mathrm{X}}$.

Proof

(1) $\mathrm{X} \cdot \mathrm{Y} = \sum_1^n x_k y_k = \sum_1^n y_k x_k = \mathrm{Y} \cdot \mathrm{X}$.
(2) $\mathrm{X} \cdot (\mathrm{Y} + \mathrm{Z}) = \mathrm{X}^T(\mathrm{Y} + \mathrm{Z}) = \mathrm{X}^T \mathrm{Y} + \mathrm{X}^T \mathrm{Z} = \mathrm{X} \cdot \mathrm{Y} + \mathrm{X} \cdot \mathrm{Z}$.
(3) See Exercise 3.
(4) $\mathrm{X} \cdot \mathrm{X} = x_1^2 + \cdots + x_n^2 = \|\mathrm{X}\|^2 \geq 0$. ■

Commutativity of the dot product implies $(\mathrm{Y} + \mathrm{Z}) \cdot \mathrm{X} = \mathrm{X} \cdot (\mathrm{Y} + \mathrm{Z})$, $\mathrm{X} \cdot \mathrm{Y} = \mathrm{Y} \cdot \mathrm{X}$, and $\mathrm{X} \cdot \mathrm{Z} = \mathrm{Z} \cdot \mathrm{X}$, so statement (2) of Theorem 3.1 is equivalent to $(\mathrm{Y} + \mathrm{Z}) \cdot \mathrm{X} = \mathrm{Y} \cdot \mathrm{X} + \mathrm{Z} \cdot \mathrm{X}$.

Given $\mathrm{X}, \mathrm{Y}_1, \ldots, \mathrm{Y}_p \in \mathcal{R}^n$ and $t_1, \ldots, t_p \in \mathcal{R}$, an inductive argument utilizing statements (2) and (3) of Theorem 3.1 shows that

$$\mathrm{X} \cdot (t_1 \mathrm{Y}_1 + \cdots + t_p \mathrm{Y}_p) = t_1(\mathrm{X} \cdot \mathrm{Y}_1) + \cdots + t_p(\mathrm{X} \cdot \mathrm{Y}_p),$$

that is,

$$\boxed{\mathrm{X} \cdot \left(\sum_1^p t_k \mathrm{Y}_k \right) = \sum_1^p t_k (\mathrm{X} \cdot \mathrm{Y}_k)}. \tag{3.2}$$

Consider now the issue of defining the angle between two vectors in $\mathcal{R}^n$. As indicated earlier, the idea is to use an n-dimensional analogue of (3.1) to determine θ. For the right side of (3.1) to be the cosine of an angle, it must be a real number that lies between -1 and 1. That conclusion, when X and Y are n-dimensional vectors, is the content of a very famous inequality.

THEOREM 3.2. **(Cauchy-Schwarz Inequality).** If X,Y $\in \mathcal{R}^n$, then

$$|\mathrm{X} \cdot \mathrm{Y}| \leq \|\mathrm{X}\|\,\|\mathrm{Y}\|, \tag{3.3}$$

with equality if and only if X and Y are linearly dependent.

Proof. If X and Y are linearly dependent, then one is a scalar multiple of the other, say $\mathrm{X} = t\mathrm{Y}$. Then

$$|\mathrm{X} \cdot \mathrm{Y}| = |(t\mathrm{Y}) \cdot \mathrm{Y}| = |t(\mathrm{Y} \cdot \mathrm{Y})| = |t|\|\mathrm{Y}\|^2 = (|t|\|\mathrm{Y}\|)\|\mathrm{Y}\| = \|t\mathrm{Y}\|\|\mathrm{Y}\| = \|\mathrm{X}\|\|\mathrm{Y}\|,$$

that is, equality occurs in (3.3). On the other hand, if X and Y are linearly independent, then neither X nor Y is 0 and, moreover, the line with equation $\mathrm{X}(t) = \mathrm{X} + t\mathrm{Y}$, $-\infty < t < \infty$, does not pass through 0 (Why?). Thus,

$$\begin{aligned} 0 < \|\mathrm{X}(t)\|^2 &= (\mathrm{X} + t\mathrm{Y}) \cdot (\mathrm{X} + t\mathrm{Y}) \\ &= (\mathrm{X} + t\mathrm{Y}) \cdot \mathrm{X} + (\mathrm{X} + t\mathrm{Y}) \cdot (t\mathrm{Y}) \\ &= \mathrm{X} \cdot \mathrm{X} + (t\mathrm{Y}) \cdot \mathrm{X} + \mathrm{X} \cdot (t\mathrm{Y}) + (t\mathrm{Y}) \cdot (t\mathrm{Y}) \\ &= \|\mathrm{X}\|^2 + 2t(\mathrm{X} \cdot \mathrm{Y}) + t^2\|\mathrm{Y}\|^2 \end{aligned}$$

for every $t \in \mathcal{R}$. The last expression is a quadratic function of t. It has no real zeros, so its discriminant is negative, that is, $[2(\mathrm{X} \cdot \mathrm{Y})]^2 - 4\|\mathrm{Y}\|^2\|\mathrm{X}\|^2 < 0$, which is equivalent to (3.3). ■

For X,Y $\in \mathcal{R}^n$, (3.3) states that $-\|\mathrm{X}\|\,\|\mathrm{Y}\| \leq \mathrm{X} \cdot \mathrm{Y} \leq \|\mathrm{X}\|\,\|\mathrm{Y}\|$. Thus, when X and Y are nonzero, $-1 \leq \dfrac{\mathrm{X} \cdot \mathrm{Y}}{\|\mathrm{X}\|\,\|\mathrm{Y}\|} \leq 1$.

DEFINITION. If X and Y are nonzero vectors in $\mathcal{R}^n$, then the *angle between* X *and* Y is the unique $\theta \in [0, \pi]$ satisfying

$$\cos(\theta) = \frac{\mathrm{X} \cdot \mathrm{Y}}{\|\mathrm{X}\|\,\|\mathrm{Y}\|}. \tag{3.4}$$

■ **EXAMPLE 3.2.** If $\mathrm{X} = \begin{bmatrix} 0 \\ -1 \\ 0 \\ 1 \end{bmatrix}$ and $\mathrm{Y} = \begin{bmatrix} 1 \\ 1 \\ 1 \\ -1 \end{bmatrix}$, then $\mathrm{X} \cdot \mathrm{Y} = -2$, $\|\mathrm{X}\| = \sqrt{2}$, $\|\mathrm{Y}\| = 2$, and $\cos(\theta) = (\mathrm{X} \cdot \mathrm{Y})/\|\mathrm{X}\|\,\|\mathrm{Y}\| = -1/\sqrt{2}$. Thus, $\theta = 3\pi/4$. □

■ **EXAMPLE 3.3.** Let $\{\mathrm{E}_1, \ldots, \mathrm{E}_n\}$ be the standard basis for $\mathcal{R}^n$ and $\mathrm{X} = [1 \quad \cdots \quad 1]^T \in \mathcal{R}^n$. For $k = 1, \ldots, n$, $\|\mathrm{E}_k\| = 1$ and $\mathrm{X} \cdot \mathrm{E}_k = 1$. Moreover, $\|\mathrm{X}\| = \sqrt{n}$, so the angle between X and E_k is the $\theta \in [0,\pi]$ satisfying

$$\cos(\theta) = \frac{X \cdot E_k}{\|X\|\|E_k\|} = \frac{1}{\sqrt{n}}.$$

Since θ is independent of k, X makes the same angle with each of the positive standard coordinate axes in $\mathcal{R}^n$. If $n = 2$ or $n = 4$, then $\theta = \pi/4$ or $\theta = \pi/3$, respectively. What happens to θ as n tends to ∞? □

Let X and Y be nonzero vectors in $\mathcal{R}^n$ and suppose s and t are positive scalars. Then sX and tY have the same directions as X and Y, respectively. Moreover, $|s| = s$ and $|t| = t$, so

$$\frac{(sX) \cdot (tY)}{\|sX\|\|tY\|} = \frac{(st)(X \cdot Y)}{|s|\|X\||t|\|Y\|} = \frac{X \cdot Y}{\|X\|\|Y\|}.$$

This calculation establishes that the angle between sX and tY is the same as the angle between X and Y (when s and t are positive), a conclusion that is consistent with our geometric instincts.

It is natural to consider nonzero vectors X,Y to be perpendicular when the angle between them is $\pi/2$. By (3.4) that occurs precisely when $X \cdot Y = 0$.

DEFINITION. Two vectors $X, Y \in \mathcal{R}^n$ are *orthogonal* or *perpendicular* if $X \cdot Y = 0$. Moreover, $X_1, \ldots, X_p \in \mathcal{R}^n$ are *mutually orthogonal* if $X_i \cdot X_j = 0$ whenever $i \neq j$. A set of mutually orthogonal vectors is called an *orthogonal set*.

REMARK. The angle between two vectors is meaningful only when both vectors are nonzero. The zero vector, however, is considered to be orthogonal to every vector because $0 \cdot X = 0$ for every $X \in \mathcal{R}^n$.

EXAMPLE 3.4. If $X = \begin{bmatrix} 1 \\ 0 \\ -2 \\ 3 \end{bmatrix}$, $Y = \begin{bmatrix} -1 \\ 2 \\ 1 \\ 1 \end{bmatrix}$, and $Z = \begin{bmatrix} 3 \\ 1 \\ 2 \\ -1 \end{bmatrix}$, then $X \cdot Y = 0 = Y \cdot Z$, so X is orthogonal to Y, and Y is orthogonal to Z. Since $X \cdot Z = -4$, $\{X, Y, Z\}$ is not an orthogonal set. □

EXAMPLE 3.5. The standard basis for $\mathcal{R}^n$ is an orthogonal set because $E_i \cdot E_j = 0$ whenever $i \neq j$. □

Given nonzero vectors, $X, Y \in \mathcal{R}^n$, let $\mathbf{x}$ and $\mathbf{y}$ be the associated elements of $\mathcal{G}^2$, and assume $\overrightarrow{PQ} \in \mathbf{x}$ and $\overrightarrow{QR} \in \mathbf{y}$. Then $\overrightarrow{PR} \in \mathbf{x} + \mathbf{y}$. If X and Y are linearly independent, then $\triangle PRQ$ indicates that $\|X + Y\| < \|X\| + \|Y\|$ (Figure 3.3.) When

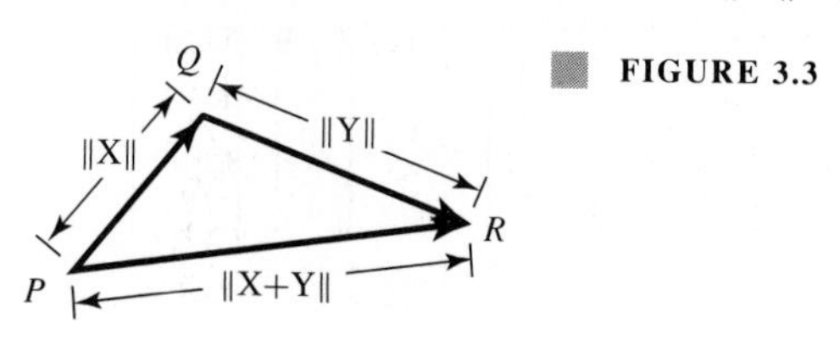

FIGURE 3.3

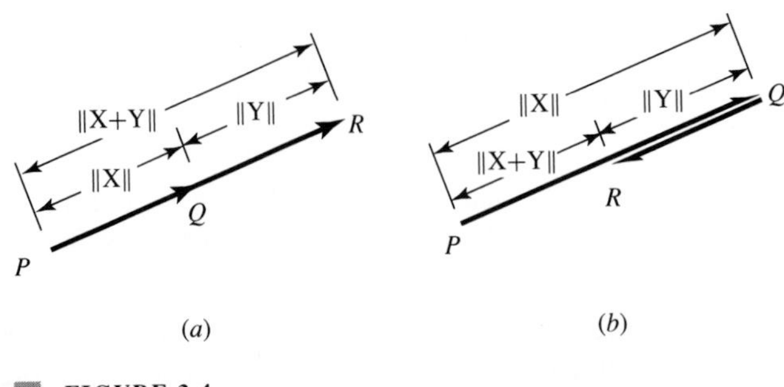

FIGURE 3.4

{X,Y} is linearly dependent, say $X = tY$, there are two alternatives, depending on the sign of t. For $t > 0$, $\|X+Y\| = \|X\| + \|Y\|$ (Figure 3.4(a)), and for $t < 0$, $\|X+Y\| < \|X\| + \|Y\|$ (Figure 3.4 (b)). In any event, $\|X+Y\| \leq \|X\| + \|Y\|$. This conclusion, known as the *triangle inequality,* furnishes an estimate for the magnitude of the sum of two vectors in terms of the magnitudes of the individual vectors. It holds for n-dimensional vectors as well, in which case the proof relies on the Cauchy-Schwarz inequality.

THEOREM 3.3. **(Triangle Inequality).** If $X,Y \in \mathcal{R}^n$, then

$$\|X+Y\| \leq \|X\| + \|Y\|. \tag{3.5}$$

Proof. By the Cauchy-Schwarz inequality, $X \cdot Y \leq |X \cdot Y| \leq \|X\|\|Y\|$, so

$$\begin{aligned}\|X+Y\|^2 &= (X+Y)\cdot(X+Y)\\ &= \|X\|^2 + 2(X\cdot Y) + \|Y\|^2\\ &\leq \|X\|^2 + 2\|X\|\|Y\| + \|Y\|^2\\ &= (\|X\| + \|Y\|)^2. \blacksquare\end{aligned}$$

Exercises 3.2

1. Assuming $X = \begin{bmatrix}1\\2\\3\end{bmatrix}$, $Y = \begin{bmatrix}-3\\2\\-1\end{bmatrix}$, and $Z = \begin{bmatrix}1\\1\\-1\end{bmatrix}$, find

   a. $X \cdot Y$. **b.** $X \cdot Z$. c. $Y \cdot Z$. **d.** $(3X + 2Y)\cdot Z$. e. $(X - Z)\cdot(Y - Z)$.

2. Find the angle between the two vectors.

   a. $\begin{bmatrix}0\\1\\1\\1\end{bmatrix}, \begin{bmatrix}1\\-1\\0\\-2\end{bmatrix}$ b. $\begin{bmatrix}1\\2\\3\end{bmatrix}, \begin{bmatrix}4\\1\\-2\end{bmatrix}$ c. $\begin{bmatrix}1\\2\\3\\4\end{bmatrix}, \begin{bmatrix}-3\\2\\-4\\-1\end{bmatrix}$ d. $\begin{bmatrix}-1\\-1\\0\\1\\1\end{bmatrix}, \begin{bmatrix}-2\\0\\2\\0\\1\end{bmatrix}$.

3. Prove property (3) of Theorem 3.1.

4. Show that $X = \begin{bmatrix} 4 \\ 3 \\ 2 \\ -1 \end{bmatrix}$, $Y = \begin{bmatrix} -3 \\ 4 \\ 1 \\ 2 \end{bmatrix}$, and $Z = \begin{bmatrix} 1 \\ -2 \\ 3 \\ 4 \end{bmatrix}$ are mutually orthogonal and find a nonzero $W \in \mathcal{R}^4$ such that {X,Y,Z,W} is an orthogonol set.

5. Suppose X and Y are nonzero vectors in $\mathcal{R}^n$ and the angle between X and Y is θ. What is the angle between X and $-Y$?

6. Suppose $X,Y \in \mathcal{R}^n$, $\|X\| = 3$, $\|Y\| = 2$, and the angle between X and Y is $\pi/3$. Find $\|X - 2Y\|$.

7. Assuming $A \in \mathcal{M}_{m\times n}$ and $B \in \mathcal{M}_{n\times p}$, show that $[AB]_{ij} = [\text{Row}_i(A)]^T \cdot \text{Col}_j(B)$.

8. Show that $(X + Y) \cdot (X - Y) = \|X\|^2 - \|Y\|^2$, where $X,Y \in \mathcal{R}^n$.

9. Show that the diagonals of a parallelogram are perpendicular if and only if the parallelogram is a rhombus.

10. Assuming $U,V \in \mathcal{R}^n$ are orthogonal vectors of length 1, find
   a. $\|U + V\|$.
   b. the angle between U and U + V.

11. Show that if $\{X_1, \ldots, X_p\} \subseteq \mathcal{R}^n$ is an orthogonal set, then

$$\|X_1 + \cdots + X_p\|^2 = \|X_1\|^2 + \cdots + \|X_p\|^2.$$

When $n = p = 2$ this is the conclusion of what famous theorem?

12. Suppose $X \in \mathcal{R}^n$ and $X \neq 0$. Assume $\{E_1, \ldots, E_n\}$ is the standard basis for $\mathcal{R}^n$ and let θ_k be the angle between X and E_k, $k = 1, \ldots, n$. Show that

$$U = \begin{bmatrix} \cos(\theta_1) \\ \vdots \\ \cos(\theta_n) \end{bmatrix}$$

is a unit vector with the same direction as X.

13. Show that if $X,Y \in \mathcal{R}^n$, $X = tY$, and $t \geq 0$, then $\|X + Y\| = \|X\| + \|Y\|$.

14. Let $\mathcal{V}$ be a subspace of $\mathcal{R}^n$ and set $\mathcal{V}^\perp = \{X \in \mathcal{R}^n : X \cdot Y = 0 \text{ for each } Y \in \mathcal{V}\}$.
   a. Show that $\mathcal{V}^\perp$ is a subspace of $\mathcal{R}^n$.
   b. What is $\{0\}^\perp$?
   c. What is $(\mathcal{R}^n)^\perp$?

15. Assuming $\mathcal{V}$ is a subspace of $\mathcal{R}^n$, show that $\mathcal{V} \cap \mathcal{V}^\perp = \{0\}$.

3.3

ORTHONORMAL BASES

If X and Y are nonzero orthogonal vectors in $\mathcal{R}^2$, then {X,Y} is linearly independent (neither vector is a scalar multiple of the other). Similarly, two or three mutually orthogonal nonzero vectors in $\mathcal{R}^3$ are linearly independent. As these observations suggest, there is a connection, in general, between orthogonality and linear independence.

THEOREM 3.4. If $X_1, \ldots, X_p$ are mutually orthogonal nonzero vectors in $\mathcal{R}^n$, then $X_1, \ldots, X_p$ are linearly independent.

Proof. Assume $\{X_1, \ldots, X_p\} \subseteq \mathcal{R}^n$ is an orthogonal set of nonzero vectors. If $t_1, \ldots, t_p$ are scalars such that

$$0 = t_1X_1 + \cdots + t_pX_p,$$

then

$$\begin{aligned} 0 = X_k \cdot 0 &= X_k \cdot (t_1X_1 + \cdots + t_kX_k + \cdots + t_pX_p) \\ &= t_1(X_k \cdot X_1) + \cdots + t_k(X_k \cdot X_k) + \cdots + t_p(X_k \cdot X_p). \end{aligned}$$

Since $X_k \cdot X_j = 0$ whenever $k \neq j$, the last equation reduces to

$$0 = t_k(X_k \cdot X_k) = t_k\|X_k\|^2,$$

and therefore $t_k = 0$ (because $X_k \neq 0$). This conclusion holds for each $k = 1, \ldots, p$, so $X_1, \ldots, X_p$ are linearly independent. ■

As a consequence of Theorem 3.4, an orthogonal set of nonzero vectors in $\mathcal{R}^n$ contains at most n vectors. Moreover, any orthogonal set of n nonzero vectors in $\mathcal{R}^n$ is a basis for $\mathcal{R}^n$.

Given a nonzero $X \in \mathcal{R}^n$, scalar multiplying X by a positive t produces a new vector with the same direction as X but whose magnitude has been adjusted by a factor of t. If $t = 1/\|X\|$ and $U = tX$, then $\|U\| = t\|X\| = 1$, and U is called the *unit vector in the direction of* X. You can think of U as being obtained by *normalizing* X, that is, by replacing X with a vector of standardized magnitude having the same direction as X.

In the previous section it was pointed out that the angle between two nonzero vectors does not depend on their magnitudes. Consequently, given nonzero $X_1, \ldots, X_p \in \mathcal{R}^n$, the normalized vectors $U_k = (1/\|X_k\|)X_k$, $k = 1, \ldots, p$, have the property that the angle between U_i and U_j is the same as the angle between X_i and X_j, $1 \leq i, j \leq p$.

DEFINITION. Mutually orthogonal unit vectors $U_1, \ldots, U_p \in \mathcal{R}_n$ are said to be *orthonormal*. Alternatively, $\{U_1, \ldots, U_p\}$ is called an *orthonormal set*.

Orthogonality of $\{X_1, \ldots, X_p\}$ is characterized by the conditions $X_i \cdot X_j = 0$, $i \neq j$. Since $X_j \cdot X_j = \|X_j\|^2$, the set is orthonormal when

$$X_i \cdot X_j = \begin{cases} 0, & \text{for } i \neq j \\ 1, & \text{for } i = j \end{cases}.$$

The usual basis for $\mathcal{R}^n$ consists of mutually orthogonal unit vectors, so the standard coordinate system for $\mathcal{R}^n$ is generated by an orthonormal basis. Certain advantages acrue to working with such a basis, one of which is the subject of the next theorem. It concludes that coordinates relative to an orthonormal basis can be calculated without having to resort to Gauss-Jordan elimination.

THEOREM 3.5. If $\mathcal{V}$ is a subspace of $\mathcal{R}^n$ with orthonormal basis $\mathcal{A} = \{U_1, \ldots, U_p\}$, then for each $X \in \mathcal{V}$,

$$X = (X \cdot U_1)U_1 + \cdots + (X \cdot U_p)U_p,$$

or equivalently,

$$[X]_{\mathcal{A}} = \begin{bmatrix} X \cdot U_1 \\ \vdots \\ X \cdot U_p \end{bmatrix}.$$

Proof. If $X \in \mathcal{V}$ and $X = t_1U_1 + \cdots + t_pU_p$, then for $k = 1, \ldots, p$,

$$\begin{aligned} X \cdot U_k &= (t_1U_1 + \cdots + t_pU_p) \cdot U_k \\ &= t_1(U_1 \cdot U_k) + \cdots + t_k(U_k \cdot U_k) + \cdots + t_p(U_p \cdot U_k) \\ &= t_k\|U_k\|^2 = t_k. \ \blacksquare \end{aligned}$$

EXAMPLE 3.6. Let $U_1 = \dfrac{1}{\sqrt{3}}\begin{bmatrix} 1 \\ 1 \\ 1 \end{bmatrix}$, $U_2 = \dfrac{1}{\sqrt{2}}\begin{bmatrix} -1 \\ 1 \\ 0 \end{bmatrix}$, and $U_3 = \dfrac{1}{\sqrt{6}}\begin{bmatrix} 1 \\ 1 \\ -2 \end{bmatrix}$ and observe that $\mathcal{A} = \{U_1, U_2, U_3\}$ is an orthonormal basis for $\mathcal{R}^3$. If $X = [1 \quad 2 \quad 3]^T$, then

$$\begin{aligned} X \cdot U_1 &= (1 + 2 + 3)(1/\sqrt{3}) = 2\sqrt{3}, \\ X \cdot U_2 &= (-1 + 2)(1/\sqrt{2}) = 1/\sqrt{2}, \\ X \cdot U_3 &= (1 + 2 - 6)(1/\sqrt{6}) = -\sqrt{3/2}, \end{aligned}$$

so

$$[X]_{\mathcal{A}} = \begin{bmatrix} 2\sqrt{3} \\ 1/\sqrt{2} \\ -\sqrt{3/2} \end{bmatrix}. \ \square$$

It is easy to see that every subspace of $\mathcal{R}^2$ or $\mathcal{R}^3$ other than $\{0\}$ has an orthonormal basis. For example, if $\mathcal{V}$ is a line through 0 and U is a unit vector in $\mathcal{V}$, then $\{U\}$ and $\{-U\}$ are orthonormal bases for $\mathcal{V}$ (Figure 3.5(a)). A plane through 0 contains infinitely

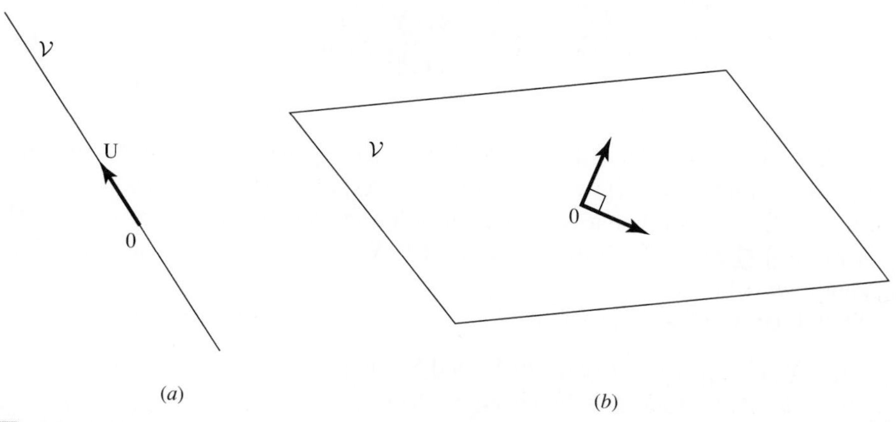

FIGURE 3.5

many pairs of orthogonal unit vectors, any one of which is an orthonormal basis for that subspace (Figure 3.5(b)). In fact, every subspace of $\mathcal{R}^n$ other than {0} has orthonormal bases, and the next task is to describe an algorithm, called the Gram-Schmidt process, for generating such bases. The algorithm is actually a procedure for replacing linearly independent vectors, $X_1, \ldots, X_p$, with mutually orthogonal vectors $Y_1, \ldots, Y_p$ such that $\mathcal{S}pan\{Y_1, \ldots, Y_p\} = \mathcal{S}pan\{X_1, \ldots, X_p\}$. It inductively generates $Y_1, \ldots, Y_p$ in such a way that for each $k = 1, \ldots, p$,

$$\mathcal{S}pan\{Y_1, \ldots, Y_k\} = \mathcal{S}pan\{X_1, \ldots, X_k\}.$$

In other words, the resulting orthogonal vectors satisfy

$$\begin{aligned} \mathcal{S}pan\{Y_1\} &= \mathcal{S}pan\{X_1\}, \\ \mathcal{S}pan\{Y_1,Y_2\} &= \mathcal{S}pan\{X_1,X_2\}, \\ \mathcal{S}pan\{Y_1,Y_2,Y_3\} &= \mathcal{S}pan\{X_1,X_2,X_3\}, \text{ and so on.} \end{aligned}$$

Turning to the process itself, assume $\{X_1, \ldots, X_p\} \subseteq \mathcal{R}^n$ is linearly independent, and begin by setting $Y_1 = X_1$. The next step is to find Y_2, such that

i. Y_2 is orthogonal to Y_1 and
ii. $\mathcal{S}pan\{Y_1,Y_2\} = \mathcal{S}pan\{X_1,X_2\}$.

To satisfy the second condition you must have $Y_2 \in \mathcal{S}pan\{X_1,X_2\}$, that is,

$$Y_2 = sX_1 + tX_2 = sY_1 + tX_2,$$

for some $s,t \in \mathcal{R}$. Orthogonality of Y_1 and Y_2 is then expressed by

$$0 = Y_2 \cdot Y_1 = (sY_1 + tX_2) \cdot Y_1 = s\|Y_1\|^2 + t(X_2 \cdot Y_1), \tag{3.6}$$

and the idea is to choose s and t to satisfy this equation. Being a single linear homogeneous equation in two unknowns, (3.6) has infinitely many solutions. This is not surprising since no restriction has been placed on $\|Y_2\|$. Setting $t = 1$ yields $s = -(X_2 \cdot Y_1)/\|Y_1\|^2$, and therefore

$$\boxed{Y_2 = X_2 - \frac{X_2 \cdot Y_1}{\|Y_1\|^2} Y_1}. \tag{3.7}$$

With this choice of Y_2, $\{Y_1,Y_2\}$ is orthogonal. Since Y_1 and Y_2 are both linear combinations of Y_1 and X_2, $\mathcal{S}pan\{Y_1,Y_2\} \subseteq \mathcal{S}pan\{Y_1,X_2\}$ (Theorem 2.12). Moreover, (3.7) also shows that X_2 as a linear combination of Y_1 and Y_2, so similar reasoning yields $\mathcal{S}pan\{Y_1,X_2\} \subseteq \mathcal{S}pan\{Y_1,Y_2\}$. Thus, $\mathcal{S}pan\{Y_1,Y_2\} = \mathcal{S}pan\{Y_1,X_2\}$, and since $Y_1 = X_1$, condition (ii) is verified.

Step three is to find Y_3, such that

i. Y_3 is orthogonal to each of Y_1 and Y_2, and
ii. $\mathcal{S}pan\{Y_1,Y_2,Y_3\} = \mathcal{S}pan\{X_1,X_2,X_3\}$.

Note that $\mathcal{S}pan\{X_1,X_2,X_3\} = \mathcal{S}pan\{Y_1,Y_2,X_3\}$, because $\mathcal{S}pan\{X_1,X_2\} = \mathcal{S}pan\{Y_1, Y_2\}$. Thus, condition (ii) amounts to $\mathcal{S}pan\{Y_1,Y_2,Y_3\} = \mathcal{S}pan\{Y_1,Y_2,X_3\}$, which is satisfied only when Y_3 is a linear combination of Y_1,Y_2, and X_3. If

$$Y_3 = rY_1 + sY_2 + tX_3,$$

then condition (i) produces two linear homogeneous equations in r, s, and t. Again, since $\|Y_3\|$ has not been specified, there are infinitely many solutions for the resulting system, and one of the scalars can be chosen arbitrarily. Setting $t = 1$, the orthogonality conditions become

$$0 = Y_3 \cdot Y_1 = r(Y_1 \cdot Y_1) + s(Y_2 \cdot Y_1) + (X_3 \cdot Y_1)$$
$$0 = Y_3 \cdot Y_2 = r(Y_1 \cdot Y_2) + s(Y_2 \cdot Y_2) + (X_3 \cdot Y_2),$$

and since $Y_1 \cdot Y_2 = Y_2 \cdot Y_1 = 0$, it follows that

$$r = -\frac{X_3 \cdot Y_1}{\|Y_1\|^2} \quad \text{and} \quad s = -\frac{X_3 \cdot Y_2}{\|Y_2\|^2}.$$

Thus, choosing

$$\boxed{Y_3 = X_3 - \frac{X_3 \cdot Y_1}{\|Y_1\|^2}Y_1 - \frac{X_3 \cdot Y_2}{\|Y_2\|^2}Y_2} \tag{3.8}$$

ensures that $\{Y_1,Y_2,Y_3\}$ is orthogonal. Observe that Y_3 is a linear combination of Y_1,Y_2, and X_3, and that X_3 is a linear combination of Y_1, Y_2, and Y_3. Thus, $\mathcal{S}pan\{Y_1,Y_2,Y_3\} = \mathcal{S}pan\{Y_1,Y_2,X_3\}$.

The subsequent steps are similar to those just completed, and the pattern emerging in (3.7) and (3.8) persists. The details involved in passing from stage k to stage $k + 1$, $1 \le k \le p - 1$, are recorded in the next theorem.

THEOREM 3.6. If $Y_1, \ldots, Y_k$ are mutually orthogonal nonzero vectors in $\mathcal{R}^n$ and $X_{k+1} \in \mathcal{R}^n$, then

$$Y_{k+1} = X_{k+1} - \frac{X_{k+1} \cdot Y_1}{\|Y_1\|^2}Y_1 - \cdots - \frac{X_{k+1} \cdot Y_k}{\|Y_k\|^2}Y_k \tag{3.9}$$

is orthogonal to each of $Y_1, \ldots, Y_k$, and

$$\mathcal{S}pan\{Y_1, \ldots, Y_k, Y_{k+1}\} = \mathcal{S}pan\{Y_1, \ldots, Y_k, X_{k+1}\}.$$

Proof. By (3.9), $Y_{k+1} \in \mathcal{S}pan\,\{Y_1, \ldots, Y_k,X_{k+1}\}$ and $X_{k+1} \in \mathcal{S}pan\,\{Y_1, \ldots, Y_k,Y_{k+1}\}$ so $\mathcal{S}pan\{Y_1, \ldots, Y_k,Y_{k+1}\} = \mathcal{S}pan\{Y_1, \ldots, Y_k,X_{k+1}\}$. For $j = 1, \ldots, k$,

$$Y_{k+1} \cdot Y_j = \left(X_{k+1} - \sum_{i=1}^{k} \frac{X_{k+1} \cdot Y_i}{\|Y_i\|^2}Y_i\right) \cdot Y_j = X_{k+1} \cdot Y_j - \sum_{i=1}^{k} \frac{X_{k+1} \cdot Y_i}{\|Y_i\|^2}(Y_i \cdot Y_j),$$

and since $Y_i \cdot Y_j = 0$ whenever $i \ne j$, the latter equation reduces to

$$Y_{k+1} \cdot Y_j = X_{k+1} \cdot Y_j - \frac{X_{k+1} \cdot Y_j}{\|Y_j\|^2}(Y_j \cdot Y_j) = 0. \blacksquare$$

■ **EXAMPLE 3.7.** Let $\mathcal{V}$ be the subspace of $\mathcal{R}^3$ represented by the plane with equation $x + y + z = 0$, and observe that

$$X_1 = \begin{bmatrix} -1 \\ 0 \\ 1 \end{bmatrix} \quad \text{and} \quad X_2 = \begin{bmatrix} -1 \\ 1 \\ 0 \end{bmatrix}$$

form a basis for $\mathcal{V}$. Since $X_1 \cdot X_2 = 1$, X_1 is not orthogonal to X_2. The Gram-Schmidt process, applied to $\{X_1,X_2\}$, produces

$$Y_1 = X_1 = \begin{bmatrix} -1 \\ 0 \\ 1 \end{bmatrix}$$

and

$$Y_2 = X_2 - \frac{X_2 \cdot Y_1}{\|Y_1\|^2} Y_1 = \begin{bmatrix} -1 \\ 1 \\ 0 \end{bmatrix} - \frac{1}{2}\begin{bmatrix} -1 \\ 0 \\ 1 \end{bmatrix} = \begin{bmatrix} -1/2 \\ 1 \\ -1/2 \end{bmatrix}.$$

Thus,

$$U_1 = \frac{Y_1}{\|Y_1\|} = \frac{1}{\sqrt{2}}\begin{bmatrix} -1 \\ 0 \\ 1 \end{bmatrix} \quad \text{and} \quad U_2 = \frac{Y_2}{\|Y_2\|} = \frac{1}{\sqrt{6}}\begin{bmatrix} -1 \\ 2 \\ -1 \end{bmatrix}$$

are orthonormal vectors that span $\mathcal{V}$. □

■ **EXAMPLE 3.8.** Suppose $\mathcal{V}$ is the subspace of $\mathcal{R}^4$ spanned by

$$X_1 = \begin{bmatrix} 2 \\ 1 \\ 0 \\ -1 \end{bmatrix}, \quad X_2 = \begin{bmatrix} 3 \\ 2 \\ 1 \\ -4 \end{bmatrix}, \quad X_3 = \begin{bmatrix} 2 \\ 2 \\ -4 \\ 0 \end{bmatrix}.$$

These vectors are linearly independent and therefore constitute a basis for $\mathcal{V}$. The Gram-Schmidt process produces $Y_1 = X_1$,

$$Y_2 = X_2 - \frac{X_2 \cdot Y_1}{\|Y_1\|^2} Y_1 = \begin{bmatrix} 3 \\ 2 \\ 1 \\ -4 \end{bmatrix} - \frac{12}{6}\begin{bmatrix} 2 \\ 1 \\ 0 \\ -1 \end{bmatrix} = \begin{bmatrix} -1 \\ 0 \\ 1 \\ -2 \end{bmatrix},$$

and

$$Y_3 = X_3 - \frac{X_3 \cdot Y_1}{\|Y_1\|^2} Y_1 - \frac{X_3 \cdot Y_2}{\|Y_2\|^2} Y_2 = \begin{bmatrix} 2 \\ 2 \\ -4 \\ 0 \end{bmatrix} - \frac{6}{6}\begin{bmatrix} 2 \\ 1 \\ 0 \\ -1 \end{bmatrix} - \frac{-6}{6}\begin{bmatrix} -1 \\ 0 \\ 1 \\ -2 \end{bmatrix} = \begin{bmatrix} -1 \\ 1 \\ -3 \\ -1 \end{bmatrix}.$$

Observe that $\|Y_1\| = \sqrt{6}$, $\|Y_2\| = \sqrt{6}$, and $\|Y_3\| = 2\sqrt{3}$, so

$$U_1 = \frac{1}{\sqrt{6}}\begin{bmatrix} 2 \\ 1 \\ 0 \\ -1 \end{bmatrix}, \quad U_2 = \frac{1}{\sqrt{6}}\begin{bmatrix} -1 \\ 0 \\ 1 \\ -2 \end{bmatrix}, \quad \text{and} \quad U_3 = \frac{1}{2\sqrt{3}}\begin{bmatrix} -1 \\ 1 \\ -3 \\ -1 \end{bmatrix}$$

form an orthonormal basis for $\mathcal{V}$. □

THEOREM 3.7. Every subspace of $\mathcal{R}^n$ other than $\{0\}$ has an orthonormal basis.

Proof. Let $\{X_1, \ldots, X_p\}$ be a basis for $\mathcal{V}$. In the Gram-Schmidt algorithm it is important that at each stage the resulting Y_k be nonzero. For $k = 1, \ldots, p-1$, linear independence of $\{X_1, \ldots, X_k, X_{k+1}\}$ implies $X_{k+1} \notin \mathcal{S}pan\{X_1, \ldots, X_k\}$. Since $\mathcal{S}pan\{X_1, \ldots, X_k\} = \mathcal{S}pan\{Y_1, \ldots, Y_k\}$, $X_{k+1} \notin \mathcal{S}pan\{Y_1, \ldots, Y_k\}$. Thus, Y_{k+1}, as given by (3.9), is not the zero vector. This conclusion holds for each k, so the lengths of $Y_1, \ldots, Y_p$ can be adjusted to obtain an orthonormal basis. ■

When the vectors under consideration are elements of $\mathcal{R}^2$ or $\mathcal{R}^3$, the steps in the Gram-Schmidt process have a geometric interpretation involving the notion of orthogonal projection. Suppose X and Y are nonzero vectors in $\mathcal{R}^3$ and set $\mathcal{V} = \mathcal{S}pan\{Y\}$. Then $\mathcal{V}$ is a line through 0, and dropping a perpendicular from X to $\mathcal{V}$ determines a unique vector in $\mathcal{V}$ called *the orthogonal projection of* X *onto* $\mathcal{V}$. It will be denoted by $\text{Proj}_{\mathcal{V}}(X)$ or $\text{Proj}_Y(X)$ (Figure 3.6). The magnitude of $\text{Proj}_Y(X)$ is

$$\|X\||\cos(\theta)| = \|X\|\frac{|X \cdot Y|}{\|X\|\|Y\|} = \frac{|X \cdot Y|}{\|Y\|},$$

where θ is the angle between X and Y. Moreover, $\text{Proj}_Y(X)$ has the same or opposite direction as Y, depending on whether $\theta \in [0,\pi/2]$ or $\theta \in [\pi/2,\pi]$, respectively. Note that $X \cdot Y = \|X\|\|Y\|\cos(\theta) \geq 0$ when $\theta \in [0,\pi/2]$ and $X \cdot Y \leq 0$ when $\theta \in [\pi/2,\pi]$. Thus, $(X \cdot Y)/\|Y\|$ has the right size and the right sign for scaling the unit vector $Y/\|Y\|$ to obtain $\text{Proj}_Y(X)$. That is,

$$\text{Proj}_Y(X) = \frac{X \cdot Y}{\|Y\|}\frac{Y}{\|Y\|} = \frac{X \cdot Y}{\|Y\|^2}Y.$$

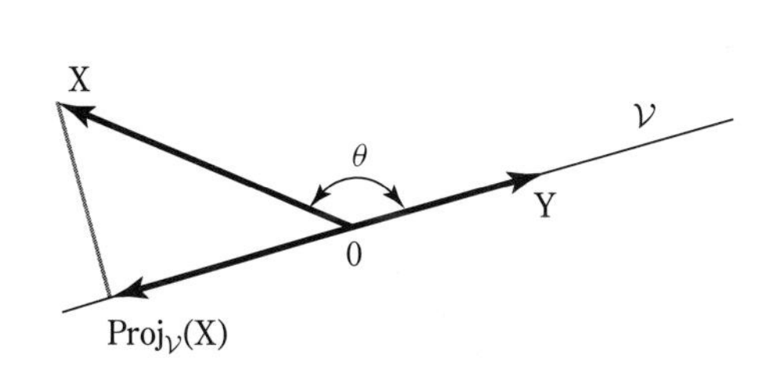

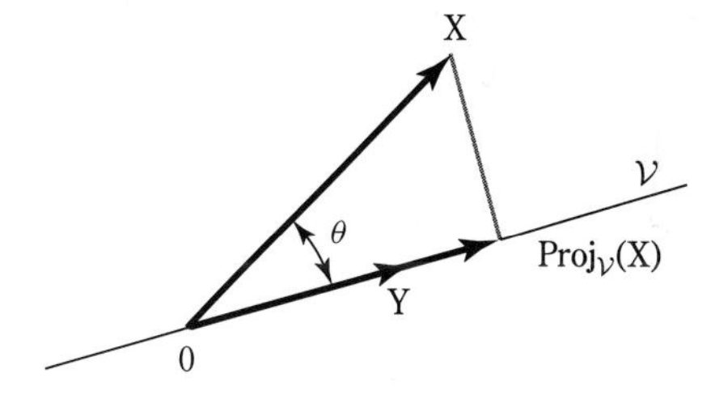

FIGURE 3.6

DEFINITION. If $X,Y \in \mathcal{R}^n$, then $\text{Proj}_Y(X) = \begin{cases} \dfrac{X \cdot Y}{\|Y\|^2}Y, & \text{if } Y \neq 0 \\ 0\ , & \text{if } Y = 0 \end{cases}$.

■ **EXAMPLE 3.9.** If $X = \begin{bmatrix} 1 \\ 2 \\ 3 \\ 4 \end{bmatrix}$ and $Y = \begin{bmatrix} 1 \\ 1 \\ 1 \\ 1 \end{bmatrix}$, then $X \cdot Y = 10$ and $\|Y\| = 2$, so

$$\text{Proj}_Y(X) = \frac{5}{2}\begin{bmatrix} 1 \\ 1 \\ 1 \\ 1 \end{bmatrix}. \ \square$$

Suppose X_1 and X_2 are linearly independent vectors in $\mathcal{R}^3$ and let $\mathcal{V} = \mathcal{S}pan\{X_1, X_2\}$. In implementing the Gram-Schmidt algorithm, you retain X_1 and replace X_2 by

$$Y_2 = X_2 - \frac{X_2 \cdot Y_1}{\|Y_1\|^2} Y_1 = X_2 - \text{Proj}_{Y_1}(X_2).$$

A geometric construction of Y_2 is illustrated in Figure 3.7.

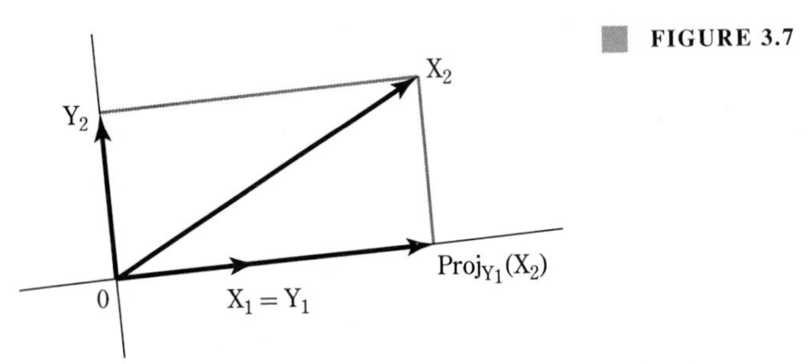

FIGURE 3.7

Assuming X_1, X_2, and X_3 are linearly independent vectors in $\mathcal{R}^3$, let Y_1 and Y_2 be the first two vectors generated by the Gram-Schmidt algorithm. Then

$$\mathcal{V} = \mathcal{S}pan\{Y_1, Y_2\} = \mathcal{S}pan\{X_1, X_2\}$$

is a plane through 0, and $X_3 \notin \mathcal{V}$. Let ℓ be the line through X_3 orthogonal to $\mathcal{V}$. The point of intersection of ℓ with $\mathcal{V}$ is *the orthogonal projection of* X_3 *onto* $\mathcal{V}$, denoted by $\text{Proj}_{\mathcal{V}}(X_3)$ (Figure 3.8). If V_1 and V_2 are projections of $\text{Proj}_{\mathcal{V}}(X_3)$ onto Y_1 and Y_2,

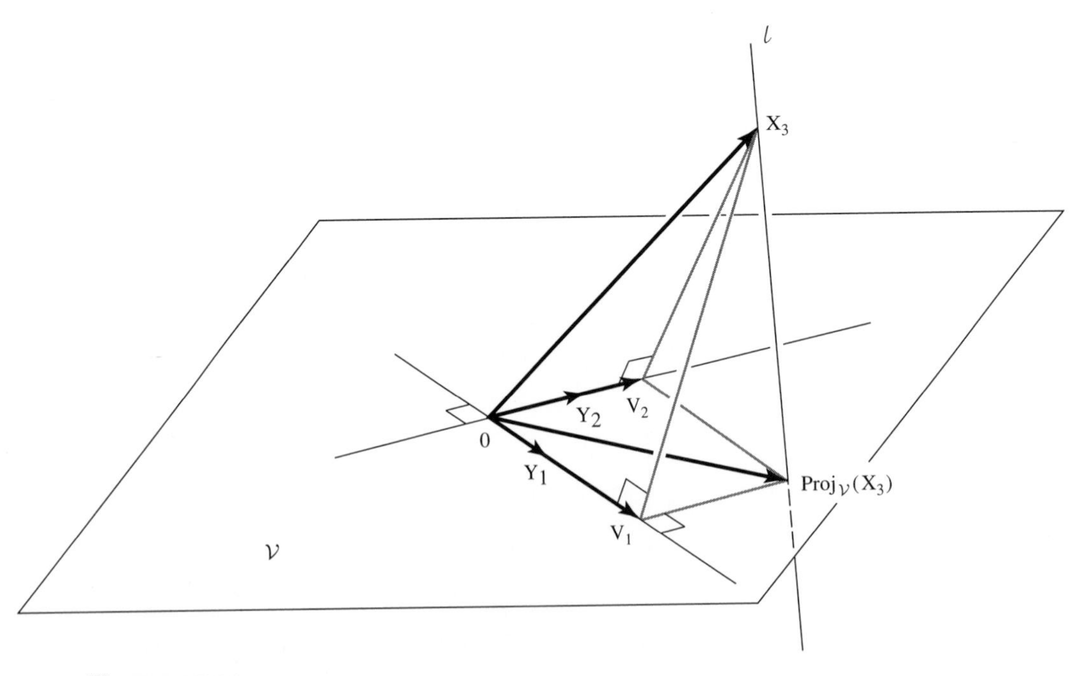

FIGURE 3.8

respectively, then 0, V_1, V_2, and $\text{Proj}_{\mathcal{V}}(X_3)$ are vertices of a rectangle that lies in $\mathcal{V}$, and $\text{Proj}_{\mathcal{V}}(X_3) = V_1 + V_2$. Note that the plane containing l and V_1 is orthogonal to the Y_1 axis. The line joining X_3 to V_1 is therefore orthogonal to the Y_1 axis, and consequently $V_1 = \text{Proj}_{Y_1}(X_3)$. Similarly, $V_2 = \text{Proj}_{Y_2}(X_3)$. Since

$$\text{Proj}_{\mathcal{V}}(X_3) = V_1 + V_2 = \text{Proj}_{Y_1}(X_3) + \text{Proj}_{Y_2}(X_3)$$

$$= \frac{X_3 \cdot Y_1}{\|Y_1\|^2} Y_1 + \frac{X_3 \cdot Y_2}{\|Y_2\|^2} Y_2,$$

the third member of the orthogonal set produced by the Gram-Schmidt process is

$$Y_3 = X_3 - \frac{X_3 \cdot Y_1}{\|Y_1\|^2} Y_1 - \frac{X_3 \cdot Y_2}{\|Y_2\|^2} Y_2 = X_3 - \text{Proj}_{\mathcal{V}}(X_3).$$

Figure 3.9 illustrates the geometric construction of Y_3.

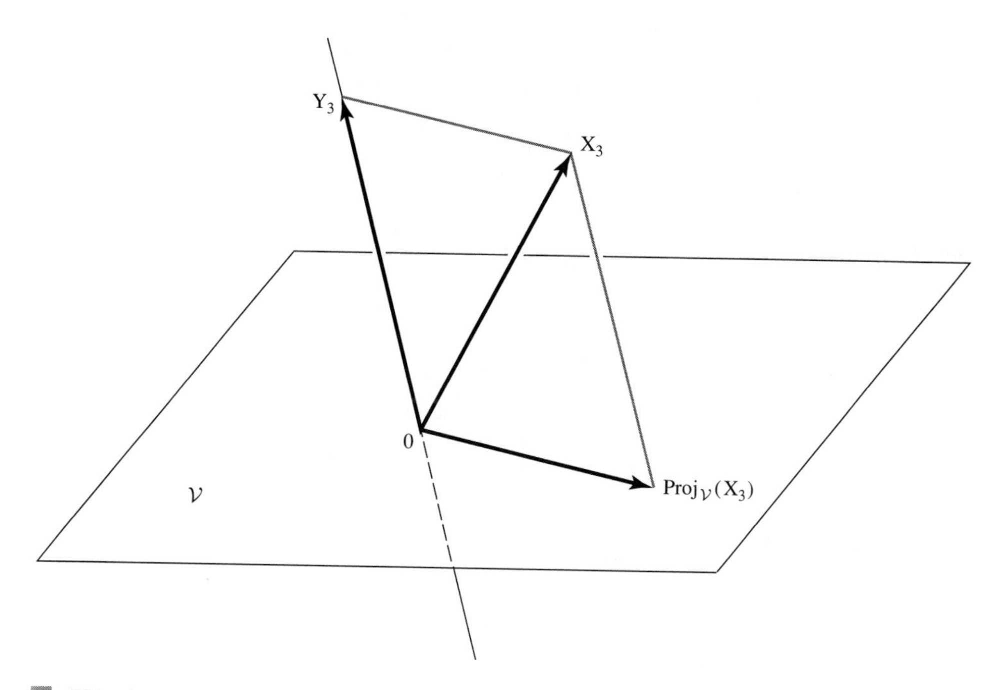

FIGURE 3.9

Exercises 3.3

1. a. Check that $\mathcal{A} = \left\{ \frac{1}{\sqrt{2}} \begin{bmatrix} 1 \\ 0 \\ -1 \end{bmatrix}, \frac{1}{3} \begin{bmatrix} 2 \\ 1 \\ 2 \end{bmatrix}, \frac{1}{3\sqrt{2}} \begin{bmatrix} 1 \\ -4 \\ 1 \end{bmatrix} \right\}$ is an orthonormal basis for $\mathcal{R}^3$.

b. Find $[X]_{\mathcal{A}}$ when $X = \begin{bmatrix} 4 \\ 1 \\ 3 \end{bmatrix}$.

c. Find $P_{\mathcal{AB}}$ when $\mathcal{B} = \left\{ \begin{bmatrix} 1 \\ 0 \\ 0 \end{bmatrix}, \begin{bmatrix} 1 \\ 1 \\ 0 \end{bmatrix}, \begin{bmatrix} 1 \\ 1 \\ 1 \end{bmatrix} \right\}$.

2. Let $\mathcal{V} = \{X \in \mathcal{R}^3 : x_1 + 2x_2 - x_3 = 0\}$.

a. Show that $\mathcal{A} = \left\{ \frac{1}{\sqrt{3}} \begin{bmatrix} -1 \\ 1 \\ 1 \end{bmatrix}, \frac{1}{\sqrt{2}} \begin{bmatrix} 1 \\ 0 \\ 1 \end{bmatrix} \right\}$ is an orthonormal basis for $\mathcal{V}$.

b. Show that $X = \begin{bmatrix} 2 \\ 1 \\ 4 \end{bmatrix} \in \mathcal{V}$ and find $[X]_{\mathcal{A}}$.

c. Show that $\mathcal{B} = \left\{ \begin{bmatrix} 0 \\ 1 \\ 2 \end{bmatrix}, \begin{bmatrix} -2 \\ 1 \\ 0 \end{bmatrix} \right\}$ is a basis for $\mathcal{V}$ and find $P_{\mathcal{AB}}$.

3. Let $\mathcal{V} = \{X \in \mathcal{R}^4 : x_2 - x_3 - x_4 = 0\}$.

a. Show that $\mathcal{A} = \left\{ \frac{1}{\sqrt{3}} \begin{bmatrix} 1 \\ 0 \\ -1 \\ 1 \end{bmatrix}, \frac{1}{\sqrt{3}} \begin{bmatrix} 1 \\ 1 \\ 1 \\ 0 \end{bmatrix}, \frac{1}{\sqrt{3}} \begin{bmatrix} 1 \\ -1 \\ 0 \\ -1 \end{bmatrix} \right\}$ is an orthonormal basis for $\mathcal{V}$.

b. Show that $X = \begin{bmatrix} 4 \\ 3 \\ 2 \\ 1 \end{bmatrix} \in \mathcal{V}$ and find $[X]_{\mathcal{A}}$.

c. Show that $\mathcal{B} = \left\{ \begin{bmatrix} 1 \\ 0 \\ 0 \\ 0 \end{bmatrix}, \begin{bmatrix} 0 \\ 1 \\ 1 \\ 0 \end{bmatrix}, \begin{bmatrix} 0 \\ 1 \\ 0 \\ 1 \end{bmatrix} \right\}$ is a basis for $\mathcal{V}$ and find $P_{\mathcal{AB}}$.

4. Assuming $\phi \in \mathcal{R}$, let $U_1 = \begin{bmatrix} \cos(\phi) \\ \sin(\phi) \end{bmatrix}$ and $U_2 = \begin{bmatrix} -\sin(\phi) \\ \cos(\phi) \end{bmatrix}$.

a. Show that $\mathcal{A} = \{U_1, U_2\}$ is an orthonormal basis for $\mathcal{R}^2$, and sketch directed line segments based at 0 representing these vectors.

b. Find $[E_1]_{\mathcal{A}}$ and $[E_2]_{\mathcal{A}}$.

5. Use the Gram-Schmidt process to find an orthonormal basis for the subspace spanned by

a. $X_1 = \begin{bmatrix} 1 \\ 2 \\ -2 \end{bmatrix}, X_2 = \begin{bmatrix} -1 \\ -3 \\ 1 \end{bmatrix}$.

b. $X_1 = \begin{bmatrix} 3 \\ 0 \\ -1 \end{bmatrix}, X_2 = \begin{bmatrix} 4 \\ 1 \\ 2 \end{bmatrix}$.

c. $X_1 = \begin{bmatrix} 1 \\ 2 \\ 0 \\ -1 \end{bmatrix}, X_2 = \begin{bmatrix} -2 \\ 0 \\ 1 \\ 2 \end{bmatrix}$.

d. $X_1 = \begin{bmatrix} 1 \\ 1 \\ 0 \\ 1 \end{bmatrix}, X_2 = \begin{bmatrix} 1 \\ 2 \\ 1 \\ 3 \end{bmatrix}, X_3 = \begin{bmatrix} -1 \\ 0 \\ 2 \\ -2 \end{bmatrix}$.

e. $X_1 = \begin{bmatrix} 1 \\ 0 \\ 0 \\ 0 \end{bmatrix}, X_2 = \begin{bmatrix} 1 \\ 1 \\ 0 \\ 0 \end{bmatrix}, X_3 = \begin{bmatrix} 1 \\ 1 \\ 1 \\ 0 \end{bmatrix}$. f. $X_1 = \begin{bmatrix} 3 \\ -2 \\ 1 \\ -1 \\ 1 \end{bmatrix}, X_2 = \begin{bmatrix} 4 \\ -1 \\ 0 \\ 0 \\ 2 \end{bmatrix}, X_3 = \begin{bmatrix} -1 \\ 1 \\ -2 \\ 2 \\ 1 \end{bmatrix}$.

g. $X_1 = \begin{bmatrix} 1 \\ 0 \\ 1 \\ 0 \\ 1 \end{bmatrix}, X_2 = \begin{bmatrix} 1 \\ 0 \\ 2 \\ 1 \\ 0 \end{bmatrix}, X_3 = \begin{bmatrix} 3 \\ 1 \\ 3 \\ 0 \\ 0 \end{bmatrix}$. h. $X_1 = \begin{bmatrix} 1 \\ 0 \\ 0 \\ 0 \\ 1 \end{bmatrix}, X_2 = \begin{bmatrix} 1 \\ 1 \\ 0 \\ 0 \\ 1 \end{bmatrix}, X_3 = \begin{bmatrix} 1 \\ 1 \\ 1 \\ 1 \\ 1 \end{bmatrix}$.

i. $X_1 = \begin{bmatrix} 1 \\ 0 \\ 1 \\ 0 \\ 0 \end{bmatrix}, X_2 = \begin{bmatrix} 1 \\ 0 \\ 1 \\ 1 \\ 1 \end{bmatrix}, X_3 = \begin{bmatrix} 3 \\ 1 \\ 1 \\ 2 \\ 0 \end{bmatrix}, X_4 \begin{bmatrix} -2 \\ 0 \\ 0 \\ -1 \\ 2 \end{bmatrix}$.

6. If $A = \begin{bmatrix} 1 & -1 & 0 & -1 \\ -1 & 1 & 2 & -3 \\ 1 & -1 & -1 & 1 \end{bmatrix}$, find

a. an orthonormal basis for $\mathcal{N}(A)$.
b. an orthonormal basis for $\mathcal{C}(A)$.

7. Assuming $X = \begin{bmatrix} -1 \\ 2 \\ 1 \end{bmatrix}$, $Y = \begin{bmatrix} -1 \\ 1 \\ -1 \end{bmatrix}$, and $Z = \begin{bmatrix} 2 \\ 1 \\ -1 \end{bmatrix}$, find

a. $\text{Proj}_Y(X)$. b. $\text{Proj}_X(Y)$. **c.** $\text{Proj}_Z(Y)$.
d. $\text{Proj}_Z(\text{Proj}_Y(X))$. **e.** $\text{Proj}_Y(\text{Proj}_X(Z))$.

8. Assuming $X = \begin{bmatrix} 2 \\ 1 \\ -1 \\ 2 \end{bmatrix}$, and $Y = \begin{bmatrix} 1 \\ 1 \\ -1 \\ -1 \end{bmatrix}$, find

a. $\text{Proj}_Y(X)$. **b.** $\text{Proj}_X(Y)$. c. $\text{Proj}_X(\text{Proj}_X(Y))$.

9. Show that if $X, Y \in \mathcal{R}^n$ and $t \neq 0$, then $\text{Proj}_{tY}(X) = \text{Proj}_Y(X)$.

10. For $X, Y \in \mathcal{R}^n$, show that $\text{Proj}_{(X+Y)}(X) + \text{Proj}_{(X+Y)}(Y) = X + Y$. Make a sketch illustrating this conclusion when $n = 2$.

11. Suppose X and Y are nonzero vectors in $\mathcal{R}^n$ and θ is the angle between X and Y. Show that $\text{Proj}_X(\text{Proj}_Y(X)) = \cos^2(\theta)\, X$.

12. Assuming X and Y are nonzero orthogonal vectors in $\mathcal{R}^n$ and $W \in \mathcal{S}pan\{X, Y\}$, show that $W = \text{Proj}_X(W) + \text{Proj}_Y(W)$.

13. Interpret the conclusion of Theorem 3.5 as a statement about projections.

14. Suppose $\mathrm{U} \in \mathcal{R}^n$ is a unit vector and $d \in \mathcal{R}$. Let $\mathcal{H} = \{\mathrm{X} \in \mathcal{R}^n : \mathrm{X} \cdot \mathrm{U} = d\}$. When $n = 2$ or $n = 3$, $\mathcal{H}$ is a line or plane, respectively, and for larger values on n, $\mathcal{H}$ is called a hyperplane.
   a. Show that $d\mathrm{U} \in \mathcal{H}$ and that $\mathrm{Proj}_{\mathrm{U}}(\mathrm{X}) = d\mathrm{U}$ for each $\mathrm{X} \in \mathcal{H}$.
   b. Show that $\mathrm{X} - d\mathrm{U}$ is orthogonal to U for each $\mathrm{X} \in \mathcal{H}$.
   c. Show that if $\mathrm{X} \in \mathcal{H}$ and $\mathrm{X} \neq d\mathrm{U}$, then $\|\mathrm{X}\| > |d|$ (that is, $d\mathrm{U}$ is the point of $\mathcal{H}$ nearest 0).
   d. Make a sketch showing directed line segment representatives of U, $d\mathrm{U}$, and $\mathrm{X} - d\mathrm{U}$ ($\mathrm{X} \in \mathcal{H}$) when $n = 2$.

3.4

ORTHOGONAL PROJECTION

In $\mathcal{R}^3$, the geometric interpretation of the Gram-Schmidt process involves dropping a perpendicular from a point to a line or plane. The line or plane contains zero, so it represents a subspace. There is no inherent notion of orthogonal projection onto a subspace of $\mathcal{R}^n$ when $n \geq 4$, but the geometry in $\mathcal{R}^3$ provides some clues as to how you might define one.

Consider a plane $\mathcal{V} \subseteq \mathcal{R}^3$ passing through 0. Let $\mathcal{W}$ be the line through 0 orthogonal to $\mathcal{V}$ and suppose $\mathrm{X} \in \mathcal{R}^3$. If $\mathrm{X} \notin \mathcal{V}$ and $\mathrm{X} \notin \mathcal{W}$, then 0, X, $\mathrm{Proj}_{\mathcal{V}}(\mathrm{X})$, and $\mathrm{Proj}_{\mathcal{W}}(\mathrm{X})$ are vertices of a rectangle, and $\mathrm{X} = \mathrm{Proj}_{\mathcal{V}}(\mathrm{X}) + \mathrm{Proj}_{\mathcal{W}}(\mathrm{X})$ (Figure 3.10). In other words, X can be decomposed into the sum of a vector in $\mathcal{V}$ and a vector orthogonal to $\mathcal{V}$. Moreover, it is geometrically clear that this decomposition is unique.

The idea of algebraically decomposing X in this manner has prospects of generalizing to $\mathcal{R}^n$ for any n. Given a subspace $\mathcal{V}$ of $\mathcal{R}^n$, if each X in $\mathcal{R}^n$ were uniquely expressible as a vector in $\mathcal{V}$ plus a vector "orthogonal to $\mathcal{V}$," then it would be reasonable

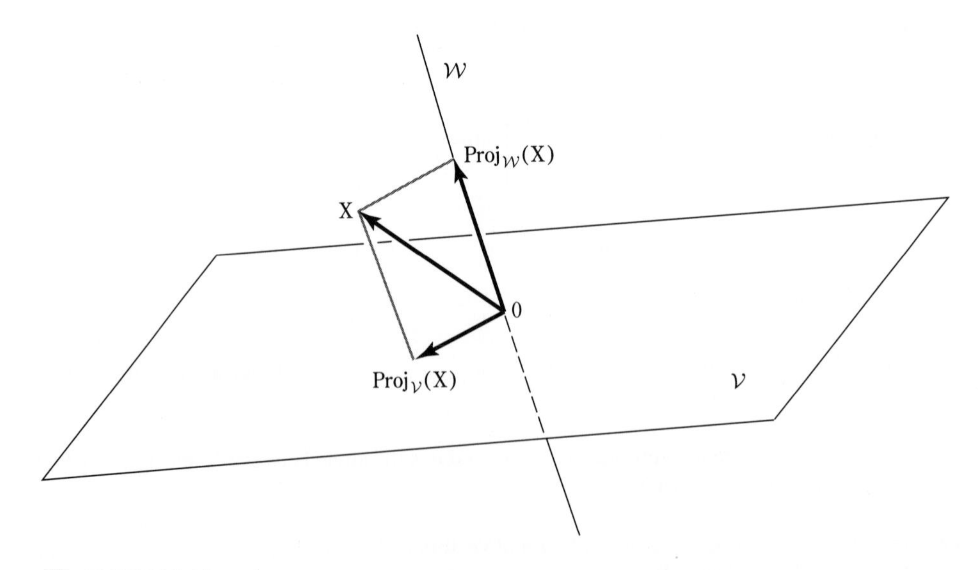

FIGURE 3.10

to consider the "component" in $\mathcal{V}$ to be the projection of X onto $\mathcal{V}$. The present section explores this line of thought, and the first step is to clarify the meaning of the phrase "orthogonal to $\mathcal{V}$."

DEFINITION. Given a subspace $\mathcal{V}$ of $\mathcal{R}^n$, $X \in \mathcal{R}^n$ is said to be *orthogonal* to $\mathcal{V}$ if $X \cdot V = 0$ for every $V \in \mathcal{V}$. The vectors orthogonal to $\mathcal{V}$ comprise a set called the *orthogonal complement* of $\mathcal{V}$, denoted by $\mathcal{V}^{\perp}$ (pronounced $\mathcal{V}$ perp).

■ **EXAMPLE 3.10.** Suppose $V = \begin{bmatrix} 1 \\ 2 \\ 3 \end{bmatrix}$ and $\mathcal{V} = \mathcal{S}pan\{V\}$. Then $X \in \mathcal{V}^{\perp}$ if and only if

$$0 = X \cdot (tV) = t(X \cdot V) = t(x_1 + 2x_2 + 3x_3)$$

for each $t \in \mathcal{R}$. Thus, $\mathcal{V}^{\perp} = \{X \in \mathcal{R}^3 : x_1 + 2x_2 + 3x_3 = 0\}$. Geometrically, $\mathcal{V}$ is the line through 0 with direction vector V, and $\mathcal{V}^{\perp}$ is the plane through 0 perpendicular to that line. □

■ **EXAMPLE 3.11.** The extreme subspaces of $\mathcal{R}^n$ are $\{0\}$ and $\mathcal{R}^n$. Since $X \cdot 0 = 0$ for every $X \in \mathcal{R}^n$, $\{0\}^{\perp} = \mathcal{R}^n$. If $X \in (\mathcal{R}^n)^{\perp}$, then $X \cdot Y = 0$ for every $Y \in \mathcal{R}^n$. In particular, $0 = X \cdot X = \|X\|^2$, so $X = 0$. Thus, $(\mathcal{R}^n)^{\perp} = \{0\}$. □

THEOREM 3.8. If $\mathcal{V} \subseteq \mathcal{R}^n$, then $\mathcal{V}^{\perp}$ is a subspace of $\mathcal{R}^n$.

Proof. If $X, Y \in \mathcal{V}^{\perp}$ and $t \in \mathcal{R}$, then for each $V \in \mathcal{V}$,

$$(X + Y) \cdot V = X \cdot V + Y \cdot V = 0 + 0 = 0,$$

and

$$(tX) \cdot V = t(X \cdot V) = t0 = 0,$$

so $X + Y \in \mathcal{V}^{\perp}$ and $tX \in \mathcal{V}^{\perp}$. ■

To conclude that X belongs to $\mathcal{V}^{\perp}$ you must check that $X \cdot V = 0$ for every V in $\mathcal{V}$. As the next theorem shows, that task can be simplified by referring to a spanning set for $\mathcal{V}$.

THEOREM 3.9. For $\mathcal{V} = \mathcal{S}pan\{X_1, \ldots, X_p\} \subseteq \mathcal{R}^n$, $X \in \mathcal{V}^{\perp}$ if and only if $X \cdot X_k = 0$, $k = 1, \ldots, p$.

Proof. If $X \in \mathcal{V}^{\perp}$, then $X \cdot V = 0$ for every $V \in \mathcal{V}$, so $X \cdot X_k = 0$, $k = 1, \ldots, p$. Conversely, suppose $X \cdot X_k = 0$, $k = 1, \ldots, p$. For each $V \in \mathcal{V}$ there are scalars $t_1, \ldots, t_p$ such that $V = t_1X_1 + \cdots + t_pX_p$, and therefore

$$X \cdot V = X \cdot (t_1X_1 + \cdots + t_pX_p) = t_1(X \cdot X_1) + \cdots + t_p(X \cdot X_p) = 0. \blacksquare$$

■ **EXAMPLE 3.12.** Let $\mathcal{V}$ be the subspace of $\mathcal{R}^5$ spanned by

$$X_1 = \begin{bmatrix} 1 \\ 0 \\ 2 \\ 0 \\ 1 \end{bmatrix}, \quad X_2 = \begin{bmatrix} 1 \\ 2 \\ -4 \\ 0 \\ 5 \end{bmatrix}, \quad X_3 = \begin{bmatrix} 1 \\ -1 \\ 5 \\ 1 \\ 2 \end{bmatrix}.$$

According to Theorem 3.9, $X \in \mathcal{V}^\perp$ if and only if $X \cdot X_1 = X \cdot X_2 = X \cdot X_3 = 0$, that is,

$$\begin{aligned} x_1 + 2x_3 + x_5 &= 0 \\ x_1 + 2x_2 - 4x_3 + 5x_5 &= 0 \\ x_1 - x_2 + 5x_3 + x_4 + 2x_5 &= 0. \end{aligned}$$

Thus,

$$\mathcal{V}^\perp = \mathcal{N}\left(\begin{bmatrix} 1 & 0 & 2 & 0 & 1 \\ 1 & 2 & -4 & 0 & 5 \\ 1 & -1 & 5 & 1 & 2 \end{bmatrix}\right).$$

Since

$$\begin{bmatrix} 1 & 0 & 2 & 0 & 1 \\ 1 & 2 & -4 & 0 & 5 \\ 1 & -1 & 5 & 1 & 2 \end{bmatrix} \longrightarrow \begin{bmatrix} 1 & 0 & 2 & 0 & 1 \\ 0 & 1 & -3 & 0 & 2 \\ 0 & 0 & 0 & 1 & 3 \end{bmatrix},$$

$\mathcal{V}^\perp = \mathcal{S}pan\{[-2 \quad 3 \quad 1 \quad 0 \quad 0]^T, [-1 \quad -2 \quad 0 \quad -3 \quad 1]^T\}$. □

THEOREM 3.10. If $\mathcal{V} = \mathcal{S}pan\{X_1, \ldots, X_p\} \subseteq \mathcal{R}^n$ and $A = [X_1 \quad \cdots \quad X_p]$, then

$$\mathcal{V}^\perp = \mathcal{N}(A^T).$$

Proof. Observe that

$$A^T X = \begin{bmatrix} X_1^T \\ \vdots \\ X_p^T \end{bmatrix} X = \begin{bmatrix} X_1^T X \\ \vdots \\ X_p^T X \end{bmatrix} = \begin{bmatrix} X_1 \cdot X \\ \vdots \\ X_p \cdot X \end{bmatrix}.$$

By Theorem 3.9, $X \in \mathcal{V}^\perp$ precisely when $X \cdot X_k = 0$, $k = 1, \ldots, p$, that is, when $A^T X = 0$. ■

Let $\mathcal{V}$ be a subspace of $\mathcal{R}^n$ with basis $\{X_1, \ldots, X_p\}$. Then $A = [X_1 \quad \cdots \quad X_p]$, being an $n \times p$ matrix with linearly independent columns, has rank p. According to Theorem 2.31, $\text{rank}(A^T) = \text{rank}(A)$, and by the rank-nullity theorem

$$\dim \mathcal{N}(A^T) + \text{rank}(A^T) = \#(\text{columns of } A^T) = n.$$

Thus, $\dim \mathcal{V}^\perp = \dim \mathcal{N}(A^T) = n - \text{rank}(A^T) = n - p$.

THEOREM 3.11. If $\mathcal{V}$ is a subspace of $\mathcal{R}^n$, then $\dim \mathcal{V} + \dim \mathcal{V}^\perp = n$.

Proof. The paragraph preceding the theorem establishes the conclusion when $\mathcal{V} \neq \{0\}$. If $\mathcal{V} = \{0\}$, then $\mathcal{V}^\perp = \mathcal{R}^n$, so $\dim \mathcal{V} + \dim \mathcal{V}^\perp = 0 + n = n$. ■

COROLLARY. If $\mathcal{V}$ is a subspace of $\mathcal{R}^n$, then $(\mathcal{V}^\perp)^\perp = \mathcal{V}$.

Proof. If $X \in \mathcal{V}$, then $X \cdot Y = 0$ for each $Y \in \mathcal{V}^\perp$, so $X \in (\mathcal{V}^\perp)^\perp$. Thus, $\mathcal{V} \subseteq (\mathcal{V}^\perp)^\perp$. Theorem 3.11, applied first to $\mathcal{V}$ and then to $\mathcal{V}^\perp$, gives

$$\dim \mathcal{V} + \dim \mathcal{V}^\perp = n = \dim \mathcal{V}^\perp + \dim (\mathcal{V}^\perp)^\perp,$$

and consequently $\dim \mathcal{V} = \dim (\mathcal{V}^\perp)^\perp$. That conclusion, together with $\mathcal{V} \subseteq (\mathcal{V}^\perp)^\perp$, implies $\mathcal{V} = (\mathcal{V}^\perp)^\perp$. ■

The stage is now set to carry out the program discussed at the beginning of this section, that is, to establish that every vector in $\mathcal{R}^n$ can be written as the sum of a vector in $\mathcal{V}$ and a vector orthogonal to $\mathcal{V}$. The vectors orthogonal to $\mathcal{V}$ constitute $\mathcal{V}^\perp$, so the idea is to show that $\mathcal{R}^n = \mathcal{V} + \mathcal{V}^\perp$. Observe that if $X \in \mathcal{V} \cap \mathcal{V}^\perp$, then X is orthogonal to itself, so $0 = X \cdot X = \|X\|^2$, and therefore $X = 0$. Thus, $\mathcal{V} \cap \mathcal{V}^\perp = \{0\}$, and by Theorem 2.36, $\mathcal{V}$ and $\mathcal{V}^\perp$ are independent.

THEOREM 3.12. If $\mathcal{V}$ is a subspace of $\mathcal{R}^n$, then $\mathcal{R}^n = \mathcal{V} \oplus \mathcal{V}^\perp$.

Proof. The comments preceding the theorem establish that $\mathcal{V} \cap \mathcal{V}^\perp = \{0\}$, so it suffices to show that $\mathcal{R}^n = \mathcal{V} + \mathcal{V}^\perp$. Let $X \in \mathcal{R}^n$. The two extreme cases will be treated first. If $\mathcal{V} = \{0\}$, then $\mathcal{V}^\perp = \mathcal{R}^n$. In this case, $X \in \mathcal{V}^\perp$, so $X = 0 + X$ expresses X as the sum of a vector in $\mathcal{V}$ and a vector in $\mathcal{V}^\perp$. When $\mathcal{V} = \mathcal{R}^n$, $\mathcal{V}^\perp = \{0\}$, and $X = X + 0$ is the desired decomposition. Now assume $\mathcal{V}$ has dimension $p \in \{1, \ldots, n-1\}$, in which case, $\dim \mathcal{V}^\perp = n - p$. Let $\mathcal{A} = \{U_1, \ldots, U_p\}$ and $\mathcal{B} = \{W_1, \ldots, W_{n-p}\}$ be orthonormal bases for $\mathcal{V}$ and $\mathcal{V}^\perp$, respectively. Then $\mathcal{A} \cup \mathcal{B}$, being a set of n mutually orthogonal nonzero vectors in $\mathcal{R}^n$, is a basis for $\mathcal{R}^n$. There are scalars $t_1, \ldots, t_p$ and $s_1, \ldots, s_{n-p}$ such that

$$X = t_1U_1 + \cdots + t_pU_p + s_1W_1 + \cdots + s_{n-p}W_{n-p},$$

and setting

$$V = t_1U_1 + \cdots + t_pU_p \quad \text{and} \quad W = s_1W_1 + \cdots + s_{n-p}W_{n-p},$$

you have $V \in \mathcal{V}$, $W \in \mathcal{V}^\perp$, and $X = V + W$. ■

DEFINITION. If $\mathcal{V}$ is a subspace of $\mathcal{R}^n$, $X \in \mathcal{R}^n$, and V and W are the unique vectors in $\mathcal{V}$ and $\mathcal{V}^\perp$, respectively, such that $X = V + W$, then V is called the *orthogonal projection* of X *onto* $\mathcal{V}$, written $V = \text{Proj}_{\mathcal{V}}(X)$.

Assume $\mathcal{V}$ is a subspace of $\mathcal{R}^n$, $X \in \mathcal{R}^n$, and $X = V + W$, where $V \in \mathcal{V}$ and $W \in \mathcal{V}^\perp$. Then $V = \text{Proj}_{\mathcal{V}}(X)$. Since $\mathcal{V} = (\mathcal{V}^\perp)^\perp$, $X = W + V$ can be interpreted as expressing X as a vector in $\mathcal{V}^\perp$ plus a vector in $(\mathcal{V}^\perp)^\perp$, so $W = \text{Proj}_{\mathcal{V}^\perp}(X)$. Thus,

$$\boxed{X = \text{Proj}_{\mathcal{V}}(X) + \text{Proj}_{\mathcal{V}^\perp}(X)}. \tag{3.10}$$

THEOREM 3.13. Suppose $\mathcal{V}$ is a subspace of $\mathcal{R}^n$ and $X \in \mathcal{R}^n$.

(1) If $\mathcal{V} = \{0\}$, then $\text{Proj}_{\mathcal{V}}(X) = 0$.
(2) If $\mathcal{V} = \mathcal{R}^n$, then $\text{Proj}_{\mathcal{V}}(X) = X$.
(3) If $\{U_1, \ldots, U_p\}$ is an orthonormal basis for $\mathcal{V}$, $1 \le p \le n-1$, then

$$\text{Proj}_{\mathcal{V}}(X) = (X \cdot U_1)U_1 + \cdots + (X \cdot U_p)U_p.$$

Proof.
(1) When $\mathcal{V} = \{0\}$, $\mathcal{V}^\perp = \mathcal{R}^n$ and $0 + X$ is the decomposition of X into the sum of a vector in $\mathcal{V}$ and a vector in $\mathcal{V}^\perp$. Thus, $\text{Proj}_{\mathcal{V}}(X) = 0$.
(2) If $\mathcal{V} = \mathcal{R}^n$, then $\mathcal{V}^\perp = \{0\}$, and part (1) together with (3.10) gives

$$X = \text{Proj}_{\mathcal{R}^n}(X) + \text{Proj}_{\{0\}}(X) = \text{Proj}_{\mathcal{R}^n}(X) + 0 = \text{Proj}_{\mathcal{R}^n}(X).$$

(3) Let $\{U_1, \ldots, U_p\}$ be an orthonormal basis for $\mathcal{V}$ and assume $X = V + W$, where $V \in \mathcal{V}$ and $W \in \mathcal{V}^\perp$. Then $\text{Proj}_{\mathcal{V}}(X) = V = (V \cdot U_1)U_1 + \cdots + (V \cdot U_p)U_p$ (Theorem 3.5). But $V \cdot U_k = X \cdot U_k$, $1 \le k \le p$, because $X \cdot U_k = (V + W) \cdot U_k = V \cdot U_k + W \cdot U_k$, and $W \cdot U_k = 0$. ■

■ **EXAMPLE 3.13.** Suppose $\mathcal{V}$ is the plane with equation $x + y + z = 0$ and set $Y = [1 \quad 1 \quad 1]^T$. For each $V = [x \quad y \quad z]^T \in \mathcal{V}$, $Y \cdot V = x + y + z = 0$, so $Y \in \mathcal{V}^\perp$. Moreover, $\dim \mathcal{V}^\perp = 1$, so

$$\{W\} = \left\{\frac{1}{\|Y\|}Y\right\} = \left\{\frac{1}{\sqrt{3}}\begin{bmatrix}1\\1\\1\end{bmatrix}\right\}$$

is an orthonormal basis for $\mathcal{V}^\perp$. As pointed out in Example 3.7,

$$U_1 = \frac{1}{\sqrt{2}}\begin{bmatrix}-1\\0\\1\end{bmatrix} \quad \text{and} \quad U_2 = \frac{1}{\sqrt{6}}\begin{bmatrix}-1\\2\\-1\end{bmatrix}$$

form an orthonormal basis for $\mathcal{V}$. If $X = \begin{bmatrix}-1\\2\\3\end{bmatrix}$, then by Theorem 3.13,

$$\text{Proj}_{\mathcal{V}}(X) = (X \cdot U_1)U_1 + (X \cdot U_2)U_2 = \frac{4}{2}\begin{bmatrix}-1\\0\\1\end{bmatrix} + \frac{2}{6}\begin{bmatrix}-1\\2\\-1\end{bmatrix} = \frac{1}{3}\begin{bmatrix}-7\\2\\5\end{bmatrix},$$

and

$$\text{Proj}_{\mathcal{V}^\perp}(X) = (X \cdot W)W = \frac{4}{3}\begin{bmatrix}1\\1\\1\end{bmatrix}.$$

Note that

$$\text{Proj}_{\mathcal{V}}(X) + \text{Proj}_{\mathcal{V}^\perp}(X) = \frac{1}{3}\begin{bmatrix}-7\\2\\5\end{bmatrix} + \frac{4}{3}\begin{bmatrix}1\\1\\1\end{bmatrix} = \begin{bmatrix}-1\\2\\3\end{bmatrix} = X. \ \square$$

Recall the kth step in the Gram-Schmidt algorithm. At that stage, k mutually orthogonal nonzero vectors, $Y_1, \ldots, Y_k$, have been generated, and the task is to construct Y_{k+1} from X_{k+1} and $Y_1, \ldots, Y_k$. If $\mathcal{V}_k = \mathcal{S}pan\{Y_1, \ldots, Y_k\}$ and $U_j = Y_j/\|Y_j\|$, $1 \le j \le k$, then $\{U_1, \ldots, U_k\}$ is an orthonormal basis for $\mathcal{V}_k$, and by Theorem 3.13,

$$\begin{aligned}\text{Proj}_{\mathcal{V}_k}(X_{k+1}) &= (X_{k+1} \cdot U_1)U_1 + \cdots + (X_{k+1} \cdot U_k)U_k \\ &= \left(X_{k+1} \cdot \frac{Y_1}{\|Y_1\|}\right)\frac{Y_1}{\|Y_1\|} + \cdots + \left(X_{k+1} \cdot \frac{Y_k}{\|Y_k\|}\right)\frac{Y_k}{\|Y_k\|} \\ &= \frac{X_{k+1} \cdot Y_1}{\|Y_1\|^2}Y_1 + \cdots + \frac{X_{k+1} \cdot Y_k}{\|Y_k\|^2}Y_k.\end{aligned}$$

Thus, the (k + 1)st vector generated by the algorithm (equation (3.9)) is

$$Y_{k+1} = X_{k+1} - \text{Proj}_{\mathcal{V}_k}(X_{k+1}).$$

The point is that this generalized notion of orthogonal projection allows us to view the

Gram-Schmidt process as successively applying a single basic operation: at stage k, project X_{k+1} orthogonally onto $\mathcal{V}_k$ and then subtract the result from X_{k+1}.

If $\mathcal{V}$ is a line or plane in $\mathcal{R}^3$ that contains 0 and X is a point of $\mathcal{R}^3$ not in $\mathcal{V}$, then the point of $\mathcal{V}$ closest of X is the orthogonal projection of X onto $\mathcal{V}$. The generalized notion of orthogonal projection in $\mathcal{R}^n$ also has this "minimal distance property."

THEOREM 3.14. If $\mathcal{V}$ is a subspace of $\mathcal{R}^n$ and $X \in \mathcal{R}^n$, then for every $Y \in \mathcal{V}$,

$$\|X - Y\| \geq \|X - \text{Proj}_{\mathcal{V}}(X)\|.$$

Equality holds only when $Y = \text{Proj}_{\mathcal{V}}(X)$.

Proof. Let $V = \text{Proj}_{\mathcal{V}}(X)$. For each $Y \in \mathcal{V}$,

$$\begin{aligned} \|X - Y\|^2 &= \|(X - V) + (V - Y)\|^2 \\ &= \|X - V\|^2 + 2(X - V) \cdot (V - Y) + \|V - Y\|^2. \end{aligned}$$

Now, $V - Y \in \mathcal{V}$ and $X - V = X - \text{Proj}_{\mathcal{V}}(X) = \text{Proj}_{\mathcal{V}^{\perp}}(X) \in \mathcal{V}^{\perp}$, so $(X - V) \cdot (V - Y) = 0$, and therefore

$$\|X - Y\|^2 = \|X - V\|^2 + \|V - Y\|^2 \geq \|X - V\|^2. \tag{3.11}$$

Thus, $\|X - Y\| \geq \|X - V\|$, with equality if and only if $Y = V$. ■

Exercises 3.4

1. Find a basis for $\mathcal{V}^{\perp}$ when $\mathcal{V}$ is the subspace spanned by

a. $\left\{\begin{bmatrix} 1 \\ 2 \\ 3 \end{bmatrix}, \begin{bmatrix} 3 \\ 2 \\ 1 \end{bmatrix}\right\}$. b. $\left\{\begin{bmatrix} 1 \\ -2 \\ 3 \\ -4 \end{bmatrix}\right\}$. **c.** $\left\{\begin{bmatrix} 1 \\ 0 \\ -1 \\ 1 \end{bmatrix}, \begin{bmatrix} 1 \\ -1 \\ -1 \\ 1 \end{bmatrix}, \begin{bmatrix} 1 \\ 1 \\ -1 \\ 1 \end{bmatrix}\right\}$.

d. $\left\{\begin{bmatrix} -1 \\ 2 \\ 1 \\ 0 \end{bmatrix}, \begin{bmatrix} 0 \\ 1 \\ 1 \\ 2 \end{bmatrix}, \begin{bmatrix} 1 \\ 0 \\ 1 \\ -2 \end{bmatrix}\right\}$. **e.** $\left\{\begin{bmatrix} 1 \\ 1 \\ -1 \\ -1 \\ 1 \end{bmatrix}, \begin{bmatrix} -1 \\ 1 \\ 1 \\ -1 \\ 1 \end{bmatrix}\right\}$. f. $\left\{\begin{bmatrix} -1 \\ 0 \\ 1 \\ 2 \\ -1 \end{bmatrix}, \begin{bmatrix} 1 \\ 1 \\ -1 \\ -2 \\ 2 \end{bmatrix}, \begin{bmatrix} -1 \\ 2 \\ 1 \\ 2 \\ 1 \end{bmatrix}\right\}$.

g. $\left\{\begin{bmatrix} 0 \\ 1 \\ -1 \\ 2 \\ 1 \end{bmatrix}, \begin{bmatrix} 0 \\ 0 \\ 2 \\ 1 \\ -1 \end{bmatrix}, \begin{bmatrix} 0 \\ 1 \\ 0 \\ 2 \\ -1 \end{bmatrix}\right\}$. h. $\left\{\begin{bmatrix} 1 \\ 0 \\ 1 \\ 0 \\ 1 \end{bmatrix}, \begin{bmatrix} 1 \\ 1 \\ 1 \\ 0 \\ 0 \end{bmatrix}, \begin{bmatrix} 0 \\ 0 \\ 1 \\ 1 \\ 1 \end{bmatrix}\right\}$. **i.** $\left\{\begin{bmatrix} -1 \\ 1 \end{bmatrix}\right\}$.

2. Suppose $\mathcal{V} = \mathcal{S}pan\left\{\begin{bmatrix} 2 \\ 1 \\ -3 \end{bmatrix}\right\}$ and $X = \begin{bmatrix} 1 \\ -1 \\ 1 \end{bmatrix}$. Find $V \in \mathcal{V}$ and $W \in \mathcal{V}^{\perp}$ such that $X = V + W$.

3. Find $\text{Proj}_{\mathcal{V}}(X)$ and $\text{Proj}_{\mathcal{V}^\perp}(X)$ when

a. $\mathcal{V} = \left\{ \begin{bmatrix} x \\ y \\ z \end{bmatrix} : x - y + z = 0 \right\}$ and $X = \begin{bmatrix} 1 \\ 2 \\ 3 \end{bmatrix}$.

b. $\mathcal{V} = \mathcal{S}pan \left\{ \begin{bmatrix} 1 \\ 1 \\ 0 \\ 0 \end{bmatrix}, \begin{bmatrix} 0 \\ 0 \\ 1 \\ 1 \end{bmatrix}, \begin{bmatrix} 0 \\ 1 \\ 1 \\ 0 \end{bmatrix} \right\}$ and $X = \begin{bmatrix} 1 \\ 2 \\ 3 \\ 4 \end{bmatrix}$.

c. $\mathcal{V} = \mathcal{S}pan \left\{ \begin{bmatrix} 1 \\ 2 \\ 1 \end{bmatrix}, \begin{bmatrix} 2 \\ 1 \\ -1 \end{bmatrix} \right\}$ and $X = \begin{bmatrix} -1 \\ 1 \\ 2 \end{bmatrix}$.

d. $\mathcal{V} = \mathcal{S}pan \left\{ \begin{bmatrix} 1 \\ 0 \\ -1 \\ 1 \end{bmatrix}, \begin{bmatrix} 1 \\ -1 \\ -1 \\ 1 \end{bmatrix}, \begin{bmatrix} 3 \\ -1 \\ -2 \\ 1 \end{bmatrix} \right\}$ and $X = \begin{bmatrix} 1 \\ 2 \\ 3 \\ 4 \end{bmatrix}$.

4. Let $A = \begin{bmatrix} 1 & -2 & 0 & -1 \\ -1 & 1 & 1 & 0 \\ 1 & -3 & 1 & -2 \end{bmatrix}$, $X = \begin{bmatrix} 1 \\ 1 \\ 1 \\ 1 \end{bmatrix}$, and $Y = \begin{bmatrix} 1 \\ 1 \\ 1 \end{bmatrix}$.

a. Find $\text{Proj}_{\mathcal{V}}(X)$ and $\text{Proj}_{\mathcal{V}^\perp}(X)$ when $\mathcal{V} = \mathcal{N}(A)$.
b. Find $\text{Proj}_{\mathcal{V}}(Y)$ and $\text{Proj}_{\mathcal{V}^\perp}(Y)$ when $\mathcal{V} = \mathcal{C}(A)$.

5. Suppose $\mathcal{V}$ is a subspace of $\mathcal{R}^n$, $X, Y \in \mathcal{R}^n$, and $t \in \mathcal{R}$. Show that
a. $\text{Proj}_{\mathcal{V}}(X + Y) = \text{Proj}_{\mathcal{V}}(X) + \text{Proj}_{\mathcal{V}}(Y)$.
b. $\text{Proj}_{\mathcal{V}}(tX) = t\,\text{Proj}_{\mathcal{V}}(X)$.

6. Assuming $\mathcal{V}$ is a subspace of $\mathcal{R}^n$ and $X \in \mathcal{V}$, show that $\text{Proj}_{\mathcal{V}}(X) = X$.

7. Suppose $\mathcal{V}$ is a subspace of $\mathcal{R}^n$ and $X \in \mathcal{R}^n$.
a. Find $\text{Proj}_{\mathcal{V}}(\text{Proj}_{\mathcal{V}}(X))$. Explain!
b. Find $\text{Proj}_{\mathcal{V}^\perp}(\text{Proj}_{\mathcal{V}}(X))$. Explain!

8. Assume $Y_1, \ldots, Y_p \in \mathcal{R}^n$, $\mathcal{V} = \mathcal{S}pan\{Y_1, \ldots, Y_p\}$, and $X \in \mathcal{R}^n$.
a. Show that if $Y_1, \ldots, Y_p$ are mutually orthogonal nonzero vectors, then

$$\text{Proj}_{\mathcal{V}}(X) = \sum_{k=1}^{p} \text{Proj}_{Y_k}(X).$$

b. Show by example in $\mathcal{R}^2$ with $p = 2$ that conclusion (a) does not hold if $Y_1, \ldots, Y_p$ are simply linearly independent. In a sketch illustrating your example label X, Y_1, Y_2, $\text{Proj}_{Y_1}(X)$, $\text{Proj}_{Y_2}(X)$, and $\text{Proj}_{\mathcal{V}}(X)$.

9. Subspaces $\mathcal{U}$ and $\mathcal{V}$ of $\mathcal{R}^n$ are said to be orthogonal if $U \cdot V = 0$ for every $U \in \mathcal{U}$ and $V \in \mathcal{V}$.
a. Show that if $\mathcal{U}$ and $\mathcal{V}$ are orthogonal, then $\mathcal{U} \cap \mathcal{V} = \{0\}$.
b. Show that if $\mathcal{U}$ and $\mathcal{V}$ are orthogonal, then for all $X \in \mathcal{R}^n$

$$\text{Proj}_{\mathcal{U} \oplus \mathcal{V}}(X) = \text{Proj}_{\mathcal{U}}(X) + \text{Proj}_{\mathcal{V}}(X).$$

10. Assuming $A \in \mathcal{M}_{m\times n}$, show that $\mathcal{N}(A^T) = \mathcal{C}(A)^\perp$.

11. Show that every subspace of $\mathcal{R}^n$ is the null space of a matrix.

3.5
ORTHOGONAL MATRICES

Suppose $A \in \mathcal{M}_{n\times n}$ and $X_j = \mathrm{Col}_j(A)$, $1 \le j \le n$. If A is nonsingular, then $X_1, \ldots, X_n$ are linearly independent, and therefore $\mathcal{A} = \{X_1, \ldots, X_n\}$ is a basis for $\mathcal{R}^n$. The present section is concerned with the special case when $\mathcal{A}$ is orthonormal.

DEFINITION. If $A \in \mathcal{M}_{n\times n}$ and the columns of A are orthonormal, then A is called an *orthogonal matrix.*

Observe that for $A = [X_1 \quad \cdots \quad X_n] \in \mathcal{M}_{n\times n}$,

$$X_i \cdot X_j = \mathrm{Row}_i(A^T)\mathrm{Col}_j(A) = [A^T A]_{ij},$$

$1 \le i,j \le n$. Thus, $X_1, \ldots, X_n$ are orthonormal precisely when

$$[A^T A]_{ij} = \begin{cases} 1, & \text{if } i = j \\ 0, & \text{if } i \ne j \end{cases},$$

that is, when $A^T A = I_n$. This establishes a useful characterization of orthogonal matrices.

THEOREM 3.15. If $A \in \mathcal{M}_{n\times n}$, then A is orthogonal if and only if $A^{-1} = A^T$.

EXAMPLE 3.14. Since $\left\{ \frac{1}{\sqrt{3}}\begin{bmatrix} 1 \\ 1 \\ 1 \end{bmatrix}, \frac{1}{\sqrt{2}}\begin{bmatrix} -1 \\ 0 \\ 1 \end{bmatrix}, \frac{1}{\sqrt{6}}\begin{bmatrix} -1 \\ 2 \\ -1 \end{bmatrix} \right\}$ is an orthonormal basis for $\mathcal{R}^3$,

$$A = \begin{bmatrix} 1/\sqrt{3} & -1/\sqrt{2} & -1/\sqrt{6} \\ 1/\sqrt{3} & 0 & 2/\sqrt{6} \\ 1/\sqrt{3} & 1/\sqrt{2} & -1/\sqrt{6} \end{bmatrix}$$

is orthogonal and

$$A^{-1} = A^T = \begin{bmatrix} 1/\sqrt{3} & 1/\sqrt{3} & 1/\sqrt{3} \\ -1/\sqrt{2} & 0 & 1/\sqrt{2} \\ -1/\sqrt{6} & 2/\sqrt{6} & -1/\sqrt{6} \end{bmatrix}. \square$$

THEOREM 3.16. If $A \in \mathcal{M}_{n\times n}$ is orthogonal, then so is A^{-1}.

Proof. Orthogonality of A implies $A^{-1} = A^T$, so $(A^{-1})^{-1} = (A^T)^{-1} = (A^{-1})^T$ (Theorem 1.19). Thus, by Theorem 3.15, A^{-1} is orthogonal. ■

Let $\mathcal{A} = \{U_1, \ldots, U_p\}$ and $\mathcal{B} = \{W_1, \ldots, W_p\}$ be orthonormal bases for a subspace $\mathcal{V} \subseteq \mathcal{R}^n$ and consider the change of coordinates matrix

$$P_{\mathcal{B}\mathcal{A}} = \begin{bmatrix} [U_1]_{\mathcal{B}} & \cdots & [U_p]_{\mathcal{B}} \end{bmatrix}.$$

By Theorem 3.5,

$$[U_j]_{\mathcal{B}} = \begin{bmatrix} U_j \cdot W_1 \\ \vdots \\ U_j \cdot W_p \end{bmatrix},$$

$1 \leq j \leq p$, so $[P_{\mathcal{B}\mathcal{A}}]_{ij} = U_j \cdot W_i$. Interchanging roles of $\mathcal{B}$ and $\mathcal{A}$ gives $[P_{\mathcal{A}\mathcal{B}}]_{ij} = W_j \cdot U_i$. Thus, for $1 \leq i, j \leq p$,

$$[P^T_{\mathcal{B}\mathcal{A}}]_{ij} = [P_{\mathcal{B}\mathcal{A}}]_{ji} = U_i \cdot W_j = [P_{\mathcal{A}\mathcal{B}}]_{ij},$$

that is,

$$P^T_{\mathcal{B}\mathcal{A}} = P_{\mathcal{A}\mathcal{B}}.$$

THEOREM 3.17. If $\mathcal{V}$ is a subspace of $\mathcal{R}^n$ with orthonormal bases $\mathcal{A}$ and $\mathcal{B}$, then $P_{\mathcal{B}\mathcal{A}}$ is orthogonal.

Proof. The discussion above establishes that $P^T_{\mathcal{B}\mathcal{A}} = P_{\mathcal{A}\mathcal{B}}$, and by Theorem 2.34, $P_{\mathcal{A}\mathcal{B}} = P^{-1}_{\mathcal{B}\mathcal{A}}$. ■

EXAMPLE 3.15. Let $\mathcal{V} = \{X \in \mathcal{R}^3 \ : \ x_1 + x_2 + x_3 = 0\}$ and set

$$U_1 = \frac{1}{\sqrt{2}}\begin{bmatrix} -1 \\ 0 \\ 1 \end{bmatrix}, U_2 = \frac{1}{\sqrt{6}}\begin{bmatrix} -1 \\ 2 \\ -1 \end{bmatrix}, W_1 = \frac{1}{\sqrt{2}}\begin{bmatrix} -1 \\ 1 \\ 0 \end{bmatrix}, \text{ and } W_2 = \frac{1}{\sqrt{6}}\begin{bmatrix} 1 \\ 1 \\ -2 \end{bmatrix}.$$

Observe that $\mathcal{A} = \{U_1, U_2\}$ and $\mathcal{B} = \{W_1, W_2\}$ are orthonormal bases for $\mathcal{V}$. Thus,

$$[U_1]_{\mathcal{B}} = \begin{bmatrix} U_1 \cdot W_1 \\ U_1 \cdot W_2 \end{bmatrix} = \begin{bmatrix} 1/2 \\ -\sqrt{3}/2 \end{bmatrix}, \qquad [U_2]_{\mathcal{B}} = \begin{bmatrix} U_2 \cdot W_1 \\ U_2 \cdot W_2 \end{bmatrix} = \begin{bmatrix} \sqrt{3}/2 \\ 1/2 \end{bmatrix},$$

and

$$P_{\mathcal{B}\mathcal{A}} = \begin{bmatrix} [U_1]_{\mathcal{B}} & [U_2]_{\mathcal{B}} \end{bmatrix} = \frac{1}{2}\begin{bmatrix} 1 & \sqrt{3} \\ -\sqrt{3} & 1 \end{bmatrix}.$$

Since $P_{\mathcal{B}\mathcal{A}}$ is orthogonal,

$$P_{\mathcal{A}\mathcal{B}} = P^{-1}_{\mathcal{B}\mathcal{A}} = P^T_{\mathcal{B}\mathcal{A}} = \frac{1}{2}\begin{bmatrix} 1 & -\sqrt{3} \\ \sqrt{3} & 1 \end{bmatrix}. \ \square$$

The standard tool for generating orthonormal bases is the Gram-Schmidt algorithm, and a careful examination of that process reveals a role played by orthogonal matrices. Suppose $X_1, \ldots, X_p$ are linearly independent vectors in $\mathcal{R}^n$ and set

$$A = [X_1 \quad \cdots \quad X_p].$$

The algorithm produces an orthogonal set $\{Y_1, \ldots, Y_p\}$ such that

$$\mathcal{V}_k = \mathcal{S}pan\{X_1, \ldots, X_k\} = \mathcal{S}pan\{Y_1, \ldots, Y_k\},$$

$k = 1, \ldots, p$. Let $U_1, \ldots, U_p$ be the associated orthonormal vectors (i.e., $U_j = Y_j/\|Y_j\|$, $1 \le j \le p$). The X's, Y's, and U's are then related as follows:

$$\|Y_1\|U_1 = Y_1 = X_1,$$

and for each $k = 1, \ldots, p-1$,

$$\|Y\|_{k+1}U_{k+1} = Y_{k+1} = X_{k+1} - \mathrm{Proj}_{\mathcal{V}_k}(X_{k+1}).$$

Since $\{U_1, \ldots, U_k\}$ is an orthonormal basis for $\mathcal{V}_k$, the last equation becomes

$$\|Y\|_{k+1}U_{k+1} = X_{k+1} - (X_{k+1} \cdot U_1)U_1 - \cdots - (X_{k+1} \cdot U_k)U_k,$$

and a little rearrangement gives

$$\begin{aligned}
X_1 &= \|Y_1\|U_1 \\
X_2 &= (X_2 \cdot U_1)U_1 + \|Y_2\|U_2 \\
X_3 &= (X_3 \cdot U_1)U_1 + (X_3 \cdot U_2)U_2 + \|Y_3\|U_3 \\
&\vdots \\
X_p &= (X_p \cdot U_1)U_1 + \cdots + (X_p \cdot U_{p-1})U_{p-1} + \|Y_p\|U_p.
\end{aligned} \tag{3.12}$$

Now, set $Q = [U_1 \quad \cdots \quad U_p]$ and observe that each line in (3.12) expresses a particular column of A as a linear combination of columns of Q. The kth equation reads

$$X_k = (X_k \cdot U_1)U_1 + \cdots + (X_k \cdot U_{k-1})U_{k-1} + \|Y_k\|U_k = [U_1 \quad \cdots \quad U_p]\begin{bmatrix} X_k \cdot U_1 \\ \vdots \\ X_k \cdot U_{k-1} \\ \|Y_k\| \\ 0 \\ \vdots \\ 0 \end{bmatrix},$$

where the last column is to be interpreted as a $p \times 1$ matrix. Thus, the relationship between $\{X_1, \ldots, X_p\}$ and $\{U_1, \ldots, U_p\}$ can be expressed concisely by the matrix equation $A = QR$, where

$$R = \begin{bmatrix} \|Y_1\| & X_2 \cdot U_1 & X_3 \cdot U_1 & \cdots & X_p \cdot U_1 \\ 0 & \|Y_2\| & X_3 \cdot U_2 & \cdots & X_p \cdot U_2 \\ 0 & 0 & \|Y_3\| & \cdots & X_p \cdot U_3 \\ \vdots & & & \ddots & \vdots \\ 0 & 0 & 0 & \cdots & \|Y_p\| \end{bmatrix}. \tag{3.13}$$

Note that R, being an upper triangular $p \times p$ matrix with positive diagonal entries, is invertible.

THEOREM 3.18. If $A \in \mathcal{M}_{n\times p}$ has linearly independent columns, then there is a $Q \in \mathcal{M}_{n\times p}$ with orthonormal columns and a nonsingular, upper triangular $R \in \mathcal{M}_{p\times p}$ such that $A = QR$. In particular, when A is square, Q is orthogonal.

■ **EXAMPLE 3.16.** Applying the Gram-Schmidt algorithm to

$$\{X_1, X_2\} = \left\{ \begin{bmatrix} 1 \\ 1 \\ -1 \\ 1 \end{bmatrix}, \begin{bmatrix} -1 \\ 1 \\ 1 \\ -1 \end{bmatrix} \right\}$$

produces $Y_1 = X_1$ and

$$Y_2 = X_2 - \frac{X_2 \cdot Y_1}{\|Y_1\|^2} Y_1 = \begin{bmatrix} -1 \\ 1 \\ 1 \\ -1 \end{bmatrix} - \frac{-2}{4} \begin{bmatrix} 1 \\ 1 \\ -1 \\ 1 \end{bmatrix} = \frac{1}{2} \begin{bmatrix} -1 \\ 3 \\ 1 \\ -1 \end{bmatrix}.$$

Since $\|Y_1\| = 2$ and $\|Y_2\| = \sqrt{3}$,

$$U_1 = \frac{1}{2} \begin{bmatrix} 1 \\ 1 \\ -1 \\ 1 \end{bmatrix} \quad \text{and} \quad U_2 = \frac{1}{2\sqrt{3}} \begin{bmatrix} -1 \\ 3 \\ 1 \\ -1 \end{bmatrix}$$

form an orthonormal basis for the subspace of $\mathcal{R}^4$ spanned by $\{X_1, X_2\}$. If $A = [X_1 \quad X_2]$ and $Q = [U_1 \quad U_2]$, then $A = QR$, where R is the upper triangular 2×2 matrix given by (3.13). That is,

$$\begin{bmatrix} 1 & -1 \\ 1 & 1 \\ -1 & 1 \\ 1 & -1 \end{bmatrix} = \frac{1}{2} \begin{bmatrix} 1 & -1/\sqrt{3} \\ 1 & \sqrt{3} \\ -1 & 1/\sqrt{3} \\ 1 & -1/\sqrt{3} \end{bmatrix} \begin{bmatrix} 2 & -1 \\ 0 & \sqrt{3} \end{bmatrix}. \quad \square$$

■ **EXAMPLE 3.17.** Let $A = [X_1 \quad X_2 \quad X_3]$, where

$$X_1 = \begin{bmatrix} 1 \\ 0 \\ 1 \end{bmatrix}, \quad X_2 = \begin{bmatrix} 1 \\ 1 \\ 0 \end{bmatrix}, \quad \text{and} \quad X_3 = \begin{bmatrix} 0 \\ 1 \\ 1 \end{bmatrix}.$$

The Gram-Schmidt process generates $Y_1 = X_1$,

$$Y_2 = X_2 - \frac{X_2 \cdot Y_1}{\|Y_1\|^2} Y_1 = \begin{bmatrix} 1 \\ 1 \\ 0 \end{bmatrix} - \frac{1}{2} \begin{bmatrix} 1 \\ 0 \\ 1 \end{bmatrix} = \frac{1}{2} \begin{bmatrix} 1 \\ 2 \\ -1 \end{bmatrix},$$

and

$$Y_3 = X_3 - \frac{X_3 \cdot Y_1}{\|Y_1\|^2} Y_1 - \frac{X_3 \cdot Y_2}{\|Y_2\|^2} Y_2 = \begin{bmatrix} 0 \\ 1 \\ 1 \end{bmatrix} - \frac{1}{2} \begin{bmatrix} 1 \\ 0 \\ 1 \end{bmatrix} - \frac{1}{6} \begin{bmatrix} 1 \\ 2 \\ -1 \end{bmatrix} = \frac{2}{3} \begin{bmatrix} -1 \\ 1 \\ 1 \end{bmatrix}.$$

Since $\|Y_1\| = \sqrt{2}$, $\|Y_2\| = \sqrt{6}/2$, and $\|Y_3\| = 2/\sqrt{3}$, the resulting orthonormal basis for $\mathcal{R}^3$ is

$$\{U_1, U_2, U_3\} = \left\{ \frac{1}{\sqrt{2}} \begin{bmatrix} 1 \\ 0 \\ 1 \end{bmatrix}, \frac{1}{\sqrt{6}} \begin{bmatrix} 1 \\ 2 \\ -1 \end{bmatrix}, \frac{1}{\sqrt{3}} \begin{bmatrix} -1 \\ 1 \\ 1 \end{bmatrix} \right\}.$$

The QR factorization of A is

$$\begin{bmatrix} 1 & 1 & 0 \\ 0 & 1 & 1 \\ 1 & 0 & 1 \end{bmatrix} = \begin{bmatrix} 1/\sqrt{2} & 1/\sqrt{6} & -1/\sqrt{3} \\ 0 & 2/\sqrt{6} & 1/\sqrt{3} \\ 1/\sqrt{2} & -1/\sqrt{6} & 1/\sqrt{3} \end{bmatrix} \begin{bmatrix} \sqrt{2} & 1/\sqrt{2} & 1/\sqrt{2} \\ 0 & \sqrt{6}/2 & 1/\sqrt{6} \\ 0 & 0 & 2/\sqrt{3} \end{bmatrix},$$

where the entries in the last matrix are computed according to (3.13). □

Calculationally, the most attractive feature of an orthogonal matrix is the ease with which it can be inverted. Thanks to Theorem 3.18, that property can be utilized in inverting an arbitrary nonsingular square matrix. If $A = QR$, where Q is orthogonal and R is upper triangular, then $A^{-1} = R^{-1}Q^{-1} = R^{-1}Q^T$. Thus, the task of inverting A reduces to that of inverting R.

Exercises 3.5

1. Check that $\mathcal{A} = \left\{ \frac{1}{\sqrt{2}} \begin{bmatrix} 1 \\ 1 \end{bmatrix}, \frac{1}{\sqrt{2}} \begin{bmatrix} 1 \\ -1 \end{bmatrix} \right\}$ and $\mathcal{B} = \left\{ \frac{1}{\sqrt{5}} \begin{bmatrix} 1 \\ 2 \end{bmatrix}, \frac{1}{\sqrt{5}} \begin{bmatrix} 2 \\ -1 \end{bmatrix} \right\}$ are orthonormal bases for $\mathcal{R}^2$ and find $P_{\mathcal{BA}}$ and $P_{\mathcal{AB}}$.

2. Let $\mathcal{V} = \{X \in \mathcal{R}^4 : x_1 - x_2 - x_3 - x_4 = 0\}$. Verify that

$$\mathcal{A} = \left\{ \frac{1}{\sqrt{2}} \begin{bmatrix} 0 \\ 1 \\ -1 \\ 0 \end{bmatrix}, \frac{1}{\sqrt{2}} \begin{bmatrix} 1 \\ 0 \\ 0 \\ 1 \end{bmatrix}, \frac{1}{2} \begin{bmatrix} 1 \\ 1 \\ 1 \\ -1 \end{bmatrix} \right\} \text{ and } \mathcal{B} = \left\{ \frac{1}{2} \begin{bmatrix} 1 \\ -1 \\ 1 \\ 1 \end{bmatrix}, \frac{1}{\sqrt{2}} \begin{bmatrix} 1 \\ 1 \\ 0 \\ 0 \end{bmatrix}, \frac{1}{\sqrt{2}} \begin{bmatrix} 0 \\ 0 \\ 1 \\ -1 \end{bmatrix} \right\}$$

are orthonormal bases for $\mathcal{V}$ and calculate $P_{\mathcal{BA}}$ and $P_{\mathcal{AB}}$.

3. Assuming $A, B \in \mathcal{M}_{n \times n}$ are orthogonal and $t \in \mathcal{R}$ prove that the indicated matrix is orthogonal or show by example that it need not be.

   a. tA, $t \neq 0$ **b.** $A + B$ c. AB

4. Show that if $A \in \mathcal{M}_{2\times 2}$ is orthogonal, then there is a $\theta \in \mathcal{R}$ such that

$$A = \begin{bmatrix} \cos(\theta) & -\sin(\theta) \\ \sin(\theta) & \cos(\theta) \end{bmatrix} \quad \text{or} \quad A = \begin{bmatrix} \cos(\theta) & \sin(\theta) \\ \sin(\theta) & -\cos(\theta) \end{bmatrix}.$$

5. Find A^2 when $A \in \mathcal{M}_{n \times n}$ is orthogonal and symmetric.

6. Show that if $A \in \mathcal{M}_{m \times m}$ and $B \in \mathcal{M}_{n \times n}$ are orthogonal, then so is $\left[\begin{array}{c|c} A & 0 \\ \hline 0 & B \end{array}\right]$.

7. Show that if $\mathcal{A} = \{U_1, \ldots, U_p\}$ and $\mathcal{B} = \{W_1, \ldots, W_p\}$ are orthonormal bases for a subspace of $\mathcal{R}^n$, $A = [U_1 \ \cdots \ U_p]$ and $B = [W_1 \ \cdots \ W_p]$, then $P_{\mathcal{BA}} = B^T A$.

8. Describe all $n \times n$, upper triangular, orthogonal matrices.

9. Suppose $\mathcal{V}$ is a subspace of $\mathcal{R}^n$ and $\mathcal{A} = \{U_1, \ldots, U_p\}$ is an orthonormal basis for $\mathcal{V}$. Show that if X,Y $\in \mathcal{V}$, then
   a. $X \cdot Y = [X]_{\mathcal{A}} \cdot [Y]_{\mathcal{A}}$.
   b. $\|X\| = \|[X]_{\mathcal{A}}\|$.

10. Use Problem 9 to obtain an alternative proof of Theorem 3.17.

11. Find a QR factorization for

    a. $\begin{bmatrix} 1 & 0 \\ 1 & 2 \end{bmatrix}$. b. $\begin{bmatrix} 1 & 0 \\ 1 & 2 \\ 1 & 1 \end{bmatrix}$. c. $\begin{bmatrix} 1 & 0 & 1 \\ 1 & 1 & 1 \\ 0 & 0 & 1 \end{bmatrix}$. **d.** $\begin{bmatrix} 1 & 1 & -1 \\ 1 & 2 & 0 \\ 0 & 1 & 4 \\ 1 & 3 & -2 \end{bmatrix}$.

12. This exercise explores the extent to which the QR factorization is unique. Assume $A \in \mathcal{M}_{n\times p}$ has linearly independent columns and $Q_1R_1 = A = Q_2R_2$, where $Q_1,Q_2 \in \mathcal{M}_{n\times p}$ have orthonormal columns and $R_1,R_2 \in \mathcal{M}_{p\times p}$ are upper triangular and nonsingular. Show that
    a. $Q_2^TQ_2 = I_p = Q_1^TQ_1$.
    b. $Q_2^TQ_1$ and $Q_1^TQ_2$ are upper triangular.
    c. there is a $p \times p$ diagonal matrix, D, such that $Q_1 = Q_2D$ and $R_2 = DR_1$.
    d. the matrix D in part (c) satisfies $|d_{kk}| = 1$, $k = 1, \ldots, p$, and then conclude that $D^{-1} = D$.

3.6
THE CROSS PRODUCT

Consider the problem of finding a $Z \in \mathcal{R}^3$ that is orthogonal to each of two given vectors, $X,Y \in \mathcal{R}^3$. Assuming

$$X = \begin{bmatrix} x_1 \\ x_2 \\ x_3 \end{bmatrix}, \quad Y = \begin{bmatrix} y_1 \\ y_2 \\ y_3 \end{bmatrix}, \quad \text{and} \quad Z = \begin{bmatrix} z_1 \\ z_2 \\ z_3 \end{bmatrix},$$

the object is to determine values of z_1, z_2, and z_3 for which $X \cdot Z = 0 = Y \cdot Z$, that is, to find a solution of the homogeneous system

$$\begin{aligned} x_1z_1 + x_2z_2 + x_3z_3 &= 0 \\ y_1z_1 + y_2z_2 + y_3z_3 &= 0. \end{aligned} \tag{3.14}$$

To that end, multiply the first equation in (3.14) by y_1, the second by x_1, and then subtract the first from the second. The result is

$$(x_1y_2 - x_2y_1)z_2 + (x_1y_3 - x_3y_1)z_3 = 0.$$

Now multiply the first equation in (3.14) by y_2, the second by x_2, and then subtract the first from the second. You get

$$-(x_1y_2 - x_2y_1)z_1 + (x_2y_3 - x_3y_2)z_3 = 0.$$

Direct substitution then confirms that the last two equations are satisfied by

$$z_3 = x_1y_2 - x_2y_1, \qquad z_1 = x_2y_3 - x_3y_2, \qquad z_2 = -(x_1y_3 - x_3y_1).$$

DEFINITION. For X,Y $\in \mathcal{R}^3$, the *cross product of* X *and* Y is $X \times Y = \begin{bmatrix} x_2y_3 - x_3y_2 \\ x_3y_1 - x_1y_3 \\ x_1y_2 - x_2y_1 \end{bmatrix}$.

THEOREM 3.19. If X,Y $\in \mathcal{R}^3$, then X $\times$ Y is orthogonal to both X and Y.

Proof. $X \cdot (X \times Y) = x_1(x_2y_3 - x_3y_2) + x_2(x_3y_1 - x_1y_3) + x_3(x_1y_2 - x_2y_1) = 0$, and similarly, $Y \cdot (X \times Y) = 0$. ■

■ **EXAMPLE 3.18.** If $X = \begin{bmatrix} 2 \\ 1 \\ 3 \end{bmatrix}$ and $Y = \begin{bmatrix} -1 \\ 3 \\ -2 \end{bmatrix}$, then

$$X \times Y = \begin{bmatrix} 1(-2) - 3(3) \\ 3(-1) - 2(-2) \\ 2(3) - 1(-1) \end{bmatrix} = \begin{bmatrix} -11 \\ 1 \\ 7 \end{bmatrix}. \square$$

System (3.14) is always satisfied by $z = 0$, but the interest lies in finding a nontrivial solution. In particular, it would be useful to know under what circumstances X $\times$ Y is nonzero.

THEOREM 3.20. If X,Y $\in \mathcal{R}^3$, then X $\times$ Y = 0 if and only if X and Y are linearly dependent.

Proof. Suppose X $\times$ Y = 0, that is,

$$x_2y_3 - x_3y_2 = 0, \qquad x_3y_1 - x_1y_3 = 0, \qquad x_1y_2 - x_2y_1 = 0. \tag{3.15}$$

Certainly X and Y are linearly dependent when Y = 0, so assume Y $\neq$ 0. Then at least one of y_1, y_2, and y_3, is different from zero. When $y_3 \neq 0$, the first two equations in (3.15) produce

$$x_2 = \left(\frac{x_3}{y_3}\right)y_2 \qquad \text{and} \qquad x_1 = \left(\frac{x_3}{y_3}\right)y_1.$$

Of course,

$$x_3 = \left(\frac{x_3}{y_3}\right)y_3,$$

so X = tY, where $t = x_3/y_3$. Similar considerations show X and Y are linearly dependent when $y_2 \neq 0$ or $y_1 \neq 0$. It is left as an exercise to verify that if X and Y are linearly dependent, then X $\times$ Y = 0. ■

■ **EXAMPLE 3.19.** Suppose X and Y are linearly independent vectors in $\mathcal{R}^3$ and let $\mathcal{V} = \mathcal{S}pan\{X,Y\}$. According to Theorems 3.19 and 3.20, X $\times$ Y is a nonzero vector that is orthogonal to both X and Y, so X $\times$ Y $\in \mathcal{V}^\perp$ (Theorem 3.9). Since $\mathcal{V}^\perp$ has dimension 1, {X $\times$ Y} is a basis for $\mathcal{V}^\perp$. □

Given that X,Y $\in \mathcal{R}^3$ are linearly independent, X $\times$ Y is just one of many nonzero vectors that are orthogonal to both X and Y. So, what makes it especially interesting? One explanation is that the cross product is useful in modeling certain physical concepts such as angular momentum and torque. There is another that is geometric in nature. If θ is the angle between X and Y, then

$$\begin{aligned}
\|X \times Y\|^2 &= (x_2y_3 - x_3y_2)^2 + (x_3y_1 - x_1y_3)^2 + (x_1y_2 - x_2y_1)^2 \\
&= (x_1^2 + x_2^2 + x_3^2)(y_1^2 + y_2^2 + y_3^2) - (x_1y_1 + x_2y_2 + x_3y_3)^2 \\
&= \|X\|^2\|Y\|^2 - (X \cdot Y)^2 \\
&= \|X\|^2\|Y\|^2 - \|X\|^2\|Y\|^2 \cos^2(\theta) \\
&= \|X\|^2\|Y\|^2 \sin^2(\theta),
\end{aligned}$$

so the magnitude of X $\times$ Y is

$$\|X \times Y\| = \|X\|\|Y\|\sin(\theta). \tag{3.16}$$

The expression on the right of (3.16) computes the area of the parallelogram with vertices 0, X, Y, and X + Y (Figure 3.11). That area, however, does not completely determine X $\times$ Y. There are two vectors of that magnitude that are orthogonal to both X and Y, namely X $\times$ Y and $-$(X $\times$ Y). It turns out X $\times$ Y is the one with the feature that the coordinate system for $\mathcal{R}^3$ determined by the basis {X,Y,X $\times$ Y} is "right-handed," a topic discussed in Chapter 5.

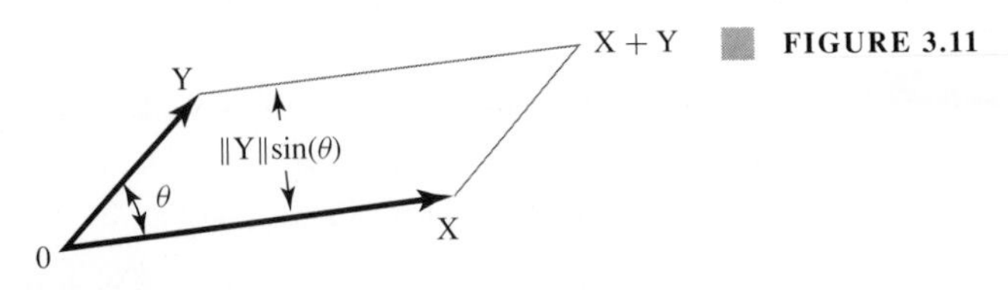

FIGURE 3.11

THEOREM 3.21. If X,Y, Z $\in \mathcal{R}^3$ and $t \in \mathcal{R}$, then.

(a) X $\times$ X = 0.
(b) X $\times$ Y = $-$(Y $\times$ X).
(c) X $\times$ (Y + Z) = X $\times$ Y + X $\times$ Z.
(d) X $\times$ (tY) = t(X $\times$ Y) = (tX) $\times$ Y.

Proof. See Exercise 4. ■

Exercises 3.6

1. Assuming $X = \begin{bmatrix} 1 \\ 2 \\ 3 \end{bmatrix}$, $Y = \begin{bmatrix} -1 \\ 0 \\ 1 \end{bmatrix}$, and $Z = \begin{bmatrix} -2 \\ 1 \\ 3 \end{bmatrix}$, find

a. X $\times$ Y. b. Y $\times$ X. **c.** (2X) $\times$ Y.
d. X $\times$ X. **e.** X $\times$ (Y + Z). f. X $\times$ Y + X $\times$ Z.
g. X $\times$ (Y $\times$ Z). h. (X $\times$ Y) $\times$ Z.

2. Calculate $E_1 \times E_2$, $E_2 \times E_3$, and $E_3 \times E_1$.

3. Show that if X and Y are linearly dependent, then $X \times Y = 0$.

4. Assuming $X, Y \in \mathcal{R}^3$ and $t \in \mathcal{R}$, show that
   a. $X \times Y = -(Y \times X)$ **b.** $X \times (Y + Z) = X \times Y + X \times Z$
   c. $X \times (tY) = t(X \times Y) = (tX) \times Y$.

5. Find the area of the parallelogram with vertices 0, X, Y, and X + Y when

$$\text{a. } X = \begin{bmatrix} 1 \\ 1 \\ 1 \end{bmatrix}, \; Y = \begin{bmatrix} 1 \\ -1 \\ 1 \end{bmatrix}, \quad \text{b. } X = \begin{bmatrix} 0 \\ 2 \\ 3 \end{bmatrix}, \; Y = \begin{bmatrix} 1 \\ -3 \\ 0 \end{bmatrix}, \quad \text{c. } X = \begin{bmatrix} 2 \\ 1 \\ 3 \end{bmatrix}, \; Y = \begin{bmatrix} 1 \\ 3 \\ 2 \end{bmatrix}.$$

6. Show that if U and V are orthogonal unit vectors in $\mathcal{R}^3$, then $\mathcal{A} = \{U, V, U \times V\}$ is an orthonormal basis for $\mathcal{R}^3$.

7. Suppose X, Y, Z are linearly independent vectors in $\mathcal{R}^3$.
   a. Explain why $X \times (Y \times Z)$ must be a linear combination of Y and Z.
   b. Verify that $X \times (Y \times Z) = (X \cdot Z)Y - (X \cdot Y)Z$.

8. If U and V are orthogonal unit vectors in $\mathcal{R}^3$, show that $V \times (U \times V) = U$.

3.7
LEAST-SQUARES APPROXIMATION

In the physical and social sciences one is often presented with the problem of inferring a formal relationship between certain variables based on experimentally obtained data. The simplest scenario involves two variables, say x and y, data points (x_k, y_k), $k = 1, \ldots, n$, and the assumption that there is a line that in some sense best fits the data. That is to say, the idea is to find a first degree polynomial, $f(x) = a + bx$, with the property that $f(x_k)$ provides a "good approximation" to y_k for each $k = 1, \ldots, n$. Figure 3.12 shows a plot of a few data points and one possible approximating line.

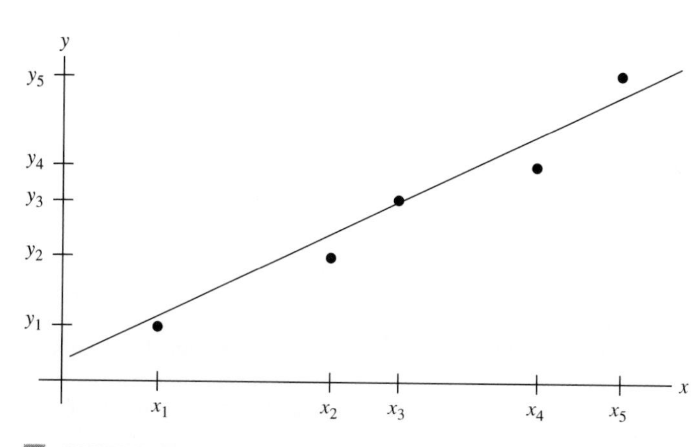

FIGURE 3.12

The quality of the fit is determined by the differences, $y_k - f(x_k)$, $k = 1, \ldots, n$, between the observed and predicted values. One natural approach is to seek a and b (and hence f) so as to minimize $\sum_1^n |y_k - f(x_k)|$. However, for statistical reasons (which won't be explored here), a preferable measure is

$$\sum_1^n [y_k - f(x_k)]^2. \tag{3.17}$$

A choice of f that minimizes (3.17) is called a *least-squares linear fit* to the data, and the resulting approximating line is referred to as the *line of regression of y on x*.

To see that this least-squares problem has a solution, it is helpful to cast the question in a vector context. Setting

$$\mathrm{Y} = \begin{bmatrix} y_1 \\ \vdots \\ y_n \end{bmatrix}, \quad \mathrm{A} = \begin{bmatrix} 1 & x_1 \\ \vdots & \vdots \\ 1 & x_n \end{bmatrix}, \quad \text{and} \quad \mathrm{Z} = \begin{bmatrix} a \\ b \end{bmatrix},$$

(3.17) becomes

$$\sum_1^n [y_k - f(x_k)]^2 = \sum_1^n [y_k - (a + bx_k)]^2 = \|\mathrm{Y} - \mathrm{AZ}\|^2,$$

so the object is to minimize $\|\mathrm{Y} - \mathrm{AZ}\|$ as Z varies over $\mathcal{R}^2$. Observe that $\mathcal{V} = \{\mathrm{AZ} : \mathrm{Z} \in \mathcal{R}^2\} = \mathcal{C}(\mathrm{A})$ is a subspace of $\mathcal{R}^n$ and

$$\min\{\|\mathrm{Y} - \mathrm{AZ}\| : \mathrm{Z} \in \mathcal{R}^2\} = \min\{\|\mathrm{Y} - \mathrm{V}\| : \mathrm{V} \in \mathcal{V}\}.$$

According to Theorem 3.14, there is a unique element of $\mathcal{V}$ that is closest to Y, namely $\mathrm{Proj}_{\mathcal{V}}(\mathrm{Y})$. Thus, a least-squares fit is obtained by solving

$$\mathrm{AZ} = \mathrm{Proj}_{\mathcal{V}}(\mathrm{Y}).$$

This system is consistent because $\mathrm{Proj}_{\mathcal{V}}(\mathrm{Y}) \in \mathcal{V} = \{\mathrm{AZ} : \mathrm{Z} \in \mathcal{R}^2\}$. It has a unique solution when the columns of A are linearly independent, a condition that occurs when at least two of $x_1, \ldots, x_n$ are distinct. These findings are summarized in the next theorem.

THEOREM 3.22. If (x_k, y_k), $1 \le k \le n$, are data points with at least two of $x_1, \ldots, x_n$ distinct, then there are unique real numbers a_0, b_0 with the property that

$$\sum_1^n [y_k - (a_0 + b_0 x_k)]^2 \le \sum_1^n [y_k - (a + bx_k)]^2$$

for all $a, b \in \mathcal{R}$. Moreover, if

$$\mathrm{Y} = \begin{bmatrix} y_1 \\ \vdots \\ y_n \end{bmatrix}, \quad \mathrm{A} = \begin{bmatrix} 1 & x_1 \\ \vdots & \vdots \\ 1 & x_n \end{bmatrix}, \quad \text{and} \quad \mathrm{Z} = \begin{bmatrix} a \\ b \end{bmatrix},$$

then $\begin{bmatrix} a_0 \\ b_0 \end{bmatrix}$ is the solution of $\mathrm{AZ} = \mathrm{Proj}_{\mathcal{V}}(\mathrm{Y})$, where $\mathcal{V} = \mathcal{C}(\mathrm{A})$.

Theorem 3.22 indicates that $\mathrm{Z}_0 = \begin{bmatrix} a_0 \\ b_0 \end{bmatrix}$ can be obtained by projecting Y onto $\mathcal{V}$ and then solving $\mathrm{AZ} = \mathrm{Proj}_{\mathcal{V}}(\mathrm{Y})$, but there is a more direct route. Note that

$$\mathrm{Y} - \mathrm{AZ}_0 = \mathrm{Y} - \mathrm{Proj}_{\mathcal{V}}(\mathrm{Y}) = \mathrm{Proj}_{\mathcal{V}^\perp}(\mathrm{Y}) \in \mathcal{V}^\perp.$$

For each $Z \in \mathcal{R}^2$, $AZ \in \mathcal{V}$, so

$$0 = (AZ) \cdot (Y - AZ_0) = (AZ)^T(Y - AZ_0) = Z^T(A^TY - A^TAZ_0)$$
$$= Z \cdot (A^TY - A^TAZ_0).$$

Thus, $A^TY - A^TAZ_0$, being orthogonal to every element of $\mathcal{R}^2$, is 0, that is, Z_0 is the solution of

$$(A^TA)Z = A^TY. \tag{3.18}$$

The equations in system (3.18) are called the *normal equations* for Z_0.

■ **EXAMPLE 3.20.** Consider data points (1,2), (2,1), (3,1), and (4, −1). In the terminology of Theorem 3.22,

$$Y = \begin{bmatrix} 2 \\ 1 \\ 1 \\ -1 \end{bmatrix} \quad \text{and} \quad A = \begin{bmatrix} 1 & 1 \\ 1 & 2 \\ 1 & 3 \\ 1 & 4 \end{bmatrix},$$

so

$$A^TA = \begin{bmatrix} 1 & 1 & 1 & 1 \\ 1 & 2 & 3 & 4 \end{bmatrix} \begin{bmatrix} 1 & 1 \\ 1 & 2 \\ 1 & 3 \\ 1 & 4 \end{bmatrix} = \begin{bmatrix} 4 & 10 \\ 10 & 30 \end{bmatrix},$$

$$A^TY = \begin{bmatrix} 1 & 1 & 1 & 1 \\ 1 & 2 & 3 & 4 \end{bmatrix} \begin{bmatrix} 2 \\ 1 \\ 1 \\ -1 \end{bmatrix} = \begin{bmatrix} 3 \\ 3 \end{bmatrix},$$

and (3.18) becomes $\begin{bmatrix} 4 & 10 \\ 10 & 30 \end{bmatrix}\begin{bmatrix} a \\ b \end{bmatrix} = \begin{bmatrix} 3 \\ 3 \end{bmatrix}$. The coefficient matrix of this system is nonsingular, and therefore

$$\begin{bmatrix} a \\ b \end{bmatrix} = \begin{bmatrix} 4 & 10 \\ 10 & 30 \end{bmatrix}^{-1} \begin{bmatrix} 3 \\ 3 \end{bmatrix} = \frac{1}{20}\begin{bmatrix} 30 & -10 \\ -10 & 4 \end{bmatrix}\begin{bmatrix} 3 \\ 3 \end{bmatrix} = \frac{1}{10}\begin{bmatrix} 30 \\ -9 \end{bmatrix}.$$

Thus, the least-squares linear fit is $f(x) = 3 - \frac{9}{10}x$ (Figure 3.13). □

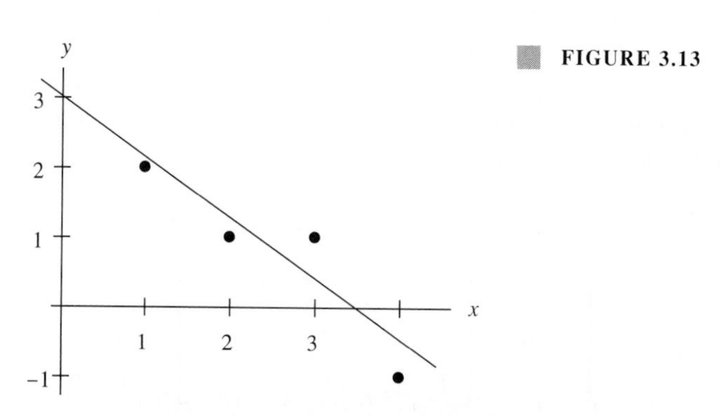

FIGURE 3.13

■ **EXAMPLE 3.21.** In an attempt to predict student success in the scientific calculus class based on performance on a calculus readiness test, the Advising Center at a certain college studies the data in the accompanying table, which relates test scores to calculus grades for a small sample of students. Grades are measured on a four-point scale and possible test scores range from 0 to 25.

Test Score	4	8	10	12	14	18	22	24
Calculus Grade	0	1	2	1	2	3	4	3

It seems reasonable to expect a linear relationship between these variables, so a least-squares linear fit is sought for the data points

$$(4,0), (8,1), (10,2), (12,1), (14,2), (18,3), (22,4), (24,3).$$

Setting

$$\mathrm{Y} = \begin{bmatrix} 0 \\ 1 \\ 2 \\ 1 \\ 2 \\ 3 \\ 4 \\ 3 \end{bmatrix} \quad \text{and} \quad \mathrm{A} = \begin{bmatrix} 1 & 4 \\ 1 & 8 \\ 1 & 10 \\ 1 & 12 \\ 1 & 14 \\ 1 & 18 \\ 1 & 22 \\ 1 & 24 \end{bmatrix},$$

yields

$$\mathrm{A}^T\mathrm{A} = \begin{bmatrix} 8 & 112 \\ 112 & 1904 \end{bmatrix} \quad \text{and} \quad \mathrm{A}^T\mathrm{Y} = \begin{bmatrix} 16 \\ 282 \end{bmatrix}.$$

The normal equations

$$\begin{bmatrix} 8 & 112 \\ 112 & 1904 \end{bmatrix} \begin{bmatrix} a \\ b \end{bmatrix} = \begin{bmatrix} 16 \\ 282 \end{bmatrix}$$

have the solution

$$\begin{bmatrix} a \\ b \end{bmatrix} = \begin{bmatrix} 8 & 112 \\ 112 & 1904 \end{bmatrix}^{-1} \begin{bmatrix} 16 \\ 282 \end{bmatrix} \approx \begin{bmatrix} -0.417 \\ 0.173 \end{bmatrix}.$$

Letting x and y denote the test score and calculus grade respectively, the line of regression of y on x is $y = -0.417 + 0.173x$. Based on this analysis, what test score predicts a grade of C or better in the calculus course? □

Sometimes theoretical considerations indicate that the curve to be fitted to the data should be of some form other than a line. Economists, for example, often use a polynomial of degree 3 to model total cost associated with production of a commodity. Consider the problem of fitting a polynomial of degree q to data points $(x_1, y_1), \ldots,$ (x_n, y_n), that is, assume the approximating function f is

$$y = f(x) = c_0 + c_1 x + \cdots + c_q x^q.$$

Set

$$\mathrm{Y} = \begin{bmatrix} y_1 \\ \vdots \\ y_n \end{bmatrix}, \quad \mathrm{A} = \begin{bmatrix} 1 & x_1 & x_1^2 & \cdots & x_1^q \\ \vdots & \vdots & \vdots & & \vdots \\ 1 & x_n & x_n^2 & \cdots & x_n^q \end{bmatrix}, \quad \text{and} \quad \mathrm{Z} = \begin{bmatrix} c_0 \\ \vdots \\ c_q \end{bmatrix},$$

and observe that

$$\sum_{1}^{n} [y_k - f(x_k)]^2 = \sum_{1}^{n} [y_k - (c_0 + c_1 x_k + \cdots + c_q x_k^q)]^2 = \|\mathrm{Y} - \mathrm{AZ}\|^2. \quad (3.19)$$

As long as the least-squares criteria is used as the measure of fit, the same sort of analysis as used to establish Theorem 3.22 shows that the Z that minimizes (3.19) is the solution of

$$(\mathrm{A}^T\mathrm{A})\mathrm{Z} = \mathrm{A}^T\mathrm{Y}.$$

■ **EXAMPLE 3.22.** Consider fitting a parabola, $y = a + bx + cx^2$, to data points $(-1,3)$, $(0,1)$, $(1,-1)$, and $(2,2)$. In the above notation,

$$\mathrm{Y} = \begin{bmatrix} 3 \\ 1 \\ -1 \\ 2 \end{bmatrix}, \quad \mathrm{A} = \begin{bmatrix} 1 & x_1 & x_1^2 \\ \vdots & \vdots & \vdots \\ 1 & x_4 & x_4^2 \end{bmatrix} = \begin{bmatrix} 1 & -1 & 1 \\ 1 & 0 & 0 \\ 1 & 1 & 1 \\ 1 & 2 & 4 \end{bmatrix}, \quad \text{and} \quad \mathrm{Z} = \begin{bmatrix} a \\ b \\ c \end{bmatrix},$$

so

$$\mathrm{A}^T\mathrm{A} = \begin{bmatrix} 1 & 1 & 1 & 1 \\ -1 & 0 & 1 & 2 \\ 1 & 0 & 1 & 4 \end{bmatrix} \begin{bmatrix} 1 & -1 & 1 \\ 1 & 0 & 0 \\ 1 & 1 & 1 \\ 1 & 2 & 4 \end{bmatrix} = \begin{bmatrix} 4 & 2 & 6 \\ 2 & 6 & 8 \\ 6 & 8 & 18 \end{bmatrix},$$

$$\mathrm{A}^T\mathrm{Y} = \begin{bmatrix} 1 & 1 & 1 & 1 \\ -1 & 0 & 1 & 2 \\ 1 & 0 & 1 & 4 \end{bmatrix} \begin{bmatrix} 3 \\ 1 \\ -1 \\ 2 \end{bmatrix} = \begin{bmatrix} 5 \\ 0 \\ 10 \end{bmatrix},$$

and the normal equations read

$$\begin{bmatrix} 4 & 2 & 6 \\ 2 & 6 & 8 \\ 6 & 8 & 18 \end{bmatrix} \begin{bmatrix} a \\ b \\ c \end{bmatrix} = \begin{bmatrix} 5 \\ 0 \\ 10 \end{bmatrix}.$$

Thus,

$$\begin{bmatrix} a \\ b \\ c \end{bmatrix} = \begin{bmatrix} 4 & 2 & 6 \\ 2 & 6 & 8 \\ 6 & 8 & 18 \end{bmatrix}^{-1} \begin{bmatrix} 5 \\ 0 \\ 10 \end{bmatrix} = \frac{1}{20} \begin{bmatrix} 11 & 3 & -5 \\ 3 & 9 & -5 \\ -5 & -5 & 5 \end{bmatrix} \begin{bmatrix} 5 \\ 0 \\ 10 \end{bmatrix} = \frac{1}{4} \begin{bmatrix} 1 \\ -7 \\ 5 \end{bmatrix},$$

and

$$f(x) = \tfrac{1}{4} - \tfrac{7}{4}x + \tfrac{5}{4}x^2$$

is the desired approximation. □

LEMMA. For $\mathrm{A} \in \mathcal{M}_{m\times n}$, $\mathcal{C}(\mathrm{A})^\perp = \mathcal{N}(\mathrm{A}^T)$ and $\mathcal{N}(\mathrm{A})^\perp = \mathcal{C}(\mathrm{A}^T)$.

Proof. The first conclusion is just a restatement of Theorem 3.10. The second follows from the first by substituting A^T for A and taking the orthogonal complement of each side. ■

THEOREM 3.23. If $\mathrm{A} \in \mathcal{M}_{m\times n}$, then

(1) $\mathcal{N}(\mathrm{A}^T\mathrm{A}) = \mathcal{N}(\mathrm{A})$, and
(2) $\mathcal{C}(\mathrm{A}^T\mathrm{A}) = \mathcal{C}(\mathrm{A}^T)$.

Proof.
(1) $AX = 0$ implies $A^TAX = A^T0 = 0$, so $\mathcal{N}(A) \subseteq \mathcal{N}(A^TA)$. On the other hand, $A^TAX = 0$ implies $X^TA^TAX = 0$, so $\|AX\|^2 = (AX)^T(AX) = X^TA^TAX = 0$, and therefore $AX = 0$. Thus, $\mathcal{N}(A^TA) \subseteq \mathcal{N}(A)$.
(2) If $Y \in \mathcal{C}(A^TA)$, then $Y = A^TAX$ for some $X \in \mathcal{R}^n$, so $Y = A^TW$, where $W = AX \in \mathcal{R}^m$. Thus, $Y \in \mathcal{C}(A^T)$. This establishes that $\mathcal{C}(A^TA) \subseteq \mathcal{C}(A^T)$. Moreover,

$$\begin{aligned} \dim \mathcal{C}(A^TA) &= n - \dim \mathcal{N}(A^TA) && \text{(rank-nullity Theorem)} \\ &= n - \dim \mathcal{N}(A) && \text{(part (1) of the present Theorem)} \\ &= \dim \mathcal{N}(A)^{\perp} && (\mathcal{R}^n = \mathcal{N}(A) \oplus \mathcal{N}(A)^{\perp}) \\ &= \dim \mathcal{C}(A^T), && \text{(preceding Lemma)} \end{aligned}$$

so $\mathcal{C}(A^TA) = \mathcal{C}(A^T)$. ■

Given $A \in \mathcal{M}_{m\times n}$, A^TA is a square matrix of size n, and by Theorems 3.23 and 2.31,

$$\text{rank}(A^TA) = \text{rank}(A^T) = \text{rank}(A).$$

If A has linearly independent columns, then A and A^TA both have rank n. In particular, A^TA is nonsingular. Under these circumstances, the coefficient matrix in the normal equations is invertible, and the solution to the least-squares linear fit problem can be written explicitly as

$$Z_0 = (A^TA)^{-1}A^TY.$$

Exercises 3.7

1. Find the least-squares linear fit to the indicated data. Plot the data and sketch the line of regression of y on x.
   a. (1,1), (2,1), (3,2)
   b. (0.5,1), (1.5,2), (2,2.5), (3,3)
   c. (−1,3), (0,2.5), (1,2), (2,1.6), (3,0.08)
   d. (−2,3), (−1, 2.5), (0,1.2), (1,−0.1), (2,−0.8)

2. The least-squares method produces infinitely many approximating lines for the data (2,1), (2,3). Find the equations of those lines and show that they all have a point in common.

3. Find the parabola that best fits the given data in the sense of least-squares approximation.
   a. (−2,1), (−1,0), (0,−1), (1,0)
   b. (−2,−1), (−1,1), (0,1), (1,0), (2,−1)

4. Find the least-squares cubic fit to the data points (−1,1), (0,0), (1,1) and (−2,0).

5. Assume $A = \begin{bmatrix} 1 & x_1 \\ \vdots & \vdots \\ 1 & x_n \end{bmatrix}$, $Y = \begin{bmatrix} y_1 \\ \vdots \\ y_n \end{bmatrix}$, and $X = \begin{bmatrix} x_1 \\ \vdots \\ x_n \end{bmatrix}$.

a. Show that the normal equations associated with (x_k, y_k), $k = 1, \ldots, n$, are

$$\begin{bmatrix} n & \sum_1^n x_k \\ \sum_1^n x_k & \|\mathrm{X}\|^2 \end{bmatrix} \begin{bmatrix} a \\ b \end{bmatrix} = \begin{bmatrix} \sum_1^n y_k \\ \mathrm{X} \cdot \mathrm{Y} \end{bmatrix}.$$

b. Assuming at least two of $x_1, \ldots, x_n$ are distinct, show that the line of regression of y on x has equation $y = a + bx$, where

$$a = \frac{\|\mathrm{X}\|^2 \sum_1^n y_k - (\mathrm{X} \cdot \mathrm{Y}) \sum_1^n x_k}{n\|\mathrm{X}\|^2 - \left(\sum_1^n x_k\right)^2} \quad \text{and} \quad b = \frac{n(\mathrm{X} \cdot \mathrm{Y}) - \sum_1^n y_k \sum_1^n x_k}{n\|\mathrm{X}\|^2 - \left(\sum_1^n x_k\right)^2}.$$

c. Show that the line of regression passes through $(\bar{x}, \bar{y})$, where

$$\bar{x} = \frac{1}{n} \sum_1^n x_k \quad \text{and} \quad \bar{y} = \frac{1}{n} \sum_1^n y_k.$$

6. The table gives values of the U.S. Consumer Price Index (CPI) for selected years since 1980.

Year	1980	1985	1990	1993	1994	1995
CPI	82.4	107.6	130.7	144.5	148.2	152.4

†

Letting x be the number of years since 1980 and y the CPI, find the line of regression of y on x. What CPI does this model predict for the year 2000?

7. Consider the following mean SAT mathematics scores in the state of Ohio.

Year	1987	1993	1994	1995	1996
Score	521	527	531	535	535

†

Based on a least-squares linear approximation to the data, predict the mean score of the year 2000.

8. The winning times in seconds at recent Summer Olympic Games for the Mens 100-Meter Run were

Year	1972	1976	1980	1984	1988	1992	1996
Time	10.14	10.06	10.25	9.99	9.92	9.96	9.84

†

What does a least-squares linear fit predict as the winning time at the 2000 Olympic Games?

9. Let $\mathrm{A} = \begin{bmatrix} 1 & x_1 & x_1^2 & \cdots & x_1^q \\ \vdots & \vdots & \vdots & & \vdots \\ 1 & x_n & x_n^2 & \cdots & x_n^q \end{bmatrix}$ and assume $t_0, \ldots, t_q \in \mathcal{R}$.

---

† Reprinted with permission from The World Almanac and Book of Facts 1997.

a. Show that $\sum_0^q t_k \operatorname{Col}_k(A) = 0$ if and only if $p(x) = t_0 + t_1 x + \cdots + t_q x^q$ has $x_1, \ldots,$ x_n as roots.
b. Show that if the sequence $x_1, \ldots, x_n$ contains $q + 1$ distinct values, then the columns of A are linearly independent.

10. Suppose $A \in \mathcal{M}_{m\times n}$ and $B \in \mathcal{R}^m$. Each solution of AX = B is a solution of $A^T AX = A^T B$, but there may be X's that satisfy the latter system without satisfying the former. Solutions of $A^T AX = A^T B$ are called *least-squares solutions* of AX = B.
    a. Show that $A^T AX = A^T B$ is consistent for every $B \in \mathcal{R}^m$, that is, that AX = B always has a least-squares solution.
    b. If A has linearly independent columns, then by Theorem 3.18 there is a $Q \in \mathcal{M}_{m\times n}$ with orthonormal columns and an upper triangular, nonsingular $R \in \mathcal{M}_{n\times n}$ such that A = QR. Show that under these circumstances, the least-squares solution of AX = B is $X = R^{-1}Q^T B$.

4

Linear Transformations

4.1 INTRODUCTION

Suppose $A \in \mathcal{M}_{m\times n}$ and consider the matrix equation

$$Y = AX. \tag{4.1}$$

If you treat $Y \in \mathcal{R}^m$ as fixed and $X \in \mathcal{R}^n$ as variable, then (4.1) describes a system of linear equations. From that perspective, interest tends to focus on the problem of solving the system (given Y, find X). Alternatively, you can think of (4.1) as a formula for transforming a given X into a corresponding Y. With that interpretation, it describes a function with domain $\mathcal{R}^n$ whose values lie in $\mathcal{R}^m$. Let T be the name of the function, that is,

$$Y = T(X) = AX.$$

If $X_1, X_2, X \in \mathcal{R}^n$ and $t \in \mathcal{R}$, then

$$T(X_1 + X_2) = A(X_1 + X_2) = AX_1 + AX_2 = T(X_1) + T(X_2)$$

and

$$T(tX) = A(tX) = t(AX) = tT(X).$$

This simple interaction between the function and the operations of addition and scalar multiplication in $\mathcal{R}^n$ and $\mathcal{R}^m$ allows T to be thoroughly analyzed using strictly algebraic tools. The business of the present chapter is to see how that is done. In the process, familiar issues regarding linear systems will acquire new interpretations.

4.2 LINEAR TRANSFORMATIONS

DEFINITION. Assume T is a function with domain $\mathcal{R}^n$ and values in $\mathcal{R}^m$ (written $T : \mathcal{R}^n \to \mathcal{R}^m$). If

$$T(\mathrm{X}_1 + \mathrm{X}_2) = T(\mathrm{X}_1) + T(\mathrm{X}_2), \tag{4.2}$$

for all $\mathrm{X}_1, \mathrm{X}_2 \in \mathcal{R}^n$, and

$$T(t\mathrm{X}) = tT(\mathrm{X}), \tag{4.3}$$

for every $\mathrm{X} \in \mathcal{R}^n$ and $t \in \mathcal{R}$, then T is said to be *linear*. Linear functions are also known as *linear transformations*.

As indicated in the introduction, the prototypes for this notion are the functions $T : \mathcal{R}^n \to \mathcal{R}^m$ defined by $T(\mathrm{X}) = \mathrm{AX}$, $\mathrm{A} \in \mathcal{M}_{m\times n}$. The next few examples show that the scope of the concept includes a variety of commonly encountered geometric operations.

EXAMPLE 4.1. For $\alpha \in \mathcal{R}$, let $T_\alpha : \mathcal{R}^2 \to \mathcal{R}^2$ be the function that rotates the plane counterclockwise about zero through an angle of α radians. Figure 4.1 illustrates the case when $\alpha > 0$. Geometrically it is easy to see that T_α is linear. Indeed, Figure 4.2(a) indicates that you can add two vectors and then rotate the result or rotate the two vectors and then add the results, and the outcome is the same. Thus, T_α satisfies (4.2). As for (4.3), whether you scale X and then rotate or rotate X and then scale, the effect is the same. Figure 4.2(b) illustrates the case when $t > 1$.

Rotations about zero are conveniently described algebraically by expressing the points of $\mathcal{R}^2$ in polar coordinates. If

$$\mathrm{X} = \begin{bmatrix} x_1 \\ x_2 \end{bmatrix} = \begin{bmatrix} r\cos(\theta) \\ r\sin(\theta) \end{bmatrix},$$

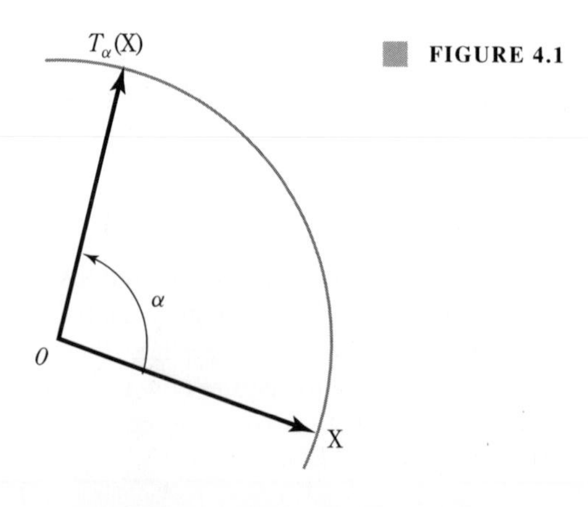

FIGURE 4.1

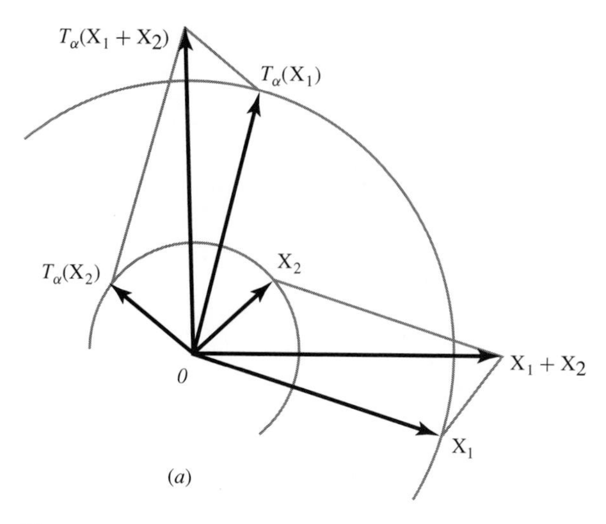

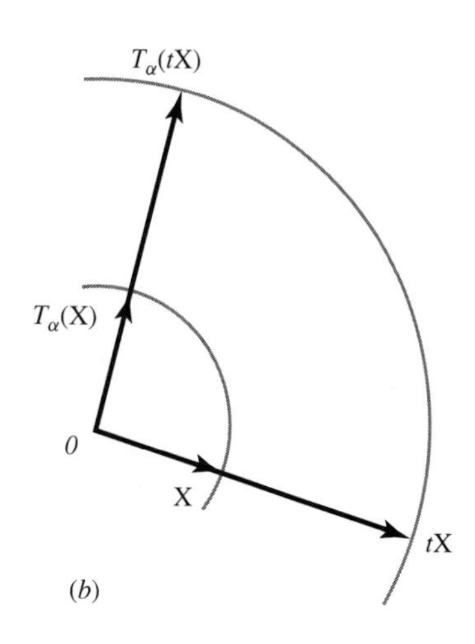

FIGURE 4.2

then

$$\begin{aligned} T_\alpha(X) &= \begin{bmatrix} r\cos(\theta+\alpha) \\ r\sin(\theta+\alpha) \end{bmatrix} \\ &= \begin{bmatrix} r\cos(\theta)\cos(\alpha) - r\sin(\theta)\sin(\alpha) \\ r\sin(\theta)\cos(\alpha) + r\cos(\theta)\sin(\alpha) \end{bmatrix} \\ &= \begin{bmatrix} x_1\cos(\alpha) - x_2\sin(\alpha) \\ x_1\sin(\alpha) + x_2\cos(\alpha) \end{bmatrix} \\ &= \begin{bmatrix} \cos(\alpha) & -\sin(\alpha) \\ \sin(\alpha) & \cos(\alpha) \end{bmatrix} \begin{bmatrix} x_1 \\ x_2 \end{bmatrix}. \end{aligned}$$

Thus, the values of T_α are generated by the matrix equation $T_\alpha(X) = AX$, where

$$A = \begin{bmatrix} \cos(\alpha) & -\sin(\alpha) \\ \sin(\alpha) & \cos(\alpha) \end{bmatrix}. \square$$

■ **EXAMPLE 4.2.** Let $T : \mathcal{R}^3 \to \mathcal{R}^3$ be the function that assigns to each $X \in \mathcal{R}^3$ its orthogonal projection onto the x_1x_2 plane. Figure 4.3 indicates that the projection of the sum of two vectors is the sum of their projections, so T satisfies (4.2). Moreover, scaling X and then projecting has the same effect as projecting X and then scaling, so T satisfies (4.3). Observe that

$$T(X) = T\left(\begin{bmatrix} x_1 \\ x_2 \\ x_3 \end{bmatrix}\right) = \begin{bmatrix} x_1 \\ x_2 \\ 0 \end{bmatrix} = \begin{bmatrix} 1 & 0 & 0 \\ 0 & 1 & 0 \\ 0 & 0 & 0 \end{bmatrix} X. \square$$

■ **EXAMPLE 4.3.** Consider the function $T : \mathcal{R}^2 \to \mathcal{R}^2$ that sends each point to its reflection in the line $\ell = \{X \in \mathcal{R}^2 : x_1 = x_2\}$. If $X \in \ell$, then $T(X) = X$, and otherwise ℓ is the perpendicular bisector of the line segment joining $T(X)$ and X. Once again it is easy to see geometrically that T is linear (Figure 4.4). Moreover,

$$T(X) = T\left(\begin{bmatrix} x_1 \\ x_2 \end{bmatrix}\right) = \begin{bmatrix} x_2 \\ x_1 \end{bmatrix} = \begin{bmatrix} 0 & 1 \\ 1 & 0 \end{bmatrix} X. \square$$

■ **EXAMPLE 4.4.** Define $\mathrm{Id} : \mathcal{R}^n \to \mathcal{R}^n$ by $\mathrm{Id}(X) = X$ (Id stands for identity). If $X_1, X_2, X \in \mathcal{R}^n$ and $t \in \mathcal{R}$, then $\mathrm{Id}(X_1 + X_2) = X_1 + X_2 = \mathrm{Id}(X_1) + \mathrm{Id}(X_2)$ and $\mathrm{Id}(tX) = tX = t\mathrm{Id}(X)$, so Id is linear. Observe that $\mathrm{Id}(X) = I_nX$. $\square$

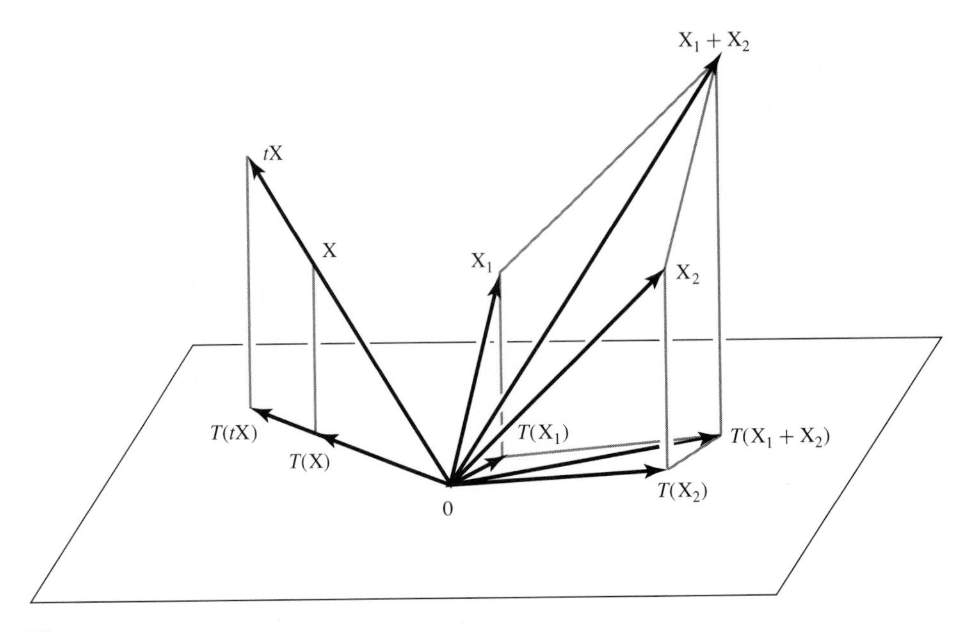

FIGURE 4.3

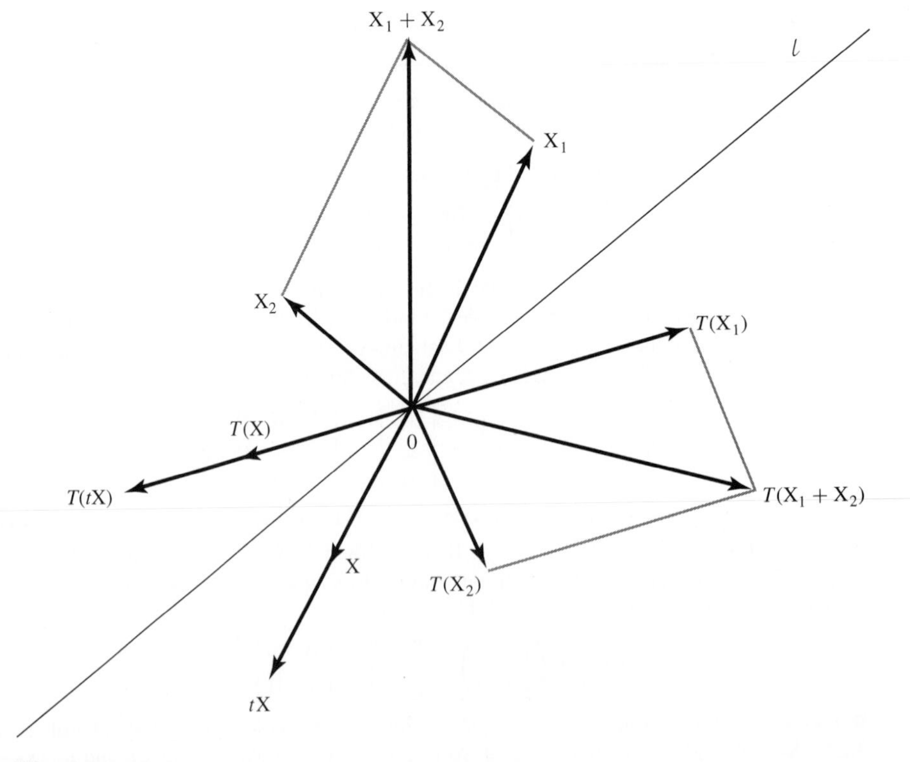

FIGURE 4.4

If $T : \mathcal{R}^n \to \mathcal{R}^m$ is linear, $X_1, X_2 \in \mathcal{R}^n$, and $s,t \in \mathcal{R}$, then (4.2) and (4.3) imply

$$T(sX_1 + tX_2) = T(sX_1) + T(tX_2) = sT(X_1) + tT(X_2).$$

More generally, given $X_1, \ldots, X_p \in \mathcal{R}^n$ and $t_1, \ldots, t_p \in \mathcal{R}$,

$$\boxed{T(t_1X_1 + \cdots + t_pX_p) = t_1T(X_1) + \cdots + t_pT(X_p)}\,. \tag{4.4}$$

Equation (4.4) is the key to understanding the workings of a linear function. It has as a direct consequence the fact that every linear $T : \mathcal{R}^n \to \mathcal{R}^m$ can be written as $T(X) = AX$ for an appropriate choice of the matrix A.

THEOREM 4.1. If $T : \mathcal{R}^n \to \mathcal{R}^m$ is linear, then there is a unique $A \in \mathcal{M}_{m\times n}$ such that

$$T(X) = AX. \tag{4.5}$$

In fact, $A = [T(E_1) \quad \cdots \quad T(E_n)]$.

Proof. If $X \in \mathcal{R}^n$, then $X = x_1E_1 + \cdots + x_nE_n$, so by (4.4),

$$\begin{aligned} T(X) &= T(x_1E_1 + \cdots + x_nE_n) = x_1T(E_1) + \cdots + x_nT(E_n) \\ &= [T(E_1) \quad \cdots \quad T(E_n)]X. \end{aligned}$$

To establish the uniqueness claim, suppose A and B are $m\times n$ matrices with the property that $AX = T(X) = BX$ for each $X \in \mathcal{R}^n$. Setting $X = E_j$ yields

$$\mathrm{Col}_j(A) = AE_j = T(E_j) = BE_j = \mathrm{Col}_j(B),$$

$j = 1, \ldots, n$, so $A = B$. ■

Thanks to Theorem 4.1, linear transformations $T : \mathcal{R}^n \to \mathcal{R}^m$ are easy to recognize. Since

$$T(X) = AX = \begin{bmatrix} a_{11}x_1 + \cdots + a_{1n}x_n \\ \vdots \\ a_{m1}x_1 + \cdots + a_{mn}x_n \end{bmatrix},$$

each coordinate of $T(X)$ is just a linear combination of the coordinates of X.

■ **EXAMPLE 4.5.** Consider $T : \mathcal{R}^2 \to \mathcal{R}^2$ given by

$$T\left(\begin{bmatrix} x_1 \\ x_2 \end{bmatrix}\right) = \begin{bmatrix} 2x_1 + 3x_2 \\ x_1^2 - x_2 \end{bmatrix}.$$

The second coordinate of $T(X)$ is not a linear combination of x_1 and x_2, so T is not linear. In fact, if $X = \begin{bmatrix} x_1 \\ x_2 \end{bmatrix}$ and $U = \begin{bmatrix} u_1 \\ u_2 \end{bmatrix}$, then

$$T(X+U) = \begin{bmatrix} 2(x_1+u_1) + 3(x_2+u_2) \\ (x_1+u_1)^2 - (x_2+u_2) \end{bmatrix} = \begin{bmatrix} (2x_1+3x_2) + (2u_1+3u_2) \\ (x_1^2 - x_2) + (u_1^2 - u_2) + 2x_1u_1 \end{bmatrix}$$

and

$$T(X) + T(U) = \begin{bmatrix} 2x_1 + 3x_2 \\ x_1^2 - x_2 \end{bmatrix} + \begin{bmatrix} 2u_1 + 3u_2 \\ u_1^2 - u_2 \end{bmatrix},$$

so $T(X+U) \neq T(X) + T(U)$ unless $x_1 = 0$ or $u_1 = 0$. Moreover, for $t \in \mathcal{R}$,

$$T(tX) = T\left(\begin{bmatrix} tx_1 \\ tx_2 \end{bmatrix}\right) = \begin{bmatrix} 2(tx_1) + 3(tx_2) \\ (tx_1)^2 - (tx_2) \end{bmatrix} = t\begin{bmatrix} 2x_1 + 3x_2 \\ tx_1^2 - x_2 \end{bmatrix},$$

whereas

$$tT(\mathrm{X}) = t\begin{bmatrix} 2x_1 + 3x_2 \\ x_1^2 - x_2 \end{bmatrix},$$

so $T(t\mathrm{X}) = tT(\mathrm{X})$ only when $t = 0$, $t = 1$, or $x_1 = 0$. □

The matrix A in Theorem 4.1 is determined by the values of T at the usual basis vectors. Those n pieces of information generate $T(\mathrm{X})$ for any $\mathrm{X} \in \mathcal{R}^n$ via matrix multiplication. In fact, given any basis $\{\mathrm{X}_1, \ldots, \mathrm{X}_n\}$ for $\mathcal{R}^n$, each X can be written as $\mathrm{X} = t_1\mathrm{X}_1 + \cdots + t_n\mathrm{X}_n$ for appropriate scalars $t_1, \ldots, t_n$, so

$$T(\mathrm{X}) = t_1 T(\mathrm{X}_1) + \cdots + t_n T(\mathrm{X}_n).$$

Since the value at X is expressed in terms of the values at $\mathrm{X}_1, \ldots, \mathrm{X}_n$, T is completely determined by its action on the basis vectors.

■ **EXAMPLE 4.6.** Let $T : \mathcal{R}^3 \to \mathcal{R}^3$ be reflection in the x_1x_3 plane. As in Example 4.3, it is easy to see geometrically that T is linear. Observe that $T(\mathrm{E}_1) = \mathrm{E}_1$, $T(\mathrm{E}_2) = -\mathrm{E}_2$ and $T(\mathrm{E}_3) = \mathrm{E}_3$. Thus, Theorem 4.1 gives

$$T(\mathrm{X}) = [\mathrm{E}_1 \quad -\mathrm{E}_2 \quad \mathrm{E}_3]\mathrm{X} = \begin{bmatrix} 1 & 0 & 0 \\ 0 & -1 & 0 \\ 0 & 0 & 1 \end{bmatrix}\begin{bmatrix} x_1 \\ x_2 \\ x_3 \end{bmatrix} = \begin{bmatrix} x_1 \\ -x_2 \\ x_3 \end{bmatrix}. \square$$

THEOREM 4.2. If $T : \mathcal{R}^n \to \mathcal{R}^m$ is linear, then $T(0) = 0$.

Proof. Since $T(\mathrm{X}) = \mathrm{AX}$, for some $\mathrm{A} \in \mathcal{M}_{m\times n}$, $T(0) = \mathrm{A}0 = 0$. ■

■ **EXAMPLE 4.7.** Assume $r,s \in \mathcal{R}$ and $\{a_k\}_0^\infty$ is a sequence of real numbers such that

$$a_{k+2} = ra_{k+1} + sa_k, \tag{4.6}$$

$k = 0, 1, \ldots$. Since $T : \mathcal{R}^2 \to \mathcal{R}$ given by $T(\mathrm{X}) = rx_1 + sx_2$ is linear and

$$a_{k+2} = T\left(\begin{bmatrix} a_{k+1} \\ a_k \end{bmatrix}\right), \quad k \geq 0,$$

(4.6) is called a *two-term linear recurrence (or difference) equation*. Note that (4.6) places no restriction on a_0 and a_1, which are called the *initial data*, but those two values uniquely determine the rest of the sequence.

When $r = s = 1$, $a_0 = 0$, and $a_1 = 1$, (4.6) generates the celebrated Fibonacci sequence, whose terms are 0, 1, 1, 2, 3, 5, 8, 13, ... It was first studied by Fibonacci in the thirteenth century, and has since played a role in the description of a surprising number of natural phenomena.

Solving (4.6) amounts to finding an explicit formula for a_k in terms of k and the initial data. One approach using matrix methods is discussed in Chapter 6. □

Exercises 4.2

1. Which functions are linear? If linear, find A such that $T(\mathrm{X}) = \mathrm{AX}$.

a. $T : \mathcal{R}^3 \to \mathcal{R}^2$, $T\left(\begin{bmatrix} x_1 \\ x_2 \\ x_3 \end{bmatrix}\right) = \begin{bmatrix} 2x_1 - 3x_2 + x_3 \\ 4x_3 - x_1 \end{bmatrix}$

b. $T : \mathcal{R}^2 \to \mathcal{R}^3, T\left(\begin{bmatrix} x_1 \\ x_2 \end{bmatrix}\right) = \begin{bmatrix} x_1 + x_2 \\ x_1x_2 \\ x_1 - x_2 \end{bmatrix}$

c. $T : \mathcal{R}^2 \to \mathcal{R}^2, T\left(\begin{bmatrix} x_1 \\ x_2 \end{bmatrix}\right) = \begin{bmatrix} x_1 - x_2 + 1 \\ x_2 - x_1 \end{bmatrix}$

d. $T : \mathcal{R}^3 \to \mathcal{R}^3$, T(X) is the projection of X onto the x_1x_3 plane.

e. $T : \mathcal{R}^2 \to \mathcal{R}^2$, T(X) is the rotation of X counterclockwise about (1,1) through $\pi/2$ radians.

f. $T : \mathcal{R}^3 \to \mathcal{R}^3$, T(X) is the reflection of X in the x_2x_3 plane.

g. $T : \mathcal{R}^3 \to \mathcal{R}^3$, T is rotation about the x_3 axis through α radians in the indicated direction.

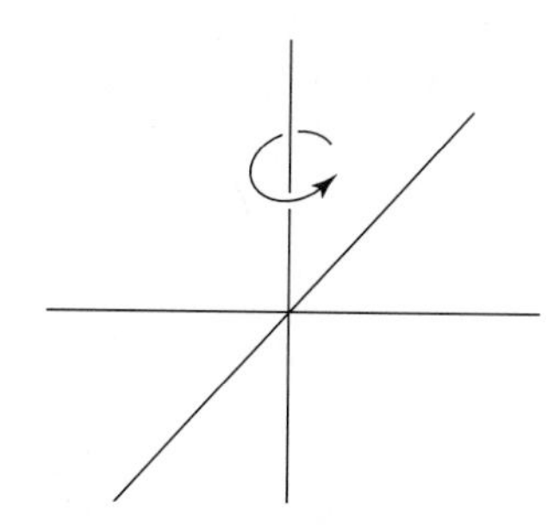

h. $T : \mathcal{R}^3 \to \mathcal{R}^3$, T(X) is the reflection of X through 0, that is, 0 is the midpoint of the line segment joining X and T(X).

i. $T : \mathcal{R}^3 \to \mathcal{R}^3$, T(X) is the reflection of X in the plane $x_1 + x_2 + x_3 = 1$.

j. $T : \mathcal{R}^3 \to \mathcal{R}^3$, $T(\text{X}) = \text{X} \times \text{C}$, where C is a fixed vector in $\mathcal{R}^3$.

k. $T : \mathcal{R}^n \to \mathcal{R}^n$, $T(\text{X}) = 5\text{X}$.

l. $T : \mathcal{R}^n \to \mathcal{R}$, $T(\text{X}) = \|\text{X}\|$.

m. $T : \mathcal{R}^n \to \mathcal{R}^n$, $T(\text{X}) = \text{X} + \text{B}$, where B is a fixed vector in $\mathcal{R}^n$.

n. $T : \mathcal{R}^n \to \mathcal{R}$, $T(\text{X}) = \text{X} \cdot \text{C}$, where C is a fixed vector in $\mathcal{R}^n$.

2. Let $\mathcal{B}$ be a basis for $\mathcal{R}^n$ and define $T : \mathcal{R}^n \to \mathcal{R}^n$ by $T(\text{X}) = [\text{X}]_{\mathcal{B}}$.
   a. Show that T is linear.
   b. Find the matrix A such that $T(\text{X}) = \text{AX}$ in the special case when $n = 2$ and $\mathcal{B} = \left\{\begin{bmatrix} 1 \\ 2 \end{bmatrix}, \begin{bmatrix} -1 \\ 3 \end{bmatrix}\right\}$.

3. Given $m,b \in \mathcal{R}$, define $T : \mathcal{R} \to \mathcal{R}$ by $T(x) = mx + b$. Is T linear? Explain!

4. If $T : \mathcal{R}^3 \to \mathcal{R}^3$ is linear, $T(\text{E}_1) = \begin{bmatrix} 2 \\ 1 \\ -3 \end{bmatrix}$, $T(\text{E}_2) = \begin{bmatrix} -1 \\ 3 \\ 2 \end{bmatrix}$, and $T(\text{E}_3) = \begin{bmatrix} 3 \\ -2 \\ 1 \end{bmatrix}$, find

   a. $T\left(\begin{bmatrix} 1 \\ 1 \\ 1 \end{bmatrix}\right)$. b. $T\left(\begin{bmatrix} 1 \\ 2 \\ 3 \end{bmatrix}\right)$. **c.** $T\left(\begin{bmatrix} -2 \\ 0 \\ 1 \end{bmatrix}\right)$.

5. If $T : \mathcal{R}^3 \to \mathcal{R}^2$ is linear, $T\left(\begin{bmatrix} 1 \\ 0 \\ 2 \end{bmatrix}\right) = \begin{bmatrix} 1 \\ 2 \end{bmatrix}$, $T\left(\begin{bmatrix} -1 \\ 1 \\ 1 \end{bmatrix}\right) = \begin{bmatrix} -2 \\ 3 \end{bmatrix}$, and

$T\left(\begin{bmatrix} 3 \\ 2 \\ -1 \end{bmatrix}\right) = \begin{bmatrix} -1 \\ 0 \end{bmatrix}$, find

a. $T\left(\begin{bmatrix} 6 \\ 1 \\ 2 \end{bmatrix}\right)$. b. $T\left(\begin{bmatrix} 1 \\ 3 \\ -2 \end{bmatrix}\right)$. **c.** $T\left(\begin{bmatrix} 3 \\ -1 \\ 3 \end{bmatrix}\right)$.

6. Find a linear $T : \mathcal{R}^3 \to \mathcal{R}^3$ such that

$$T\left(\begin{bmatrix} 1 \\ 1 \\ 1 \end{bmatrix}\right) = \begin{bmatrix} 1 \\ 2 \\ 3 \end{bmatrix}, \quad T\left(\begin{bmatrix} -1 \\ 0 \\ 1 \end{bmatrix}\right) = \begin{bmatrix} 3 \\ -2 \\ 1 \end{bmatrix}, \quad \text{and} \quad T\left(\begin{bmatrix} 0 \\ 1 \\ 1 \end{bmatrix}\right) = \begin{bmatrix} 2 \\ 3 \\ -1 \end{bmatrix}.$$

7. Suppose $\mathcal{V}$ is a subspace of $\mathcal{R}^n$ and $T : \mathcal{R}^n \to \mathcal{R}^n$ is orthogonal projection onto $\mathcal{V}$. Show that T is linear.

8. Find the matrix A such that $\text{Proj}_{\mathcal{V}}(\text{X}) = \text{AX}$ when

a. $\mathcal{V} = \mathcal{S}pan\left\{\begin{bmatrix} 1 \\ -1 \end{bmatrix}\right\}$. b. $\mathcal{V} = \mathcal{S}pan\left\{\begin{bmatrix} 1 \\ 1 \\ 0 \\ 1 \end{bmatrix}, \begin{bmatrix} 1 \\ 0 \\ 1 \\ -1 \end{bmatrix}\right\}$.

c. $\mathcal{V} = \left\{\text{X} \in \mathcal{R}^3 : x_1 + x_2 + x_3 = 0\right\}$.

9. Suppose $\{a_k\}_0^\infty$ is generated by (4.6).

a. Show that $\begin{bmatrix} a_{k+1} \\ a_{k+2} \end{bmatrix} = \begin{bmatrix} 0 & 1 \\ s & r \end{bmatrix}\begin{bmatrix} a_k \\ a_{k+1} \end{bmatrix}$, $k = 0, 1, 2, \ldots$.

b. Setting $\text{A} = \begin{bmatrix} 0 & 1 \\ s & r \end{bmatrix}$ and $\text{X}_k = \begin{bmatrix} a_k \\ a_{k+1} \end{bmatrix}$, show that $\text{X}_k = \text{A}^k\text{X}_0$, $k \geq 0$.

4.3
KERNEL AND RANGE

The next definition introduces terminology commonly used in describing the behavior of functions. Our interest is in examining these ideas in the context of linear functions.

DEFINITION. Let T be a function with domain $\mathcal{R}^n$ and values in $\mathcal{R}^m$. For $\text{X} \in \mathcal{R}^n$, $T(\text{X})$ is called the *image of* X. The *range* of T, denoted by $\mathcal{R}ng(T)$, is the set of all images, that is,

$$\mathcal{R}ng(T) = \{T(\text{X}) : \text{X} \in \mathcal{R}^n\}.$$

Note that $\mathcal{R}ng(T) \subseteq \mathcal{R}^m$. For $\mathcal{V} \subseteq \mathcal{R}^n$, the *image of* $\mathcal{V}$ is

$$T(\mathcal{V}) = \{T(\text{X}) : \text{X} \in \mathcal{V}\}.$$

In particular, $T(\mathcal{R}^n) = \mathcal{R}ng(T)$. T is said to be *onto* when $\mathcal{R}ng(T) = \mathcal{R}^m$, that is, when for each $Y \in \mathcal{R}^m$ there is an $X \in \mathcal{R}^n$ such that $T(X) = Y$.

T is *one-to-one* if different elements of $\mathcal{R}^n$ always have different images, that is, if

$$X_1 \neq X_2 \Rightarrow T(X_1) \neq T(X_2),$$

or equivalently,

$$T(X_1) = T(X_2) \Rightarrow X_1 = X_2.$$

■ **EXAMPLE 4.8.** Let $T : \mathcal{R}^3 \to \mathcal{R}^3$ be projection onto the x_1x_2 plane (Example 4.2), that is,

$$T\left(\begin{bmatrix} x_1 \\ x_2 \\ x_3 \end{bmatrix}\right) = \begin{bmatrix} x_1 \\ x_2 \\ 0 \end{bmatrix}.$$

The range of T is the x_1x_2 plane. It is a proper subset of $\mathcal{R}^3$, so T is not onto. For each $Y = [a \;\; b \;\; 0]^T \in \mathcal{R}ng(T)$, there are infinitely many X such that $T(X) = Y$, namely, $X = [a \;\; b \;\; t]^T$, $t \in \mathcal{R}$, so T is not one-to-one. Geometrically, $\{X : T(X) = Y\}$ is the vertical line intersecting the x_1x_2 plane at Y. □

■ **EXAMPLE 4.9.** Let $T : \mathcal{R}^2 \to \mathcal{R}^2$ be reflection in the line with equation $x_2 = x_1$ (Example 4.3). Since each point in $\mathcal{R}^2$ is the reflection of exactly one point in $\mathcal{R}^2$, T is both one-to-one and onto. □

THEOREM 4.3. If $A \in \mathcal{M}_{m\times n}$ and $T : \mathcal{R}^n \to \mathcal{R}^m$ is given by $T(X) = AX$, then $\mathcal{R}ng(T) = \mathcal{C}(A)$. In particular, $\mathcal{R}ng(T)$ is a subspace of $\mathcal{R}^m$.

Proof. $\mathcal{R}ng(T) = \{AX : X \in \mathcal{R}^n\} = \left\{\sum_1^n x_k \text{Col}_k(A) : x_1, \ldots, x_n \in \mathcal{R}\right\} = \mathcal{C}(A)$. ■

If $T : \mathcal{R}^n \to \mathcal{R}^m$ is linear, then $\dim \mathcal{R}ng(T) \leq m$. The only subspace of $\mathcal{R}^m$ with dimension m is $\mathcal{R}^m$ itself, so T is onto if and only if $\dim \mathcal{R}ng(T) = m$. Moreover, if A is the $m \times n$ matrix such that $T(X) = AX$, then $\mathcal{R}ng(T) = \mathcal{C}(A)$, so $\dim \mathcal{R}ng(T) = \text{rank}(A)$. Thus, T is onto precisely when A has rank m.

THEOREM 4.4. If $A \in \mathcal{M}_{m\times n}$ and $T : \mathcal{R}^n \to \mathcal{R}^m$ is defined by $T(X) = AX$, then T is onto if and only if $\text{rank}(A) = m$.

■ **EXAMPLE 4.10.** Consider $T : \mathcal{R}^4 \to \mathcal{R}^3$ given by $T(X) = \begin{bmatrix} 1 & 1 & 0 & -1 \\ -2 & -2 & 1 & 0 \\ -1 & -1 & 2 & -3 \end{bmatrix} X$.

Since

$$\begin{bmatrix} 1 & 1 & 0 & -1 \\ -2 & -2 & 1 & 0 \\ -1 & -1 & 2 & -3 \end{bmatrix} \longrightarrow \begin{bmatrix} 1 & 1 & 0 & -1 \\ 0 & 0 & 1 & -2 \\ 0 & 0 & 0 & 0 \end{bmatrix},$$

$\text{rank}(A) = 2$, $\mathcal{R}ng(T)$ is a two-dimensional subspace of $\mathcal{R}^3$, and T is not onto. □

■ **EXAMPLE 4.11.** Let $T_\alpha : \mathcal{R}^2 \to \mathcal{R}^2$ be counterclockwise rotation about 0 through α radians (Example 4.1). Then $T_\alpha(X) = AX$, where

$$A = \begin{bmatrix} \cos(\alpha) & -\sin(\alpha) \\ \sin(\alpha) & \cos(\alpha) \end{bmatrix}.$$

Since $\cos^2(\alpha) + \sin^2(\alpha) = 1 \neq 0$, A is nonsingular (Example 1.36). Thus, rank(A) = 2, and T_α is onto. (You can also convince yourself easily through geometric considerations that T_α is onto.) □

THEOREM 4.5. If $T : \mathcal{R}^n \to \mathcal{R}^m$ is linear and $\{X_1, \ldots, X_n\}$ is a basis for $\mathcal{R}^n$, then $\mathcal{Rng}(T) = \mathcal{Span}\{T(X_1), \ldots, T(X_n)\}$.

Proof. Since $T(X_1), \ldots, T(X_n) \in \mathcal{Rng}(T)$, each vector in $\mathcal{Span}\{T(X_1), \ldots, T(X_n)\}$ is a linear combination of vectors in $\mathcal{Rng}(T)$ and, as such, is an element of $\mathcal{Rng}(T)$ ($\mathcal{Rng}(T)$ is a subspace). Thus, $\mathcal{Span}\{T(X_1), \ldots, T(X_n)\} \subseteq \mathcal{Rng}(T)$. To establish the opposite inclusion, suppose $Y \in \mathcal{Rng}(T)$. Then $Y = T(X)$ for some $X \in \mathcal{R}^n$, and $X = \sum_1^n t_k X_k$ for some $t_1, \ldots, t_n \in \mathcal{R}$, so $Y = T\left(\sum_1^n t_k X_k\right) = \sum_1^n t_k T(X_k) \in \mathcal{Span}\{T(X_1), \ldots, T(X_n)\}$. ■

REMARK. Theorem 4.5 does not claim that $\{T(X_1), \ldots, T(X_n)\}$ is a basis for $\mathcal{Rng}(T)$. There is no guarantee that $\{T(X_1), \ldots, T(X_n)\}$ is linearly independent.

Every linear $T : \mathcal{R}^n \to \mathcal{R}^m$ sends 0 to 0 (Theorem 4.2). Moreover, if T is one-to-one, then 0 is the only vector in $\mathcal{R}^n$ with image 0. On the other hand, if it is known that $T(X) = 0$ only when $X = 0$, then for all $X_1, X_2 \in \mathcal{R}^n$,

$$T(X_1) = T(X_2) \Rightarrow T(X_1 - X_2) = T(X_1) - T(X_2) = 0 \Rightarrow X_1 - X_2 = 0,$$

that is, T is one-to-one.

DEFINITION. If $T : \mathcal{R}^n \to \mathcal{R}^m$ is linear, then the *kernel* of T, denoted by $\mathcal{Ker}(T)$, is the set of vectors in $\mathcal{R}^n$ having image 0, that is, $\mathcal{Ker}(T) = \{X \in \mathcal{R}^n : T(X) = 0\}$.

The discussion preceding the definition establishes the following characterization of one-to-one linear functions.

THEOREM 4.6. If $T : \mathcal{R}^n \to \mathcal{R}^m$ is linear, then T is one-to-one if and only if $\mathcal{Ker}(T) = \{0\}$.

THEOREM 4.7. If $A \in \mathcal{M}_{m\times n}$ and $T : \mathcal{R}^n \to \mathcal{R}^m$ is given by $T(X) = AX$, then $\mathcal{Ker}(T) = \mathcal{N}(A)$. In particular, $\mathcal{Ker}(T)$ is a subspace of $\mathcal{R}^n$.

Proof. $\mathcal{Ker}(T) = \{X \in \mathcal{R}^n : AX = 0\} = \mathcal{N}(A)$. ■

■ EXAMPLE 4.12. Consider $T : \mathcal{R}^3 \to \mathcal{R}^2$ defined by $T(X) = \begin{bmatrix} 1 & 3 & 5 \\ 2 & 4 & 6 \end{bmatrix} X$. Since

$$A = \begin{bmatrix} 1 & 3 & 5 \\ 2 & 4 & 6 \end{bmatrix} \longrightarrow \begin{bmatrix} 1 & 0 & -1 \\ 0 & 1 & 2 \end{bmatrix},$$

$\dim \mathcal{N}(A) = 1$ and $\dim \mathcal{C}(A) = 2$. It follows that $\mathcal{Ker}(T) = \mathcal{N}(A) \neq \{0\}$, so T is not one-to-one, and $\mathcal{Rng}(T) = \mathcal{C}(A) = \mathcal{R}^2$, so T is onto. □

■ EXAMPLE 4.13. Let $T(X) = \begin{bmatrix} 1 & -3 \\ -1 & 2 \\ 2 & -1 \end{bmatrix} X$. Since $\begin{bmatrix} 1 & -3 \\ -1 & 2 \\ 2 & -1 \end{bmatrix} \longrightarrow \begin{bmatrix} 1 & 0 \\ 0 & 1 \\ 0 & 0 \end{bmatrix}$,

$\mathcal{Ker}(T) = \{0\}$ and $\mathcal{Rng}(T)$ is a two-dimensional subspace of $\mathcal{R}^3$. Thus, T is one-to-one but not onto. □

THEOREM 4.8. If $T : \mathcal{R}^n \to \mathcal{R}^m$ is linear, then $\dim \mathcal{R}ng(T) + \dim \mathcal{K}er(T) = n$.

Proof. Let A be the $m \times n$ matrix such that $T(\mathrm{X}) = \mathrm{AX}$. Then $\mathcal{C}(\mathrm{A}) = \mathcal{R}ng(T)$, $\mathcal{N}(\mathrm{A}) = \mathcal{K}er(T)$, and the rank-nullity theorem gives

$$n = \dim \mathcal{C}(\mathrm{A}) + \dim \mathcal{N}(\mathrm{A}) = \dim \mathcal{R}ng(T) + \dim \mathcal{K}er(T). \blacksquare$$

Notice that the relationship between $\dim \mathcal{R}ng(T)$ and $\dim \mathcal{K}er(T)$ is complementary; the larger one is, the smaller the other must be.

THEOREM 4.9. Assume $T : \mathcal{R}^n \to \mathcal{R}^m$ is linear.

(1) If $n < m$, then T is not onto.
(2) If $n > m$, then T is not one-to-one.
(3) If $n = m$, then T is onto if and only if T is one-to-one.

Proof.
(1) It follows from Theorem 4.8 that $\dim \mathcal{R}ng(T) \leq n$. In this case $n < m$, so $\mathcal{R}ng(T)$ is a proper subspace of $\mathcal{R}^m$. Thus, T is not onto.
(2) Since $\mathcal{R}ng(T)$ is a subspace of $\mathcal{R}^m$, $\dim \mathcal{R}ng(T) \leq m$. In this case, $m < n$, so by Theorem 4.8, $\dim \mathcal{K}er(T) > 0$. Thus, T is not one-to-one.
(3) Here, $\mathcal{R}ng(T) \subseteq \mathcal{R}^n$. T is onto if and only if $\dim \mathcal{R}ng(T) = n$, which (by Theorem 4.8) occurs when $\dim \mathcal{K}er(T) = 0$, that is, when T is one-to-one. $\blacksquare$

As a result of Theorem 4.9, a linear $T : \mathcal{R}^n \to \mathcal{R}^m$ cannot be both one-to-one and onto unless $n = m$. When $n = m$, T has either both properties or neither.

Exercises 4.3

1. Assuming $T(\mathrm{X}) = \mathrm{AX}$, find a basis for $\mathcal{K}er(T)$ and a basis for $\mathcal{R}ng(T)$. Determine if T is one-to-one and/or onto.

a. $T : \mathcal{R}^3 \to \mathcal{R}^4$, $\mathrm{A} = \begin{bmatrix} 1 & 1 & 0 \\ -1 & 1 & 2 \\ 0 & 2 & 2 \\ -1 & 3 & 4 \end{bmatrix}$

b. $T : \mathcal{R}^3 \to \mathcal{R}^3$, $\mathrm{A} = \begin{bmatrix} 1 & -1 & 0 \\ 0 & 1 & -1 \\ -1 & 0 & 1 \end{bmatrix}$

c. $T : \mathcal{R}^2 \to \mathcal{R}^3$, $\mathrm{A} = \begin{bmatrix} 1 & 4 \\ 2 & 5 \\ 3 & 6 \end{bmatrix}$

d. $T : \mathcal{R}^6 \to \mathcal{R}^4$, $\mathrm{A} = \begin{bmatrix} 1 & 1 & 2 & 1 & 3 & 1 \\ 0 & 0 & 1 & 2 & 1 & -2 \\ 1 & 1 & 2 & 2 & 5 & 1 \\ 0 & 0 & 0 & 0 & 1 & 1 \end{bmatrix}$

e. $T : \mathcal{R}^2 \to \mathcal{R}^2$, $\mathrm{A} = \begin{bmatrix} 0 & 0 \\ 0 & 2 \end{bmatrix}$

f. $T : \mathcal{R}^4 \to \mathcal{R}^3$, $\mathrm{A} = \begin{bmatrix} 1 & 1 & 0 & 0 \\ 0 & 1 & 1 & 0 \\ 0 & 0 & 1 & 1 \end{bmatrix}$

g. $T : \mathcal{R}^3 \to \mathcal{R}^3$, $\mathrm{A} = \begin{bmatrix} 1 & 0 & 1 \\ 1 & 1 & 0 \\ 0 & 1 & 1 \end{bmatrix}$

h. $T : \mathcal{R}^3 \to \mathcal{R}^4$, $\mathrm{A} = \begin{bmatrix} 1 & 0 & -1 \\ 0 & 2 & 0 \\ 3 & 0 & 1 \\ -1 & 2 & 1 \end{bmatrix}$

2. Suppose C is a nonzero vector in $\mathcal{R}^n$ and define $T : \mathcal{R}^n \to \mathcal{R}$ by $T(\mathrm{X}) = \mathrm{X} \cdot \mathrm{C}$ (Problem 1(n), Exercises 4.2).
   a. Find the dimension of $\mathcal{K}er(T)$.
   b. What is the range of T?

3. Suppose $\mathcal{B}$ is a basis for $\mathcal{R}^n$ and $T : \mathcal{R}^n \to \mathcal{R}^n$ is defined by $T(\mathrm{X}) = [\mathrm{X}]_{\mathcal{B}}$ (Problem 2, Exercises 4.2). Show that T is one-to-one and onto.

4. Suppose M,B $\in \mathcal{R}^3$ and M $\neq 0$. Let $\mathcal{l}$ be the line in $\mathcal{R}^3$ described by $\mathrm{X}(t) = t\mathrm{M} + \mathrm{B}$, $-\infty < t < \infty$, and assume $T : \mathcal{R}^3 \to \mathcal{R}^3$ is linear.
   a. Show that $T(\mathcal{l})$ is either a line or a point.
   b. Show that if T is one-to-one, then $T(\mathcal{l})$ is a line.

5. Show that if A $\in \mathcal{M}_{n\times n}$ and $T : \mathcal{R}^n \to \mathcal{R}^n$ is given by $T(\mathrm{X}) = \mathrm{AX}$, then T is onto if and only if A is nonsingular.

6. Suppose A $\in \mathcal{M}_{4\times 5}$ and $T(\mathrm{X}) = \mathrm{AX}$. Assuming T is onto, what is the dimension of $\mathcal{K}er(T)$?

7. Let $\mathcal{V}$ be a subspace of $\mathcal{R}^n$ and $T(\mathrm{X}) = \mathrm{Proj}_{\mathcal{V}}(\mathrm{X})$ (Problem 7, Exercises 4.2).
   a. What is the range of T?
   b. What is the kernel of T?

8. Assume C is a nonzero vector in $\mathcal{R}^3$ and define $T : \mathcal{R}^3 \to \mathcal{R}^3$ by $T(\mathrm{X}) = \mathrm{X} \times \mathrm{C}$ (Problem 1(j), Exercises 4.2).
   a. What is the kernel of T?
   b. What is the range of T?

9. Suppose $T : \mathcal{R}^n \to \mathcal{R}^m$ is linear and $\mathcal{B} = \{\mathrm{X}_1, \ldots, \mathrm{X}_p\} \subseteq \mathcal{R}^n$. Show that
   a. if $T(\mathcal{B})$ is linearly independent, then $\mathcal{B}$ is linearly independent.
   b. if T is one-to-one and $\mathcal{B}$ is linearly independent, then $T(\mathcal{B})$ is linearly independent.

4.4
AFFINE SUBSETS

The distinguishing feature of a linear transformation is the manner in which it interacts with linear combinations, that is, the behavior expressed by

$$T(t_1\mathrm{X}_1 + \cdots + t_p\mathrm{X}_p) = t_1T(\mathrm{X}_1) + \cdots + t_pT(\mathrm{X}_p). \tag{4.7}$$

Thanks to (4.7), subsets of the domain with a structure based on the notion of linear combination have images that are particularly easy to identify. The principal examples of such subsets are subspaces.

THEOREM 4.10. Suppose $T : \mathcal{R}^n \to \mathcal{R}^m$ is linear and $\mathcal{V}$ is a subspace of $\mathcal{R}^n$. Then $T(\mathcal{V})$ is a subspace of $\mathcal{R}^m$, and $\dim T(\mathcal{V}) \leq \dim \mathcal{V}$. Moreover, if $\mathcal{B}$ is a basis for $\mathcal{V}$, then $T(\mathcal{V}) = \mathcal{S}pan\{T(\mathcal{B})\}$.

Proof. Assuming $\mathcal{V} \neq \{0\}$ and $\mathcal{B} = \{X_1, \ldots, X_p\}$ is a basis for $\mathcal{V}$,

$$\begin{aligned} T(\mathcal{V}) &= \{T(X) : X \in \mathcal{V}\} \\ &= \left\{ T\left(\sum_1^p t_k X_k\right) : t_1, \ldots, t_p \in \mathcal{R} \right\} \\ &= \left\{ \sum_1^p t_k T(X_k) : t_1, \ldots, t_p \in \mathcal{R} \right\} \\ &= \mathcal{S}pan\{T(X_1), \ldots, T(X_p)\} \\ &= \mathcal{S}pan\{T(\mathcal{B})\}. \end{aligned}$$

Thus, $T(\mathcal{V})$ is a subspace of $\mathcal{R}^m$ (Theorem 2.11). If $T(\mathcal{V}) \neq \{0\}$, then $T(\mathcal{B})$ has a linearly independent subset that spans $T(\mathcal{V})$, and therefore $T(\mathcal{V})$ has a basis consisting of at most p vectors. Of course when $T(\mathcal{V}) = \{0\}$, dim $T(\mathcal{V}) = 0$. In either event, dim $T(\mathcal{V}) \leq p = \dim \mathcal{V}$. Finally, $\mathcal{V} = \{0\}$ implies $T(\mathcal{V}) = \{0\}$, and in this case the conclusions follow trivially. ■

Suppose $T : \mathcal{R}^3 \to \mathcal{R}^3$ is linear and $\mathcal{V}$ is a plane through 0. By Theorem 4.10, $T(\mathcal{V})$ is a subspace of $\mathcal{R}^3$ of dimension at most 2, so the image of $\mathcal{V}$ is either a plane through 0, a line through 0, or $\{0\}$. Similarly, the image of a line through 0 is either a line through 0 or $\{0\}$.

■ **EXAMPLE 4.14.** Consider $T : \mathcal{R}^3 \to \mathcal{R}^3$ given by $T(X) = \begin{bmatrix} 2 & 2 & 2 \\ -1 & -2 & 0 \\ 0 & 1 & -1 \end{bmatrix} X$. Let

$$X_1 = \begin{bmatrix} 1 \\ -1 \\ 0 \end{bmatrix}, \quad X_2 = \begin{bmatrix} -1 \\ 1 \\ 1 \end{bmatrix}, \quad X_3 = \begin{bmatrix} 0 \\ 1 \\ -1 \end{bmatrix},$$

and observe that

$$T(X_1) = \begin{bmatrix} 0 \\ 1 \\ -1 \end{bmatrix}, \quad T(X_2) = \begin{bmatrix} 2 \\ -1 \\ 0 \end{bmatrix}, \quad T(X_3) = \begin{bmatrix} 0 \\ -2 \\ 2 \end{bmatrix}.$$

Both $\mathcal{V} = \mathcal{S}pan\{X_1, X_2\}$ and $\mathcal{W} = \mathcal{S}pan\{X_1, X_3\}$ have dimension 2. Since $\{T(X_1), T(X_2)\}$ is linearly independent, dim $T(\mathcal{V}) = \dim \mathcal{S}pan\{T(X_1), T(X_2)\} = 2$. On the other hand, $T(X_3) = -2T(X_1)$, so $\mathcal{S}pan\{T(X_1), T(X_3)\} = \mathcal{S}pan\{T(X_1)\}$, and $T(\mathcal{W})$ is one-dimensional (Figure 4.5). □

Lines and planes in space that do not pass through 0 are not subspaces of $\mathcal{R}^3$, but they certainly are geometrically similar to subspaces. A line ℓ, for example, is described parametrically by

$$X(t) = tM + B, \quad -\infty < t < \infty,$$

where $M, B \in \mathcal{R}^3$ and $M \neq 0$. Since $M \neq 0$, $\mathcal{V} = \mathcal{S}pan\{M\}$ is a one-dimensional subspace of $\mathcal{R}^3$. It is, in fact, the line through 0 parallel to ℓ. Note that ℓ is obtained by adding B to each vector in $\mathcal{V}$, that is,

$$\ell = \{B + X : X \in \mathcal{V}\}.$$

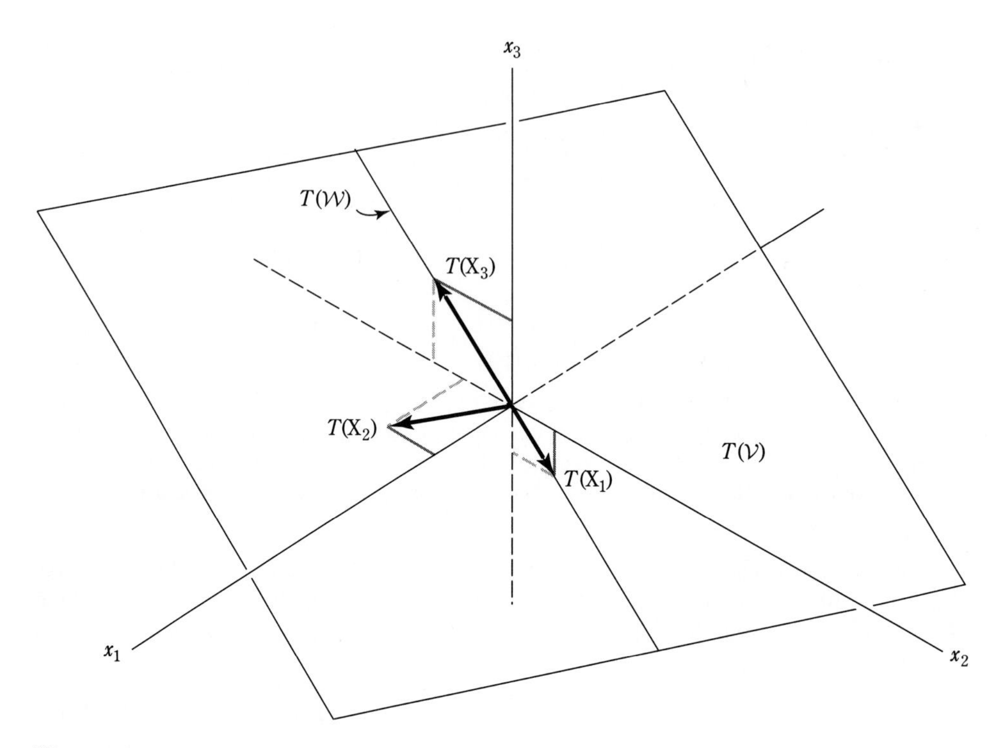

FIGURE 4.5

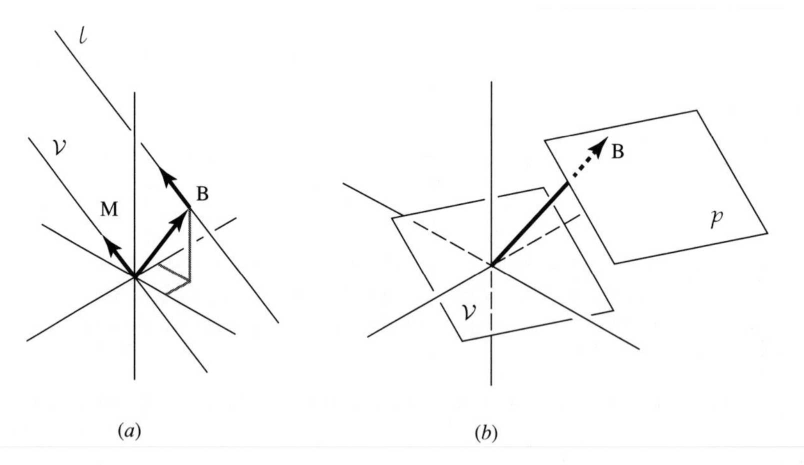

FIGURE 4.6

Similarly, if $p \subseteq \mathcal{R}^3$ is a plane, $B \in p$, and $\mathcal{V}$ is the plane through 0 parallel to p (Figure 4.6(b)), then $\mathcal{V}$ is a two-dimensional subspace of $\mathcal{R}^3$, and

$$p = \{B + X \,:\, X \in \mathcal{V}\}.$$

The geometric effect of adding B to each element of a set is to translate that set by $\|B\|$ units in the direction of B. Therein lies the motivation for the following terminology.

DEFINITION. For $\mathcal{V} \subseteq \mathcal{R}^n$ and $B \in \mathcal{R}^n$, $B + \mathcal{V} = \{B + X : X \in \mathcal{V}\}$ is the *translate of* $\mathcal{V}$ *by* B. A translate of a subspace is called an *affine subset*, and if $\mathcal{V}$ is a subspace of dimension p, then $B + \mathcal{V}$ is a *p-dimensional affine subset.*

■ **EXAMPLE 4.15.** The zero-dimensional affine subsets of $\mathcal{R}^3$ are the translates of $\mathcal{V} = \{0\}$, that is, the subsets of $\mathcal{R}^3$ that consist of a single point. The affine subsets of $\mathcal{R}^3$ of dimensions 1 and 2 are the lines and planes, respectively. The only three-dimensional affine subset of $\mathcal{R}^3$ is $\mathcal{R}^3$ itself. □

■ **EXAMPLE 4.16.** The solution set for a consistent system of linear equations in n unknowns is an affine subset of $\mathcal{R}^n$. Indeed, if $A \in \mathcal{M}_{m\times n}$, $B \in \mathcal{R}^m$, and X_0 is any solution of $AX = B$, then by Theorem 2.10,

$$\{X : AX = B\} = \{X_0 + Y : Y \in \mathcal{N}(A)\} = X_0 + \mathcal{N}(A). \ \square$$

THEOREM 4.11. If $\mathcal{V} \subseteq \mathcal{R}^n$, $B \in \mathcal{R}^n$, and $T : \mathcal{R}^n \to \mathcal{R}^m$ is linear, then

$$T(B + \mathcal{V}) = T(B) + T(\mathcal{V}).$$

In particular, the image under T of an affine subset is an affine subset.

Proof. Since T is linear,

$$\begin{aligned} T(B + \mathcal{V}) &= \{T(B + X) : X \in \mathcal{V}\} \\ &= \{T(B) + T(X) : X \in \mathcal{V}\} \\ &= \{T(B) + Y : Y \in T(\mathcal{V})\} \\ &= T(B) + T(\mathcal{V}). \end{aligned}$$

If $\mathcal{V}$ is a subspace of $\mathcal{R}^n$, then $T(\mathcal{V})$ is a subspace of $\mathcal{R}^m$ (Theorem 4.10), and therefore $T(B + \mathcal{V})$ is an affine subset of $\mathcal{R}^m$. ■

■ **EXAMPLE 4.17.** Let $d \in \mathcal{R}$ and consider the line ℓ in $\mathcal{R}^2$ with equation

$$x_1 + x_2 = d.$$

Solving this "system" produces a parametric description of ℓ, namely,

$$X(t) = \begin{bmatrix} d - t \\ t \end{bmatrix} = t\begin{bmatrix} -1 \\ 1 \end{bmatrix} + \begin{bmatrix} d \\ 0 \end{bmatrix}, \quad -\infty < t < \infty,$$

so ℓ has $M = \begin{bmatrix} -1 \\ 1 \end{bmatrix}$ and $B = \begin{bmatrix} d \\ 0 \end{bmatrix}$ as direction and position vectors, respectively. Let $\mathcal{V} = \mathcal{S}pan\{M\}$. Then $\mathcal{V}$ is the line through 0 parallel to ℓ, and $\ell = B + \mathcal{V}$. Now, consider $T : \mathcal{R}^2 \to \mathcal{R}^2$ given by

$$T(X) = \begin{bmatrix} 2 & 3 \\ 1 & 4 \end{bmatrix} X.$$

$T(\ell)$ is the affine subset of $\mathcal{R}^2$ described parametrically by

$$Y(t) = T(X(t)) = tT\left(\begin{bmatrix} -1 \\ 1 \end{bmatrix}\right) + T\left(\begin{bmatrix} d \\ 0 \end{bmatrix}\right) = t\begin{bmatrix} 1 \\ 3 \end{bmatrix} + d\begin{bmatrix} 2 \\ 1 \end{bmatrix},$$

$-\infty < t < \infty$, that is, the line with direction vector $T(M) = \begin{bmatrix} 1 \\ 3 \end{bmatrix}$ and position vector

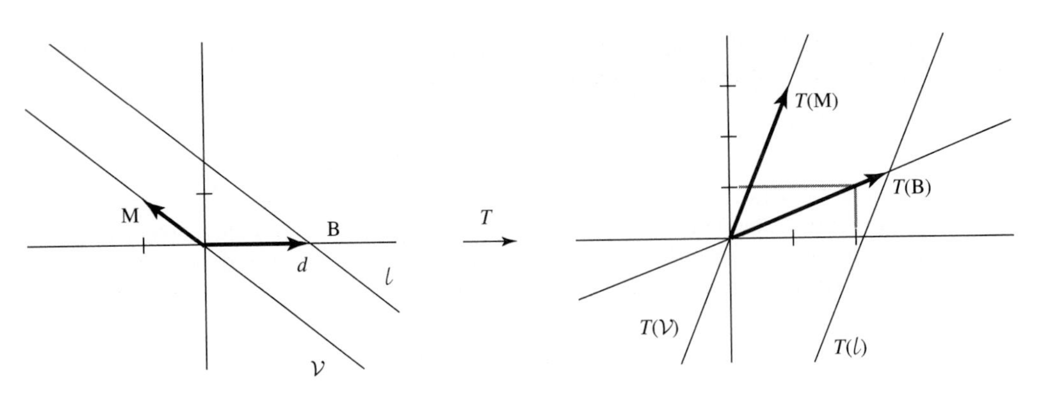

FIGURE 4.7

$T(\mathrm{B}) = d\begin{bmatrix} 2 \\ 1 \end{bmatrix}$. Figure 4.7 illustrates the action of T on ℓ when $d > 0$. Note that the direction vectors for ℓ and $T(\ell)$ do not depend on d. As d varies from $-\infty$ to ∞, the associated ℓ's are parallel and sweep out the entire plane, with different values of d corresponding to different x_1 intercepts. Such a collection of lines is called a *ruling* of the plane. The image lines also *rule* the plane, the various choices for d determining the points of intersection of the ruling lines, $T(\ell)$, with $\mathcal{S}pan\left\{\begin{bmatrix} 2 \\ 1 \end{bmatrix}\right\}$. □

The next example illustrates the use of the foregoing ideas in analyzing a general linear function from $\mathcal{R}^2$ to $\mathcal{R}^2$.

■ **EXAMPLE 4.18.** Suppose $T : \mathcal{R}^2 \to \mathcal{R}^2$ is linear and $\mathcal{A} = \{\mathrm{X}_1, \mathrm{X}_2\}$ is a basis for $\mathcal{R}^2$. The $\mathcal{A}$ coordinate axes are $\mathcal{X}_1 = \mathcal{S}pan\{\mathrm{X}_1\}$ and $\mathcal{X}_2 = \mathcal{S}pan\{\mathrm{X}_2\}$, and their images are $\mathcal{Y}_1 = \mathcal{S}pan\{\mathrm{Y}_1\}$ and $\mathcal{Y}_2 = \mathcal{S}pan\{\mathrm{Y}_2\}$, where $\mathrm{Y}_k = T(\mathrm{X}_k)$, $k = 1, 2$. Remember that $\mathcal{R}ng(T) = \mathcal{S}pan\{\mathrm{Y}_1, \mathrm{Y}_2\}$. There are three cases to consider.

First, suppose $\mathcal{B} = \{\mathrm{Y}_1, \mathrm{Y}_2\}$ is linearly independent. Then $\mathcal{B}$ is a basis for $\mathcal{R}^2$, $\mathcal{R}ng(T) = \mathcal{R}^2$, and T is both onto and one-to-one (Theorem 4.9(3)). For $\mathrm{X} \in \mathcal{R}^2$,

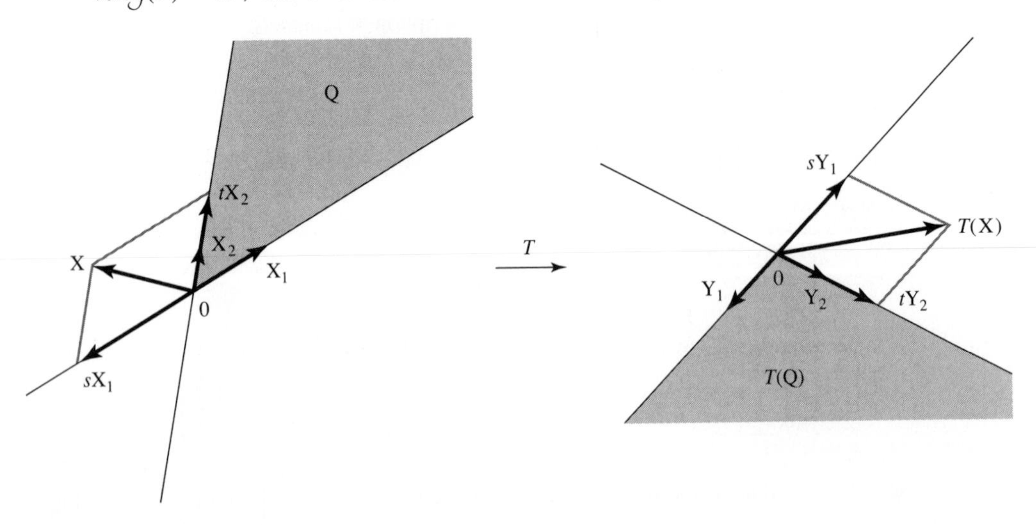

FIGURE 4.8

$X = sX_1 + tX_2$ for some choice of s and t, and consequently $T(X) = sY_1 + tY_2$. Notice that the scalars determining X relative to $\mathcal{A}$ are the same as those determining $T(X)$ relative to $\mathcal{B}$, that is, that

$$[T(X)]_{\mathcal{B}} = \begin{bmatrix} s \\ t \end{bmatrix} = [X]_{\mathcal{A}}.$$

If you think of $Q = \{sX_1 + tX_2 : s,t > 0\}$ as the first "quadrant" in the $\mathcal{A}$ system, then its image, $T(Q) = \{sY_1 + tY_2 : s,t > 0\}$, is the first "quadrant" in the $\mathcal{B}$ system.

Consider now a typical element of $\mathcal{R}^2$, written as $sX_1 + tX_2$, $s,t \in \mathcal{R}$. Holding s fixed and letting t vary,

$$X(t) = sX_1 + tX_2, \quad t \in \mathcal{R},$$

describes the line ℓ_s passing through sX_1 parallel to the X_2 axis. Its image,

$$Y(t) = T(X(t)) = sY_1 + tY_2, \quad t \in \mathcal{R},$$

is the line passing through sY_1 parallel to the Y_2 axis. Think of $\mathcal{R}^2$ as being ruled by the parallel lines ℓ_s, $-\infty < s < \infty$. T transforms them into parallel lines $T(\ell_s)$, which are "strung" along the Y_1 axis in a manner analogous to that of ℓ_s along the X_1 axis. That is, ℓ_s crosses the X_1 axis at sX_1, and $T(\ell_s)$ crosses the Y_1 axis at sY_1, $-\infty < s < \infty$ (Figure 4.9).

In the remaining two cases, $\{Y_1,Y_2\}$ is linearly dependent, so $\mathcal{Rng}(T)$ is a line through 0 or $\{0\}$. In either event, T is neither onto nor one-to-one.

Suppose dim $\mathcal{Rng}(T) = 1$. Then dim $\mathcal{Ker}(T) = 1$ (Theorem 4.8). Under these circumstances it helps to use a basis $\mathcal{A} = \{X_1,X_2\}$ with the property that one of its vectors, say X_2, is in $\mathcal{Ker}(T)$. Then $Y_2 = T(X_2) = 0$, $Y_1 = T(X_1) \neq 0$, and $\mathcal{Rng}(T) = \mathcal{Span}\{Y_1\} = \mathcal{Y}_1$. For $s,t \in \mathcal{R}$,

$$T(sX_1 + tX_2) = sY_1 + tY_2 = sY_1 + 0 = sY_1,$$

so $\mathcal{R}^2$ is "collapsed" by T onto the line $\mathcal{Y}_1$. The last equation is very precise about how the collapsing takes place. All the points on

$$\ell_s : X(t) = sX_1 + tX_2, \quad -\infty < t < \infty,$$

are sent to sY_1 (Figure 4.10), and this is the case for each s.

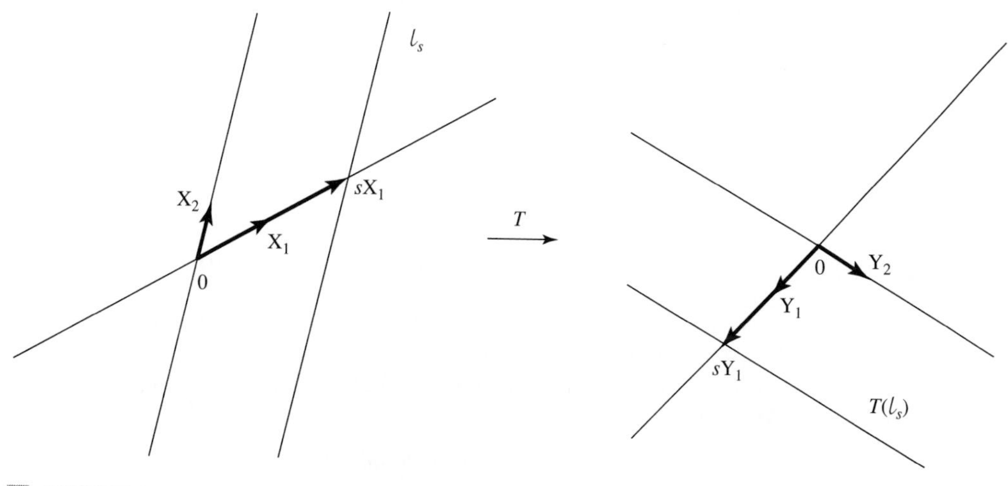

FIGURE 4.9

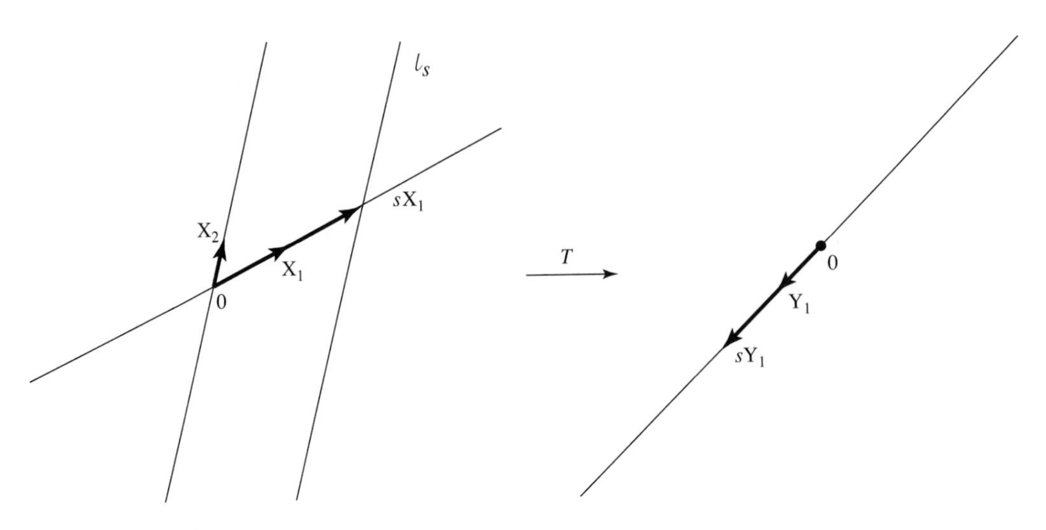

FIGURE 4.10

Finally, if dim $\mathcal{Rng}(T) = 0$, then $\mathcal{Rng}(T) = \{0\}$, so T collapses the entire plane onto 0. □

The analysis in the last example revolves around some issues that are of considerable importance in understanding the workings of a linear function $T : \mathcal{R}^n \to \mathcal{R}^m$. Given a basis $\mathcal{A}$ for $\mathcal{R}^n$, $T(\mathcal{A})$ is always a spanning set for $\mathcal{Rng}(T)$. If it is linearly independent, then $T(\mathcal{A})$ is a basis for $\mathcal{Rng}(T)$. In that event, the range has the same dimension as the domain, and no collapsing takes place. Alternatively, dim $\mathcal{Rng}(T) < n$, and then interest tends to focus on the nature of the collapsing. The next result shows that it is the one-to-one linear functions that preserve independence, and consequently, T being one-to-one implies $T(\mathcal{A})$ is a basis for $\mathcal{Rng}(T)$.

THEOREM 4.12. If $T : \mathcal{R}^n \to \mathcal{R}^m$ is linear, then the following are equivalent:

(1) T is one-to-one.
(2) $T(\mathcal{A})$ is linearly independent whenever $\mathcal{A}$ is linearly independent.

Proof.
(1) ⇒ (2). Suppose T is one-to-one and $\mathcal{A} = \{X_1, \ldots, X_p\}$ is linearly independent. If $t_1, \ldots, t_p$ are scalars such that

$$0 = t_1 T(X_1) + \cdots + t_p T(X_p) = T(t_1 X_1 + \cdots + t_p X_p),$$

then $t_1 X_1 + \cdots + t_p X_p \in \mathcal{Ker}(T)$. Since $\mathcal{Ker}(T) = \{0\}$ (Theorem 4.6),

$$t_1 X_1 + \cdots + t_p X_p = 0,$$

and the linear independence of $\mathcal{A}$ then implies that $t_1, \ldots, t_p$ are all 0's. Thus, $T(\mathcal{A})$ is linearly independent.

(2) ⇒ (1). Suppose the image of each linearly independent set is linearly independent. If $X \in \mathcal{R}^n$ and $X \neq 0$, then $\mathcal{A} = \{X\}$ is linearly independent, so $T(\mathcal{A}) = \{T(X)\}$ is linearly independent. Thus, $T(X) \neq 0$. It follows that $\mathcal{Ker}(T) = \{0\}$ and, therefore, that T is one-to-one. ■

COROLLARY. If $T : \mathcal{R}^n \to \mathcal{R}^m$ is linear and one-to-one, then dim $\mathcal{Rng}(T) = n$.

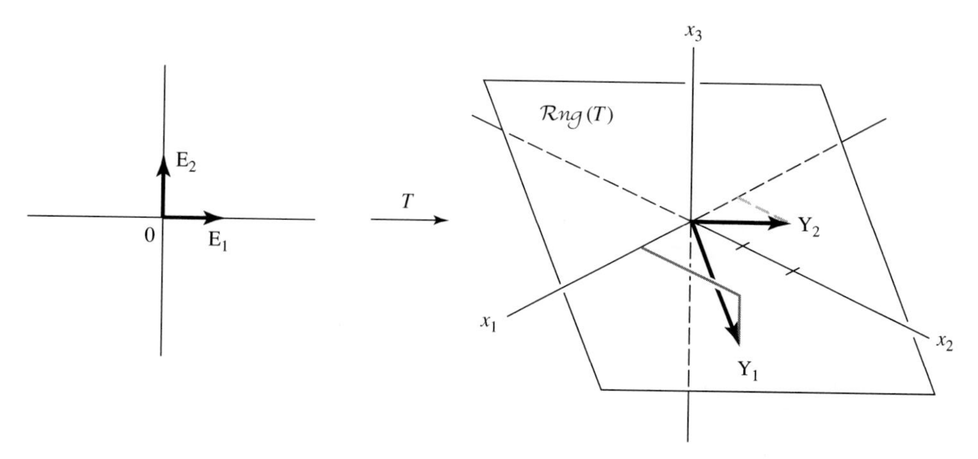

FIGURE 4.11

■ **EXAMPLE 4.19.** Define $T : \mathcal{R}^2 \to \mathcal{R}^3$ by $T(X) = AX$, where

$$A = \begin{bmatrix} 1 & -1 \\ 2 & 1 \\ -1 & 0 \end{bmatrix}.$$

Since the columns of A are linearly independent, dim $\mathcal{R}ng(T) = 2$ and dim $\mathcal{K}er(T) = 0$. Thus, T is one-to-one but not onto. The range of T, being a two-dimensional subspace of $\mathcal{R}^3$, is a plane through zero, and by Theorem 4.12, the image of any basis for $\mathcal{R}^2$ is a basis for $\mathcal{R}ng(T)$. In particular,

$$Y_1 = T(E_1) = \begin{bmatrix} 1 \\ 2 \\ -1 \end{bmatrix} \quad \text{and} \quad Y_2 = T(E_2) = \begin{bmatrix} -1 \\ 1 \\ 0 \end{bmatrix}$$

are linearly independent spanning vectors for $\mathcal{R}ng(T)$ (Figure 4.11). □

When $T : \mathcal{R}^n \to \mathcal{R}^m$ is not one-to-one, there is at least one Y in $\mathcal{R}ng(T)$ that is the image of more than one X in $\mathcal{R}^n$, and $\{X \in \mathcal{R}^n : T(X) = Y\}$ is the set of all points in the domain that are collapsed by T onto Y. To understand the action of T it helps to know the structure of these sets.

DEFINITION. Assume $T : \mathcal{R}^n \to \mathcal{R}^m$ and $Y \in \mathcal{R}ng(T)$. Each $X \in \mathcal{R}^n$ such that $T(X) = Y$ is called a *preimage of* Y. The set of all preimages of Y is

$$T^{-1}(Y) = \{X \in \mathcal{R}^n : T(X) = Y\}.$$

More generally, if $\mathcal{Y} \subseteq \mathcal{R}ng(T)$, then $T^{-1}(\mathcal{Y}) = \{X \in \mathcal{R}^n : T(X) \in \mathcal{Y}\}$.

■ **EXAMPLE 4.20.** Let

$$A = \begin{bmatrix} 1 & 2 & 1 \\ -1 & -1 & 0 \end{bmatrix},$$

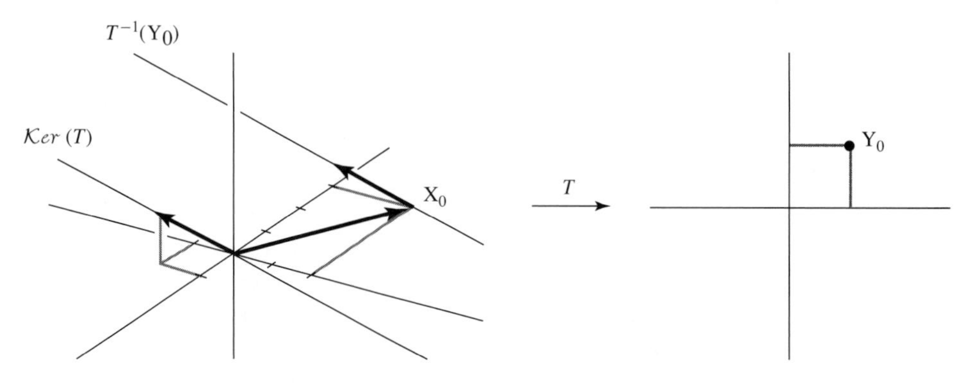

FIGURE 4.12

and consider $T : \mathcal{R}^3 \to \mathcal{R}^2$ given by $T(X) = AX$. Since

$$A \longrightarrow \begin{bmatrix} 1 & 0 & -1 \\ 0 & 1 & 1 \end{bmatrix},$$

$\mathcal{K}er(T) = \mathcal{S}pan\{[1 \quad -1 \quad 1]^T\}$ and $\dim \mathcal{R}ng(T) = \text{rank}(A) = 2$. Thus, T is onto but not one-to-one. Given $Y \in \mathcal{R}^2$, $T^{-1}(Y)$ is the set of solutions of $Y = AX$, so you calculate $T^{-1}(Y)$ by row reducing $[A \mid Y]$. In particular, when $Y_0 = \begin{bmatrix} 1 \\ 1 \end{bmatrix}$,

$$[A \mid Y_0] = \left[\begin{array}{ccc|c} 1 & 2 & 1 & 1 \\ -1 & -1 & 0 & 1 \end{array}\right] \longrightarrow \left[\begin{array}{ccc|c} 1 & 0 & -1 & -3 \\ 0 & 1 & 1 & 2 \end{array}\right].$$

The solutions of $AX = Y_0$ are therefore

$$X = \begin{bmatrix} t-3 \\ -t+2 \\ t \end{bmatrix} = t\begin{bmatrix} 1 \\ -1 \\ 1 \end{bmatrix} + \begin{bmatrix} -3 \\ 2 \\ 0 \end{bmatrix}, \qquad t \in \mathcal{R}.$$

Each is the sum of $X_0 = [-3 \quad 2 \quad 0]^T$ and a scalar multiple of $M = [1 \quad -1 \quad 1]^T$. Note that $\mathcal{K}er(T) = \mathcal{S}pan\{M\}$, so

$$T^{-1}(Y_0) = \{X_0 + X : X \in \mathcal{K}er(T)\} = X_0 + \mathcal{K}er(T).$$

Different choices of Y_0 generate different values of X_0, but in each case $T^{-1}(Y_0)$ is a translate of $\mathcal{K}er(T)$ (Figure 4.12). □

The behavior exhibited in Example 4.20 is typical of linear functions. If $T : \mathcal{R}^n \to \mathcal{R}^m$ is linear and $Y \in \mathcal{R}ng(T)$, then $T^{-1}(Y)$ is a translate of $\mathcal{K}er(T)$.

THEOREM 4.13. Suppose $T : \mathcal{R}^n \to \mathcal{R}^m$ is linear, $Y_0 \in \mathcal{R}ng(T)$, and $T(X_0) = Y_0$. Then

$$T^{-1}(Y_0) = X_0 + \mathcal{K}er(T).$$

Proof. If $X \in T^{-1}(Y_0)$, then $T(X) = Y_0 = T(X_0)$, so $T(X - X_0) = T(X) - T(X_0) = 0$. Thus, $V = X - X_0 \in \mathcal{K}er(T)$, and $X = X_0 + V \in X_0 + \mathcal{K}er(T)$. On the other hand, if $X \in X_0 + \mathcal{K}er(T)$, then $X = X_0 + V$, $V \in \mathcal{K}er(T)$, and $T(X) = T(X_0 + V) = T(X_0) + T(V) = Y_0 + 0 = Y_0$. Thus, $X \in T^{-1}(Y_0)$. ■

REMARK. If A is the $m \times n$ matrix such that $T(X) = AX$, then $T^{-1}(Y_0)$ is the solution set for the system $AX = Y_0$ and $\mathcal{K}er(T) = \mathcal{N}(A)$. A glance back at Theorem 2.10 reveals that

Theorem 4.13 is just a reformulation of the previous result using the language of linear transformations and affine subsets.

■ **EXAMPLE 4.21.** Let $T : \mathcal{R}^3 \to \mathcal{R}^2$ be the linear function with equation

$$T(\mathrm{X}) = \begin{bmatrix} 1 & -1 & 1 \\ -2 & 2 & -2 \end{bmatrix} \mathrm{X}.$$

Since

$$\mathrm{A} = \begin{bmatrix} 1 & -1 & 1 \\ -2 & 2 & -2 \end{bmatrix} \longrightarrow \begin{bmatrix} 1 & -1 & 1 \\ 0 & 0 & 0 \end{bmatrix},$$

$$\mathcal{K}er(T) = \mathcal{S}pan\{\mathrm{X}_2, \mathrm{X}_3\} \quad \text{and} \quad \mathcal{R}ng(T) = \mathcal{S}pan\{\mathrm{Y}_1\},$$

where

$$\mathrm{X}_2 = \begin{bmatrix} 1 \\ 1 \\ 0 \end{bmatrix}, \quad \mathrm{X}_3 = \begin{bmatrix} -1 \\ 0 \\ 1 \end{bmatrix}, \quad \text{and} \quad \mathrm{Y}_1 = \begin{bmatrix} 1 \\ -2 \end{bmatrix}.$$

Thus, $\mathcal{R}ng(T)$ is the line through 0 with direction vector $\mathrm{Y}_1 = T(\mathrm{E}_1)$, and $\mathcal{K}er(T)$ is the plane through 0, X_2, and X_3 (Figure 4.13). According to Theorem 4.13, each point of

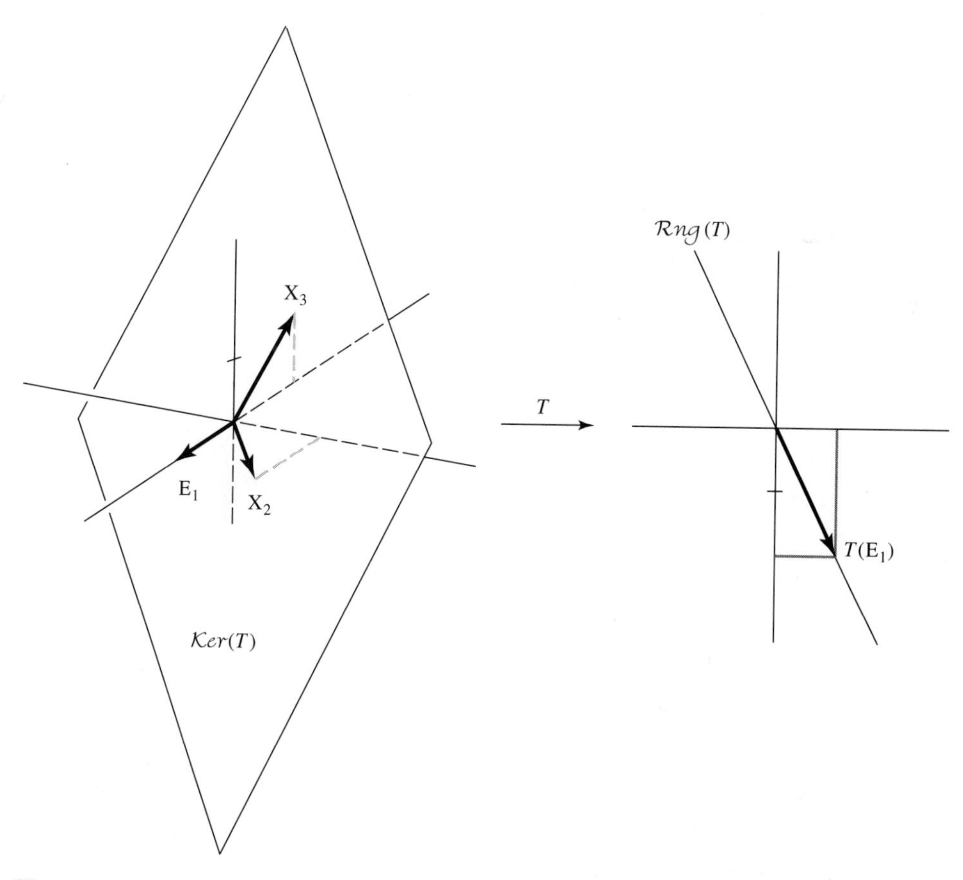

FIGURE 4.13

$\mathcal{Rng}(T)$ is the image of a translate of $\mathcal{Ker}(T)$. For example, $Y_0 = \begin{bmatrix} -3 \\ 6 \end{bmatrix} = -3\begin{bmatrix} 1 \\ -2 \end{bmatrix}$ $\in \mathcal{Rng}(T)$, and

$$[A \mid Y_0] = \left[\begin{array}{rrr|r} 1 & -1 & 1 & -3 \\ -2 & 2 & -2 & 6 \end{array}\right] \longrightarrow \left[\begin{array}{rrr|r} 1 & -1 & 1 & -3 \\ 0 & 0 & 0 & 0 \end{array}\right],$$

so

$$T^{-1}(Y_0) = \left\{ \begin{bmatrix} -3 \\ 0 \\ 0 \end{bmatrix} + s\begin{bmatrix} 1 \\ 1 \\ 0 \end{bmatrix} + t\begin{bmatrix} -1 \\ 0 \\ 1 \end{bmatrix} : s,t \in \mathcal{R} \right\}$$

$$= \begin{bmatrix} -3 \\ 0 \\ 0 \end{bmatrix} + \mathcal{Span}\{X_2, X_3\}.$$

Thus, $T^{-1}(Y_0) = X_0 + \mathcal{Ker}(T)$, where $X_0 = [-3 \quad 0 \quad 0]^T$ (Figure 4.14).

There is an alternative way to see the action of T. Suppose $B \in \mathcal{Ker}(T)$ and ℓ is the line through B parallel to E_1. Then

$$\ell \ : \ X(t) = tE_1 + B, \quad -\infty < t < \infty,$$

and

$$T(\ell) \ : \ Y(t) = tT(E_1) + T(B) = tY_1, \quad -\infty < t < \infty.$$

Thus, $T(\ell) = \mathcal{Rng}(T)$. As B travels through $\mathcal{Ker}(T)$, ℓ sweeps out all of $\mathcal{R}^3$, and each of these lines has $\mathcal{Rng}(T)$ as its image (Figure 4.15). The action of T can therefore be viewed as either collapsing each plane parallel to $\mathcal{Ker}(T)$ to a point of $\mathcal{Rng}(T)$ or as sending each line parallel to E_1 onto $\mathcal{Rng}(T)$. □

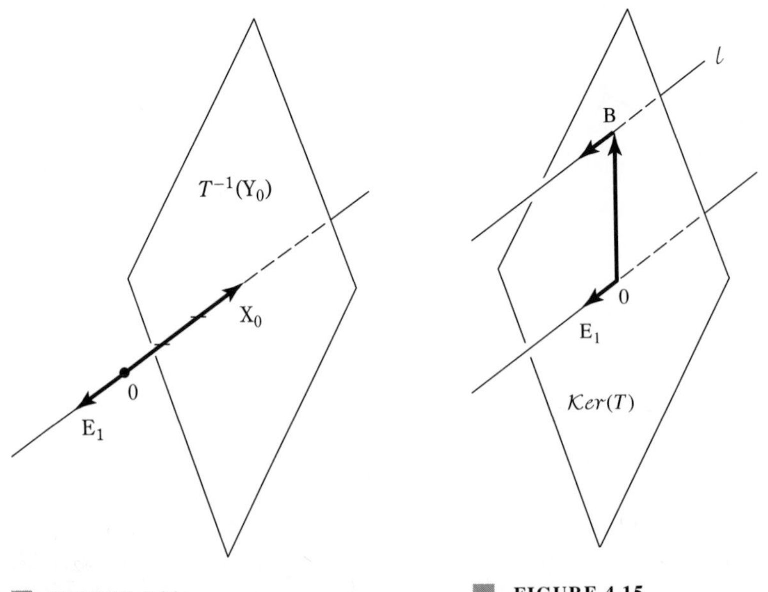

FIGURE 4.14 **FIGURE 4.15**

Exercises 4.4

1. Suppose $T(X) = \begin{bmatrix} -6 & 8 & -5 \\ 0 & 1 & -1 \\ 2 & 0 & -1 \end{bmatrix} X, X \in \mathcal{R}^3$. Find a spanning set for $T(\mathcal{V})$ and the dimension of $T(\mathcal{V})$ when

   a. $\mathcal{V} = Span\left\{\begin{bmatrix} 1 \\ 0 \\ -1 \end{bmatrix}, \begin{bmatrix} 1 \\ -1 \\ -3 \end{bmatrix}\right\}$.

   b. $\mathcal{V} = Span\left\{\begin{bmatrix} -1 \\ 1 \\ 2 \end{bmatrix}, \begin{bmatrix} 1 \\ 2 \\ 2 \end{bmatrix}\right\}$.

   c. $\mathcal{V} = Span\left\{\begin{bmatrix} 1 \\ 1 \\ 1 \end{bmatrix}, \begin{bmatrix} 0 \\ 2 \\ 2 \end{bmatrix}\right\}$.

   d. $\mathcal{V} = \mathcal{R}^3$.

2. Define $T : \mathcal{R}^4 \to \mathcal{R}^3$ by $T(X) = \begin{bmatrix} 1 & 1 & 1 & 3 \\ -1 & 0 & 1 & -2 \\ 1 & 2 & 3 & 4 \end{bmatrix} X$. Find a spanning set for $T(\mathcal{V})$ and the dimension of $T(\mathcal{V})$ when

   a. $\mathcal{V} = Span\left\{\begin{bmatrix} 1 \\ -1 \\ -1 \\ 1 \end{bmatrix}, \begin{bmatrix} 0 \\ 1 \\ 0 \\ -1 \end{bmatrix}, \begin{bmatrix} 1 \\ 1 \\ 1 \\ -1 \end{bmatrix}\right\}$.

   b. $\mathcal{V} = Span\left\{\begin{bmatrix} 0 \\ 1 \\ 1 \\ -1 \end{bmatrix}, \begin{bmatrix} 2 \\ -1 \\ -2 \\ 1 \end{bmatrix}, \begin{bmatrix} 1 \\ -3 \\ 0 \\ 1 \end{bmatrix}\right\}$.

   c. $\mathcal{V} = Span\left\{\begin{bmatrix} -1 \\ -3 \\ 1 \\ 1 \end{bmatrix}, \begin{bmatrix} 5 \\ 0 \\ 1 \\ -2 \end{bmatrix}\right\}$.

   d. $\mathcal{V} = \mathcal{R}^4$.

3. Find a subspace $\mathcal{V}$ and vector B such that $\mathcal{S} = B + \mathcal{V}$.

   a. $\mathcal{S} = \{X \in \mathcal{R}^2 : x_1 = \frac{1}{3}x_2 - 2\}$

   b. $\mathcal{S} = \{X \in \mathcal{R}^3 : x_1 + 2x_2 - 3x_3 = 4\}$

   c. $\mathcal{S} = \left\{X \in \mathcal{R}^4 : \begin{bmatrix} 1 & -1 & 2 & -3 \\ -1 & 2 & -3 & 4 \end{bmatrix} X = \begin{bmatrix} 5 \\ -6 \end{bmatrix}\right\}$

4. Suppose p_1 and p_2 are parallel planes in $\mathcal{R}^3$, $T : \mathcal{R}^3 \to \mathcal{R}^3$ is linear, and $\mathcal{K}er(T) \neq \mathcal{R}^3$. Explain why $T(p_1)$ and $T(p_2)$ are either parallel planes, parallel lines, or two points.

5. Define $T : \mathcal{R}^2 \to \mathcal{R}^2$ by $T(X) = \begin{bmatrix} 1 & -1 \\ -1 & 2 \end{bmatrix} X$. For $X_1 = \begin{bmatrix} 3 \\ 1 \end{bmatrix}$ and $X_2 = \begin{bmatrix} -1 \\ 1 \end{bmatrix}$, sketch

   a. $\mathcal{S} = \{sX_1 + tX_2 : s \in \mathcal{R}, -2 \le t \le 1\}$.

   b. $\mathcal{P} = \{sX_1 + tX_2 : -1 \le s \le 1, 0 \le t \le 2\}$.

   c. $T(\mathcal{S})$.

   d. $T(\mathcal{P})$.

6. Consider the following subsets of $\mathcal{R}^2$: $\mathcal{X}_1 = Span\{E_1\}$, $\mathcal{X}_2 = Span\{E_2\}$, $Q = \{X : x_1, x_2 > 0\}$, $H = \{X : x_2 < 0\}$, $\mathcal{U} = \{X : x_1 = x_2\}$, $\mathcal{L} = \{X : x_1 = x_2 + 2\}$, and $C = \{X : -1 \le x_1 \le 0, 0 \le x_2 \le 1\}$. Sketch $T(\mathcal{X}_1)$, $T(\mathcal{X}_2)$, $T(Q)$, $T(H)$, $T(\mathcal{U})$, $T(\mathcal{L})$, and

$T(\mathrm{C})$ when

a. $T(\mathrm{X}) = \begin{bmatrix} 1 & -3 \\ 1 & 1 \end{bmatrix}\mathrm{X}$. b. $T(\mathrm{X}) = \begin{bmatrix} 2 & 0 \\ 0 & -3 \end{bmatrix}\mathrm{X}$.

7. Consider the following subsets of $\mathcal{R}^3$: $\mathcal{X}_1 = \mathcal{S}pan\{\mathrm{E}_1\}$, $\mathcal{X}_2 = \mathcal{S}pan\{\mathrm{E}_2\}$, $\mathcal{X}_3 = \mathcal{S}pan\{\mathrm{E}_3\}$,

$\mathcal{U} = \{\mathrm{X} : x_2 = 0\}$, $p = \{\mathrm{X} : x_2 = 1\}$, $\mathrm{H} = \{\mathrm{X} : x_2 < 0\}$,

$\mathcal{l} = \begin{bmatrix} 1 \\ 2 \\ 3 \end{bmatrix} + \mathcal{V}$. Sketch $T(\mathcal{X}_1)$, $T(\mathcal{X}_2)$, $T(\mathcal{X}_3)$, $T(\mathcal{U})$, $T(p)$, $T(\mathrm{H})$, $T(\mathcal{V})$, and $T(\mathcal{l})$ when

a. $\mathrm{A} = \begin{bmatrix} 1 & 0 & -1 \\ -2 & 2 & 1 \\ 0 & 0 & 1 \end{bmatrix}$. b. $\mathrm{A} = \begin{bmatrix} 1 & 0 & -1 \\ 0 & -1 & 0 \\ 0 & 1 & 0 \end{bmatrix}$.

8. Suppose $T : \mathcal{R}^n \to \mathcal{R}^m$ is linear, $\mathcal{V}$ is a subspace of $\mathcal{R}^n$, and $\mathcal{V} \cap \mathcal{K}er(T) = \{0\}$.
   a. Show that $\dim T(\mathcal{V}) = \dim \mathcal{V}$. (Hint: Assuming $\{\mathrm{X}_1, \ldots, \mathrm{X}_p\}$ is a basis for $\mathcal{V}$, show that $\{T(\mathrm{X}_1), \ldots, T(\mathrm{X}_p)\}$ is a basis for $T(\mathcal{V})$.)
   b. Show that if $\mathrm{V}_1, \mathrm{V}_2 \in \mathcal{V}$ and $T(\mathrm{V}_1) = T(\mathrm{V}_2)$, then $\mathrm{V}_1 = \mathrm{V}_2$. (That is, T acts as a one-to-one function on $\mathcal{V}$.)

9. Suppose $T : \mathcal{R}^n \to \mathcal{R}^m$ is linear and $\mathcal{V}$ is a subspace of $\mathcal{R}^n$. Show that $\dim T(\mathcal{V}) = \dim \mathcal{V} - \dim \mathcal{V} \cap \mathcal{K}er(T)$.

10. Let $T : \mathcal{R}^3 \to \mathcal{R}^3$ be the linear function with equation $T(X) = \begin{bmatrix} -1 & 1 & 0 \\ 2 & 0 & 1 \\ 0 & 2 & 1 \end{bmatrix}\mathrm{X}$.
   a. Sketch $\mathcal{K}er(T)$.
   b. Sketch $\mathcal{R}ng(T)$.
   c. Let $\mathcal{U} = \mathcal{S}pan\{\mathrm{E}_1, \mathrm{E}_2\}$ and suppose $\mathrm{B} \in \mathcal{R}^3$. Show that $T(\mathrm{B} + \mathcal{U}) = \mathcal{R}ng(T)$.

11. Define $T : \mathcal{R}^2 \to \mathcal{R}^2$ by $T(\mathrm{X}) = \begin{bmatrix} a & 0 \\ 0 & b \end{bmatrix}\mathrm{X}$. Describe the action of T geometrically when

a. $a > 0$, $b = 1$. b. $a > 0$, $b > 0$. c. $a < 0$, $b > 0$. d. $a > 0$, $b = 0$.

12. Define $T : \mathcal{R}^3 \to \mathcal{R}^2$ by $T(\mathrm{X}) = \begin{bmatrix} 1 & 1 & 0 \\ -1 & 1 & -2 \end{bmatrix}\mathrm{X}$. Find

a. $\mathcal{K}er(T)$. b. $T^{-1}\left(\begin{bmatrix} 3 \\ 1 \end{bmatrix}\right)$. **c.** $T^{-1}\left(\begin{bmatrix} -1 \\ 5 \end{bmatrix}\right)$. d. $T^{-1}\left(\mathcal{S}pan\left\{\begin{bmatrix} 3 \\ 1 \end{bmatrix}\right\}\right)$.

13. Define $T : \mathcal{R}^3 \to \mathcal{R}^3$ by $T(\mathrm{X}) = \begin{bmatrix} -2 & 1 & 0 \\ 2 & 2 & 6 \\ 1 & -1 & -1 \end{bmatrix}\mathrm{X}$. Find

a. $\mathcal{K}er(T)$. b. $T^{-1}\left(\begin{bmatrix} 5 \\ 1 \\ -2 \end{bmatrix}\right)$.

c. $T^{-1}\left(\begin{bmatrix} 4 \\ 2 \\ -3 \end{bmatrix}\right)$. d. $T^{-1}\left(\mathcal{S}pan\left\{\begin{bmatrix} 4 \\ 2 \\ -3 \end{bmatrix}\right\}\right)$.

14. Suppose $T : \mathcal{R}^2 \to \mathcal{R}^2$ is given by $T(X) = \begin{bmatrix} 1 & 1 \\ 2 & 1 \end{bmatrix} X$ and $Q = \{Y \in \mathcal{R}^2 : y_1, y_2 > 0\}$. Find $T^{-1}(Q)$.

15. Suppose $T : \mathcal{R}^n \to \mathcal{R}^m$ is linear, $\mathcal{W}$ is a subspace of $\mathcal{R}^m$, and $\mathcal{W} \subseteq \mathcal{Rng}(T)$.
    a. Show that $\mathcal{V} = T^{-1}(\mathcal{W})$ is a subspace of $\mathcal{R}^n$.
    b. Assuming $\{W_1, \ldots, W_q\}$ is a basis for $\mathcal{W}$, $T^{-1}(W_k) = B_k + \mathcal{Ker}(T)$, $1 \le k \le q$, and $\{X_1, \ldots, X_p\}$ is a basis for $\mathcal{Ker}(T)$, show that $\{B_1, \ldots, B_q, X_1, \ldots, X_p\}$ is a basis for $\mathcal{V}$.

16. Assuming $\mathcal{V}$ is a subspace of $\mathcal{R}^n$ and $B \in \mathcal{R}^n$, describe the circumstances under which $B + \mathcal{V} = \mathcal{V}$.

17. a. Show that for $X,Y \in \mathcal{R}^3$, $\mathcal{L}(X,Y) = \{tX + (1-t)Y : t \in [0,1]\}$ is the line segment joining X and Y.
    b. A subset $\mathcal{S}$ of $\mathcal{R}^n$ is said to be *convex* if $\mathcal{L}(X,Y) \subseteq \mathcal{S}$ whenever $X,Y \in \mathcal{S}$. Show that every affine subset of $\mathcal{R}^n$ is convex.

4.5
MATRIX REPRESENTATIONS

For each linear $T : \mathcal{R}^n \to \mathcal{R}^m$, there is an $m \times n$ matrix A such that

$$Y = T(X) = AX. \tag{4.8}$$

This equation calculates values of T in terms of the usual coordinate systems for $\mathcal{R}^n$ and $\mathcal{R}^m$. That is, X in (4.8) is the description relative to the standard basis for $\mathcal{R}^n$ of a point in the domain, and Y is the description of its image relative to the standard basis for $\mathcal{R}^m$. There are other choices of coordinate systems for describing the domain and range variables, and each generates a "formula" for T. The present section explores these "formulas."

THEOREM 4.14. Assume $T : \mathcal{R}^n \to \mathcal{R}^m$ is linear. If $\mathcal{A} = \{X_1, \ldots, X_n\}$ is a basis for $\mathcal{R}^n$ and $\mathcal{B} = \{Y_1, \ldots, Y_m\}$ is a basis for $\mathcal{R}^m$, then there is a unique $m \times n$ matrix, $[T]_{\mathcal{BA}}$, with the property that for all $X \in \mathcal{R}^n$,

$$[T(X)]_{\mathcal{B}} = [T]_{\mathcal{BA}}[X]_{\mathcal{A}}. \tag{4.9}$$

Moreover,

$$[T]_{\mathcal{BA}} = \big[[T(X_1)]_{\mathcal{B}} \quad \cdots \quad [T(X_n)]_{\mathcal{B}}\big]. \tag{4.10}$$

Proof. If $X \in \mathcal{R}^n$ and $[X]_{\mathcal{A}} = [t_1 \ \cdots \ t_n]^T$, then $X = \sum_1^n t_k X_k$, $T(X) = \sum_1^n t_k T(X_k)$, and

$$\begin{aligned}
[T(X)]_{\mathcal{B}} &= [t_1 T(X_1) + \cdots + t_n T(X_n)]_{\mathcal{B}} \\
&= t_1[T(X_1)]_{\mathcal{B}} + \cdots + t_n[T(X_n)]_{\mathcal{B}} \\
&= \big[[T(X_1)]_{\mathcal{B}} \quad \cdots \quad [T(X_n)]_{\mathcal{B}}\big]\begin{bmatrix} t_1 \\ \vdots \\ t_n \end{bmatrix} \\
&= \big[[T(X_1)]_{\mathcal{B}} \quad \cdots \quad [T(X_n)]_{\mathcal{B}}\big][X]_{\mathcal{A}}.
\end{aligned}$$

Thus, $[T]_{\mathcal{BA}} = \big[[T(X_1)]_{\mathcal{B}} \quad \cdots \quad [T(X_n)]_{\mathcal{B}}\big]$ has property (4.9). For the uniqueness claim, suppose A and B are two such matrices, that is, $A[X]_{\mathcal{A}} = [T(X)]_{\mathcal{B}} = B[X]_{\mathcal{A}}$, for all $X \in \mathcal{R}^n$. If $X = X_j$, then $[X]_{\mathcal{A}} = E_j$, so

$$\mathrm{Col}_j(A) = AE_j = A[X]_{\mathcal{A}} = B[X]_{\mathcal{A}} = BE_j = \mathrm{Col}_j(B),$$

$j = 1, \ldots, n$. Thus, A = B. ■

■ **EXAMPLE 4.22.** Consider $T : \mathcal{R}^3 \to \mathcal{R}^2$ given by

$$T(X) = \begin{bmatrix} 1 & -2 & 3 \\ 2 & 1 & -1 \end{bmatrix} X. \tag{4.11}$$

Observe that $\mathcal{A} = \{X_1, X_2, X_3\} = \left\{ \begin{bmatrix} 0 \\ 0 \\ 1 \end{bmatrix}, \begin{bmatrix} 0 \\ 1 \\ 1 \end{bmatrix}, \begin{bmatrix} 1 \\ 1 \\ 1 \end{bmatrix} \right\}$ and $\mathcal{B} = \{Y_1, Y_2\} = \left\{ \begin{bmatrix} 1 \\ 1 \end{bmatrix}, \begin{bmatrix} -1 \\ 1 \end{bmatrix} \right\}$

are bases for $\mathcal{R}^3$ and $\mathcal{R}^2$, respectively. The columns of $[T]_{\mathcal{BA}}$, namely $[T(X_1)]_{\mathcal{B}}$, $[T(X_2)]_{\mathcal{B}}$, and $[T(X_3)]_{\mathcal{B}}$, are computed simultaneously by the row-reduction

$$[Y_1 \quad Y_2 \mid T(X_1) \quad T(X_2) \quad T(X_3)] \longrightarrow \big[I_2 \mid [T(X_1)]_{\mathcal{B}} \quad [T(X_2)]_{\mathcal{B}} \quad [T(X_3)]_{\mathcal{B}}\big].$$

From (4.11),

$$T(X_1) = \begin{bmatrix} 3 \\ -1 \end{bmatrix}, \quad T(X_2) = \begin{bmatrix} 1 \\ 0 \end{bmatrix}, \quad \text{and} \quad T(X_3) = \begin{bmatrix} 2 \\ 2 \end{bmatrix},$$

so

$$[Y_1 \quad Y_2 \mid T(X_1) \quad T(X_2) \quad T(X_3)] = \left[\begin{array}{cc|ccc} 1 & -1 & 3 & 1 & 2 \\ 1 & 1 & -1 & 0 & 2 \end{array}\right]$$
$$\longrightarrow \left[\begin{array}{cc|ccc} 1 & 0 & 1 & 1/2 & 2 \\ 0 & 1 & -2 & -1/2 & 0 \end{array}\right].$$

Thus,

$$[T]_{\mathcal{BA}} = \frac{1}{2}\begin{bmatrix} 2 & 1 & 4 \\ -4 & -1 & 0 \end{bmatrix}.$$

To illustrate how $[T]_{\mathcal{BA}}$ calculates values of T, consider

$$X = 2X_1 - 3X_2 + X_3 = 2\begin{bmatrix} 0 \\ 0 \\ 1 \end{bmatrix} - 3\begin{bmatrix} 0 \\ 1 \\ 1 \end{bmatrix} + \begin{bmatrix} 1 \\ 1 \\ 1 \end{bmatrix} = \begin{bmatrix} 1 \\ -2 \\ 0 \end{bmatrix}.$$

Then $[X]_{\mathcal{A}} = [2 \quad -3 \quad 1]^T$, and by (4.9),

$$[T(X)]_{\mathcal{B}} = [T]_{\mathcal{BA}}[X]_{\mathcal{A}} = \frac{1}{2}\begin{bmatrix} 2 & 1 & 4 \\ -4 & -1 & 0 \end{bmatrix}\begin{bmatrix} 2 \\ -3 \\ 1 \end{bmatrix} = \frac{1}{2}\begin{bmatrix} 5 \\ -5 \end{bmatrix}.$$

Thus,

$$T(X) = \frac{5}{2}Y_1 - \frac{5}{2}Y_2 = \begin{bmatrix} 5 \\ 0 \end{bmatrix}. \quad \square$$

In the notation of Theorem 4.14, the algorithm for calculating $[T]_{\mathcal{BA}}$ is

$$[Y_1 \quad \cdots \quad Y_m \mid T(X_1) \quad \cdots \quad T(X_n)] \longrightarrow [I_m \mid [T]_{\mathcal{BA}}].$$

■ **EXAMPLE 4.23.** Define $T : \mathcal{R}^3 \to \mathcal{R}^4$ by $T(\mathrm{X}) = \begin{bmatrix} -6 & 2 & 2 \\ -7 & 3 & 5 \\ 9 & -5 & -5 \\ -6 & 4 & 8 \end{bmatrix} \mathrm{X}$, and consider the bases

$$\mathcal{A} = \{\mathrm{X}_1, \mathrm{X}_2, \mathrm{X}_3\} = \left\{ \begin{bmatrix} 1 \\ 2 \\ 0 \end{bmatrix}, \begin{bmatrix} 0 \\ 2 \\ -1 \end{bmatrix}, \begin{bmatrix} -1 \\ -3 \\ 1 \end{bmatrix} \right\}$$

and

$$\mathcal{B} = \{\mathrm{Y}_1, \mathrm{Y}_2, \mathrm{Y}_3, \mathrm{Y}_4\} = \left\{ \begin{bmatrix} 1 \\ 0 \\ -3 \\ -1 \end{bmatrix}, \begin{bmatrix} 1 \\ 1 \\ 0 \\ 0 \end{bmatrix}, \begin{bmatrix} 3 \\ 2 \\ -2 \\ -1 \end{bmatrix}, \begin{bmatrix} -1 \\ 0 \\ 0 \\ 2 \end{bmatrix} \right\}$$

for $\mathcal{R}^3$ and $\mathcal{R}^4$, respectively. Observe that

$$[\mathrm{Y}_1 \quad \mathrm{Y}_2 \quad \mathrm{Y}_3 \quad \mathrm{Y}_4 \mid T(\mathrm{X}_1) \quad T(\mathrm{X}_2) \quad T(\mathrm{X}_3)] = \left[\begin{array}{rrrr|rrr} 1 & 1 & 3 & -1 & -2 & 2 & 2 \\ 0 & 1 & 2 & 0 & -1 & 1 & 3 \\ -3 & 0 & -2 & 0 & -1 & -5 & 1 \\ -1 & 0 & -1 & 2 & 2 & 0 & 2 \end{array}\right]$$

$$\to \left[\begin{array}{rrrr|rrr} 1 & 0 & 0 & 0 & 1 & 1 & -1 \\ 0 & 1 & 0 & 0 & 1 & -1 & 1 \\ 0 & 0 & 1 & 0 & -1 & 1 & 1 \\ 0 & 0 & 0 & 1 & 1 & 1 & 1 \end{array}\right],$$

so

$$[T]_{\mathcal{B}\mathcal{A}} = \begin{bmatrix} 1 & 1 & -1 \\ 1 & -1 & 1 \\ -1 & 1 & 1 \\ 1 & 1 & 1 \end{bmatrix}. \square$$

The standard basis for $\mathcal{R}^n$ has been denoted by $\mathcal{E}$. Sometimes, for the sake of clarity, it is helpful to have the notation reflect the dimension of $\mathcal{R}^n$, so for that purpose the symbol $\mathcal{E}_n$ will be used. If $T : \mathcal{R}^n \to \mathcal{R}^m$ is linear and the standard bases have been selected for describing both the domain and range variables, then in the notation of Theorem 4.14, $\mathcal{A} = \mathcal{E}_n$, $\mathcal{B} = \mathcal{E}_m$, and

$$\mathrm{Col}_j([T]_{\mathcal{E}_m\mathcal{E}_n}) = [T(\mathrm{E}_j)]_{\mathcal{E}_m} = T(\mathrm{E}_j),$$

$1 \leq j \leq n$. Thus,

$$[T]_{\mathcal{E}_m\mathcal{E}_n} = [T(\mathrm{E}_1) \quad \cdots \quad T(\mathrm{E}_n)],$$

which is the matrix A of Theorem 4.1. Since $[\mathrm{X}]_{\mathcal{E}_n} = \mathrm{X}$ and $[T(\mathrm{X})]_{\mathcal{E}_m} = T(\mathrm{X})$, (4.5) is just a special case of (4.9).

When $n = m$, there is the option of using the same basis $\mathcal{B}$ to describe both the domain and range variables. The notation can then be simplified by writing $[T]_{\mathcal{B}}$ instead of $[T]_{\mathcal{B}\mathcal{B}}$, and (4.9) becomes $[T(\mathrm{X})]_{\mathcal{B}} = [T]_{\mathcal{B}}[\mathrm{X}]_{\mathcal{B}}$.

■ **EXAMPLE 4.24.** Suppose $\mathcal{V} = \{X \in \mathcal{R}^3 : x_1 = x_2\}$, $T : \mathcal{R}^3 \to \mathcal{R}^3$ is projection onto $\mathcal{V}$, and

$$\mathcal{B} = \{X_1, X_2, X_3\} = \left\{ \begin{bmatrix} 1 \\ 1 \\ 0 \end{bmatrix}, \begin{bmatrix} -1 \\ -1 \\ 1 \end{bmatrix}, \begin{bmatrix} 1 \\ -1 \\ 0 \end{bmatrix} \right\}.$$

Observe that $X_1, X_2 \in \mathcal{V}$ and $X_3 \in \mathcal{V}^\perp$ ($X_3 \cdot X_1 = 0 = X_3 \cdot X_2$). Thus,

$$\begin{aligned} T(X_1) &= X_1 = 1X_1 + 0X_2 + 0X_3, \\ T(X_2) &= X_2 = 0X_1 + 1X_2 + 0X_3, \\ T(X_3) &= \ 0 \ = 0X_1 + 0X_2 + 0X_3, \end{aligned}$$

and

$$[T]_{\mathcal{B}} = \begin{bmatrix} [T(X_1)]_{\mathcal{B}} & [T(X_2)]_{\mathcal{B}} & [T(X_3)]_{\mathcal{B}} \end{bmatrix} = \begin{bmatrix} 1 & 0 & 0 \\ 0 & 1 & 0 \\ 0 & 0 & 0 \end{bmatrix}. \ \square$$

A linear transformation, T, has as many matrix representatives, $[T]_{\mathcal{BA}}$, as there are choices for $\mathcal{A}$ and $\mathcal{B}$. Any two such matrices are related, and it should come as no surprise that the connection between them involves changing coordinates in the domain and range.

THEOREM 4.15. If $T : \mathcal{R}^n \to \mathcal{R}^m$ is linear, $\mathcal{A}$ and $\mathcal{C}$ are bases for $\mathcal{R}^n$, and $\mathcal{B}$ and $\mathcal{D}$ are bases for $\mathcal{R}^m$, then

$$[T]_{\mathcal{DC}} = P_{\mathcal{DB}}[T]_{\mathcal{BA}}P_{\mathcal{AC}}.$$

Proof. $[T]_{\mathcal{DC}}$ stands for the unique $m \times n$ matrix satisfying $[T(X)]_{\mathcal{D}} = [T]_{\mathcal{DC}}[X]_{\mathcal{C}}$, $X \in \mathcal{R}^n$, so it suffices to show that $P_{\mathcal{DB}}[T]_{\mathcal{BA}}P_{\mathcal{AC}}$ has this property. Indeed,

$$\begin{aligned} (P_{\mathcal{DB}}[T]_{\mathcal{BA}}P_{\mathcal{AC}})[X]_{\mathcal{C}} &= (P_{\mathcal{DB}}[T]_{\mathcal{BA}})(P_{\mathcal{AC}}[X]_{\mathcal{C}}) && \text{(matrix algebra)} \\ &= (P_{\mathcal{DB}}[T]_{\mathcal{BA}})[X]_{\mathcal{A}} && \text{(Theorem 2.33)} \\ &= P_{\mathcal{DB}}([T]_{\mathcal{BA}}[X]_{\mathcal{A}}) && \text{(matrix algebra)} \\ &= P_{\mathcal{DB}}[T(X)]_{\mathcal{B}} && \text{(Theorem 4.14)} \\ &= [T(X)]_{\mathcal{D}}. && \text{(Theorem 2.33)} \ \blacksquare \end{aligned}$$

■ **EXAMPLE 4.25.** Recall $T : \mathcal{R}^3 \to \mathcal{R}^2$ in Example 4.22 given by

$$T(X) = \begin{bmatrix} 1 & -2 & 3 \\ 2 & 1 & -1 \end{bmatrix} X.$$

Of course,

$$[T]_{\mathcal{E}_2\mathcal{E}_3} = [T(E_1) \quad T(E_2) \quad T(E_3)] = \begin{bmatrix} 1 & -2 & 3 \\ 2 & 1 & -1 \end{bmatrix}.$$

If

$$\mathcal{A} = \left\{ \begin{bmatrix} 0 \\ 0 \\ 1 \end{bmatrix}, \begin{bmatrix} 0 \\ 1 \\ 1 \end{bmatrix}, \begin{bmatrix} 1 \\ 1 \\ 1 \end{bmatrix} \right\} \quad \text{and} \quad \mathcal{B} = \left\{ \begin{bmatrix} 1 \\ 1 \end{bmatrix}, \begin{bmatrix} -1 \\ 1 \end{bmatrix} \right\},$$

then $[T]_{\mathcal{BA}} = \mathrm{P}_{\mathcal{BE}_2}[T]_{\mathcal{E}_2\mathcal{E}_3}\mathrm{P}_{\mathcal{E}_3\mathcal{A}}$ (Theorem 4.15). According to Example 2.46,

$$\mathrm{P}_{\mathcal{E}_3\mathcal{A}} = \begin{bmatrix} 0 & 0 & 1 \\ 0 & 1 & 1 \\ 1 & 1 & 1 \end{bmatrix} \quad \text{and} \quad \mathrm{P}_{\mathcal{E}_2\mathcal{B}} = \begin{bmatrix} 1 & -1 \\ 1 & 1 \end{bmatrix}.$$

Then

$$\mathrm{P}_{\mathcal{BE}_2} = \mathrm{P}_{\mathcal{E}_2\mathcal{B}}^{-1} = \begin{bmatrix} 1 & -1 \\ 1 & 1 \end{bmatrix}^{-1} = \frac{1}{2}\begin{bmatrix} 1 & 1 \\ -1 & 1 \end{bmatrix},$$

and

$$\mathrm{P}_{\mathcal{BE}_2}[T]_{\mathcal{E}_2\mathcal{E}_3}\mathrm{P}_{\mathcal{E}_3\mathcal{A}} = \frac{1}{2}\begin{bmatrix} 1 & 1 \\ -1 & 1 \end{bmatrix}\begin{bmatrix} 1 & -2 & 3 \\ 2 & 1 & -1 \end{bmatrix}\begin{bmatrix} 0 & 0 & 1 \\ 0 & 1 & 1 \\ 1 & 1 & 1 \end{bmatrix} = \frac{1}{2}\begin{bmatrix} 2 & 1 & 4 \\ -4 & -1 & 0 \end{bmatrix}.$$

Note that this agrees with the conclusion of Example 4.22. □

In the terminology of Theorem 4.15, there are four matrix representations for T involving bases $\mathcal{A}$, $\mathcal{B}$, $\mathcal{C}$, and $\mathcal{D}$, namely, $[T]_{\mathcal{DC}}$, $[T]_{\mathcal{DA}}$, $[T]_{\mathcal{BC}}$, and $[T]_{\mathcal{BA}}$. Any two are related as indicated in that theorem. For example,

$$[T]_{\mathcal{DA}} = \mathrm{P}_{\mathcal{DB}}[T]_{\mathcal{BC}}\mathrm{P}_{\mathcal{CA}},$$

and

$$[T]_{\mathcal{DA}} = \mathrm{P}_{\mathcal{DB}}[T]_{\mathcal{BA}}\mathrm{P}_{\mathcal{AA}} = \mathrm{P}_{\mathcal{DB}}[T]_{\mathcal{BA}},$$

where the last equality utilizes the fact that $\mathrm{P}_{\mathcal{AA}} = \mathrm{I}_n$. Similarly, $[T]_{\mathcal{BA}} = [T]_{\mathcal{BC}}\mathrm{P}_{\mathcal{CA}}$. Note how the order of the subscripts indicates the placement of the factors.

THEOREM 4.16. If $T : \mathcal{R}^n \to \mathcal{R}^n$ is linear and $\mathcal{A}$ and $\mathcal{B}$ are bases for $\mathcal{R}^n$, then

$$[T]_{\mathcal{B}} = \mathrm{P}_{\mathcal{AB}}^{-1}[T]_{\mathcal{A}}\mathrm{P}_{\mathcal{AB}}.$$

Proof. $[T]_{\mathcal{B}} = \mathrm{P}_{\mathcal{BA}}[T]_{\mathcal{A}}\mathrm{P}_{\mathcal{AB}}$ (Theorem 4.15), and $\mathrm{P}_{\mathcal{BA}} = \mathrm{P}_{\mathcal{AB}}^{-1}$ (Theorem 2.34). ■

■ **EXAMPLE 4.26.** Suppose $\mathcal{L}$ is the line in $\mathcal{R}^2$ with equation $y = mx$ and $T : \mathcal{R}^2 \to \mathcal{R}^2$ is reflection in $\mathcal{L}$. It is easy to see geometrically (as in Example 4.3) that T is linear. If $\mathrm{X} \in \mathcal{L}$, then $T(\mathrm{X}) = \mathrm{X}$, and if X lies on the line through 0 perpendicular to $\mathcal{L}$, then $T(\mathrm{X}) = -\mathrm{X}$. The last two observations suggest a coordinate system for $\mathcal{R}^2$ with respect to which the matrix representative for T is especially simple. Set

$$\mathrm{X}_1 = \begin{bmatrix} 1 \\ m \end{bmatrix}, \quad \mathrm{X}_2 = \begin{bmatrix} -m \\ 1 \end{bmatrix}, \quad \text{and} \quad \mathcal{B} = \{\mathrm{X}_1, \mathrm{X}_2\}.$$

Note that $\mathrm{X}_1 \in \mathcal{L}$ and X_2 is perpendicular to $\mathcal{L}$ (Figure 4.16). Since $T(\mathrm{X}_1) = \mathrm{X}_1 = \mathrm{X}_1 + 0\mathrm{X}_2$ and $T(\mathrm{X}_2) = -\mathrm{X}_2 = 0\mathrm{X}_1 + (-1)\mathrm{X}_2$,

$$[T]_{\mathcal{B}} = \begin{bmatrix} [T(\mathrm{X}_1)]_{\mathcal{B}} & [T(\mathrm{X}_2)]_{\mathcal{B}} \end{bmatrix} = \begin{bmatrix} 1 & 0 \\ 0 & -1 \end{bmatrix}.$$

Having obtained one matrix for T, Theorem 4.16 can now be used to get the formula for T relative to the standard coordinate system. Indeed,

$$\mathrm{P}_{\mathcal{EB}} = \begin{bmatrix} 1 & -m \\ m & 1 \end{bmatrix}, \quad \text{so} \quad \mathrm{P}_{\mathcal{BE}} = \mathrm{P}_{\mathcal{EB}}^{-1} = \frac{1}{1+m^2}\begin{bmatrix} 1 & m \\ -m & 1 \end{bmatrix},$$

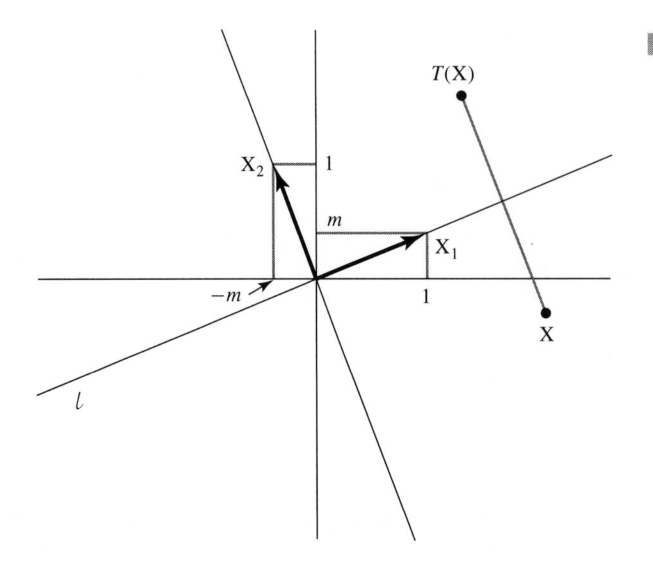

FIGURE 4.16

and therefore

$$\begin{aligned}[T]_{\mathcal{E}} &= \mathrm{P}_{\mathcal{E}\mathcal{B}}[T]_{\mathcal{B}}\mathrm{P}_{\mathcal{B}\mathcal{E}} \\ &= \begin{bmatrix} 1 & -m \\ m & 1 \end{bmatrix}\begin{bmatrix} 1 & 0 \\ 0 & -1 \end{bmatrix}\left(\frac{1}{1+m^2}\begin{bmatrix} 1 & m \\ -m & 1 \end{bmatrix}\right) \\ &= \frac{1}{1+m^2}\begin{bmatrix} 1-m^2 & 2m \\ 2m & m^2-1 \end{bmatrix}. \square\end{aligned}$$

Exercises 4.5

1. If $T(\mathrm{X}) = \begin{bmatrix} -2 & 1 \\ 0 & -1 \\ 1 & 2 \end{bmatrix}\mathrm{X}$, $\mathcal{A} = \left\{\begin{bmatrix} 2 \\ -1 \end{bmatrix}, \begin{bmatrix} -1 \\ 2 \end{bmatrix}\right\}$, and $\mathcal{B} = \left\{\begin{bmatrix} 1 \\ 0 \\ -1 \end{bmatrix}, \begin{bmatrix} -1 \\ 1 \\ 0 \end{bmatrix}, \begin{bmatrix} -1 \\ 1 \\ -1 \end{bmatrix}\right\}$ find

   a. $[T]_{\mathcal{B}\mathcal{A}}$. b. $[T]_{\mathcal{E}\mathcal{A}}$. **c.** $[T]_{\mathcal{B}\mathcal{E}}$.

2. If $T(\mathrm{X}) = \begin{bmatrix} 1 & 2 & 2 & 1 \\ 3 & -1 & 1 & -3 \\ 0 & -2 & 2 & 0 \end{bmatrix}\mathrm{X}$, $\mathcal{A} = \left\{\begin{bmatrix} 1 \\ 1 \\ 0 \\ 0 \end{bmatrix}, \begin{bmatrix} -1 \\ 1 \\ 0 \\ 0 \end{bmatrix}, \begin{bmatrix} 0 \\ 0 \\ 1 \\ 1 \end{bmatrix}, \begin{bmatrix} 0 \\ 0 \\ -1 \\ 1 \end{bmatrix}\right\}$, and

   $\mathcal{B} = \left\{\begin{bmatrix} 1 \\ 0 \\ 1 \end{bmatrix}, \begin{bmatrix} 1 \\ 1 \\ 0 \end{bmatrix}, \begin{bmatrix} 0 \\ 0 \\ 1 \end{bmatrix}\right\}$, find

   a. $[T]_{\mathcal{B}\mathcal{A}}$. **b.** $[T]_{\mathcal{E}\mathcal{A}}$. c. $[T]_{\mathcal{B}\mathcal{E}}$.

3. If $T(\mathrm{X}) = \begin{bmatrix} 2 & -1 & 1 \\ 0 & -1 & 1 \\ 2 & -2 & 2 \end{bmatrix}\mathrm{X}$, $\mathcal{A} = \left\{ \begin{bmatrix} 1 \\ 0 \\ 0 \end{bmatrix}, \begin{bmatrix} 1 \\ 1 \\ 0 \end{bmatrix}, \begin{bmatrix} 1 \\ 1 \\ 1 \end{bmatrix} \right\}$, and $\mathcal{B} = \left\{ \begin{bmatrix} 2 \\ 0 \\ 1 \end{bmatrix}, \begin{bmatrix} 1 \\ -1 \\ 0 \end{bmatrix}, \begin{bmatrix} 0 \\ 0 \\ 1 \end{bmatrix} \right\}$, find

a. $[T]_{\mathcal{BA}}$. b. $[T]_{\mathcal{A}}$. **c.** $[T]_{\mathcal{B}}$.

4. If $T(\mathrm{X}) = \begin{bmatrix} 1 & 1 & 1 & 0 \\ 0 & 1 & 1 & 1 \\ 1 & 0 & 1 & 1 \\ 1 & 1 & 0 & 1 \end{bmatrix}\mathrm{X}$ and $\mathcal{B} = \left\{ \begin{bmatrix} 1 \\ 0 \\ 0 \\ 0 \end{bmatrix}, \begin{bmatrix} 2 \\ 1 \\ 0 \\ 0 \end{bmatrix}, \begin{bmatrix} 3 \\ 2 \\ 1 \\ 0 \end{bmatrix}, \begin{bmatrix} 4 \\ 3 \\ 2 \\ 1 \end{bmatrix} \right\}$, find

a. $[T]_{\mathcal{EB}}$. **b.** $[T]_{\mathcal{BE}}$. c. $[T]_{\mathcal{B}}$.

5. If $T(\mathrm{X}) = \begin{bmatrix} 7 & 4 & 5 & 6 \\ 5 & 0 & 4 & 8 \\ -1 & -5 & 0 & 5 \end{bmatrix}\mathrm{X}$, $\mathcal{A} = \left\{ \begin{bmatrix} 1 \\ 0 \\ 0 \\ 0 \end{bmatrix}, \begin{bmatrix} 1 \\ -1 \\ 0 \\ 0 \end{bmatrix}, \begin{bmatrix} 1 \\ -1 \\ 1 \\ 0 \end{bmatrix}, \begin{bmatrix} 1 \\ -1 \\ 1 \\ -1 \end{bmatrix} \right\}$, and

$\mathcal{B} = \left\{ \begin{bmatrix} 1 \\ 2 \\ 2 \end{bmatrix}, \begin{bmatrix} 1 \\ 1 \\ 0 \end{bmatrix}, \begin{bmatrix} 2 \\ 1 \\ -1 \end{bmatrix} \right\}$, find

a. $[T]_{\mathcal{BA}}$. **b.** $[T]_{\mathcal{EA}}$. c. $[T]_{\mathcal{BE}}$.

6. If $T : \mathcal{R}^2 \to \mathcal{R}^2$ is linear, $\mathcal{B} = \left\{ \begin{bmatrix} 1 \\ -3 \end{bmatrix}, \begin{bmatrix} -1 \\ 1 \end{bmatrix} \right\}$, and $[T]_{\mathcal{B}} = \begin{bmatrix} 1 & 0 \\ 0 & 2 \end{bmatrix}$, find $[T]_{\mathcal{E}}$.

7. Suppose $T : \mathcal{R}^2 \to \mathcal{R}^3$ is linear, $\mathcal{A} = \left\{ \begin{bmatrix} 1 \\ 1 \end{bmatrix}, \begin{bmatrix} -1 \\ 1 \end{bmatrix} \right\}$, $\mathcal{B} = \left\{ \begin{bmatrix} 1 \\ 0 \\ 2 \end{bmatrix}, \begin{bmatrix} 0 \\ 2 \\ -1 \end{bmatrix}, \begin{bmatrix} 2 \\ 1 \\ 0 \end{bmatrix} \right\}$,

and $[T]_{\mathcal{BA}} = \begin{bmatrix} 1 & -6 \\ -2 & 5 \\ 3 & -4 \end{bmatrix}$. Find $T(\mathrm{X})$ when

a. $\mathrm{X} = \begin{bmatrix} 0 \\ 2 \end{bmatrix}$. **b.** $\mathrm{X} = \begin{bmatrix} 2 \\ 0 \end{bmatrix}$. c. $\mathrm{X} = \begin{bmatrix} -1 \\ 5 \end{bmatrix}$.

8. Suppose $\mathcal{A} = \{\mathrm{X}_1, \ldots, \mathrm{X}_n\}$ and $\mathcal{B} = \{\mathrm{W}_1, \ldots, \mathrm{W}_n\}$ are bases for $\mathcal{R}^n$.
a. Find $[\mathrm{Id}]_{\mathcal{A}}$ and $[\mathrm{Id}]_{\mathcal{B}}$.
b. What are the columns of $[\mathrm{Id}]_{\mathcal{BA}}$? Do you recognize this matrix?

9. Suppose $\mathcal{A} = \{\mathrm{X}_1, \mathrm{X}_2\}$ is a basis for $\mathcal{R}^2$ and $T : \mathcal{R}^2 \to \mathcal{R}^2$ is one-to-one. Let $\mathrm{Y}_1 = T(\mathrm{X}_1)$, $\mathrm{Y}_2 = T(\mathrm{X}_2)$, and $\mathcal{B} = \{\mathrm{Y}_1, \mathrm{Y}_2\}$.
a. Explain why $\mathcal{B}$ is a basis for $\mathcal{R}^2$.
b. Find $[T]_{\mathcal{BA}}$.

10. Suppose $\mathcal{V}$ is the plane in $\mathcal{R}^3$ with equation $x_1 + x_2 + x_3 = 0$ and $T : \mathcal{R}^3 \to \mathcal{R}^3$ is projection onto $\mathcal{V}$. Let $\mathcal{B} = \{\mathrm{X}_1, \mathrm{X}_2, \mathrm{X}_3\} = \left\{ \begin{bmatrix} 1 \\ 1 \\ 1 \end{bmatrix}, \begin{bmatrix} -1 \\ 1 \\ 0 \end{bmatrix}, \begin{bmatrix} -1 \\ 0 \\ 1 \end{bmatrix} \right\}$, and observe that X_2, $\mathrm{X}_3 \in \mathcal{V}$ and $\mathrm{X}_1 \in \mathcal{V}^{\perp}$. Find
a. $[T]_{\mathcal{B}}$. b. $[T]_{\mathcal{E}}$. **c.** $[T]_{\mathcal{BE}}$.

11. Find $[T]_{\mathcal{E}}$ when $\mathcal{V} \subseteq \mathcal{R}^n$ is the line through 0 with slope m and $T : \mathcal{R}^2 \to \mathcal{R}^2$ is projection onto $\mathcal{V}$.

12. Suppose $\mathcal{A} = \{X_1, X_2, X_3\}$ is a basis for $\mathcal{R}^3$, $\mathcal{B} = \{X_3, X_2, X_1\}$, and $T : \mathcal{R}^3 \to \mathcal{R}^3$ is the linear function such that $[T]_{\mathcal{A}} = \begin{bmatrix} 1 & 2 & 3 \\ 4 & 5 & 6 \\ 7 & 8 & 9 \end{bmatrix}$. Find

a. $[T]_{\mathcal{B}}$. b. $[T]_{\mathcal{BA}}$. c. $[T]_{\mathcal{AB}}$.

13. Suppose $\mathcal{B} = \{X_1, X_2\}$ is the basis for $\mathcal{R}^2$ indicated in the figure and $Q = \{sX_1 + tX_2 : s,t > 0\}$. Find $T(Q)$ when $T : \mathcal{R}^2 \to \mathcal{R}^2$ is the linear function such that

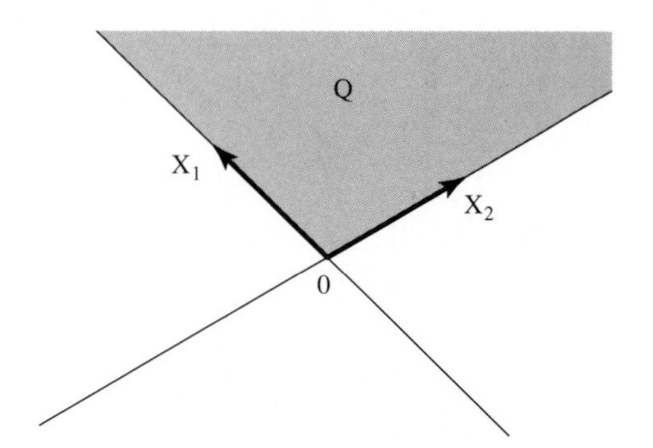

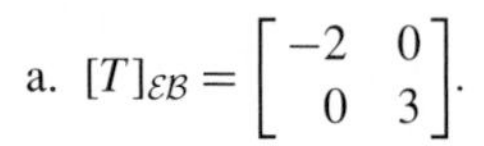
a. $[T]_{\mathcal{EB}} = \begin{bmatrix} -2 & 0 \\ 0 & 3 \end{bmatrix}$.

b. $[T]_{\mathcal{B}} = \begin{bmatrix} -2 & 0 \\ 0 & 3 \end{bmatrix}$.

14. Assume $\mathcal{V}$ is a p-dimensional subspace of $\mathcal{R}^n$, $1 \leq p \leq n-1$, and $T : \mathcal{R}^n \to \mathcal{R}^n$ is projection onto $\mathcal{V}$. Let $\{X_1, \ldots, X_p\}$ and $\{W_1, \ldots, W_{n-p}\}$ be bases for $\mathcal{V}$ and $\mathcal{V}^\perp$, respectively, and set $\mathcal{B} = \{X_1, \ldots, X_p, W_1, \ldots, W_{n-p}\}$. Find $[T]_{\mathcal{B}}$.

15. Assume $\mathcal{V}$ is a p-dimensional subspace of $\mathcal{R}^n$, $1 \leq p \leq n-1$, and $T : \mathcal{R}^n \to \mathcal{R}^n$ is projection onto $\mathcal{V}$. Let $\{U_1, \ldots, U_p\}$ be an orthonormal basis for $\mathcal{V}$ and set $A = [U_1 \cdots U_p]$. Show that $[T]_{\mathcal{E}} = AA^T$.

16. Suppose $T : \mathcal{R}^n \to \mathcal{R}^m$ is linear, $\mathcal{A}$ is a basis for $\mathcal{R}^n$, and $\mathcal{B}$ is a basis for $\mathcal{R}^m$. Show that
a. $X \in \mathcal{K}er(T)$ if and only if $[X]_{\mathcal{A}} \in \mathcal{N}([T]_{\mathcal{BA}})$.
b. $Y \in \mathcal{R}ng(T)$ if and only if $[Y]_{\mathcal{B}} \in \mathcal{C}([T]_{\mathcal{BA}})$.

4.6

ALGEBRA OF LINEAR TRANSFORMATIONS

There is a natural way to add two functions $S,T : \mathcal{R}^n \to \mathcal{R}^m$ to obtain $S + T : \mathcal{R}^n \to \mathcal{R}^m$, called the *sum* of S and T. The value of $S + T$ at X is

$$(S + T)(X) = S(X) + T(X). \tag{4.12}$$

Note that the plus sign is used ambiguously. On the right it indicates addition of vectors, and on the left it stands for addition of functions, but the proper interpretation in each case is clear from the context.

If S and T are linear, then $S(X) = AX$ and $T(X) = BX$ for some $A,B \in \mathcal{M}_{m\times n}$, and consequently $(S + T)(X) = S(X) + T(X) = AX + BX = (A + B)X$. Thus, the sum of two linear functions is linear.

There is also a natural way to scalar multiply $T : \mathcal{R}^n \to \mathcal{R}^m$. For $t \in \mathcal{R}$, $tT : \mathcal{R}^n \to \mathcal{R}^m$ is defined by

$$(tT)(\mathrm{X}) = tT(\mathrm{X}). \tag{4.13}$$

If T is linear, with $T(\mathrm{X}) = \mathrm{BX}$, then $(tT)(\mathrm{X}) = tT(\mathrm{X}) = t(\mathrm{BX}) = (t\mathrm{B})\mathrm{X}$, so tT is linear as well.

THEOREM 4.17. If $S,T : \mathcal{R}^n \to \mathcal{R}^m$ are linear and $t \in \mathcal{R}$, then $S + T$ and tT are linear. Moreover, given bases $\mathcal{A}$ and $\mathcal{B}$ for $\mathcal{R}^n$ and $\mathcal{R}^m$, respectively,

(1) $[S + T]_{\mathcal{BA}} = [S]_{\mathcal{BA}} + [T]_{\mathcal{BA}}$, and
(2) $[tT]_{\mathcal{BA}} = t[T]_{\mathcal{BA}}$.

Proof. The earlier discussion establishes the linearity of $S + T$ and tT. For $\mathrm{X} \in \mathcal{R}^n$,

$$\begin{aligned}
[(S+T)(\mathrm{X})]_{\mathcal{B}} &= [S(\mathrm{X}) + T(\mathrm{X})]_{\mathcal{B}} && \text{(addition of functions)}\\
&= [S(\mathrm{X})]_{\mathcal{B}} + [T(\mathrm{X})]_{\mathcal{B}} && \text{(Theorem 2.32)}\\
&= [S]_{\mathcal{BA}}[\mathrm{X}]_{\mathcal{A}} + [T]_{\mathcal{BA}}[\mathrm{X}]_{\mathcal{A}} && \text{(Theorem 4.14)}\\
&= ([S]_{\mathcal{BA}} + [T]_{\mathcal{BA}})[\mathrm{X}]_{\mathcal{A}}, && \text{(matrix algebra)}
\end{aligned}$$

so $[S + T]_{\mathcal{BA}} = [S]_{\mathcal{BA}} + [T]_{\mathcal{BA}}$. Justification of (2) is left as an exercise. ■

■ **EXAMPLE 4.27.** Let $S,T : \mathcal{R}^3 \to \mathcal{R}^3$ be rotations through $\pi/2$ radians about the x_1 and x_2 axes, respectively, in the directions indicated in Figure 4.17. Observe that $S(\mathrm{E}_1) = \mathrm{E}_1$, $S(\mathrm{E}_2) = \mathrm{E}_3$, $S(\mathrm{E}_3) = -\mathrm{E}_2$, and $T(\mathrm{E}_1) = \mathrm{E}_3$, $T(\mathrm{E}_2) = \mathrm{E}_2$, $T(\mathrm{E}_3) = -\mathrm{E}_1$. Thus,

$$S(\mathrm{X}) = \mathrm{AX} = \begin{bmatrix} 1 & 0 & 0 \\ 0 & 0 & -1 \\ 0 & 1 & 0 \end{bmatrix}\mathrm{X}, \quad T(\mathrm{X}) = \mathrm{BX} = \begin{bmatrix} 0 & 0 & -1 \\ 0 & 1 & 0 \\ 1 & 0 & 0 \end{bmatrix}\mathrm{X},$$

$$(S+T)(\mathrm{X}) = (\mathrm{A}+\mathrm{B})\mathrm{X} = \begin{bmatrix} 1 & 0 & -1 \\ 0 & 1 & -1 \\ 1 & 1 & 0 \end{bmatrix}\mathrm{X},$$

and

$$(3T)(\mathrm{X}) = (3\mathrm{B})\mathrm{X} = \begin{bmatrix} 0 & 0 & -3 \\ 0 & 3 & 0 \\ 3 & 0 & 0 \end{bmatrix}\mathrm{X}. \ \square$$

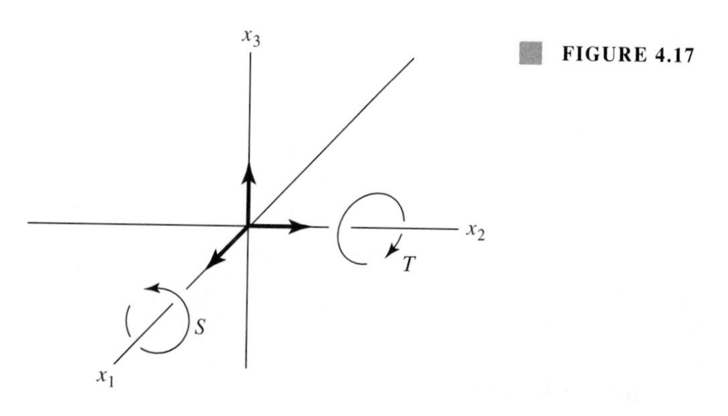

FIGURE 4.17

Subtraction for functions $S,T : \mathcal{R}^n \to \mathcal{R}^m$ is defined by

$$(S - T)(\mathrm{X}) = S(\mathrm{X}) - T(\mathrm{X}).$$

Note that $S - T = S + (-1)T$. If S and T are linear and $\mathcal{A}$ and $\mathcal{B}$ are bases for $\mathcal{R}^n$ and $\mathcal{R}^m$, respectively, then Theorem 4.17 implies $S - T$ is linear and

$$[S - T]_{\mathcal{BA}} = [S + (-1)T]_{\mathcal{BA}} = [S]_{\mathcal{BA}} + (-1)[T]_{\mathcal{BA}} = [S]_{\mathcal{BA}} - [T]_{\mathcal{BA}}.$$

■ **EXAMPLE 4.28.** Suppose $\mathcal{V}$ is a plane in $\mathcal{R}^3$ that passes through 0 and assume $T : \mathcal{R}^3 \to \mathcal{R}^3$ is reflection in $\mathcal{V}$. If $\mathrm{X} \in \mathcal{V}$, then $T(\mathrm{X}) = \mathrm{X}$; otherwise X and $T(\mathrm{X})$ are endpoints of a line segment perpendicular to $\mathcal{V}$ which is bisected by $\mathcal{V}$ (See Figure 4.18). The midpoint of that line segment is the projection of X onto $\mathcal{V}$, that is,

$$\tfrac{1}{2}[\mathrm{X} + T(\mathrm{X})] = \mathrm{Proj}_{\mathcal{V}}(\mathrm{X}), \tag{4.14}$$

so

$$T(\mathrm{X}) = 2\mathrm{Proj}_{\mathcal{V}}(\mathrm{X}) - \mathrm{X}. \tag{4.15}$$

Setting $S(\mathrm{X}) = \mathrm{Proj}_{\mathcal{V}}(\mathrm{X})$, (4.15) becomes $T(\mathrm{X}) = 2S(\mathrm{X}) - \mathrm{X} = (2S)(\mathrm{X}) - \mathrm{Id}(\mathrm{X}) = (2S - \mathrm{Id})(\mathrm{X})$, that is, $T = 2S - \mathrm{Id}$. Since S and Id are both linear, T is linear (Theorem 4.17). Now, suppose $\{\mathrm{X}_1,\mathrm{X}_2\}$ is a basis for $\mathcal{V}$ and X_3 is any nonzero vector in $\mathcal{R}^3$ orthogonal to $\mathcal{V}$. Then $\mathcal{B} = \{\mathrm{X}_1,\mathrm{X}_2,\mathrm{X}_3\}$ is a basis for $\mathcal{R}^3$, $S(\mathrm{X}_1) = \mathrm{X}_1$, $S(\mathrm{X}_2) = \mathrm{X}_2$, $S(\mathrm{X}_3) = 0$, and

$$[S]_{\mathcal{B}} = \begin{bmatrix}[S(\mathrm{X}_1)]_{\mathcal{B}} & [S(\mathrm{X}_2)]_{\mathcal{B}} & [S(\mathrm{X}_3)]_{\mathcal{B}}\end{bmatrix} = \begin{bmatrix}1 & 0 & 0\\ 0 & 1 & 0\\ 0 & 0 & 0\end{bmatrix}.$$

Thus,

$$[T]_{\mathcal{B}} = [2S - \mathrm{Id}]_{\mathcal{B}} = 2[S]_{\mathcal{B}} - [\mathrm{Id}]_{\mathcal{B}} = 2\begin{bmatrix}1 & 0 & 0\\ 0 & 1 & 0\\ 0 & 0 & 0\end{bmatrix} - \begin{bmatrix}1 & 0 & 0\\ 0 & 1 & 0\\ 0 & 0 & 1\end{bmatrix}$$

$$= \begin{bmatrix}1 & 0 & 0\\ 0 & 1 & 0\\ 0 & 0 & -1\end{bmatrix}. \ \square$$

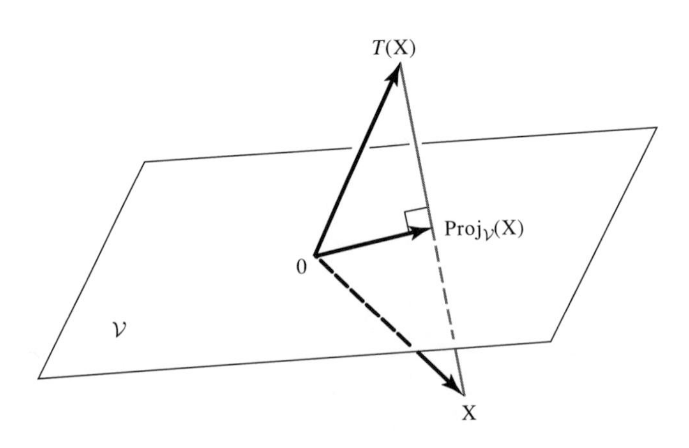

■ **FIGURE 4.18**

Each nontrivial subspace $\mathcal{V}$ of $\mathcal{R}^3$ has an associated geometric notion of reflection. The case when dim $\mathcal{V} = 2$ is discussed in the previous example. If $\mathcal{V}$ is a line through 0, then the reflection of X in $\mathcal{V}$ is the point $T(\mathrm{X})$ such that $\mathcal{V}$ is the perpendicular bisector of the line segment joining X and $T(\mathrm{X})$. In both cases, (4.14) expresses the connection between reflection in $\mathcal{V}$ and projection onto $\mathcal{V}$, which suggests that (4.15) is an appropriate algebraic vehicle for extending the notion of reflection to higher dimensional settings.

DEFINITION. If $\mathcal{V}$ is a subspace of $\mathcal{R}^n$ and $\mathrm{X} \in \mathcal{R}^n$, then the *reflection of* X *in* $\mathcal{V}$ is $\mathrm{Refl}_{\mathcal{V}}(\mathrm{X}) = 2\mathrm{Proj}_{\mathcal{V}}(\mathrm{X}) - \mathrm{X}$.

Reflection in $\mathcal{V}$ defines a function $T : \mathcal{R}^n \to \mathcal{R}^n$, and since $T = 2S - \mathrm{Id}$, where S is projection onto $\mathcal{V}$, T is linear (Theorem 4.17).

EXAMPLE 4.29. Consider the subspace $\mathcal{V}$ of $\mathcal{R}^4$ spanned by

$$\mathrm{U}_1 = \frac{1}{\sqrt{2}}\begin{bmatrix} 1 \\ 0 \\ 1 \\ 0 \end{bmatrix}, \quad \mathrm{U}_2 = \frac{1}{\sqrt{2}}\begin{bmatrix} 0 \\ 1 \\ 0 \\ 1 \end{bmatrix}, \quad \mathrm{U}_3 = \frac{1}{2}\begin{bmatrix} 1 \\ -1 \\ -1 \\ 1 \end{bmatrix},$$

and let $T(\mathrm{X}) = \mathrm{Refl}_{\mathcal{V}}(\mathrm{X})$. Note that $\mathcal{B} = \{\mathrm{U}_1, \mathrm{U}_2, \mathrm{U}_3\}$ is orthonormal. If $S(\mathrm{X}) = \mathrm{Proj}_{\mathcal{V}}(\mathrm{X})$, then $S(\mathrm{X}) = (\mathrm{X} \cdot \mathrm{U}_1)\mathrm{U}_1 + (\mathrm{X} \cdot \mathrm{U}_2)\mathrm{U}_2 + (\mathrm{X} \cdot \mathrm{U}_3)\mathrm{U}_3$ (Theorem 3.13), and in particular,

$$S(\mathrm{E}_1) = \frac{1}{2}\begin{bmatrix} 1 \\ 0 \\ 1 \\ 0 \end{bmatrix} + 0\begin{bmatrix} 0 \\ 1 \\ 0 \\ 1 \end{bmatrix} + \frac{1}{4}\begin{bmatrix} 1 \\ -1 \\ -1 \\ 1 \end{bmatrix} = \frac{1}{4}\begin{bmatrix} 3 \\ -1 \\ 1 \\ 1 \end{bmatrix},$$

$$S(\mathrm{E}_2) = 0\begin{bmatrix} 1 \\ 0 \\ 1 \\ 0 \end{bmatrix} + \frac{1}{2}\begin{bmatrix} 0 \\ 1 \\ 0 \\ 1 \end{bmatrix} - \frac{1}{4}\begin{bmatrix} 1 \\ -1 \\ -1 \\ 1 \end{bmatrix} = \frac{1}{4}\begin{bmatrix} -1 \\ 3 \\ 1 \\ 1 \end{bmatrix},$$

$$S(\mathrm{E}_3) = \frac{1}{2}\begin{bmatrix} 1 \\ 0 \\ 1 \\ 0 \end{bmatrix} + 0\begin{bmatrix} 0 \\ 1 \\ 0 \\ 1 \end{bmatrix} - \frac{1}{4}\begin{bmatrix} 1 \\ -1 \\ -1 \\ 1 \end{bmatrix} = \frac{1}{4}\begin{bmatrix} 1 \\ 1 \\ 3 \\ -1 \end{bmatrix},$$

$$S(\mathrm{E}_4) = 0\begin{bmatrix} 1 \\ 0 \\ 1 \\ 0 \end{bmatrix} + \frac{1}{2}\begin{bmatrix} 0 \\ 1 \\ 0 \\ 1 \end{bmatrix} + \frac{1}{4}\begin{bmatrix} 1 \\ -1 \\ -1 \\ 1 \end{bmatrix} = \frac{1}{4}\begin{bmatrix} 1 \\ 1 \\ -1 \\ 3 \end{bmatrix}.$$

Thus,

$$[S]_{\mathcal{E}} = \frac{1}{4}\begin{bmatrix} 3 & -1 & 1 & 1 \\ -1 & 3 & 1 & 1 \\ 1 & 1 & 3 & -1 \\ 1 & 1 & -1 & 3 \end{bmatrix},$$

and

$$[T]_{\mathcal{E}} = 2[S]_{\mathcal{E}} - [\mathrm{Id}]_{\mathcal{E}} = \frac{1}{2}\begin{bmatrix} 3 & -1 & 1 & 1 \\ -1 & 3 & 1 & 1 \\ 1 & 1 & 3 & -1 \\ 1 & 1 & -1 & 3 \end{bmatrix} - \begin{bmatrix} 1 & 0 & 0 & 0 \\ 0 & 1 & 0 & 0 \\ 0 & 0 & 1 & 0 \\ 0 & 0 & 0 & 1 \end{bmatrix}$$

$$= \frac{1}{2}\begin{bmatrix} 1 & -1 & 1 & 1 \\ -1 & 1 & 1 & 1 \\ 1 & 1 & 1 & -1 \\ 1 & 1 & -1 & 1 \end{bmatrix}. \square$$

■ **EXAMPLE 4.30.** Suppose $\mathcal{V}$ is a subspace of $\mathcal{R}^n$ and $T(\mathrm{X}) = \mathrm{Refl}_{\mathcal{V}}(\mathrm{X})$. If $\mathcal{V} = \mathcal{R}^n$, then

$$T(\mathrm{X}) = 2\mathrm{Proj}_{\mathcal{R}^n}(\mathrm{X}) - \mathrm{Id}(\mathrm{X}) = 2\mathrm{X} - \mathrm{X} = \mathrm{X},$$

that is, reflection in $\mathcal{R}^n$ is the identity transformation. When $\mathcal{V} = \{0\}$,

$$T(\mathrm{X}) = 2\mathrm{Proj}_{\{0\}}(\mathrm{X}) - \mathrm{Id}(\mathrm{X}) = 2(0) - \mathrm{X} = -\mathrm{X},$$

so reflection in $\{0\}$ is $-\mathrm{Id}$. $\square$

When $T : \mathcal{R}^n \to \mathcal{R}^m$ is one-to-one and onto, its action can be reversed by a function $T^{-1} : \mathcal{R}^m \to \mathcal{R}^n$, called the *inverse* of T. The onto assumption ensures that for each $\mathrm{Y} \in \mathcal{R}^m$ there is an $\mathrm{X} \in \mathcal{R}^n$ such that $T(\mathrm{X}) = \mathrm{Y}$, the one-to-oneness guarantees that the X is unique, and the inverse sends Y to X. That is,

$$T^{-1}(\mathrm{Y}) = \mathrm{X} \quad \text{if and only if} \quad T(\mathrm{X}) = \mathrm{Y}.$$

When T is linear, invertibility can occur only when $n = m$, for otherwise T cannot be both one-to-one and onto. In fact, when $n = m$, T is one-to-one if and only if T is onto (Theorem 4.9).

Assume $T : \mathcal{R}^n \to \mathcal{R}^n$ is linear and let A be the $n \times n$ matrix such that

$$\mathrm{Y} = T(\mathrm{X}) = \mathrm{AX}. \tag{4.16}$$

According to Theorem 4.6, T is one-to-one if and only if $\mathcal{K}er(T) = \mathcal{N}(\mathrm{A}) = \{0\}$, and the latter condition is equivalent to A being nonsingular (Theorem 1.14). Thus, T has an inverse precisely when A has an inverse. As long as A is nonsingular, (4.16) can be solved for X to obtain

$$\mathrm{X} = T^{-1}(\mathrm{Y}) = \mathrm{A}^{-1}\mathrm{Y}. \tag{4.17}$$

Besides furnishing a formula for T^{-1}, (4.17) shows that T^{-1} is linear.

THEOREM 4.18. Suppose $T : \mathcal{R}^n \to \mathcal{R}^n$ is linear and $\mathcal{B}$ is a basis for $\mathcal{R}^n$. Then T is one-to-one if and only if $[T]_{\mathcal{B}}$ is nonsingular. If T is one-to-one, then T^{-1} is linear, and

$$[T^{-1}]_{\mathcal{B}} = [T]_{\mathcal{B}}^{-1}.$$

Proof. Set $\mathrm{A} = [T]_{\mathcal{E}}$. Then $T(\mathrm{X}) = \mathrm{AX}$, and the discussion preceding the theorem shows that T is one-to-one if and only if A is invertible. By Theorem 4.16,

$$[T]_{\mathcal{B}} = \mathrm{P}_{\mathcal{EB}}^{-1}[T]_{\mathcal{E}}\mathrm{P}_{\mathcal{EB}} = \mathrm{P}_{\mathcal{EB}}^{-1}\mathrm{AP}_{\mathcal{EB}}.$$

Since a product of nonsingular matrices is nonsingular, A is invertible whenever $[T]_{\mathcal{B}}$ is invertible. Assuming T is one-to-one, the earlier discussion established the linearity of T^{-1}.

If $Y \in \mathcal{R}^n$ and $X = T^{-1}(Y)$, then $T(X) = Y$ and $[Y]_\mathcal{B} = [T(X)]_\mathcal{B} = [T]_\mathcal{B}[X]_\mathcal{B}$. Solving the last equation for $[X]_\mathcal{B}$ produces

$$[T^{-1}(Y)]_\mathcal{B} = [X]_\mathcal{B} = [T]_\mathcal{B}^{-1}[Y]_\mathcal{B},$$

so $[T^{-1}]_\mathcal{B} = [T]_\mathcal{B}^{-1}$. ■

■ **EXAMPLE 4.31.** Let $A = \begin{bmatrix} 1 & 2 \\ 3 & 4 \end{bmatrix}$ and define $T : \mathcal{R}^2 \to \mathcal{R}^2$ by $T(X) = AX$. The columns of A are linearly independent, so A is nonsingular and T has an inverse. Moreover,

$$T^{-1}(Y) = A^{-1}Y = -\frac{1}{2}\begin{bmatrix} 4 & -2 \\ -3 & 1 \end{bmatrix} Y. \square$$

■ **EXAMPLE 4.32.** Suppose $\mathcal{V}$ is a plane in $\mathcal{R}^3$ that contains 0 and $T : \mathcal{R}^3 \to \mathcal{R}^3$ is reflection in $\mathcal{V}$. If Y is the reflection of X, then X is the reflection of Y, so T is invertible and $T^{-1} = T$. Example 4.27 showed that if $\mathcal{B} = \{X_1, X_2, X_3\}$ is a basis for $\mathcal{R}^3$ such that $X_1, X_2 \in \mathcal{V}$ and $X_3 \in \mathcal{V}^\perp$, then

$$[T]_\mathcal{B} = \begin{bmatrix} 1 & 0 & 0 \\ 0 & 1 & 0 \\ 0 & 0 & -1 \end{bmatrix}.$$

Note that $[T]_\mathcal{B}$ is nonsingular. □

Given $S : \mathcal{R}^n \to \mathcal{R}^m$ and $T : \mathcal{R}^m \to \mathcal{R}^p$, the *composition of T with S* is $T \circ S : \mathcal{R}^n \to \mathcal{R}^p$ defined by

$$(T \circ S)(X) = T(S(X)).$$

When S and T are linear, there are $m \times n$ and $p \times m$ matrices A and B, respectively, such that $S(X) = AX$ and $T(Y) = BY$, so $(T \circ S)(X) = T(S(X)) = BS(X) = B(AX) = (BA)X$. Thus, $T \circ S$ is linear and $[T \circ S]_{\mathcal{E}_p\mathcal{E}_n} = BA = [T]_{\mathcal{E}_p\mathcal{E}_m}[S]_{\mathcal{E}_m\mathcal{E}_n}$.

THEOREM 4.19. If $S : \mathcal{R}^n \to \mathcal{R}^m$ and $T : \mathcal{R}^m \to \mathcal{R}^p$ are linear, then $T \circ S$ is linear. Given bases $\mathcal{A}$, $\mathcal{B}$, and $\mathcal{C}$ for $\mathcal{R}^n$, $\mathcal{R}^m$, and $\mathcal{R}^p$, respectively,

$$[T \circ S]_{\mathcal{C}\mathcal{A}} = [T]_{\mathcal{C}\mathcal{B}}[S]_{\mathcal{B}\mathcal{A}}. \tag{4.18}$$

Proof. The discussion above establishes the linearity of $T \circ S$ and shows that (4.18) holds in the special case when $\mathcal{A} = \mathcal{E}_n$, $\mathcal{B} = \mathcal{E}_m$, and $\mathcal{C} = \mathcal{E}_p$. The proof of (4.18) for arbitrary $\mathcal{A}$, $\mathcal{B}$, and $\mathcal{C}$ is left as an exercise. ■

■ **EXAMPLE 4.33.** Suppose $S : \mathcal{R}^2 \to \mathcal{R}^2$ is rotation by $\pi/2$ radians clockwise about 0 and $T : \mathcal{R}^2 \to \mathcal{R}^2$ is reflection in the line $x_2 = -x_1$. Observe that $S(E_1) = -E_2$, $S(E_2) = E_1$, $T(E_1) = -E_2$, and $T(E_2) = -E_1$ (see Figure 4.19). Thus,

$$[S]_\mathcal{E} = \begin{bmatrix} 0 & 1 \\ -1 & 0 \end{bmatrix} \quad \text{and} \quad [T]_\mathcal{E} = \begin{bmatrix} 0 & -1 \\ -1 & 0 \end{bmatrix}.$$

Moreover, $(T \circ S)(E_1) = T(S(E_1)) = T(-E_2) = E_1$ and $(T \circ S)(E_2) = T(S(E_2)) = T(E_1) = -E_2$, so

$$[T \circ S]_\mathcal{E} = \begin{bmatrix} 1 & 0 \\ 0 & -1 \end{bmatrix}.$$

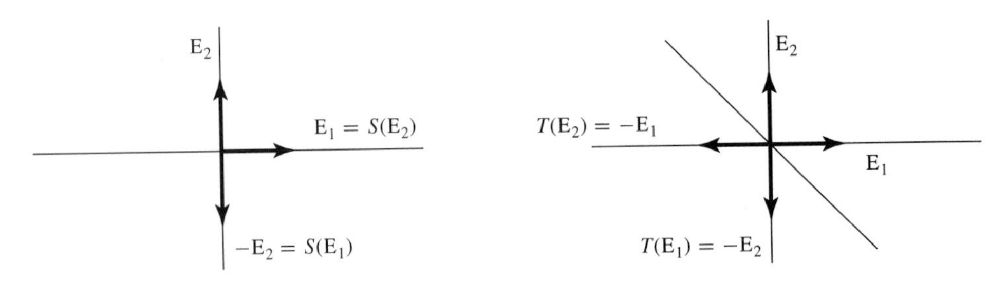

FIGURE 4.19

Note that

$$[T]_{\mathcal{E}}[S]_{\mathcal{E}} = \begin{bmatrix} 0 & -1 \\ -1 & 0 \end{bmatrix}\begin{bmatrix} 0 & 1 \\ -1 & 0 \end{bmatrix} = \begin{bmatrix} 1 & 0 \\ 0 & -1 \end{bmatrix} = [T \circ S]_{\mathcal{E}}. \square$$

■ **EXAMPLE 4.34.** Suppose $\mathcal{V}$ is the plane in $\mathcal{R}^3$ with equation $x_1 - 2x_2 + x_3 = 0$, $\mathcal{W}$ is the x_2x_3 plane, $T(\mathrm{X}) = \mathrm{Proj}_{\mathcal{W}}(\mathrm{X})$, and $S(\mathrm{X}) = \mathrm{Refl}_{\mathcal{V}}(\mathrm{X})$. Consider $T \circ S$. The action of S is most easily expressed in terms of a coordinate system in which two of the axes lie in $\mathcal{V}$ and the third is orthogonal to $\mathcal{V}$, such as $\mathcal{B} = \{\mathrm{X}_1, \mathrm{X}_2, \mathrm{X}_3\}$, where

$$\mathrm{X}_1 = \begin{bmatrix} 1 \\ 0 \\ -1 \end{bmatrix}, \quad \mathrm{X}_2 = \begin{bmatrix} 0 \\ 1 \\ 2 \end{bmatrix}, \quad \text{and} \quad \mathrm{X}_3 = \begin{bmatrix} 1 \\ -2 \\ 1 \end{bmatrix}.$$

Note that $\mathrm{X}_1, \mathrm{X}_2 \in \mathcal{V}$ and $\mathrm{X}_3 \cdot \mathrm{X}_1 = 0 = \mathrm{X}_3 \cdot \mathrm{X}_2$ (so $\mathrm{X}_3 \in \mathcal{V}^{\perp}$). Since $S(\mathrm{X}_1) = \mathrm{X}_1$, $S(\mathrm{X}_2) = \mathrm{X}_2$, and $S(\mathrm{X}_3) = -\mathrm{X}_3$,

$$[S]_{\mathcal{B}} = \big[[S(\mathrm{X}_1)]_{\mathcal{B}} \quad [S(\mathrm{X}_2)]_{\mathcal{B}} \quad [S(\mathrm{X}_3)]_{\mathcal{B}}\big] = \begin{bmatrix} 1 & 0 & 0 \\ 0 & 1 & 0 \\ 0 & 0 & -1 \end{bmatrix}.$$

Moreover,

$$[T]_{\mathcal{E}\mathcal{B}} = [T(\mathrm{X}_1) \quad T(\mathrm{X}_2) \quad T(\mathrm{X}_3)] = \begin{bmatrix} 0 & 0 & 0 \\ 0 & 1 & -2 \\ -1 & 2 & 1 \end{bmatrix},$$

so Theorem 4.19 gives

$$[T \circ S]_{\mathcal{E}\mathcal{B}} = [T]_{\mathcal{E}\mathcal{B}}[S]_{\mathcal{B}} = \begin{bmatrix} 0 & 0 & 0 \\ 0 & 1 & -2 \\ -1 & 2 & 1 \end{bmatrix}\begin{bmatrix} 1 & 0 & 0 \\ 0 & 1 & 0 \\ 0 & 0 & -1 \end{bmatrix} = \begin{bmatrix} 0 & 0 & 0 \\ 0 & 1 & 2 \\ -1 & 2 & -1 \end{bmatrix}.$$

The matrix representative for $T \circ S$ relative to the standard basis can now be obtained by making a change of coordinates in the domain, that is, $[T \circ S]_{\mathcal{E}} = [T \circ S]_{\mathcal{E}\mathcal{B}}\mathrm{P}_{\mathcal{B}\mathcal{E}}$. Since

$$\mathrm{P}_{\mathcal{B}\mathcal{E}} = \mathrm{P}_{\mathcal{E}\mathcal{B}}^{-1} = \begin{bmatrix} 1 & 0 & 1 \\ 0 & 1 & -2 \\ -1 & 2 & 1 \end{bmatrix}^{-1} = \frac{1}{6}\begin{bmatrix} 5 & 2 & -1 \\ 2 & 2 & 2 \\ 1 & -2 & 1 \end{bmatrix},$$

$$[T \circ S]_{\mathcal{E}} = \begin{bmatrix} 0 & 0 & 0 \\ 0 & 1 & 2 \\ -1 & 2 & -1 \end{bmatrix}\left(\frac{1}{6}\begin{bmatrix} 5 & 2 & -1 \\ 2 & 2 & 2 \\ 1 & -2 & 1 \end{bmatrix}\right) = \frac{1}{3}\begin{bmatrix} 0 & 0 & 0 \\ 2 & -1 & 2 \\ -1 & 2 & 2 \end{bmatrix}. \square$$

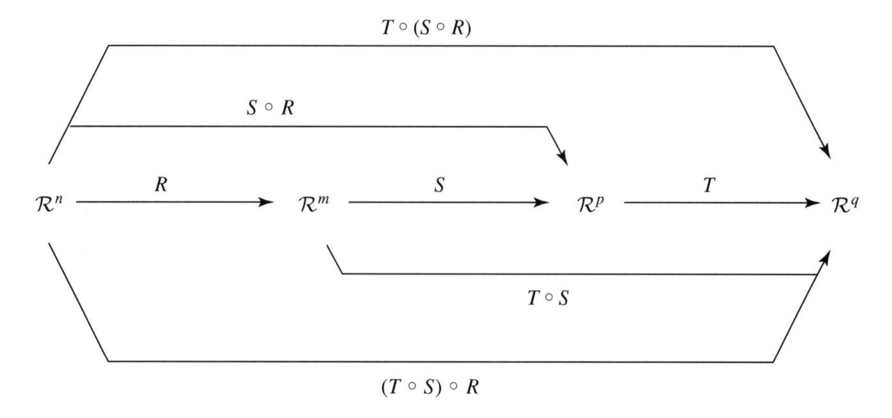

FIGURE 4.20

Suppose $R : \mathcal{R}^n \to \mathcal{R}^m$, $S : \mathcal{R}^m \to \mathcal{R}^p$, and $T : \mathcal{R}^p \to \mathcal{R}^q$ are linear and assume

$$R(\mathrm{X}) = \mathrm{AX}, \quad S(\mathrm{Y}) = \mathrm{BY}, \quad \text{and} \quad T(\mathrm{Z}) = \mathrm{CZ},$$

where $\mathrm{A} \in \mathcal{M}_{m\times n}$, $\mathrm{B} \in \mathcal{M}_{p\times m}$, $\mathrm{C} \in \mathcal{M}_{q\times p}$. Various compositions involving R, S, and T are indicated schematically in Figure 4.20. Note that $S \circ R$, $T \circ S$, $T \circ (S \circ R)$, and $(T \circ S) \circ R$ are linear (Theorem 4.19), and that

$$(S \circ R)(\mathrm{X}) = (\mathrm{BA})\mathrm{X} \quad \text{and} \quad (T \circ S)(\mathrm{Y}) = (\mathrm{CB})\mathrm{Y}.$$

Moreover,

$$[(T \circ S) \circ R](\mathrm{X}) = [(\mathrm{CB})\mathrm{A}]\mathrm{X} = [\mathrm{C}(\mathrm{BA})]\mathrm{X} = (T \circ (S \circ R))(\mathrm{X}),$$

so $T \circ (S \circ R) = (T \circ S) \circ R$. Since the placement of parentheses has no bearing on the outcome, it is customary to write $T \circ S \circ R$ for the composition of T, S, and R.

When $m = n = p$, that is, $R, S : \mathcal{R}^n \to \mathcal{R}^n$, there is the possibility of composing R with S or S with R. Note that $R \circ S$ is usually different from $S \circ R$, because

$$(R \circ S)(\mathrm{X}) = (\mathrm{AB})\mathrm{X}, \quad (S \circ R)(\mathrm{X}) = (\mathrm{BA})\mathrm{X},$$

and matrix multiplication is not commutative. In this context, either function can be composed with itself, and that process can be iterated. The notations for $R \circ R$ and $R \circ R \circ R$ are R^2 and R^3, respectively, and the k-fold composition of R with itself is abbreviated R^k, $k = 1, 2, \ldots$. Also, $R^k(\mathrm{X}) = \mathrm{A}^k\mathrm{X}$.

Finally, if $R : \mathcal{R}^n \to \mathcal{R}^n$ is assumed to be one-to-one, then for each $\mathrm{X} \in \mathcal{R}^n$,

$$(R^{-1} \circ R)(\mathrm{X}) = (\mathrm{A}^{-1}\mathrm{A})\mathrm{X} = \mathrm{I}_n\mathrm{X} \qquad \text{and} \qquad (R \circ R^{-1})(\mathrm{X}) = (\mathrm{AA}^{-1})\mathrm{X} = \mathrm{I}_n\mathrm{X}.$$

Thus,

$$R^{-1} \circ R = \mathrm{Id} = R \circ R^{-1}.$$

Exercises 4.6

1. Assuming $T : \mathcal{R}^n \to \mathcal{R}^m$ is linear and $\mathcal{A}$ and $\mathcal{B}$ are bases for $\mathcal{R}^n$ and $\mathcal{R}^m$, respectively, show that $[tT]_{\mathcal{BA}} = t[T]_{\mathcal{BA}}$.

2. Define $R, S, T : \mathcal{R}^3 \to \mathcal{R}^3$ as follows:

   R = rotation by $\pi/2$ radians about the x_2 axis in the indicated direction,
   S = projection onto the plane $x_2 = -x_1$,
   T = reflection in the plane $x_2 = x_1$.

   Find

   a.  $[R]_{\mathcal{E}}$. b. $[S]_{\mathcal{E}}$. **c.** $[T]_{\mathcal{E}}$.
   d. $[R \circ T]_{\mathcal{E}}$. **e.** $[T \circ R]_{\mathcal{E}}$.
   f. $[2S - R + 3T]_{\mathcal{E}}$. **g.** $[S^2]_{\mathcal{E}}$.

3. Find $[T]_{\mathcal{E}}$ when $T : \mathcal{R}^3 \to \mathcal{R}^3$ is reflection in $\mathcal{V}$ and

   a. $\mathcal{V} = Span\left\{\begin{bmatrix}1\\1\\1\end{bmatrix}\right\}$. b. $\mathcal{V} = Span\left\{\begin{bmatrix}-1\\1\\0\end{bmatrix}, \begin{bmatrix}-1\\0\\1\end{bmatrix}\right\}$.

   c. $\mathcal{V} = Span\left\{\begin{bmatrix}-1\\-1\\2\end{bmatrix}, \begin{bmatrix}-2\\1\\1\end{bmatrix}\right\}$. d. $\mathcal{V} = \{X : x_1 - x_2 + 2x_3 = 0\}$.

4. Find $T_1 \circ T_2 \circ T_3$ when $T_k : \mathcal{R}^3 \to \mathcal{R}^3$ is reflection in $\mathcal{V}_k$, $k = 1, 2, 3$, and
   a. $\mathcal{V}_1 = Span\{E_1, E_2\}$, $\mathcal{V}_2 = Span\{E_2, E_3\}$, and $\mathcal{V}_3 = Span\{E_1, E_3\}$.
   b. $\mathcal{V}_1 = Span\{E_1\}$, $\mathcal{V}_2 = Span\{E_2\}$, and $\mathcal{V}_3 = Span\{E_3\}$.
   Do your conclusions hold when E_1, E_2, and E_3 are replaced by arbitrary orthonormal vectors U_1, U_2, and U_3?

5. Let S and T be rotations about the x_1 and x_2 axes, respectively, by $\pi/4$ radians in the indicated directions. Find $[T \circ S]_{\mathcal{E}}$.

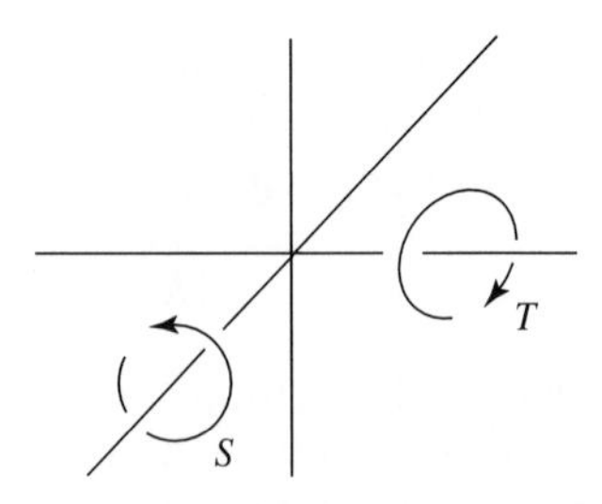

6. Assume $\mathcal{V}$ is a subspace of $\mathcal{R}^n$, $X \in \mathcal{R}^n$, and V, W are the unique vectors in $\mathcal{V}$ and $\mathcal{V}^\perp$, respectively, such that $X = V + W$. Show that $\text{Refl}_{\mathcal{V}}(X) = V - W$. Illustrate the conclusion with a sketch in $\mathcal{R}^2$ for the case when $\mathcal{V}$ is a line through 0.

7. Suppose $\mathcal{V}$ is a subspace of $\mathcal{R}^n$ and $T(X) = \text{Refl}_{\mathcal{V}}(X)$.
   a. Show that if $X \in \mathcal{V}$, then $T(X) = X$.
   b. Show that if $X \in \mathcal{V}^\perp$, then $T(X) = -X$.

c. Let $\{X_1, \ldots, X_p\}$ and $\{W_1, \ldots, W_{n-p}\}$ be bases for $\mathcal{V}$ and $\mathcal{V}^\perp$, respectively, and set $\mathcal{B} = \{X_1, \ldots, X_p, W_1, \ldots, W_{n-p}\}$. Find $[T]_\mathcal{B}$.
d. Show that T is one-to-one and onto.

8. Verify that T is one-to-one and find $[T^{-1}]_\mathcal{E}$ when

a. $T : \mathcal{R}^3 \to \mathcal{R}^3, \quad T(X) = \begin{bmatrix} 2x_1 + x_2 + 3x_3 \\ x_1 + 2x_3 \\ x_2 + x_3 \end{bmatrix}.$

b. $T : \mathcal{R}^3 \to \mathcal{R}^3, \quad T(X) = \begin{bmatrix} x_1 + x_2 + x_3 \\ x_2 - x_3 \\ -x_1 + x_3 \end{bmatrix}.$

c. $T : \mathcal{R}^2 \to \mathcal{R}^2$ is rotation by α radians counterclockwise about 0.

d. $T : \mathcal{R}^2 \to \mathcal{R}^2, \quad \mathcal{B} = \left\{ \begin{bmatrix} 0 \\ 1 \end{bmatrix}, \begin{bmatrix} 1 \\ 1 \end{bmatrix} \right\}, \quad$ and $\quad [T]_\mathcal{B} = \begin{bmatrix} 1 & -1 \\ -1 & 2 \end{bmatrix}.$

9. Assume $T : \mathcal{R}^n \to \mathcal{R}^n$ is a one-to-one linear function, and let $\mathcal{A}$ and $\mathcal{B}$ be bases for $\mathcal{R}^n$. Show that $[T^{-1}]_{\mathcal{AB}} = [T]_{\mathcal{BA}}^{-1}$.

10. If $\mathcal{V}$ is a subspace of $\mathcal{R}^n$, $T(X) = \text{Refl}_\mathcal{V}(X)$, and $S(X) = \text{Proj}_\mathcal{V}(X)$, show that
a. $S^2 = S$.
b. $T^2 = \text{Id}$.

11. Assume $\mathcal{V}$ and $\mathcal{W}$ are subspaces of $\mathcal{R}^n$ such that $\mathcal{R}^n = \mathcal{V} \oplus \mathcal{W}$. Define $T : \mathcal{R}^n \to \mathcal{R}^n$ as follows: for $X \in \mathcal{R}^n$, $X = V + W$, $V \in \mathcal{V}$, $W \in \mathcal{W}$, and $T(X) = V$.
a. Show that T is linear.
b. Show that $T^2 = T$.
c. Show that $\mathcal{R}ng(T) = \mathcal{V}$ and $\mathcal{K}er(T) = \mathcal{W}$.
d. In the case when $n = 2$ and $\mathcal{V}$ and $\mathcal{W}$ are lines through 0, make a sketch illustrating $\mathcal{V}$, $\mathcal{W}$, X, V, and W. Does it seem reasonable to call $T(X)$ the *projection of X onto $\mathcal{V}$ parallel to $\mathcal{W}$*?

12. Suppose $T : \mathcal{R}^n \to \mathcal{R}^n$ is linear and $T^2 = T$. Show that $\mathcal{R}^n = \mathcal{R}ng(T) \oplus \mathcal{K}er(T)$ (Hint: $X = T(X) + (X - T(X))$ and conclude that T is projection onto $\mathcal{R}ng(T)$ parallel to $\mathcal{K}er(T)$.

13. Suppose $\mathcal{V}$ is a subspace of $\mathcal{R}^n$ and S and T are reflections in $\mathcal{V}$ and $\mathcal{V}^\perp$, respectively.
a. Show that $S + T = 0$.
b. Show that $S \circ T = -\text{Id}$.
c. Make a sketch illustrating these conclusions when $n = 2$ and $\mathcal{V}$ is a line through 0.

14. Assume $S : \mathcal{R}^n \to \mathcal{R}^m$ and $T : \mathcal{R}^m \to \mathcal{R}^p$ are linear functions.
a. Show that if $\mathcal{A}$, $\mathcal{B}$, and $\mathcal{C}$ are bases for $\mathcal{R}^n$, $\mathcal{R}^m$, and $\mathcal{R}^p$, respectively, then $[T \circ S]_{\mathcal{CA}} = [T]_{\mathcal{CB}}[S]_{\mathcal{BA}}$.
b. Show that if S and T are both one-to-one, then $T \circ S$ is one-to-one.
c. Show that if S and T are both onto, then $T \circ S$ is onto.
d. Assuming S and T are one-to-one and onto, show that $(T \circ S)^{-1} = S^{-1} \circ T^{-1}$.

4.7
ISOMETRIES AND SIMILARITIES

In the study of Euclidean geometry one encounters certain functions from $\mathcal{R}^2$ to $\mathcal{R}^2$ known as rigid motions of the plane. There are three types:

i. translation,
ii. rotation about a particular point, and
iii. reflection in a given line.

A rigid motion $F : \mathcal{R}^2 \to \mathcal{R}^2$ has two geometrically significant properties. First, it "preserves distances." That is, for all X,Y $\in \mathcal{R}^2$, the distance between $F(\mathrm{X})$ and $F(\mathrm{Y})$ is the same as the distance between X and Y. Second, it "preserves angles." If ℓ_1 and ℓ_2 are lines that intersect to form an angle, then $F(\ell_1)$ and $F(\ell_2)$ are lines that intersect to form the same angle.

Although rigid motions need not be linear (a translation, for example, does not send 0 to 0), it can be shown that for each rigid motion $F : \mathcal{R}^2 \to \mathcal{R}^2$ there is a linear function $T : \mathcal{R}^2 \to \mathcal{R}^2$ and a constant B $\in \mathcal{R}^2$ such that

$$F(\mathrm{X}) = T(\mathrm{X}) + \mathrm{B} \tag{4.19}$$

(see Exercise 7). Functions F and T, related as in (4.19), are translations of one another. If one preserves distances and/or angles, so does the other. The present section formulates and explores these distance and angle preserving ideas for linear transformations from $\mathcal{R}^n$ to $\mathcal{R}^n$.

DEFINITION. The *Euclidean distance between* X,Y$\in \mathcal{R}^n$ is

$$\|\mathrm{X} - \mathrm{Y}\| = \sqrt{\sum_1^n (x_k - y_k)^2},$$

and $F : \mathcal{R}^n \to \mathcal{R}^n$ is *distance preserving* if

$$\|F(\mathrm{X}) - F(\mathrm{Y})\| = \|\mathrm{X} - \mathrm{Y}\| \tag{4.20}$$

for all X,Y $\in \mathcal{R}^n$. Distance preserving functions are called *isometries*.

If $F : \mathcal{R}^n \to \mathcal{R}^n$ is distance preserving, X,Y $\in \mathcal{R}^n$, and X $\neq$ Y, then (4.20) implies $F(\mathrm{X}) \neq F(\mathrm{Y})$, so isometries are necessarily one-to-one.

Suppose $T : \mathcal{R}^n \to \mathcal{R}^n$ is linear. If T is an isometry, then $\|T(\mathrm{X}) - T(\mathrm{Y})\| = \|\mathrm{X} - \mathrm{Y}\|$ for all X,Y $\in \mathcal{R}^n$, and when Y $= 0$ this equation reduces to

$$\|T(\mathrm{X})\| = \|\mathrm{X}\|. \tag{4.21}$$

On the other hand, if T is known to satisfy (4.21), then for all X,Y $\in \mathcal{R}^n$, $\|T(\mathrm{X}) - T(\mathrm{Y})\| = \|T(\mathrm{X} - \mathrm{Y})\| = \|\mathrm{X} - \mathrm{Y}\|$, so T is an isometry. These observations are recorded in the next theorem.

THEOREM 4.20. If $T : \mathcal{R}^n \to \mathcal{R}^n$ is linear, then T is an isometry if and only if $\|T(\mathrm{X})\| = \|\mathrm{X}\|$ for all X $\in \mathcal{R}^n$.

EXAMPLE 4.35. Suppose $\mathcal{V}$ is a subspace of $\mathcal{R}^3$ and $T(\mathrm{X}) = \mathrm{Refl}_{\mathcal{V}}(\mathrm{X})$. It is clear from geometric considerations that T satisfies (4.21) when $\mathcal{V}$ is a plane or line through 0. For $\mathcal{V} = \{0\}$ or $\mathcal{V} = \mathcal{R}^3$, $T = -\mathrm{Id}$ or $T = \mathrm{Id}$, respectively, so T satisfies (4.21) in these cases as well. Thus, reflection in a subspace of $\mathcal{R}^3$ is an isometry. □

EXAMPLE 4.36. If $T : \mathcal{R}^2 \to \mathcal{R}^2$ is given by

$$T(\mathrm{X}) = \frac{1}{2}\begin{bmatrix} \sqrt{3} & -1 \\ 1 & \sqrt{3} \end{bmatrix}\mathrm{X} = \frac{1}{2}\begin{bmatrix} \sqrt{3}x_1 - \quad x_2 \\ x_1 + \sqrt{3}x_2 \end{bmatrix},$$

then

$$\|T(\mathrm{X})\|^2 = \tfrac{1}{4}\big((\sqrt{3}x_1 - x_2)^2 + (x_1 + \sqrt{3}x_2)^2\big) = \|\mathrm{X}\|^2,$$

so T is an isometry (Theorem 4.20). □

Suppose $T : \mathcal{R}^3 \to \mathcal{R}^3$ is linear and consider a line ℓ with equation

$$\mathrm{X}(t) = tM + \mathrm{B}, \quad -\infty < t < \infty,$$

where M,B $\in \mathcal{R}^3$ and M $\neq 0$. The image of ℓ under T is described by

$$\mathrm{Y}(t) = T(\mathrm{X}(t)) = tT(M) + T(\mathrm{B}), \quad -\infty < t < \infty.$$

If $T(\mathrm{M}) \neq 0$, then $T(\ell)$ is a line, and otherwise $T(\ell) = \{T(\mathrm{B})\}$. For $T(\ell)$ to be a line whenever ℓ is a line, it is necessary that $T(\mathrm{M}) \neq 0$ whenever $\mathrm{M} \neq 0$, that is, that T is one-to-one. Assume then that T is one-to-one and suppose

$$\ell_1 : \mathrm{X}_1(t) = t\mathrm{M}_1 + \mathrm{B}_1 \qquad \text{and} \qquad \ell_2 : \mathrm{X}_2(t) = t\mathrm{M}_2 + \mathrm{B}_2$$

intersect. The angle between ℓ_1 and ℓ_2 is the angle between their direction vectors, that is, the unique $\theta \in [0,\pi]$ satisfying

$$\cos(\theta) = \frac{\mathrm{M}_1 \cdot \mathrm{M}_2}{\|\mathrm{M}_1\|\,\|\mathrm{M}_2\|}.$$

The images of ℓ_1 and ℓ_2 are

$$\mathcal{L}_1 : \mathrm{Y}_1(t) = tT(\mathrm{M}_1) + T(\mathrm{B}_1) \qquad \text{and} \qquad \mathcal{L}_2 : \mathrm{Y}_2(t) = tT(\mathrm{M}_2) + T(\mathrm{B}_2),$$

$-\infty < t < \infty$, and the angle between them is the unique $\phi \in [0,\pi]$ such that

$$\cos(\phi) = \frac{T(\mathrm{M}_1) \cdot T(\mathrm{M}_2)}{\|T(\mathrm{M}_1)\|\,\|T(\mathrm{M}_2)\|}.$$

Of course T preserves the angle θ when $\phi = \theta$.

DEFINITION. A one-to-one linear function $T : \mathcal{R}^n \to \mathcal{R}^n$ is *angle preserving* if

$$\frac{T(\mathrm{X}) \cdot T(\mathrm{Y})}{\|T(\mathrm{X})\|\,\|T(\mathrm{Y})\|} = \frac{\mathrm{X} \cdot \mathrm{Y}}{\|\mathrm{X}\|\,\|\mathrm{Y}\|} \tag{4.22}$$

for all nonzero X,Y $\in \mathcal{R}^n$. Angle preserving functions are called *similarity transformations*, or simply *similarities*.

■ **EXAMPLE 4.37.** Let t be a nonzero scalar and define $T : \mathcal{R}^n \to \mathcal{R}^n$ by $T(\mathrm{X}) = t\mathrm{X}$ ($T = t\mathrm{Id}$). For all $\mathrm{X,Y} \in \mathcal{R}^n$,

$$\|T(\mathrm{X}) - T(\mathrm{Y})\| = \|t\mathrm{X} - t\mathrm{Y}\| = |t|\|\mathrm{X} - \mathrm{Y}\|.$$

The action of T alters distances by a factor of $|t|$, so T is an isometry if and only if $t = \pm 1$. However, for nonzero $\mathrm{X,Y} \in \mathcal{R}^n$,

$$\frac{T(\mathrm{X}) \cdot T(\mathrm{Y})}{\|T(\mathrm{X})\|\|T(\mathrm{Y})\|} = \frac{(t\mathrm{X}) \cdot (t\mathrm{Y})}{\|t\mathrm{X}\|\|t\mathrm{Y}\|} = \frac{t^2(\mathrm{X} \cdot \mathrm{Y})}{|t|\|\mathrm{X}\||t|\|\mathrm{Y}\|} = \frac{\mathrm{X} \cdot \mathrm{Y}}{\|\mathrm{X}\|\|\mathrm{Y}\|},$$

so T preserves angles regardless of the size of t. When $t > 0$, $t\mathrm{Id}$ is called a *pure dilation*, or a *pure change of scale*. For $t < 0$, $t\mathrm{Id} = -|t|\mathrm{Id}$, in which case T is the composition of a pure dilation and reflection in $\{0\}$. □

Let Δ be a triangle in $\mathcal{R}^2$ with vertices U, V, and W, and assume $T : \mathcal{R}^2 \to \mathcal{R}^2$ is linear and one-to-one. The vertices of Δ determine three distinct lines, and since T is one-to-one, their images are also distinct lines. Thus, $T(\Delta)$ is the triangle with vertices $T(\mathrm{U})$, $T(\mathrm{V})$, and $T(\mathrm{W})$ (Figure 4.21).

Suppose T is angle preserving. The interior angle of Δ at U is the angle between $\mathrm{W} - \mathrm{U}$ and $\mathrm{V} - \mathrm{U}$, and the interior angle of $T(\Delta)$ at $T(\mathrm{U})$ is the angle between $T(\mathrm{W}) - T(\mathrm{U})$ and $T(\mathrm{V}) - T(\mathrm{U})$. Since

$$T(\mathrm{W}) - T(\mathrm{U}) = T(\mathrm{W} - \mathrm{U}) \qquad \text{and} \qquad T(\mathrm{V}) - T(\mathrm{U}) = T(\mathrm{V} - \mathrm{U}),$$

the angle preserving assumption ensures that these two angles are the same. This sort of analysis applies at each vertex, so $T(\Delta)$ is similar to Δ. Thus, similarities on $\mathcal{R}^2$ send triangles to similar triangles.

If you assume instead that T is distance preserving, then Δ and $T(\Delta)$ have corresponding sides of equal length and are therefore congruent. Congruent triangles have equal corresponding interior angles, so you might suspect that linear isometries preserve angles as well as distances. In fact, every linear isometry is a similarity, a statement whose justification rests on the following result.

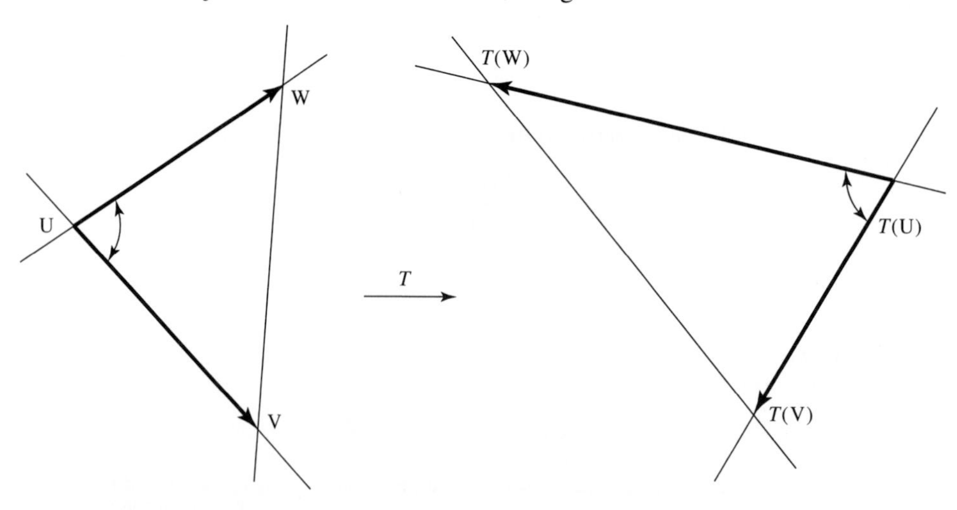

FIGURE 4.21

THEOREM 4.21. If $T : \mathcal{R}^n \to \mathcal{R}^n$ is linear, then the following are equivalent:

(1) $T(X) \cdot T(Y) = X \cdot Y$ for all $X, Y \in \mathcal{R}^n$,
(2) $\|T(X)\| = \|X\|$ for all $X \in \mathcal{R}^n$.

Proof.
(1) $\Rightarrow$ (2). Setting $Y = X$ in statement (1) yields $\|T(X)\|^2 = \|X\|^2$, so $\|T(X)\| = \|X\|$.

(2) $\Rightarrow$ (1). For $X, Y \in \mathcal{R}^n$, $\|X + Y\| = \|T(X + Y)\|$ implies $(X + Y) \cdot (X + Y) = \|X + Y\|^2 = \|T(X + Y)\|^2 = \|T(X) + T(Y)\|^2 = [T(X) + T(Y)] \cdot [T(X) + T(Y)]$. Expanding the dot products leads to

$$\|X\|^2 + 2(X \cdot Y) + \|Y\|^2 = \|T(X)\|^2 + 2[T(X) \cdot T(Y)] + \|T(Y)\|^2.$$

Since $\|T(X)\| = \|X\|$ and $\|T(Y)\| = \|Y\|$, it follows that $T(X) \cdot T(Y) = X \cdot Y$. ■

THEOREM 4.22. Every linear isometry on $\mathcal{R}^n$ is a similarity transformation.

Proof. If T is a linear isometry on $\mathcal{R}^n$, then $\|T(X)\| = \|X\|$, $X \in \mathcal{R}^n$, so by Theorem 4.21, $T(X) \cdot T(Y) = X \cdot Y$, $X, Y \in \mathcal{R}^n$. Thus, for all nonzero $X, Y \in \mathcal{R}^n$,

$$\frac{T(X) \cdot T(Y)}{\|T(X)\| \|T(Y)\|} = \frac{X \cdot Y}{\|X\| \|Y\|}. \ ■$$

Similarities need not be isometries (Example 4.36), but they are very closely related to isometries. As the next theorem indicates, each similarity on $\mathcal{R}^n$ is the composition of a linear isometry and a change of scale.

THEOREM 4.23. If $T : \mathcal{R}^n \to \mathcal{R}^n$ is linear, then T is a similarity if and only if there is a linear isometry $S : \mathcal{R}^n \to \mathcal{R}^n$ and a $t > 0$ such that $T = tS$.

Proof. Suppose T is a similarity. If X and Y are nonzero vectors in $\mathcal{R}^n$ such that $\|X\| = \|Y\|$, then

$$(X + Y) \cdot (X - Y) = \|X\|^2 + Y \cdot X - X \cdot Y - \|Y\|^2 = \|X\|^2 - \|Y\|^2 = 0,$$

so $X + Y$ is orthogonal to $X - Y$. Since T preserves angles, $T(X + Y)$ is orthogonal to $T(X - Y)$, and therefore

$$\begin{aligned} 0 = T(X + Y) \cdot T(X - Y) &= [T(X) + T(Y)] \cdot [T(X) - T(Y)] \\ &= \|T(X)\|^2 - \|T(Y)\|^2. \end{aligned}$$

Thus, $\|T(X)\| = \|T(Y)\|$. This shows that the images under a similarity of two vectors of equal length are vectors of equal length. In particular, all unit vectors in $\mathcal{R}^n$ have images of some fixed length, say t. Now, for each nonzero $X \in \mathcal{R}^n$,

$$T(X) = T\left(\|X\| \frac{X}{\|X\|}\right) = \|X\| T\left(\frac{X}{\|X\|}\right).$$

Since $X/\|X\|$ is a unit vector, $\|T(X/\|X\|)\| = t$, and therefore $\|T(X)\| = t\|X\|$. Observe that $S = (1/t)T$ is linear and

$$\|S(X)\| = (1/t)\|T(X)\| = (1/t)t\|X\| = \|X\|,$$

$X \in \mathcal{R}^n$. Thus, S is an isometry and $T = tS$. This completes the first half of the proof. The other half is left as an exercise. ■

If $T : \mathcal{R}^n \to \mathcal{R}^n$ is a linear isometry, then T preserves angles as well as lengths, and therefore the image of an orthonormal basis is an orthonormal basis. Thus, $A = [T(E_1) \quad \cdots \quad T(E_n)]$ is an orthogonal matrix. Conversely, if A is orthogonal and $T(X) = AX$, then $A^T = A^{-1}$, and for all $X, Y \in \mathcal{R}^n$,

$$T(X) \cdot T(Y) = (AX)^T(AY) = X^T(A^TA)Y = X^TI_nY = X^TY = X \cdot Y.$$

Thus, $\|T(X)\| = \|X\|$ for all $X \in \mathcal{R}^n$ (Theorem 4.21), and T is an isometry.

THEOREM 4.24. If $T : \mathcal{R}^n \to \mathcal{R}^n$ is linear and $\mathcal{B}$ is an orthonormal basis for $\mathcal{R}^n$, then T is an isometry if and only if $[T]_{\mathcal{B}}$ is orthogonal.

Proof. The discussion above shows that T is an isometry if and only if $A = [T]_{\mathcal{E}}$ is orthogonal. Moreover, $P = P_{\mathcal{EB}}$ is orthogonal (Theorem 3.17), so $PP^T = I$, and therefore $[T]_{\mathcal{B}} = P^{-1}AP = P^TAP$. Thus,

$$[T]_{\mathcal{B}}^T[T]_{\mathcal{B}} = (P^TAP)^TP^TAP = P^TA^TPP^TAP = P^TA^TAP.$$

If follows from the last equation that $A^TA = I$ if and only if $[T]_{\mathcal{B}}^T[T]_{\mathcal{B}} = I$. ■

■ **EXAMPLE 4.38.** Consider $T : \mathcal{R}^4 \to \mathcal{R}^4$ with equation $T(X) = AX$, where

$$A = \begin{bmatrix} -1 & \sqrt{2} & 0 & -1 \\ \sqrt{2} & 0 & 0 & -\sqrt{2} \\ 1 & \sqrt{2} & 0 & 1 \\ 0 & 0 & 2 & 0 \end{bmatrix}.$$

The columns of A are mutually orthogonal vectors of length 2, so $B = \frac{1}{2}A$ is an orthogonal matrix. According to Theorem 4.24, $S(X) = BX$ is an isometry, and therefore $T = 2S$ is a similarity (Theorem 4.23). ■

■ **EXAMPLE 4.39.** Orthogonal 2×2 matrices come in two forms, namely

$$A = \begin{bmatrix} \cos(\theta) & -\sin(\theta) \\ \sin(\theta) & \cos(\theta) \end{bmatrix} \quad \text{and} \quad B = \begin{bmatrix} \cos(\theta) & \sin(\theta) \\ \sin(\theta) & -\cos(\theta) \end{bmatrix},$$

$\theta \in \mathcal{R}$ (Problem 4, Exercises 3.5). Consider the isometries T and S defined by $T(X) = AX$ and $S(X) = BX$. The geometric action of T has already been discussed; it rotates the plane counterclockwise about 0 through angle θ (Example 4.1). Since

$$B = \begin{bmatrix} \cos(\theta) & -\sin(\theta) \\ \sin(\theta) & \cos(\theta) \end{bmatrix}\begin{bmatrix} 1 & 0 \\ 0 & -1 \end{bmatrix} = A\begin{bmatrix} 1 & 0 \\ 0 & -1 \end{bmatrix},$$

$S = T \circ R$, where $R : \mathcal{R}^2 \to \mathcal{R}^2$ is given by

$$R(X) = \begin{bmatrix} 1 & 0 \\ 0 & -1 \end{bmatrix}\begin{bmatrix} x_1 \\ x_2 \end{bmatrix} = \begin{bmatrix} x_1 \\ -x_2 \end{bmatrix}.$$

Thus, the action of S can be viewed as a reflection in the x_1 axis followed by counterclockwise rotation by θ radians (Figure 4.22). Note that the perpendicular bisector of the segment joining X and $S(X)$ is the line through 0 making angle $\theta/2$ with the positive x_1 axis. In other words, $S(X)$ appears to be the reflection of X in the line $y = mx$, where $m = \tan(\theta/2)$. The matrix representing this reflection relative to the standard coordinate system was found in Example 4.26 to be

$$\frac{1}{1+m^2}\begin{bmatrix} 1-m^2 & 2m \\ 2m & m^2-1 \end{bmatrix}.$$

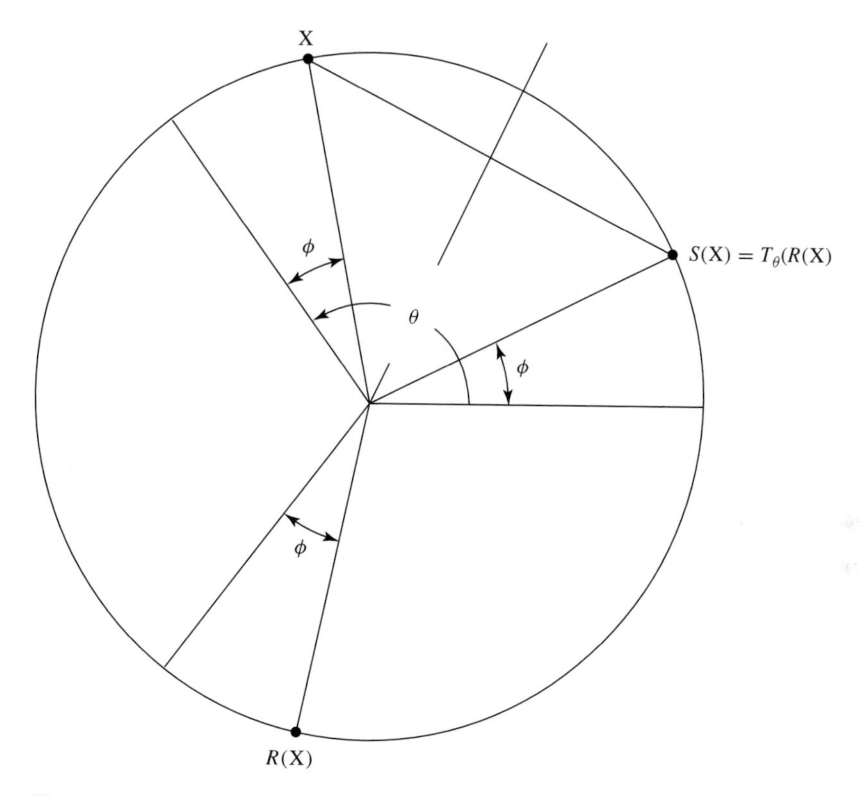

FIGURE 4.22

Note that

$$\frac{1-m^2}{1+m^2} = \frac{1-\tan^2(\theta/2)}{1+\tan^2(\theta/2)} = \frac{\cos^2(\theta/2)-\sin^2(\theta/2)}{\cos^2(\theta/2)+\sin^2(\theta/2)} = \cos(\theta),$$

and

$$\frac{2m}{1+m^2} = \frac{2\tan(\theta/2)}{1+\tan^2(\theta/2)} = \frac{2\cos(\theta/2)\sin(\theta/2)}{\cos^2(\theta/2)+\sin^2(\theta/2)} = \sin(\theta),$$

so

$$\frac{1}{1+m^2}\begin{bmatrix} 1-m^2 & 2m \\ 2m & m^2-1 \end{bmatrix} = \begin{bmatrix} \cos(\theta) & \sin(\theta) \\ \sin(\theta) & -\cos(\theta) \end{bmatrix}. \ \square$$

Exercises 4.7

1. Determine whether $T : \mathcal{R}^2 \to \mathcal{R}^2$ given by $T(X) = \begin{bmatrix} 1 & -1 \\ 1 & 1 \end{bmatrix} X$ is

a. a similarity. b. an isometry.

2. Show that if $S,T : \mathcal{R}^n \to \mathcal{R}^n$ are linear isometries, then so is
   a. $S \circ T$. b. T^{-1}.

3. Suppose $S : \mathcal{R}^n \to \mathcal{R}^n$ is an isometry and $t \in \mathcal{R}$, $t \neq 0$. Show that $T = tS$ is a similarity. (This completes the proof of Theorem 4.23.)

4. Let $[T]_{\mathcal{E}}$ be the given matrix. Is T an isometry? A similarity?

   a. $\frac{1}{2}\begin{bmatrix} \sqrt{2} & -1 & 1 \\ \sqrt{2} & 1 & -1 \\ 0 & \sqrt{2} & \sqrt{2} \end{bmatrix}$ **b.** $\begin{bmatrix} 2 & -1 & 1 & 0 \\ 1 & 0 & -2 & 1 \\ 0 & 1 & 1 & 2 \\ -1 & -2 & 0 & 1 \end{bmatrix}$ **c.** $\begin{bmatrix} 6 & -4 \\ 8 & 3 \end{bmatrix}$

   d. $\begin{bmatrix} 1 & -1 & 0 & 1 \\ 0 & 1 & 1 & 1 \\ 1 & 0 & 1 & -1 \\ 1 & 1 & -1 & 0 \end{bmatrix}$ e. $\begin{bmatrix} 0 & -3 & 0 \\ 2 & 0 & 0 \\ 0 & 0 & 4 \end{bmatrix}$ f. $\begin{bmatrix} 0 & -1 \\ 1 & 0 \end{bmatrix}$

5. Assuming $\mathcal{V}$ is a subspace of $\mathcal{R}^n$, show that $T(\mathrm{X}) = \mathrm{Refl}_{\mathcal{V}}(\mathrm{X})$ is an isometry.

6. Let $T : \mathcal{R}^n \to \mathcal{R}^n$ be a similarity and suppose $\mathcal{V}$ is a subspace of $\mathcal{R}^n$. Show that $T(\mathcal{V}^{\perp}) = T(\mathcal{V})^{\perp}$.

7. Assume $\mathrm{C} \in \mathcal{R}^2$ and $G : \mathcal{R}^2 \to \mathcal{R}^2$ is translation by C, that is, $G(\mathrm{X}) = \mathrm{X} + \mathrm{C}$. Note that G is one-to-one and onto, and $G^{-1}(\mathrm{X}) = \mathrm{X} - \mathrm{C}$.
   a. Suppose F and T are rotations of the plane counterclockwise through α radians about C and 0, respectively. Convince yourself geometrically that $F = G \circ T \circ G^{-1}$ and find $\mathrm{B} \in \mathcal{R}^2$ such that $F(\mathrm{X}) = T(\mathrm{X}) + \mathrm{B}$.
   b. Suppose $\mathcal{L}$ is a line passing through C and let F be reflection in $\mathcal{L}$. Let R be reflection in the line through 0 parallel to $\mathcal{L}$. Convince yourself that $F = G \circ R \circ G^{-1}$ and find $\mathrm{B} \in \mathcal{R}^2$ such that $F(\mathrm{X}) = R(\mathrm{X}) + \mathrm{B}$.
   c. Suppose $R,S : \mathcal{R}^2 \to \mathcal{R}^2$ are linear and $\mathrm{D},\mathrm{E} \in \mathcal{R}^2$. Let $F(\mathrm{X}) = S(\mathrm{X}) + \mathrm{D}$ and $H(\mathrm{X}) = R(\mathrm{X}) + \mathrm{E}$, $\mathrm{X} \in \mathcal{R}^2$. Find a linear function $T : \mathcal{R}^2 \to \mathcal{R}^2$ and $\mathrm{B} \in \mathcal{R}^2$ such that $(F \circ H)(\mathrm{X}) = T(\mathrm{X}) + \mathrm{B}$.

5

The Determinant

5.1 INTRODUCTION

Associated with

$$A = \begin{bmatrix} a & b \\ c & d \end{bmatrix}$$

is a real number, called the determinant of A, defined by

$$\det(A) = ad - bc.$$

You may recall that this number provides a simple test for invertibility of A. According to Example 1.36, A is nonsingular if and only if $\det(A) \neq 0$, and as long as $\det(A) \neq 0$,

$$A^{-1} = \frac{1}{(ad - bc)} \begin{bmatrix} d & -b \\ -c & a \end{bmatrix}.$$

There is also a geometric interpretation for det(A). When A is nonsingular, its columns,

$$X_1 = \begin{bmatrix} a \\ c \end{bmatrix} \quad \text{and} \quad X_2 = \begin{bmatrix} b \\ d \end{bmatrix},$$

are linearly independent and therefore determine a parallelogram Π with vertices 0, X_1, X_2, and $X_1 + X_2$ (Figure 5.1). The area of Π can be expressed in terms of the lengths of its sides and the angle θ between X_1 and X_2 by

$$\text{Area}(\Pi) = \|X_1\|\|X_2\|\sin(\theta).$$

Moreover,

$$\sin(\theta) = \sqrt{1 - \cos^2(\theta)} = \sqrt{1 - \left(\frac{X_1 \cdot X_2}{\|X_1\|\|X_2\|}\right)^2},$$

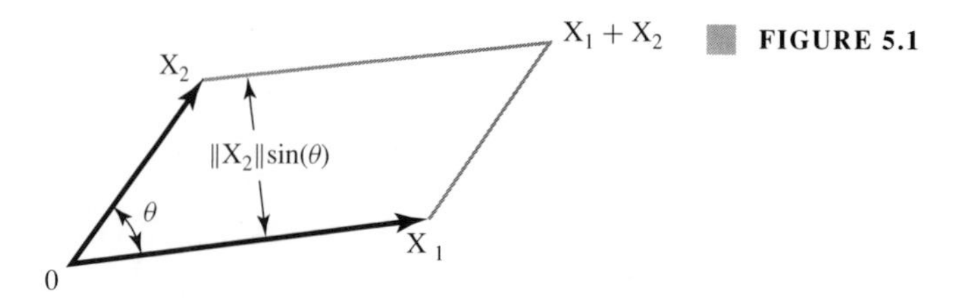

FIGURE 5.1

so

$$\begin{aligned}\|X_1\|\,\|X_2\|\sin(\theta) &= \sqrt{\|X_1\|^2\|X_2\|^2 - (X_1 \cdot X_2)^2} \\ &= \sqrt{(a^2+c^2)(b^2+d^2) - (ab+cd)^2} \\ &= \sqrt{(ad-bc)^2} \\ &= |ad-bc|.\end{aligned}$$

Thus,

$$\text{Area}(\Pi) = |\det(A)|. \tag{5.1}$$

When A is singular, det(A) = 0. In that event, the columns of A are linearly dependent, and Π degenerates to a segment of the line through 0, X_1, and X_2. Line segments have zero area, so (5.1) holds in that case as well.

The present chapter extends these ideas to $n \times n$ matrices and discusses the interpretation of the resulting function

$$\det : \mathcal{M}_{n\times n} \to \mathcal{R}.$$

A careful development of the determinant function involves technical issues regarding permutations that aren't used elsewhere in this text. Consequently, the presentation that follows glosses over those details and focuses instead on how to calculate det(A) and on the properties of the determinant function.

5.2 THE DETERMINANT FUNCTION

For $A \in \mathcal{M}_{n\times n}$, the $(n-1)\times(n-1)$ matrix that remains after deleting $\text{Row}_i(A)$ and $\text{Col}_j(A)$ is denoted by $M_{ij}(A)$, or simply M_{ij}.

EXAMPLE 5.1. If

$$A = \begin{bmatrix} 1 & -2 & 0 \\ 3 & 1 & -4 \\ 0 & 2 & 3 \end{bmatrix}, \text{ then } M_{23} = \begin{bmatrix} 1 & -2 \\ 0 & 2 \end{bmatrix} \text{ and } M_{11} = \begin{bmatrix} 1 & -4 \\ 2 & 3 \end{bmatrix}. \square$$

EXAMPLE 5.2. For

$$A = \begin{bmatrix} -2 & 1 & 0 & -1 \\ 3 & 0 & 1 & 2 \\ -1 & 4 & 2 & 0 \\ 2 & 1 & 1 & -3 \end{bmatrix}, \; M_{43} = \begin{bmatrix} -2 & 1 & -1 \\ 3 & 0 & 2 \\ -1 & 4 & 0 \end{bmatrix} \text{ and } M_{32} = \begin{bmatrix} -2 & 0 & -1 \\ 3 & 1 & 2 \\ 2 & 1 & -3 \end{bmatrix}. \square$$

DEFINITION. The *determinant* of a square matrix A, abbreviated det(A), is defined inductively as follows:

1. If $A = \begin{bmatrix} a & b \\ c & d \end{bmatrix}$, then $\det(A) = ad - bc$.
2. If $n \geq 3$ and det(A) has been defined for each $A \in \mathcal{M}_{(n-1)\times(n-1)}$, then for $A \in \mathcal{M}_{n\times n}$,

$$\det(A) = \sum_{k=1}^{n} a_{1k}(-1)^{1+k}\det(M_{1k}). \tag{5.2}$$
$$= a_{11}\det(M_{11}) - a_{12}\det(M_{12}) + \cdots + (-1)^{1+n}a_{1n}\det(M_{1n}).$$

■ **EXAMPLE 5.3.** $\det\begin{bmatrix} 2 & -3 \\ 4 & 1 \end{bmatrix} = 2 \cdot 1 - (-3)4 = 14.$ □

■ **EXAMPLE 5.4.**

$$\det\begin{bmatrix} 1 & -2 & 0 \\ 3 & 1 & -4 \\ 0 & 2 & 3 \end{bmatrix} = 1\det\begin{bmatrix} 1 & -4 \\ 2 & 3 \end{bmatrix} - (-2)\det\begin{bmatrix} 3 & -4 \\ 0 & 3 \end{bmatrix} + 0\det\begin{bmatrix} 3 & 1 \\ 0 & 2 \end{bmatrix}$$
$$= 1(3-(-8)) - (-2)(9-0) + 0(6-0) = 29. \ \square$$

■ **EXAMPLE 5.5.**

$$\det\begin{bmatrix} -2 & 1 & 0 & -1 \\ 3 & 0 & 1 & 2 \\ -1 & 4 & 2 & 0 \\ 2 & 1 & 1 & -3 \end{bmatrix} = (-2)\det\begin{bmatrix} 0 & 1 & 2 \\ 4 & 2 & 0 \\ 1 & 1 & -3 \end{bmatrix} - 1\det\begin{bmatrix} 3 & 1 & 2 \\ -1 & 2 & 0 \\ 2 & 1 & -3 \end{bmatrix}$$
$$+ 0\det\begin{bmatrix} 3 & 0 & 2 \\ -1 & 4 & 0 \\ 2 & 1 & -3 \end{bmatrix} - (-1)\det\begin{bmatrix} 3 & 0 & 1 \\ -1 & 4 & 2 \\ 2 & 1 & 1 \end{bmatrix},$$
$$= -2\left\{0\det\begin{bmatrix} 2 & 0 \\ 1 & -3 \end{bmatrix} - 1\det\begin{bmatrix} 4 & 0 \\ 1 & -3 \end{bmatrix} + 2\det\begin{bmatrix} 4 & 2 \\ 1 & 1 \end{bmatrix}\right\}$$
$$-1\left\{3\det\begin{bmatrix} 2 & 0 \\ 1 & -3 \end{bmatrix} - 1\det\begin{bmatrix} -1 & 0 \\ 2 & -3 \end{bmatrix} + 2\det\begin{bmatrix} -1 & 2 \\ 2 & 1 \end{bmatrix}\right\}$$
$$+\left\{3\det\begin{bmatrix} 4 & 2 \\ 1 & 1 \end{bmatrix} - 0\det\begin{bmatrix} -1 & 2 \\ 2 & 1 \end{bmatrix} + 1\det\begin{bmatrix} -1 & 4 \\ 2 & 1 \end{bmatrix}\right\}$$
$$= -2\big(0-(-12)+2(2)\big) - \big(3(-6)-(3)+2(-5)\big)$$
$$+\big(3(2)-0+(-9)\big)$$
$$= -4. \ \square$$

Equation (5.2) is just one of several expressions for det(A). Each displays det(A) as a sum whose terms involve the elements of a particular row or column and the determinants of their associated $(n-1)\times(n-1)$ matrices. The various expressions are recorded in the next theorem.

THEOREM 5.1. If $A \in \mathcal{M}_{n\times n}$ and $1 \le i, j \le n$, then

$$\sum_{k=1}^{n} a_{ik}(-1)^{i+k}\det(M_{ik}) = \det(A) = \sum_{k=1}^{n} a_{kj}(-1)^{k+j}\det(M_{kj}). \tag{5.3}$$

Proof. Omitted. ■

■ **EXAMPLE 5.6.** Consider $A = \begin{bmatrix} 1 & -2 & 0 \\ 3 & 1 & -4 \\ 0 & 2 & 3 \end{bmatrix}$ from Example 5.4. When $i = 3$, the left side of (5.3) is

$$\begin{aligned} &a_{31}(-1)^{3+1}\det(M_{31}) + a_{32}(-1)^{3+2}\det(M_{32}) + a_{33}(-1)^{3+3}\det(M_{33}) \\ &= 0\det\begin{bmatrix} -2 & 0 \\ 1 & -4 \end{bmatrix} - 2\det\begin{bmatrix} 1 & 0 \\ 3 & -4 \end{bmatrix} + 3\det\begin{bmatrix} 1 & -2 \\ 3 & 1 \end{bmatrix} \\ &= 0 - 2(-4-0) + 3(1-(-6)) = 29, \end{aligned}$$

and when $j = 2$, the right side of (5.3) is

$$\begin{aligned} &a_{12}(-1)^{1+2}\det(M_{12}) + a_{22}(-1)^{2+2}\det(M_{22}) + a_{32}(-1)^{3+2}\det(M_{32}) \\ &= -(-2)\det\begin{bmatrix} 3 & -4 \\ 0 & 3 \end{bmatrix} + 1\det\begin{bmatrix} 1 & 0 \\ 0 & 3 \end{bmatrix} - 2\det\begin{bmatrix} 1 & 0 \\ 3 & -4 \end{bmatrix} \\ &= 2(9-0) + 1(3-0) - 2(-4-0) = 29. \end{aligned}$$

There are six of these expressions for det(A), one for each row and column. You might check that they all produce the value 29. □

DEFINITION. If $A \in \mathcal{M}_{n\times n}$ and $1 \le i, j \le n$, then $\det(M_{ij})$ and $(-1)^{i+j}\det(M_{ij})$ are called the *minor* and *cofactor of the i,j position,* respectively. The left and right sides of (5.3) are referred to as *Laplace expansions* of det(A) by the ith row and jth column, respectively.

Observe that $(-1)^{i+j}$, occurring in the cofactor of the i,j position, associates a plus or minus sign with each position in the matrix according to the pattern

$$\begin{bmatrix} + & - & + & - & \cdot & \cdot & \cdot \\ - & + & - & + & \cdot & \cdot & \cdot \\ + & - & + & - & \cdot & \cdot & \cdot \\ - & + & - & + & \cdot & \cdot & \cdot \\ \cdot & \cdot & \cdot & \cdot & \cdot & \cdot & \cdot \\ \cdot & \cdot & \cdot & \cdot & \cdot & \cdot & \cdot \\ \cdot & \cdot & \cdot & \cdot & \cdot & \cdot & \cdot \end{bmatrix}.$$

Given $A \in \mathcal{M}_{n\times n}$, Laplace expansion of det(A) requires you to calculate the determinants of n matrices of size $(n-1) \times (n-1)$. Each of those, in turn, involves the determinants of $n-1$ matrices of size $(n-2) \times (n-2)$. Continuing in this manner, you ultimately compute the determinants of $n(n-1)\cdots 4\cdot 3 = n!/2$ matrices of size 2×2. This is an imposing task when n is a large integer. Consequently, Laplace expansion by itself is usually not a very efficient procedure for finding det(A). However, in some special circumstances it quickly produces the desired result.

■ **EXAMPLE 5.7.** Let $A = \begin{bmatrix} 1 & 2 & 0 & -1 \\ 4 & -1 & 2 & 6 \\ -3 & 0 & 0 & -2 \\ 2 & -4 & 0 & 5 \end{bmatrix}$. Observe that Laplace expansion of det(A) by column 3 makes efficient use of the zeros there. Indeed,

$$\begin{aligned} \det(A) &= 0(-1)^{1+3}\det(M_{13}) + 2(-1)^{2+3}\det(M_{23}) + 0(-1)^{3+3}\det(M_{33}) \\ &\quad + 0(-1)^{4+3}\det(M_{43}) \\ &= (-2)\det\begin{bmatrix} 1 & 2 & -1 \\ -3 & 0 & -2 \\ 2 & -4 & 5 \end{bmatrix}. \end{aligned}$$

Then, expansion by either the second row or the second column exploits the 0 in the 2,2 position of the last matrix. Expanding by the second row yields

$$\begin{aligned} \det(A) &= (-2)\left((-3)(-1)^{2+1}\det\begin{bmatrix} 2 & -1 \\ -4 & 5 \end{bmatrix} + 0 + (-2)(-1)^{2+3}\det\begin{bmatrix} 1 & 2 \\ 2 & -4 \end{bmatrix}\right) \\ &= (-2)\big(3(10-4) + 2(-4-4)\big) = -4. \ \square \end{aligned}$$

■ **EXAMPLE 5.8.** If $A \in \mathcal{M}_{n \times n}$ and each entry in a row or column of A is zero, then $\det(A) = 0$. Indeed, when $\text{Row}_i(A)$ consists of zeros, the left side of (5.3) becomes

$$\det(A) = \sum_{k=1}^{n} a_{ik}(-1)^{i+k}\det(M_{ik}) = \sum_{k=1}^{n} 0(-1)^{i+k}\det(M_{ik}) = 0. \ \square$$

■ **EXAMPLE 5.9.** Successive expansions by the first column give

$$\det\begin{bmatrix} a & \cdot & \cdot & \cdot \\ 0 & b & \cdot & \cdot \\ 0 & 0 & c & \cdot \\ 0 & 0 & 0 & d \end{bmatrix} = a\det\begin{bmatrix} b & \cdot & \cdot \\ 0 & c & \cdot \\ 0 & 0 & d \end{bmatrix} = (ab)\det\begin{bmatrix} c & \cdot \\ 0 & d \end{bmatrix} = abcd.$$

In fact, you can see inductively that the determinant of any $n \times n$ upper triangular matrix is the product of its diagonal entries. In particular, I_n is upper triangular, and its diagonal entries are 1's, so $\det(I_n) = 1$. □

THEOREM 5.2. If $A \in \mathcal{M}_{n \times n}$ has two identical rows or columns, then $\det(A) = 0$.

Proof. Suppose A has two identical rows. Laplace expansion will show that if the conclusion holds for $(n-1) \times (n-1)$ matrices, then it also holds for $n \times n$ matrices. Assume $A \in \mathcal{M}_{n \times n}$, $n \geq 3$, and $\text{Row}_i(A) = \text{Row}_j(A)$, $1 \leq i < j \leq n$. Laplace expansion by row r produces

$$\det(A) = \sum_{k=1}^{n} a_{rk}(-1)^{r+k}\det(M_{rk}).$$

As long as $r \neq i$ and $r \neq j$, M_{rk} has two identical rows. If the conclusion holds for all $(n-1) \times (n-1)$ matrices, then $\det(M_{rk}) = 0$, $k = 1, \ldots, n$, and consequently $\det(A) = 0$. To start the inductive process you must check the first value of n. When $n = 2$,

$$A = \begin{bmatrix} a & b \\ a & b \end{bmatrix},$$

and $\det(A) = ab - ab = 0$. A similar argument utilizing a column expansion addresses the case when A has two identical columns. ■

REMARK. Although the determinant was defined for matrices of size $n \times n$, where $n \geq 2$, it is sometimes helpful to have such a notion for 1×1 matrices. Consequently, $[a]$ is considered to have determinant a. With this convention, $\det\begin{bmatrix} a & b \\ c & d \end{bmatrix} = ad - bc$ can be interpreted as a Laplace expansion.

Assuming $A \in \mathcal{M}_{n\times n}$, the Laplace expansion of det(A) by the first row is

$$\det(A) = a_{11}\det\begin{bmatrix} a_{22} & \cdots & a_{2n} \\ \vdots & & \vdots \\ a_{n2} & \cdots & a_{nn} \end{bmatrix} - a_{12}\det\begin{bmatrix} a_{21} & a_{23} & \cdots & a_{2n} \\ \vdots & & & \vdots \\ a_{n1} & a_{n3} & \cdots & a_{nn} \end{bmatrix} + \cdots$$

$$+ (-1)^{1+n}a_{1n}\det\begin{bmatrix} a_{21} & \cdots & a_{2,n-1} \\ \vdots & & \vdots \\ a_{n1} & \cdots & a_{n,n-1} \end{bmatrix}. \tag{5.4}$$

Each term on the right has the form

$$\pm a_{1k_1}\det(M),$$

where $1 \leq k_1 \leq n$ and M represents an $(n-1) \times (n-1)$ matrix. Observe that

i. the entries in the first row of M come from $\text{Row}_2(A)$, and
ii. M contains no entries from $\text{Col}_{k_1}(A)$.

Expanding the determinants on the right of (5.4) by the first row expresses det(A) as a sum of $n(n-1)$ new terms of the form

$$\pm a_{1k_1}a_{2k_2}\det(M),$$

where $1 \leq k_2 \leq n$, $k_2 \neq k_1$. This time M denotes an $(n-2) \times (n-2)$ matrix such that

i. the entries in the first row of M come from $\text{Row}_3(A)$, and
ii. M contains no entries from $\text{Col}_{k_1}(A)$ or $\text{Col}_{k_2}(A)$.

Continuing in this manner produces

$$\det(A) = \sum \pm a_{1k_1}a_{2k_2}\cdots a_{nk_n}, \tag{5.5}$$

where $k_1, k_2, \ldots, k_n$ is a permutation of $1, 2, \ldots, n$, and the summation is understood to be taken over all $n!$ permutations. The plus or minus sign in

$$\pm a_{1k_1}a_{2k_2}\cdots a_{nk_n}$$

depends in some manner on the permutation $k_1, k_2, \ldots, k_n$. Notice that (5.5) expresses det(A) as a sum of $n!$ terms each of which, apart from its sign, is a product of n entries in A such that no two factors come from the same row or column.

Formal development of the determinant often uses (5.5) to define det(A). Of course, that approach requires an explanation as to how a given permutation of

$1, 2, \ldots, n$ determines the plus or minus sign in the corresponding term in the sum. In this book, the only term in (5.5) whose sign is needed is the one corresponding to $k_1 = 1, k_2 = 2, \ldots, k_n = n$, that is,

$$\pm a_{11}a_{22}\cdots a_{nn}.$$

Referring to (5.4), you can see that this product arises in the calculation of

$$a_{11}\det\begin{bmatrix} a_{22} & \cdots & a_{2n} \\ \vdots & & \vdots \\ a_{n2} & \cdots & a_{nn} \end{bmatrix}.$$

In expanding the last determinant by the first row, the term in question appears in

$$a_{11}a_{22}\det\begin{bmatrix} a_{33} & \cdots & a_{3n} \\ \vdots & & \vdots \\ a_{n3} & \cdots & a_{nn} \end{bmatrix}.$$

The pattern continues. At each stage the expression is preceded by $+$, so the term in (5.5) involving the product of the diagonal entries of A appears with a plus sign.

Exercises 5.2

1. Find the determinant of each matrix using the Laplace expansion by the (i) second row, (ii) third column, and (iii) first column.

a. $\begin{bmatrix} 1 & 2 & -1 \\ 0 & 3 & 2 \\ -1 & 1 & 2 \end{bmatrix}$ b. $\begin{bmatrix} 3 & 1 & -1 \\ 2 & 3 & 0 \\ 2 & 0 & 0 \end{bmatrix}$ **c.** $\begin{bmatrix} 1 & -3 & -2 \\ 2 & 1 & 3 \\ -3 & 2 & -1 \end{bmatrix}$

2. Find the determinant of

a. $\begin{bmatrix} -1 & 0 & 2 \\ 2 & 3 & 0 \\ 1 & 1 & 4 \end{bmatrix}$. b. $\begin{bmatrix} 1 & -1 & -3 \\ 2 & 2 & 2 \\ 3 & 0 & -2 \end{bmatrix}$. **c.** $\begin{bmatrix} 1 & 0 & 0 \\ 2 & 3 & 0 \\ 4 & 5 & 6 \end{bmatrix}$.

d. $\begin{bmatrix} 1 & 2 & 3 \\ 3 & 2 & 1 \\ 1 & 2 & 3 \end{bmatrix}$. **e.** $\begin{bmatrix} 0 & 0 & a \\ 0 & b & 1 \\ c & 1 & 1 \end{bmatrix}$. f. $\begin{bmatrix} 0 & a & 1 \\ b & 0 & b \\ 1 & a & 0 \end{bmatrix}$.

g. $\begin{bmatrix} 4 & -1 & -3 & 0 \\ 0 & 0 & 2 & 0 \\ 1 & 0 & 2 & 3 \\ -2 & 1 & 0 & 1 \end{bmatrix}$ h. $\begin{bmatrix} 1 & 2 & 0 & 3 \\ 0 & -1 & 1 & 2 \\ 1 & -2 & 1 & 0 \\ 1 & 1 & 0 & 2 \end{bmatrix}$ **i.** $\begin{bmatrix} 1 & 1 & 1 & 1 \\ 0 & 2 & 0 & 1 \\ 0 & 0 & 3 & 1 \\ 5 & 0 & 0 & 4 \end{bmatrix}$

j. $\begin{bmatrix} \cos(\alpha) & -\sin(\alpha) \\ \sin(\alpha) & \cos(\alpha) \end{bmatrix}$ **k.** $\begin{bmatrix} 0 & 0 & 0 & 1 \\ 0 & a & b & 0 \\ 0 & c & d & 0 \\ 1 & 0 & 0 & 0 \end{bmatrix}$ l. $\begin{bmatrix} 0 & 1 & 0 \\ 1 & 1 & 1 \\ 0 & 1 & 0 \end{bmatrix}$

3. Find the determinant of the $n \times n$ elementary matrix $E_i(t)$. How about $E_{ij}(t)$?

4. Find det(A) when $A = \begin{bmatrix} 1 & 0 & 0 & \cdots & 0 & 0 & 1 \\ 1 & 1 & 0 & & & 0 & 0 \\ 0 & 1 & 1 & \ddots & & & \vdots \\ \vdots & & & \ddots & 1 & 0 & 0 \\ 0 & 0 & & & 1 & 1 & 0 \\ 0 & 0 & 0 & \cdots & 0 & 1 & 1 \end{bmatrix} \in \mathcal{M}_{n\times n}$.

5. Show that if $A = \begin{bmatrix} x & 0 & 0 & \cdots & 0 & -1 \\ 0 & x & & & & 0 \\ \vdots & & x & & & \vdots \\ 0 & & & \ddots & x & 0 \\ -1 & 0 & 0 & \cdots & 0 & x \end{bmatrix} \in \mathcal{M}_{n\times n}$, then $\det(A) = x^{n-2}(x^2 - 1)$.

6. a. Show that $\begin{bmatrix} 0 & 0 & \cdots & 0 & 1 \\ \vdots & 1 & \ddots & & 0 \\ 0 & & \ddots & 1 & \vdots \\ 1 & 0 & \cdots & 0 & 0 \end{bmatrix} \in \mathcal{M}_{p\times p}$ has determinant -1.

   b. Show that $\det(E_{ij}) = -1$.

7. **a.** Show that if $A \in \mathcal{M}_{2\times 2}$ and $t \in \mathcal{R}$, then $\det(tA) = t^2\det(A)$.
   b. What is the corresponding result when $A \in \mathcal{M}_{3\times 3}$?
   c. State the relationship between $\det(tA)$ and $\det(A)$ when $A \in \mathcal{M}_{n\times n}$ and give an inductive argument to support your statement.

8. Let $A = \begin{bmatrix} a & b \\ c & d \end{bmatrix}$. Calculate $\det(A^2)$ and relate your answer to det(A).

5.3

PROPERTIES OF DET(A)

Evaluating det(A) by Laplace expansion is a tedious job when A does not have an ample number of strategically placed zero entries. However, it is always possible to row reduce A, and since rref(A) is upper triangular, its determinant is the product of its diagonal entries. Thus, understanding how the determinant interacts with Gaussian elimination might lead to an efficient way of calculating det(A). Of course row reduction involves a sequence of elementary row operations, so the crucial question is: given an elementary matrix E, how are det(A) and det(EA) related?

THEOREM 5.3. If $A \in \mathcal{M}_{n\times n}$, $t \in \mathcal{R}$, $1 \le i,j \le n$, and $i \ne j$, then

(1) $\det(E_{ij}A) = -\det(A)$.
(2) $\det(E_i(t)A) = t\det(A)$.
(3) $\det(E_{ij}(t)A) = \det(A)$.

Proof.
(1) Since $E_{ij} = E_{ji}$, you may assume $i < j$. The proof is inductive. If

$$A = \begin{bmatrix} a & b \\ c & d \end{bmatrix},$$

then $i = 1, j = 2$, and

$$\det(E_{12}A) = \det\left(\begin{bmatrix} c & d \\ a & b \end{bmatrix}\right) = cb - ad = -(ad - bc) = -\det(A).$$

Thus, (1) holds when $n = 2$. Assume $n \geq 3$ and (1) holds for all $(n-1) \times (n-1)$ matrices. For $A \in \mathcal{M}_{n\times n}$, set $B = E_{ij}A$, and choose $m \in \{1, \ldots, n\}$ such that $m \neq i, j$. Expanding det(B) by row m produces

$$\begin{aligned} \det(B) &= \sum_{k=1}^{n} b_{mk}(-1)^{m+k}\det(M_{mk}(B)) \\ &= \sum_{k=1}^{n} a_{mk}(-1)^{m+k}\det(M_{mk}(B)). \end{aligned} \tag{5.6}$$

Since $m \neq i, j$, $M_{mk}(B)$ is the result of interchanging two rows of $M_{mk}(A)$, and by the induction assumption, $\det(M_{mk}(B)) = -\det(M_{mk}(A))$. Thus, (5.6) becomes

$$\det(B) = -\sum_{k=1}^{n} a_{mk}(-1)^{m+k}\det(M_{mk}(A)) = -\det(A).$$

(2) Let $B = E_i(t)A$. Then $b_{ik} = ta_{ik}$, $M_{ik}(B) = M_{ik}(A)$, and expansion by row i produces

$$\det(B) = \sum_{k=1}^{n} b_{ik}(-1)^{i+k}\det(M_{ik}(B)) = \sum_{k=1}^{n} ta_{ik}(-1)^{i+k}\det(M_{ik}(A)) = t\det(A).$$

(3) Let $B = E_{ij}(t)A$. Then $b_{jk} = a_{jk} + ta_{ik}$ and $M_{jk}(B) = M_{jk}(A)$. The Laplace expansion of det(B) by row j is

$$\begin{aligned} \det(B) &= \sum_{k=1}^{n} b_{jk}(-1)^{j+k}\det(M_{jk}(B)) = \sum_{k=1}^{n} (a_{jk} + ta_{ik})(-1)^{j+k}\det(M_{jk}(A)) \\ &= \sum_{k=1}^{n} a_{jk}(-1)^{j+k}\det(M_{jk}(A)) + t\sum_{k=1}^{n} a_{ik}(-1)^{j+k}\det(M_{jk}(A)) \\ &= \det(A) + t\sum_{k=1}^{n} a_{ik}(-1)^{j+k}\det(M_{jk}(A)). \end{aligned}$$

To evaluate the last sum, consider the matrix C that results from replacing $\text{Row}_j(A)$ by a copy of $\text{Row}_i(A)$. Observe that $c_{jk} = a_{ik}$ and $M_{jk}(C) = M_{jk}(A)$, so

$$\det(C) = \sum_{k=1}^{n} c_{jk}(-1)^{j+k}\det(M_{jk}(C)) = \sum_{k=1}^{n} a_{ik}(-1)^{j+k}\det(M_{jk}(A)).$$

Thus, $\det(B) = \det(A) + t\det(C)$. Since C has two identical rows, $\det(C) = 0$, and therefore $\det(B) = \det(A)$. ■

During Gaussian elimination, the type III operations generate the zero entries in the row reduced matrix, and by Theorem 5.3, those operations have no effect on the value of the determinant. An interchange of rows, however, changes the sign of the determinant, and the effect of a type II operation is indicated by

$$\det\begin{bmatrix} \vdots & \cdots & \vdots \\ ta_{i1} & & ta_{in} \\ \vdots & \cdots & \vdots \end{bmatrix} = t\det\begin{bmatrix} \vdots & \cdots & \vdots \\ a_{i1} & & a_{in} \\ \vdots & \cdots & \vdots \end{bmatrix}. \tag{5.7}$$

The next example illustrates the use of these properties.

EXAMPLE 5.10. Consider

$$A = \begin{bmatrix} 2 & -1 & 3 & 1 \\ 2 & 0 & 2 & 4 \\ 3 & 1 & 1 & 1 \\ -2 & 3 & 1 & 0 \end{bmatrix}.$$

Factoring 2 from the second row produces a leading one in the first column without introducing fractional entries elsewhere, and by (5.7),

$$\det\begin{bmatrix} 2 & -1 & 3 & 1 \\ 2 & 0 & 2 & 4 \\ 3 & 1 & 1 & 1 \\ -2 & 3 & 1 & 0 \end{bmatrix} = 2\det\begin{bmatrix} 2 & -1 & 3 & 1 \\ 1 & 0 & 1 & 2 \\ 3 & 1 & 1 & 1 \\ -2 & 3 & 1 & 0 \end{bmatrix}.$$

Then operations $r_{21}(-2)$, $r_{23}(-3)$, and $r_{24}(2)$ give

$$\det(A) = 2\det\begin{bmatrix} 0 & -1 & 1 & -3 \\ 1 & 0 & 1 & 2 \\ 0 & 1 & -2 & -5 \\ 0 & 3 & 3 & 4 \end{bmatrix}.$$

Now expand by the first column to obtain

$$\det(A) = 2(-1)\det\begin{bmatrix} -1 & 1 & -3 \\ 1 & -2 & -5 \\ 3 & 3 & 4 \end{bmatrix}.$$

Finally, $r_{12}(1)$, $r_{13}(3)$, and expansion by column 1 yield

$$\det(A) = -2\det\begin{bmatrix} -1 & 1 & -3 \\ 0 & -1 & -8 \\ 0 & 6 & -5 \end{bmatrix} = (-2)(-1)\det\begin{bmatrix} -1 & -8 \\ 6 & -5 \end{bmatrix} = 106. \ \square$$

EXAMPLE 5.11. You can obtain the determinants of the elementary matrices by setting $A = I_n$ in Theorem 5.3. Indeed,

$$\det(E_{ij}) = \det(E_{ij}I_n) = -\det(I_n) = -1,$$
$$\det(E_i(t)) = \det(E_i(t)I_n) = t\det(I_n) = t,$$
$$\det(E_{ij}(t)) = \det(E_{ij}(t)I_n) = \det(I_n) = 1.$$

Note that in each case, the determinant is not zero. $\square$

The conclusions in Example 5.11 allow Theorem 5.3 to be reformulated in the following concise manner.

COROLLARY. If $A \in \mathcal{M}_{n\times n}$ and E is an $n \times n$ elementary matrix, then

$$\det(EA) = \det(E)\det(A). \tag{5.8}$$

Given $A \in \mathcal{M}_{n\times n}$ and elementary $E_1, \ldots, E_q \in \mathcal{M}_{n\times n}$, successive applications of (5.8) give

$$\begin{aligned}\det(E_q \cdots E_1 A) &= \det(E_q(E_{q-1} \cdots E_1 A))\\ &= \det(E_q)\det(E_{q-1}(E_{q-2} \cdots E_1 A))\\ &\ \ \vdots\\ &= \det(E_q)\det(E_{q-1}) \cdots \det(E_1)\det(A).\end{aligned} \tag{5.9}$$

In particular, when $A = I_n$, (5.9) becomes

$$\det(E_q \cdots E_1) = \det(E_q)\det(E_{q-1}) \cdots \det(E_1),$$

and consequently (5.9) can be rewritten as

$$\det(E_q \cdots E_1 A) = \det(E_q \cdots E_1)\det(A). \tag{5.10}$$

Now, recall that an $n \times n$ matrix is nonsingular if and only if it can be expressed as a product of elementary matrices (Theorem 1.14). Thus, (5.10) can be interpreted as

$$\det(PA) = \det(P)\det(A),$$

where P is nonsingular. These observations suggest the possibility of a special interaction between the determinant and matrix multiplication.

THEOREM 5.4. If $A, B \in \mathcal{M}_{n\times n}$, then $\det(BA) = \det(B)\det(A)$.

Proof. The discussion preceding the theorem establishes the conclusion when B is nonsingular. Suppose B is singular and let $C = \mathrm{rref}(B)$. Then C has a row of zeros, so $\det(C) = 0$ (Example 5.8). Moreover, there is a nonsingular P such that $B = PC$ (Theorem 1.16), and the previously established case yields

$$\det(B) = \det(PC) = \det(P)\det(C) = 0.$$

Now, $BA = (PC)A = P(CA)$, so $\det(BA) = \det(P)\det(CA)$. Moreover, C having a row of zeros implies CA has a row of zeros, so $\det(CA) = 0$. Thus,

$$\det(BA) = 0 = 0\det(A) = \det(B)\det(A). \blacksquare$$

The conclusion of Theorem 5.4 extends to products of more than two matrices by a standard inductive argument. If $A_1, \ldots, A_q \in \mathcal{M}_{n\times n}$, then

$$\det(A_1 \cdots A_q) = \det(A_1) \cdots \det(A_q). \tag{5.11}$$

THEOREM 5.5. $A \in \mathcal{M}_{n\times n}$ is nonsingular if and only if $\det(A) \neq 0$, in which case

$$\det(A^{-1}) = \frac{1}{\det(A)}.$$

Proof. If A is nonsingular, then $I_n = AA^{-1}$, and by Theorem 5.4,

$$1 = \det(I_n) = \det(AA^{-1}) = \det(A)\det(A^{-1}).$$

Thus, $\det(A) \neq 0$ and $\det(A^{-1}) = [\det(A)]^{-1}$. On the other hand, the proof of Theorem 5.4 established that singular matrices have determinant 0, so $\det(A) \neq 0$ implies A is nonsingular. $\blacksquare$

■ **EXAMPLE 5.12.** If P $\in \mathcal{M}_{n\times n}$ is nonsingular, A $\in \mathcal{M}_{n\times n}$, and B $=$ P^{-1}AP, then

$$\det(\mathrm{B}) = \det(\mathrm{P}^{-1})\det(\mathrm{A})\det(\mathrm{P}) = \frac{1}{\det(\mathrm{P})}\det(\mathrm{A})\det(\mathrm{P}) = \det(\mathrm{A}).$$

In particular, when $T : \mathcal{R}^n \to \mathcal{R}^n$ is linear and $\mathcal{A}$ and $\mathcal{B}$ are bases for $\mathcal{R}^n$,

$$\det[T]_{\mathcal{B}} = \det(\mathrm{P}_{\mathcal{AB}}^{-1}[T]_{\mathcal{A}}\mathrm{P}_{\mathcal{AB}}) = \det[T]_{\mathcal{A}}. \ \square$$

THEOREM 5.6. For A $\in \mathcal{M}_{n\times n}$, $\det(\mathrm{A}^T) = \det(\mathrm{A})$.

Proof. If A is singular, then A^T is also singular, so $\det(\mathrm{A}^T) = 0 = \det(\mathrm{A})$. When A is nonsingular, there are elementary matrices $\mathrm{E}_1, \ldots, \mathrm{E}_q$ such that $\mathrm{A} = \mathrm{E}_1 \cdots \mathrm{E}_q$, so

$$\det(\mathrm{A}) = \det(\mathrm{E}_1) \cdots \det(\mathrm{E}_q),$$

and

$$\det(\mathrm{A}^T) = \det(\mathrm{E}_q^T \cdots \mathrm{E}_1^T) = \det(\mathrm{E}_q^T) \cdots \det(\mathrm{E}_1^T).$$

Thus, it suffices to show that $\det(\mathrm{E}^T) = \det(\mathrm{E})$ whenever E is elementary. That conclusion follows from the observations: $\mathrm{E}_{ij}^T = \mathrm{E}_{ij}$, $\mathrm{E}_i(t)^T = \mathrm{E}_i(t)$, and $\mathrm{E}_{ij}(t)^T = \mathrm{E}_{ji}(t)$. ■

■ **EXAMPLE 5.13.** If A is orthogonal, then $\mathrm{A}^{-1} = \mathrm{A}^T$, and by Theorems 5.5 and 5.6,

$$\frac{1}{\det(\mathrm{A})} = \det(\mathrm{A}^{-1}) = \det(\mathrm{A}^T) = \det(\mathrm{A}).$$

Thus, $[\det(\mathrm{A})]^2 = 1$, that is, $\det(\mathrm{A}) = \pm 1$. □

THEOREM 5.7. If A $\in \mathcal{M}_{n\times n}$ and $t \in \mathcal{R}$, then $\det(t\mathrm{A}) = t^n \det(\mathrm{A})$.

Proof. See Exercise 3. ■

■ **EXAMPLE 5.14.** Suppose A and B are 3×3 matrices such that $\det(\mathrm{A}) = 18$ and $\det(\mathrm{B}) = 3$. By Theorems 5.7, 5.4, and 5.5,

$$\det(6\mathrm{A}^{-1}\mathrm{B}) = 6^3\det(\mathrm{A}^{-1})\det(\mathrm{B}) = 6^3\frac{\det(\mathrm{B})}{\det(\mathrm{A})} = 6^3\left(\frac{3}{18}\right) = 36. \ \square$$

Exercises 5.3

1. Use the technique in Example 5.10 to evaluate the determinant of

a. $\begin{bmatrix} 1 & 1 & 0 & -1 \\ 2 & 4 & 1 & -3 \\ -1 & 0 & 2 & 4 \\ 0 & 2 & 3 & 2 \end{bmatrix}$. b. $\begin{bmatrix} 1 & 1 & 1 & 1 \\ 1 & 2 & 2 & 2 \\ 2 & 2 & 3 & 3 \\ 3 & 3 & 3 & 4 \end{bmatrix}$. c. $\begin{bmatrix} 1 & -1 & 1 & -1 \\ -1 & 1 & 1 & -1 \\ 1 & -1 & -1 & -1 \\ -1 & -1 & 1 & 1 \end{bmatrix}$.

2. Show that $\det\begin{bmatrix} 1 & 1 & 1 & 1 \\ 1 & a & 1 & 1 \\ 1 & 1 & b & 1 \\ 1 & 1 & 1 & c \end{bmatrix} = (a-1)(b-1)(c-1)$.

3. Prove Theorem 5.7.

4. Assuming $A = \begin{bmatrix} a & b & c \\ d & e & f \\ g & h & i \end{bmatrix}$ and $\det(A) = -3$, find

a. $\det(-A)$. b. $\det(2A^T)$. c. $\det \begin{bmatrix} i & h & g \\ f & e & d \\ c & b & a \end{bmatrix}$.

5. Assuming $A, B \in \mathcal{M}_{3\times 3}$, $\det(A) = -1/2$, and $\det(B) = 6$, find
a. $\det[(2A)B^{-1}]$. **b.** $\det[A^2(-B)^T]$. c. $\det\left[\frac{1}{2}(A^{-1}B)^2\right]$.

6. Use Theorem 5.3(1) to show that if A is a matrix with two identical rows, then $\det(A) = 0$.

7. Find the determinant of $A = \begin{bmatrix} 1 & \cdots & 1 \\ \vdots & \cdot^{\cdot^{\cdot}} & \\ 1 & & 0 \end{bmatrix} \in \mathcal{M}_{n\times n}$.

8. Assuming $A, B \in \mathcal{M}_{n\times n}$, show the statement is true or find a counterexample.
a. $\det(A + B) = \det(A) + \det(B)$.
b. $\det(AB) = \det(BA)$.
c. $\det(A + B^T) = \det(A^T + B)$.

9. Suppose $A \in \mathcal{M}_{n\times n}$ and $A^3 = A$. What are the possible values for $\det(A)$?

10. Show that an $n \times n$ skew-symmetric matrix is singular if n is odd.

11. Suppose $A \in \mathcal{M}_{n\times n}$ and let F_{ij}, $F_i(t)$, and $F_{ij}(t)$ be the elementary matrices that result from performing column operations c_{ij}, $c_i(t)$, and $c_{ij}(t)$ on I_n, respectively. Show that
a. $\det(AF_{ij}) = -\det(A)$. b. $\det(AF_i(t)) = t\det(A)$. **c.** $\det(AF_{ij}(t)) = \det(A)$.

12. Suppose A is a 5×5 matrix and $\text{Row}_4(A) = -2\text{Row}_1(A) + 3\text{Row}_2(A) - \text{Row}_5(A)$. Show that A is singular.

13. Show that $\det \left[\begin{array}{c|c} A & C \\ \hline 0 & B \end{array}\right] = \det(A)\det(B)$, where $A \in \mathcal{M}_{n\times n}$, $B \in \mathcal{M}_{m\times m}$, and $C \in \mathcal{M}_{n\times m}$.

14. If $P = (a,b)$ and $Q = (c,d)$ are distinct points in the plane, show that

$$\det \begin{bmatrix} x & a & c \\ y & b & d \\ 1 & 1 & 1 \end{bmatrix} = 0$$

is the equation of the line through P and Q.

15. Suppose $A \in \mathcal{M}_{n\times n}$, $X \in \mathcal{R}^n$, and $1 \le j \le n$. Let $A_j(X)$ be the matrix obtained by replacing $\text{Col}_j(A)$ by X, that is,

$$A_j(X) = [\text{Col}_1(A) \quad \cdots \quad \text{Col}_{j-1}(A) \quad X \quad \text{Col}_{j+1}(A) \quad \cdots \quad \text{Col}_n(A)].$$

Show that $T : \mathcal{R}^n \to \mathcal{R}$ defined by $T(X) = \det(A_j(X))$ is linear.

16. Assume $A \in \mathcal{M}_{n\times n}$ is nonsingular and all the entries in A and A^{-1} are integers. Show that $\det(A) = \pm 1$. Find a 2×2 matrix with no zero entries that has these properties.

5.4 THE ADJOINT AND CRAMER'S RULE

DEFINITION. For $A \in \mathcal{M}_{n\times n}$ and $1 \le i,j \le n$, the *cofactor of* a_{ij} is

$$\mathrm{cof}_{ij}(A) = (-1)^{i+j}\det(M_{ij}(A)),$$

and the *classical adjoint* of A, denoted by Adj(A), is the $n \times n$ matrix with entries

$$[\mathrm{Adj}(A)]_{ij} = \mathrm{cof}_{ji}(A).$$

In other words, Adj(A) is the transpose of the matrix obtained by replacing each entry in A by its cofactor.

■ **EXAMPLE 5.15.** If $A = \begin{bmatrix} 1 & 2 & 3 \\ 3 & 1 & 2 \\ 2 & 1 & 3 \end{bmatrix}$, then

$$\mathrm{cof}_{11}(A) = (-1)^{1+1}\det\begin{bmatrix} 1 & 2 \\ 1 & 3 \end{bmatrix} = 1, \qquad \mathrm{cof}_{12}(A) = (-1)^{1+2}\det\begin{bmatrix} 3 & 2 \\ 2 & 3 \end{bmatrix} = -5,$$

$$\mathrm{cof}_{13}(A) = (-1)^{1+3}\det\begin{bmatrix} 3 & 1 \\ 2 & 1 \end{bmatrix} = 1, \qquad \mathrm{cof}_{21}(A) = (-1)^{2+1}\det\begin{bmatrix} 2 & 3 \\ 1 & 3 \end{bmatrix} = -3,$$

$$\mathrm{cof}_{22}(A) = (-1)^{2+2}\det\begin{bmatrix} 1 & 3 \\ 2 & 3 \end{bmatrix} = -3, \qquad \mathrm{cof}_{23}(A) = (-1)^{2+3}\det\begin{bmatrix} 1 & 2 \\ 2 & 1 \end{bmatrix} = 3,$$

$$\mathrm{cof}_{31}(A) = (-1)^{3+1}\det\begin{bmatrix} 2 & 3 \\ 1 & 2 \end{bmatrix} = 1, \qquad \mathrm{cof}_{32}(A) = (-1)^{3+2}\det\begin{bmatrix} 1 & 3 \\ 3 & 2 \end{bmatrix} = 7,$$

$$\mathrm{cof}_{33}(A) = (-1)^{3+3}\det\begin{bmatrix} 1 & 2 \\ 3 & 1 \end{bmatrix} = -5,$$

so

$$\mathrm{Adj}(A) = \begin{bmatrix} 1 & -5 & 1 \\ -3 & -3 & 3 \\ 1 & 7 & -5 \end{bmatrix}^T = \begin{bmatrix} 1 & -3 & 1 \\ -5 & -3 & 7 \\ 1 & 3 & -5 \end{bmatrix}. \ \square$$

Interest in the adjoint stems from its appearance in the following explicit "formula" for the inverse of a nonsingular matrix.

THEOREM 5.8. If $A \in \mathcal{M}_{n\times n}$, then

$$A\ \mathrm{Adj}(A) = \det(A)\, I_n = \mathrm{Adj}(A)\, A. \tag{5.12}$$

In particular, when A is nonsingular,

$$A^{-1} = \frac{1}{\det(A)}\,\mathrm{Adj}(A). \tag{5.13}$$

Proof. For $1 \le i,j \le n$,

$$\begin{aligned} [A\ \mathrm{Adj}(A)]_{ij} &= \sum_{k=1}^{n} a_{ik}[\mathrm{Adj}(A)]_{kj} = \sum_{k=1}^{n} a_{ik}\mathrm{cof}_{jk}(A) \\ &= \sum_{k=1}^{n} a_{ik}(-1)^{j+k}\det(M_{jk}(A)). \end{aligned} \tag{5.14}$$

When $i = j$, (5.14) is the Laplace expansion of det(A) by the ith row. Suppose $i \neq j$ and let C be the $n \times n$ matrix obtained by replacing $\text{Row}_j(\text{A})$ by a copy of $\text{Row}_i(\text{A})$. Then (5.14) is the Laplace expansion of det(C) by the jth row, and since rows i and j of C are identical, det(C) = 0. Thus,

$$[\text{A Adj(A)}]_{ij} = \begin{cases} \det(\text{A}), & \text{if } i = j \\ 0, & \text{if } i \neq j \end{cases},$$

that is, $\text{A Adj(A)} = \det(\text{A})\, \text{I}_n$. A similar argument utilizing column expansions shows that $\text{Adj(A) A} = \text{I}_n$. Finally, when A is nonsingular, (5.12) yields

$$\text{I}_n = \frac{1}{\det(\text{A})}\left(\text{A Adj(A)}\right) = \text{A}\left(\frac{1}{\det(\text{A})}\text{Adj(A)}\right),$$

which establishes (5.13). ■

■ **EXAMPLE 5.16.** Consider $\text{A} = \begin{bmatrix} 1 & 2 & 3 \\ 3 & 1 & 2 \\ 2 & 1 & 3 \end{bmatrix}$. Previous calculations in Example 5.15 produced

$$\text{Adj(A)} = \begin{bmatrix} 1 & -3 & 1 \\ -5 & -3 & 7 \\ 1 & 3 & -5 \end{bmatrix},$$

so

$$\text{A Adj(A)} = \begin{bmatrix} 1 & 2 & 3 \\ 3 & 1 & 2 \\ 2 & 1 & 3 \end{bmatrix}\begin{bmatrix} 1 & -3 & 1 \\ -5 & -3 & 7 \\ 1 & 3 & -5 \end{bmatrix} = \begin{bmatrix} -6 & 0 & 0 \\ 0 & -6 & 0 \\ 0 & 0 & -6 \end{bmatrix} = -6\text{I}_3.$$

According to (5.12) and (5.13), det(A) = −6, and

$$\text{A}^{-1} = \frac{1}{\det(\text{A})}\text{Adj(A)} = -\frac{1}{6}\begin{bmatrix} 1 & -3 & 1 \\ -5 & -3 & 7 \\ 1 & 3 & -5 \end{bmatrix}. \ \square$$

■ **EXAMPLE 5.17.** Assume $\text{A} \in \mathcal{M}_{n\times n}$. By Theorems 5.4, 5.8, and 5.7,

$$\det(\text{A})\det(\text{Adj(A)}) = \det(\text{A Adj(A)}) = \det(\det(\text{A})\text{I}_n) = [\det(\text{A})]^n.$$

If A is nonsingular, then $\det(\text{A}) \neq 0$, and consequently

$$\det(\text{Adj(A)}) = [\det(\text{A})]^{n-1}. \tag{5.15}$$

In particular, $\det(\text{Adj(A)}) \neq 0$, so Adj(A) is also nonsingular. It can be shown that Adj(A) is singular when A is singular (Problem 3), in which case both A and Adj(A) have determinant 0. Thus, (5.15) holds for any $\text{A} \in \mathcal{M}_{n\times n}$. □

If $\text{A} \in \mathcal{M}_{n\times n}$ is nonsingular and $\text{B} \in \mathcal{R}^n$, then AX = B has unique solution $\text{X} = \text{A}^{-1}\text{B}$, and (5.13) leads to the following description of X.

THEOREM 5.9. **(Cramer's Rule).** Assume $\text{A} \in \mathcal{M}_{n\times n}$ is nonsingular and $\text{B} \in \mathcal{R}^n$. Let A_j be the matrix obtained by replacing the jth column of A by B, that is,

$$\text{A}_j = [\text{Col}_1(\text{A}) \ \cdots \ \text{Col}_{j-1}(\text{A}) \ \ \text{B} \ \ \text{Col}_{j+1}(\text{A}) \ \cdots \ \text{Col}_n(\text{A})].$$

The unique solution of AX = B is $[x_1 \ \cdots \ x_n]^T$, where

$$x_j = \frac{\det(\text{A}_j)}{\det(\text{A})}, \quad j = 1, \ldots, n. \tag{5.16}$$

Proof. By (5.13), the solution of $AX = B$ is $X = A^{-1}B = \frac{1}{\det(A)}\left(\mathrm{Adj}(A)B\right)$, so

$$\begin{aligned} x_j = [X]_{j1} &= \frac{1}{\det(A)}[\mathrm{Adj}(A)B]_{j1} \\ &= \frac{1}{\det(A)}\sum_{k=1}^{n}[\mathrm{Adj}(A)]_{jk}[B]_{k1} = \frac{1}{\det(A)}\sum_{k=1}^{n} b_k \mathrm{cof}_{kj}(A) \\ &= \frac{1}{\det(A)}\sum_{k=1}^{n} b_k(-1)^{k+j}\det(M_{kj}(A)). \end{aligned}$$

Observe that $b_k = [A_j]_{kj}$ and $M_{kj}(A) = M_{kj}(A_j)$, $k = 1, \ldots, n$, so the last sum is the Laplace expansion of $\det(A_j)$ by column j. Thus, $x_j = \det(A_j)/\det(A)$. ■

■ **EXAMPLE 5.18.** For $A = \begin{bmatrix} 1 & 2 & 3 \\ 3 & 1 & 2 \\ 2 & 1 & 3 \end{bmatrix}$ and $B = \begin{bmatrix} -1 \\ 0 \\ 2 \end{bmatrix}$,

$$A_1 = \begin{bmatrix} -1 & 2 & 3 \\ 0 & 1 & 2 \\ 2 & 1 & 3 \end{bmatrix}, \quad A_2 = \begin{bmatrix} 1 & -1 & 3 \\ 3 & 0 & 2 \\ 2 & 2 & 3 \end{bmatrix}, \quad A_3 = \begin{bmatrix} 1 & 2 & -1 \\ 3 & 1 & 0 \\ 2 & 1 & 2 \end{bmatrix},$$

and a little calculation produces

$$\det(A_1) = 1, \quad \det(A_2) = 19, \quad \text{and} \quad \det(A_3) = -11.$$

Since $\det(A) = -6$ (Example 5.16), the solution of $AX = B$ has coordinates

$$x_1 = -\tfrac{1}{6}, \quad x_2 = -\tfrac{19}{6}, \quad \text{and} \quad x_3 = \tfrac{11}{6}. \ \square$$

REMARK. Equation (5.13) is not an efficient means of calculating A^{-1}. It is interesting mostly because it gives an explicit formula (albeit a complicated one) for the entries in A^{-1} in terms of the entries in A. A similar comment applies to (5.16).

Exercises 5.4

1. Find Adj(A) and use it, if possible, to find A^{-1}.

a. $\begin{bmatrix} 1 & 1 & 2 \\ 2 & 3 & 3 \\ 4 & 4 & 5 \end{bmatrix}$ b. $\begin{bmatrix} 1 & 2 & 3 \\ 0 & 4 & 5 \\ 0 & 0 & 6 \end{bmatrix}$ **c.** $\begin{bmatrix} 1 & 2 & 3 \\ 4 & 5 & 6 \\ 7 & 8 & 9 \end{bmatrix}$ d. $\begin{bmatrix} a & 0 & 0 \\ 0 & b & 0 \\ 0 & 0 & c \end{bmatrix}$

e. $\begin{bmatrix} 1 & 0 & 1 & 0 \\ 0 & 2 & 0 & 1 \\ 0 & 0 & 3 & 0 \\ 0 & 0 & 0 & 4 \end{bmatrix}$ f. $\begin{bmatrix} 1 & -1 & 0 & 1 \\ 0 & 1 & 1 & -1 \\ -1 & 0 & 1 & 1 \\ 1 & 1 & -1 & 0 \end{bmatrix}$ **g.** $\begin{bmatrix} 1 & 1 & 1 & 0 \\ 1 & 1 & 0 & 1 \\ 1 & 0 & 1 & 1 \\ 0 & 1 & 1 & 1 \end{bmatrix}$.

2. Assuming $A = \begin{bmatrix} a & b \\ c & d \end{bmatrix}$, find Adj(A) and Adj(Adj(A)).

3. Show that if $A \in \mathcal{M}_{n\times n}$ is singular, then Adj(A) is singular.

4. Suppose $A \in \mathcal{M}_{n\times n}$ and $t \in \mathcal{R}$. Show that $\mathrm{Adj}(tA) = t^{n-1}\mathrm{Adj}(A)$.

5. Show that if A has integer entries and $\det(A) = \pm 1$, then A^{-1} has integer entries.

6. Assuming A is nonsingular, show that $\operatorname{Adj}(A^{-1}) = \dfrac{1}{\det(A)} A = [\operatorname{Adj}(A)]^{-1}$.

7. Show that if $A \in \mathcal{M}_{n \times n}$ is nonsingular, then $\operatorname{Adj}(\operatorname{Adj}(A)) = [\det(A)]^{n-2} A$.

8. **a.** How is $\operatorname{Adj}(A^T)$ related to $\operatorname{Adj}(A)$?
   b. Show that if A is symmetric, then Adj(A) is symmetric.

9. For nonsingular $A, B \in \mathcal{M}_{n \times n}$, show that $\operatorname{Adj}(AB) = \operatorname{Adj}(B)\operatorname{Adj}(A)$.

10. Use Cramer's rule to solve $AX = B$ when

a. $A = \begin{bmatrix} 2 & 5 \\ 1 & -3 \end{bmatrix}$, $B = \begin{bmatrix} 1 \\ 2 \end{bmatrix}$.

b. $A = \begin{bmatrix} 1 & 0 & 1 \\ -1 & 1 & 0 \\ 0 & -1 & 1 \end{bmatrix}$, $B = \begin{bmatrix} 1 \\ 2 \\ 3 \end{bmatrix}$.

c. $A = \begin{bmatrix} 1 & 0 & 2 & 1 \\ 0 & 2 & -1 & 0 \\ 0 & 0 & -1 & 2 \\ 0 & 0 & 0 & 2 \end{bmatrix}$, $B = \begin{bmatrix} 1 \\ 2 \\ 3 \\ 4 \end{bmatrix}$.

5.5
ORIENTATION AND VOLUME

Given a nonzero $X \in \mathcal{R}^2$, the line ℓ through 0 and X divides $\mathcal{R}^2$ into two half-planes. Imagine traveling along ℓ in the direction indicated by X. The half-planes to your left and right are called the *upper* and *lower half-planes,* respectively, *relative to* X and are denoted by $H^+(X)$ and $H^-(X)$ (Figure 5.2). When $X = E_1$,

$$H^+(E_1) = \left\{ \begin{bmatrix} x_1 \\ x_2 \end{bmatrix} : x_2 > 0 \right\},$$

in which case, "upper" has its customary meaning.

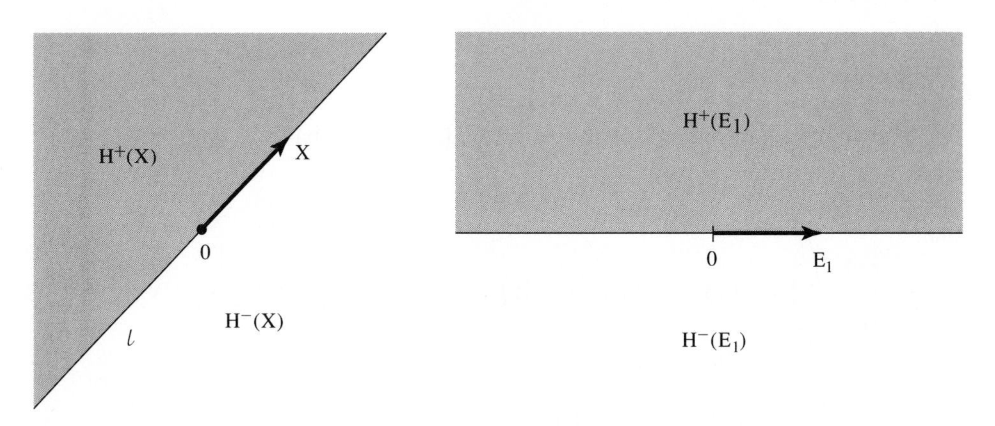

FIGURE 5.2

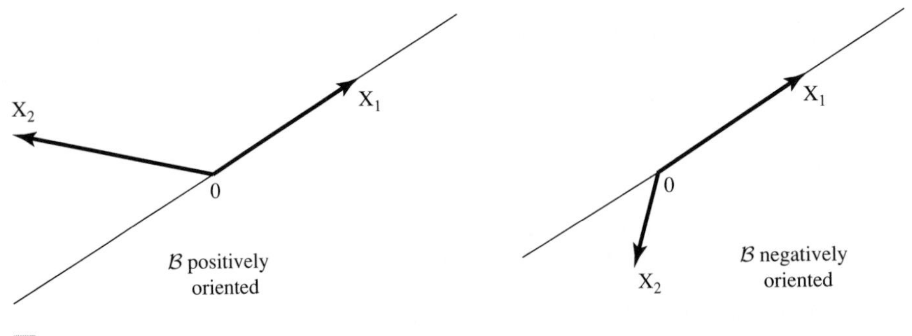

FIGURE 5.3

Now suppose $X_1, X_2 \in \mathcal{R}^2$ are linearly independent. Since 0, X_1, and X_2 are noncollinear, X_2 lies in one of the half-planes determined by X_1. The basis $\mathcal{B} = \{X_1, X_2\}$ is said to be *positively* or *negatively oriented* according as $X_2 \in H^+(X_1)$ or $X_2 \in H^-(X_1)$, respectively (Figure 5.3).

Let T be a rotation about 0 that sends X_1 to the positive E_1 axis (Figure 5.4). Then

$$T(X_1) = \begin{bmatrix} a \\ 0 \end{bmatrix} \quad \text{and} \quad T(X_2) = \begin{bmatrix} b \\ c \end{bmatrix},$$

where $a = \|X_1\| > 0$. The images of $H^+(X_1)$ and $H^-(X_1)$ are $H^+(E_1)$ and $H^-(E_1)$, respectively. Observe that $\mathcal{B}$ is positively oriented when $T(X_2) \in H^+(E_1)$, that is, when $c > 0$.

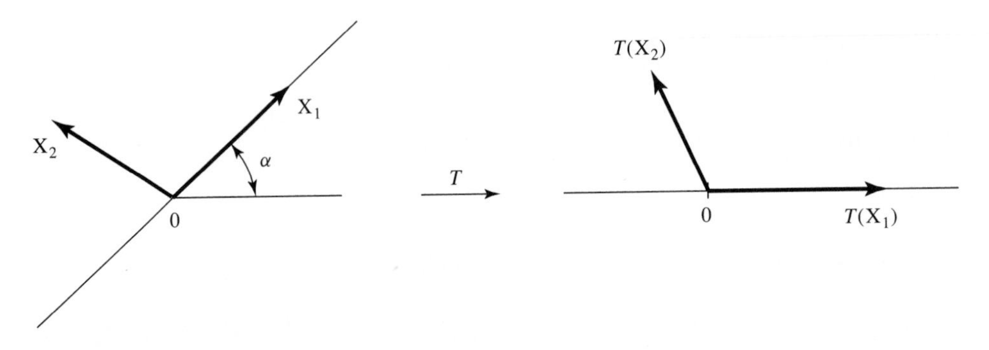

FIGURE 5.4

The matrix representing T with respect to the standard basis is

$$[T]_{\mathcal{E}} = \begin{bmatrix} \cos(\alpha) & \sin(\alpha) \\ -\sin(\alpha) & \cos(\alpha) \end{bmatrix},$$

and

$$\begin{bmatrix} a & b \\ 0 & c \end{bmatrix} = [T(X_1) \quad T(X_2)] = \big[[T]_{\mathcal{E}}X_1 \quad [T]_{\mathcal{E}}X_2\big] = [T]_{\mathcal{E}}[X_1 \quad X_2].$$

Since $\det[T]_{\mathcal{E}} = \cos^2(\alpha) + \sin^2(\alpha) = 1$,

$$ac = \det\begin{bmatrix} a & b \\ 0 & c \end{bmatrix} = \det[T]_{\mathcal{E}}\det[X_1 \quad X_2] = \det[X_1 \quad X_2].$$

The orientation of $\mathcal{B}$ is determined by the sign of c, and since $a > 0$, c has the same sign as $\det[X_1 \quad X_2]$. Thus, $\mathcal{B}$ is positively or negatively oriented according as $\det[X_1 \quad X_2]$ is positive or negative, respectively.

■ **EXAMPLE 5.19.** If $X_1 = \begin{bmatrix} 1 \\ -2 \end{bmatrix}$, $X_2 = \begin{bmatrix} -3 \\ 1 \end{bmatrix}$, and $\mathcal{B} = \{X_1, X_2\}$, then

$$\det[X_1 \quad X_2] = \det \begin{bmatrix} 1 & -3 \\ -2 & 1 \end{bmatrix} = -5 < 0,$$

so $\mathcal{B}$ is negatively oriented. On the other hand, $\mathcal{C} = \{X_2, X_1\}$ is positively oriented because $\det[X_2 \quad X_1] = -\det[X_1 \quad X_2] = 5 > 0$. □

The prototypical examples of positively and negatively oriented bases for $\mathcal{R}^2$ are $\mathcal{E} = \{E_1, E_2\}$ and $\mathcal{D} = \{E_1, -E_2\}$, respectively. Basis $\{X_1, X_2\}$ is positively oriented when the location of X_2 relative to X_1 more closely resembles that of E_2 to E_1 than that of $-E_2$ to E_1.

Consider now a basis $\mathcal{B} = \{X_1, X_2, X_3\}$ for $\mathcal{R}^3$. Let p be the plane spanned by X_1 and X_2, and observe that p divides $\mathcal{R}^3$ into two half-spaces, one of which contains X_3. The plan is to designate one of them as "upper" relative to $\{X_1, X_2\}$ and declare $\mathcal{B}$ positively oriented when X_3 lies in that half-space. The choice is based on a comparison with $\mathcal{E} = \{E_1, E_2, E_3\}$, in which case the usual upper half-space is

$$H^+(E_1, E_2) = \left\{ \begin{bmatrix} x_1 \\ x_2 \\ x_3 \end{bmatrix} : x_3 > 0 \right\}.$$

To make the comparison, you transform p into $\mathcal{S}pan\{E_1, E_2\}$ via a linear function L that is composed of rotations about various coordinate axes. This can be done so that $L(X_1)$ is on the positive E_1 axis and $L(X_2)$ is on the same side of the E_1 axis as E_2, that is, so that $\{L(X_1), L(X_2)\}$ resembles $\{E_1, E_2\}$ as much as possible. The half-space determined by p that gets transformed into $H^+(E_1, E_2)$ is called the *upper half-space relative to* $\{X_1, X_2\}$, denoted by $H^+\{X_1, X_2\}$, and $H^-(X_1, X_2)$ is the other (*lower*) half-space relative to $\{X_1, X_2\}$. The next task is to exhibit the transformation L. A practical test for determining the orientation of $\mathcal{B}$ will then be forthcoming.

Let T be a rotation about the E_1 axis such that $T(X_1) \in \mathcal{S}pan\{E_1, E_3\}$ (Figure 5.5(a)). Then choose a rotation S about the E_2 axis that sends $T(X_1)$ to the positive E_1 axis (Figure 5.5(b)). The matrices representing T and S relative to the standard coordinate system are

$$[T]_{\mathcal{E}} = \begin{bmatrix} 1 & 0 & 0 \\ 0 & \cos(\alpha) & -\sin(\alpha) \\ 0 & \sin(\alpha) & \cos(\alpha) \end{bmatrix} \quad \text{and} \quad [S]_{\mathcal{E}} = \begin{bmatrix} \cos(\beta) & 0 & \sin(\beta) \\ 0 & 1 & 0 \\ -\sin(\beta) & 0 & \cos(\beta) \end{bmatrix},$$

where α and β are the angles indicated in Figure 5.5. Now, $S \circ T : \mathcal{R}^3 \to \mathcal{R}^3$ is linear, and $W_1 = (S \circ T)(X_1)$ lies on the positive E_1 axis. Set $W_2 = (S \circ T)(X_2)$. Let R be a rotation about the E_1 axis such that $R(W_2)$ lies in $\mathcal{S}pan\{E_1, E_2\}$, on the same side of the E_1 axis as E_2, that is, $R(W_2) = bE_1 + cE_2$, and $c > 0$. Then

$$L = R \circ S \circ T \tag{5.17}$$

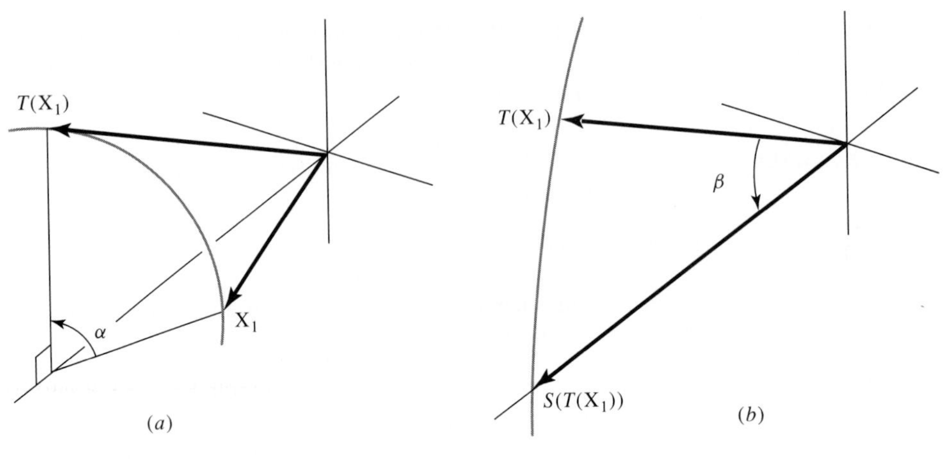

FIGURE 5.5

is the desired transformation. It satisfies

$$L(X_1) = \begin{bmatrix} a \\ 0 \\ 0 \end{bmatrix}, \quad L(X_2) = \begin{bmatrix} b \\ c \\ 0 \end{bmatrix}, \quad \text{and} \quad L(X_3) = \begin{bmatrix} d \\ e \\ f \end{bmatrix},$$

where $a = \|X_1\| > 0$ and $c > 0$, so

$$\begin{bmatrix} a & b & d \\ 0 & c & e \\ 0 & 0 & f \end{bmatrix} = [L(X_1) \quad L(X_2) \quad L(X_3)] = \big[[L]_{\mathcal{E}}X_1 \quad [L]_{\mathcal{E}}X_2 \quad [L]_{\mathcal{E}}X_3\big]$$

$$= [L]_{\mathcal{E}}[X_1 \quad X_2 \quad X_3].$$

Note that $[L]_{\mathcal{E}} = [R]_{\mathcal{E}}[S]_{\mathcal{E}}[T]_{\mathcal{E}}$. Since $[T]_{\mathcal{E}}$, $[S]_{\mathcal{E}}$, and $[R]_{\mathcal{E}}$ all have determinant 1, $\det[L]_{\mathcal{E}} = \det[R]_{\mathcal{E}}\det[S]_{\mathcal{E}}\det[T]_{\mathcal{E}} = 1$, and therefore

$$acf = \det\begin{bmatrix} a & b & d \\ 0 & c & e \\ 0 & 0 & f \end{bmatrix} = \det[L]_{\mathcal{E}}\det[X_1 \quad X_2 \quad X_3] = \det[X_1 \quad X_2 \quad X_3].$$

Now, $\mathcal{B}$ is positively oriented when $L(X_3) \in H^+(E_1,E_2)$, that is, when $f > 0$. Since a and c are positive, f and $\det[X_1 \quad X_2 \quad X_3]$ have the same sign. Thus, $\mathcal{B}$ is positively oriented when $\det[X_1 \quad X_2 \quad X_3] > 0$.

The preceding discussions in $\mathcal{R}^2$ and $\mathcal{R}^3$ suggest the following notion of orientation of bases for $\mathcal{R}^n$.

DEFINITION. A basis $\mathcal{B} = \{X_1, \ldots, X_n\}$ for $\mathcal{R}^n$ is *positively oriented* when $\det[X_1 \quad \cdots \quad X_n] > 0$ and *negatively oriented* when $\det[X_1 \quad \cdots \quad X_n] < 0$.

■ **EXAMPLE 5.20.** If $X_1 = \begin{bmatrix} 1 \\ 1 \\ 0 \end{bmatrix}$, $X_2 = \begin{bmatrix} 0 \\ 1 \\ 1 \end{bmatrix}$, $X_3 = \begin{bmatrix} 1 \\ 0 \\ 1 \end{bmatrix}$, and $\mathcal{B} = \{X_1, X_2, X_3\}$, then

$$\det[X_1 \quad X_2 \quad X_3] = \det \begin{bmatrix} 1 & 0 & 1 \\ 1 & 1 & 0 \\ 0 & 1 & 1 \end{bmatrix} = 2 > 0,$$

so $\mathcal{B}$ is positively oriented. On the other hand,

$$\det[X_1 \quad X_3 \quad X_2] = -\det[X_1 \quad X_2 \quad X_3] = -2 < 0$$

implies $\{X_1, X_3, X_2\}$ is negatively oriented. There are six permutations of X_1, X_2, and X_3, each yielding a basis for $\mathcal{R}^3$. You might check that three of these bases are positively oriented and the others are negatively oriented. □

■ **EXAMPLE 5.21.** The standard basis for $\mathcal{R}^n$ is positively oriented, because $\det[E_1 \cdots E_n] = \det(I_n) = 1$. □

■ **EXAMPLE 5.22.** If X and Y are linearly independent vectors in $\mathcal{R}^3$, then

$$X \times Y = \begin{bmatrix} x_2y_3 - x_3y_2 \\ x_3y_1 - x_1y_3 \\ x_1y_2 - x_2y_1 \end{bmatrix} \neq 0,$$

and therefore $\mathcal{B} = \{X, Y, X \times Y\}$ is a basis for $\mathcal{R}^3$. Expanding $\det[X \quad Y \quad X \times Y]$ by the third column gives

$$\det \begin{bmatrix} x_1 & y_1 & x_2y_3 - x_3y_2 \\ x_2 & y_2 & x_3y_1 - x_1y_3 \\ x_3 & y_3 & x_1y_2 - x_2y_1 \end{bmatrix} = (x_2y_3 - x_3y_2)^2 + (x_3y_1 - x_1y_3)^2 + (x_1y_2 - x_2y_1)^2 > 0,$$

so $\mathcal{B}$ is positively oriented. □

There is a well-known rule of thumb, called the right-hand rule, for checking the orientation of a basis $\mathcal{B} = \{X_1, X_2, X_3\}$ for $\mathcal{R}^3$. When the first and second fingers of your right hand are positioned to point in the directions of X_1 and X_2, respectively, the thumb of that hand points into $H^+(X_1, X_2)$. Thus, $\mathcal{B}$ is positively oriented if the first finger, second finger, and thumb of the right hand can be aligned with X_1, X_2, and X_3, respectively. A positively oriented basis for $\mathcal{R}^3$ is said to establish a *right-handed coordinate system.*

Suppose X_1 and X_2 are linearly independent vectors in $\mathcal{R}^2$ and let Π be the parallelogram with vertices 0, X_1, X_2, and $X_1 + X_2$. As indicated in the introductory section

$$\text{Area}(\Pi) = |\det[X_1 \quad X_2]|.$$

Since the sign of $\det[X_1 \quad X_2]$ determines the orientation of $\mathcal{B} = \{X_1, X_2\}$,

$$\det[X_1 \quad X_2] = \begin{cases} \text{Area}(\Pi), & \text{if } \mathcal{B} \text{ is positively oriented} \\ -\text{Area}(\Pi), & \text{if } \mathcal{B} \text{ is negatively oriented} \end{cases}.$$

A similar connection exists between the determinant of a 3×3 matrix and the volume of an associated three-dimensional solid determined by its columns. The parallelepiped generated by linearly independent vectors $X_1, X_2, X_3 \in \mathcal{R}^3$ is $\Pi = \{rX_1 + sX_2 + tX_3 : 0 \leq r,s,t \leq 1\}$ (Figure 5.6). It is a polyhedron with six faces, occurring in three opposite pairs of parallel congruent parallelgrams. To compute the volume of Π it is convenient to first "rotate" space so that one of the faces lies in the coordinate plane spanned by E_1 and E_2. The volume is then the product of the area of that face and the vertical distance between it and its opposite face. The "rotation" is accomplished by the function L in (5.17). Recall that

$$L(X_1) = \begin{bmatrix} a \\ 0 \\ 0 \end{bmatrix}, \quad L(X_2) = \begin{bmatrix} b \\ c \\ 0 \end{bmatrix}, \quad \text{and } L(X_3) = \begin{bmatrix} d \\ e \\ f \end{bmatrix},$$

where $a = \|X_1\| > 0, c > 0,$ and

$$acf = \det \begin{bmatrix} a & b & d \\ 0 & c & e \\ 0 & 0 & f \end{bmatrix} = \det[X_1 \quad X_2 \quad X_3].$$

Now, $L(\Pi)$ lies in $H^+(E_1,E_2)$ or $H^-(E_1,E_2)$, depending on whether $\mathcal{B}$ is positively or negatively oriented, respectively. In either case, its altitude is $|f|$, and the face lying in the E_1E_2 coordinate plane has area ac (see Figure 5.7 for the case when $\mathcal{B}$ is positively oriented). Thus,

$$\text{Vol}(\Pi) = \text{Vol}(L(\Pi)) = ac|f| = |\det[X_1 \quad X_2 \quad X_3]|,$$

and therefore

$$\det[X_1 \quad X_2 \quad X_3] = \begin{cases} \text{Vol}(\Pi), & \text{if } \mathcal{B} \text{ is positively oriented} \\ -\text{Vol}(\Pi), & \text{if } \mathcal{B} \text{ is negatively oriented} \end{cases}.$$

DEFINITION. The *parallelepiped generated by* $X_1, \ldots, X_n \in \mathcal{R}^n$ is

$$\Pi(X_1, \ldots, X_n) = \left\{ \sum_1^n t_k X_k \; : \; 0 \leq t_1, \ldots, t_n \leq 1 \right\},$$

and its *n-dimensional volume* is

$$\text{Vol}(\Pi(X_1, \ldots, X_n)) = |\det[X_1 \quad \cdots \quad X_n]|. \tag{5.18}$$

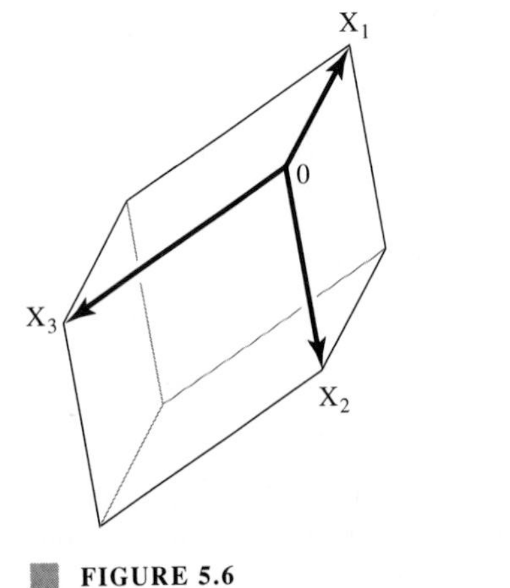

FIGURE 5.6

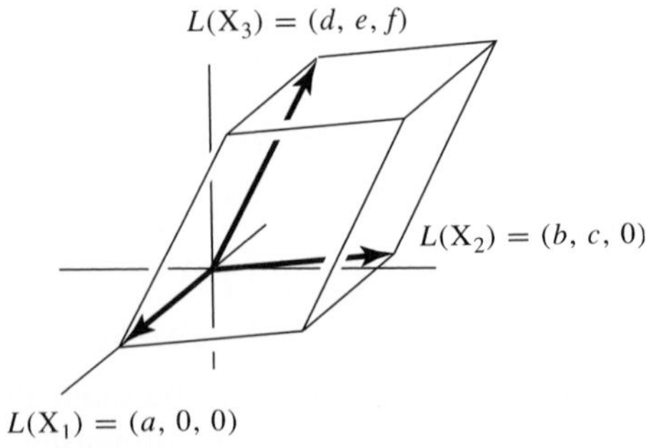

FIGURE 5.7

■ **EXAMPLE 5.23.** If $X_1 = \begin{bmatrix} 2 \\ -1 \\ 1 \end{bmatrix}$, $X_2 = \begin{bmatrix} 3 \\ 0 \\ -2 \end{bmatrix}$, and $X_3 = \begin{bmatrix} 1 \\ -1 \\ 1 \end{bmatrix}$, then

$$\det[X_1 \quad X_2 \quad X_3] = \det \begin{bmatrix} 2 & 3 & 1 \\ -1 & 0 & -1 \\ 1 & -2 & 1 \end{bmatrix} = -2,$$

so $\mathcal{B} = \{X_1, X_2, X_3\}$ is negatively oriented, and

$$\mathrm{Vol}(\Pi(X_1, X_2, X_3)) = |\det[X_1 \quad X_2 \quad X_3]| = 2. \ \square$$

The definition of $\Pi(X_1, \ldots, X_n)$ does not require $X_1, \ldots, X_n$ to be linearly independent. When they are linearly dependent, the parallelepiped is in some sense degenerate, and that condition is reflected in the value assigned by (5.18). Indeed, in this case $[X_1 \ \cdots \ X_n]$ is singular, so its determinant is 0. This outcome is consistent with your geometric intuition regarding area in $\mathcal{R}^2$ and volume in $\mathcal{R}^3$. For example, if X_1 and X_2 are linearly dependent vectors in $\mathcal{R}^2$, then $\mathcal{S}pan\{X_1, X_2\}$ is either a line through 0 or $\{0\}$, and $\Pi(X_1, X_2)$, being a subset of $\mathcal{S}pan\{X_1, X_2\}$, is either a line segment or a point. In either event, $\Pi(X_1, X_2)$ has area 0. Similarly, when $X_1, X_2, X_3 \in \mathcal{R}^3$ are linearly dependent, $\Pi(X_1, X_2, X_3)$ is either a piece of a plane, a segment of a line, or $\{0\}$, so the volume of $\Pi(X_1, X_2, X_3)$ is 0.

Suppose $T : \mathcal{R}^n \to \mathcal{R}^n$ is linear and $\{X_1, \ldots, X_n\} \subseteq \mathcal{R}^n$. Since

$$\left\{T\left(\sum_1^n t_k X_k\right) : 0 \le t_1, \ldots, t_n \le 1\right\} = \left\{\sum_1^n t_k T(X_k) : 0 \le t_1, \ldots, t_n \le 1\right\},$$

the image of the parallelepiped generated by $X_1, \ldots, X_n$ is the parallelepiped generated by $T(X_1), \ldots, T(X_n)$. That is,

$$T(\Pi(X_1, \ldots, X_n)) = \Pi(T(X_1), \ldots, T(X_n)). \tag{5.19}$$

Moreover, $T(X_k) = [T]_{\mathcal{E}} X_k$, $k = 1, \ldots, n$, so

$$[T(X_1) \quad \cdots \quad T(X_n)] = \big[[T]_{\mathcal{E}} X_1 \quad \cdots \quad [T]_{\mathcal{E}} X_n\big] = [T]_{\mathcal{E}}[X_1 \quad \cdots \quad X_n]. \tag{5.20}$$

These observations lead to the following conclusion regarding the effect of the action of T on the volume of a parallelepiped.

THEOREM 5.10. If $T : \mathcal{R}^n \to \mathcal{R}^n$ is linear, $X_1, \ldots, X_n \in \mathcal{R}^n$, and $\Pi = \Pi(X_1, \ldots, X_n)$, then

$$\mathrm{Vol}(T(\Pi)) = |\det([T]_{\mathcal{E}})| \ \mathrm{Vol}(\Pi).$$

Proof. From (5.20), $\det[T(X_1) \ \cdots \ T(X_n)] = \det\,[T]_{\mathcal{E}} \det[X_1 \ \cdots \ X_n]$, so

$$\begin{aligned} \mathrm{Vol}(T(\Pi)) &= |\det[T(X_1) \quad \cdots \quad T(X_n)]| \\ &= |\det[T]_{\mathcal{E}}| \, |\det[X_1 \quad \cdots \quad X_n]| \\ &= |\det[T]_{\mathcal{E}}| \mathrm{Vol}(\Pi). \ ■ \end{aligned}$$

■ **EXAMPLE 5.24.** Define $T : \mathcal{R}^3 \to \mathcal{R}^3$ by

$$T(X) = \begin{bmatrix} 1 & -1 & 2 \\ 0 & 2 & -3 \\ 2 & 0 & -1 \end{bmatrix} X,$$

and consider the parallelepiped Π generated by $X_1 = \begin{bmatrix} 1 \\ 1 \\ 1 \end{bmatrix}$, $X_2 = \begin{bmatrix} -1 \\ 2 \\ 1 \end{bmatrix}$, and

$X_3 = \begin{bmatrix} 2 \\ 1 \\ -2 \end{bmatrix}$. Observe that

$$\mathrm{Vol}(\Pi) = |\det[X_1 \quad X_2 \quad X_3]| = \left| \det \begin{bmatrix} 1 & -1 & 2 \\ 1 & 2 & 1 \\ 1 & 1 & -2 \end{bmatrix} \right| = 10,$$

and

$$\det[T]_{\mathcal{E}} = \det \begin{bmatrix} 1 & -1 & 2 \\ 0 & 2 & -3 \\ 2 & 0 & -1 \end{bmatrix} = -4.$$

By Theorem 5.10,

$$\mathrm{Vol}(T(\Pi)) = |\det[T]_{\mathcal{E}}| \; \mathrm{Vol}(\Pi) = 40.$$

The volume of $T(\Pi)$ can also be calculated directly. Indeed,

$$T(X_1) = \begin{bmatrix} 2 \\ -1 \\ 1 \end{bmatrix}, \quad T(X_2) = \begin{bmatrix} -1 \\ 1 \\ -3 \end{bmatrix}, \quad \text{and} \quad T(X_3) = \begin{bmatrix} -3 \\ 8 \\ 6 \end{bmatrix},$$

so

$$\mathrm{Vol}(T(\Pi)) = |\det[T(X_1) \quad T(X_2) \quad T(X_3)]| = \left| \det \begin{bmatrix} 2 & -1 & -3 \\ -1 & 1 & 8 \\ 1 & -3 & 6 \end{bmatrix} \right| = 40. \;\square$$

The conclusion of Theorem 5.10 applies to more general sets than parallelepipeds, but to make the extension one has to discuss the meaning of "n-dimensional volume" for arbitrary subsets of $\mathcal{R}^n$. Such an inquiry involves limiting processes that properly belong to a course in analysis. Nevertheless, if S is a subset of $\mathcal{R}^n$ for which n-dimensional volume is meaningful, then

$$\mathrm{Vol}(T(S)) = |\det[T]_{\mathcal{E}}| \; \mathrm{Vol}(S). \tag{5.21}$$

As S varies, (5.21) states that the volume of $T(S)$ remains proportional to that of S, with proportionality constant $|\det[T]_{\mathcal{E}}|$. The next example is a particularly elegant application of these ideas.

■ **EXAMPLE 5.25.** Let $S = \left\{ \begin{bmatrix} x \\ y \end{bmatrix} : x^2 + y^2 \le 1 \right\}$, and consider $T : \mathcal{R}^2 \to \mathcal{R}^2$ with equation

$$\begin{bmatrix} u \\ v \end{bmatrix} = T(X) = \begin{bmatrix} a & 0 \\ 0 & b \end{bmatrix} \begin{bmatrix} x \\ y \end{bmatrix},$$

$0 < b < a$. Then $u = ax$, $v = by$, and

$$T(S) = \left\{ \begin{bmatrix} ax \\ by \end{bmatrix} : x^2 + y^2 \le 1 \right\} = \left\{ \begin{bmatrix} u \\ v \end{bmatrix} : \frac{u^2}{a^2} + \frac{v^2}{b^2} \le 1 \right\}.$$

Thus, the image of the disk S is the ellipse centered at the origin with major and minor axes of lengths $2a$ and $2b$, respectively. Since Area $(S) = \pi$ and

$$|\det[T]_{\mathcal{E}}| = \left|\det \begin{bmatrix} a & 0 \\ 0 & b \end{bmatrix}\right| = ab,$$

Area $(T(S)) = ab\pi$. □

A good deal of information about T is conveyed by $\det[T]_{\mathcal{E}}$. Indeed, T is one-to-one if and only if $\det[T]_{\mathcal{E}} \neq 0$, and the size of $\det[T]_{\mathcal{E}}$ governs the change in n-dimensional volume under the action of T. The sign of $\det[T]_{\mathcal{E}}$ also has a useful interpretation. If T is one-to-one and $\mathcal{B} = \{X_1, \ldots, X_n\}$ is a basis for $\mathcal{R}^n$, then $T(\mathcal{B})$ is a basis for $\mathcal{R}^n$, and by (5.20),

$$\det[T(X_1) \quad \cdots \quad T(X_n)] = \det[T]_{\mathcal{E}} \det[X_1 \quad \cdots \quad X_n].$$

Thus, $T(\mathcal{B})$ and $\mathcal{B}$ have the same or opposite orientations, depending on whether $\det[T]_{\mathcal{E}}$ is positive or negative, respectively.

DEFINITION. A one-to-one linear transformation $T : \mathcal{R}^n \to \mathcal{R}^n$ is said to be *orientation preserving* or *orientation reversing* according as $\det[T]_{\mathcal{E}} > 0$ or $\det[T]_{\mathcal{E}} < 0$, respectively.

■ **EXAMPLE 5.26.** Suppose $T : \mathcal{R}^2 \to \mathcal{R}^2$ is one-to-one and linear. Assume $U \in \mathcal{R}^2$ is nonzero, and let ℓ be the line through 0 and U. Since T is one-to-one, $T(U) \neq 0$, so $T(\ell)$ is the line through 0 and T(U). Now, ℓ divides $\mathcal{R}^2$ into two half-planes, which are transformed by T into the half-planes determined by $T(\ell)$. Suppose T is orientation preserving. If $X \in H^+(U)$, then {U,X} is positively oriented, so $\{T(U), T(X)\}$ is positively oriented, that is, $T(X) \in H^+(T(U))$. Thus, $T(H^+(U)) = H^+(T(U))$. In other words, the half-plane to your left as you travel ℓ in the direction of U corresponds under T to the half-plane to your left when you travel $T(\ell)$ in the direction of T(U) (Figure 5.8). Similarly, when T is orientation reversing, $T(H^+(U)) = H^-(T(U))$. □

If $T : \mathcal{R}^n \to \mathcal{R}^n$ is linear and $\mathcal{B}$ is a basis for $\mathcal{R}^n$, then by Example 5.12, $\det[T]_{\mathcal{B}} = \det[T]_{\mathcal{E}}$. Thus, regardless of the choice of $\mathcal{B}$, the sign of $\det[T]_{\mathcal{B}}$ determines whether T is orientation preserving, and $|\det[T]_{\mathcal{B}}|$ is the "change of volume" factor for T.

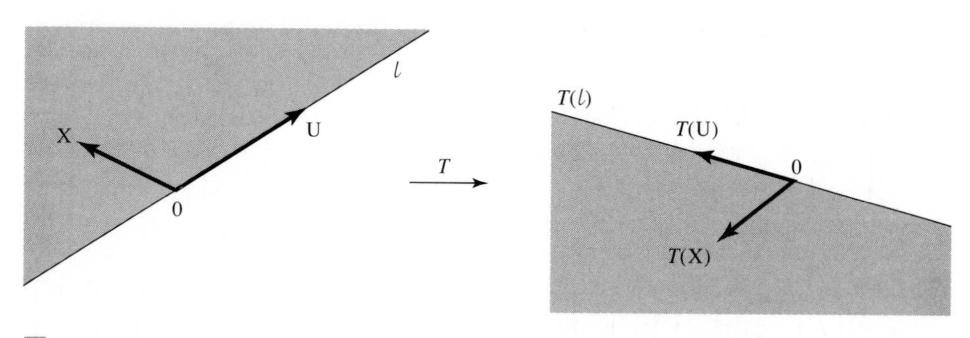

FIGURE 5.8

■ **EXAMPLE 5.27.** Suppose $\mathcal{V}$ is a two-dimensional subspace of $\mathcal{R}^3$ and let T be reflection in $\mathcal{V}$. If $\{U_1,U_2\}$ is an orthonormal basis for $\mathcal{V}$ and $U_3 = U_1 \times U_2$, then $\mathcal{B} = \{U_1,U_2,U_3\}$ is an orthonormal basis for $\mathcal{R}^3$ such that $T(U_1) = U_1$, $T(U_2) = U_2$, and $T(U_3) = -U_3$. Since

$$\det[T]_{\mathcal{B}} = \det\begin{bmatrix} 1 & 0 & 0 \\ 0 & 1 & 0 \\ 0 & 0 & -1 \end{bmatrix} = -1,$$

T is orientation reversing. But $|\det[T]_{\mathcal{B}}| = 1$, so $\mathrm{Vol}(T(\Pi)) = \mathrm{Vol}(\Pi)$ for any parallelepiped $\Pi \subseteq \mathcal{R}^3$. □

Exercises 5.5

1. Determine the orientation of each basis.

   a. $\left\{\begin{bmatrix} 1 \\ 1 \end{bmatrix}, \begin{bmatrix} 1 \\ -1 \end{bmatrix}\right\}$ b. $\left\{\begin{bmatrix} -2 \\ 3 \end{bmatrix}, \begin{bmatrix} -4 \\ 5 \end{bmatrix}\right\}$ c. $\left\{\begin{bmatrix} 1 \\ 1 \\ 0 \end{bmatrix}, \begin{bmatrix} -1 \\ 0 \\ 2 \end{bmatrix}, \begin{bmatrix} 3 \\ 1 \\ 0 \end{bmatrix}\right\}$

   d. $\left\{\begin{bmatrix} 1 \\ 2 \\ 0 \\ 0 \end{bmatrix}, \begin{bmatrix} 0 \\ 2 \\ 1 \\ 0 \end{bmatrix}, \begin{bmatrix} 0 \\ 0 \\ 1 \\ 2 \end{bmatrix}, \begin{bmatrix} 2 \\ 0 \\ 0 \\ 1 \end{bmatrix}\right\}$ e. $\left\{\begin{bmatrix} 0 \\ 0 \\ 0 \\ 1 \end{bmatrix}, \begin{bmatrix} 0 \\ 1 \\ 0 \\ 0 \end{bmatrix}, \begin{bmatrix} 0 \\ 0 \\ 1 \\ 0 \end{bmatrix}, \begin{bmatrix} 1 \\ 0 \\ 0 \\ 0 \end{bmatrix}\right\}$

   f. $\left\{\begin{bmatrix} 0 \\ 0 \\ 1 \end{bmatrix}, \begin{bmatrix} 0 \\ -2 \\ 2 \end{bmatrix}, \begin{bmatrix} 3 \\ 1 \\ 3 \end{bmatrix}\right\}$.

2. Assuming X_1 and X_2 are linearly independent vectors in $\mathcal{R}^2$, illustrate with appropriate sketches the fact that $X_2 \in H^+(X_1)$ if and only if $X_1 \in H^-(X_2)$. What does this say about the orientations of $\{X_1,X_2\}$ and $\{X_2,X_1\}$?

3. Let $X_1 = \begin{bmatrix} 1 \\ 1 \\ 0 \end{bmatrix}$, $X_2 = \begin{bmatrix} 1 \\ -1 \\ \sqrt{2} \end{bmatrix}$, and $X_3 = \begin{bmatrix} 0 \\ \sqrt{2} \\ 1 \end{bmatrix}$ and consider $L : \mathcal{R}^3 \to \mathcal{R}^3$ described in (5.17).

   a. Find $[L]_{\mathcal{E}}$.
   b. Calculate $L(X_3)$ and use it to decide whether $\mathcal{B} = \{X_1,X_2,X_3\}$ is positively or negatively oriented.

4. Sketch the parallelogram generated by the given vectors and find its area.

   a. $\begin{bmatrix} 3 \\ -1 \end{bmatrix}, \begin{bmatrix} -4 \\ 2 \end{bmatrix}$ b. $\begin{bmatrix} -1 \\ -2 \end{bmatrix}, \begin{bmatrix} 0 \\ 3 \end{bmatrix}$ c. $\begin{bmatrix} 2 \\ 1 \end{bmatrix}, \begin{bmatrix} -6 \\ -3 \end{bmatrix}$

5. Find the volume of the polyhedron generated by

   a. $\begin{bmatrix} 1 \\ 1 \\ 1 \end{bmatrix}, \begin{bmatrix} 1 \\ -1 \\ 1 \end{bmatrix}, \begin{bmatrix} -1 \\ 1 \\ 1 \end{bmatrix}$. b. $\begin{bmatrix} -1 \\ 2 \\ -3 \end{bmatrix}, \begin{bmatrix} 3 \\ -2 \\ 1 \end{bmatrix}, \begin{bmatrix} 2 \\ 3 \\ -1 \end{bmatrix}$. c. $\begin{bmatrix} -1 \\ 1 \\ 0 \end{bmatrix}, \begin{bmatrix} 1 \\ 0 \\ -1 \end{bmatrix}, \begin{bmatrix} 0 \\ -1 \\ 1 \end{bmatrix}$.

6. Find the four-dimensional volume of the parallelepiped generated by

a. $\begin{bmatrix}1\\0\\0\\1\end{bmatrix}, \begin{bmatrix}0\\2\\3\\0\end{bmatrix}, \begin{bmatrix}0\\-1\\4\\0\end{bmatrix}, \begin{bmatrix}-1\\0\\0\\2\end{bmatrix}.$ b. $\begin{bmatrix}1\\-1\\0\\0\end{bmatrix}, \begin{bmatrix}0\\-1\\1\\0\end{bmatrix}, \begin{bmatrix}0\\0\\1\\-1\end{bmatrix}, \begin{bmatrix}-1\\0\\0\\1\end{bmatrix}.$

7. Let $T(\mathrm{X}) = \begin{bmatrix}1 & 2\\3 & 0\end{bmatrix}\mathrm{X}$ and $\Pi = \Pi\left(\begin{bmatrix}-1\\1\end{bmatrix}, \begin{bmatrix}1\\1\end{bmatrix}\right)$. Sketch Π and $T(\Pi)$ and verify the conclusion of Theorem 5.10.

8. Let $\Pi = \Pi\left(\begin{bmatrix}1\\-1\\-1\end{bmatrix}, \begin{bmatrix}-1\\1\\-1\end{bmatrix}, \begin{bmatrix}-1\\-1\\1\end{bmatrix}\right)$. Find $\mathrm{Vol}(\Pi)$ and $\mathrm{Vol}(T(\Pi))$ when $T(\mathrm{X}) = \mathrm{AX}$ and A is

a. $\begin{bmatrix}1 & 0 & -2\\0 & -1 & 1\\-2 & 1 & 0\end{bmatrix}.$ b. $\begin{bmatrix}1 & 2 & -1\\1 & -1 & 2\\2 & 1 & 1\end{bmatrix}.$ **c.** $\begin{bmatrix}0 & 0 & 2\\0 & -1 & 3\\3 & 2 & 1\end{bmatrix}.$

9. Suppose $\{\mathrm{X}_1, \ldots, \mathrm{X}_n\}$ is a basis for $\mathcal{R}^n$, $\Pi = \Pi(\mathrm{X}_1, \ldots, \mathrm{X}_n)$, and $T : \mathcal{R}^n \to \mathcal{R}^n$ is linear but not one-to-one. Show that $\mathrm{Vol}(T(\Pi)) = 0$.

10. Assuming $\mathrm{X}_1, \ldots, \mathrm{X}_n \in \mathcal{R}^n$ and $k_1, \ldots, k_n$ is a permutation of $1, \ldots, n$, show that $\mathrm{Vol}(\Pi(\mathrm{X}_{k_1}, \ldots, \mathrm{X}_{k_n})) = \mathrm{Vol}(\Pi(\mathrm{X}_1, \ldots, \mathrm{X}_n))$.

11. Suppose $\mathrm{X}_1, \ldots, \mathrm{X}_n \in \mathcal{R}^n$, $\Pi = \Pi(\mathrm{X}_1, \ldots, \mathrm{X}_n)$, and $T : \mathcal{R}^n \to \mathcal{R}^n$ is an isometry. Show that $\mathrm{Vol}(T(\Pi)) = \mathrm{Vol}(\Pi)$.

12. Find $[T]_{\mathcal{E}}$ and decide whether T is orientation preserving or reversing.

a. $T : \mathcal{R}^2 \to \mathcal{R}^2$ is reflection in the line $x_2 = x_1$.

b. $T : \mathcal{R}^3 \to \mathcal{R}^3, T(\mathrm{X}) = \begin{bmatrix}x_1 - x_2\\x_2 - x_3\\x_3 + x_1\end{bmatrix}.$

c. $T : \mathcal{R}^4 \to \mathcal{R}^4, T(\mathrm{X}) = [x_1 + x_4 \quad x_3 \quad x_2 \quad x_1]^T.$

13. Determine whether $\mathrm{Refl}_{\mathcal{V}} : \mathcal{R}^3 \to \mathcal{R}^3$ is orientation preserving or reversing when $\mathcal{V}$ is

a. $\mathcal{S}pan\left\{\begin{bmatrix}1\\1\\0\end{bmatrix}\right\}.$ **b.** $\mathcal{S}pan\left\{\begin{bmatrix}1\\1\\0\end{bmatrix}, \begin{bmatrix}-1\\1\\0\end{bmatrix}\right\}.$ c. $\{0\}$.

14. Suppose $\mathcal{V}$ is a p-dimensional subspace of $\mathcal{R}^n$ and $T : \mathcal{R}^n \to \mathcal{R}^n$ is reflection in $\mathcal{V}$. Is T orientation preserving? Explain!

15. Assume $T : \mathcal{R}^n \to \mathcal{R}^n$ is linear and orientation preserving. Are the following functions orientation preserving?

a. T^{-1} **b.** $tT, t \neq 0$ c. T^2

16. Suppose $S,T : \mathcal{R}^n \to \mathcal{R}^n$ are orientation preserving linear functions.
    a. Show that $S \circ T$ is orientation preserving.
    b. What if S and T are orientation reversing?

17. Let $\mathcal{B} = \{X_1, X_2\}$ be a basis for $\mathcal{R}^2$. The Gram-Schmidt process, applied to $\mathcal{B}$, produces orthogonal vectors Y_1, Y_2, which, when normalized, constitute an orthonormal basis $\mathcal{A} = \{U_1, U_2\}$.
    a. Make two sketches indicating X_1, X_2 and Y_1, Y_2, one for the case when $\mathcal{B}$ is positively oriented, the other for $\mathcal{B}$ negatively oriented. Do $\mathcal{A}$ and $\mathcal{B}$ have the same orientation?
    b. Show that there is a $t \in \mathcal{R}$ such that $[Y_1 \quad Y_2] = [X_1 \quad X_2] \begin{bmatrix} 1 & t \\ 0 & 1 \end{bmatrix}$.
    c. Show that $\det[U_1 \quad U_2]$ and $\det[X_1 \quad X_2]$ have the same sign.
    d. Formulate and prove similar statements for bases for $\mathcal{R}^3$.

6

Diagonalization

6.1 INTRODUCTION

A linear function from $\mathcal{R}^n$ to $\mathcal{R}^n$ is often called a *linear operator*. It operates on n-dimensional vectors to produce n-dimensional image vectors, so its action can be thought of as a repositioning of the elements of $\mathcal{R}^n$, a repositioning that is consistent with the constraints of linearity.

Suppose T is a linear operator on $\mathcal{R}^2$, $\mathrm{X} \in \mathcal{R}^2$ and $\mathrm{X} \neq 0$. Then $\mathcal{V} = \mathcal{S}pan\{\mathrm{X}\}$ is the line through zero with direction X, and as long as $T(\mathrm{X}) \neq 0, T(\mathcal{V}) = \mathcal{S}pan\{T(\mathrm{X})\}$ is the line through 0 with direction $T(\mathrm{X})$. It is sometimes possible to select X so the direction of $T(\mathcal{V})$ is parallel to that of $\mathcal{V}$, that is, so

$$T(\mathrm{X}) = \lambda \mathrm{X},$$

for some $\lambda \in \mathcal{R}$. When this occurs, $\mathcal{V}$ remains intact under the action of T $(T(\mathcal{V}) = \mathcal{V})$, but its individual points get repositioned in accordance with linearity $(T(t\mathrm{X}) = tT(\mathrm{X}))$. If you have two such vectors, X_1 and X_2, which are linearly independent, then $\mathcal{B} = \{\mathrm{X}_1,\mathrm{X}_2\}$ generates a coordinate system for $\mathcal{R}^2$ whose axes remain "invariant" under T. In that event,

$$T(\mathrm{X}_1) = \lambda_1 \mathrm{X}_1 \qquad \text{and} \qquad T(\mathrm{X}_2) = \lambda_2 \mathrm{X}_2,$$

$\lambda_1,\lambda_2 \in \mathcal{R}$, and the matrix representing T relative to $\mathcal{B}$ is

$$[T]_{\mathcal{B}} = \big[[T(\mathrm{X}_1)]_{\mathcal{B}}\ [T(\mathrm{X}_2)]_{\mathcal{B}}\big] = \begin{bmatrix} \lambda_1 & 0 \\ 0 & \lambda_2 \end{bmatrix}.$$

An operator of this sort is geometrically easy to understand. Consider the $\mathcal{B}$ coordinate axes, $\mathcal{V}_k = \mathcal{S}pan\{\mathrm{X}_k\}, k = 1, 2$. If $\mathrm{X} \in \mathcal{V}_k$, then $\mathrm{X} = t\mathrm{X}_k$ for some $t \in \mathcal{R}$, and therefore

$$T(\mathrm{X}) = T(t\mathrm{X}_k) = tT(\mathrm{X}_k) = t(\lambda_k \mathrm{X}_k) = \lambda_k(t\mathrm{X}_k) = \lambda_k \mathrm{X}.$$

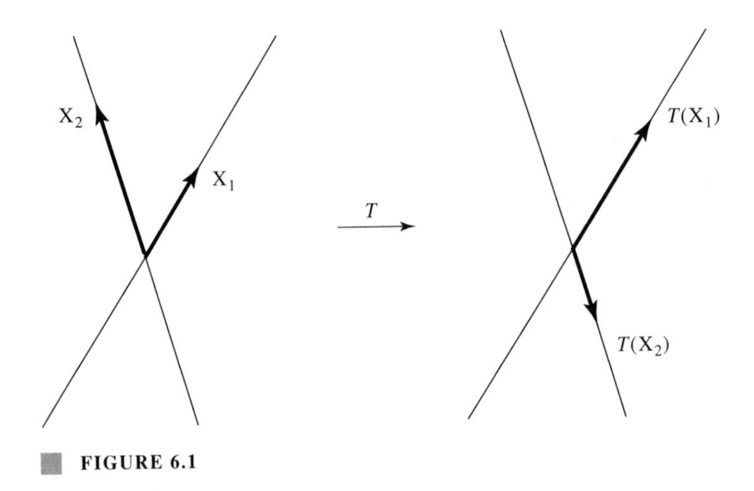

FIGURE 6.1

Thus, for each $X \in \mathcal{V}_k$,

$$\|T(X)\| = |\lambda_k|\|X\|,$$

and the direction of $T(X)$ is the same as or opposite that of X, depending on the sign of λ_k. In other words, along each $\mathcal{B}$ coordinate axis, T acts either as a pure dilation (when $\lambda_k > 0$) or as a dilation followed by a reflection in $\{0\}$ (when $\lambda_k < 0$). Figure 6.1 depicts the case when $\lambda_1 > 1$ and $-1 < \lambda_2 < 0$.

For a given linear operator T on $\mathcal{R}^n$, there is no guarantee that $\mathcal{R}^n$ has a basis $\mathcal{B}$ such that $[T]_{\mathcal{B}}$ is diagonal, but when one exists it certainly provides a natural coordinate system in which to describe the action of the function. The present chapter considers the issues that arise in trying to find such a basis and explores some of the advantages that acrue to "diagonalizing" T in this manner.

6.2 EIGENVALUES AND EIGENVECTORS

DEFINITION. Suppose T is a linear operator on $\mathcal{R}^n$. If there is a $\lambda \in \mathcal{R}$ and a nonzero $X \in \mathcal{R}^n$ such that

$$T(X) = \lambda X,$$

then λ is called an *eigenvalue* (or *characteristic value*) of T, and X is an *eigenvector* (or *characteristic vector*) associated with λ.

Note that the definition of eigenvector requires X to be nonzero. As long as T is linear, $T(X) = \lambda X$ is satisfied by $X = 0$ regardless of the value of λ, so it is only the nonzero solutions of such equations that provide information about the function.

EXAMPLE 6.1. Suppose $\mathcal{V}$ is a plane through 0 in $\mathcal{R}^3$ and T is reflection in $\mathcal{V}$. If $X \in \mathcal{V}$, then $T(X) = X = 1X$, and therefore each element of $\mathcal{V}$ except 0 is an eigenvector of T with eigenvalue 1. When $X \in \mathcal{V}^{\perp}$, $T(X) = -X = (-1)X$, so the nonzero elements of $\mathcal{V}^{\perp}$ are eigenvectors of T with eigenvalue -1. T has no other eigenvalues. Indeed, if $X \notin \mathcal{V}$ and $X \notin \mathcal{V}^{\perp}$, then 0, X, and $T(X)$ are noncollinear, so there is no scalar λ such that $T(X) = \lambda X$. □

■ **EXAMPLE 6.2.** Given $\alpha \in [0,2\pi)$, let $T_\alpha : \mathcal{R}^2 \to \mathcal{R}^2$ be counterclockwise rotation about 0 through α radians. Observe that $T_\pi(X) = -X = (-1)X$, so each nonzero $X \in \mathcal{R}^2$ is an eigenvector of T_π with eigenvalue $\lambda = -1$. When $\alpha \in (0,\pi) \cup (\pi,2\pi)$, 0, X, and $T_\alpha(X)$ are noncollinear (provided $X \neq 0$), and therefore T_α has no eigenvalues. Finally, since $T_0(X) = X = 1X$, $\lambda = 1$ is an eigenvalue of T_0, and each $X \in \mathcal{R}^2$ except 0 is an associated eigenvector. □

Assuming T is a linear operator on $\mathcal{R}^n$ and $\lambda \in \mathcal{R}$, let

$$\mathcal{V}(\lambda) = \{X \in \mathcal{R}^n : T(X) = \lambda X\}.$$

Observe that $0 \in \mathcal{V}(\lambda)$. Moreover, λ is an eigenvalue of T when $\mathcal{V}(\lambda) \neq \{0\}$, in which case $\mathcal{V}(\lambda)$ is called the *eigenspace* associated with λ.

THEOREM 6.1. If $T : \mathcal{R}^n \to \mathcal{R}^n$ is linear and $\lambda \in \mathcal{R}$, then $\mathcal{V}(\lambda) = \mathcal{K}er(\lambda \mathrm{Id} - T)$. In particular, $\mathcal{V}(\lambda)$ is a subspace of $\mathcal{R}^n$.

Proof. $T(X) = \lambda X$ if and only if $0 = \lambda X - T(X) = \lambda \mathrm{Id}(X) - T(X) = (\lambda \mathrm{Id} - T)(X)$. Thus, $\mathcal{V}(\lambda) = \mathcal{K}er(\lambda \mathrm{Id} - T)$. By Theorem 4.7, $\mathcal{V}(\lambda)$ is a subspace of $\mathcal{R}^n$. ■

■ **EXAMPLE 6.3.** The eigenvalues of T in Example 6.1 are 1 and -1, with $\mathcal{V}(1) = \mathcal{V}$ and $\mathcal{V}(-1) = \mathcal{V}^{\perp}$. □

■ **EXAMPLE 6.4.** Recall T_α from Example 6.2. When $\alpha = 0$, the only eigenvalue is 1, and $\mathcal{V}(1) = \mathcal{R}^2$. When $\alpha = \pi$, -1 is the only eigenvalue, and $\mathcal{V}(-1) = \mathcal{R}^2$. In the remaining cases, there are no eigenvalues and hence no eigenspaces. □

■ **EXAMPLE 6.5.** Suppose $\mathcal{V}$ is a line through 0 in $\mathcal{R}^2$ and T is projection onto $\mathcal{V}$. Then $T(X) = X = 1X$ if and only if $X \in \mathcal{V}$, and $T(X) = 0 = 0X$ if and only if $X \in \mathcal{V}^{\perp}$. Thus, 1 and 0 are eigenvalues of T, with $\mathcal{V}(1) = \mathcal{V}$ and $\mathcal{V}(0) = \mathcal{V}^{\perp}$. If $X \notin \mathcal{V}$ and $X \notin \mathcal{V}^{\perp}$, then $T(X)$ is not a scalar multiple of X, so T has no other eigenvalues. □

Let T be a linear operator on $\mathcal{R}^n$ with eigenvalue λ and suppose $\mathcal{B} = \{X_1, \ldots, X_p\}$ is a basis for $\mathcal{V}(\lambda)$. Then

$$T(\mathcal{V}(\lambda)) = \mathcal{S}pan\{T(X_1), \ldots, T(X_p)\} = \mathcal{S}pan\{\lambda X_1, \ldots, \lambda X_p\}.$$

Since

$$\sum_1^p t_k(\lambda X_k) = \sum_1^p (\lambda t_k)X_k,$$

each linear combination of $\lambda X_1, \ldots, \lambda X_p$ is a linear combination of $X_1, \ldots, X_p$. Thus,

$$T(\mathcal{V}(\lambda)) = \mathcal{S}pan\{\lambda X_1, \ldots, \lambda X_p\} \subseteq \mathcal{S}pan\{X_1, \ldots, X_p\} = \mathcal{V}(\lambda).$$

When $\lambda \neq 0$ the opposite inclusion also holds, because each linear combination of $X_1, \ldots, X_p$ can then be expressed as a linear combination of $\lambda X_1, \ldots, \lambda X_p$ by writing

$$\sum_1^p t_k X_k = \sum_1^p (t_k/\lambda)(\lambda X_k).$$

There are, therefore, exactly two possibilities for the image of $\mathcal{V}(\lambda)$, namely

$$T(\mathcal{V}(\lambda)) = \begin{cases} \mathcal{V}(\lambda), & \text{if } \lambda \neq 0 \\ \{0\}, & \text{if } \lambda = 0 \end{cases}.$$

In other words, $\mathcal{V}(\lambda)$ remains intact under the action of T when $\lambda \neq 0$, but it gets collapsed onto $\{0\}$ when $\lambda = 0$.

DEFINITION. If T is a linear operator on $\mathcal{R}^n$ and $\mathcal{V}$ is a subspace of $\mathcal{R}^n$ such that $T(\mathcal{V}) \subseteq \mathcal{V}$, then $\mathcal{V}$ is said to be *invariant* under T or *T-invariant*.

The trivial subspaces of $\mathcal{R}^n$ are invariant under any linear operator, because $T(\{0\}) = \{T(0)\} = \{0\}$ and $T(\mathcal{R}^n) \subseteq \mathcal{R}^n$. The discussion above shows that each eigenspace of T is T-invariant.

■ **EXAMPLE 6.6.** Let T be a linear operator on $\mathcal{R}^n$ and observe that

$$\mathcal{V}(0) = \{\mathrm{X} \in \mathcal{R}^n : T(\mathrm{X}) = 0\mathrm{X} = 0\} = \mathcal{K}er(T).$$

Thus, T has eigenvalue 0 if and only if $\mathcal{K}er(T) \neq \{0\}$. The latter condition is equivalent to T not being one-to-one (Theorem 4.6). □

Each linear $T : \mathcal{R}^n \to \mathcal{R}^n$ has an associated $\mathrm{A} \in \mathcal{M}_{n\times n}$ such that $T(\mathrm{X}) = \mathrm{AX}$, so the statement $T(\mathrm{X}) = \lambda\mathrm{X}$ can be reformulated as the matrix equation $\mathrm{AX} = \lambda\mathrm{X}$. Therein lies the motivation for the notions of eigenvalue and eigenvector for a square matrix.

DEFINITION. Suppose $\mathrm{A} \in \mathcal{M}_{n\times n}$ and $\lambda \in \mathcal{R}$. If there is a nonzero $\mathrm{X} \in \mathcal{R}^n$ such that $\mathrm{AX} = \lambda\mathrm{X}$, then X is said to be an *eigenvector* of A with *eigenvalue* λ, and $\{\mathrm{X} \in \mathcal{R}^n : \mathrm{AX} = \lambda\mathrm{X}\}$ is the *eigenspace* of A associated with λ.

Given $\mathrm{A} \in \mathcal{M}_{n\times n}$, $\mathrm{X} \in \mathcal{R}^n$, and $\lambda \in \mathcal{R}$, $\mathrm{AX} = \lambda\mathrm{X}$ if and only if

$$0 = \lambda\mathrm{X} - \mathrm{AX} = (\lambda\mathrm{I}_n - \mathrm{A})\mathrm{X}.$$

Thus, the eigenspace of A associated with λ is the null space of $\lambda\mathrm{I}_n - \mathrm{A}$. If T is the linear operator on $\mathcal{R}^n$ represented by A (relative to $\mathcal{E}$), then $(\lambda\mathrm{Id} - T)(\mathrm{X}) = (\lambda\mathrm{I}_n - \mathrm{A})\mathrm{X}$, so

$$\mathcal{V}(\lambda) = \mathcal{K}er(\lambda\mathrm{Id} - T) = \mathcal{N}(\lambda\mathrm{I}_n - \mathrm{A}). \tag{6.1}$$

Exercises 6.2

1. Assuming $\mathcal{V} = \mathcal{S}pan\left\{\begin{bmatrix} 1 \\ 2 \\ 1 \end{bmatrix}\right\}$, find the eigenvalues and associated eigenspaces for $T : \mathcal{R}^3 \to \mathcal{R}^3$ when

   a. $T(\mathrm{X}) = \mathrm{Refl}_{\mathcal{V}}(\mathrm{X})$.
   b. $T(\mathrm{X}) = \mathrm{Proj}_{\mathcal{V}}(\mathrm{X})$.

2. Suppose S and T are linear operators on $\mathcal{R}^n$, $t \in \mathcal{R}$, and X is an eigenvector of both S and T with eigenvalues λ and μ, respectively. Show that X is an eigenvector of the indicated operator and find the associated eigenvalue.
   a. $S + T$ **b.** tT c. $T \circ S$.

3. Let C be a nonzero vector in $\mathcal{R}^3$ and define $T : \mathcal{R}^3 \to \mathcal{R}^3$ by $T(\mathrm{X}) = \mathrm{X} \times \mathrm{C}$. Find the eigenvalues and associated eigenspaces of T.

4. Suppose T is a one-to-one linear operator on $\mathcal{R}^n$ and X is an eigenvector of T with eigenvalue λ. Show that X is an eigenvector of T^{-1} and find the associated eigenvalue.

5. a. What are the possible eigenvalues for an isometry $T : \mathcal{R}^n \to \mathcal{R}^n$?
   b. Do all isometries have eigenvalues?

6. **a.** Show that E_1, E_2, and E_3 are eigenvectors of $\begin{bmatrix} 1 & 0 & 0 \\ 0 & 2 & 0 \\ 0 & 0 & 3 \end{bmatrix}$. What are the associated eigenvalues?
   b. Show that 1, 2, and 3 are eigenvalues of $\begin{bmatrix} 1 & 1 & 1 \\ 0 & 2 & 2 \\ 0 & 0 & 3 \end{bmatrix}$.

7. Suppose $\mathrm{A} \in \mathcal{M}_{n\times n}$ and X is an eigenvector of A with eigenvalue λ. Show that X is an eigenvector of A^2. What's the eigenvalue?

8. What are the possible eigenvalues of A when $\mathrm{A} \in \mathcal{M}_{n\times n}$ and $\mathrm{A}^2 = \mathrm{I}$? How about when $\mathrm{A}^k = \mathrm{I}$, k a positive integer?

9. Suppose $\mathrm{A} \in \mathcal{M}_{n\times n}$ and $\mathrm{A}^k = 0$ for some positive integer k.
   a. Show that A is singular. b. Show that the only eigenvalue of A is $\lambda = 0$.

10. Suppose T is a linear operator on $\mathcal{R}^n$ with distinct eigenvalues λ and μ. Show that $\mathcal{V}(\lambda) \cap \mathcal{V}(\mu) = \{0\}$.

11. Show that if T is a linear operator on $\mathcal{R}^n$, then $\mathcal{K}er(T)$ and $\mathcal{R}ng(T)$ are invariant under T.

12. Suppose $T : \mathcal{R}^n \to \mathcal{R}^n$ is linear and $\mathcal{V}$ and $\mathcal{W}$ are T-invariant subspaces of $\mathcal{R}^n$.
   a. Show that $\mathcal{V} + \mathcal{W}$ is invariant under T. b. Show that $\mathcal{V} \cap \mathcal{W}$ is invariant under T.

13. If $\mathcal{V}$ is a nontrivial subspace of $\mathcal{R}^n$, $T(\mathrm{X}) = \mathrm{Refl}_{\mathcal{V}}(\mathrm{X})$ and $S(\mathrm{X}) = \mathrm{Proj}_{\mathcal{V}}(\mathrm{X})$, show that
   a. X is an eigenvector of S with eigenvalue λ if and only if X is an eigenvector of T with eigenvalue $2\lambda - 1$.
   b. The eigenvalues of S are 1 and 0, with $\mathcal{V}(1) = \mathcal{V}$ and $\mathcal{V}(0) = \mathcal{V}^{\perp}$.
   c. The eigenvalues of T are 1 and -1, with $\mathcal{V}(1) = \mathcal{V}$ and $\mathcal{V}(-1) = \mathcal{V}^{\perp}$.

14. Suppose $\mathrm{A} \in \mathcal{M}_{m\times m}$, $\mathrm{B} \in \mathcal{M}_{n\times n}$, $\mathrm{U} \in \mathcal{R}^m$, $\mathrm{V} \in \mathcal{R}^n$, and $\lambda \in \mathcal{R}$. Let

$$\mathrm{C} = \left[\begin{array}{c|c} \mathrm{A} & 0 \\ \hline 0 & \mathrm{B} \end{array}\right] \quad \text{and} \quad \mathrm{X} = \left[\begin{array}{c} \mathrm{U} \\ \hline \mathrm{V} \end{array}\right].$$

   a. Show that $\mathrm{CX} = \lambda\mathrm{X}$ if and only if $\mathrm{AU} = \lambda\mathrm{U}$ and $\mathrm{BV} = \lambda\mathrm{V}$.
   b. Show that λ is an eigenvalue of C if and only if λ is an eigenvalue of either A or B.

6.3
THE CHARACTERISTIC POLYNOMIAL

The examples in the last section featured simple geometric operators whose eigenvalues and eigenvectors could be found by inspection. We now turn to an algebraic procedure for identifying these items. Since T and $[T]_{\mathcal{E}}$ have the same eigenvalues and eigenspaces, it suffices to consider the problem for $n \times n$ matrices.

Let $\mathrm{A} \in \mathcal{M}_{n\times n}$. The real number λ is an eigenvalue of A when $\mathrm{AX} = \lambda\mathrm{X}$ for some nonzero $\mathrm{X} \in \mathcal{R}^n$, that is, when the homogeneous system $(\lambda\mathrm{I} - \mathrm{A})\mathrm{X} = 0$ has a nontrivial solution. That, according to Theorem 1.14, occurs when $\lambda\mathrm{I} - \mathrm{A}$ is singular. Thus, the eigenvalues of A are the λ's for which $\det(\lambda\mathrm{I} - \mathrm{A}) = 0$, with corresponding eigenspaces $\mathcal{V}(\lambda) = \mathcal{N}(\lambda\mathrm{I} - \mathrm{A})$.

THEOREM 6.2. The eigenvalues of $\mathrm{A} \in \mathcal{M}_{n\times n}$ are the roots of $\det(\lambda\mathrm{I} - \mathrm{A}) = 0$, and for each such λ, $\mathcal{V}(\lambda) = \mathcal{N}(\lambda\mathrm{I} - \mathrm{A})$.

■ **EXAMPLE 6.7.** If $\mathrm{A} = \begin{bmatrix} 4 & 1 \\ -2 & 1 \end{bmatrix}$, then

$$0 = \det(\lambda\mathrm{I} - \mathrm{A}) = \det\begin{bmatrix} \lambda - 4 & -1 \\ 2 & \lambda - 1 \end{bmatrix} = \lambda^2 - 5\lambda + 6 = (\lambda - 3)(\lambda - 2)$$

if and only if $\lambda = 3$ or $\lambda = 2$. Thus, A has eigenvalues 3 and 2. Moreover,

$$3\mathrm{I} - \mathrm{A} = \begin{bmatrix} -1 & -1 \\ 2 & 2 \end{bmatrix} \longrightarrow \begin{bmatrix} 1 & 1 \\ 0 & 0 \end{bmatrix} \quad \text{and} \quad 2\mathrm{I} - \mathrm{A} = \begin{bmatrix} -2 & -1 \\ 2 & 1 \end{bmatrix} \longrightarrow \begin{bmatrix} 1 & 1/2 \\ 0 & 0 \end{bmatrix},$$

so $\mathcal{V}(3) = \mathcal{N}(3\mathrm{I} - \mathrm{A}) = \mathcal{S}pan\left\{\begin{bmatrix} -1 \\ 1 \end{bmatrix}\right\}$, and $\mathcal{V}(2) = \mathcal{N}(2\mathrm{I} - \mathrm{A}) = \mathcal{S}pan\left\{\begin{bmatrix} -1 \\ 2 \end{bmatrix}\right\}$. □

■ **EXAMPLE 6.8.** For $\mathrm{A} = \begin{bmatrix} 0 & 1 & 1 \\ 1 & 0 & 1 \\ 1 & 1 & 0 \end{bmatrix}$, Laplace expansion of $\det(\lambda\mathrm{I} - \mathrm{A})$ by the first row produces

$$\begin{aligned} \det\begin{bmatrix} \lambda & -1 & -1 \\ -1 & \lambda & -1 \\ -1 & -1 & \lambda \end{bmatrix} &= \lambda(\lambda^2 - 1) - (-1)(-\lambda - 1) + (-1)(1 + \lambda) \\ &= (\lambda + 1)[\lambda(\lambda - 1) - 1 - 1] \\ &= (\lambda + 1)(\lambda^2 - \lambda - 2) \\ &= (\lambda + 1)^2(\lambda - 2), \end{aligned}$$

so -1 and 2 are the eigenvalues of A. Since

$$(-1)\mathrm{I} - \mathrm{A} = \begin{bmatrix} -1 & -1 & -1 \\ -1 & -1 & -1 \\ -1 & -1 & -1 \end{bmatrix} \longrightarrow \begin{bmatrix} 1 & 1 & 1 \\ 0 & 0 & 0 \\ 0 & 0 & 0 \end{bmatrix}$$

and

$$2\mathrm{I} - \mathrm{A} = \begin{bmatrix} 2 & -1 & -1 \\ -1 & 2 & -1 \\ -1 & -1 & 2 \end{bmatrix} \longrightarrow \begin{bmatrix} 1 & 0 & -1 \\ 0 & 1 & -1 \\ 0 & 0 & 0 \end{bmatrix},$$

$$\mathcal{V}(-1) = \mathcal{N}((-1)\mathrm{I} - \mathrm{A}) = \mathcal{S}pan\left\{\begin{bmatrix} -1 \\ 1 \\ 0 \end{bmatrix}, \begin{bmatrix} -1 \\ 0 \\ 1 \end{bmatrix}\right\},$$

and

$$\mathcal{V}(2) = \mathcal{N}(2\mathrm{I} - \mathrm{A}) = \mathcal{S}pan\left\{\begin{bmatrix} 1 \\ 1 \\ 1 \end{bmatrix}\right\}. \square$$

Note that in the previous two examples $c(\lambda) = \det(\lambda \mathrm{I} - \mathrm{A})$ is a polynomial in λ whose degree is the size of A.

THEOREM 6.3. If $\mathrm{A} \in \mathcal{M}_{n \times n}$, then $c(\lambda) = \det(\lambda \mathrm{I} - \mathrm{A})$ is a polynomial of degree n.

Proof. The determinant of an $n \times n$ matrix is a sum of $n!$ terms, each, apart from its sign, being a product of n entries in the matrix. Moreover, no two factors in a given term come from the same row or column. Consider

$$\lambda \mathrm{I} - \mathrm{A} = \begin{bmatrix} \lambda - a_{11} & -a_{12} & -a_{13} & \cdots & -a_{1n} \\ -a_{21} & \lambda - a_{22} & -a_{23} & \cdots & -a_{2n} \\ -a_{31} & -a_{32} & \lambda - a_{33} & \cdots & -a_{3n} \\ \vdots & \vdots & \vdots & & \vdots \\ -a_{n1} & -a_{n2} & -a_{n3} & \cdots & \lambda - a_{nn} \end{bmatrix}.$$

The only term in $\det(\lambda \mathrm{I} - \mathrm{A})$ involving all the diagonal entries is

$$(\lambda - a_{11})(\lambda - a_{22}) \cdots (\lambda - a_{nn}), \tag{6.2}$$

which expands to

$$\lambda^n - (a_{11} + a_{22} + \cdots + a_{nn})\lambda^{n-1} + (\text{polynomial of degree} \leq n - 2).$$

Each remaining term has at least one nondiagonal entry as a factor. If $-a_{ij}$, $i \neq j$, is a factor, then $\lambda - a_{ii}$ and $\lambda - a_{jj}$ are not, because a term never has two factors from the same row or column. Thus, each term in $\det(\lambda \mathrm{I} - \mathrm{A})$ except (6.2) has at most $n - 2$ factors containing λ and therefore expands to a polynomial of degree at most $n - 2$. Summing the terms produces a polynomial of degree n. ■

The coefficients of λ^n and λ^{n-1} in $c(\lambda)$ are 1 and $-(a_{11} + \cdots + a_{nn})$, respectively. Both come from (6.2). The sum of the diagonal entries of an $n \times n$ matrix is called the *trace* of the matrix, so

$$c(\lambda) = \lambda^n - [\text{trace}(\mathrm{A})]\lambda^{n-1} + \cdots + a_0. \tag{6.3}$$

Moreover,

$$a_0 = c(0) = \det(0\mathrm{I} - \mathrm{A}) = \det(-\mathrm{A}) = (-1)^n \det(\mathrm{A}).$$

DEFINITION. For $\mathrm{A} \in \mathcal{M}_{n \times n}$, $c_{\mathrm{A}}(\lambda) = \det(\lambda \mathrm{I} - \mathrm{A})$ is called the *characteristic polynomial* of A. If there is no danger of confusion, $c_{\mathrm{A}}(\lambda)$ will be shortened to $c(\lambda)$.

Theorems 6.2 and 6.3 convert the task of finding eigenvalues for an $n \times n$ matrix into the problem of determining roots of a polynomial of degree n. The fundamental theorem of algebra states that such a polynomial has n roots, counting multiplicity, as long as both real and complex numbers are allowed. Complex numbers can be viewed as eigenvalues of a matrix, but in doing so one is led to consider eigenvectors with complex coordinates, and the discussion then takes place in the set $\mathcal{C}^n$ of ordered n-tuples of complex numbers. In keeping with our focus on $\mathcal{R}^n$, we are concerned with only the real roots of the characteristic polynomial, so for our purposes, an $n \times n$ matrix has at most n eigenvalues, and it may possibly have none.

■ **EXAMPLE 6.9.** For $\mathrm{A} = \begin{bmatrix} 1 & 0 & 1 \\ 0 & 3 & 0 \\ -1 & 0 & 1 \end{bmatrix}$, $\lambda\mathrm{I} - \mathrm{A} = \begin{bmatrix} \lambda - 1 & 0 & -1 \\ 0 & \lambda - 3 & 0 \\ 1 & 0 & \lambda - 1 \end{bmatrix}$, and Laplace expansion of $\det(\lambda\mathrm{I} - \mathrm{A})$ by the second row produces

$$c(\lambda) = (\lambda - 3)\det\begin{bmatrix} \lambda - 1 & -1 \\ 1 & \lambda - 1 \end{bmatrix} = (\lambda - 3)(\lambda^2 - 2\lambda + 2).$$

The quadratic factor has no real roots, so 3 is the only eigenvalue. □

■ **EXAMPLE 6.10.** If $\mathrm{A} = \begin{bmatrix} 0 & 0 & 1 & 0 \\ 0 & 3 & 0 & 0 \\ 0 & 0 & 0 & -2 \\ 4 & 0 & 0 & 0 \end{bmatrix}$, then

$$\det(\lambda\mathrm{I} - \mathrm{A}) = \det\begin{bmatrix} \lambda & 0 & -1 & 0 \\ 0 & \lambda - 3 & 0 & 0 \\ 0 & 0 & \lambda & 2 \\ -4 & 0 & 0 & \lambda \end{bmatrix} = (\lambda - 3)\det\begin{bmatrix} \lambda & -1 & 0 \\ 0 & \lambda & 2 \\ -4 & 0 & \lambda \end{bmatrix},$$

and expanding the last determinant by the first row yields

$$c(\lambda) = (\lambda - 3)(\lambda^3 + 8) = (\lambda - 3)(\lambda + 2)(\lambda^2 - 2\lambda + 4).$$

Since the quadratic factor is irreducible, the eigenvalues are 3 and -2. □

The foregoing development shows that you can study the eigenvalue-eigenvector problem for a linear operator T on $\mathcal{R}^n$ by analyzing $\mathrm{A} = [T]_{\mathcal{E}}$ via its characteristic polynomial. Consider now what information might be obtained from other matrix representatives of T. Given a basis $\mathcal{B}$ for $\mathcal{R}^n$, $T(\mathrm{X}) = \lambda\mathrm{X}$ if and only if $[T(\mathrm{X})]_{\mathcal{B}} = [\lambda\mathrm{X}]_{\mathcal{B}}$, or equivalently,

$$[T]_{\mathcal{B}}[\mathrm{X}]_{\mathcal{B}} = \lambda[\mathrm{X}]_{\mathcal{B}}.$$

Thus, the eigenvalues and eigenvectors of T are related to those of $[T]_{\mathcal{B}}$ as follows.

THEOREM 6.4. If T is a linear operator on $\mathcal{R}^n$ and $\mathcal{B}$ is a basis for $\mathcal{R}^n$, then λ is an eigenvalue of T with eigenvector X if and only if λ is an eigenvalue of $[T]_{\mathcal{B}}$ with eigenvector $[\mathrm{X}]_{\mathcal{B}}$.

■ **EXAMPLE 6.11.** Let $\mathcal{B} = \{\mathrm{X}_1, \mathrm{X}_2, \mathrm{X}_3\} = \left\{ \begin{bmatrix} 1 \\ 1 \\ 1 \end{bmatrix}, \begin{bmatrix} 1 \\ 1 \\ 0 \end{bmatrix}, \begin{bmatrix} 0 \\ 1 \\ 1 \end{bmatrix} \right\}$ and suppose T is the linear operator on $\mathcal{R}^3$ such that

$$[T]_{\mathcal{B}} = \begin{bmatrix} 2 & 1 & -3 \\ 0 & -1 & 0 \\ 0 & 0 & -1 \end{bmatrix}.$$

Since

$$\det(\lambda \mathrm{I} - [\mathrm{T}]_{\mathcal{B}}) = \det \begin{bmatrix} \lambda - 2 & -1 & 3 \\ 0 & \lambda + 1 & 0 \\ 0 & 0 & \lambda + 1 \end{bmatrix} = (\lambda - 2)(\lambda + 1)^2,$$

$[T]_{\mathcal{B}}$ has eigenvalues 2 and -1. According to Theorem 6.4, 2 and -1 are also the eigenvalues of T. Now,

$$2\mathrm{I} - [T]_{\mathcal{B}} = \begin{bmatrix} 0 & -1 & 3 \\ 0 & 3 & 0 \\ 0 & 0 & 3 \end{bmatrix} \longrightarrow \begin{bmatrix} 0 & 1 & 0 \\ 0 & 0 & 1 \\ 0 & 0 & 0 \end{bmatrix},$$

so

$$\mathcal{N}(2\mathrm{I} - [T]_{\mathcal{B}}) = \mathcal{S}pan \left\{ \begin{bmatrix} 1 \\ 0 \\ 0 \end{bmatrix} \right\}.$$

Although $\begin{bmatrix} 1 \\ 0 \\ 0 \end{bmatrix}$ is an eigenvector of $[T]_{\mathcal{B}}$ with eigenvalue 2, the corresponding eigenvector for T is the $\mathrm{X} \in \mathcal{R}^3$ satisfying $[\mathrm{X}]_{\mathcal{B}} = \begin{bmatrix} 1 \\ 0 \\ 0 \end{bmatrix}$, that is, $\mathrm{X} = 1\mathrm{X}_1 + 0\mathrm{X}_2 + 0\mathrm{X}_3 = \begin{bmatrix} 1 \\ 1 \\ 1 \end{bmatrix}$.

Similarly,

$$(-1)\mathrm{I} - [T]_{\mathcal{B}} = \begin{bmatrix} -3 & -1 & 3 \\ 0 & 0 & 0 \\ 0 & 0 & 0 \end{bmatrix} \longrightarrow \begin{bmatrix} 1 & 1/3 & -1 \\ 0 & 0 & 0 \\ 0 & 0 & 0 \end{bmatrix},$$

so

$$\mathcal{N}((-1)\mathrm{I} - [T]_{\mathcal{B}}) = \mathcal{S}pan \left\{ \begin{bmatrix} -1 \\ 3 \\ 0 \end{bmatrix}, \begin{bmatrix} 1 \\ 0 \\ 1 \end{bmatrix} \right\},$$

and

$$(-1)\mathrm{X}_1 + 3\mathrm{X}_2 + 0\mathrm{X}_3 = \begin{bmatrix} 2 \\ 2 \\ -1 \end{bmatrix} \quad \text{and} \quad 1\mathrm{X}_1 + 0\mathrm{X}_2 + 1\mathrm{X}_3 = \begin{bmatrix} 1 \\ 2 \\ 2 \end{bmatrix}$$

are eigenvectors of T with eigenvalue -1. □

If $\mathcal{A}$ and $\mathcal{B}$ are two bases for $\mathcal{R}^n$, then by Theorem 6.4, $[T]_{\mathcal{A}}$ and $[T]_{\mathcal{B}}$ have the same eigenvalues (those of T). In fact,

$$[T]_{\mathcal{B}} = \mathrm{P}_{\mathcal{AB}}^{-1}[T]_{\mathcal{A}}\mathrm{P}_{\mathcal{AB}},$$

(Theorem 4.16) so

$$\lambda \mathrm{I} - [T]_{\mathcal{B}} = \lambda \mathrm{P}_{\mathcal{AB}}^{-1}\mathrm{I}\mathrm{P}_{\mathcal{AB}} - \mathrm{P}_{\mathcal{AB}}^{-1}[T]_{\mathcal{A}}\mathrm{P}_{\mathcal{AB}} = \mathrm{P}_{\mathcal{AB}}^{-1}(\lambda \mathrm{I} - [\mathrm{T}]_{\mathcal{A}})\mathrm{P}_{\mathcal{AB}},$$

and therefore (Example 5.12)

$$\det(\lambda I - [T]_{\mathcal{B}}) = \det(\lambda I - [T]_{\mathcal{A}}).$$

Thus, $[T]_{\mathcal{A}}$ and $[T]_{\mathcal{B}}$ have the same characteristic polynomial. This allows us to formulate a notion of characteristic polynomial for a linear operator.

DEFINITION. If T is a linear operator on $\mathcal{R}^n$ and $\mathcal{B}$ is any basis for $\mathcal{R}^n$, then $c(\lambda) = \det(\lambda I - [T]_{\mathcal{B}})$ is called the *characteristic polynomial* of T.

Although $[T]_{\mathcal{A}}$ and $[T]_{\mathcal{B}}$ have the same eigenvalues, they do not have the same eigenspaces. According to Theorem 6.4, the eigenvectors of $[T]_{\mathcal{A}}$ and $[T]_{\mathcal{B}}$ are the $\mathcal{A}$ and $\mathcal{B}$ coordinates, respectively, of the eigenvectors of T, that is,

$$\mathcal{N}(\lambda I - [T]_{\mathcal{A}}) = \{[X]_{\mathcal{A}} : X \in \mathcal{V}(\lambda)\},$$

whereas

$$\mathcal{N}(\lambda I - [T]_{\mathcal{B}}) = \{[X]_{\mathcal{B}} : X \in \mathcal{V}(\lambda)\}.$$

Recall that the linear isometries on $\mathcal{R}^n$ are the functions $T(X) = AX$ where A is an orthogonal $n \times n$ matrix. The analysis of orthogonal 2×2 matrices in Example 4.39 showed that a linear isometry on $\mathcal{R}^2$ is either rotation about 0 or reflection in a line through 0. The next example treats the case when $n = 3$. It shows that a linear isometry on $\mathcal{R}^3$ is either rotation (about an axis through 0) or rotation followed by reflection in the plane through 0 orthogonal to the axis of rotation. The argument is a particularly nice application of the ideas we have developed.

■ **EXAMPLE 6.12.** Suppose A is an orthogonal 3×3 matrix and $T(X) = AX$. The only possible eigenvalues for an isometry are ± 1, because $T(X) = \lambda X$ implies $\|X\| = \|T(X)\| = |\lambda|\|X\|$. Now $c(\lambda)$, being a polynomial of degree 3, has at least one real root, so T has either 1 or -1 as an eigenvalue. Note also that $\det(A) = \pm 1$ (Example 5.13). We first consider the case when $\det(A) = 1$ and show that under those circumstances 1 is an eigenvalue of T. Indeed, since $I - A = A(A^T - I) = A(A - I)^T$,

$$\det(I - A) = \det(A)\det(A - I)^T = \det(A - I) = (-1)^3\det(I - A),$$

and therefore $c(1) = \det(I - A) = 0$. Now, choose a unit vector $W \in \mathcal{V}(1)$ and set $\mathcal{W} = \mathcal{S}pan\{W\}$. Then $\mathcal{W}$ is an axis through 0 whose points are fixed under T ($T(X) = 1X = X$ for each $X \in \mathcal{W}$). For each $X \in \mathcal{W}^\perp$,

$$T(X) \cdot W = T(X) \cdot T(W) = X \cdot W = 0$$

(Theorem 4.21), so $T(X) \in \mathcal{W}^\perp$. Thus, $T(\mathcal{W}^\perp) \subseteq \mathcal{W}^\perp$. But $\mathcal{W}^\perp$ and $T(\mathcal{W}^\perp)$ both have dimension 2 (T is one-to-one), so $T(\mathcal{W}^\perp) = \mathcal{W}^\perp$. To understand the action of T on $\mathcal{W}^\perp$, consider an orthonormal basis {U,V} for $\mathcal{W}^\perp$ and observe that $\mathcal{B} = \{U,V,W\}$ is an orthonormal basis for $\mathcal{R}^3$. Since $T(U), T(V) \in \mathcal{W}^\perp$, the third coordinates of $[T(U)]_{\mathcal{B}}$ and $[T(V)]_{\mathcal{B}}$ are zeros. Moreover, $[T(W)]_{\mathcal{B}} = [W]_{\mathcal{B}} = E_3$, and therefore

$$[T]_{\mathcal{B}} = \left[\begin{array}{cc|c} \multicolumn{2}{c|}{\multirow{2}{*}{C}} & 0 \\ & & 0 \\ \hline 0 & 0 & 1 \end{array}\right],$$

for some $C \in \mathcal{M}_{2\times 2}$. Theorem 4.24 states that $[T]_{\mathcal{B}}$ is orthogonal, so C is orthogonal. Also, $\det(C) = \det[T]_{\mathcal{B}}$, and by Example 5.12, $\det[T]_{\mathcal{B}} = \det(A) = 1$. Orthogonal 2×2 matrices with determinant 1 have the form

$$C = \begin{bmatrix} \cos(\theta) & -\sin(\theta) \\ \sin(\theta) & \cos(\theta) \end{bmatrix},$$

$\theta \in \mathcal{R}$ (Example 4.39). Thus, T performs a rotation of $\mathcal{R}^3$ about $\mathcal{W}$. A similar argument shows that if $\det(A) = -1$, then -1 is an eigenvalue of T, and

$$[T]_{\mathcal{B}} = \left[\begin{array}{cc|c} \multicolumn{2}{c|}{\mathrm{C}} & 0 \\ & & 0 \\ \hline 0 & 0 & -1 \end{array}\right] = \begin{bmatrix} 1 & 0 & 0 \\ 0 & 1 & 0 \\ 0 & 0 & -1 \end{bmatrix} \left[\begin{array}{cc|c} \multicolumn{2}{c|}{\mathrm{C}} & 0 \\ & & 0 \\ \hline 0 & 0 & 1 \end{array}\right].$$

The first factor on the right represents reflection in $\mathcal{W}^{\perp}$, relative to $\mathcal{B}$, so in this case T amounts to a rotation about $\mathcal{W}$ followed by reflection in $\mathcal{W}^{\perp}$. □

Exercises 6.3

1. Find the eigenvalues and associated eigenspaces for

a. $\begin{bmatrix} 1 & 2 \\ 4 & 3 \end{bmatrix}$. b. $\begin{bmatrix} 7 & -3 \\ 4 & -1 \end{bmatrix}$. **c.** $\begin{bmatrix} 6 & -3 \\ 2 & 1 \end{bmatrix}$.

d. $\begin{bmatrix} -2 & 1 \\ -4 & 3 \end{bmatrix}$. **e.** $\begin{bmatrix} -5 & 7 & -4 \\ -2 & 4 & -2 \\ 4 & -4 & 3 \end{bmatrix}$. f. $\begin{bmatrix} 0 & 0 & -3 \\ 0 & 2 & 0 \\ 1 & 0 & 0 \end{bmatrix}$.

g. $\begin{bmatrix} -1 & 4 & 2 \\ 0 & 1 & 0 \\ -4 & 8 & 5 \end{bmatrix}$. h. $\begin{bmatrix} 3 & 1 & -1 \\ 1 & 3 & -1 \\ -1 & -1 & 5 \end{bmatrix}$. **i.** $\begin{bmatrix} 2 & 1 & 0 \\ 0 & 3 & 0 \\ 0 & 0 & 3 \end{bmatrix}$.

j. $\begin{bmatrix} 0 & 1 & 1 & 1 \\ 1 & 0 & 1 & 1 \\ 1 & 1 & 0 & 1 \\ 1 & 1 & 1 & 0 \end{bmatrix}$. **k.** $\begin{bmatrix} 2 & 0 & 0 & 0 \\ 0 & 0 & -1 & 0 \\ 0 & 0 & 0 & -1 \\ 0 & 1 & 0 & 0 \end{bmatrix}$. l. $\begin{bmatrix} -1 & 1 & 0 & 0 \\ 1 & -1 & 0 & 0 \\ 0 & 0 & -2 & 0 \\ 0 & 0 & 0 & -2 \end{bmatrix}$.

2. Find the characteristic polynomial for $\begin{bmatrix} a & b \\ c & d \end{bmatrix}$ and confirm that it has the form indicated in (6.3).

3. Show that the characteristic polynomial for $\begin{bmatrix} \cos(\alpha) & -\sin(\alpha) \\ \sin(\alpha) & \cos(\alpha) \end{bmatrix}$ has no real roots when $\alpha \in (0,\pi) \cup (\pi,2\pi)$.

4. a. Find the characteristic polynomial for $\begin{bmatrix} a & d & e \\ 0 & b & f \\ 0 & 0 & c \end{bmatrix}$.

b. What are the eigenvalues of an upper triangular $n \times n$ matrix?

5. If $A \in \mathcal{M}_{n\times n}$, show that A and A^T have the same characteristic polynomial.

6. If $A \in \mathcal{M}_{n\times n}$ is nonsingular and $B = A^{-1}$, show that $c_B(\lambda) = \dfrac{(-\lambda)^n}{\det(A)} c_A(1/\lambda)$.

7. If $A \in \mathcal{M}_{n\times n}$ and $B = A^2$, show that $c_B(\lambda^2) = (-1)^n c_A(\lambda) c_A(-\lambda)$.

8. Suppose $A \in \mathcal{M}_{n\times n}$, $t \neq 0$, and $B = tA$. Show that $c_B(\lambda) = t^n c_A(\lambda/t)$.

9. Let $\mathcal{B} = \left\{ \begin{bmatrix} 1 \\ 1 \\ 0 \end{bmatrix}, \begin{bmatrix} 0 \\ 1 \\ 1 \end{bmatrix}, \begin{bmatrix} 1 \\ 0 \\ 1 \end{bmatrix} \right\}$ and assume T is a linear operator on $\mathcal{R}^3$. Find a basis for each eigenspace of T when

a. $[T]_{\mathcal{B}} = \begin{bmatrix} -5 & 7 & -4 \\ -2 & 4 & -2 \\ 4 & -4 & 3 \end{bmatrix}$. b. $[T]_{\mathcal{B}} = \begin{bmatrix} 3 & 1 & -1 \\ 1 & 3 & -1 \\ -1 & -1 & 5 \end{bmatrix}$. c. $[T]_{\mathcal{B}} = \begin{bmatrix} -1 & 4 & 2 \\ 0 & 1 & 0 \\ -4 & 8 & 5 \end{bmatrix}$.

(Hint: Refer to Problem 1, parts (e), (h), and (g).)

10. If $A \in \mathcal{M}_{m\times m}$, $B \in \mathcal{M}_{n\times n}$ and $C = \left[\begin{array}{c|c} A & 0 \\ \hline 0 & B \end{array}\right]$, show that $c_C(\lambda) = c_A(\lambda) c_B(\lambda)$.

11. Show that if $f(x) = a_0 + a_1 x + a_2 x^2 + \cdots + a_{n-1}x^{n-1} + x^n$, $n \geq 2$, and

$$A = \begin{bmatrix} 0 & 1 & 0 & \cdots & 0 \\ 0 & 0 & 1 & & 0 \\ 0 & 0 & 0 & & \vdots \\ \vdots & \vdots & \vdots & & 0 \\ 0 & 0 & 0 & & 1 \\ -a_0 & -a_1 & -a_2 & \cdots & -a_{n-1} \end{bmatrix},$$

then $\det(\lambda I - A) = f(\lambda)$. (In other words, every polynomial of degree 2 or larger is the characteristic polynomial of a matrix.)

12. Suppose $\mathcal{V}$ is a subspace of $\mathcal{R}^n$ of dimension p, $0 < p < n$, and S and T are reflection in $\mathcal{V}$ and projection onto $\mathcal{V}$, respectively. Find the characteristic polynomials for S and T.

13. Show that if $S,T : \mathcal{R}^3 \to \mathcal{R}^3$ are rotations (about axes through 0), then $S \circ T$ is a rotation.

6.4
DIAGONALIZATION OF LINEAR OPERATORS

The time has come to resolve the issue raised in the introduction to this chapter. Given a linear operator T on $\mathcal{R}^n$, under what circumstances will there be a coordinate system for $\mathcal{R}^n$ whose axes are invariant under T? A basis $\mathcal{B} = \{X_1, \ldots, X_n\}$ generates such a coordinate system when its axes $\mathcal{V}_k = \mathcal{S}pan\{X_k\}$ satisfy

$$T(\mathcal{V}_k) \subseteq \mathcal{V}_k,$$

that is, when $T(X_k) \in \mathcal{S}pan\{X_k\}$, $k = 1, \ldots, n$. This condition is equivalent to $T(X_k) = \lambda_k X_k$, $\lambda_k \in \mathcal{R}$, $k = 1, \ldots, n$, in which case

$$[T]_{\mathcal{B}} = \begin{bmatrix} [T(X_1)]_{\mathcal{B}} & \cdots & [T(X_n)]_{\mathcal{B}} \end{bmatrix} = \begin{bmatrix} \lambda_1 & & 0 \\ & \ddots & \\ 0 & & \lambda_n \end{bmatrix}.$$

Thus, a coordinate system with T-invariant axes is associated with a matrix representative of T that is diagonal.

DEFINITION. A linear operator T on $\mathcal{R}^n$ is said to be *diagonalizable* if there is a basis $\mathcal{B}$ for $\mathcal{R}^n$ such that $[T]_{\mathcal{B}}$ is diagonal. The $n \times n$ diagonal matrix with diagonal entries $\lambda_1, \ldots, \lambda_n$ will be denoted by $D[\lambda_1, \ldots, \lambda_n]$, that is,

$$D[\lambda_1, \ldots, \lambda_n] = \begin{bmatrix} \lambda_1 & & 0 \\ & \ddots & \\ 0 & & \lambda_n \end{bmatrix}.$$

Diagonalizability of T depends on what eigenvalues and eigenvectors it has. The next theorem spells out the connection between these notions.

THEOREM 6.5. If $T : \mathcal{R}^n \to \mathcal{R}^n$ is linear and $\mathcal{B} = \{X_1, \ldots, X_n\}$ is a basis for $\mathcal{R}^n$, then $[T]_{\mathcal{B}} = D[\lambda_1, \ldots, \lambda_n]$ if and only if X_k is an eigenvector of T with eigenvalue λ_k, $1 \leq k \leq n$.

Proof. The linear independence of $\mathcal{B}$ ensures that $X_k \neq 0$, $k = 1, \ldots, n$. Now, $\text{Col}_k([T]_{\mathcal{B}}) = [T(X_k)]_{\mathcal{B}}$, and $\text{Col}_k(D[\lambda_1, \ldots, \lambda_n]) = \lambda_k E_k = \lambda_k [X_k]_{\mathcal{B}}$, so $[T]_{\mathcal{B}} = D[\lambda_1, \ldots, \lambda_n]$ if and only if $[T(X_k)]_{\mathcal{B}} = \lambda_k [X_k]_{\mathcal{B}}$, which is equivalent to $T(X_k) = \lambda_k X_k$, $k = 1, \ldots, n$. ■

Theorem 6.5 states that the linear operator T on $\mathcal{R}^n$ is diagonalizable if and only if there is a basis for $\mathcal{R}^n$ consisting of eigenvectors of T. Moreover, when $[T]_{\mathcal{B}}$ is diagonal, its diagonal entries are the eigenvalues of T, and the elements of $\mathcal{B}$ are associated eigenvectors.

■ **EXAMPLE 6.13.** Let T be the linear operator on $\mathcal{R}^3$ given by $T(X) = AX$, where

$$A = \begin{bmatrix} 0 & 1 & 1 \\ 1 & 0 & 1 \\ 1 & 1 & 0 \end{bmatrix}.$$

Example 6.8 established that A has eigenvalues -1 and 2, with eigenspaces

$$\mathcal{V}(-1) = \mathcal{S}pan\left\{ \begin{bmatrix} -1 \\ 1 \\ 0 \end{bmatrix}, \begin{bmatrix} -1 \\ 0 \\ 1 \end{bmatrix} \right\} \quad \text{and} \quad \mathcal{V}(2) = \mathcal{S}pan\left\{ \begin{bmatrix} 1 \\ 1 \\ 1 \end{bmatrix} \right\}.$$

Set $X_1 = \begin{bmatrix} -1 \\ 1 \\ 0 \end{bmatrix}$, $X_2 = \begin{bmatrix} -1 \\ 0 \\ 1 \end{bmatrix}$, and $X_3 = \begin{bmatrix} 1 \\ 1 \\ 1 \end{bmatrix}$. Then $\mathcal{B} = \{X_1, X_2, X_3\}$ is a basis for

$\mathcal{R}^3$, T is diagonalizable, and

$$[T]_{\mathcal{B}} = \begin{bmatrix} -1 & 0 & 0 \\ 0 & -1 & 0 \\ 0 & 0 & 2 \end{bmatrix}.$$

Note that $\mathcal{A} = \{X_1, X_3, X_2\}$ is also a basis for $\mathcal{R}^3$ and that

$$[T]_{\mathcal{A}} = \begin{bmatrix} -1 & 0 & 0 \\ 0 & 2 & 0 \\ 0 & 0 & -1 \end{bmatrix}. \square$$

■ **EXAMPLE 6.14.** The operator $T(X) = \begin{bmatrix} 3 & 2 \\ 0 & 3 \end{bmatrix} X$ has characteristic polynomial

$$c(\lambda) = \det \begin{bmatrix} \lambda - 3 & -2 \\ 0 & \lambda - 3 \end{bmatrix} = (\lambda - 3)^2.$$

The only eigenspace,

$$\mathcal{V}(3) = \mathcal{N}\left(\begin{bmatrix} 0 & -1 \\ 0 & 0 \end{bmatrix}\right) = Span\left\{\begin{bmatrix} 1 \\ 0 \end{bmatrix}\right\},$$

is one-dimensional. Since T does not have two linearly independent eigenvectors, T is not diagonalizable. $\square$

Diagonalizability, like any other feature of a linear operator, is reflected in some manner in the matrices that represent that operator. Suppose $A \in \mathcal{M}_{n\times n}$ and $T(X) = AX$. Assuming T is diagonalizable, there are real numbers $\lambda_1, \ldots, \lambda_n$ and a basis $\mathcal{B} = \{X_1, \ldots, X_n\}$ such that $[T]_{\mathcal{B}} = D[\lambda_1, \ldots, \lambda_n]$. Setting $P = P_{\mathcal{E}\mathcal{B}} = [X_1 \ \cdots \ X_n]$, you have

$$D[\lambda_1, \ldots, \lambda_n] = [T]_{\mathcal{B}} = P_{\mathcal{E}\mathcal{B}}^{-1}[T]_{\mathcal{E}}P_{\mathcal{E}\mathcal{B}} = P^{-1}AP. \tag{6.4}$$

Therein lies the motivation for the notion of diagonalizability for a square matrix.

DEFINITION. $A \in \mathcal{M}_{n\times n}$ is *diagonalizable* if there is a nonsingular $P \in \mathcal{M}_{n\times n}$ and a diagonal $D \in \mathcal{M}_{n\times n}$ such that $D = P^{-1}AP$.

The discussion above indicates that if T is diagonalizable, then A is diagonalizable, where $T(X) = AX$. Consider the reverse implication. If P is nonsingular and $P^{-1}AP = D[\lambda_1, \ldots, \lambda_n] = D$, then

$$AP = PD. \tag{6.5}$$

Moreover, $Col_k(AP) = ACol_k(P)$, and $Col_k(PD) = PCol_k(D) = P(\lambda_k E_k) = \lambda_k Col_k(P)$, so (6.5) is equivalent to

$$ACol_k(P) = \lambda_k Col_k(P),$$

$1 \le k \le n$. Thus, $X_k = Col_k(P)$ is an eigenvector of A with eigenvalue λ_k. The columns of P are linearly independent (P is nonsingular), so $\{X_1, \ldots, X_n\}$ is a basis for $\mathcal{R}^n$

consisting of eigenvectors of A. Since A and T have the same eigenvectors, it follows from Theorem 6.5 that T is diagonalizable.

These observations are summarized in the next theorem.

THEOREM 6.6. $A \in \mathcal{M}_{n\times n}$ is diagonalizable if and only if there is a basis for $\mathcal{R}^n$ consisting of eigenvectors of A. If $X_1, \ldots, X_n$ are linearly independent eigenvectors with eigenvalues $\lambda_1, \ldots, \lambda_n$, respectively, and $P = [X_1 \ \cdots \ X_n]$, then

$$P^{-1}AP = \begin{bmatrix} \lambda_1 & & 0 \\ & \ddots & \\ 0 & & \lambda_n \end{bmatrix}.$$

■ **EXAMPLE 6.15.** $A = \begin{bmatrix} 4 & 1 \\ -2 & 1 \end{bmatrix}$ from Example 6.7 has eigenvalues 3 and 2, with associated eigenvectors

$$X_1 = \begin{bmatrix} -1 \\ 1 \end{bmatrix} \quad \text{and} \quad X_2 = \begin{bmatrix} -1 \\ 2 \end{bmatrix},$$

respectively. Since $\{X_1, X_2\}$ is a basis for $\mathcal{R}^2$, A is diagonalizable. If $P = [X_1 \ \ X_2]$, then

$$P^{-1}AP = \begin{bmatrix} -2 & -1 \\ 1 & 1 \end{bmatrix}\begin{bmatrix} 4 & 1 \\ -2 & 1 \end{bmatrix}\begin{bmatrix} -1 & -1 \\ 1 & 2 \end{bmatrix} = \begin{bmatrix} 3 & 0 \\ 0 & 2 \end{bmatrix}. \ \square$$

■ **EXAMPLE 6.16.** The only eigenvalue of $A = \begin{bmatrix} 1 & 0 & 1 \\ 0 & 3 & 0 \\ -1 & 0 & 1 \end{bmatrix}$ is $\lambda = 3$ (Example 6.9), and

$$3I - A = \begin{bmatrix} 2 & 0 & -1 \\ 0 & 0 & 0 \\ 1 & 0 & 2 \end{bmatrix} \longrightarrow \begin{bmatrix} 1 & 0 & 0 \\ 0 & 0 & 1 \\ 0 & 0 & 0 \end{bmatrix},$$

so $\mathcal{V}(3) = \mathcal{S}pan\left\{\begin{bmatrix} 0 \\ 1 \\ 0 \end{bmatrix}\right\}$. There is no basis for $\mathcal{R}^3$ consisting of eigenvectors of A, and therefore A is not diagonalizable. $\square$

For a given linear operator, the maximal number of linearly independent eigenvectors associated with a particular eigenvalue is the dimension of the eigenspace. To decide if the operator is diagonalizable you might obtain a basis for each eigenspace and, from those vectors, try to assemble a basis for $\mathcal{R}^n$. This raises the following questions:

> Is a set consisting of basis vectors for the eigenspaces necessarily linearly independent? (A)

> Does such a set contain n vectors? (B)

We'll start with question (A). Suppose T is a linear operator on $\mathcal{R}^n$ with two different eigenvalues, λ and μ. Observe that $\mathcal{V}(\lambda) \cap \mathcal{V}(\mu) = \{0\}$. Indeed, if $X \in \mathcal{V}(\lambda) \cap \mathcal{V}(\mu)$, then $\lambda X = T(X) = \mu X$, and therefore $(\lambda - \mu)X = 0$. Since $\lambda \neq \mu$, $X = 0$. Now, assume $\{X_1, \ldots, X_p\} \subseteq \mathcal{V}(\lambda)$ and $\{W_1, \ldots, W_q\} \subseteq \mathcal{V}(\mu)$ are linearly independent and set $\mathcal{C} = \{X_1, \ldots, X_p, W_1, \ldots, W_q\}$. The object is to check $\mathcal{C}$ for linear independence, so suppose there are scalars $t_1, \ldots, t_p$ and $s_1, \ldots, s_q$ such that

$$0 = t_1X_1 + \cdots + t_pX_p + s_1W_1 + \cdots + s_qW_q.$$

Setting $X = \sum_1^p t_kX_k$ and $W = \sum_1^q s_kW_k$, you have $X \in \mathcal{V}(\lambda)$, $W \in \mathcal{V}(\mu)$, and

$$0 = X + W. \tag{6.6}$$

Since $\mathcal{V}(\lambda) \cap \mathcal{V}(\mu) = \{0\}$, $\mathcal{V}(\lambda)$ and $\mathcal{V}(\mu)$ are independent subspaces (Theorem 2.36 and its Corollary), and therefore $X = 0 = W$. Thus,

$$t_1X_1 + \cdots + t_pX_p = 0 = s_1W_1 + \cdots + s_qW_q.$$

Linear independence of $\{X_1, \ldots, X_p\}$ and $\{W_1, \ldots, W_q\}$ then implies $t_1 = \cdots = t_p = 0$ and $s_1 = \cdots = s_q = 0$, so $\mathcal{C}$ is linearly independent.

The key step in the argument above is the conclusion that $X = 0 = W$ whenever $X \in \mathcal{V}(\lambda)$, $W \in \mathcal{V}(\mu)$, and $0 = X + W$. The next theorem extends that result to the case of more than two eigenvalues.

THEOREM 6.7. Suppose T is a linear operator on $\mathcal{R}^n$ and $\lambda_1, \ldots, \lambda_q$ are distinct eigenvalues of T. If $V_k \in \mathcal{V}(\lambda_k)$, $k = 1, \ldots, q$, then

$$0 = V_1 + V_2 + \cdots + V_q \quad \text{implies} \quad V_1 = \cdots = V_q = 0. \tag{6.7}$$

Proof. Let $S_k = \lambda_k \mathrm{Id} - T$, $1 \leq k \leq q$. Then $\mathcal{K}er(S_k) = \mathcal{V}(\lambda_k)$ (Theorem 6.1). In particular, $S_k(V_k) = 0$. Moreover,

$$S_k(V_j) = \lambda_k \mathrm{Id}(V_j) - T(V_j) = \lambda_k V_j - \lambda_j V_j = (\lambda_k - \lambda_j)V_j,$$

$j = 1, \ldots, q$. Thus, applying S_1 to both sides of (6.7) produces

$$\begin{aligned} 0 = S_1(0) &= S_1(V_1) + S_1(V_2) + \cdots + S_1(V_q) \\ &= 0 + (\lambda_1 - \lambda_2)V_2 + \cdots + (\lambda_1 - \lambda_q)V_q. \end{aligned}$$

Applying S_2 to both sides of the last equation gives

$$\begin{aligned} 0 &= S_2((\lambda_1 - \lambda_2)V_2) + S_2((\lambda_1 - \lambda_3)V_3) + \cdots + S_2((\lambda_1 - \lambda_q)V_q) \\ &= 0 + (\lambda_1 - \lambda_3)S_2(V_3) + \cdots + (\lambda_1 - \lambda_q)S_2(V_q) \\ &= (\lambda_1 - \lambda_3)(\lambda_2 - \lambda_3)V_3 + \cdots + (\lambda_1 - \lambda_q)(\lambda_2 - \lambda_q)V_q, \end{aligned}$$

after which application of $S_4, S_5, \ldots, S_{q-1}$ in succession results in

$$0 = (\lambda_1 - \lambda_q)(\lambda_2 - \lambda_q)\cdots(\lambda_{q-1} - \lambda_q)V_q.$$

Since the eigenvalues are distinct, $V_q = 0$, and (6.7) reduces to

$$0 = V_1 + \cdots + V_{q-1}. \tag{6.8}$$

Now, applying $S_1, \ldots, S_{q-2}$ in succession to (6.8), the same sort of analysis produces $V_{q-1} = 0$ and $0 = V_1 + \cdots + V_{q-2}$. Continuing in this manner leads to the conclusion $V_j = 0$, $j = 1, \ldots, q$. ■

COROLLARY. If $T : \mathcal{R}^n \to \mathcal{R}^n$ is linear, with distinct eigenvalues $\lambda_1, \ldots, \lambda_q$, then $\mathcal{V}(\lambda_1), \mathcal{V}(\lambda_2), \ldots, \mathcal{V}(\lambda_q)$ are independent subspaces.

Proof. If $\sum_1^q \mathrm{V}_k = \mathrm{X} = \sum_1^q \mathrm{W}_k$, where $\mathrm{V}_k, \mathrm{W}_k \in \mathcal{V}(\lambda_k)$, $k = 1, \ldots, q$, then

$$0 = (\mathrm{W}_1 - \mathrm{V}_1) + \cdots + (\mathrm{W}_q - \mathrm{V}_q).$$

Since $\mathrm{W}_k - \mathrm{V}_k \in \mathcal{V}(\lambda_k)$, Theorem 6.7 gives $\mathrm{W}_k - \mathrm{V}_k = 0$, $1 \le k \le q$, that is, X has a unique a representation as the sum of vectors in $\mathcal{V}(\lambda_1), \mathcal{V}(\lambda_2), \ldots, \mathcal{V}(\lambda_q)$. ■

Question (A) can now be settled.

THEOREM 6.8. Suppose T is a linear operator on $\mathcal{R}^n$ and $\lambda_1, \ldots, \lambda_q$ are distinct eigenvalues of T. If $\mathcal{B}_j$ is a linearly independent subset of $\mathcal{V}(\lambda_j)$, $1 \le j \le q$, then

$$\mathcal{B} = \bigcup_1^q \mathcal{B}_j$$

is linearly independent.

Proof. Suppose $\mathcal{B}_j = \{\mathrm{X}_{j1}, \ldots, \mathrm{X}_{jm_j}\}$, $1 \le j \le q$, and $t_{11}, \ldots, t_{1m_1}, t_{21}, \ldots, t_{2m_2}, \ldots, t_{q1}, \ldots, t_{qm_q}$ are scalars such that

$$0 = \sum_{k=1}^{m_1} t_{1k}\mathrm{X}_{1k} + \sum_{k=1}^{m_2} t_{2k}\mathrm{X}_{2k} + \cdots + \sum_{k=1}^{m_q} t_{qk}\mathrm{X}_{qk}. \tag{6.9}$$

Then $\mathrm{V}_j = \sum_{k=1}^{m_j} t_{jk}\mathrm{X}_{jk} \in \mathcal{V}(\lambda_j)$, and $0 = \mathrm{V}_1 + \mathrm{V}_2 + \cdots + \mathrm{V}_q$, so by Theorem 6.7,

$$0 = \mathrm{V}_j = \sum_{k=1}^{m_j} t_{jk}\mathrm{X}_{jk},$$

$j = 1, \ldots, q$. Linear independence of $\mathcal{B}_j$ then implies that $t_{j1}, \ldots, t_{jm_j}$ are all 0's. This conclusion holds for each j, so the scalars in (6.9) are all zeros, and $\mathcal{B}$ therefore is linearly independent. ■

COROLLARY. A linear operator on $\mathcal{R}^n$ with n distinct eigenvalues is diagonalizable.

Proof. Suppose T is a linear operator on $\mathcal{R}^n$ with n distinct eigenvalues. Choosing one eigenvector from each eigenspace produces a set of n linearly independent vectors and, therefore, a basis for $\mathcal{R}^n$. By Theorem 6.5, T is diagonalizable. ■

■ **EXAMPLE 6.17.** Suppose $\mathrm{A} = \begin{bmatrix} 1 & 1 & 1 \\ 0 & 2 & 1 \\ 0 & 0 & 3 \end{bmatrix}$ and $T(\mathrm{X}) = \mathrm{AX}$. T has characteristic polynomial

$$c(\lambda) = \det \begin{bmatrix} \lambda - 1 & -1 & -1 \\ 0 & \lambda - 2 & -1 \\ 0 & 0 & \lambda - 3 \end{bmatrix} = (\lambda - 1)(\lambda - 2)(\lambda - 3),$$

so its eigenvalues are 1, 2, and 3. According to the previous Corollary, T is diagonalizable. Since

$$1\mathrm{I} - \mathrm{A} = \begin{bmatrix} 0 & -1 & -1 \\ 0 & -1 & -1 \\ 0 & 0 & -2 \end{bmatrix} \longrightarrow \begin{bmatrix} 0 & 1 & 0 \\ 0 & 0 & 1 \\ 0 & 0 & 0 \end{bmatrix},$$

$$2\mathrm{I} - \mathrm{A} = \begin{bmatrix} 1 & -1 & -1 \\ 0 & 0 & -1 \\ 0 & 0 & -1 \end{bmatrix} \longrightarrow \begin{bmatrix} 1 & -1 & 0 \\ 0 & 0 & 1 \\ 0 & 0 & 0 \end{bmatrix},$$

and

$$3\mathrm{I} - \mathrm{A} = \begin{bmatrix} 2 & -1 & -1 \\ 0 & 1 & -1 \\ 0 & 0 & 0 \end{bmatrix} \longrightarrow \begin{bmatrix} 1 & 0 & -1 \\ 0 & 1 & -1 \\ 0 & 0 & 0 \end{bmatrix},$$

$$\mathcal{V}(1) = Span\left\{\begin{bmatrix} 1 \\ 0 \\ 0 \end{bmatrix}\right\}, \quad \mathcal{V}(2) = Span\left\{\begin{bmatrix} 1 \\ 1 \\ 0 \end{bmatrix}\right\}, \quad \text{and} \quad \mathcal{V}(3) = Span\left\{\begin{bmatrix} 1 \\ 1 \\ 1 \end{bmatrix}\right\}.$$

If

$$\mathrm{P} = \begin{bmatrix} 1 & 1 & 1 \\ 0 & 1 & 1 \\ 0 & 0 & 1 \end{bmatrix}, \quad \text{then} \quad \mathrm{P}^{-1}\mathrm{AP} = \begin{bmatrix} 1 & 0 & 0 \\ 0 & 2 & 0 \\ 0 & 0 & 3 \end{bmatrix}. \square$$

Now consider question (B). Suppose T is a linear operator on $\mathcal{R}^n$ whose distinct eigenvalues are $\lambda_1, \ldots, \lambda_q$. The maximal number of linearly independent eigenvectors associated with λ_j is dim $\mathcal{V}(\lambda_j)$. If $\mathcal{B}_j$ is a basis for $\mathcal{V}(\lambda_j)$, $1 \leq j \leq q$, then $\mathcal{B} = \cup_1^q \mathcal{B}_j$ is a linearly independent set (Theorem 6.8) containing

$$N = \dim \mathcal{V}(\lambda_1) + \dim \mathcal{V}(\lambda_2) + \cdots + \dim \mathcal{V}(\lambda_q)$$

eigenvectors of T. Of course, $\mathcal{B} \subseteq \mathcal{R}^n$ implies $N \leq n$. Whether or not equality occurs depends in part on the size of dim $\mathcal{V}(\lambda_j)$, $1 \leq k \leq q$, which in turn is governed by the multiplicity of λ_j as a zero of the characteristic polynomial.

DEFINITION. Suppose f is a polynomial of degree n and $f(x_0) = 0$. The *algebraic multiplicity* of x_0 is the integer m satisfying

$$f(x) = (x - x_0)^m g(x),$$

where g is a polynomial of degree $n - m$ and $g(x_0) \neq 0$.

■ **EXAMPLE 6.18.** The real zeros of $f(x) = x^3(x + 4)^2(x - 2)(x^4 + 3)$ are 0, -4, and 2, and their algebraic multiplicities are 3, 2, and 1, respectively. □

THEOREM 6.9. If T is a linear operator on $\mathcal{R}^n$ with characteristic polynomial c and μ is a zero of c of multiplicity m, then dim $\mathcal{V}(\mu) \leq m$.

Proof. Let $\{\mathrm{X}_1, \ldots, \mathrm{X}_r\}$ be a basis for $\mathcal{V}(\mu)$. Assuming first that $r < n$, choose $\mathrm{W}_1, \ldots, \mathrm{W}_{n-r}$ so that $\mathcal{B} = \{\mathrm{X}_1, \ldots, \mathrm{X}_r, \mathrm{W}_1, \ldots, \mathrm{W}_{n-r}\}$ is a basis for $\mathcal{R}^n$. For $j = 1, \ldots, r$, $T(\mathrm{X}_j) = \mu\mathrm{X}_j$, and therefore $[T(\mathrm{X}_j)]_{\mathcal{B}} = \mu\mathrm{E}_j$. Thus,

$$[T]_{\mathcal{B}} = \left[\begin{array}{c|c} \mu\mathrm{I}_r & \mathrm{A} \\ \hline 0 & \mathrm{B} \end{array}\right],$$

where A and B are $r \times (n - r)$ and $(n - r) \times (n - r)$ matrices, respectively. Then

$$c(\lambda) = \det(\lambda\mathrm{I} - [T]_{\mathcal{B}}) = \det\left[\begin{array}{c|c} (\lambda - \mu)\mathrm{I}_r & -\mathrm{A} \\ \hline 0 & \lambda\mathrm{I}_{n-r} - \mathrm{B} \end{array}\right]$$

$$= (\lambda - \mu)^r g(\lambda),$$

where $g(\lambda) = \det(\lambda I_{n-r} - B)$ is a polynomial of degree $n - r$. The last equation states that $(\lambda - \mu)^r$ divides $c(\lambda)$. Thus, $\dim \mathcal{V}(\mu) = r \leq m$. When $r = n$, a basis $\mathcal{B}$ for $\mathcal{V}(\mu)$ is also a basis for $\mathcal{R}^n$, and $[T]_{\mathcal{B}} = \mu I_n$. In this case,

$$c(\lambda) = \det(\lambda I_n - [T]_{\mathcal{B}}) = \det((\lambda - \mu)I_n) = (\lambda - \mu)^n,$$

so $m = n = r$. ■

The stage is now set to complete the discussion of diagonalizability.

THEOREM 6.10. Suppose T is a linear operator on $\mathcal{R}^n$ with characteristic polynomial c. If the distinct zeros of c are $\lambda_1, \ldots, \lambda_q$, with multiplicities $m_1, \ldots, m_q$, respectively, then the following are equivalent:

(1) T is diagonalizable
(2) $m_1 + \cdots + m_q = n$ and $\dim \mathcal{V}(\lambda_j) = m_j, j = 1, \ldots, q$.

Proof.
(1) $\Rightarrow$ (2). The factorization of c is

$$c(\lambda) = (\lambda - \lambda_1)^{m_1}(\lambda - \lambda_2)^{m_2} \cdots (\lambda - \lambda_q)^{m_q} g(\lambda),$$

where g is a polynomial of degree $n - \sum_1^q m_j$ and g has no real zeros. Of course $\sum_1^q m_j \leq n$. By Theorem 6.9, $\dim \mathcal{V}(\lambda_j) \leq m_j$, $1 \leq j \leq q$, so

$$\sum_1^q \dim \mathcal{V}(\lambda_j) \leq \sum_1^q m_j \leq n. \tag{6.10}$$

The left side of (6.10) is the maximal number of independent eigenvectors. If $\sum_1^q m_j < n$ or $\dim \mathcal{V}(\lambda_j) < m_j$ for some j, then (6.10) implies T has fewer than n linearly independent eigenvectors, contrary to the assumption that T is diagonalizable. Thus, $\sum_1^q m_j = n$, and $\dim \mathcal{V}(\lambda_j) = m_j, j = 1, \ldots, q$.

(2) $\Rightarrow$ (1). Let $\mathcal{B}_j$ be a basis for $\mathcal{V}(\lambda_j)$, $1 \leq j \leq q$, and set $\mathcal{B} = \cup_1^q \mathcal{B}_j$. By Theorem 6.8, $\mathcal{B}$ is linearly independent, and by (2), $\mathcal{B}$ contains n vectors, so $\mathcal{B}$ is a basis for $\mathcal{R}^n$ consisting of eigenvectors of T. ■

■ **EXAMPLE 6.19.** Consider $T(X) = AX = \begin{bmatrix} 2 & 0 & -1 & 1 \\ 0 & 3 & 0 & 0 \\ -1 & 0 & 2 & 1 \\ 1 & 0 & 1 & 2 \end{bmatrix} X$. Since

$$\det(\lambda I - A) = \det \begin{bmatrix} \lambda - 2 & 0 & 1 & -1 \\ 0 & \lambda - 3 & 0 & 0 \\ 1 & 0 & \lambda - 2 & -1 \\ -1 & 0 & -1 & \lambda - 2 \end{bmatrix} = (\lambda - 3)\det \begin{bmatrix} \lambda - 2 & 1 & -1 \\ 1 & \lambda - 2 & -1 \\ -1 & -1 & \lambda - 2 \end{bmatrix}$$

$$= (\lambda - 3)\left\{(\lambda - 2)[\lambda^2 - 4\lambda + 3] - (\lambda - 3) - (\lambda - 3)\right\} \quad \begin{pmatrix}\text{expansion} \\ \text{by first row}\end{pmatrix}$$
$$= (\lambda - 3)^2\{(\lambda - 2)(\lambda - 1) - 2\}$$
$$= \lambda(\lambda - 3)^3,$$

the characteristic polynomial has zeros $\lambda_1 = 0$ and $\lambda_2 = 3$, with multiplicities $m_1 = 1$ and $m_2 = 3$. Moreover,

$$0I - A = \begin{bmatrix} -2 & 0 & 1 & -1 \\ 0 & -3 & 0 & 0 \\ 1 & 0 & -2 & -1 \\ -1 & 0 & -1 & -2 \end{bmatrix} \longrightarrow \begin{bmatrix} 1 & 0 & 0 & 1 \\ 0 & 1 & 0 & 0 \\ 0 & 0 & 1 & 1 \\ 0 & 0 & 0 & 0 \end{bmatrix}$$

and

$$3\mathrm{I}-\mathrm{A}=\begin{bmatrix}1&0&1&-1\\0&0&0&0\\1&0&1&-1\\-1&0&-1&1\end{bmatrix}\longrightarrow\begin{bmatrix}1&0&1&-1\\0&0&0&0\\0&0&0&0\\0&0&0&0\end{bmatrix},$$

so

$$\mathcal{V}(0)=\mathcal{S}pan\left\{\begin{bmatrix}-1\\0\\-1\\1\end{bmatrix}\right\},\quad\text{and}\quad\mathcal{V}(3)=\mathcal{S}pan\left\{\begin{bmatrix}-1\\0\\1\\0\end{bmatrix},\begin{bmatrix}0\\1\\0\\0\end{bmatrix},\begin{bmatrix}1\\0\\0\\1\end{bmatrix}\right\}.$$

Since dim $\mathcal{V}(0)=m_1$, dim $\mathcal{V}(3)=m_2$, and $m_1+m_2=4$, T is diagonalizable. If

$$\mathcal{B}=\left\{\begin{bmatrix}-1\\0\\1\\0\end{bmatrix},\begin{bmatrix}0\\1\\0\\0\end{bmatrix},\begin{bmatrix}1\\0\\0\\1\end{bmatrix},\begin{bmatrix}-1\\0\\-1\\1\end{bmatrix}\right\},\quad\text{then}\quad[T]_{\mathcal{B}}=\begin{bmatrix}3&0&0&0\\0&3&0&0\\0&0&3&0\\0&0&0&0\end{bmatrix},$$

and if

$$\mathrm{P}=\begin{bmatrix}-1&0&1&-1\\0&1&0&0\\1&0&0&-1\\0&0&1&1\end{bmatrix},\quad\text{then}\quad\mathrm{P}^{-1}\mathrm{AP}=\begin{bmatrix}3&0&0&0\\0&3&0&0\\0&0&3&0\\0&0&0&0\end{bmatrix}.\ \square$$

For a linear operator T with eigenvalue λ, the dimension of $\mathcal{V}(\lambda)$ is referred to as the *geometric multiplicity* of λ. It can be no larger than the algebraic multiplicity of λ (Theorem 6.9), and the two are the same when T is diagonalizable (Theorem 6.10).

Exercises 6.4

1. Determine whether A is diagonalizable. If it is, find a nonsingular P and diagonal D such that $\mathrm{P}^{-1}\mathrm{AP}=\mathrm{D}$.

a. $\begin{bmatrix}6&2\\-3&1\end{bmatrix}$ b. $\begin{bmatrix}0&1\\-9&6\end{bmatrix}$ **c.** $\begin{bmatrix}9&-25\\4&-11\end{bmatrix}$

d. $\begin{bmatrix}-1&0&1\\0&-1&1\\1&1&0\end{bmatrix}$ **e.** $\begin{bmatrix}3&-1&4\\1&2&1\\-2&1&-3\end{bmatrix}$ f. $\begin{bmatrix}1&0&1\\0&1&0\\0&0&1\end{bmatrix}$

g. $\begin{bmatrix}5&0&2\\-2&1&3\\0&0&5\end{bmatrix}$ h. $\begin{bmatrix}0&1&0\\0&0&1\\2&0&-1\end{bmatrix}$ **i.** $\begin{bmatrix}3&1&3\\1&3&3\\-1&-1&-1\end{bmatrix}$

j. $\begin{bmatrix}2&0&0&0\\0&1&1&0\\0&1&1&0\\0&0&0&2\end{bmatrix}$ **k.** $\begin{bmatrix}0&-1&-1&0\\1&0&0&1\\1&0&0&1\\0&-1&-1&0\end{bmatrix}$ l. $\begin{bmatrix}0&0&1&0\\0&1&0&0\\1&0&0&1\\0&0&0&1\end{bmatrix}$

m. $\begin{bmatrix} 2 & 0 & 1 & 0 & 0 \\ 0 & 2 & 0 & 1 & 0 \\ 0 & 0 & 3 & 0 & 1 \\ 0 & 0 & 0 & 3 & 0 \\ 0 & 0 & 0 & 0 & 3 \end{bmatrix}$ n. $\begin{bmatrix} 3 & -1 & 0 & 0 & 0 \\ -1 & 3 & 0 & 0 & 0 \\ 0 & 0 & 2 & 0 & 0 \\ 0 & 0 & -3 & -1 & 0 \\ 0 & 0 & -3 & 0 & -1 \end{bmatrix}$

o. $\begin{bmatrix} 4 & -2 & 0 & 0 & 0 \\ -3 & 9 & 0 & 0 & 0 \\ 0 & 0 & 3 & 0 & 0 \\ 0 & 0 & 0 & -1 & 0 \\ 0 & 0 & 0 & 2 & 3 \end{bmatrix}$

2. **a.** Show that $A = \begin{bmatrix} 2 & 0 & 0 \\ 0 & 2 & 0 \\ 0 & 0 & 2 \end{bmatrix}$, $B = \begin{bmatrix} 2 & 1 & 0 \\ 0 & 2 & 0 \\ 0 & 0 & 2 \end{bmatrix}$, and $C = \begin{bmatrix} 2 & 1 & 0 \\ 0 & 2 & 1 \\ 0 & 0 & 2 \end{bmatrix}$ all have characteristic polynomial $c(\lambda) = (\lambda - 2)^3$ and in each case find the dimension of $\mathcal{N}(2I - A)$.
   b. Find a 7×7 matrix with characteristic polynomial $c(\lambda) = (\lambda - 2)^7$ such that $\dim \mathcal{N}(2I - A) = 4$.

3. Show that $A = \begin{bmatrix} a & b \\ b & c \end{bmatrix}$ is diagonalizable for all $a,b,c \in \mathcal{R}$.

4. Assuming A is a diagonalizable $n \times n$ matrix, show that
   a. A^T is diagonalizable.
   b. A^2 is diagonalizable.
   c. A^{-1} is diagonalizable (provided A is nonsingular).

5. Show that if A is a diagonalizable $n \times n$ matrix with only one eigenvalue, λ, then $A = \lambda I_n$.

6. Show that if $A \in \mathcal{M}_{m\times m}$ and $B \in \mathcal{M}_{n\times n}$ are both diagonalizable, then $\left[\begin{array}{c|c} A & 0 \\ \hline 0 & B \end{array}\right]$ is diagonalizable.

7. Suppose U and V are nonzero vectors in $\mathcal{R}^n$ and $A = UV^T$.
   a. Show that U is an eigenvector of A with eigenvalue $\lambda = U \cdot V = \text{trace}(A)$.
   b. Show that $\mathcal{N}(A)$ has dimension $n - 1$. Interpret this conclusion in terms of eigenvalues and associated eigenspaces for A.
   c. Show that if $U \cdot V \neq 0$, then A is diagonalizable.
   d. Show by example that if $U \cdot V = 0$, then A need not be diagonalizable.

6.5
SIMILARITY AND SYMMETRY

Suppose T is a linear operator on $\mathcal{R}^n$ and $\mathcal{A}$ and $\mathcal{B}$ are bases for $\mathcal{R}^n$. Since $[T]_\mathcal{A}$ and $[T]_\mathcal{B}$ represent the same function, you might expect various features of T to present themselves as matrix properties that $[T]_\mathcal{A}$ and $[T]_\mathcal{B}$ have in common. For example, $[T]_\mathcal{A}$ is nonsingular if and only if $[T]_\mathcal{B}$ is nonsingular, because either occurrence is equivalent to T being one-to-one. Moreover, $[T]_\mathcal{A}$ and $[T]_\mathcal{B}$ have the same eigenvalues (those of T). The two matrices are connected algebraically by

$$[T]_{\mathcal{B}} = P_{\mathcal{AB}}^{-1}[T]_{\mathcal{A}}P_{\mathcal{AB}},$$

which serves as motivation for a notion of similarity for square matrices.

DEFINITION. Given A,B $\in \mathcal{M}_{n\times n}$, A is *similar* to B if there is a nonsingular P $\in \mathcal{M}_{n\times n}$ such that $B = P^{-1}AP$.

Formally, the order of appearance of A and B in this definition makes a difference, but it is easy to see that if A is similar to B, then B is similar to A. Indeed, if $B = P^{-1}AP$, then $A = PBP^{-1} = (P^{-1})^{-1}BP^{-1}$, so there is a nonsingular $Q \in \mathcal{M}_{n\times n}$ ($Q = P^{-1}$) such that $A = Q^{-1}BQ$.

THEOREM 6.11. For A,B $\in \mathcal{M}_{n\times n}$, the following statements are equivalent:

(1) A is similar to B.
(2) There is a linear operator T on $\mathcal{R}^n$ and bases, $\mathcal{A}$ and $\mathcal{B}$, for $\mathcal{R}^n$ such that $A = [T]_{\mathcal{A}}$ and $B = [T]_{\mathcal{B}}$.

Proof. That (2) implies (1) was noted above. For the reverse implication, suppose $P \in \mathcal{M}_{n\times n}$ is nonsingular and $B = P^{-1}AP$. Consider $T : \mathcal{R}^n \to \mathcal{R}^n$ given by $T(X) = AX$, that is, suppose $A = [T]_{\mathcal{E}}$. Observe that $\mathcal{B} = \{\mathrm{Col}_1(P), \ldots, \mathrm{Col}_n(P)\}$ is a basis for $\mathcal{R}^n$ (P is nonsingular). Moreover, $P_{\mathcal{EB}} = P$, so

$$B = P^{-1}AP = P_{\mathcal{EB}}^{-1}[T]_{\mathcal{E}}P_{\mathcal{EB}} = [T]_{\mathcal{B}}. \blacksquare$$

THEOREM 6.12. If A and B are similar $n \times n$ matrices, then

(1) $\det(A) = \det(B)$.
(2) $\mathrm{rank}(A) = \mathrm{rank}(B)$ and $\mathrm{nul}(A) = \mathrm{nul}(B)$.
(3) $c_A(\lambda) = c_B(\lambda)$.
(4) $\mathrm{trace}(A) = \mathrm{trace}(B)$.

Proof. Assume P is nonsingular and $B = P^{-1}AP$. Part (1) was established in Example 5.12. That $\mathrm{rank}(A) = \mathrm{rank}(B)$ is a direct consequence of Theorem 2.29, and then $\mathrm{nul}(A) = \mathrm{nul}(B)$ follows from the rank-nullity theorem. Since

$$\lambda I - B = \lambda P^{-1}IP - P^{-1}AP = P^{-1}(\lambda I - A)P,$$

$\lambda I - A$ is similar to $\lambda I - B$. Thus, (3) is a consequence of (1). Finally, the coefficients of λ^{n-1} in $c_A(\lambda)$ and $c_B(\lambda)$ are -trace(A) and -trace(B), respectively, (recall (6.3)) so (3) implies (4). $\blacksquare$

■ **EXAMPLE 6.20.** If A and B are similar $n \times n$ matrices and P is a nonsingular matrix such that $B = P^{-1}AP$, then

$$B^2 = (P^{-1}AP)(P^{-1}AP) = P^{-1}AI_nAP = P^{-1}A^2P,$$

$$B^3 = B^2B = (P^{-1}A^2P)(P^{-1}AP) = P^{-1}A^3P,$$

and (by induction), $B^k = P^{-1}A^kP$ for each positive integer k. □

If $A \in \mathcal{M}_{n\times n}$ is diagonalizable and P is a nonsingular $n \times n$ matrix such that $P^{-1}AP = D[\lambda_1, \ldots, \lambda_n] = D$, then Example 6.20 gives $P^{-1}A^kP = D^k$, so

$$A^k = PD^kP^{-1}, \tag{6.11}$$

$k = 1, 2, \ldots$. Since $D^k = D[\lambda_1^k, \ldots, \lambda_n^k]$, (6.11) provides an approach to the problem of calculating powers of A.

■ **EXAMPLE 6.21.** Consider the two-term linear recurrence relation

$$a_{k+2} = ra_{k+1} + sa_k, \quad k = 0, 1, 2, \ldots \tag{6.12}$$

discussed in Example 4.7. Observe that $\begin{bmatrix} a_{k+1} \\ a_{k+2} \end{bmatrix} = \begin{bmatrix} 0 & 1 \\ s & r \end{bmatrix} \begin{bmatrix} a_k \\ a_{k+1} \end{bmatrix}$. Setting $X_k = \begin{bmatrix} a_k \\ a_{k+1} \end{bmatrix}$ and $A = \begin{bmatrix} 0 & 1 \\ s & r \end{bmatrix}$ produces the one-term relation

$$X_{k+1} = AX_k, \quad k = 0, 1, 2, \ldots,$$

whose solution can be easily expressed in terms of the initial data $X_0 = \begin{bmatrix} a_0 \\ a_1 \end{bmatrix}$. Indeed, $X_1 = AX_0$, $X_2 = AX_1 = A(AX_0) = A^2X_0$, and in general,

$$X_k = A^kX_0, \quad k = 1, 2, \ldots. \tag{6.13}$$

To find the powers of A, consider the case when A is diagonalizable, with independent eigenvectors U and V and associated eigenvalues μ and ν, respectively. If $P = [U \quad V]$, then $P^{-1}AP = D = \begin{bmatrix} \mu & 0 \\ 0 & \nu \end{bmatrix}$, and (6.11) converts (6.13) to $X_k = PD^kP^{-1}X_0$. After multiplying the last equation on the left by P^{-1}, the change of variables $Y = P^{-1}X$ yields

$$Y_k = \begin{bmatrix} \mu^k & 0 \\ 0 & \nu^k \end{bmatrix} Y_0.$$

If $Y_0 = \begin{bmatrix} c \\ d \end{bmatrix}$, then $Y_k = \begin{bmatrix} c\mu^k \\ d\nu^k \end{bmatrix}$, and therefore

$$X_k = PY_k = [U \quad V] \begin{bmatrix} c\mu^k \\ d\nu^k \end{bmatrix} = c\mu^kU + d\nu^kV. \tag{6.14}$$

The solution of (6.12) is the first coordinate of X_k. If u and v are the first coordinates of U and V, respectively, then

$$a_k = cu\mu^k + dv\nu^k, \quad k = 0, 1, 2, \ldots.$$

For the Fibonacci sequence (Example 4.7), $A = \begin{bmatrix} 0 & 1 \\ 1 & 1 \end{bmatrix}$ and $X_0 = \begin{bmatrix} 0 \\ 1 \end{bmatrix}$. The eigenvalues of A are $\mu = (1 - \sqrt{5})/2$ and $\nu = (1 + \sqrt{5})/2$, and a little calculation produces

$$\mathcal{V}(\mu) = \mathcal{S}pan\left\{\begin{bmatrix} 1 \\ \mu \end{bmatrix}\right\} \quad \text{and} \quad \mathcal{V}(\nu) = \mathcal{S}pan\left\{\begin{bmatrix} 1 \\ \nu \end{bmatrix}\right\}$$

(see Exercise 3). Thus,

$$P = \begin{bmatrix} 1 & 1 \\ \mu & \nu \end{bmatrix}, Y_0 = P^{-1}X_0 = \frac{1}{\sqrt{5}} \begin{bmatrix} \nu & -1 \\ -\mu & 1 \end{bmatrix} \begin{bmatrix} 0 \\ 1 \end{bmatrix} = \frac{1}{\sqrt{5}} \begin{bmatrix} -1 \\ 1 \end{bmatrix},$$

and by (6.14),

$$\begin{bmatrix} a_k \\ a_{k+1} \end{bmatrix} = X_k = \frac{-1}{\sqrt{5}}\mu^k \begin{bmatrix} 1 \\ \mu \end{bmatrix} + \frac{1}{\sqrt{5}}\nu^k \begin{bmatrix} 1 \\ \nu \end{bmatrix}.$$

The terms of the Fibonacci sequence are therefore given explicitly by

$$a_k = \frac{1}{\sqrt{5}}\left(\frac{1+\sqrt{5}}{2}\right)^k - \frac{1}{\sqrt{5}}\left(\frac{1-\sqrt{5}}{2}\right)^k, \quad k = 0, 1, 2, \ldots . \square$$

For a square matrix to be diagonalizable, its characteristic polynomial must factor completely into linear factors (its zeros must all be real) and the dimension of each eigenspace must be as large as possible (the geometric multiplicity of the eigenvalue must equal its algebraic multiplicity). Checking these criteria is not a simple task. Often the roots of $c(\lambda)$ cannot be obtained explicitly, and one must resort to numerical techniques to approximate them. Faced with this reality, it is natural to seek sufficient conditions for diagonalizability that are verifiable. The next theorem is the best known result of this type.

THEOREM 6.13. If $A \in \mathcal{M}_{n\times n}$ is symmetric, then A is diagonalizable.

Proof. Omitted. ■

Symmetry not only ensures diagonalizability, it also places orthogonality restrictions on the eigenvectors.

THEOREM 6.14. If $A \in \mathcal{M}_{n\times n}$ is symmetric, λ and μ are distinct eigenvalues of A, $X \in \mathcal{V}(\lambda)$, and $W \in \mathcal{V}(\mu)$, then X is orthogonal to W.

Proof. For $X \in \mathcal{V}(\lambda)$ and $W \in \mathcal{V}(\mu)$, $\lambda(X \cdot W) = (\lambda X) \cdot W = (AX)^T W = X^T A^T W = X^T(AW) = X \cdot (\mu W) = \mu(X \cdot W)$, so $(\lambda - \mu)(X \cdot W) = 0$. Since $\lambda \neq \mu$, $X \cdot W = 0$. ■

■ **EXAMPLE 6.22.** As established in Example 6.8, $\begin{bmatrix} 0 & 1 & 1 \\ 1 & 0 & 1 \\ 1 & 1 & 0 \end{bmatrix}$ has eigenvalues -1 and 2, with eigenspaces $\mathcal{V}(-1) = \mathcal{S}pan\{X_1, X_2\}$ and $\mathcal{V}(2) = \mathcal{S}pan\{X_3\}$, where

$$X_1 = \begin{bmatrix} -1 \\ 1 \\ 0 \end{bmatrix}, \quad X_2 = \begin{bmatrix} -1 \\ 0 \\ 1 \end{bmatrix}, \quad \text{and} \quad X_3 = \begin{bmatrix} 1 \\ 1 \\ 1 \end{bmatrix}.$$

Note that $X_1 \cdot X_3 = 0 = X_2 \cdot X_3$. If $X \in \mathcal{V}(-1)$ and $W \in \mathcal{V}(2)$, then $X = rX_1 + sX_2$ and $W = tX_3$, for appropriate r, s, and t, so

$$X \cdot W = (rX_1 + sX_2) \cdot (tX_3) = rt(X_1 \cdot X_3) + st(X_2 \cdot X_3) = 0. \square$$

THEOREM 6.15. For $A \in \mathcal{M}_{n\times n}$, the following statements are equivalent:

(1) A is symmetric.
(2) There is an orthonormal basis for $\mathcal{R}^n$ consisting of eigenvectors of A.
(3) There is an orthogonal P and diagonal D such that $P^T AP = D$.

Proof.

(1) $\Rightarrow$ (2) Suppose the distinct eigenvalues of A are $\lambda_1, \ldots, \lambda_q$, with multiplicities $m_1, \ldots, m_q$, respectively. Since A is diagonalizable,

$$\sum_1^q m_j = n \qquad \text{and} \qquad \dim\mathcal{V}(\lambda_j) = m_j, \quad 1 \le j \le q.$$

Let $\mathcal{A}_j$ be an orthonormal basis for $\mathcal{V}(\lambda_j)$. When $i \neq j$, the vectors in $\mathcal{A}_i$ are orthogonal to those in $\mathcal{A}_j$ (Theorem 6.14), so $\mathcal{A} = \cup_1^q \mathcal{A}_j$ is an orthonormal basis for $\mathcal{R}^n$.

(2) $\Rightarrow$ (3) Suppose $\mathcal{A} = \{U_1, \ldots, U_n\}$ is an orthonormal basis of eigenvectors of A, with associated eigenvalues $\lambda_1, \ldots, \lambda_n$. Then $P = [U_1 \ \cdots \ U_n]$ is orthogonal, and $P^TAP = P^{-1}AP = D[\lambda_1, \ldots, \lambda_n]$.

(3) $\Rightarrow$ (1) If P is orthogonal, D is diagonal, and $P^TAP = D$, then $A = PDP^T$, so $A^T = (P^T)^TD^TP^T = PDP^T = A$. ■

■ **EXAMPLE 6.23.** The characteristic polynomial for $A = \begin{bmatrix} 2 & 2 \\ 2 & -1 \end{bmatrix}$ is

$$\det\begin{bmatrix} \lambda - 2 & -2 \\ -2 & \lambda + 1 \end{bmatrix} = \lambda^2 - \lambda - 6 = (\lambda - 3)(\lambda + 2),$$

so A has eigenvalues 3 and -2. Since

$$3I - A = \begin{bmatrix} 1 & -2 \\ -2 & 4 \end{bmatrix} \longrightarrow \begin{bmatrix} 1 & -2 \\ 0 & 0 \end{bmatrix} \text{ and } -2I - A = \begin{bmatrix} -4 & -2 \\ -2 & -1 \end{bmatrix} \longrightarrow \begin{bmatrix} 1 & 1/2 \\ 0 & 0 \end{bmatrix},$$

$$\mathcal{V}(3) = \mathcal{S}pan\left\{\begin{bmatrix} 2 \\ 1 \end{bmatrix}\right\} \quad \text{and} \quad \mathcal{V}(-2) = \mathcal{S}pan\left\{\begin{bmatrix} -1 \\ 2 \end{bmatrix}\right\}.$$

Observe that $P = \dfrac{1}{\sqrt{5}}\begin{bmatrix} 2 & -1 \\ 1 & 2 \end{bmatrix}$ is orthogonal and $P^TAP = \begin{bmatrix} 3 & 0 \\ 0 & -2 \end{bmatrix}$. □

■ **EXAMPLE 6.24.** For $A = \begin{bmatrix} 2 & 1 & 1 \\ 1 & 2 & 1 \\ 1 & 1 & 2 \end{bmatrix}$, $\lambda I - A = \begin{bmatrix} \lambda - 2 & -1 & -1 \\ -1 & \lambda - 2 & -1 \\ -1 & -1 & \lambda - 2 \end{bmatrix}$, and expansion by the first row produces

$$\begin{aligned} c(\lambda) &= (\lambda - 2)(\lambda^2 - 4\lambda + 3) + (-\lambda + 1) - (\lambda - 1) \\ &= (\lambda - 1)[(\lambda - 2)(\lambda - 3) - 2] \\ &= (\lambda - 1)(\lambda^2 - 5\lambda + 4) \\ &= (\lambda - 1)^2(\lambda - 4). \end{aligned}$$

Thus, A has eigenvalues 1 and 4. Moreover,

$$1I - A = \begin{bmatrix} -1 & -1 & -1 \\ -1 & -1 & -1 \\ -1 & -1 & -1 \end{bmatrix} \longrightarrow \begin{bmatrix} 1 & 1 & 1 \\ 0 & 0 & 0 \\ 0 & 0 & 0 \end{bmatrix}$$

and

$$4I - A = \begin{bmatrix} 2 & -1 & -1 \\ -1 & 2 & -1 \\ -1 & -1 & 2 \end{bmatrix} \longrightarrow \begin{bmatrix} 1 & 0 & -1 \\ 0 & 1 & -1 \\ 0 & 0 & 0 \end{bmatrix}$$

so

$$\mathcal{V}(1) = \mathcal{S}pan\left\{\begin{bmatrix} -1 \\ 0 \\ 1 \end{bmatrix}, \begin{bmatrix} -1 \\ 1 \\ 0 \end{bmatrix}\right\} \quad \text{and} \quad \mathcal{V}(4) = \mathcal{S}pan\left\{\begin{bmatrix} 1 \\ 1 \\ 1 \end{bmatrix}\right\}.$$

Set

$$X_1 = \begin{bmatrix} -1 \\ 0 \\ 1 \end{bmatrix}, \quad X_2 = \begin{bmatrix} -1 \\ 1 \\ 0 \end{bmatrix}, \quad X_3 = \begin{bmatrix} 1 \\ 1 \\ 1 \end{bmatrix},$$

and note that $X_1 \cdot X_3 = 0 = X_2 \cdot X_3$, but $X_1 \cdot X_2 \neq 0$. The Gram-Schmidt algorithm, applied to $\{X_1, X_2\}$, yields $Y_1 = X_1$ and

$$Y_2 = X_2 - \frac{X_2 \cdot Y_1}{\|Y_1\|^2} Y_1 = \begin{bmatrix} -1 \\ 1 \\ 0 \end{bmatrix} - \frac{1}{2}\begin{bmatrix} -1 \\ 0 \\ 1 \end{bmatrix} = \begin{bmatrix} -1/2 \\ 1 \\ -1/2 \end{bmatrix} = \frac{1}{2}\begin{bmatrix} -1 \\ 2 \\ -1 \end{bmatrix}.$$

Thus,

$$U_1 = Y_1/\|Y_1\| = \frac{1}{\sqrt{2}}\begin{bmatrix} -1 \\ 0 \\ 1 \end{bmatrix} \quad \text{and} \quad U_2 = Y_2/\|Y_2\| = \frac{1}{\sqrt{6}}\begin{bmatrix} -1 \\ 2 \\ -1 \end{bmatrix}$$

form an orthonormal basis for $\mathcal{V}(1)$. Also, $U_3 = X_3/\sqrt{3}$ is a unit vector spanning $\mathcal{V}(4)$, so $\{U_1, U_2, U_3\}$ is an orthonormal basis for $\mathcal{R}^3$ consisting of eigenvectors of A. Thus,

$$P = [\,U_1 \quad U_2 \quad U_3\,] = \begin{bmatrix} -1/\sqrt{2} & -1/\sqrt{6} & 1/\sqrt{3} \\ 0 & 2/\sqrt{6} & 1/\sqrt{3} \\ -1/\sqrt{2} & -1/\sqrt{6} & 1/\sqrt{3} \end{bmatrix}$$

is orthogonal and

$$P^T AP = \begin{bmatrix} 1 & 0 & 0 \\ 0 & 1 & 0 \\ 0 & 0 & 4 \end{bmatrix}. \ \square$$

Exercises 6.5

1. For $A, B, C \in \mathcal{M}_{n\times n}$, show that if A is similar to B and B is similar to C, then A is similar to C.

2. Suppose $A, B \in \mathcal{M}_{n\times n}$ and A is similar to B. Show that
 a. A^T similar to B^T.
 b. A nonsingular implies B is nonsingular and A^{-1} is similar to B^{-1}.

3. Verify the claims made about the eigenvalues and eigenvectors of $\begin{bmatrix} 0 & 1 \\ 1 & 1 \end{bmatrix}$ in Example 6.21. (Hint: $\mu + \nu = 1$ and $\mu\nu = -1$).

4. Find an orthogonal P and diagonal D such that $P^T AP = D$ when A is

a. $\begin{bmatrix} 0 & 1 \\ 1 & 0 \end{bmatrix}$. b. $\begin{bmatrix} 1 & 2 \\ 2 & 1 \end{bmatrix}$. **c.** $\begin{bmatrix} -1 & 2 \\ 2 & 2 \end{bmatrix}$.

d. $\begin{bmatrix} 1 & 0 & 1 \\ 0 & 1 & 0 \\ 1 & 0 & 1 \end{bmatrix}$. **e.** $\begin{bmatrix} 3 & 2 & 4 \\ 2 & 0 & 2 \\ 4 & 2 & 3 \end{bmatrix}$. f. $\begin{bmatrix} 3 & 1 & 1 \\ 1 & 3 & 1 \\ 1 & 1 & 3 \end{bmatrix}$.

g. $\begin{bmatrix} 5 & 0 & 1 & 0 \\ 0 & 2 & 0 & -4 \\ 1 & 0 & 5 & 0 \\ 0 & -4 & 0 & 2 \end{bmatrix}$. h. $\begin{bmatrix} 2 & 0 & 1 & -1 \\ 0 & 2 & 1 & 1 \\ 1 & 1 & 1 & 0 \\ -1 & 1 & 0 & 1 \end{bmatrix}$. **i.** $\begin{bmatrix} 1 & 0 & 2 & 0 \\ 0 & 2 & 0 & 1 \\ 2 & 0 & 1 & 0 \\ 0 & 1 & 0 & 2 \end{bmatrix}$.

5. Suppose A is a diagonalizable $n \times n$ matrix whose distinct eigenvalues are $\lambda_1, \ldots, \lambda_q$, with multiplicities $m_1, \ldots, m_q$. Show that
a. $\det(A) = (\lambda_1)^{m_1} \cdots (\lambda_q)^{m_q}$.
b. $\operatorname{trace}(A) = \sum_1^q m_k \lambda_k$.

6. Suppose $A \in \mathcal{M}_{n \times n}$ is symmetric, with eigenvalues $\lambda_1, \ldots, \lambda_n$. Show that if $\lambda_k = \pm 1$, $k = 1, \ldots, n$, then A is orthogonal.

7. Given $g(x) = a_0 + a_1 x + a_2 x^2 + \cdots + a_q x^q$ and $A \in \mathcal{M}_{n \times n}$, $g(A)$ denotes the $n \times n$ matrix $a_0 I_n + a_1 A + a_2 A^2 + \cdots + a_q A^q$.
a. Show that if A is similar to B, then $g(A)$ is similar to $g(B)$.
b. Show that $g(D[\lambda_1, \ldots, \lambda_n]) = D[g(\lambda_1), \ldots, g(\lambda_n)]$.
c. Show that if A is diagonalizable and c is the characteristic polynomial for A, then $c(A) = 0$.

8. Suppose $\mathcal{V}$ is a nontrivial subspace of $\mathcal{R}^n$, $T(X) = \text{Proj}_{\mathcal{V}}(X)$, and $S(X) = \text{Refl}_{\mathcal{V}}(X)$. Show that $[T]_{\mathcal{E}}$ and $[S]_{\mathcal{E}}$ are symmetric.

9. Solve the following recurrence relations:
a. $a_{k+2} = a_{k+1} + 2a_k, \quad a_0 = a_1 = 1$.
b. $a_{k+2} = -2a_{k+1} + 3a_k, \quad a_0 = 0, a_1 = 1$.

6.6

QUADRATIC FORMS

A quadratic polynomial in two variables looks like

$$F(X) = a + b_1 x_1 + b_2 x_2 + c_1 x_1^2 + c_2 x_1 x_2 + c_3 x_2^2,$$

where $X = [x_1 \quad x_2]^T$ and $a, b_1, b_2, c_1, c_2, c_3 \in \mathcal{R}$. Its terms of degree 2 generate a function $Q : \mathcal{R}^2 \to \mathcal{R}$ with values

$$Q(X) = c_1 x_2^2 + c_2 x_1 x_2 + c_3 x_2^2,$$

called a quadratic form in x_1 and x_2. You may recall that the level curves of Q, that is, the sets

$$\Gamma_d = \{X \in \mathcal{R}^2 : Q(X) = d\},$$

$d \in \mathcal{R}$, are conic sections "centered" at 0 (provided $\Gamma_d \neq \phi$). When $c_2 = 0$ the lines of symmetry of Γ_d are the standard coordinate axes, and otherwise they are obtained by a rotation of those axes.

A quadratic form in three variables is a function $Q : \mathcal{R}^3 \to \mathcal{R}$ defined by

$$Q(X) = c_1 x_1^2 + c_2 x_1 x_2 + c_3 x_2^2 + c_4 x_1 x_3 + c_5 x_3^2 + c_6 x_2 x_3,$$

where $c_1, \ldots, c_6 \in \mathcal{R}$. The expression on the right is simply a linear combination of all possible terms of degree 2 that can be formed from the variables x_1, x_2, and x_3. Again, the level sets

$$\Gamma_d = \{X \in \mathcal{R}^3 : Q(X) = d\}$$

(called quadric surfaces) are familiar geometric objects, including ellipsoids and hyperboloids.

DEFINITION. A *quadratic form* on $\mathcal{R}^n$ is a function $Q : \mathcal{R}^n \to \mathcal{R}$ such that

$$Q(\mathrm{X}) = \sum_{i=1}^{n}\sum_{j=1}^{n} a_{ij}x_i x_j, \tag{6.15}$$

where $a_{ij} \in \mathcal{R}$, $1 \leq i, j \leq n$.

Given Q as in (6.15), set $\mathrm{A} = \begin{bmatrix} a_{11} & \cdots & a_{1n} \\ \vdots & & \vdots \\ a_{n1} & \cdots & a_{nn} \end{bmatrix}$ and observe that

$$\begin{aligned}
\mathrm{X}^T\mathrm{AX} &= [x_1 \quad \cdots \quad x_n]\begin{bmatrix} a_{11} & \cdots & a_{1n} \\ \vdots & & \vdots \\ a_{n1} & \cdots & a_{nn} \end{bmatrix}\begin{bmatrix} x_1 \\ \vdots \\ x_n \end{bmatrix} = [x_1 \quad \cdots \quad x_n]\begin{bmatrix} \sum_{j=1}^{n} a_{1j}x_j \\ \vdots \\ \sum_{j=1}^{n} a_{nj}x_j \end{bmatrix} \\
&= \left[\sum_{j=1}^{n} a_{1j}x_1x_j + \cdots + \sum_{j=1}^{n} a_{nj}x_nx_j\right] \\
&= \left[\sum_{i=1}^{n}\sum_{j=1}^{n} a_{ij}x_ix_j\right].
\end{aligned}$$

That is,

$$Q(\mathrm{X}) = \mathrm{X}^T\mathrm{AX}.$$

■ **EXAMPLE 6.25.** For $\mathrm{X} \in \mathcal{R}^2$,

$$\begin{aligned}
\mathrm{X}^T\begin{bmatrix} 1 & 3 \\ 1 & 6 \end{bmatrix}\mathrm{X} &= [x_1 \quad x_2]\begin{bmatrix} x_1 + 3x_2 \\ x_1 + 6x_2 \end{bmatrix} = x_1^2 + 3x_1x_2 + x_2x_1 + 6x_2^2 \\
&= x_1^2 + 4x_1x_2 + 6x_2^2,
\end{aligned}$$

and

$$\begin{aligned}
\mathrm{X}^T\begin{bmatrix} 1 & 2 \\ 2 & 6 \end{bmatrix}\mathrm{X} &= [x_1 \quad x_2]\begin{bmatrix} x_1 + 2x_2 \\ 2x_1 + 6x_2 \end{bmatrix} = x_1^2 + 2x_1x_2 + 2x_2x_1 + 6x_2^2 \\
&= x_1^2 + 4x_1x_2 + 6x_2^2.
\end{aligned}$$

In fact,

$$Q(\mathrm{X}) = x_1^2 + 4x_1x_2 + 6x_2^2 = \mathrm{X}^T\begin{bmatrix} 1 & a \\ b & 6 \end{bmatrix}\mathrm{X}$$

for any $a,b \in \mathcal{R}$ satisfying $a + b = 4$. □

Let $Q : \mathcal{R}^n \to \mathcal{R}$ be a quadratic form and assume A is an $n \times n$ matrix such that

$$Q(\mathrm{X}) = \mathrm{X}^T\mathrm{AX} = \sum_{i=1}^{n}\sum_{j=1}^{n} a_{ij}x_ix_j. \tag{6.16}$$

When $i \neq j$, $a_{ij}x_ix_j$ and $a_{ji}x_jx_i$ combine to form a single term of degree 2 in x_i and x_j, namely,

$$(a_{ij} + a_{ji})x_ix_j,$$

and there are many ways to choose the entries a_{ij} and a_{ji} without changing the value of $a_{ij} + a_{ji}$. However, by insisting that $a_{ij} = a_{ji}$, $i \neq j$, you obtain a unique symmetric matrix that generates the values of Q.

■ **EXAMPLE 6.26.** If $Q(\mathrm{X}) = x_1^2 + x_1x_2 + 2x_2^2 + 6x_1x_3 + 4x_3^2 - 5x_2x_3$, then the symmetric 3×3 matrix satisfying $Q(\mathrm{X}) = \mathrm{X}^T\mathrm{AX}$ is

$$\mathrm{A} = \begin{bmatrix} 1 & 1/2 & 3 \\ 1/2 & 2 & -5/2 \\ 3 & -5/2 & 4 \end{bmatrix}.$$

Note that for $i \neq j$ the entry in the i, j position of A is half the coefficient of x_ix_j in Q. □

■ **EXAMPLE 6.27.** If $Q : \mathcal{R}^4 \to \mathcal{R}$ is the quadratic form with equation

$$Q(\mathrm{X}) = -x_1^2 + 2x_1x_2 + 2x_2^2 - 10x_2x_3 - 4x_3x_4 + 3x_4^2 + 2\sqrt{2}x_1x_4,$$

then

$$Q(\mathrm{X}) = \mathrm{X}^T \begin{bmatrix} -1 & 1 & 0 & \sqrt{2} \\ 1 & 2 & -5 & 0 \\ 0 & -5 & 0 & -2 \\ \sqrt{2} & 0 & -2 & 3 \end{bmatrix} \mathrm{X}. \square$$

Suppose A is a symmetric $n \times n$ matrix and $Q(\mathrm{X}) = \mathrm{X}^T\mathrm{AX}$, $\mathrm{X} \in \mathcal{R}^n$. According to Theorem 6.15, $\mathcal{R}^n$ has an orthonormal basis $\mathcal{A} = \{\mathrm{U}_1, \ldots, \mathrm{U}_n\}$ consisting of eigenvectors of A. If $\mathrm{AU}_k = \lambda_k\mathrm{U}_k$, $k = 1, \ldots, n$, and $\mathrm{P} = [\mathrm{U}_1 \ \cdots \ \mathrm{U}_n]$, then

$$\mathrm{P}^T\mathrm{AP} = \begin{bmatrix} \lambda_1 & & 0 \\ & \ddots & \\ 0 & & \lambda_n \end{bmatrix}.$$

Recall that $\mathrm{P} = \mathrm{P}_{\mathcal{E}\mathcal{A}}$. Setting $\mathrm{Y} = [\mathrm{X}]_{\mathcal{A}}$, you have $\mathrm{PY} = \mathrm{P}_{\mathcal{E}\mathcal{A}}[\mathrm{X}]_{\mathcal{A}} = [\mathrm{X}]_{\mathcal{E}} = \mathrm{X}$, and therefore

$$\begin{aligned} \mathrm{X}^T\mathrm{AX} = (\mathrm{PY})^T\mathrm{A}(\mathrm{PY}) = \mathrm{Y}^T(\mathrm{P}^T\mathrm{AP})\mathrm{Y} = \mathrm{Y}^T \begin{bmatrix} \lambda_1 & & 0 \\ & \ddots & \\ 0 & & \lambda_n \end{bmatrix} \mathrm{Y} \\ = \lambda_1 y_1^2 + \cdots + \lambda_n y_n^2. \end{aligned} \tag{6.17}$$

The last expression gives the value of $Q(\mathrm{X})$ in terms of the $\mathcal{A}$ coordinates of X, and you will note that it contains no cross terms (terms of the form cy_iy_j, $i \neq j$). Thus, the behavior of Q is easier to analyze when viewed from the $\mathcal{A}$ coordinate system.

■ **EXAMPLE 6.28.** Suppose $\mathrm{A} = \begin{bmatrix} 1 & 2 \\ 2 & 1 \end{bmatrix}$ and $Q(\mathrm{X}) = \mathrm{X}^T\mathrm{AX} = x_1^2 + 4x_1x_2 + x_2^2$. Since

$$c(\lambda) = \det \begin{bmatrix} \lambda - 1 & -2 \\ -2 & \lambda - 1 \end{bmatrix} = \lambda^2 - 2\lambda - 3 = (\lambda - 3)(\lambda + 1),$$

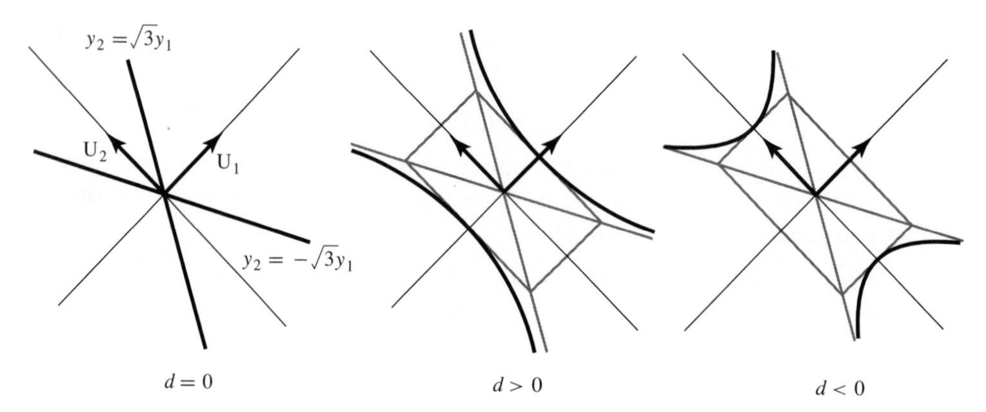

FIGURE 6.2

A has eigenvalues 3 and -1. Moreover,

$$3\mathrm{I} - \mathrm{A} = \begin{bmatrix} 2 & -2 \\ -2 & 2 \end{bmatrix} \longrightarrow \begin{bmatrix} 1 & -1 \\ 0 & 0 \end{bmatrix}$$

and

$$(-1)\mathrm{I} - \mathrm{A} = \begin{bmatrix} -2 & -2 \\ -2 & -2 \end{bmatrix} \longrightarrow \begin{bmatrix} 1 & 1 \\ 0 & 0 \end{bmatrix},$$

so

$$\mathcal{V}(3) = \mathcal{S}pan\left\{\begin{bmatrix} 1 \\ 1 \end{bmatrix}\right\} \quad \text{and} \quad \mathcal{V}(-1) = \mathcal{S}pan\left\{\begin{bmatrix} -1 \\ 1 \end{bmatrix}\right\}.$$

Thus, $\mathcal{A} = \{\mathrm{U}_1, \mathrm{U}_2\} = \left\{\frac{1}{\sqrt{2}}\begin{bmatrix} 1 \\ 1 \end{bmatrix}, \frac{1}{\sqrt{2}}\begin{bmatrix} -1 \\ 1 \end{bmatrix}\right\}$ is an orthonormal basis for $\mathcal{R}^2$ consisting of eigenvectors for A. Setting $\mathrm{P} = [\mathrm{U}_1\ \mathrm{U}_2]$ and $\mathrm{X} = \mathrm{PY}$ gives

$$\mathrm{P}^T\mathrm{AP} = \begin{bmatrix} 3 & 0 \\ 0 & -1 \end{bmatrix}$$

and

$$\mathrm{X}^T\mathrm{AX} = \mathrm{Y}^T\begin{bmatrix} 3 & 0 \\ 0 & -1 \end{bmatrix}\mathrm{Y} = 3y_1^2 - y_2^2.$$

Now, for $d \in \mathcal{R}$, $\Gamma_d = \{\mathrm{X} \in \mathcal{R}^2 : Q(\mathrm{X}) = d\}$ is the conic section whose equation in the $\mathcal{A}$ coordinate system is

$$3y_1^2 - y_2^2 = d. \tag{6.18}$$

When $d = 0$, (6.18) becomes $|y_2| = \sqrt{3}\,|y_1|$, that is, Γ_0 consists of two lines intersecting at 0 with slopes $\pm\sqrt{3}$. If $d \neq 0$, then (6.18) is the equation of a hyperbola whose foci lie on the U_1 axis or U_2 axis, depending on whether $d > 0$ or $d < 0$, respectively. Figure 6.2 illustrates the three cases. □

■ **EXAMPLE 6.29.** Consider $Q : \mathcal{R}^3 \to \mathcal{R}$ with equation

$$Q(\mathrm{X}) = 2x_1^2 + 2x_1x_2 + 2x_2^2 + 2x_1x_3 + 2x_3^2 + 2x_2x_3.$$

Observe that $Q(X) = X^T AX$, where $A = \begin{bmatrix} 2 & 1 & 1 \\ 1 & 2 & 1 \\ 1 & 1 & 2 \end{bmatrix}$. Example 6.24 showed that

$$U_1 = \frac{1}{\sqrt{2}}\begin{bmatrix} -1 \\ 0 \\ 1 \end{bmatrix}, \quad U_2 = \frac{1}{\sqrt{6}}\begin{bmatrix} -1 \\ 2 \\ -1 \end{bmatrix}, \quad \text{and} \quad U_3 = \frac{1}{\sqrt{3}}\begin{bmatrix} 1 \\ 1 \\ 1 \end{bmatrix}$$

are orthonormal eigenvectors of A with eigenvalues 4, 4, and 1, respectively. If $P = [U_1 \quad U_2 \quad U_3]$ and $X = PY$, then

$$X^T AX = Y^T \begin{bmatrix} 4 & 0 & 0 \\ 0 & 4 & 0 \\ 0 & 0 & 1 \end{bmatrix} Y = 4y_1^2 + 4y_2^2 + y_3^2.$$

The equation of $\Gamma_d = \{X \in \mathcal{R}^3 : Q(X) = d\}$ in the $\mathcal{A}$ coordinate system is

$$4y_1^2 + 4y_2^2 + y_3^2 = d. \tag{6.19}$$

It has no solutions when $d < 0$ (i.e., $\Gamma_d = \varnothing$). For $d = 0$, the only solution is $Y = 0$, so $\Gamma_0 = \{0\}$. If $d > 0$, then Γ_d is an ellipsoid. It has circular cross sections orthogonal to the U_3 axis and elliptical cross sections orthogonal to the U_1 or U_2 axes. □

Exercises 6.6

1. Find the symmetric $n \times n$ matrix A such that $Q(X) = X^T AX$ when
   a. $Q(X) = 2x_1^2 - 2x_1x_2 - 3x_2^2, \quad n = 2.$
   b. $Q(X) = -4x_1x_2, \quad n = 2.$
   c. $Q(X) = x_1^2 + x_1x_2 - x_2^2 - 3x_1x_3 + 2x_3^2 - 2x_2x_3, \quad n = 3.$
   d. $Q(X) = 4x_2x_3 + 6x_1x_2 - 8x_1x_3, \quad n = 3.$
   e. $Q(X) = (x_1 - x_2)^2 + (x_2 - x_3)^2 + (x_1 - x_3)^2, \quad n = 3.$
   f. $Q(X) = (x_1 + 2x_2)(x_2 - x_3), \quad n = 3.$
   g. $Q(X) = ax_1^2 + bx_2^2 + cx_3^2, \quad n = 3.$

2. For the given A, find an orthonormal basis $\mathcal{A}$ for $\mathcal{R}^n$ with the property that the expression for $Q(X) = X^T AX$ in $\mathcal{A}$ coordinates has no cross terms.

   a. $\begin{bmatrix} 11 & 3 \\ 3 & 19 \end{bmatrix}$ b. $\begin{bmatrix} 0 & 1 \\ 1 & 0 \end{bmatrix}$ **c.** $\begin{bmatrix} 3 & 0 & -1 \\ 0 & 4 & 0 \\ -1 & 0 & 3 \end{bmatrix}$

   d. $\begin{bmatrix} -3 & 1 & 2 \\ 1 & -3 & 2 \\ 2 & 2 & -4 \end{bmatrix}$ **e.** $\begin{bmatrix} 0 & 0 & -1 \\ 0 & 0 & 1 \\ -1 & 1 & 0 \end{bmatrix}$ f. $\begin{bmatrix} 2 & 1 & 1 & 1 \\ 1 & 2 & 1 & 1 \\ 1 & 1 & 2 & 1 \\ 1 & 1 & 1 & 2 \end{bmatrix}$.

3. If $Q(X) = X^T \begin{bmatrix} 0 & 4 \\ 4 & 6 \end{bmatrix} X, \ X \in \mathcal{R}^2$, sketch

   a. $\Gamma_0 = \{X : Q(X) = 0\}$. b. $\Gamma_8 = \{X : Q(X) = 8\}$.

4. Assuming $Q(\mathrm{X}) = \mathrm{X}^T \begin{bmatrix} -11 & 12 \\ 12 & -4 \end{bmatrix} \mathrm{X}$, $\mathrm{X} \in \mathcal{R}^2$, sketch

   a. $\{\mathrm{X} : Q(\mathrm{X}) < 0\}$. b. $\Gamma_{20} = \{\mathrm{X} : Q(\mathrm{X}) = 20\}$.

5. Let $Q(\mathrm{X}) = \mathrm{X}^T \begin{bmatrix} 11 & 3 \\ 3 & 19 \end{bmatrix} \mathrm{X}$, $\mathrm{X} \in \mathcal{R}^2$. Sketch

   a. $\Gamma_0 = \{\mathrm{X} : Q(\mathrm{X}) = 0\}$. b. $\Gamma_{100} = \{\mathrm{X} : Q(\mathrm{X}) = 100\}$.

6. Assuming $Q(\mathrm{X}) = \mathrm{X}^T \begin{bmatrix} 1 & -4 & 8 \\ -4 & 7 & 4 \\ 8 & 4 & 1 \end{bmatrix} \mathrm{X}$, $\mathrm{X} \in \mathcal{R}^3$, describe

   a. $\Gamma_0 = \{\mathrm{X} : Q(\mathrm{X}) = 0\}$. b. $\Gamma_9 = \{\mathrm{X} : Q(\mathrm{X}) = 9\}$.

7. If $Q(\mathrm{X}) = \mathrm{X}^T \begin{bmatrix} 1 & -1 & 0 \\ -1 & 1 & 0 \\ 0 & 0 & 4 \end{bmatrix} \mathrm{X}$, $\mathrm{X} \in \mathcal{R}^3$, describe

   a. $\Gamma_0 = \{\mathrm{X} : Q(\mathrm{X}) = 0\}$. b. $\Gamma_4 = \{\mathrm{X} : Q(\mathrm{X}) = 4\}$.

8. Suppose $\mathrm{A,B} \in \mathcal{M}_{n\times n}$ and $\mathrm{X}^T\mathrm{AX} = \mathrm{X}^T\mathrm{BX}$ for all $\mathrm{X} \in \mathcal{R}^n$.
   a. Show that $a_{kk} = b_{kk}$, $k = 1, \ldots, n$.
   b. Show that if A and B are both symmetric, then $\mathrm{A} = \mathrm{B}$.

6.7
POSITIVE DEFINITE FORMS

The present section is concerned with analyzing the sign of a quadratic form. This topic arises, for example, in the classification of critical points of a real-valued function. A careful treatment of that material belongs to calculus, but the following brief description of it is very helpful in motivating the material that follows.

Assume $\mathrm{J} \subseteq \mathcal{R}$ is an open interval centered at 0 and consider a function $f : \mathrm{J} \to \mathcal{R}$ that has a third derivative at each point of J. Taylor's theorem states that for each $x \in \mathrm{J}$ there is a number $t(x)$ between 0 and x such that

$$f(x) = f(0) + f'(0)x + \frac{f''(0)}{2!}x^2 + \frac{f'''(t(x))}{3!}x^3. \tag{6.20}$$

The Taylor polynomial of order 2 for f is

$$p(x) = f(0) + f'(0)x + \frac{f''(0)}{2!}x^2,$$

and

$$E(x) = \frac{f'''(t(x))}{3!}x^3$$

is the error committed when using $p(x)$ as an approximation to $f(x)$. Recall that p is the unique polynomial of degree less than or equal to 2 such that

$$p(0) = f(0), \quad p'(0) = f'(0), \quad \text{and} \quad p''(0) = f''(0).$$

You might expect $p(x)$ to provide a good approximation to $f(x)$ when x is near 0, and indeed it does. It can be shown under mild assumptions on f that

$$\lim_{x \to 0} \frac{E(x)}{x^2} = \lim_{x \to 0} \frac{f(x) - p(x)}{x^2} = 0. \tag{6.21}$$

In other words, the error term approaches 0 more quickly than does x^2 as x tends to 0.

Suppose f has a critical point at $x = 0$. Then $f'(0) = 0$, and (6.20) reduces to

$$f(x) = f(0) + \frac{f''(0)}{2!}x^2 + \mathrm{E}(x). \tag{6.22}$$

Now, f has a local maximum or minimum at 0 when $f(x) \le f(0)$ or $f(x) \ge f(0)$, respectively, for all x sufficiently near 0. Whether one of these comparisons holds depends, by (6.22), on the sign of

$$\frac{f''(0)}{2!}x^2 + E(x) = x^2\left(\frac{f''(0)}{2!} + \frac{E(x)}{x^2}\right), \tag{6.23}$$

and (6.21) shows that (6.23) has the same sign as $f''(0)$ when x is sufficiently near 0. Thus, $f''(0) < 0$ or $f''(0) > 0$ implies that $f(0)$ is a local maximum or local minimum, respectively. These results are known as the second derivative test for local extrema. This particular derivation is not the one you usually find in calculus texts, but it provides an approach to classifying local extrema that generalizes to functions of several variables.

Suppose $\Delta \subseteq \mathcal{R}^2$ is a disk centered at 0 and $f : \Delta \to \mathcal{R}$ has partial derivatives of order 3 at each point of Δ. The Taylor polynomial for f of order 2 is

$$p(\mathrm{X}) = f(0) + f_x(0)x + f_y(0)y + \tfrac{1}{2}f_{xx}(0)x^2 + f_{xy}(0)xy + \tfrac{1}{2}f_{yy}(0)y^2,$$

where $\mathrm{X} = [x \quad y]^T$. It is the unique polynomial in two variables of degree at most 2 satisfying

$$p(0) = f(0), \quad p_x(0) = f_x(0), \quad p_y(0) = f_y(0),$$

$$p_{xx}(0) = f_{xx}(0), \quad p_{xy}(0) = f_{xy}(0), \quad \text{and} \quad p_{yy}(0) = f_{yy}(0).$$

Taylor's theorem for functions of two variables states that

$$f(\mathrm{X}) = p(\mathrm{X}) + E(\mathrm{X}), \tag{6.24}$$

where $E(\mathrm{X})$ is an error term with the property that

$$\lim_{\mathrm{X} \to 0} \frac{E(\mathrm{X})}{\|\mathrm{X}\|^2} = 0. \tag{6.25}$$

If f has a critical point at $\mathrm{X} = 0$, then $f_x(0) = 0 = f_y(0)$, and (6.24) becomes

$$f(\mathrm{X}) = f(0) + \tfrac{1}{2}\left(f_{xx}(0)x^2 + 2f_{xy}(0)xy + f_{yy}(0)y^2\right) + E(\mathrm{X}).$$

Note that,

$$Q(\mathrm{X}) = f_{xx}(0)x^2 + 2f_{xy}(0)xy + f_{yy}(0)y^2$$

is a quadratic form on $\mathcal{R}^2$ and

$$f(\mathrm{X}) = f(0) + \left(\tfrac{1}{2}Q(\mathrm{X}) + \mathrm{E}(\mathrm{X})\right).$$

Comparing $f(\mathrm{X})$ to $f(0)$ amounts to examining the sign of $\frac{1}{2}Q(\mathrm{X}) + \mathrm{E}(\mathrm{X})$, and it can be shown, thanks to (6.25), that when X is near 0 the dominant term in the last expression is $\frac{1}{2}Q(\mathrm{X})$. Thus, whether $f(0)$ is a local maximum or local minimum depends on the sign of $Q(\mathrm{X})$ when X is close to 0.

Let $\mathrm{A} \in \mathcal{M}_{n\times n}$ and $Q(\mathrm{X}) = \mathrm{X}^T\mathrm{A}\mathrm{X}$, $\mathrm{X} \in \mathcal{R}^n$. Note that $Q(0) = 0$. For $\mathrm{X}_0 \neq 0$,

$$Q(t\mathrm{X}_0) = (t\mathrm{X}_0)^T\mathrm{A}(t\mathrm{X}_0) = t^2(\mathrm{X}_0^T\mathrm{A}\mathrm{X}_0) = t^2Q(\mathrm{X}_0).$$

Thus, either $Q(\mathrm{X})$ and $Q(\mathrm{X}_0)$ have the same sign for all nonzero $\mathrm{X} \in \mathcal{S}pan\{\mathrm{X}_0\}$, or $Q(\mathrm{X})$ is identically 0 along $\mathcal{S}pan\{\mathrm{X}_0\}$.

DEFINITION. Suppose Q is a quadratic form on $\mathcal{R}^n$.

1. If $Q(\mathrm{X}) > 0$ (< 0) for all $\mathrm{X} \neq 0$, then Q is said to be *positive* (*negative*) *definite*.
2. If $Q(\mathrm{X}) \geq 0$ (≤ 0) for all X, then Q is *positive* (*negative*) *semidefinite*.
3. When Q has both positive and negative values, Q is *not semidefinite*.

■ **EXAMPLE 6.30.** Since $Q(\mathrm{X}) = 4x_1^2 - 4x_1x_2 + x_2^2 = (2x_1 - x_2)^2 \geq 0$ on $\mathcal{R}^2$, Q is positive semidefinite. It is not positive definite because $Q(\mathrm{X}) = 0$ along the line $x_2 = 2x_1$. □

■ **EXAMPLE 6.31.** If $Q : \mathcal{R}^2 \to \mathcal{R}$ is given by

$$Q(\mathrm{X}) = x_1^2 + 2x_1x_2 - 3x_2^2 = (x_1 + 3x_2)(x_1 - x_2),$$

then $Q(\mathrm{X}) = 0$ when $x_1 = -3x_2$ or $x_1 = x_2$. These lines divide $\mathcal{R}^2$ into four sectors. In two of them $Q(\mathrm{X})$ is positive; in the other two $Q(\mathrm{X})$ is negative (Figure 6.3). Thus, Q is not semidefinite. □

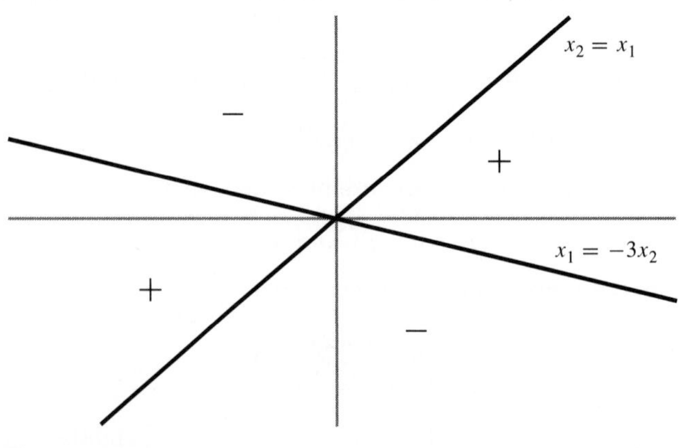

FIGURE 6.3

■ **EXAMPLE 6.32.** $Q(\mathrm{X}) = 3x_1^2 + x_2^2 + 2x_3^2$ is positive definite on $\mathcal{R}^3$; it is 0 at $\mathrm{X} = 0$ and positive elsewhere. $\mathrm{P(X)} = 3x_1^2 + x_2^2$ is positive semidefinite. Its values are nonnegative, but $\mathrm{P}(t\mathrm{E}_3) = 0, t \in \mathcal{R}$. $R(\mathrm{X}) = 3x_1^2 + x_2^2 - 2x_3^2$ is not semidefinite. If $\mathrm{X} = s\mathrm{E}_1 + t\mathrm{E}_2$ and s and t are not both zero, then $R(\mathrm{X}) = 3s^2 + t^2 > 0$, but $R(t\mathrm{E}_3) = -2t^2 < 0$ when $t \neq 0$. □

The last example illustrates how the lack of cross terms in a quadratic form makes its status with regard to definiteness easy to establish. Assuming A is symmetric, $Q(\mathrm{X}) = \mathrm{X}^T\mathrm{AX}$ lacks cross terms precisely when A is diagonal, a rather special case. But symmetric matrices are diagonalizable, so the same sort of analysis can be applied to any quadratic form as long as you express its values in terms of coordinates with respect to an orthonormal basis of eigenvectors.

THEOREM 6.16. If $\mathrm{A} \in \mathcal{M}_{n\times n}$ is symmetric, with eigenvalues $\lambda_1, \ldots, \lambda_n$, and $Q(\mathrm{X}) = \mathrm{X}^T\mathrm{AX}$, then Q is

(1) positive (negative) definite if and only if $\lambda_k > 0\ (< 0)$, $1 \leq k \leq n$.
(2) positive (negative) semidefinite if and only if $\lambda_k \geq 0\ (\leq 0)$, $1 \leq k \leq n$.
(3) not semidefinite if and only if A has at least one positive and one negative eigenvalue.

Proof. Let $\mathcal{A} = \{\mathrm{U}_1, \ldots, \mathrm{U}_n\}$ be an orthonormal basis for $\mathcal{R}^n$ such that $\mathrm{AU}_k = \lambda_k\mathrm{U}_k$, $k = 1, \ldots, n$. Set $\mathrm{P} = [\mathrm{U}_1 \ \cdots \ \mathrm{U}_n] = \mathrm{P}_{\mathcal{E}\mathcal{A}}$ and $\mathrm{X} = \mathrm{PY}$. Then

$$Q(\mathrm{X}) = \mathrm{X}^T\mathrm{AX} = \mathrm{Y}^T\begin{bmatrix} \lambda_1 & & 0 \\ & \ddots & \\ 0 & & \lambda_n \end{bmatrix}\mathrm{Y} = \lambda_1 y_1^2 + \cdots + \lambda_n y_n^2. \tag{6.26}$$

(1) If $\lambda_k > 0$, $1 \leq k \leq n$, then $\lambda_1 y_1^2 + \cdots + \lambda_n y_n^2 > 0$ for $\mathrm{Y} \neq 0$. Moreover, $\mathrm{X} \neq 0$ implies $\mathrm{Y} \neq 0$ $(\mathrm{X} = \mathrm{PY})$, so $Q(\mathrm{X}) > 0$ whenever $\mathrm{X} \neq 0$. Thus, Q is positive definite. Conversely, if Q is positive definite, then for $k = 1, \ldots, n$,

$$0 < Q(\mathrm{U}_k) = \mathrm{U}_k^T\mathrm{AU}_k = \mathrm{U}_k^T(\lambda_k\mathrm{U}_k) = \lambda_k\|\mathrm{U}_k\|^2 = \lambda_k.$$

The conclusion relating negative definiteness with the conditions $\lambda_k < 0$, $k = 1, \ldots, n$, is obtained in a completely analogous manner.

The proof of (2) involves only minor modifications in the proof of (1), and (3) is left as an exercise. ■

■ **EXAMPLE 6.33.** The eigenvalues of $\begin{bmatrix} 1 & 2 \\ 2 & 1 \end{bmatrix}$ are 3 and -1 (Example 6.28), so $Q(\mathrm{X}) = \mathrm{X}^T\begin{bmatrix} 1 & 2 \\ 2 & 1 \end{bmatrix}\mathrm{X}$ is not semidefinite. □

■ **EXAMPLE 6.34.** If $\mathrm{A} = \begin{bmatrix} 2 & 1 \\ 1 & 2 \end{bmatrix}$, then $\det(\lambda\mathrm{I} - \mathrm{A}) = \lambda^2 - 4\lambda + 3 = (\lambda - 3)(\lambda - 1)$.

Both eigenvalues are positive, so $Q(\mathrm{X}) = \mathrm{X}^T\begin{bmatrix} 2 & 1 \\ 1 & 2 \end{bmatrix}\mathrm{X}$ is positive definite. □

■ **EXAMPLE 6.35.** Consider the general quadratic form on $\mathcal{R}^2$ given by $Q(\mathrm{X}) = \mathrm{X}^T\mathrm{AX}$, where

$$\mathrm{A} = \begin{bmatrix} a & b \\ b & c \end{bmatrix}.$$

Suppose $c(\lambda) = \det(\lambda I - A) = (\lambda - \lambda_1)(\lambda - \lambda_2)$. Since $(-1)^2\det(A) = \det(-A) = c(0) = \lambda_1\lambda_2$,

$$\det(A) = \lambda_1\lambda_2,$$

and therefore $\det(A) < 0$ if and only if λ_1 and λ_2 have opposite signs. Thus, $\det(A) < 0$ indicates Q is not semidefinite (Theorem 6.16). When $\det(A) > 0$, λ_1 and λ_2 are both positive or both negative, so Q is either positive or negative definite. You can decide which is the case by evaluating Q at any nonzero X. Since

$$Q(E_1) = E_1^T \begin{bmatrix} a & b \\ b & c \end{bmatrix} E_1 = a,$$

the outcome depends on the sign of a. In summary,

$$Q \text{ is positive (negative) definite} \iff \det(A) > 0 \text{ and } a > 0\ (a < 0),$$

$$Q \text{ is not semidefinite} \iff \det(A) < 0.$$

□

■ **EXAMPLE 6.36.** Suppose $\Delta \subseteq \mathcal{R}^2$ is a disk centered at 0 and $f : \Delta \to \mathcal{R}$ satisfies the hypotheses of Taylor's theorem. Let Q be the quadratic form occuring in the Taylor polynomial for f of order 2, that is,

$$Q(X) = f_{xx}(0)x^2 + 2f_{xy}(0)xy + f_{yy}(0)y^2.$$

Then $Q(X) = X^TAX$, where $X = [x \quad y]^T$ and

$$A = \begin{bmatrix} f_{xx}(0) & f_{xy}(0) \\ f_{xy}(0) & f_{yy}(0) \end{bmatrix}.$$

Assuming 0 is a critical point of f, $f(0)$ is a local maximum or minimum when Q is negative definite or positive definite, respectively, and $f(0)$ is a saddle point when Q is not semidefinite. According to Example 6.35 these findings can be summarized as follows:

1. $f(0)$ is a local maximum when $f_{xx}(0)f_{yy}(0) - f_{xy}(0)^2 > 0$ and $f_{xx}(0) < 0$.
2. $f(0)$ is a local minimum when $f_{xx}(0)f_{yy}(0) - f_{xy}(0)^2 > 0$ and $f_{xx}(0) > 0$.
3. $f(0)$ is a saddle point when $f_{xx}(0)f_{yy}(0) - f_{xy}(0)^2 < 0$.

Collectively these conclusions are known as the second derivative test for local extrema of a real-valued function of two variables. □

DEFINITION. A symmetric $n \times n$ matrix A is said to be *positive (negative) definite*, *positive (negative) semidefinite*, or *not semidefinite* according as the $Q(X) = X^TAX$, $X \in \mathcal{R}^n$, is positive (negative) definite, positive (negative) semidefinite, or not semidefinite, respectively.

Definiteness for a symmetric 2×2 matrix A is determined by the signs of a_{11} and $\det(A)$ (Example 6.35). The next result is a similar test for larger matrices. Parts of its proof can be found in the exercises.

THEOREM 6.17. If $A \in \mathcal{M}_{n\times n}$ is symmetric and $A_k = \begin{bmatrix} a_{11} & \cdots & a_{1k} \\ \vdots & & \vdots \\ a_{k1} & \cdots & a_{kk} \end{bmatrix}$, $k = 1, \ldots, n$, then

(1) A is positive definite $\Leftrightarrow \det(A_k) > 0$, for each $k = 1, \ldots, n$, and
(2) A is negative definite $\Leftrightarrow (-1)^k \det(A_k) > 0$, for each $k = 1, \ldots, n$.

■ **EXAMPLE 6.37.** If $A = \begin{bmatrix} 2 & -1 & 1 \\ -1 & 2 & -2 \\ 1 & -2 & 5 \end{bmatrix}$, then $A_1 = [2]$, $A_2 = \begin{bmatrix} 2 & -1 \\ -1 & 2 \end{bmatrix}$, and

$A_3 = A$. Since $\det(A_1) = 2 > 0$, $\det(A_2) = 3 > 0$, and $\det(A_3) = 9 > 0$, A is positive definite. □

■ **EXAMPLE 6.38.** When $A = \begin{bmatrix} -4 & 0 & -2 & 0 \\ 0 & -2 & 0 & 1 \\ -2 & 0 & -3 & 1 \\ 0 & 1 & 1 & -2 \end{bmatrix}$,

$$A_1 = [-4], \quad A_2 = \begin{bmatrix} -4 & 0 \\ 0 & -2 \end{bmatrix}, \quad A_3 = \begin{bmatrix} -4 & 0 & -2 \\ 0 & -2 & 0 \\ -2 & 0 & -3 \end{bmatrix}, \text{ and } A_4 = A.$$

Since $\det(A_1) = -4$, $\det(A_2) = 8$, $\det(A_3) = -16$, and $\det(A_4) = 16$, A is negative definite. □

Exercises 6.7

1. Classify the quadratic forms in Problem 2 of Exercises 6.6 as positive or negative definite, positive or negative semidefinite, or not semidefinite.

2. Describe geometrically the set $\{X \in \mathcal{R}^3 : Q(X) > 0\}$ when
   a. $Q(X) = x_1^2 + x_2^2 - x_3^2$.
   b. $Q(X) = 3x_1^2 - 2x_2^2 + x_3^2$.

3. Prove part (3) of Theorem 6.16.

4. Check $Q(X) = X^T A X$ for positive or negative definiteness when A is

a. $\begin{bmatrix} 2 & 1 & 0 \\ 1 & 1 & 1 \\ 0 & 1 & 3 \end{bmatrix}$. b. $\begin{bmatrix} -1 & 1 & 0 \\ 1 & -2 & 1 \\ 0 & 1 & -3 \end{bmatrix}$. **c.** $\begin{bmatrix} -1 & 1 & -1 \\ 1 & -3 & 2 \\ -1 & 2 & -3 \end{bmatrix}$.

d. $\begin{bmatrix} 3 & 0 & -1 & 0 \\ 0 & 3 & 0 & 1 \\ -1 & 0 & 3 & 0 \\ 0 & 1 & 0 & 3 \end{bmatrix}$. **e.** $\begin{bmatrix} 2 & -1 & 1 & 0 \\ -1 & 1 & 0 & 1 \\ 1 & 0 & 2 & 1 \\ 0 & 1 & 1 & 3 \end{bmatrix}$. f. $\begin{bmatrix} -10 & -2 & 0 & 2 \\ -2 & -13 & 0 & 4 \\ 0 & 0 & -9 & 0 \\ 2 & 4 & 0 & -13 \end{bmatrix}$.

5. Assuming A and B are symmetric, positive definite, $n \times n$ matrices, determine which if any of the following matrices are positive definite.
   a. $A + B$ b. tA, $t \in \mathcal{R}$ c. A^{-1} (A nonsingular)

6. Find symmetric 2×2 matrices A and B such that $\det(\mathrm{A}) = 0 = \det(\mathrm{B})$ and $Q(\mathrm{X}) = \mathrm{X}^T\mathrm{AX}$ is positive semidefinite but $R(\mathrm{X}) = \mathrm{X}^T\mathrm{BX}$ is negative semidefinite.

7. Assuming $\mathrm{A} \in \mathcal{M}_{n\times n}$ is nonsingular, show that $\mathrm{B} = \mathrm{A}^T\mathrm{A}$ is positive definite.

8. a. Show A is positive definite if and only if $-\mathrm{A}$ is negative definite.
b. Use part (a) to show that (1) $\Rightarrow$ (2) in Theorem 6.17.

9. Suppose $\mathrm{A} \in \mathcal{M}_{n\times n}$ is symmetric. Show that A positive definite implies $a_{kk} > 0$, $k = 1, \ldots, n$. What are the corresponding conclusions when A is positive semidefinite, negative definite, or negative semidefinite?

10. Show that if $\mathrm{A} \in \mathcal{M}_{n\times n}$ is symmetric and positive definite, then $\det(\mathrm{A}) > 0$.

11. Assume $\mathrm{A} \in \mathcal{M}_{n\times n}$ is symmetric and positive definite, $\mathrm{A}_k = \begin{bmatrix} a_{11} & \cdots & a_{1k} \\ \vdots & & \vdots \\ a_{k1} & \cdots & a_{kk} \end{bmatrix}$, and Q_k is the quadratic form on $\mathcal{R}^k$ given by $Q_k(\mathrm{Z}) = \mathrm{Z}^T\mathrm{A}_k\mathrm{Z}$, $1 \le k \le n$.

a. Show that Q_k is positive definite. (Hint: If $\mathrm{Z} \in \mathcal{R}^k$, $0 \in \mathcal{R}^{n-\mathrm{k}}$, and $\mathrm{X} = \begin{bmatrix} \mathrm{Z} \\ \hline 0 \end{bmatrix}$, then $Q(\mathrm{X}) = Q_k(\mathrm{Z})$.)
b. Show that $\det(\mathrm{A}_k) > 0$, $1 \le k \le n$.

12. Suppose $\mathrm{A} \in \mathcal{M}_{n\times n}$ is symmetric and positive definite. Show that there is a symmetric B such that $\mathrm{B}^2 = \mathrm{A}$ (i.e., A has a square root).

13. Let $Q(\mathrm{X}) = ax_1^2 + bx_1x_2 + cx_2^2$, where $\mathrm{X} = \begin{bmatrix} x_1 \\ x_2 \end{bmatrix}$ and $a,b,c \in \mathcal{R}$, and assume $\mathrm{P} \in \mathcal{M}_{2\times 2}$ is orthogonal. Substituting $\mathrm{X} = \mathrm{PY}$ produces an expression for Q of the form $\alpha y_1^2 + \beta y_1 y_2 + \gamma y_2^2$, where $\mathrm{Y} = \begin{bmatrix} y_1 \\ y_2 \end{bmatrix}$ and $\alpha,\beta,\gamma \in \mathcal{R}$. Show that $b^2 - 4ac = \beta^2 - 4\alpha\gamma$.
($b^2 - 4ac$ is called the discriminant of Q. It is invariant under an orthogonal change of coordinates.)

7

Abstract Vector Spaces

7.1 INTRODUCTION

The development of Euclidean n-space, as presented in Chapter 2, was largely motivated by the need to create a mathematical context in which to study and interpret the solutions of linear equations. That effort produced an algebraic system whose essential ingredients are

i. a set, $\mathcal{R}^n$ (whose elements are called vectors),
ii. a procedure (called addition) for combining two elements of the set to obtain a third element of the set, and
iii. a procedure (called scalar multiplication) for combining a real number with an element of the set to obtain another element of the set.

The system has certain algebraic features, as determined by the rules governing the use of the operations (e.g., commutativity and associativity of addition).

You have seen that the key ideas involved in describing the solutions of a linear system are purely algebraic. That is, notions like subspace, spanning set, linear independence, and the like are formulated strictly in terms of addition and scalar multiplication. Moreover, careful perusal of the proofs of the theorems in Chapter 2 reveals that for the most part they depend only on the properties of the algebraic operations. The actual nature of the vectors plays little or no role in the arguments, and the symbols X, Y, . . . could just as well be place holders for purely abstract objects as place holders for n-tuples of real numbers. Admittedly, the same cannot be said for the algorithms developed in Chapter 2. Listing vectors as columns in a matrix and row reducing the result certainly relies on the convention of writing elements of $\mathcal{R}^n$ as $n \times 1$ matrices. Nevertheless, the basic facts about $\mathcal{R}^n$ and its subspaces are consequences of the algebraic rules for operating with n-dimensional vectors.

In the present chapter, the essential algebraic features of $\mathcal{R}^n$ are used as axioms to define an abstract mathematical system called a vector space. The idea is to capture the pure algebraic structure of $\mathcal{R}^n$ by focusing on the rules the operations satisfy. With arguments based only on those rules, you can establish general results that are independent of the idiosyncrasies of the individual vectors. This achieves a certain economy, in as much as the conclusions are valid in any context where the objects satisfy the specified rules. Besides, it is very satisfying aesthetically to recognize that a small collection of basic principles can provide a unified means of structuring information in a variety of different settings.

Most of the theorems in $\mathcal{R}^n$ translate directly into results about abstract vector spaces. Sometimes the proofs must be modified to suit the new circumstances, but often they are identical to the previous ones. You might view this entire chapter as an introduction to abstraction (a common theme in mathematics). Along the way you will encounter linearity and its interpretations in important contexts other than $\mathcal{R}^n$, and you will have an opportunity to experience the power and beauty of the axiomatic method.

7.2 VECTOR SPACES AND SUBSPACES

DEFINITION. Let $\mathcal{V}$ be a nonempty set. Suppose there is a means of combining any two elements of $\mathcal{V}$, say $\mathbf{u}$ and $\mathbf{v}$, to obtain another element of $\mathcal{V}$, denoted by $\mathbf{u} + \mathbf{v}$, and suppose there is a means of combining any $t \in \mathcal{R}$ with any $\mathbf{v} \in \mathcal{V}$ to obtain an element, $t\mathbf{v}$, of $\mathcal{V}$. The set $\mathcal{V}$, together with these operations of *addition* and *scalar multiplication*, is said to be a *vector space over* $\mathcal{R}$ if for all $\mathbf{u},\mathbf{v},\mathbf{w} \in \mathcal{V}$ and for all $s,t \in \mathcal{R}$,

(1) $\mathbf{u} + \mathbf{v} = \mathbf{v} + \mathbf{u}$, (commutativity of $+$)
(2) $\mathbf{u} + (\mathbf{v} + \mathbf{w}) = (\mathbf{u} + \mathbf{v}) + \mathbf{w}$, (associativity of $+$)
(3) there is a unique element of $\mathcal{V}$, denoted by $\mathbf{0}_{\mathcal{V}}$, such that $\mathbf{v} + \mathbf{0}_{\mathcal{V}} = \mathbf{v}$ for each $\mathbf{v} \in \mathcal{V}$, (additive identity)
(4) for each $\mathbf{v} \in \mathcal{V}$ there is a unique element of $\mathcal{V}$, denoted by $-\mathbf{v}$, such that $\mathbf{v} + (-\mathbf{v}) = \mathbf{0}_{\mathcal{V}}$, (additive inverse)
(5) $(s + t)\mathbf{v} = s\mathbf{v} + t\mathbf{v}$,
(6) $s(\mathbf{u} + \mathbf{v}) = s\mathbf{u} + s\mathbf{v}$,
(7) $s(t\mathbf{v}) = (st)\mathbf{v}$,
(8) $1\mathbf{v} = \mathbf{v}$.

The elements of $\mathcal{V}$ are called *vectors*, and $\mathbf{0}_{\mathcal{V}}$ is known as the *zero vector*.

These axioms are the first eight rules, listed in Theorem 2.5, governing algebraic operations with elements of $\mathcal{R}^n$. Here, however, $+$ stands for an abstract notion of addition, and the juxtaposition $t\mathbf{v}$ indicates an abstract notion of scalar multiplication. Bold lower case letters like $\mathbf{u}$, $\mathbf{v}$, . . . will denote elements of an abstract vector space, but the practice of representing elements of Euclidean n-space by capital letters like X and Y will be retained when $\mathcal{V} = \mathcal{R}^n$.

In Euclidean n-space, the algebraic structure of the system evolved from the form of the vectors. That is, the notion of n-tuple was the primitive concept, and $\mathcal{R}^n$ was just the collection of all such objects, together with operations of addition and scalar multiplication that arose in the process of manipulating solutions of a linear system. In the abstract setting, the opposite point of view prevails. Here the fundamental notion is that of a vector space, based as it is on the rules governing the operations. An abstract vector is just an element of a vector space. The next few examples introduce some vector spaces other than $\mathcal{R}^n$ that are of particular interest.

■ **EXAMPLE 7.1.** Let $\mathcal{V} = \mathcal{M}_{m \times n}$. Here, the objects under consideration are $m \times n$ matrices. As long as addition and scalar multiplication are defined in the usual manner, Axioms (1)–(8) of the definition of a vector space are just restatements of familiar facts from matrix algebra. Thus, $\mathcal{M}_{m \times n}$ can be viewed as a vector space. Its additive identity is the $m \times n$ matrix whose entries are all 0's, and the additive inverse of $A \in \mathcal{M}_{m \times n}$ is the $m \times n$ matrix, $-A$, with entries $[-A]_{ij} = -[A]_{ij}$. □

■ **EXAMPLE 7.2.** Assume $J \subseteq \mathcal{R}$ is an interval and let $\mathcal{F}(J)$ be the collection of real-valued functions defined on J. Elements of $\mathcal{F}(J)$ will be denoted by lower case letters like f and g, and the fact that f is defined on J and has values in $\mathcal{R}$ is indicated by writing $f : J \rightarrow \mathcal{R}$. If $f \in \mathcal{F}(J)$ and $x \in J$, then $f(x)$ is the value of f at x. Two functions, $f,g \in \mathcal{F}(J)$, are considered to be the same when they have the same values, that is, when $f(x) = g(x)$, $x \in J$. Addition and scalar multiplication for functions are defined in terms of the addition and scalar multiplication of their values, that is, for $f,g \in \mathcal{F}(J)$ and $t \in \mathcal{R}$,

$$(f+g)(x) = f(x) + g(x),$$

and

$$(tf)(x) = tf(x).$$

To show that $\mathcal{F}(J)$ is a vector space you must check that these operations satisfy the eight defining axioms. For example, Axiom (2) reads

$$f + (g+h) = (f+g) + h,$$

for any choice of $f,g,h \in \mathcal{F}(J)$. Its verification amounts to comparing the values of $f + (g+h)$ and $(f+g)+h$. Indeed, for each $x \in J$,

$$\begin{aligned}
(f+(g+h))(x) &= f(x) + (g+h)(x) && \text{(defn. of + in } \mathcal{F}(J)) \\
&= f(x) + (g(x) + h(x)) && \text{(defn. of + in } \mathcal{F}(J)) \\
&= (f(x) + g(x)) + h(x) && \text{(associative law of + in } \mathcal{R}) \\
&= (f+g)(x) + h(x) && \text{(defn. of + in } \mathcal{F}(J)) \\
&= ((f+g)+h)(x). && \text{(defn. of + in } \mathcal{F}(J))
\end{aligned}$$

The additive identity in $\mathcal{F}(J)$ is the function, 0, whose values are all zeros, that is, the function $0 : J \rightarrow \mathcal{R}$ such that $0(x) = 0$ for every $x \in J$. Here the symbol 0 is used to denote both the function and its value, a practice that is tolerated because the meaning in each case can be inferred from its use in the sentence. For $f \in \mathcal{F}(J)$ and $x \in J$,

$$\begin{aligned}
(f+0)(x) &= f(x) + 0(x) && \text{(defn. of + in } \mathcal{F}(J)) \\
&= f(x) + 0 && \text{(defn. of } 0 : J \rightarrow \mathcal{R}) \\
&= f(x), && \text{(algebra in } \mathcal{R})
\end{aligned}$$

so $f + 0 = f$.

The additive inverse of $f \in \mathcal{F}(\mathrm{J})$ is $-f : \mathrm{J} \to \mathcal{R}$ defined by $(-f)(x) = -f(x)$, $x \in \mathrm{J}$. For each $x \in \mathrm{J}$,

$$\begin{aligned}
(f + (-f))(x) &= f(x) + (-f)(x) && \text{(defn. of + in } \mathcal{F}(\mathrm{J})) \\
&= f(x) + (-f(x)) && \text{(defn. of } -f) \\
&= 0 && \text{(algebra in } \mathcal{R}) \\
&= 0(x), && \text{(defn. of } 0 : \mathrm{J} \to \mathcal{R})
\end{aligned}$$

so $f + (-f) = 0$.

The remaining axioms can be checked by the reader. In each case, the argument uses only the definitions of addition and scalar multiplication of functions and standard algebraic properties of $\mathcal{R}$. □

Notice that verification of Axioms (1)–(8) in the last example did not rely on properties of J. Any nonempty subset of $\mathcal{R}$ would have sufficed as a domain for the functions.

■ **EXAMPLE 7.3.** Let $\mathcal{R}^\infty$ denote the set of all sequences of real numbers, that is,

$$\mathcal{R}^\infty = \left\{ \{x_k\}_0^\infty \ : \ \text{where } x_k \in \mathcal{R}, \quad k = 0, 1, 2, \ldots \right\}.$$

A sequence is an infinite analogue of an ordered n-tuple, that is, $\{x_k\}_0^\infty$ can be thought of as $(x_0, x_1, x_2, \ldots)$. Addition and scalar multiplication are defined as follows. If $\mathrm{X} = \{x_k\}_0^\infty$, $\mathrm{Y} = \{y_k\}_0^\infty$, and $t \in \mathcal{R}$, then

$$\mathrm{X} + \mathrm{Y} = \{x_k + y_k\}_0^\infty \qquad \text{and} \qquad t\mathrm{X} = \{tx_k\}_0^\infty,$$

or equivalently

$$(x_0, x_1, x_2, \ldots) + (y_0, y_1, y_2, \ldots) = (x_0 + y_0, x_1 + y_1, x_2 + y_2, \ldots)$$

and

$$t(x_0, x_1, x_2, \ldots) = (tx_0, tx_1, tx_2, \ldots).$$

Checking that these operations satisfy Axioms (1)–(8) is no more difficult than doing so for $\mathcal{R}^n$. □

REMARK. Formally, a sequence is a real-valued function with domain $\{0, 1, \ldots\}$. That is, the terms of $\{x_k\}_0^\infty$ can be interpreted as values of $f : \{0, 1, 2, \ldots\} \to \mathcal{R}$ defined by $f(k) = x_k$. If $t \in \mathcal{R}$ and $f, g : \{0, 1, 2, \ldots\} \to \mathcal{R}$, with $f(k) = x_k$ and $g(k) = y_k$, then

$$(f + g)(k) = f(k) + g(k) = x_k + y_k$$

and

$$(tf)(k) = tf(k) = tx_k,$$

so addition and scalar multiplication of sequences are just special cases of addition and scalar multiplication of functions. In other words, $\mathcal{R}^\infty = \mathcal{F}(\mathrm{J})$, where $\mathrm{J} = \{0, 1, 2, \ldots\}$.

The next theorem lists some elementary properties of abstract vectors. In Euclidean n-space, they followed easily from the nature of the vectors and standard algebraic features of the real number system. In the abstract case, however, they must be established as consequences of the axioms defining a vector space.

THEOREM 7.1. If $\mathcal{V}$ is a vector space, $\mathbf{v} \in \mathcal{V}$, and $t \in \mathcal{R}$, then

(1) $0\mathbf{v} = \mathbf{0}_{\mathcal{V}}$,
(2) $t\mathbf{0}_{\mathcal{V}} = \mathbf{0}_{\mathcal{V}}$,
(3) $(-1)\mathbf{v} = -\mathbf{v}$, and
(4) $t\mathbf{v} = \mathbf{0}_{\mathcal{V}}$ if and only if $t = 0$ or $\mathbf{v} = \mathbf{0}_{\mathcal{V}}$.

Proof.
(1) Since $0 = 0 + 0$, it follows from Axiom (5) that

$$0\mathbf{v} = (0 + 0)\mathbf{v} = 0\mathbf{v} + 0\mathbf{v}. \tag{7.1}$$

Adding the additive inverse of $0\mathbf{v}$ to both sides of (7.1) produces

$$\begin{aligned} \mathbf{0}_{\mathcal{V}} &= 0\mathbf{v} + (-0\mathbf{v}) && \text{(Axiom (4))} \\ &= (0\mathbf{v} + 0\mathbf{v}) + (-0\mathbf{v}) && \text{(equation (7.1))} \\ &= 0\mathbf{v} + (0\mathbf{v} + (-0\mathbf{v})) && \text{(Axiom (2))} \\ &= 0\mathbf{v} + \mathbf{0}_{\mathcal{V}} && \text{(Axiom (4))} \\ &= 0\mathbf{v}. && \text{(Axiom (3))} \end{aligned}$$

(2) Since $\mathbf{0}_{\mathcal{V}} = \mathbf{0}_{\mathcal{V}} + \mathbf{0}_{\mathcal{V}}$ (Axiom (3)), it follows from Axiom (6) that

$$t\mathbf{0}_{\mathcal{V}} = t(\mathbf{0}_{\mathcal{V}} + \mathbf{0}_{\mathcal{V}}) = t\mathbf{0}_{\mathcal{V}} + t\mathbf{0}_{\mathcal{V}}. \tag{7.2}$$

Adding the additive inverse of $t\mathbf{0}_{\mathcal{V}}$ to both sides of (7.2) produces

$$\begin{aligned} \mathbf{0}_{\mathcal{V}} &= t\mathbf{0}_{\mathcal{V}} + (-t\mathbf{0}_{\mathcal{V}}) && \text{(Axiom (4))} \\ &= (t\mathbf{0}_{\mathcal{V}} + t\mathbf{0}_{\mathcal{V}}) + (-t\mathbf{0}_{\mathcal{V}}) && \text{(equation (7.2))} \\ &= t\mathbf{0}_{\mathcal{V}} + (t\mathbf{0}_{\mathcal{V}} + (-t\mathbf{0}_{\mathcal{V}})) && \text{(Axiom (2))} \\ &= t\mathbf{0}_{\mathcal{V}} + \mathbf{0}_{\mathcal{V}} && \text{(Axiom (4))} \\ &= t\mathbf{0}_{\mathcal{V}}. && \text{(Axiom (3))} \end{aligned}$$

(3) Since $-\mathbf{v}$ is the unique $\mathbf{x} \in \mathcal{V}$ satisfying $\mathbf{v} + \mathbf{x} = \mathbf{0}_{\mathcal{V}}$, it suffices to show that $\mathbf{x} = (-1)\mathbf{v}$ has this property. Indeed,

$$\begin{aligned} \mathbf{v} + (-1)\mathbf{v} &= 1\mathbf{v} + (-1)\mathbf{v} && \text{(Axiom (8))} \\ &= (1 + (-1))\mathbf{v} && \text{(Axiom (5))} \\ &= 0\mathbf{v} && \text{(algebra in } \mathcal{R}) \\ &= \mathbf{0}_{\mathcal{V}}. && \text{((1) of the present theorem)} \end{aligned}$$

(4) Suppose $t\mathbf{v} = \mathbf{0}_{\mathcal{V}}$. When $t = 0$ there is nothing to do. If $t \neq 0$, then

$$\begin{aligned} \mathbf{v} &= 1\mathbf{v} && \text{(Axiom (8))} \\ &= (t^{-1}t)\mathbf{v} && \text{(algebra in } \mathcal{R}) \\ &= t^{-1}(t\mathbf{v}) && \text{(Axiom (7))} \\ &= t^{-1}\mathbf{0}_{\mathcal{V}} && \text{(hypothesis)} \\ &= \mathbf{0}_{\mathcal{V}}. && \text{((2) of the present theorem)} \end{aligned}$$

On the other hand, if $t = 0$ or $\mathbf{v} = \mathbf{0}_{\mathcal{V}}$, then (1) or (2) implies $t\mathbf{v} = \mathbf{0}_{\mathcal{V}}$. ∎

Suppose $\mathcal{V}$ is a vector space and $\mathcal{W} \subseteq \mathcal{V}$. The elements of $\mathcal{W}$, being vectors in $\mathcal{V}$, can be algebraically manipulated using the operations of addition and scalar multiplication on $\mathcal{V}$, but there is no guarantee that such activity will lead to vectors in $\mathcal{W}$. For the operations on $\mathcal{V}$ to be operations on $\mathcal{W}$, it is necessary that the sum of any two elements of $\mathcal{W}$ be an element of $\mathcal{W}$ and that any scalar multiple of an element of $\mathcal{W}$ be an element of $\mathcal{W}$. As in $\mathcal{R}^n$, subsets with these two features are of special interest.

DEFINITION. Assuming $\mathcal{V}$ is a vector space, $\mathcal{W} \subseteq \mathcal{V}$ is *closed under addition* if $\mathbf{w} + \mathbf{v} \in \mathcal{W}$ whenever $\mathbf{w} \in \mathcal{W}$ and $\mathbf{v} \in \mathcal{W}$. If $t\mathbf{w} \in \mathcal{W}$ for each $t \in \mathcal{R}$ and $\mathbf{w} \in \mathcal{W}$, then $\mathcal{W}$ is *closed under scalar multiplication.* A subset of $\mathcal{V}$ that is closed under both addition and scalar multiplication is called a *subspace* of $\mathcal{V}$ and is said to *inherit* its algebraic operations from $\mathcal{V}$.

The next result shows that a subspace of $\mathcal{V}$, together with the algebraic operations it inherits from $\mathcal{V}$, is itself a vector space.

THEOREM 7.2. If $\mathcal{V}$ is a vector space and $\mathcal{W}$ is a subspace of $\mathcal{V}$, then $\mathcal{W}$ is a vector space.

Proof. The object is to show that Axioms (1)–(8) hold in $\mathcal{W}$. Axioms (1), (2), (5), (6), (7), and (8) are automatically satisfied by elements of $\mathcal{W}$ because they hold for all vectors in $\mathcal{V}$.

Axiom (3): To see that $\mathcal{W}$ has a unique additive identity, first observe that $\mathbf{0}_{\mathcal{V}}$, being the additive identity in $\mathcal{V}$, satisfies $\mathbf{w} + \mathbf{0}_{\mathcal{V}} = \mathbf{w}$ for all $\mathbf{w} \in \mathcal{W}$. However, it has not yet been established that $\mathbf{0}_{\mathcal{V}} \in \mathcal{W}$. The latter conclusion follows from the fact that $\mathcal{W}$ is closed under scalar multiplication. Indeed, $0\mathbf{w} \in \mathcal{W}$ for any $\mathbf{w} \in \mathcal{W}$, and $0\mathbf{w} = \mathbf{0}_{\mathcal{V}}$ (Theorem 7.1), so $\mathbf{0}_{\mathcal{V}} \in \mathcal{W}$. It remains to show that no other element of $\mathcal{W}$ acts as an additive identity. To that end, suppose $\mathbf{z} \in \mathcal{W}$ and $\mathbf{w} + \mathbf{z} = \mathbf{w}$ for each $\mathbf{w} \in \mathcal{W}$. Setting $\mathbf{w} = \mathbf{0}_{\mathcal{V}}$ yields $\mathbf{0}_{\mathcal{V}} + \mathbf{z} = \mathbf{0}_{\mathcal{V}}$. But $\mathbf{0}_{\mathcal{V}} + \mathbf{z} = \mathbf{z}$ (because $\mathbf{0}_{\mathcal{V}}$ is the additive identity in $\mathcal{V}$), so $\mathbf{z} = \mathbf{0}_{\mathcal{V}}$.

Axiom (4): Let $\mathbf{w} \in \mathcal{W}$. Being a vector in $\mathcal{V}$, $\mathbf{w}$ has an additive inverse in $\mathcal{V}$, denoted by $-\mathbf{w}$. Since $-\mathbf{w} = (-1)\mathbf{w}$ (Theorem 7.1), closure of $\mathcal{W}$ under scalar multiplication implies $-\mathbf{w} \in \mathcal{W}$. Thus, each element of $\mathcal{W}$ has an additive inverse in $\mathcal{W}$. That inverse is unique, for otherwise $\mathbf{w}$ would have more than one additive inverse in $\mathcal{V}$. ■

Given a vector space $\mathcal{V}$, $\{\mathbf{0}_{\mathcal{V}}\}$ and $\mathcal{V}$ itself are closed under addition and scalar multiplication, so both are subspaces of $\mathcal{V}$. They are referred to as the trivial subspaces. For any subspace $\mathcal{W}$ of $\mathcal{V}$, $\{\mathbf{0}_{\mathcal{V}}\} \subseteq \mathcal{W} \subseteq \mathcal{V}$, so $\{\mathbf{0}_{\mathcal{V}}\}$ and $\mathcal{V}$ are the smallest and largest subspaces, respectively.

■ **EXAMPLE 7.4.** Let $\mathcal{V} = \mathcal{M}_{n\times n}$ and consider $\mathcal{S}_{n\times n} = \{\mathrm{A} \in \mathcal{V} : \mathrm{A} = \mathrm{A}^T\}$. For A,B $\in \mathcal{S}_{n\times n}$ and $t \in \mathcal{R}$,

$$(\mathrm{A}+\mathrm{B})^T = \mathrm{A}^T + \mathrm{B}^T = \mathrm{A}+\mathrm{B}$$

and

$$(t\mathrm{A})^T = t\mathrm{A}^T = t\mathrm{A},$$

so $\mathcal{S}_{n\times n}$ is closed under addition and scalar multiplication. Thus, the set of $n \times n$ symmetric matrices forms a subspace of $\mathcal{M}_{n\times n}$. □

■ **EXAMPLE 7.5.** Let $\mathcal{P}$ be the set of polynomials. Then $\mathcal{P} \subseteq \mathcal{F}(-\infty,\infty)$, and a typical $f \in \mathcal{P}$ has values

$$f(x) = a_0 + a_1 x + \cdots + a_n x^n,$$

where n is some nonnegative integer and $a_0, a_1, \ldots, a_n \in \mathcal{R}$. For $t \in \mathcal{R}$,

$$(tf)(x) = (ta_0) + (ta_1)x + \cdots + (ta_n)x^n,$$

so $tf \in \mathcal{P}$. If g is another element of $\mathcal{P}$, say $g(x) = b_0 + b_1 x + \cdots + b_m x^m$, then combining like terms in $f(x) + g(x)$ confirms that $f + g$ is also a polynomial. Thus $\mathcal{P}$, being closed under addition and scalar multiplication, is a subspace of $\mathcal{F}(-\infty,\infty)$. □

■ **EXAMPLE 7.6.** Given a fixed positive integer n, let

$$\mathcal{P}_n = \{f \in \mathcal{P} : \deg(f) \leq n\}.$$

If $f,g \in \mathcal{P}_n$ and $t \in \mathcal{R}$, then $f(x) = \sum_0^n a_k x^k$ and $g(x) = \sum_0^n b_k x^k$, where $a_k, b_k \in \mathcal{R}$, $k = 0, 1, \ldots, n$, and therefore

$$(f + g)(x) = \sum_0^n (a_k + b_k)x^k \quad \text{and} \quad (tf)(x) = \sum_0^n (ta_k)x^k.$$

Since $\deg(f + g) \leq n$ and $\deg(tf) \leq n$, $\mathcal{P}_n$ is closed under addition and scalar multiplication. Thus, $\mathcal{P}_n$ is a subspace of $\mathcal{P}$. □

■ **EXAMPLE 7.7.** Let $\mathcal{C} \subseteq \mathcal{R}^\infty$ be the collection of convergent sequences. Standard calculus results state that the sum of two convergent sequences is convergent and that a scalar multiple of a convergent sequence is convergent, so $\mathcal{C}$ is a subspace of $\mathcal{R}^\infty$. □

DEFINITION. Assume $\mathcal{V}$ is a vector space and $\mathcal{A} \subseteq \mathcal{V}$. A *linear combination of elements of* $\mathcal{A}$ is a vector

$$t_1\mathbf{v}_1 + \cdots + t_p\mathbf{v}_p = \sum_1^p t_k\mathbf{v}_k,$$

where p is a positive integer, $t_1, \ldots, t_p \in \mathcal{R}$, and $\mathbf{v}_1, \ldots, \mathbf{v}_p \in \mathcal{A}$. The set of all linear combinations of elements of $\mathcal{A}$ is denoted by $\mathcal{S}pan\{\mathcal{A}\}$ and if $\mathcal{W} = \mathcal{S}pan\{\mathcal{A}\}$, then $\mathcal{A}$ is said to be a *spanning set* for $\mathcal{W}$.

This notion of spanning set is a bit more subtle than that discussed in Chapter 2. The difference here is that $\mathcal{A}$ need not be a finite set. Each linear combination of elements of $\mathcal{A}$ involves only a finite number of vectors in $\mathcal{A}$, but the members of $\mathcal{S}pan\{\mathcal{A}\}$ need not all be generated by the same finite set of vectors. The next example illustrates this point.

■ **EXAMPLE 7.8.** Let $\mathcal{V} = \mathcal{F}(-\infty,\infty)$ and consider $\mathcal{A} = \{1, x, x^2, x^3, \ldots\}$. If $f \in \mathcal{S}pan\{\mathcal{A}\}$, then there are integers $k_1, \ldots, k_p$ satisfying $0 \leq k_1 < k_2 < \cdots < k_p$ such that

$$f(x) = a_1 x^{k_1} + \cdots + a_p x^{k_p},$$

where $a_1, \ldots, a_p \in \mathcal{R}$. Thus, $f \in \mathcal{P}$. Conversely, every polynomial is a linear combination of elements of $\mathcal{A}$, so $\mathcal{S}pan\{\mathcal{A}\} = \mathcal{P}$. Note that there is no finite subset of $\mathcal{A}$ that spans $\mathcal{P}$.

Indeed, if $\mathcal{B} \subseteq \mathcal{A}$ is finite and m is the largest integer such that $x^m \in \mathcal{B}$, then $\deg f \leq m$ for each $f \in Span\{\mathcal{B}\}$. □

If $\mathcal{A}$ consists of p vectors, then each $\mathbf{v} \in Span\{\mathcal{A}\}$ is expressible in terms of the same p vectors, and $Span\{\mathcal{A}\}$ has the same meaning as in Chapter 2.

■ **EXAMPLE 7.9.** If n is a positive integer and $\mathcal{A} = \{1, x, x^2, \ldots, x^n\}$, then

$$Span\{\mathcal{A}\} = \left\{\sum_0^n a_k x^k : a_0, \ldots, a_n \in \mathcal{R}\right\} = \mathcal{P}_n. \square$$

■ **EXAMPLE 7.10.** Let $\mathcal{A} = \{A_{11}, A_{12}, A_{21}, A_{22}\} \subseteq \mathcal{M}_{2\times2}$, where

$$A_{11} = \begin{bmatrix} 1 & 0 \\ 0 & 0 \end{bmatrix}, \quad A_{12} = \begin{bmatrix} 0 & 1 \\ 0 & 0 \end{bmatrix}, \quad A_{21} = \begin{bmatrix} 0 & 0 \\ 1 & 0 \end{bmatrix}, \quad \text{and} \quad A_{22} = \begin{bmatrix} 0 & 0 \\ 0 & 1 \end{bmatrix}.$$

For $a, b, c, d \in \mathcal{R}$,

$$aA_{11} + bA_{12} + cA_{21} + dA_{22} = \begin{bmatrix} a & b \\ c & d \end{bmatrix},$$

so $Span\{\mathcal{A}\} = \mathcal{M}_{2\times2}$. More generally, $\mathcal{M}_{m\times n} = Span\{A_{ij} : 1 \leq i \leq m,\ 1 \leq j \leq n\}$, where A_{ij} is the $m \times n$ matrix with i, j entry 1 and other entries 0's. □

THEOREM 7.3. If $\mathcal{A}$ is a nonempty subset of a vector space $\mathcal{V}$, then $Span\{\mathcal{A}\}$ is a subspace of $\mathcal{V}$.

Proof. Suppose $\mathbf{w}, \mathbf{v} \in Span\{\mathcal{A}\}$ and $t \in \mathcal{R}$. There are positive integers p and q, vectors $\mathbf{w}_1, \ldots, \mathbf{w}_p$ and $\mathbf{v}_1, \ldots, \mathbf{v}_q$ in $\mathcal{A}$, and scalars $t_1, \ldots, t_p$ and $s_1, \ldots, s_q$ such that

$$\mathbf{w} = t_1\mathbf{w}_1 + \cdots + t_p\mathbf{w}_p \qquad \text{and} \qquad \mathbf{v} = s_1\mathbf{v}_1 + \cdots + s_q\mathbf{v}_q.$$

Thus,

$$\mathbf{w} + \mathbf{v} = \sum_1^p t_k\mathbf{w}_k + \sum_1^q s_k\mathbf{v}_k \in Span\{\mathcal{A}\}$$

and

$$t\mathbf{w} = t\left(\sum_1^p t_k\mathbf{w}_k\right) = \sum_1^p (tt_k)\mathbf{w}_k \in Span\{\mathcal{A}\}. \blacksquare$$

■ **EXAMPLE 7.11.** Consider $\mathcal{V} = \{f \in \mathcal{P}_2 : f(1) = f(2)\}$. A member of $\mathcal{P}_2$ looks like $f(x) = a + bx + cx^2$, and it satisfies the membership criterion for $\mathcal{V}$ when

$$a + b + c = f(1) = f(2) = a + 2b + 4c,$$

that is, when $b = -3c$. Thus,

$$\mathcal{V} = \{a + c(x^2 - 3x) : a, c \in \mathcal{R}\} = Span\{1, x^2 - 3x\}.$$

By Theorem 7.3, $\mathcal{V}$ is a subspace of $\mathcal{P}_2$. □

Exercises 7.2

1. Which of Axioms (1)–(8) are satisfied by the following sets and their operations of "addition" and "scalar multiplication"?

   a. $\mathcal{V}$ is the collection of all subsets of $\mathcal{R}$. For $A, B \in \mathcal{V}$ and $t \in \mathcal{R}$,

$$A + B = A \cup B \qquad \text{and} \qquad tA = A.$$

b. $\mathcal{V} = \{x \in \mathcal{R} : x > 0\}$. For $x,y \in \mathcal{V}$ and $t \in \mathcal{R}$,

$$x + y = xy \qquad \text{and} \qquad tx = x^t.$$

c. $\mathcal{V}$ is the collection of subspaces of $\mathcal{R}^n$. For $\mathcal{U}, \mathcal{W} \in \mathcal{V}$ and $t \in \mathcal{R}$,

$$\mathcal{U} + \mathcal{W} = \{\mathbf{u} + \mathbf{w} : \mathbf{u} \in \mathcal{U}, \mathbf{w} \in \mathcal{W}\} \qquad \text{and} \qquad t\mathcal{U} = \{t\mathbf{u} : \mathbf{u} \in \mathcal{U}\}.$$

2. Which of the following subsets of $\mathcal{M}_{n\times n}$ are closed under addition? Closed under scalar multiplication? Subspaces of $\mathcal{M}_{n\times n}$?
   a. $\mathcal{U} = \{\text{A} : \text{A is upper triangular}\}$
   b. $\mathcal{D} = \{\text{A} : \text{A is diagonal}\}$
   c. $\mathcal{T} = \{\text{A} : \text{trace(A)} = 0\}$ $\quad$ $(\text{trace(A)} = a_{11} + a_{22} + \cdots + a_{nn})$
   d. $\mathcal{W} = \{\text{A} : \text{A is singular}\}$
   e. $\mathcal{W} = \{\text{A} : \text{AC} = \text{CA, where C is a fixed } n \times n \text{ matrix}\}$
   f. $\mathcal{W} = \{\text{A} : \text{A is symmetric and positive semidefinite}\}$
   g. $\mathcal{W} = \{\text{A} : \text{X}_0 \in \mathcal{R}^n \text{ is an eigenvector of A}\}$

3. Which of the following subsets of $\mathcal{P}$ are closed under addition? Closed under scalar multiplication? Subspaces of $\mathcal{P}$?
   a. $\mathcal{W} = \{f : \deg(f) \text{ is an even integer}\}$
   b. $\mathcal{W} = \{f : f(-x) = f(x) \text{ for all } x \in \mathcal{R}\}$
   c. $\mathcal{W} = \{f : f \text{ is divisible by } q, \text{ where } q \text{ is a fixed polynomial}\}$

4. Which of the following subsets of $\mathcal{R}^\infty$ are closed under addition? Closed under scalar multiplication? Subspaces of $\mathcal{R}^\infty$?
   a. $\mathcal{W} = \left\{\{x_k\}_0^\infty : \sum_0^\infty x_k < \infty\right\}$
   b. $\mathcal{W} = \left\{\{x_k\}_0^\infty : x_{k+1} \geq x_k, \quad k = 0, 1, 2, \ldots\right\}$ (the increasing sequences)
   c. $\mathcal{W} = \left\{\{x_k\}_0^\infty : x_k = 0 \text{ for all but finitely many values of } k\right\}$
   d. $\mathcal{W} = \left\{\{x_k\}_0^\infty : x_{k+2} = x_{k+1} + x_k, \quad k = 0, 1, 2, \ldots\right\}$

5. Sketch the graph of the additive identity of $\mathcal{F}[0,1]$. How is the graph of $f \in \mathcal{F}[0,1]$ related to that of its additive inverse?

6. Assuming $a,b \in \mathcal{R}$, with $a < b$, show that the following subsets of $\mathcal{F}[a,b]$ are subspaces.
   a. $\mathcal{C}[a,b] = \{f \in \mathcal{F}[a,b] : f \text{ is continuous}\}$
   b. $\mathcal{D}[a,b] = \{f \in \mathcal{F}[a,b] : f \text{ is differentiable}\}$
   c. $\mathcal{W} = \{f \in \mathcal{C}[a,b] : \int_a^b f(x)\, dx = 0\}$
   d. $\mathcal{W} = \{f \in \mathcal{D}[a,b] : f' + f = 0\}$
   e. $\mathcal{W} = \{f \in \mathcal{F}[a,b] : f(c) = 0, \text{ where } c \text{ is a fixed point in } [a,b]\}$

7. Find a spanning set for each of the following subspaces.
   a. $\{\text{A} \in \mathcal{M}_{3\times 3} : \text{A is diagonal}\}$
   b. $\{\text{A} \in \mathcal{M}_{3\times 3} : \text{A is symmetric}\}$
   c. $\{\text{A} \in \mathcal{M}_{2\times 2} : \text{trace(A)} = 0\}$
   d. $\{f \in \mathcal{P}_2 : f(1) = 0\}$
   e. $\{f \in \mathcal{P}_2 : \int_0^1 f(x)\, dx = 0\}$
   f. $\{f \in \mathcal{P} : f(-x) = f(x) \text{ for all } x \in \mathcal{R}\}$
   g. $\{(x_0, x_1, \ldots) \in \mathcal{R}^\infty : x_k = 0 \text{ for all } k > n\}$ (n a fixed positive integer)
   h. $\{(x_0, x_1, \ldots) \in \mathcal{R}^\infty : x_k = 0 \text{ for all but finitely many values of } k\}$.

7.3
INDEPENDENCE AND DIMENSION

In an abstract vector space, as in $\mathcal{R}^n$, the notion of linear independence for a subset $\mathcal{A}$ addresses the issue of how many ways the additive identity can be expressed as a linear combination of elements of $\mathcal{A}$. The only new wrinkle is that the set may now contain infinitely many vectors. Keep in mind, however, that a particular linear combination involves only a finite number of vectors.

DEFINITION. Suppose $\mathcal{V}$ is a vector space and $\mathcal{A} \subseteq \mathcal{V}$. If $\mathbf{0}_\mathcal{V}$ can be expressed as a linear combination of elements of $\mathcal{A}$ in a nontrivial manner (using at least one nonzero scalar), then $\mathcal{A}$ is *linearly dependent.* A subset of $\mathcal{V}$ that is not linearly dependent is *linearly independent.*

For finite sets $\{\mathbf{v}_1, \ldots, \mathbf{v}_p\} \subseteq \mathcal{V}$, the meaning of linear dependence is the same as in Chapter 2, that is, there is a nontrivial choice of $t_1, \ldots, t_p$ such that

$$t_1\mathbf{v}_1 + \cdots + t_p\mathbf{v}_p = \mathbf{0}_\mathcal{V}.$$

An infinite $\mathcal{A} \subseteq \mathcal{V}$ has many finite subsets with which to build linear combinations. Such a set is linearly dependent when $\mathbf{0}_\mathcal{V}$ can be written as a nontrivial linear combination of the vectors in any one of those subsets, that is, when $\mathcal{A}$ has a finite subset that is linearly dependent. Alternatively, linear independence occurs when every finite subset of $\mathcal{A}$ is linearly independent. Notice that $\{\mathbf{0}_\mathcal{V}\}$ is linearly dependent (since $t\mathbf{0}_\mathcal{V} = \mathbf{0}_\mathcal{V}$ for any $t \in \mathcal{R}$), so any subset of $\mathcal{V}$ containing $\mathbf{0}_\mathcal{V}$ is linearly dependent.

■ **EXAMPLE 7.12.** Consider $\mathcal{A} = \{1, x, x^2, \ldots, x^n\} \subseteq \mathcal{P}$. A linear combination of the elements in $\mathcal{A}$ produces a polynomial

$$f(x) = a_0 + a_1x + a_2x^2 + \cdots + a_nx^n.$$

To say that f is the additive identity of $\mathcal{P}$ is to say that $f(x) = 0$ for every $x \in \mathcal{R}$, which occurs only when $a_0 = \cdots = a_n = 0$. Thus, $\mathcal{A}$ is linearly independent. Essentially the same argument shows that $\mathcal{B} = \{1, x, x^2, \ldots\}$ is linearly independent. Indeed, a linear combination of elements of $\mathcal{B}$ is a polynomial

$$g(x) = a_1x^{k_1} + \cdots + a_px^{k_p},$$

where $k_1, \ldots, k_p$ are integers satisfying $0 \le k_1 < k_2 < \cdots < k_p$, and $g = 0$ if and only if $a_1, \ldots, a_p$ are all zeros. □

■ **EXAMPLE 7.13.** Let $\mathcal{A} = \left\{ \begin{bmatrix} 1 & 2 \\ -3 & 4 \end{bmatrix}, \begin{bmatrix} -3 & -1 \\ -4 & 2 \end{bmatrix}, \begin{bmatrix} 4 & 3 \\ 1 & 2 \end{bmatrix} \right\} \subseteq \mathcal{M}_{2\times 2}$. Observe that

$$r\begin{bmatrix} 1 & 2 \\ -3 & 4 \end{bmatrix} + s\begin{bmatrix} -3 & -1 \\ -4 & 2 \end{bmatrix} + t\begin{bmatrix} 4 & 3 \\ 1 & 2 \end{bmatrix} = \begin{bmatrix} 0 & 0 \\ 0 & 0 \end{bmatrix}$$

if and only if

$$\begin{bmatrix} 0 \\ 0 \\ 0 \\ 0 \end{bmatrix} = \begin{bmatrix} r - 3s + 4t \\ 2r - s + 3t \\ -3r - 4s + t \\ 4r + 2s + 2t \end{bmatrix} = \begin{bmatrix} 1 & -3 & 4 \\ 2 & -1 & 3 \\ -3 & -4 & 1 \\ 4 & 2 & 2 \end{bmatrix} \begin{bmatrix} r \\ s \\ t \end{bmatrix}.$$

Since

$$\begin{bmatrix} 1 & -3 & 4 \\ 2 & -1 & 3 \\ -3 & -4 & 1 \\ 4 & 2 & 2 \end{bmatrix} \longrightarrow \begin{bmatrix} 1 & 0 & 1 \\ 0 & 1 & -1 \\ 0 & 0 & 0 \\ 0 & 0 & 0 \end{bmatrix},$$

r, s, and t need not all be zeros, and $\mathcal{A}$ therefore is linearly dependent. □

■ **EXAMPLE 7.14.** Let $\mathcal{A} = \{1 - x^2, -1 + x, 1 - x + x^2\} \subseteq \mathcal{P}_2$. If $r,s,t \in \mathcal{R}$ and

$$\begin{aligned} 0 &= r(1 - x^2) + s(-1 + x) + t(1 - x + x^2) \\ &= (r - s + t) + (s - t)x + (-r + t)x^2, \end{aligned}$$

then

$$\begin{bmatrix} 0 \\ 0 \\ 0 \end{bmatrix} = \begin{bmatrix} r - s + t \\ s - t \\ -r + t \end{bmatrix} = \begin{bmatrix} 1 & -1 & 1 \\ 0 & 1 & -1 \\ -1 & 0 & 1 \end{bmatrix} \begin{bmatrix} r \\ s \\ t \end{bmatrix}.$$

Since

$$\begin{bmatrix} 1 & -1 & 1 \\ 0 & 1 & -1 \\ -1 & 0 & 1 \end{bmatrix} \longrightarrow \begin{bmatrix} 1 & 0 & 0 \\ 0 & 1 & 0 \\ 0 & 0 & 1 \end{bmatrix},$$

$r = s = t = 0$, that is, $\mathcal{A}$ is linearly independent. □

DEFINITION. A *basis* for a vector space $\mathcal{V}$ is a subset of $\mathcal{V}$ that is linearly independent and spans $\mathcal{V}$.

In Euclidean n-space the concept of dimension rests on the fact that any two bases for a subspace have the same number of elements. To establish a corresponding result for abstract vector spaces one must contend with the possibility that a basis contains infinitely many vectors. With minor modifications to accommodate that contingency, the previous line of reasoning still produces the desired conclusion. The argument relies on the following analogue of Theorem 2.20.

THEOREM 7.4. Suppose p is a positive integer and $\mathcal{V}$ is a vector space with a basis $\mathcal{B}$ consisting of p vectors. If $\mathcal{A} \subseteq \mathcal{V}$ and $\mathcal{A}$ has more than p elements, then $\mathcal{A}$ is linearly dependent.

Proof. If $\mathcal{A}$ is finite, then aside from notational differences, the proof is identical to that of Theorem 2.20. If $\mathcal{A}$ is infinite and $q > p$, then $\mathcal{A}$ has a finite subset consisting of q elements, and the proof in the previous case shows that this finite subset is linearly dependent. Having a linearly dependent finite subset, $\mathcal{A}$ itself is linearly dependent. ■

THEOREM 7.5. If $\mathcal{A}$ and $\mathcal{B}$ are bases for a vector space $\mathcal{V}$, then $\mathcal{A}$ and $\mathcal{B}$ are either both infinite or both finite. In the latter case, $\mathcal{A}$ and $\mathcal{B}$ have the same number of elements.

Proof. When $\mathcal{A}$ and $\mathcal{B}$ are both infinite there is nothing to do, so assume at least one is finite, say $\mathcal{A}$. Being linearly independent, $\mathcal{B}$ cannot contain more elements than $\mathcal{A}$ (Theorem 7.4), so $\mathcal{B}$ is finite. Switching roles of $\mathcal{A}$ and $\mathcal{B}$ then leads to the conclusion that $\mathcal{A}$ and $\mathcal{B}$ have the same number of elements. ■

It can be shown that every nontrivial vector space has a basis, but the tools required to do so lie beyond the scope of this book.

THEOREM 7.6. Every vector space containing more than one element has a basis. Moreover, if $\mathcal{V}$ is a vector space and $\mathcal{A}$ is a linearly independent subset of $\mathcal{V}$, then there is a basis $\mathcal{B}$ for $\mathcal{V}$ such that $\mathcal{A} \subseteq \mathcal{B}$.

Proof. Omitted. ■

DEFINITION. A vector space $\mathcal{V}$ having a finite basis is said to be *finite dimensional,* and the number of elements in the basis is the *dimension* of $\mathcal{V}$. Otherwise, $\mathcal{V}$ is *infinite dimensional,* written dim $\mathcal{V} = \infty$. Although $\{\mathbf{0}_{\mathcal{V}}\}$ has no basis, it is considered to be finite dimensional with dimension 0.

■ **EXAMPLE 7.15.** Examples 7.8, 7.9, and 7.12 indicate that $\mathcal{B} = \{1, x, x^2, \ldots\}$ and $\mathcal{A} = \{1, x, x^2, \ldots, x^n\}$ are bases for $\mathcal{P}$ and $\mathcal{P}_n$, respectively, so $\mathcal{P}$ is infinite dimensional and dim $\mathcal{P}_n = n + 1$. □

■ **EXAMPLE 7.16.** Example 7.10 established that $\mathcal{A} = \{A_{11}, A_{12}, A_{21}, A_{22}\}$ spans $\mathcal{M}_{2\times 2}$, where

$$A_{11} = \begin{bmatrix} 1 & 0 \\ 0 & 0 \end{bmatrix}, \quad A_{12} = \begin{bmatrix} 0 & 1 \\ 0 & 0 \end{bmatrix}, \quad A_{21} = \begin{bmatrix} 0 & 0 \\ 1 & 0 \end{bmatrix}, \quad \text{and} \quad A_{22} = \begin{bmatrix} 0 & 0 \\ 0 & 1 \end{bmatrix}.$$

Since

$$\begin{bmatrix} 0 & 0 \\ 0 & 0 \end{bmatrix} = aA_{11} + bA_{12} + cA_{21} + dA_{22} = \begin{bmatrix} a & b \\ c & d \end{bmatrix}$$

implies $a = b = c = d = 0$, $\mathcal{A}$ is also linearly independent. Thus, $\mathcal{A}$ is a basis for $\mathcal{M}_{2\times 2}$, and dim $\mathcal{M}_{2\times 2} = 4$. Similar reasoning shows that if A_{ij} is the $m \times n$ matrix whose i, j entry is 1 and whose other entries are zeros, then $\mathcal{A} = \{A_{ij} : 1 \le i \le m, 1 \le j \le n\}$ is a basis for $\mathcal{M}_{m\times n}$. Thus, dim $\mathcal{M}_{m\times n} = mn$. □

■ **EXAMPLE 7.17.** According to Example 7.11, $\mathcal{V} = \{f \in \mathcal{P}_2 : f(1) = f(2)\}$ has $\mathcal{A} = \{1, x^2 - 3x\}$ as a spanning set. Since

$$0 = s1 + t(x^2 - 3x) = s - 3tx + tx^2$$

implies $s = t = 0$, $\mathcal{A}$ is linearly independent. Thus, $\mathcal{A}$ is a basis for $\mathcal{V}$ and dim $\mathcal{V} = 2$. □

An infinite dimensional vector space always has a linearly independent subset containing infinitely many vectors (e.g., any basis). Conversely, if a vector space has an infinite linearly independent subset, then by Theorem 7.4, it cannot have a finite basis, so its dimension is ∞.

■ **EXAMPLE 7.18.** Let X_k be the element of $\mathcal{R}^\infty$ that has a 1 in the kth position and zeros elsewhere, that is,

$$\begin{aligned} X_1 &= (1, 0, 0, 0, \cdots) \\ X_2 &= (0, 1, 0, 0, \cdots) \\ X_3 &= (0, 0, 1, 0, \cdots) \\ &\vdots \end{aligned}$$

To check $\mathcal{A} = \{X_1, X_2, X_3, \dots\}$ for linear independence, consider a finite subset, $\mathcal{B} = \{X_{k_1}, \dots, X_{k_p}\}$. If $t_1, t_2, \dots, t_p \in \mathcal{R}$, then

$$t_1 X_{k_1} + \cdots + t_p X_{k_p}$$

is the element of $\mathcal{R}^\infty$ having $t_1, t_2, \dots, t_p$ in positions $k_1, k_2, \dots, k_p$, respectively, and zeros in the remaining positions. This linear combination represents the zero vector only when $t_1 = \cdots = t_p = 0$, so $\mathcal{B}$ is linearly independent. Since each finite subset of $\mathcal{A}$ is linearly independent, $\mathcal{A}$ itself is linearly independent. Note, however, that $\mathcal{A}$ does not span $\mathcal{R}^\infty$. A linear combination of elements of $\mathcal{A}$ has nonzero entries in only a finite number of positions, so for example $(1, 1, 1, \dots) \notin \mathcal{S}pan\{\mathcal{A}\}$. Although $\mathcal{A}$ is not a basis for $\mathcal{R}^\infty$, it is an infinite linearly independent subset of $\mathcal{R}^\infty$, so $\dim \mathcal{R}^\infty = \infty$. □

The next three results are abstract vector space analogues of Theorems 2.15, 2.21, and 2.22. Their proofs are virtually identical to those of their counterparts in Chapter 2 (minor modifications being needed when the subset in question is infinite) and will therefore be omitted.

THEOREM 7.7. If $\mathcal{V}$ is a vector space and $\mathcal{A}$ is a subset of $\mathcal{V}$ containing at least two elements, then the following statements are equivalent:

(1) One of the vectors in $\mathcal{A}$ is a linear combination of the others.
(2) $\mathcal{A}$ has a proper subset, $\mathcal{C}$, such that $\mathcal{S}pan\{\mathcal{C}\} = \mathcal{S}pan\{\mathcal{A}\}$.
(3) $\mathcal{A}$ is linearly dependent.

THEOREM 7.8. If $\mathcal{V}$ is a vector space, $\mathcal{A} \subseteq \mathcal{V}$ is linearly independent, and $\mathbf{v} \in \mathcal{V}$, then $\mathcal{A} \cup \{\mathbf{v}\}$ is linearly independent if and only if $\mathbf{v} \notin \mathcal{S}pan\{\mathcal{A}\}$.

THEOREM 7.9. If p is a positive integer, $\mathcal{V}$ is a vector space of dimension p, and $\mathcal{A} \subseteq \mathcal{V}$ consists of p vectors, then the following are equivalent:

(1) $\mathcal{A}$ is linearly independent.
(2) $\mathcal{S}pan\{\mathcal{A}\} = \mathcal{V}$.

■ **EXAMPLE 7.19.** Let $\mathcal{A} = \{f, g, h\} \subseteq \mathcal{P}_2$, where

$$f(x) = 1 + x + x^2, \quad g(x) = x - x^2, \quad \text{and} \quad h(x) = 1 + 2x,$$

and set $\mathcal{V} = \mathcal{S}pan\{\mathcal{A}\}$, Observe that

$$\begin{aligned} 0 &= rf(x) + sg(x) + th(x) \\ &= r(1 + x + x^2) + s(x - x^2) + t(1 + 2x) \\ &= (r + t) + (r + s + 2t)x + (r - s)x^2 \end{aligned}$$

if and only if

$$\begin{bmatrix} 0 \\ 0 \\ 0 \end{bmatrix} = \begin{bmatrix} r + t \\ r + s + 2t \\ r - s \end{bmatrix} = \begin{bmatrix} 1 & 0 & 1 \\ 1 & 1 & 2 \\ 1 & -1 & 0 \end{bmatrix} \begin{bmatrix} r \\ s \\ t \end{bmatrix}.$$

Since

$$\begin{bmatrix} 1 & 0 & 1 \\ 1 & 1 & 2 \\ 1 & -1 & 0 \end{bmatrix} \longrightarrow \begin{bmatrix} 1 & 0 & 1 \\ 0 & 1 & 1 \\ 0 & 0 & 0 \end{bmatrix},$$

$r = s = -t$, that is,

$$0 = (-t)f + (-t)g + th.$$

Thus, $\mathcal{A}$ is linearly dependent. As long as $t \neq 0$, the latter equation can be solved for f, g, or h, expressing that function as a linear combination of the others, so $\{f,g\}$, $\{f,h\}$, and $\{g,h\}$ are proper subsets of $\mathcal{A}$ that span $\mathcal{V}$. Testing $\{f,g\}$ for linear independence amounts to repeating the earlier calculations with the terms contributed by h deleted. The resulting system is

$$\begin{bmatrix} 0 \\ 0 \\ 0 \end{bmatrix} = \begin{bmatrix} 1 & 0 \\ 1 & 1 \\ 1 & -1 \end{bmatrix} \begin{bmatrix} r \\ s \end{bmatrix}, \quad \text{and} \quad \begin{bmatrix} 1 & 0 \\ 1 & 1 \\ 1 & -1 \end{bmatrix} \longrightarrow \begin{bmatrix} 1 & 0 \\ 0 & 1 \\ 0 & 0 \end{bmatrix},$$

so $r = s = 0$. Thus, $\{f,g\}$ is a basis for $\mathcal{V}$, and $\dim \mathcal{V} = 2$.

Now, $a + bx + cx^2 \in \mathcal{S}pan\{f, g\}$ if and only if there are scalars r and s such that

$$a + bx + cx^2 = r(1 + x + x^2) + s(x - x^2),$$

or equivalently,

$$\begin{bmatrix} a \\ b \\ c \end{bmatrix} = \begin{bmatrix} 1 & 0 \\ 1 & 1 \\ 1 & -1 \end{bmatrix} \begin{bmatrix} r \\ s \end{bmatrix}.$$

Since

$$\left[\begin{array}{cc|c} 1 & 0 & a \\ 1 & 1 & b \\ 1 & -1 & c \end{array}\right] \longrightarrow \left[\begin{array}{cc|c} 1 & 0 & a \\ 0 & 1 & -a + b \\ 0 & 0 & -2a + b + c \end{array}\right],$$

the membership criterion is $-2a + b + c = 0$. Note that $q(x) = 1 - 3x + 4x^2$ does not satisfy this condition, so $q \notin \mathcal{V}$. Thus, $\mathcal{B} = \{f,g,q\}$ is linearly independent (Theorem 7.8), and since $\dim \mathcal{P}_2 = 3$, $\mathcal{B}$ is a basis for $\mathcal{P}_2$. □

■ **EXAMPLE 7.20.** Assume $a,b \in \mathcal{R}$, with $a \neq b$. Let $\mathcal{A} = \{e^{ax}, e^{bx}\}$ and consider $\mathcal{V} = \mathcal{S}pan\{\mathcal{A}\} \subseteq \mathcal{F}(-\infty,\infty)$. If s and t are scalars such that

$$0 = se^{ax} + te^{bx} \tag{7.3}$$

for all $x \in \mathcal{R}$, then differentiating both sides of (7.3) produces

$$0 = ase^{ax} + bte^{bx}, \tag{7.4}$$

and substituting $x = 0$ in (7.3) and (7.4) leads to

$$\begin{bmatrix} 0 \\ 0 \end{bmatrix} = \begin{bmatrix} s + t \\ as + bt \end{bmatrix} = \begin{bmatrix} 1 & 1 \\ a & b \end{bmatrix} \begin{bmatrix} s \\ t \end{bmatrix}.$$

Since $a \neq b$, $\begin{bmatrix} 1 & 1 \\ a & b \end{bmatrix}$ is nonsingular, and therefore $s = t = 0$. Thus, $\mathcal{A}$ is linearly independent and $\dim \mathcal{V} = 2$. Now consider the particular case $a = 1$ and $b = -1$, that is,

$$\mathcal{V} = \mathcal{S}pan\{e^x, e^{-x}\}.$$

Recall that

$$\cosh(x) = \tfrac{1}{2}(e^x + e^{-x}) \quad \text{and} \quad \sinh(x) = \tfrac{1}{2}(e^x - e^{-x}),$$

so $\mathcal{B} = \{\cosh(x), \sinh(x)\} \subseteq \mathcal{V}$. Suppose s and t are scalars such that

$$0 = s\cosh(x) + t\sinh(x). \tag{7.5}$$

When $x = 0$, you obtain $0 = s\cosh(0) + t\sinh(0) = s \cdot 1 + t \cdot 0 = s$, so (7.5) reduces to $0 = t\sinh(x)$. Moreover, $\sinh(x) \neq 0$ when $x \neq 0$, so t must be 0. Thus, $\mathcal{B}$ is linearly independent. By Theorem 7.9, $\mathcal{B}$ is a basis for $\mathcal{V}$. □

The key hypothesis in Theorem 7.9 is the assumption that the number of vectors in $\mathcal{A}$ is the same as the dimension of $\mathcal{V}$. If that number is finite, then $\mathcal{A}$ cannot have one of the basis properties without the other. The situation is quite different when $\dim \mathcal{V} = \infty$. The subset of $\mathcal{R}^\infty$ discussed in Example 7.18 is linearly independent and has infinitely many elements, but it doesn't span $\mathcal{R}^\infty$.

These remarks serve as a warning that there are subtle differences between finite and infinite dimensional vector spaces. If $\mathcal{V}$ has dimension n, then as far as algebraic structure is concerned, $\mathcal{V}$ is indistinguishable from $\mathcal{R}^n$ (a point discussed carefully in Section 7.5). The infinite dimensional case is complicated by the fact that infinite sets come in different sizes, that is, two sets being infinite does not necessarily mean they are considered to have the same number of elements. It is difficult to appreciate the impact of that statement without a thorough understanding of cardinality, a topic that lies beyond the scope of this book.

THEOREM 7.10. If $\mathcal{V}$ is a finite dimensional vector space with basis $\mathcal{A}$, then each $\mathbf{v} \in \mathcal{V}$ is uniquely expressible as a linear combination of elements of $\mathcal{A}$.

Proof. See Theorem 2.18. ■

Exercises 7.3

1. Check each set for linear independence. If linearly dependent, express one of the elements as a linear combination of the others and find a linearly independent subset which has the same span.
   a. $\{1 - 2x + x^2, 2 - x + x^2\} \subseteq \mathcal{P}_2$
   b. $\{1 + 2x^2 - x^3, -1 + x - x^2 + 2x^3, -1 + 3x + x^2 + 4x^3\} \subseteq \mathcal{P}_3$
   c. $\{\sin(x), \cos(x)\} \subseteq \mathcal{F}(-\infty,\infty)$
   d. $\{1, \ln(x), \ln(2x)\} \subseteq \mathcal{F}(0,\infty)$
   e. $\{e^x, xe^x, x^2e^x, x^3e^x\} \subseteq \mathcal{F}(-\infty,\infty)$
   f. $\{\sin^2(x), \cos^2(x), \cos(2x)\} \subseteq \mathcal{F}(-\infty,\infty)$
   g. $\{1, \sqrt{x}, \sqrt[3]{x}\} \subseteq \mathcal{F}[0,\infty)$
   h. $\left\{\frac{1}{x}, 1, \frac{2-3x}{x}\right\} \subseteq \mathcal{F}(\mathrm{J})$, where $\mathrm{J} = (-\infty,0) \cup (0,\infty)$
   i. $\left\{\begin{bmatrix} 1 & 1 \\ 0 & 1 \end{bmatrix}, \begin{bmatrix} 1 & 0 \\ 1 & 1 \end{bmatrix}, \begin{bmatrix} 0 & 1 \\ 1 & 0 \end{bmatrix}, \begin{bmatrix} 1 & 0 \\ 0 & 1 \end{bmatrix}\right\} \subseteq \mathcal{M}_{2\times 2}$

j. $\left\{\begin{bmatrix}1 & 2\\0 & 3\end{bmatrix}, \begin{bmatrix}1 & 0\\2 & 3\end{bmatrix}, \begin{bmatrix}0 & 1\\2 & 3\end{bmatrix}, \begin{bmatrix}1 & 2\\3 & 0\end{bmatrix}\right\} \subseteq \mathcal{M}_{2\times 2}$

k. $\left\{\begin{bmatrix}1 & 1 & 0\\0 & 1 & 1\end{bmatrix}, \begin{bmatrix}0 & 1 & 1\\1 & 0 & 1\end{bmatrix}, \begin{bmatrix}1 & 0 & 1\\1 & 1 & 0\end{bmatrix}, \begin{bmatrix}1 & 1 & 1\\0 & 1 & 0\end{bmatrix}\right\} \subseteq \mathcal{M}_{2\times 3}$

l. $\{[1\ \ -1\ \ 1\ \ 0], [0\ \ 1\ \ -1\ \ 1], [-1\ \ 1\ \ -1\ \ 0], [0\ \ -1\ \ 1\ \ 1]\} \subseteq \mathcal{M}_{1\times 4}$

2. Suppose $a,b,c \in \mathcal{R}$ and $\mathcal{A} = \{e^{ax}, e^{bx}, e^{cx}\}$.
   a. Using the ideas in Example 7.20, show that $\mathcal{A}$ is linearly independent if and only if

$$A = \begin{bmatrix}1 & 1 & 1\\a & b & c\\a^2 & b^2 & c^2\end{bmatrix}$$

   is nonsingular.
   b. Show that A is nonsingular when a, b, and c are distinct.

3. Assuming $J = (-\infty,0) \cup (0,\infty)$ and n is a positive integer, show that the following subsets of $\mathcal{F}(J)$ are linearly independent.
   a. $\{1, x^{-1}, x^{-2}, \ldots, x^{-n}\}$
   b. $\{1, x^{-1}, x^{-2}, x^{-3}, \ldots\}$

4. Check $\mathcal{A} = \{Y_1, Y_2, Y_3, \ldots\} \subseteq \mathcal{R}^\infty$ for linear independence when
   a. $Y_1 = (1,0,0,0,0,\ldots)$, $Y_2 = (1,1,0,0,0,\ldots)$, $Y_3 = (1,1,1,0,0,\ldots)$, ...
   b. $Y_1 = (1,1,1,1,1,\ldots)$, $Y_2 = (0,1,1,1,1,\ldots)$, $Y_3 = (0,0,1,1,1,\ldots)$, ...
   c. $Y_1 = (0,1,1,1,1,\ldots)$, $Y_2 = (1,0,1,1,1,\ldots)$, $Y_3 = (1,1,0,1,1,\ldots)$, ...

5. Is $\mathcal{B}$ a basis for $\mathcal{V}$? Explain!
   a. $\mathcal{B} = \{1 + x^2, 1 + x, x + x^2\}$, $\mathcal{V} = \mathcal{P}_2$
   b. $\mathcal{B} = \left\{\begin{bmatrix}1 & 1\\0 & 0\end{bmatrix}, \begin{bmatrix}0 & 1\\-1 & 0\end{bmatrix}, \begin{bmatrix}-1 & 1\\0 & 1\end{bmatrix}\right\}$, $\mathcal{V} = \mathcal{M}_{2\times 2}$
   c. $\mathcal{B} = \{1 - x^3, 2x^3 - x, 1 + 2x, x^3 + x - 2\}$, $\mathcal{V} = \mathcal{P}_3$
   d. $\mathcal{B} = \{(1,1,1,1,1,\ldots), (0,1,1,1,1,\ldots), (0,0,1,1,1,\ldots), \ldots\}$, $\mathcal{V} = \mathcal{R}^\infty$

6. Find the dimension of each subspace.
   a. $\mathcal{W} = \{A \in \mathcal{M}_{3\times 3} : A \text{ is symmetric}\}$
   b. $\mathcal{W} = \{A \in \mathcal{M}_{2\times 2} : \text{trace}(A) = 0\}$
   c. $\mathcal{W} = \{f \in \mathcal{P} : f(-x) = f(x) \text{ for all } x \in \mathcal{R}\}$
   d. $\mathcal{W} = \{f \in \mathcal{P}_3 : \int_0^1 f(x)\,dx = 0\}$
   e. $\mathcal{W} = \left\{A \in \mathcal{M}_{2\times 2} : A\begin{bmatrix}0 & 1\\1 & 0\end{bmatrix} = \begin{bmatrix}0 & 1\\1 & 0\end{bmatrix}A\right\}$
   f. $\mathcal{W} = \{X \in \mathcal{R}^\infty : x_{2k} = 0, k = 0, 1, 2, \ldots\}$

7. Show that the sequences $\{a_k\}_0^\infty$ satisfying the linear recurrence relation $a_{k+2} = ra_{k+1} + sa_k$ constitute a two-dimensional subspace of $\mathcal{R}^\infty$. (Hint: each solution is determined by its initial data a_0 and a_1.)

7.4
LINEAR FUNCTIONS

DEFINITION. Assuming $\mathcal{V}$ and $\mathcal{W}$ are vector spaces, $T : \mathcal{V} \to \mathcal{W}$ is *linear* if for all $\mathbf{u},\mathbf{v} \in \mathcal{V}$ and for all $t \in \mathcal{R}$,

$$T(\mathbf{u}+\mathbf{v}) = T(\mathbf{u}) + T(\mathbf{v}) \tag{7.6}$$

and

$$T(t\mathbf{v}) = tT(\mathbf{v}) \tag{7.7}$$

Aside from the abstract vector space setting, this notion of linearity is identical to the one discussed in Chapter 4. Of course the + signs on the left and right of (7.6) refer to the addition operations in $\mathcal{V}$ and $\mathcal{W}$, respectively, and a similar comment pertains to the scalar multiplications in (7.7). If $T : \mathcal{V} \to \mathcal{W}$ is linear, then (7.6), (7.7), and a standard induction argument yield

$$T\left(\sum_1^p t_k\mathbf{v}_k\right) = \sum_1^p t_kT(\mathbf{v}_k), \tag{7.8}$$

whenever $\mathbf{v}_1, \ldots, \mathbf{v}_p \in \mathcal{V}$ and $t_1, \ldots, t_p \in \mathcal{R}$.

■ **EXAMPLE 7.21.** Consider $T : \mathcal{M}_{m\times n} \to \mathcal{M}_{n\times m}$ defined by $T(\mathrm{A}) = \mathrm{A}^T$. For $\mathrm{A},\mathrm{B} \in \mathcal{M}_{m\times n}$ and $t \in \mathcal{R}$,

$$T(\mathrm{A}+\mathrm{B}) = (\mathrm{A}+\mathrm{B})^T = \mathrm{A}^T + \mathrm{B}^T = T(\mathrm{A}) + T(\mathrm{B})$$

and

$$T(t\mathrm{A}) = (t\mathrm{A})^T = t\mathrm{A}^T = tT(\mathrm{A}),$$

so T is linear. □

The next two examples show that differentiation and integration are linear operations. To facilitate their discussion, some additional notation will be useful. Given an interval $\mathrm{J} \subseteq \mathcal{R}$, $\mathcal{C}(\mathrm{J})$ and $\mathcal{D}(\mathrm{J})$ denote the subsets of $\mathcal{F}(\mathrm{J})$ consisting of the continuous and differentiable functions, respectively. The sum of two continuous functions is continuous, and a constant multiple of a continuous function is continuous, so $\mathcal{C}(\mathrm{J})$ is a subspace. Similarly, $\mathcal{D}(\mathrm{J})$ is a subspace. In particular cases like $\mathrm{J} = [a,b]$, $\mathcal{C}([a,b])$ and $\mathcal{D}([a,b])$ are shortened to $\mathcal{C}[a,b]$ and $\mathcal{D}[a,b]$.

■ **EXAMPLE 7.22.** Define $D : \mathcal{D}(\mathrm{J}) \to \mathcal{F}(\mathrm{J})$ by $D(f) = f'$. Since

$$D(f+g) = (f+g)' = f' + g' = D(f) + D(g)$$

and

$$D(tf) = (tf)' = tf' = tD(f),$$

for all $f,g \in \mathcal{D}(\mathrm{J})$ and $t \in \mathcal{R}$, D is linear. □

■ **EXAMPLE 7.23.** Let $a,b \in \mathcal{R}$, with $a < b$, and define $T : \mathcal{C}[a,b] \to \mathcal{R}$ by

$$T(f) = \int_a^b f(x)\,dx.$$

If $f,g \in \mathcal{C}[a,b]$ and $t \in \mathcal{R}$, then

$$T(f+g) = \int_a^b [f(x)+g(x)]\,dx = \int_a^b f(x)\,dx + \int_a^b g(x)\,dx = T(f)+T(g)$$

and

$$T(tf) = \int_a^b tf(x)\,dx = t\int_a^b f(x)\,dx = tT(f),$$

so T is linear. □

The next two theorems are abstract vector space analogues of earlier results in Euclidean n-space. In Chapter 4 they were simple consequences of the fact that each linear $T : \mathcal{R}^n \to \mathcal{R}^m$ has an $m \times n$ matrix A such that $T(\mathrm{X}) = \mathrm{AX}$, but here they must be established without the aid of a matrix representation.

THEOREM 7.11. If $\mathcal{V}$ and $\mathcal{W}$ are vector spaces and $T : \mathcal{V} \to \mathcal{W}$ is linear, then

(1) $T(\mathbf{0}_{\mathcal{V}}) = \mathbf{0}_{\mathcal{W}}$, and
(2) $T(-\mathbf{v}) = -T(\mathbf{v})$, $\mathbf{v} \in \mathcal{V}$.

Proof. For $\mathbf{v} \in \mathcal{V}$,

$$\begin{aligned} T(\mathbf{0}_{\mathcal{V}}) &= T(0\mathbf{v}) && \text{(Theorem 7.1)} \\ &= 0T(\mathbf{v}) && \text{(linearity of } T) \\ &= \mathbf{0}_{\mathcal{W}}, && \text{(Theorem 7.1)} \end{aligned}$$

and

$$\begin{aligned} T(-\mathbf{v}) &= T((-1)\mathbf{v}) && \text{(Theorem 7.1)} \\ &= (-1)T(\mathbf{v}) && \text{(linearity of } T) \\ &= -T(\mathbf{v}). && \text{(Theorem 7.1)} \ ■ \end{aligned}$$

DEFINITION. Assuming $\mathcal{V}$ and $\mathcal{W}$ are vector spaces and $T : \mathcal{V} \to \mathcal{W}$,

$$\mathcal{Rng}(T) = \{\mathbf{w} \in \mathcal{W} : \mathbf{w} = T(\mathbf{v}) \quad \text{for some} \quad \mathbf{v} \in \mathcal{V}\}$$

and

$$\mathcal{Ker}(T) = \{\mathbf{v} \in \mathcal{V} : T(\mathbf{v}) = \mathbf{0}_{\mathcal{W}}\}.$$

THEOREM 7.12. If $\mathcal{V}$ and $\mathcal{W}$ are vector spaces and $T : \mathcal{V} \to \mathcal{W}$ is linear, then

(1) $\mathcal{Rng}(T)$ is a subspace of $\mathcal{W}$, and
(2) $\mathcal{Ker}(T)$ is a subspace of $\mathcal{V}$.

Proof. Given $\mathbf{w}_1,\mathbf{w}_2,\mathbf{w} \in \mathcal{Rng}(T)$, there are $\mathbf{v}_1,\mathbf{v}_2,\mathbf{v} \in \mathcal{V}$ such that $T(\mathbf{v}_1) = \mathbf{w}_1$, $T(\mathbf{v}_2) = \mathbf{w}_2$, and $T(\mathbf{v}) = \mathbf{w}$. Since $\mathbf{w}_1 + \mathbf{w}_2 = T(\mathbf{v}_1) + T(\mathbf{v}_2) = T(\mathbf{v}_1 + \mathbf{v}_2)$, and $t\mathbf{w} = tT(\mathbf{v}) = T(t\mathbf{v})$,

$\mathbf{w}_1 + \mathbf{w}_2 \in \mathcal{R}ng(T)$ and $t\mathbf{w} \in \mathcal{R}ng(T)$. Thus, $\mathcal{R}ng(T)$ is a subspace. Part (2) is left as an exercise. ■

Suppose $\mathcal{V}$ and $\mathcal{W}$ are vector spaces, $T : \mathcal{V} \to \mathcal{W}$ is linear, and $\mathcal{U}$ is a subspace of $\mathcal{V}$. The *restriction of T to $\mathcal{U}$*, denoted by $T|_{\mathcal{U}}$, is the function from $\mathcal{U}$ to $\mathcal{W}$ defined by

$$(T|_{\mathcal{U}})(\mathbf{u}) = T(\mathbf{u}), \quad \mathbf{u} \in \mathcal{U}.$$

Formally, $T|_{\mathcal{U}}$ is different from T because it has a different domain, but its values are just the values of T on that domain. Now, $\mathcal{U}$ itself is a vector space (Theorem 7.2), and T satisfies linearity conditions (7.6) and (7.7) for all $\mathbf{u}, \mathbf{v} \in \mathcal{V}$ and $t \in \mathcal{R}$, so it certainly does so for all $\mathbf{u}, \mathbf{v} \in \mathcal{U}$ and $t \in \mathcal{R}$. Thus, $T|_{\mathcal{U}} : \mathcal{U} \to \mathcal{W}$ is linear.

THEOREM 7.13. Suppose $\mathcal{V}$ and $\mathcal{W}$ are vector spaces and $T : \mathcal{V} \to \mathcal{W}$ is linear. If $\mathcal{U}$ is a subspace of $\mathcal{V}$ and $\mathcal{A}$ is a spanning set for $\mathcal{U}$, then

(1) $T(\mathcal{U})$ is a subspace of $\mathcal{W}$, and
(2) $T(\mathcal{A})$ is a spanning set for $T(\mathcal{U})$.

In particular, if $\mathcal{V} = \mathcal{S}pan\{\mathcal{A}\}$, then $\mathcal{R}ng(T) = \mathcal{S}pan\{T(\mathcal{A})\}$.

Proof. Since $T(\mathcal{U}) = \{T(\mathbf{u}) : \mathbf{u} \in \mathcal{U}\} = \mathcal{R}ng(T|_{\mathcal{U}})$ and $T|_{\mathcal{U}} : \mathcal{U} \to \mathcal{W}$ is linear, $T(\mathcal{U})$ is a subspace of $\mathcal{W}$ (Theorem 7.12). Suppose $\mathcal{A}$ is a spanning set for $\mathcal{U}$. Of course $T(\mathcal{A}) \subseteq T(\mathcal{U})$. Moreover, $T(\mathcal{U})$, being closed under addition and scalar multiplication, contains all linear combinations of elements of $T(\mathcal{A})$, that is, $\mathcal{S}pan\{T(\mathcal{A})\} \subseteq T(\mathcal{U})$. For the reverse containment, suppose $\mathbf{w} \in T(\mathcal{U})$ and choose $\mathbf{u} \in \mathcal{U}$ such that $\mathbf{w} = T(\mathbf{u})$. Since $\mathcal{U} = \mathcal{S}pan\{\mathcal{A}\}$, there is a positive integer p such that

$$\mathbf{u} = \sum_{1}^{p} t_k \mathbf{u}_k,$$

where $\mathbf{u}_1, \ldots, \mathbf{u}_p \in \mathcal{A}$ and $t_1, \ldots, t_p \in \mathcal{R}$, and therefore

$$\mathbf{w} = T(\mathbf{u}) = T\left(\sum_{1}^{p} t_k \mathbf{u}_k\right) = \sum_{1}^{p} t_k T(\mathbf{u}_k).$$

Thus, $\mathbf{w} \in \mathcal{S}pan\{T(\mathcal{A})\}$. The final statement follows from (2) and the fact that $T(\mathcal{V}) = \mathcal{R}ng(T)$. ■

■ **EXAMPLE 7.24.** Consider $T : \mathcal{M}_{2\times2} \to \mathcal{M}_{2\times2}$ defined by $T(\mathrm{A}) = \mathrm{A} - \mathrm{A}^T$. Set $\mathcal{A} = \{\mathrm{A}_{11}, \mathrm{A}_{12}, \mathrm{A}_{21}, \mathrm{A}_{22}\}$, where

$$\mathrm{A}_{11} = \begin{bmatrix} 1 & 0 \\ 0 & 0 \end{bmatrix}, \quad \mathrm{A}_{12} = \begin{bmatrix} 0 & 1 \\ 0 & 0 \end{bmatrix}, \quad \mathrm{A}_{21} = \begin{bmatrix} 0 & 0 \\ 1 & 0 \end{bmatrix}, \quad \text{and} \quad \mathrm{A}_{22} = \begin{bmatrix} 0 & 0 \\ 0 & 1 \end{bmatrix}.$$

Then $\mathcal{M}_{2\times2} = \mathcal{S}pan\{\mathcal{A}\}$, and

$$\begin{aligned} \mathcal{R}ng(T) &= \mathcal{S}pan\{T(\mathrm{A}_{11}), T(\mathrm{A}_{12}), T(\mathrm{A}_{21}), T(\mathrm{A}_{22})\} \\ &= \mathcal{S}pan\left\{\begin{bmatrix} 0 & 0 \\ 0 & 0 \end{bmatrix}, \begin{bmatrix} 0 & 1 \\ -1 & 0 \end{bmatrix}, \begin{bmatrix} 0 & -1 \\ 1 & 0 \end{bmatrix}, \begin{bmatrix} 0 & 0 \\ 0 & 0 \end{bmatrix}\right\}. \end{aligned}$$

Note that $T(\mathrm{A}_{21}) = -T(\mathrm{A}_{12})$. Deleting $T(\mathrm{A}_{21})$ and the two copies of the zero vector leaves a linearly independent subset of $T(\mathcal{A})$ that spans $\mathcal{R}ng(T)$, namely,

$$\left\{\begin{bmatrix} 0 & 1 \\ -1 & 0 \end{bmatrix}\right\}.$$

Thus, $\mathcal{Rng}(T)$ is the one-dimensional subspace of $\mathcal{M}_{2\times 2}$ consisting of the matrices $\begin{bmatrix} 0 & c \\ -c & 0 \end{bmatrix}$, $c \in \mathcal{R}$, that is, the 2×2 skew-symmetric matrices. Moreover, $A \in \mathcal{Ker}(T)$ if and only if $0 = T(A) = A - A^T$, so $\mathcal{Ker}(T) = \mathcal{S}_{2\times 2}$ (the 2×2 symmetric matrices). □

THEOREM 7.14. If $\mathcal{V}$ and $\mathcal{W}$ are vector spaces and $T : \mathcal{V} \to \mathcal{W}$ is linear, then T is one-to-one if and only if $\mathcal{Ker}(T) = \{\mathbf{0}_\mathcal{V}\}$.

Proof. Suppose T is one-to-one. If $\mathbf{v} \in \mathcal{Ker}(T)$, then $T(\mathbf{v}) = \mathbf{0}_\mathcal{W} = T(\mathbf{0}_\mathcal{V})$, and therefore $\mathbf{v} = \mathbf{0}_\mathcal{V}$. Thus, $\mathcal{Ker}(T) = \{\mathbf{0}_\mathcal{V}\}$. On the other hand, if $\mathcal{Ker}(T) = \{\mathbf{0}_\mathcal{V}\}$ and $\mathbf{v}_1, \mathbf{v}_2 \in \mathcal{V}$ such that $T(\mathbf{v}_1) = T(\mathbf{v}_2)$, then

$$T(\mathbf{v}_1 - \mathbf{v}_2) = T(\mathbf{v}_1) - T(\mathbf{v}_2) = \mathbf{0}_\mathcal{W},$$

so $\mathbf{v}_1 - \mathbf{v}_2 \in \mathcal{Ker}(T)$, and therefore $\mathbf{v}_1 - \mathbf{v}_2 = \mathbf{0}_\mathcal{V}$. Thus, T is one-to-one. ■

EXAMPLE 7.25. Consider $D : \mathcal{D}(\mathrm{J}) \to \mathcal{F}(\mathrm{J})$ (Example 7.22). The kernel of D consists of those $f : \mathrm{J} \to \mathcal{R}$ satisfying $f'(x) = 0, x \in \mathrm{J}$, that is, the constant functions. Since $\mathcal{Ker}(D)$ contains nonzero vectors, D is not one-to-one. Neither is D onto. Consider, for example, $\mathrm{J} = [0,1]$ and $g \in \mathcal{F}[0,1]$ defined by

$$g(x) = \begin{cases} 0, & \text{if } x \in [0, 1/2] \\ 1, & \text{if } x \in (1/2, 1] \end{cases}.$$

An $f \in \mathcal{D}[0,1]$ such that $f' = g$ would be constant on [0,1/2], and its graph on (1/2,1] would be a line with slope 1. There are continuous functions with those two properties (Figure 7.1), but none that are differentiable at $x = 1/2$. Thus, there is no $f \in \mathcal{D}[0,1]$ such that $D(f) = g$.

Now consider the action of D on the subspace $\mathcal{P}_n$. Let $f \in \mathcal{P}_n$ and set $g = D(f)$. Then

$$f(x) = a_0 + a_1x + a_2x^2 + \cdots + a_nx^n$$

and

$$g(x) = f'(x) = a_1 + 2a_2x + \cdots + na_nx^{n-1},$$

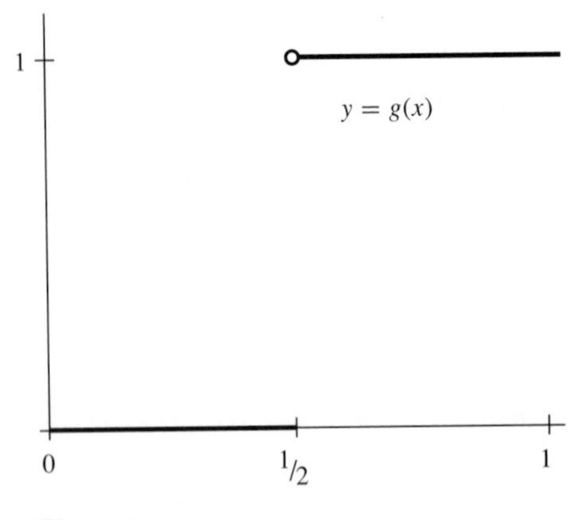

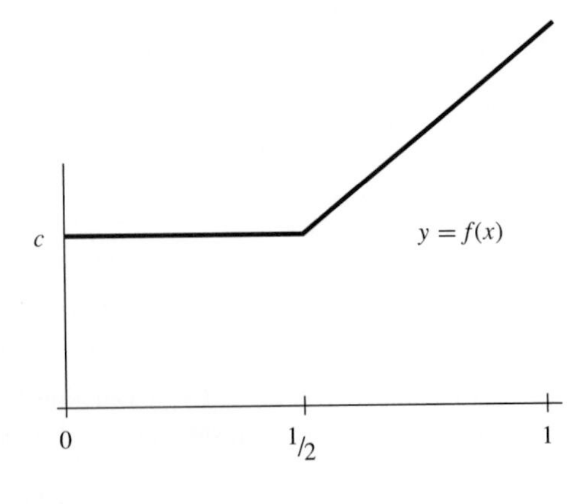

FIGURE 7.1

so $g \in \mathcal{P}_{n-1}$. Moreover, each element of $\mathcal{P}_{n-1}$ is the image of an element of $\mathcal{P}_n$. Indeed, if

$$g(x) = b_0 + b_1 x + \cdots + b_{n-1} x^{n-1}$$

and

$$f(x) = c + b_0 x + (b_1/2)x^2 + \cdots + (b_{n-1}/n)x^n,$$

then $f \in \mathcal{P}_n$ and $D(f) = g$. Thus, $D|_{\mathcal{P}_n} : \mathcal{P}_n \to \mathcal{P}_{n-1}$ is onto. Since $\mathcal{K}er(D|_{\mathcal{P}_n})$ contains the constant functions, $D|_{\mathcal{P}_n}$ is not one-to-one. □

THEOREM 7.15. Suppose $\mathcal{V}$ and $\mathcal{W}$ are vector spaces, $T : \mathcal{V} \to \mathcal{W}$ is linear, and $\mathcal{A} \subseteq \mathcal{V}$. If T is one-to-one and $\mathcal{A}$ is linearly independent, then $T(\mathcal{A})$ is linearly independent.

Proof. Consider a finite subset $\{\mathbf{w}_1, \ldots, \mathbf{w}_p\}$ of $T(\mathcal{A})$ and let $\mathbf{v}_1, \ldots, \mathbf{v}_p \in \mathcal{A}$ such that $T(\mathbf{v}_k) = \mathbf{w}_k$, $1 \le k \le p$. If

$$\mathbf{0}_{\mathcal{W}} = \sum_1^p t_k \mathbf{w}_k = \sum_1^p t_k T(\mathbf{v}_k) = T\left(\sum_1^p t_k \mathbf{v}_k\right),$$

then $t_1\mathbf{v}_1 + \cdots + t_p\mathbf{v}_p \in \mathcal{K}er(T)$, so

$$\mathbf{0}_{\mathcal{V}} = t_1\mathbf{v}_1 + \cdots + t_p\mathbf{v}_p.$$

Since $\mathcal{A}$ is linearly independent, $t_1 = \cdots = t_p = 0$, and therefore $\{\mathbf{w}_1, \ldots, \mathbf{w}_p\}$ is linearly independent. Thus, each finite subset of $T(\mathcal{A})$ is linearly independent. ■

COROLLARY. If $\mathcal{V}$ and $\mathcal{W}$ are vector spaces and $T : \mathcal{V} \to \mathcal{W}$ is linear and one-to-one, then $\dim \mathcal{R}ng(T) = \dim \mathcal{V}$.

Proof. If $\mathcal{A}$ is a basis for $\mathcal{V}$, then $T(\mathcal{A})$ is linearly independent (Theorem 7.15) and $T(\mathcal{A})$ spans $\mathcal{R}ng(T)$ (Theorem 7.13), so $T(\mathcal{A})$ is a basis for $\mathcal{R}ng(T)$. Since T is one-to-one, $\mathcal{A}$ and $T(\mathcal{A})$ have the same number of elements. ■

If $T : \mathcal{V} \to \mathcal{W}$ is one-to-one and onto, then for each $\mathbf{w} \in \mathcal{W}$ there is a unique $\mathbf{v} \in \mathcal{V}$ such that $T(\mathbf{v}) = \mathbf{w}$. Consequently there is a function $T^{-1} : \mathcal{W} \to \mathcal{V}$, called the inverse of T, defined by $T^{-1}(\mathbf{w}) = \mathbf{v}$ if and only if $T(\mathbf{v}) = \mathbf{w}$.

THEOREM 7.16. If $\mathcal{V}$ and $\mathcal{W}$ are vector spaces and $T : \mathcal{V} \to \mathcal{W}$ is linear, one-to-one, and onto, then $T^{-1} : \mathcal{W} \to \mathcal{V}$ is linear.

Proof. See Exercise 8. ■

Given vector spaces $\mathcal{U}$, $\mathcal{V}$, and $\mathcal{W}$, the composition of $T : \mathcal{V} \to \mathcal{W}$ with $S : \mathcal{U} \to \mathcal{V}$ is the function $T \circ S : \mathcal{U} \to \mathcal{W}$ defined by $(T \circ S)(\mathbf{u}) = T(S(\mathbf{u}))$, $\mathbf{u} \in \mathcal{U}$.

THEOREM 7.17. If $\mathcal{U}$, $\mathcal{V}$, and $\mathcal{W}$ are vector spaces and $S : \mathcal{U} \to \mathcal{V}$ and $T : \mathcal{V} \to \mathcal{W}$ are linear, then $T \circ S : \mathcal{U} \to \mathcal{W}$ is linear.

Proof. For all $\mathbf{u}, \mathbf{v} \in \mathcal{U}$ and $t \in \mathcal{R}$,

$$\begin{aligned}
(T \circ S)(\mathbf{u} + \mathbf{v}) &= T(S(\mathbf{u} + \mathbf{v})) && \text{(defn. of } T \circ S\text{)}\\
&= T(S(\mathbf{u}) + S(\mathbf{v})) && \text{(linearity of } S\text{)}\\
&= T(S(\mathbf{u})) + T(S(\mathbf{v})) && \text{(linearity of } T\text{)}\\
&= (T \circ S)(\mathbf{u}) + (T \circ S)(\mathbf{v}), && \text{(defn. of } T \circ S\text{)}
\end{aligned}$$

and

$$\begin{aligned}(T \circ S)(t\mathbf{u}) &= T(S(t\mathbf{u})) && \text{(defn. of } T \circ S\text{)}\\ &= T(tS(\mathbf{u})) && \text{(linearity of } S\text{)}\\ &= tT(S(\mathbf{u})) && \text{(linearity of } T\text{)}\\ &= t(T \circ S)(\mathbf{u}). && \text{(defn. of } T \circ S\text{)} \blacksquare\end{aligned}$$

If $\mathcal{U}$, $\mathcal{V}$, $\mathcal{W}$, and $\mathcal{Z}$ are vector spaces, $R : \mathcal{U} \to \mathcal{V}$, $S : \mathcal{V} \to \mathcal{W}$, and $T : \mathcal{W} \to \mathcal{Z}$, then $T \circ (S \circ R) = (T \circ S) \circ R$. Indeed, for each $\mathbf{u} \in \mathcal{U}$,

$$\begin{aligned}[T \circ (S \circ R)](\mathbf{u}) &= T((S \circ R)(\mathbf{u})) && \text{(defn. of } T \circ (S \circ R)\text{)}\\ &= T(S(R(\mathbf{u}))) && \text{(defn. of } S \circ R\text{)}\\ &= (T \circ S)(R(\mathbf{u})) && \text{(defn. of } T \circ S\text{)}\\ &= [(T \circ S) \circ R](\mathbf{u}). && \text{(defn. of } (T \circ S) \circ R\text{)}\end{aligned}$$

When $\mathcal{U} = \mathcal{V} = \mathcal{W}$, $S \circ T$ and $T \circ S$ are both meaningful but are not usually the same function. If $S \circ T = T \circ S$, then S and T are said to commute. Each operator on $\mathcal{V}$ commutes with itself ($T \circ T = T \circ T$), and T^2 is the customary abbreviation for $T \circ T$. For each positive integer k, T^k denotes the composition of k copies of T. By Theorem 7.17, T^k is linear whenever T is linear.

DEFINITION. Assume $\mathcal{V}$ is a vector space, $T : \mathcal{V} \to \mathcal{V}$ is linear, $\lambda \in \mathcal{R}$, and $\mathbf{v} \in \mathcal{V}$. If $\mathbf{v} \neq \mathbf{0}_{\mathcal{V}}$ and $T(\mathbf{v}) = \lambda\mathbf{v}$, then $\mathbf{v}$ is an *eigenvector* of T with *eigenvalue* λ. The *eigenspace* of T associated with eigenvalue λ is $\mathcal{V}(\lambda) = \{\mathbf{v} \in \mathcal{V} : T(\mathbf{v}) = \lambda\mathbf{v}\}$.

THEOREM 7.18. If $\mathcal{V}$ is a vector space, $T : \mathcal{V} \to \mathcal{V}$ is linear, and $\lambda \in \mathcal{R}$ is an eigenvalue of T, then $\mathcal{V}(\lambda)$ is a subspace of $\mathcal{V}$.

Proof. See Exercise 13. ■

■ **EXAMPLE 7.26.** Consider $T : \mathcal{M}_{n\times n} \to \mathcal{R}_{n\times n}$ defined by $T(\mathrm{A}) = \mathrm{A}^T$. If λ is an eigenvalue of T, then there is an $\mathrm{A} \in \mathcal{M}_{n\times n}$ such that $\mathrm{A} \neq 0$ and

$$\mathrm{A}^T = \lambda\mathrm{A}. \tag{7.9}$$

Being nonzero, A has at least one nonzero entry, say, a_{ij}. If $i = j$, then by (7.9),

$$a_{ii} = [\mathrm{A}^T]_{ii} = [\lambda\mathrm{A}]_{ii} = \lambda a_{ii},$$

and therefore $\lambda = 1$. When $i \neq j$, the i,j and j,i positions in (7.9) yield

$$a_{ji} = \lambda a_{ij} \qquad \text{and} \qquad a_{ij} = \lambda a_{ji},$$

respectively, so $a_{ij} = \lambda(\lambda a_{ij}) = \lambda^2 a_{ij}$. Thus, $\lambda^2 = 1$. These observations show that ± 1 are the only possible eigenvalues. In fact, both are eigenvalues, with $\mathcal{V}(1)$ and $\mathcal{V}(-1)$ consisting of the symmetric and skew-symmetric matrices, respectively. □

Let $\mathcal{C}^\infty(\mathrm{J})$ be the subspace of $\mathcal{F}(\mathrm{J})$ consisting of infinitely differentiable functions. Observe that $D : \mathcal{C}^\infty(\mathrm{J}) \to \mathcal{C}^\infty(\mathrm{J})$ is a linear operator. The next example identifies its eigenvalues.

■ **EXAMPLE 7.27.** Assume $\lambda \in \mathcal{R}$ and set $y = f(x)$, where $f \in \mathcal{C}^{\infty}(J)$. Suppose

$$f' = D(f) = \lambda f,$$

or equivalently,

$$\frac{dy}{dx} = \lambda y. \tag{7.10}$$

You may recall that (7.10) is the differential equation governing exponential growth or decay. A quick calculation verifies that $y = e^{\lambda x}$ is a solution of (7.10), and since the latter function in not the zero element of $\mathcal{C}^{\infty}(J)$, λ is an eigenvalue of D. If $f \in \mathcal{V}(\lambda)$ (i.e., $f' = \lambda f$) and $g(x) = e^{-\lambda x} f(x)$, then

$$\begin{aligned} g'(x) &= e^{-\lambda x} f'(x) - \lambda e^{-\lambda x} f(x) \\ &= e^{-\lambda x}[f'(x) - \lambda f(x)] \\ &= 0, \end{aligned}$$

$x \in [a,b]$, so $g(x) \equiv c$ for some $c \in \mathcal{R}$, and therefore $f(x) = ce^{\lambda x}$. Thus,

$$\mathcal{V}(\lambda) = \{ce^{\lambda x} \ : \ c \in \mathcal{R}\} = \mathcal{S}pan\{e^{\lambda x}\}.$$

These conclusions establish that each real number is an eigenvalue for D and that the corresponding eigenspace has dimension 1.

Exercises 7.4

1. Determine whether T is linear.
   a. $T : \mathcal{P}_n \to \mathcal{P}_n, T(f)(x) = xf'(x)$
   b. $T : \mathcal{M}_{n\times n} \to \mathcal{R}, T(A) = \text{trace}(A)$
   c. $T : \mathcal{F}(-\infty,\infty), \to \mathcal{F}(-\infty,\infty), T(f)(x) = f(x - c)$, c a fixed real number
   d. $T : \mathcal{M}_{n\times n} \to \mathcal{R}, T(A) = \det(A)$
   e. $T : \mathcal{F}[a,b] \to \mathcal{R}, T(f) = f(c)$, c a fixed number in $[a,b]$
   f. $T : \mathcal{P} \to \mathcal{P}, T(f)(x) = x^2 f(x)$

2. Show that T is linear. Find a basis for $\mathcal{K}er(T)$ and a basis for $\mathcal{R}ng(T)$. Is T one-to-one and/or onto?

   a. $T : \mathcal{M}_{2\times 2} \to \mathcal{R}^3, T\left(\begin{bmatrix} a & b \\ c & d \end{bmatrix}\right) = \begin{bmatrix} a+b \\ b+c \\ c+d \end{bmatrix}$

   b. $T : \mathcal{P}_2 \to \mathcal{P}_2, T(f) = xf' + f$

   c. $T : \mathcal{P}_3 \to \mathcal{R}, T(f) = f(1)$

   d. $T : \mathcal{R}^{\infty} \to \mathcal{R}^{\infty}, T(X) = (x_1 - x_0, x_2 - x_1, x_3 - x_2, \dots)$.

3. Assuming $T : \mathcal{V} \to \mathcal{W}$ is linear, show that $\mathcal{K}er(T)$ is a subspace of $\mathcal{V}$.

4. Show that $T : \mathcal{R}^{\infty} \to \mathcal{R}^{\infty}$ defined by $T(x_0, x_1, x_2, \dots) = (0, x_0, x_1, x_2, \dots)$ is linear. Is T one-to-one? Is T onto?

5. Show that $T : \mathcal{R}^{\infty} \to \mathcal{R}^{\infty}$ given by $T(x_0, x_1, x_2, \dots) = (x_1, x_2, x_3, \dots)$ is linear. Is T one-to-one? Is T onto?

6. Assuming $P \in \mathcal{M}_{m\times m}$ and $Q \in \mathcal{M}_{n\times n}$ are nonsingular, define $T : \mathcal{M}_{m\times n} \to \mathcal{M}_{m\times n}$ by $T(A) = PAQ$. Show that T is linear. Is T one-to-one? Is T onto?

7. Find the kernel of $D^2 : \mathcal{C}^\infty(-\infty,\infty) \to \mathcal{C}^\infty(-\infty,\infty)$.

8. Prove Theorem 7.16.

9. Define $T : \mathcal{P} \to \mathcal{P}$ by $T(f) = xf$ and assume $D : \mathcal{P} \to \mathcal{P}$ is the differentiation operator.
   a. Show that $T \circ D \neq D \circ T$.
   b. Calculate $(D \circ T)^2(f)$ when $f(x) = a_0 + a_1x + a_2x^2 + \cdots + a_nx^n$.

10. Assume $\mathcal{V}$, $\mathcal{W}$, and $\mathcal{Z}$ are vector spaces and $S : \mathcal{V} \to \mathcal{W}$ and $T : \mathcal{W} \to \mathcal{Z}$ are linear.
   a. Show that $\mathcal{K}er(S) \subseteq \mathcal{K}er(T \circ S)$ and $\mathcal{R}ng(T \circ S) \subseteq \mathcal{R}ng(T)$.
   b. Show that $T \circ S$ one-to-one $\Rightarrow$ S is one-to-one, and $T \circ S$ onto $\Rightarrow$ T is onto.

11. Given a vector space $\mathcal{V}$, define $\mathrm{Id}_\mathcal{V} : \mathcal{V} \to \mathcal{V}$ by $\mathrm{Id}_\mathcal{V}(\mathbf{v}) = \mathbf{v}$. If $S,T : \mathcal{V} \to \mathcal{V}$, show that
   a. $\mathrm{Id}_\mathcal{V}$ is linear. b. $\mathrm{Id}_\mathcal{V} \circ S = S = S \circ \mathrm{Id}_\mathcal{V}$.
   c. $S \circ T = \mathrm{Id}_\mathcal{V} = T \circ S$ implies S is one-to-one and onto and $T = S^{-1}$.

12. Assuming $T : \mathcal{V} \to \mathcal{V}$ is linear, show that $\lambda = 0$ is an eigenvalue of T if and only if T is not one-to-one.

13. Prove Theorem 7.18.

14. Show that $T : \mathcal{R}^\infty \to \mathcal{R}^\infty$ defined by $T(x_0, x_1, x_2, \ldots) = (0, x_0, x_1, \ldots)$ has no eigenvalues.

15. Define $T : \mathcal{R}^\infty \to \mathcal{R}^\infty$ by $T(x_0, x_1, x_2, \ldots) = (x_1, x_2, x_3, \ldots)$. Show that every $\lambda \in \mathcal{R}$ is an eigenvalue for T and find a basis for $\mathcal{V}(\lambda)$.

16. Show that every real number is an eigenvalue of D^2.

7.5
COORDINATES

Let $\mathcal{V}$ be a vector space with basis $\mathcal{A} = \{\mathbf{v}_1, \ldots, \mathbf{v}_p\}$. Each element of $\mathcal{V}$ can be uniquely written as a linear combination of $\mathbf{v}_1, \ldots, \mathbf{v}_p$ and can therefore be identified with a p-tuple of real numbers. In describing the vectors in $\mathcal{V}$ by reference to a particular basis one has in effect established a coordinate system for $\mathcal{V}$.

DEFINITION. Assume $\mathcal{V}$ is a finite dimensional vector space with basis $\mathcal{A} = \{\mathbf{v}_1, \ldots, \mathbf{v}_p\}$. The *coordinates* of $\mathbf{v} \in \mathcal{V}$ relative to $\mathcal{A}$ are the scalars $t_1, \ldots, t_p$ satisfying

$$\mathbf{v} = t_1\mathbf{v}_1 + \cdots + t_p\mathbf{v}_p,$$

and $[\mathbf{v}]_\mathcal{A} = \begin{bmatrix} t_1 \\ \vdots \\ t_p \end{bmatrix}$ is the $\mathcal{A}$ *coordinate vector* for $\mathbf{v}$. Note that $[\mathbf{v}]_\mathcal{A} \in \mathcal{R}^p$.

■ **EXAMPLE 7.28.** Consider $f \in \mathcal{P}_2$, where $f(x) = 1 + 2x + 3x^2$. The usual basis for $\mathcal{P}_2$ is $\mathcal{A} = \{1, x, x^2\}$, and

$$[f]_{\mathcal{A}} = \begin{bmatrix} 1 \\ 2 \\ 3 \end{bmatrix}.$$

A second basis for $\mathcal{P}_2$ is $\mathcal{B} = \{1 + x^2, 1 + x, x + x^2\}$. The coordinates of f relative to $\mathcal{B}$, are the scalars r, s, and t satisfying

$$1 + 2x + 3x^2 = r(1 + x^2) + s(1 + x) + t(x + x^2),$$

$x \in \mathcal{R}$. Comparing coefficients of like powers of x on both sides produces

$$\begin{bmatrix} 1 \\ 2 \\ 3 \end{bmatrix} = \begin{bmatrix} r + s \\ s + t \\ r + t \end{bmatrix} = \begin{bmatrix} 1 & 1 & 0 \\ 0 & 1 & 1 \\ 1 & 0 & 1 \end{bmatrix} \begin{bmatrix} r \\ s \\ t \end{bmatrix}.$$

Since

$$\left[\begin{array}{ccc|c} 1 & 1 & 0 & 1 \\ 0 & 1 & 1 & 2 \\ 1 & 0 & 1 & 3 \end{array}\right] \longrightarrow \left[\begin{array}{ccc|c} 1 & 0 & 0 & 1 \\ 0 & 1 & 0 & 0 \\ 0 & 0 & 1 & 2 \end{array}\right],$$

$r = 1$, $s = 0$, and $t = 2$, that is, $[f]_{\mathcal{B}} = \begin{bmatrix} 1 \\ 0 \\ 2 \end{bmatrix}$. □

■ **EXAMPLE 7.29.** The usual basis for $\mathcal{S}_{2\times 2}$ (the 2×2 symmetric matrices) is $\mathcal{A} = \left\{\begin{bmatrix} 1 & 0 \\ 0 & 0 \end{bmatrix}, \begin{bmatrix} 0 & 1 \\ 1 & 0 \end{bmatrix}, \begin{bmatrix} 0 & 0 \\ 0 & 1 \end{bmatrix}\right\}$. If $\mathrm{A} \in \mathcal{S}_{2\times 2}$, then there are real numbers a, b, and c such that $\mathrm{A} = \begin{bmatrix} a & b \\ b & c \end{bmatrix} = a\begin{bmatrix} 1 & 0 \\ 0 & 0 \end{bmatrix} + b\begin{bmatrix} 0 & 1 \\ 1 & 0 \end{bmatrix} + c\begin{bmatrix} 0 & 0 \\ 0 & 1 \end{bmatrix}$, so $[\mathrm{A}]_{\mathcal{A}} = \begin{bmatrix} a \\ b \\ c \end{bmatrix}$. □

■ **EXAMPLE 7.30.** If $\mathcal{V}$ is a vector space with basis $\mathcal{A} = \{\mathbf{v}_1, \ldots, \mathbf{v}_p\}$, then for each $k = 1, \ldots, p$, $[\mathbf{v}_k]_{\mathcal{A}} = \mathrm{E}_k$, where E_k is the kth member of the standard basis for $\mathcal{R}^p$. Also, $\mathbf{0}_{\mathcal{V}} = 0\mathbf{v}_1 + \cdots + 0\mathbf{v}_p$, so

$$[\mathbf{0}_{\mathcal{V}}]_{\mathcal{A}} = \begin{bmatrix} 0 \\ \vdots \\ 0 \end{bmatrix} \in \mathcal{R}^p. \ \square$$

The next theorem records the interplay between the algebraic operations in a finite dimensional vector space and the coordinates, relative to a fixed basis, of the vectors in the space. Its proof, being identical to that of Theorem 2.32, is not repeated.

THEOREM 7.19. If $\mathcal{V}$ is a finite dimensional vector space with basis $\mathcal{A}$, $\mathbf{u},\mathbf{v} \in \mathcal{V}$ and $t \in \mathcal{R}$, then

(1) $[\mathbf{u} + \mathbf{v}]_{\mathcal{A}} = [\mathbf{u}]_{\mathcal{A}} + [\mathbf{v}]_{\mathcal{A}}$, and
(2) $[t\mathbf{v}]_{\mathcal{A}} = t[\mathbf{v}]_{\mathcal{A}}$.

An inductive argument employing (1) and (2) of Theorem 7.19 gives

$$[t_1\mathbf{v}_1 + \cdots + t_k\mathbf{v}_k]_{\mathcal{A}} = t_1[\mathbf{v}_1]_{\mathcal{A}} + \cdots + t_k[\mathbf{v}_k]_{\mathcal{A}} \tag{7.11}$$

whenever $\mathbf{v}_1, \ldots, \mathbf{v}_k \in \mathcal{V}$ and $t_1, \ldots, t_k \in \mathcal{R}$.

An algebraic manipulation in $\mathcal{V}$ is a finite sequence of additions and/or scalar multiplications involving elements of $\mathcal{V}$, the net effect of which is a linear combination $t_1\mathbf{v}_1 + \cdots + t_k\mathbf{v}_k$, where $\mathbf{v}_1, \ldots, \mathbf{v}_k \in \mathcal{V}$ and $t_1, \ldots, t_k \in \mathcal{R}$. Equation (7.11) states that if you replace each vector by its coordinates relative to a fixed basis and perform the same manipulation on the coordinate vectors, you will obtain the coordinates of $t_1\mathbf{v}_1 + \cdots + t_k\mathbf{v}_k$. In other words, you can transform the abstract vectors into p-tuples, manipulate the p-tuples, and then use the resulting coordinate vector to recover the element of $\mathcal{V}$ that would have been produced had the manipulation been performed on the original vectors in $\mathcal{V}$. The advantage in doing so is that $\mathcal{R}^p$ is familiar territory. Algorithms developed previously for making decisions in Euclidean space can be applied to the coordinate vectors, and the outcomes can be interpreted as conclusions about the abstract vectors.

The transition from vectors to coordinates is effected as follows. Assuming $\mathcal{V}$ is a finite dimensional vector space with basis $\mathcal{A} = \{\mathbf{v}_1, \ldots, \mathbf{v}_p\}$, define $C_{\mathcal{A}} : \mathcal{V} \rightarrow \mathcal{R}^p$ by

$$C_{\mathcal{A}}(\mathbf{v}) = [\mathbf{v}]_{\mathcal{A}}.$$

According to Theorem 7.19, if $\mathbf{u},\mathbf{v} \in \mathcal{V}$ and $t \in \mathcal{R}$, then

$$C_{\mathcal{A}}(\mathbf{u}+\mathbf{v}) = [\mathbf{u}+\mathbf{v}]_{\mathcal{A}} = [\mathbf{u}]_{\mathcal{A}} + [\mathbf{v}]_{\mathcal{A}} = C_{\mathcal{A}}(\mathbf{u}) + C_{\mathcal{A}}(\mathbf{v})$$

and

$$C_{\mathcal{A}}(t\mathbf{v}) = [t\mathbf{v}]_{\mathcal{A}} = t[\mathbf{v}]_{\mathcal{A}} = tC_{\mathcal{A}}(\mathbf{v}),$$

so $C_{\mathcal{A}}$ is linear. Since coordinates are unique, $[\mathbf{u}]_{\mathcal{A}} = [\mathbf{v}]_{\mathcal{A}}$ implies $\mathbf{u} = \mathbf{v}$, that is, $C_{\mathcal{A}}$ is one-to-one. It then follows from the Corollary to Theorem 7.15 that $\dim \mathcal{Rng}(C_{\mathcal{A}}) = \dim \mathcal{V} = p = \dim \mathcal{R}^p$, and therefore $C_{\mathcal{A}}$ is onto.

THEOREM 7.20. Let $\mathcal{V}$ be a finite dimensional vector space with basis $\mathcal{A}$ and suppose $\mathbf{u}_1, \ldots, \mathbf{u}_q \in \mathcal{V}$. Then

(1) $\mathbf{u}_1, \ldots, \mathbf{u}_q$ are linearly independent if and only if $[\mathbf{u}_1]_{\mathcal{A}}, \ldots, [\mathbf{u}_q]_{\mathcal{A}}$ are linearly independent, and

(2) $\mathcal{U} = \mathcal{Span}\{\mathbf{u}_1, \ldots, \mathbf{u}_q\}$ if and only if $C_{\mathcal{A}}(\mathcal{U}) = \mathcal{Span}\{[\mathbf{u}_1]_{\mathcal{A}}, \ldots, [\mathbf{u}_q]_{\mathcal{A}}\}$.

Proof. Conclusion (1) is a direct consequence of Theorem 7.15 and the fact that $C_{\mathcal{A}} : \mathcal{V} \rightarrow \mathcal{R}^p$ and $C_{\mathcal{A}}^{-1} : \mathcal{R}^p \rightarrow \mathcal{V}$ are both one-to-one and linear. Conclusion (2) follows similarly from Theorem 7.13. ■

■ **EXAMPLE 7.31.** Let $\mathcal{U}$ be the subspace of $\mathcal{P}_3$ spanned by $\mathcal{B} = \{f_1, f_2, f_3\}$, where

$$f_1(x) = -1 + x - x^2, \quad f_2(x) = 2x - x^2 + x^3, \quad \text{and} \quad f_3(x) = 1 + x + x^3.$$

The usual basis for $\mathcal{P}_3$ is $\mathcal{A} = \{1, x, x^2, x^3\}$, and

$$[f_1]_{\mathcal{A}} = \begin{bmatrix} -1 \\ 1 \\ -1 \\ 0 \end{bmatrix}, \quad [f_2]_{\mathcal{A}} = \begin{bmatrix} 0 \\ 2 \\ -1 \\ 1 \end{bmatrix}, \quad \text{and} \quad [f_3]_{\mathcal{A}} = \begin{bmatrix} 1 \\ 1 \\ 0 \\ 1 \end{bmatrix}.$$

The row reduction

$$\mathrm{A} = \begin{bmatrix} -1 & 0 & 1 \\ 1 & 2 & 1 \\ -1 & -1 & 0 \\ 0 & 1 & 1 \end{bmatrix} \longrightarrow \begin{bmatrix} 1 & 0 & -1 \\ 0 & 1 & 1 \\ 0 & 0 & 0 \\ 0 & 0 & 0 \end{bmatrix} = \mathrm{B}$$

reveals that $\{[f_1]_{\mathcal{A}}, [f_2]_{\mathcal{A}}, [f_3]_{\mathcal{A}}\}$ is linearly dependent, so by Theorem 7.20(1), $\mathcal{B}$ is linearly dependent. Moreover,

$$C_{\mathcal{A}}(\mathcal{U}) = \mathcal{S}pan\{[f_1]_{\mathcal{A}}, [f_2]_{\mathcal{A}}, [f_3]_{\mathcal{A}}\} = \mathcal{C}(\mathrm{A}),$$

and the leading ones in B indicate that $\{[f_1]_{\mathcal{A}}, [f_2]_{\mathcal{A}}\}$ is a basis for $\mathcal{C}(\mathrm{A})$, so $\{f_1, f_2\}$ is a basis for $\mathcal{U}$ (Theorem 7.20). □

■ **EXAMPLE 7.32.** Consider $\mathcal{B} = \{\mathrm{A,B,C,D}\} \subseteq \mathcal{M}_{2\times 2}$, where

$$\mathrm{A} = \begin{bmatrix} 1 & 2 \\ -1 & 3 \end{bmatrix}, \quad \mathrm{B} = \begin{bmatrix} 0 & -1 \\ 2 & 0 \end{bmatrix}, \quad \mathrm{C} = \begin{bmatrix} 2 & 1 \\ 0 & -1 \end{bmatrix}, \quad \text{and} \quad \mathrm{D} = \begin{bmatrix} 1 & 1 \\ 2 & 2 \end{bmatrix}.$$

The usual basis for $\mathcal{M}_{2\times 2}$, is

$$\mathcal{A} = \left\{ \begin{bmatrix} 1 & 0 \\ 0 & 0 \end{bmatrix}, \begin{bmatrix} 0 & 1 \\ 0 & 0 \end{bmatrix}, \begin{bmatrix} 0 & 0 \\ 1 & 0 \end{bmatrix}, \begin{bmatrix} 0 & 0 \\ 0 & 1 \end{bmatrix} \right\},$$

and

$$[\mathrm{A}]_{\mathcal{A}} = \begin{bmatrix} 1 \\ 2 \\ -1 \\ 3 \end{bmatrix}, \quad [\mathrm{B}]_{\mathcal{A}} = \begin{bmatrix} 0 \\ -1 \\ 2 \\ 0 \end{bmatrix}, \quad [\mathrm{C}]_{\mathcal{A}} = \begin{bmatrix} 2 \\ 1 \\ 0 \\ -1 \end{bmatrix}, \quad \text{and} \quad [\mathrm{D}]_{\mathcal{A}} = \begin{bmatrix} 1 \\ 1 \\ 2 \\ 2 \end{bmatrix}.$$

Since

$$\begin{bmatrix} 1 & 0 & 2 & 1 \\ 2 & -1 & 1 & 1 \\ -1 & 2 & 0 & 2 \\ 3 & 0 & -1 & 2 \end{bmatrix} \longrightarrow \begin{bmatrix} 1 & 0 & 0 & 0 \\ 0 & 1 & 0 & 0 \\ 0 & 0 & 1 & 0 \\ 0 & 0 & 0 & 1 \end{bmatrix},$$

$\{[\mathrm{A}]_{\mathcal{A}}, [\mathrm{B}]_{\mathcal{A}}, [\mathrm{C}]_{\mathcal{A}}, [\mathrm{D}]_{\mathcal{A}}\}$ is linearly independent, and therefore $\mathcal{B}$ is linearly independent (Theorem 7.20). Since $\mathcal{B}$ contains four vectors and dim $\mathcal{M}_{2\times 2} = 4$, $\mathcal{B}$ is a basis for $\mathcal{M}_{2\times 2}$. □

DEFINITION. Vector spaces $\mathcal{V}$ and $\mathcal{W}$ are said to be *isomorphic* if there is a linear $T : \mathcal{V} \to \mathcal{W}$ that is one-to-one and onto. Such a function is called an *isomorphism*.

Suppose $T : \mathcal{V} \to \mathcal{W}$ is an isomorphism. Since T is one-to-one and onto, the correspondence

$$\mathbf{v} \longleftrightarrow T(\mathbf{v})$$

is a one-to-one pairing of the elements of $\mathcal{V}$ with those of $\mathcal{W}$. The linearity of T ensures that if $\mathbf{v}_1, \ldots, \mathbf{v}_p \in \mathcal{V}$ and $t_1, \ldots, t_p \in \mathcal{R}$, then

$$t_1\mathbf{v}_1 + \cdots + t_p\mathbf{v}_p \longleftrightarrow t_1T(\mathbf{v}_1) + \cdots + t_pT(\mathbf{v}_p),$$

that is, that the outcome of an algebraic manipulation of elements of $\mathcal{V}$ is always associated under this pairing with the vector that results from performing the same algebraic manipulation on the corresponding elements of $\mathcal{W}$. Although the vectors in $\mathcal{V}$ may not resemble those in $\mathcal{W}$, the isomorphism associates them in such a way that the algebraic behavior of those in the one set matches that of their associates in the other. In that sense, the two vector spaces are algebraically indistinguishable.

■ **EXAMPLE 7.33.** The correspondence $\mathbf{x} \longleftrightarrow \begin{bmatrix} x_1 \\ x_2 \end{bmatrix}$ in Section 2.2 of Chapter 2 defines a function from $\mathcal{G}^2$ to $\mathcal{R}^2$ that is one-to-one and onto. Theorems 2.2 and 2.3 state that this function is linear, so $\mathcal{G}^2$ is isomorphic to $\mathcal{R}^2$. □

THEOREM 7.21. Every vector space of dimension n is isomorphic to $\mathcal{R}^n$.

Proof. If $\mathcal{V}$ is an n-dimensional vector space and $\mathcal{A}$ is a basis for $\mathcal{V}$, then $C_{\mathcal{A}} : \mathcal{V} \to \mathcal{R}^n$ is an isomorphism. ■

Suppose $\mathcal{V}$ and $\mathcal{W}$ are finite dimensional vector spaces with respective bases $\mathcal{A} = \{\mathbf{v}_1, \ldots, \mathbf{v}_n\}$ and $\mathcal{B} = \{\mathbf{w}_1, \ldots, \mathbf{w}_m\}$ and assume $T : \mathcal{V} \to \mathcal{W}$ is linear. By resorting to $\mathcal{A}$ coordinates for $\mathbf{v} \in \mathcal{V}$ and $\mathcal{B}$ coordinates for $T(\mathbf{v})$, you can get a matrix representative for T. Indeed, consider the diagram

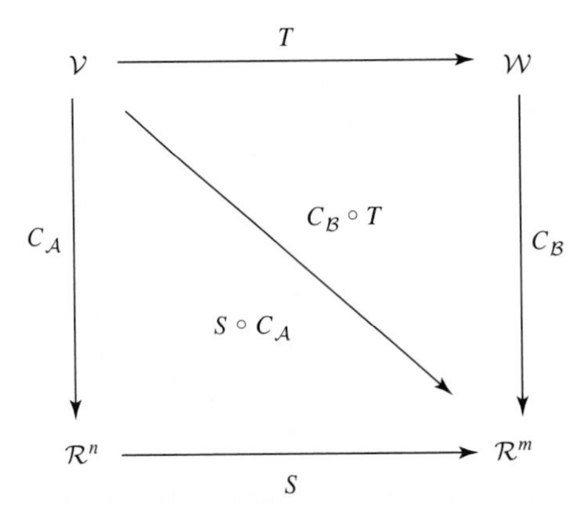

where $C_{\mathcal{A}}(\mathbf{v}) = [\mathbf{v}]_{\mathcal{A}}$, $C_{\mathcal{B}}(\mathbf{w}) = [\mathbf{w}]_{\mathcal{B}}$, and $S : \mathcal{R}^n \to \mathcal{R}^m$ is defined by

$$S = C_{\mathcal{B}} \circ T \circ C_{\mathcal{A}}^{-1}.$$

Being a composition of linear functions, S is linear, and therefore

$$S(\mathrm{X}) = \mathrm{AX}, \tag{7.12}$$

where $\mathrm{A} = [S(\mathrm{E}_1) \quad \cdots \quad S(\mathrm{E}_n)]$ (Theorem 4.1). Observe that

$$\begin{aligned} S \circ C_{\mathcal{A}} &= (C_{\mathcal{B}} \circ T \circ C_{\mathcal{A}}^{-1}) \circ C_{\mathcal{A}} = (C_{\mathcal{B}} \circ T) \circ (C_{\mathcal{A}}^{-1} \circ C_{\mathcal{A}}) \\ &= (C_{\mathcal{B}} \circ T) \circ \mathrm{Id}_{\mathcal{V}} = C_{\mathcal{B}} \circ T. \end{aligned}$$

Thus, for each $\mathbf{v} \in \mathcal{V}$,

$$S([\mathbf{v}]_{\mathcal{A}}) = (S \circ C_{\mathcal{A}})(\mathbf{v}) = (C_{\mathcal{B}} \circ T)(\mathbf{v}) = [T(\mathbf{v})]_{\mathcal{B}}, \tag{7.13}$$

and substituting $\mathrm{X} = [\mathbf{v}]_{\mathcal{A}}$ in (7.12) produces

$$[T(\mathbf{v})]_{\mathcal{B}} = \mathrm{A}[\mathbf{v}]_{\mathcal{A}}.$$

Finally, since $[\mathbf{v}_k]_{\mathcal{A}} = \mathrm{E}_k$, (7.13) gives $S(\mathrm{E}_k) = S([\mathbf{v}_k]_{\mathcal{A}}) = [T(\mathbf{v}_k)]_{\mathcal{B}}$, $k = 1, \ldots, n$, and therefore

$$A = [S(E_1) \quad \cdots \quad S(E_n)] = \big[[T(\mathbf{v}_1)]_{\mathcal{B}} \quad \cdots \quad [T(\mathbf{v}_n)]_{\mathcal{B}}\big].$$

THEOREM 7.22. Let $\mathcal{V}$ and $\mathcal{W}$ be finite dimensional vector spaces and suppose $T : \mathcal{V} \rightarrow \mathcal{W}$ is linear. If $\mathcal{A} = \{\mathbf{v}_1, \ldots, \mathbf{v}_n\}$ and $\mathcal{B} = \{\mathbf{w}_1, \ldots, \mathbf{w}_m\}$ are bases for $\mathcal{V}$ and $\mathcal{W}$, respectively, then there is a unique $m \times n$ matrix, $[T]_{\mathcal{BA}}$, such that

$$[T(\mathbf{v})]_{\mathcal{B}} = [T]_{\mathcal{BA}}[\mathbf{v}]_{\mathcal{A}}, \tag{7.14}$$

for all $\mathbf{v} \in \mathcal{V}$. Moreover,

$$[T]_{\mathcal{BA}} = \big[[T(\mathbf{v}_1)]_{\mathcal{B}} \quad \cdots \quad [T(\mathbf{v}_n)]_{\mathcal{B}}\big]. \tag{7.15}$$

Proof. Only the uniqueness of $[T]_{\mathcal{BA}}$ remains to be checked, and that task is left as an exercise. ■

■ **EXAMPLE 7.34.** If $D : \mathcal{P}_3 \rightarrow \mathcal{P}_2$ is differentiation, $\mathcal{A} = \{1,x,x^2,x^3\}$ and $\mathcal{B} = \{1,x,x^2\}$, then

$$[D(1)]_{\mathcal{B}} = [0]_{\mathcal{B}} = \begin{bmatrix} 0 \\ 0 \\ 0 \end{bmatrix}, \quad [D(x)]_{\mathcal{B}} = [1]_{\mathcal{B}} = \begin{bmatrix} 1 \\ 0 \\ 0 \end{bmatrix},$$

$$[D(x^2)]_{\mathcal{B}} = [2x]_{\mathcal{B}} = \begin{bmatrix} 0 \\ 2 \\ 0 \end{bmatrix}, \quad \text{and} \quad [D(x^3)]_{\mathcal{B}} = [3x^2]_{\mathcal{B}} = \begin{bmatrix} 0 \\ 0 \\ 3 \end{bmatrix},$$

so

$$[D]_{\mathcal{BA}} = \begin{bmatrix} 0 & 1 & 0 & 0 \\ 0 & 0 & 2 & 0 \\ 0 & 0 & 0 & 3 \end{bmatrix}.$$

For $f(x) = 4 + 3x + 2x^2 + x^3$, $[f]_{\mathcal{A}} = \begin{bmatrix} 4 \\ 3 \\ 2 \\ 1 \end{bmatrix}$, and by (7.14),

$$[D(f)]_{\mathcal{B}} = [D]_{\mathcal{BA}}[f]_{\mathcal{A}} = \begin{bmatrix} 0 & 1 & 0 & 0 \\ 0 & 0 & 2 & 0 \\ 0 & 0 & 0 & 3 \end{bmatrix} \begin{bmatrix} 4 \\ 3 \\ 2 \\ 1 \end{bmatrix} = \begin{bmatrix} 3 \\ 4 \\ 3 \end{bmatrix}.$$

Thus,

$$D(f)(x) = 3 + 4x + 3x^2.$$

Note that the last function is the derivative of f. □

If $\mathcal{V}$ is a finite dimensional vector space with basis $\mathcal{A}$ and $T : \mathcal{V} \rightarrow \mathcal{V}$ is linear, then in keeping with previous practice, $[T]_{\mathcal{AA}}$ is shortened to $[T]_{\mathcal{A}}$.

■ **EXAMPLE 7.35.** Define $T : \mathcal{M}_{2\times2} \rightarrow \mathcal{M}_{2\times2}$ by $T(A) = 3A - 2A^T$ and let

$$\mathcal{A} = \left\{ \begin{bmatrix} 1 & 0 \\ 0 & 0 \end{bmatrix}, \begin{bmatrix} 0 & 1 \\ 0 & 0 \end{bmatrix}, \begin{bmatrix} 0 & 0 \\ 1 & 0 \end{bmatrix}, \begin{bmatrix} 0 & 0 \\ 0 & 1 \end{bmatrix} \right\}.$$

Then

$$T\left(\begin{bmatrix}1 & 0\\0 & 0\end{bmatrix}\right)=\begin{bmatrix}1 & 0\\0 & 0\end{bmatrix},\quad T\left(\begin{bmatrix}0 & 1\\0 & 0\end{bmatrix}\right)=\begin{bmatrix}0 & 3\\-2 & 0\end{bmatrix},$$

$$T\left(\begin{bmatrix}0 & 0\\1 & 0\end{bmatrix}\right)=\begin{bmatrix}0 & -2\\3 & 0\end{bmatrix},\quad T\left(\begin{bmatrix}0 & 0\\1 & 0\end{bmatrix}\right)=\begin{bmatrix}0 & 0\\0 & 1\end{bmatrix},$$

and

$$[T]_{\mathcal{A}}=\begin{bmatrix}1 & 0 & 0 & 0\\0 & 3 & -2 & 0\\0 & -2 & 3 & 0\\0 & 0 & 0 & 1\end{bmatrix}.$$

If

$$\mathrm{A}=\begin{bmatrix}1 & 2\\3 & 4\end{bmatrix},\quad \text{then}\quad [\mathrm{A}]_{\mathcal{A}}=\begin{bmatrix}1\\2\\3\\4\end{bmatrix},$$

and by (7.14)

$$[T(\mathrm{A})]_{\mathcal{A}}=\begin{bmatrix}1 & 0 & 0 & 0\\0 & 3 & -2 & 0\\0 & -2 & 3 & 0\\0 & 0 & 0 & 1\end{bmatrix}\begin{bmatrix}1\\2\\3\\4\end{bmatrix}=\begin{bmatrix}1\\0\\5\\4\end{bmatrix}.$$

Thus, $T(\mathrm{A})=\begin{bmatrix}1 & 0\\5 & 4\end{bmatrix}$. Of course, $T(\mathrm{A})$ can also be calculated directly by

$$3\begin{bmatrix}1 & 2\\3 & 4\end{bmatrix}-2\begin{bmatrix}1 & 2\\3 & 4\end{bmatrix}^T=\begin{bmatrix}3 & 6\\9 & 12\end{bmatrix}-\begin{bmatrix}2 & 6\\4 & 8\end{bmatrix}=\begin{bmatrix}1 & 0\\5 & 4\end{bmatrix}.\ \square$$

When $\mathcal{V}$ and $\mathcal{W}$ are finite dimensional, $\mathcal{Ker}(T)$ and $\mathcal{Rng}(T)$ can be identified by referring to coordinate systems for $\mathcal{V}$ and $\mathcal{W}$ and analyzing the corresponding matrix for T. The next theorem spells out the connections between these items.

THEOREM 7.23. If $\mathcal{V}$ and $\mathcal{W}$ are finite dimensional vector spaces with bases $\mathcal{A}$ and $\mathcal{B}$, respectively, and if $T : \mathcal{V} \to \mathcal{W}$ is linear, then

$$C_{\mathcal{A}}(\mathcal{Ker}(T))=\mathcal{N}([T]_{\mathcal{BA}})\qquad\text{and}\qquad C_{\mathcal{B}}(\mathcal{Rng}(T))=\mathcal{C}([T]_{\mathcal{BA}}).$$

Proof. Assume dim $\mathcal{V}=n$ and note that $\mathcal{R}^n=C_{\mathcal{A}}(\mathcal{V})=\{[\mathbf{v}]_{\mathcal{A}} : \mathbf{v}\in\mathcal{V}\}$. Thus,

$$\begin{aligned}
\mathcal{N}([T]_{\mathcal{BA}}) &= \{\mathrm{X}\in\mathcal{R}^n : [T]_{\mathcal{BA}}\mathrm{X}=0\} && \text{(defn. of null space)}\\
&= \{[\mathbf{v}]_{\mathcal{A}} : [T]_{\mathcal{BA}}[\mathbf{v}]_{\mathcal{A}}=0,\ \mathbf{v}\in\mathcal{V}\} && (\mathcal{R}^n=C_{\mathcal{A}}(\mathcal{V}))\\
&= \{[\mathbf{v}]_{\mathcal{A}} : [T(\mathbf{v})]_{\mathcal{B}}=0,\ \mathbf{v}\in\mathcal{V}\} && \text{(Theorem 7.22)}\\
&= \{C_{\mathcal{A}}(\mathbf{v}) : T(\mathbf{v})=\mathbf{0}_{\mathcal{W}},\ \mathbf{v}\in\mathcal{V}\} && ([\mathbf{w}]_{\mathcal{B}}=0 \Leftrightarrow \mathbf{w}=\mathbf{0}_{\mathcal{W}})\\
&= \{C_{\mathcal{A}}(\mathbf{v}) : \mathbf{v}\in\mathcal{Ker}(T)\} && \text{(defn. of } \mathcal{Ker}(T))\\
&= C_{\mathcal{A}}(\mathcal{Ker}(T)),
\end{aligned}$$

and

$$\begin{aligned}
\mathcal{C}([T]_{\mathcal{BA}}) &= \{[T]_{\mathcal{BA}}\mathrm{X} \ : \ \mathrm{X} \in \mathcal{R}^n\} && \text{(equation (2.1))}\\
&= \{[T]_{\mathcal{BA}}[\mathbf{v}]_{\mathcal{A}} \ : \ \mathbf{v} \in \mathcal{V}\} && (\mathcal{R}^n = C_{\mathcal{A}}(\mathcal{V}))\\
&= \{[T(\mathbf{v})]_{\mathcal{B}} \ : \ \mathbf{v} \in \mathcal{V}\} && \text{(Theorem 7.22)}\\
&= \{[\mathbf{w}]_{\mathcal{B}} \ : \ \mathbf{w} \in \mathcal{R}ng(T) && \text{(defn. of } \mathcal{R}ng(T))\\
&= C_{\mathcal{B}}(\mathcal{R}ng(T)). \ \blacksquare
\end{aligned}$$

■ **EXAMPLE 7.36.** Suppose $T : \mathcal{P}_2 \to \mathcal{M}_{2\times 2}$ is given by

$$T(f) = \begin{bmatrix} f(0) & f(1) \\ f(1) & f(0) \end{bmatrix}.$$

Let $\mathcal{A} = \{f_1, f_2, f_3\}$, where $f_1(x) = 1, f_2(x) = x, f_3(x) = x^2$, and set

$$\mathcal{B} = \left\{ \begin{bmatrix} 1 & 0 \\ 0 & 0 \end{bmatrix}, \begin{bmatrix} 0 & 1 \\ 0 & 0 \end{bmatrix}, \begin{bmatrix} 0 & 0 \\ 1 & 0 \end{bmatrix}, \begin{bmatrix} 0 & 0 \\ 0 & 1 \end{bmatrix} \right\}.$$

Then $T(f_1) = \begin{bmatrix} 1 & 1 \\ 1 & 1 \end{bmatrix}$, $T(f_2) = T(f_3) = \begin{bmatrix} 0 & 1 \\ 1 & 0 \end{bmatrix}$, and

$$[T]_{\mathcal{BA}} = \begin{bmatrix} [T(f_1)]_{\mathcal{B}} & [T(f_2)]_{\mathcal{B}} & [T(f_3)]_{\mathcal{B}} \end{bmatrix} = \begin{bmatrix} 1 & 0 & 0 \\ 1 & 1 & 1 \\ 1 & 1 & 1 \\ 1 & 0 & 0 \end{bmatrix}.$$

Since

$$\begin{bmatrix} 1 & 0 & 0 \\ 1 & 1 & 1 \\ 1 & 1 & 1 \\ 1 & 0 & 0 \end{bmatrix} \longrightarrow \begin{bmatrix} 1 & 0 & 0 \\ 0 & 1 & 1 \\ 0 & 0 & 0 \\ 0 & 0 & 0 \end{bmatrix},$$

$$\mathcal{N}([T]_{\mathcal{BA}}) = \mathcal{S}pan\left\{ \begin{bmatrix} 0 \\ -1 \\ 1 \end{bmatrix} \right\} \quad \text{and} \quad \mathcal{C}([T]_{\mathcal{BA}}) = \mathcal{S}pan\left\{ \begin{bmatrix} 1 \\ 1 \\ 1 \\ 1 \end{bmatrix}, \begin{bmatrix} 0 \\ 1 \\ 1 \\ 0 \end{bmatrix} \right\}.$$

By Theorems 7.23 and 7.20,

$$\{-x + x^2\} \quad \text{and} \quad \left\{ \begin{bmatrix} 1 & 1 \\ 1 & 1 \end{bmatrix}, \begin{bmatrix} 0 & 1 \\ 1 & 0 \end{bmatrix} \right\}$$

are bases for $\mathcal{K}er(T)$ and $\mathcal{R}ng(T)$, respectively. Note that $\dim \mathcal{K}er(T) = 1 > 0$, so T is not one-to-one, and $\dim \mathcal{R}ng(T) = 2 < \dim \mathcal{M}_{2\times 2}$, so T is not onto. □

THEOREM 7.24. If $\mathcal{V}$ and $\mathcal{W}$ are finite dimensional vector spaces and $T : \mathcal{V} \to \mathcal{W}$ is linear, then

$$\dim \mathcal{R}ng(T) + \dim \mathcal{K}er(T) = \dim \mathcal{V}.$$

Proof. Suppose $\dim \mathcal{V} = n$, $\dim \mathcal{W} = m$, and $\mathcal{A}$ and $\mathcal{B}$ are bases for $\mathcal{V}$ and $\mathcal{W}$, respectively. Let $C_{\mathcal{A}} : \mathcal{V} \to \mathcal{R}^n$ and $C_{\mathcal{B}} : \mathcal{W} \to \mathcal{R}^m$ be the corresponding coordinate isomorphisms. By Theorem 7.23 and the Corollary to Theorem 7.15,

$$\dim \mathcal{K}er(T) = \dim \mathrm{C}_{\mathcal{A}}(\mathcal{K}er(T)) = \dim \mathcal{N}([T]_{\mathcal{BA}})$$

and

$$\dim \mathcal{R}ng(T) = \dim \mathrm{C}_{\mathcal{B}}(\mathcal{R}ng(T)) = \dim \mathcal{C}([T]_{\mathcal{BA}})$$

Since dim $\mathcal{V} = n = \#$ of columns in $[T]_{\mathcal{BA}}$, the conclusion follows from the rank-nullity theorem (Theorem 2.26). ■

THEOREM 7.25. Suppose $\mathcal{V}$ and $\mathcal{W}$ are n-dimensional vector spaces with bases $\mathcal{A}$ and $\mathcal{B}$, respectively, and $T : \mathcal{V} \to \mathcal{W}$ is linear. Then T is one-to-one if and only if $[T]_{\mathcal{BA}}$ is nonsingular. Moreover, if T is one-to-one, then T is onto, and

$$[T^{-1}]_{\mathcal{AB}} = [T]_{\mathcal{BA}}^{-1}.$$

In particular, when $\mathcal{V} = \mathcal{W}$ and $\mathcal{A} = \mathcal{B}$, $[T^{-1}]_{\mathcal{A}} = [T]_{\mathcal{A}}^{-1}$.

Proof. If $[T]_{\mathcal{BA}}$ is nonsingular, then $\mathcal{N}([T]_{\mathcal{BA}}) = \{0\}$, and from Theorem 7.23,

$$\mathcal{K}er(T) = C_{\mathcal{A}}^{-1}(\mathcal{N}([T]_{\mathcal{BA}}) = C_{\mathcal{A}}^{-1}(\{0\}) = \{\mathbf{0}_{\mathcal{V}}\}.$$

Thus, T is one-to-one. Now, suppose T is one-to-one and let $\mathcal{A} = \{\mathbf{v}_1, \ldots, \mathbf{v}_n\}$. By Theorems 7.15 and 7.13, $T(\mathcal{A}) = \{T(\mathbf{v}_1), \ldots, T(\mathbf{v}_n)\}$ is a basis for $\mathcal{R}ng(T)$, so dim $\mathcal{R}ng(T) = n =$ dim $\mathcal{W}$. Thus, T is onto. Also, the linear independence of $\{T(\mathbf{v}_1), \ldots, T(\mathbf{v}_n)\}$ implies $\{[T(\mathbf{v}_1)]_{\mathcal{B}}, \ldots, [T(\mathbf{v}_n)]_{\mathcal{B}}\}$ is linearly independent. The latter coordinate vectors are the columns of $[T]_{\mathcal{BA}}$, so $[T]_{\mathcal{BA}}$ is nonsingular. Finally, for each $\mathbf{w} \in \mathcal{W}$ there is a $\mathbf{v} \in \mathcal{V}$ such that $T(\mathbf{v}) = \mathbf{w}$, and by (7.14), $[\mathbf{w}]_{\mathcal{B}} = [T]_{\mathcal{BA}}[\mathbf{v}]_{\mathcal{A}}$. Thus,

$$[T^{-1}(\mathbf{w})]_{\mathcal{A}} = [\mathbf{v}]_{\mathcal{A}} = [T]_{\mathcal{BA}}^{-1}[\mathbf{w}]_{\mathcal{B}},$$

that is, $[T^{-1}]_{\mathcal{AB}} = [T]_{\mathcal{BA}}^{-1}$. ■

■ **EXAMPLE 7.37.** Assume $T : \mathcal{P}_2 \to \mathcal{P}_2$ is the linear operator given by

$$T(f) = f' + 2f,$$

and let $\mathcal{A} = \{f_1, f_2, f_3\}$, where $f_1(x) = 1$, $f_2(x) = x$, and $f_3(x) = x^2$. Then

$$T(f_1)(x) = 2, \quad T(f_2)(x) = 1 + 2x, \quad \text{and} \quad T(f_3)(x) = 2x + 2x^2,$$

and therefore

$$[T]_{\mathcal{A}} = \begin{bmatrix} 2 & 1 & 0 \\ 0 & 2 & 2 \\ 0 & 0 & 2 \end{bmatrix}.$$

Since $[T]_{\mathcal{A}}$ is nonsingular, T is one-to-one and onto and

$$[T^{-1}]_{\mathcal{A}} = [T]_{\mathcal{A}}^{-1} = \frac{1}{4}\begin{bmatrix} 2 & -1 & 1 \\ 0 & 2 & -2 \\ 0 & 0 & 2 \end{bmatrix}.$$

Suppose $g(x) = 3 + 2x + x^2$ and $f = T^{-1}(g)$. Then $[g]_{\mathcal{A}} = [3 \quad 2 \quad 1]^T$, and

$$[f]_{\mathcal{A}} = [T^{-1}(g)]_{\mathcal{A}} = \frac{1}{4}\begin{bmatrix} 2 & -1 & 1 \\ 0 & 2 & -2 \\ 0 & 0 & 2 \end{bmatrix}\begin{bmatrix} 3 \\ 2 \\ 1 \end{bmatrix} = \frac{1}{4}\begin{bmatrix} 5 \\ 2 \\ 2 \end{bmatrix}.$$

Thus, $f(x) = 5/4 + (1/2)x + (1/2)x^2$. Note that $f'(x) + 2f(x) = [(1/2) + x] + [(5/2) + x + x^2] = 3 + 2x + x^2$. □

THEOREM 7.26. If $\mathcal{U}$, $\mathcal{V}$, and $\mathcal{W}$ are finite dimensional vector spaces with bases $\mathcal{A}$, $\mathcal{B}$, and $\mathcal{C}$, respectively, and $T : \mathcal{U} \to \mathcal{V}$ and $S : \mathcal{V} \to \mathcal{W}$ are linear, then

$$[S \circ T]_{\mathcal{CA}} = [S]_{\mathcal{CB}}[T]_{\mathcal{BA}}.$$

In particular, if $\mathcal{U} = \mathcal{V} = \mathcal{W}$ and $\mathcal{A} = \mathcal{B} = \mathcal{C}$, then $[S \circ T]_{\mathcal{A}} = [S]_{\mathcal{A}}[T]_{\mathcal{A}}$.

Proof. Omitted. ■

Suppose $\mathcal{V}$ is a finite dimensional vector space, $\mathcal{A}$ is a basis for $\mathcal{V}$, and $T : \mathcal{V} \to \mathcal{V}$ is linear. For $\mathbf{v} \in \mathcal{V}$ and $\lambda \in \mathcal{R}$, $T(\mathbf{v}) = \lambda\mathbf{v}$ if and only if

$$\lambda[\mathbf{v}]_{\mathcal{A}} = [\lambda\mathbf{v}]_{\mathcal{A}} = [T(\mathbf{v})]_{\mathcal{A}} = [T]_{\mathcal{A}}[\mathbf{v}]_{\mathcal{A}}.$$

Thus, you can investigate the eigenvalue-eigenspace problem for T by referring to a coordinate system for $\mathcal{V}$ and studying $[T]_{\mathcal{A}}$. Note that T and $[T]_{\mathcal{A}}$ have the same eigenvalues, but that $\mathbf{v} \in \mathcal{V}(\lambda)$ if and only if $[\mathbf{v}]_{\mathcal{A}} \in \mathcal{N}(\lambda\mathrm{I} - [T]_{\mathcal{A}})$.

■ **EXAMPLE 7.38.** Define $T : \mathcal{M}_{2\times 2} \to \mathcal{M}_{2\times 2}$ by $T(\mathrm{A}) = 3\mathrm{A} - 2\mathrm{A}^T$. Example 7.35 showed that if $\mathcal{A}$ is the standard basis for $\mathcal{M}_{2\times 2}$, then

$$[T]_{\mathcal{A}} = \begin{bmatrix} 1 & 0 & 0 & 0 \\ 0 & 3 & -2 & 0 \\ 0 & -2 & 3 & 0 \\ 0 & 0 & 0 & 1 \end{bmatrix}.$$

Since

$$\det(\lambda\mathrm{I} - [T]_{\mathcal{A}}) = \det \begin{bmatrix} \lambda - 1 & 0 & 0 & 0 \\ 0 & \lambda - 3 & 2 & 0 \\ 0 & 2 & \lambda - 3 & 0 \\ 0 & 0 & 0 & \lambda - 1 \end{bmatrix} = (\lambda - 1)^3(\lambda - 5),$$

1 and 5 are the eigenvalues of T. Observe that

$$1\mathrm{I} - [T]_{\mathcal{A}} = \begin{bmatrix} 0 & 0 & 0 & 0 \\ 0 & -2 & 2 & 0 \\ 0 & 2 & -2 & 0 \\ 0 & 0 & 0 & 0 \end{bmatrix} \longrightarrow \begin{bmatrix} 0 & 1 & -1 & 0 \\ 0 & 0 & 0 & 0 \\ 0 & 0 & 0 & 0 \\ 0 & 0 & 0 & 0 \end{bmatrix}$$

and

$$5\mathrm{I} - [T]_{\mathcal{A}} = \begin{bmatrix} 4 & 0 & 0 & 0 \\ 0 & 2 & 2 & 0 \\ 0 & 2 & 2 & 0 \\ 0 & 0 & 0 & 4 \end{bmatrix} \longrightarrow \begin{bmatrix} 1 & 0 & 0 & 0 \\ 0 & 1 & 1 & 0 \\ 0 & 0 & 0 & 1 \\ 0 & 0 & 0 & 0 \end{bmatrix},$$

so

$$\mathcal{N}(1\mathrm{I} - [T]_{\mathcal{A}}) = \mathcal{S}pan\left\{ \begin{bmatrix} 1 \\ 0 \\ 0 \\ 0 \end{bmatrix}, \begin{bmatrix} 0 \\ 1 \\ 1 \\ 0 \end{bmatrix}, \begin{bmatrix} 0 \\ 0 \\ 0 \\ 1 \end{bmatrix} \right\}$$

and

$$\mathcal{N}(5\mathrm{I} - [T]_{\mathcal{A}}) = \mathcal{S}pan\left\{ \begin{bmatrix} 0 \\ -1 \\ 1 \\ 0 \end{bmatrix} \right\}.$$

Thus,

$$\mathcal{V}(1) = \mathcal{S}pan\left\{ \begin{bmatrix} 1 & 0 \\ 0 & 0 \end{bmatrix}, \begin{bmatrix} 0 & 1 \\ 1 & 0 \end{bmatrix}, \begin{bmatrix} 0 & 0 \\ 0 & 1 \end{bmatrix} \right\},$$

and

$$\mathcal{V}(5) = \mathcal{S}pan\left\{ \begin{bmatrix} 0 & -1 \\ 1 & 0 \end{bmatrix} \right\}. \square$$

Exercises 7.5

1. $\mathcal{A} = \{1 + x^2, 2x + x^2, 1 + x + x^2\}$ is a basis for $\mathcal{P}_2$. Find

a. f when $[f]_{\mathcal{A}} = \begin{bmatrix} 3 \\ -2 \\ 1 \end{bmatrix}$. b. $[f]_{\mathcal{A}}$ when $f(x) = -4 + x - 2x^2$.

2. $\mathcal{B} = \left\{ \begin{bmatrix} 1 & 1 \\ 1 & 0 \end{bmatrix}, \begin{bmatrix} 0 & 1 \\ -1 & 0 \end{bmatrix}, \begin{bmatrix} 1 & -1 \\ 0 & 1 \end{bmatrix}, \begin{bmatrix} -1 & 0 \\ 0 & 1 \end{bmatrix} \right\}$ is a basis for $\mathcal{M}_{2\times 2}$. Find

a. A when $[\mathrm{A}]_{\mathcal{B}} = \begin{bmatrix} 1 \\ -2 \\ 3 \\ -4 \end{bmatrix}$. b. $[\mathrm{A}]_{\mathcal{B}}$ when $\mathrm{A} = \begin{bmatrix} 1 & 2 \\ 3 & 4 \end{bmatrix}$.

3. Show that $\mathcal{B} = \left\{ \begin{bmatrix} 1 & 1 \\ 1 & 0 \end{bmatrix}, \begin{bmatrix} 1 & 1 \\ 1 & 1 \end{bmatrix}, \begin{bmatrix} 0 & 1 \\ 1 & 1 \end{bmatrix} \right\}$ is a basis for $\mathcal{S}_{2\times 2}$ and find $[\mathrm{I}_2]_{\mathcal{B}}$.

4. Find the dimension of $\mathcal{S}pan\{1 + x + x^2 - x^3, -1 - x + x^2 + x^3, -1 - x + 5x^2 + x^3\}$.

5. Let $\mathcal{A} = \{1,x,x^2\}$, $\mathcal{B} = \{1 + x, 1 - x, x + x^2\}$ and $T : \mathcal{P}_2 \to \mathcal{P}_2$ be the linear function such that $[T]_{\mathcal{BA}} = \begin{bmatrix} 1 & -2 & 3 \\ -4 & 5 & -6 \\ 7 & -8 & 9 \end{bmatrix}$. Find $T(f)$ when $f(x) = 1 + x + x^2$.

6. **a.** $T : \mathcal{M}_{2\times 2} \to \mathcal{P}_2$, $T\left(\begin{bmatrix} a & b \\ c & d \end{bmatrix}\right) = (a + b - c) + (b + c + d)x + (a - 2c - d)x^2$.

Find $[T]_{\mathcal{BA}}$ when $\mathcal{A} = \left\{ \begin{bmatrix} 1 & 0 \\ 0 & 0 \end{bmatrix}, \begin{bmatrix} 0 & 1 \\ 0 & 0 \end{bmatrix}, \begin{bmatrix} 0 & 0 \\ 1 & 0 \end{bmatrix}, \begin{bmatrix} 0 & 0 \\ 0 & 1 \end{bmatrix} \right\}$ and $\mathcal{B} = \{1,x,x^2\}$.

b. $T : \mathcal{M}_{2\times 2} \to \mathcal{M}_{1\times 3}$, $T(\mathrm{A}) = [1 \quad -1]\mathrm{A}\begin{bmatrix} 1 & 1 & 0 \\ 0 & 1 & 1 \end{bmatrix}$. Find $[T]_{\mathcal{BA}}$ when $\mathcal{A} = \left\{ \begin{bmatrix} 1 & 0 \\ 0 & 0 \end{bmatrix}, \begin{bmatrix} 0 & 1 \\ 0 & 0 \end{bmatrix}, \begin{bmatrix} 0 & 0 \\ 1 & 0 \end{bmatrix}, \begin{bmatrix} 0 & 0 \\ 0 & 1 \end{bmatrix} \right\}$ and $\mathcal{B} = \{[1 \quad 0 \quad 0], [0 \quad 1 \quad 0], [0 \quad 0 \quad 1]\}$.

c. $T : \mathcal{P}_3 \to \mathcal{P}_3$, $T(f) = x^2 f'' + xf' + f$. Find $[T]_{\mathcal{A}}$ when $\mathcal{A} = \{1,x,x^2,x^3\}$.

d. $T : \mathcal{P}_2 \to \mathcal{P}_2$, $T(f)(x) = f(x + 1)$. Find $[T]_{\mathcal{A}}$ when $\mathcal{A} = \{1,x,x^2\}$.

e. $T : \mathcal{P}_2 \to \mathcal{R}$, $T(f) = \int_0^1 f(x)\,dx$. Find $[T]_{\mathcal{BA}}$ when $\mathcal{A} = \{1,x,x^2\}$ and $\mathcal{B} = \{1\}$.

f. $T : \mathcal{M}_{2\times 2} \to \mathcal{M}_{2\times 2}$, $T\left(\begin{bmatrix} a & b \\ c & d \end{bmatrix}\right) = \begin{bmatrix} a - b + c & b - c + d \\ c - d + a & d - a + b \end{bmatrix}$. Find $[T]_{\mathcal{A}}$ when $\mathcal{A} = \left\{ \begin{bmatrix} 1 & 0 \\ 0 & 0 \end{bmatrix}, \begin{bmatrix} 0 & 1 \\ 0 & 0 \end{bmatrix}, \begin{bmatrix} 0 & 0 \\ 1 & 0 \end{bmatrix}, \begin{bmatrix} 0 & 0 \\ 0 & 1 \end{bmatrix} \right\}$.

g. $T : \mathcal{P}_2 \to \mathcal{P}_2$, $T(f)(x) = \frac{1}{2}[f(x + 1) - f(x - 1)]$. Find $[T]_{\mathcal{A}}$ when $\mathcal{A} = \{1,x,x^2\}$.

h. $T : \mathcal{M}_{2\times 2} \to \mathcal{R}^4, T\left(\begin{bmatrix} a & b \\ c & d \end{bmatrix}\right) = \begin{bmatrix} a+b+c \\ b+c+d \\ c+d+a \\ d+a+b \end{bmatrix}$. Find $[T]_{\mathcal{BA}}$ when

$$\mathcal{B} = \left\{ \begin{bmatrix} 1 \\ 0 \\ 0 \\ 0 \end{bmatrix}, \begin{bmatrix} 2 \\ 1 \\ 0 \\ 0 \end{bmatrix}, \begin{bmatrix} 3 \\ 2 \\ 1 \\ 0 \end{bmatrix}, \begin{bmatrix} 4 \\ 3 \\ 2 \\ 1 \end{bmatrix} \right\} \text{ and } \mathcal{A} = \left\{ \begin{bmatrix} 1 & 0 \\ 0 & 0 \end{bmatrix}, \begin{bmatrix} 0 & 1 \\ 0 & 0 \end{bmatrix}, \begin{bmatrix} 0 & 0 \\ 1 & 0 \end{bmatrix}, \begin{bmatrix} 0 & 0 \\ 0 & 1 \end{bmatrix} \right\}.$$

i. $T : \mathcal{P}_3 \to \mathcal{R}^3, T(f) = \begin{bmatrix} f(1) \\ f'(1) \\ f''(1) \end{bmatrix}$. Find $[T]_{\mathcal{EA}}$ when $\mathcal{A} = \{1, x, x^2, x^3\}$.

7. For each function in Problem 6, find a basis for $\mathcal{Ker}(T)$, find a basis for $\mathcal{Rng}(T)$, and determine whether T is one-to-one and/or onto.

8. Suppose $\mathcal{V}$ is a vector space with basis $\mathcal{A} = \{\mathbf{v}_1, \ldots, \mathbf{v}_n\}$. Show that if A and B are $m \times n$ matrices satisfying $\mathrm{A}[\mathbf{v}]_{\mathcal{A}} = \mathrm{B}[\mathbf{v}]_{\mathcal{A}}$ for all $\mathbf{v} \in \mathcal{V}$, then $\mathrm{A} = \mathrm{B}$.

9. Let T be the function in Problem 6(f). Find $T^{-1}\left(\begin{bmatrix} a & b \\ c & d \end{bmatrix}\right)$.

10. Define $T : \mathcal{P}_2 \to \mathcal{P}_2$ by $T(f)(x) = \frac{1}{2}[f(x+1) + f(x-1)]$, and let $\mathcal{A} = \{1, x, x^2\}$.

a. Show T is linear.
b. Find $[T]_{\mathcal{A}}$.
c. Show T is one-to-one and onto.
d. Find $[T^{-1}]_{\mathcal{A}}$.

11. Let $D : \mathcal{P}_3 \to \mathcal{P}_2$ be the differentiation operator and define $T : \mathcal{P}_2 \to \mathcal{P}_3$ by $T(f)(x) = \int_0^x f(u)\,du$. Assuming $\mathcal{A} = \{1, x, x^2, x^3\}$ and $\mathcal{B} = \{1, x, x^2\}$, find $[T]_{\mathcal{AB}}$ and show that $[D]_{\mathcal{BA}}[T]_{\mathcal{AB}} = \mathrm{I}_3$.

12. Suppose $\mathcal{V}$ is a p-dimensional vector space with basis $\mathcal{A} = \{\mathbf{v}_1, \ldots, \mathbf{v}_p\}$ and $C_{\mathcal{A}} : \mathcal{V} \to \mathcal{R}^p$ is the associated coordinate isomorphism. Find $[C_{\mathcal{A}}]_{\mathcal{EA}}$.

13. If $\mathcal{V}$ and $\mathcal{W}$ are vector spaces and $T : \mathcal{V} \to \mathcal{W}$ is linear, then the rank of T is defined as the dimension of $\mathcal{Rng}(T)$. Assuming $\mathcal{A}$ and $\mathcal{B}$ are bases for $\mathcal{V}$ and $\mathcal{W}$, respectively, show that $\mathrm{rank}(T) = \mathrm{rank}([T]_{\mathcal{BA}})$.

14. Let $\mathcal{V}$ be a vector space with bases $\mathcal{A} = \{\mathbf{v}_1, \ldots, \mathbf{v}_n\}$ and $\mathcal{B} = \{\mathbf{w}_1, \ldots, \mathbf{w}_n\}$ and consider $\mathrm{Id}_{\mathcal{V}} : \mathcal{V} \to \mathcal{V}$ defined by $\mathrm{Id}_{\mathcal{V}}(\mathbf{v}) = \mathbf{v}$, $\mathbf{v} \in \mathcal{V}$. Show that

a. $[\mathrm{Id}_{\mathcal{V}}]_{\mathcal{BA}} = \begin{bmatrix} [\mathbf{v}_1]_{\mathcal{B}} & \cdots & [\mathbf{v}_n]_{\mathcal{B}} \end{bmatrix}$.
b. $\mathrm{P}_{\mathcal{BA}} = [\mathrm{Id}_{\mathcal{V}}]_{\mathcal{BA}}$ is nonsingular.
c. $[\mathbf{v}]_{\mathcal{B}} = \mathrm{P}_{\mathcal{BA}}[\mathbf{v}]_{\mathcal{A}}$ for all $\mathbf{v} \in \mathcal{V}$,
d. $\mathrm{P}_{\mathcal{BA}} = \mathrm{P}_{\mathcal{AB}}^{-1}$.

15. Suppose $\mathcal{V}$ is a three-dimensional vector space with basis $\mathcal{A} = \{\mathbf{v}_1, \mathbf{v}_2, \mathbf{v}_3\}$ and $\mathcal{B} = \{\mathbf{v}_1 + 2\mathbf{v}_2, \mathbf{v}_1 - \mathbf{v}_2 + \mathbf{v}_3, 2\mathbf{v}_1 + \mathbf{v}_3\}$. Show that $\mathcal{B}$ is a basis for $\mathcal{V}$. Find $\mathrm{P}_{\mathcal{BA}}$ and $\mathrm{P}_{\mathcal{AB}}$.

16. Assume $a, b \in \mathcal{R}$, $a \neq b$. Let $\mathcal{B} = \{(a+x)^2, (a+x)(b+x), (b+x)^2\}$ and $\mathcal{A} = \{1, x, x^2\}$. Show that $\mathcal{B}$ is a basis for $\mathcal{P}_2$. Find $\mathrm{P}_{\mathcal{AB}}$ and $\mathrm{P}_{\mathcal{BA}}$.

17. Show that if $\mathcal{V}$ is a finite dimensional vector space with bases $\mathcal{A}$ and $\mathcal{B}$ and $T : \mathcal{V} \to \mathcal{V}$ is linear, then $[T]_{\mathcal{B}} = \mathrm{P}_{\mathcal{AB}}^{-1}[T]_{\mathcal{A}}\mathrm{P}_{\mathcal{AB}}$.

18. Find the eigenvalues of T and a basis for each eigenspace.
    a. $T : \mathcal{M}_{2\times 2} \to \mathcal{M}_{2\times 2}, T(\mathrm{A}) = \mathrm{A} + \mathrm{A}^T$
    b. $D : \mathcal{P}_2 \to \mathcal{P}_2, D(f) = f'$
    c. $T : \mathcal{P}_2 \to \mathcal{P}_2$, where $T(f)(x) = \frac{1}{2}[f(x+1) + f(x-1)]$
    d. $T : \mathcal{P}_2 \to \mathcal{P}_2$, for $f(x) = a + bx + cx^2$, $T(f)(x) = (a + 2c) + bx + (2a + c)x^2$
    e. $T : \mathcal{M}_{2\times 2} \to \mathcal{M}_{2\times 2}, T\left(\begin{bmatrix} a & b \\ c & d \end{bmatrix}\right) = \begin{bmatrix} a+b & b+c \\ c+d & d+a \end{bmatrix}$
    f. $T : \mathcal{M}_{2\times 2} \to \mathcal{M}_{2\times 2}, T\left(\begin{bmatrix} a & b \\ c & d \end{bmatrix}\right) = \begin{bmatrix} d & -c \\ -b & a \end{bmatrix}$

7.6
NORMS AND INNER PRODUCTS

The vector concept in $\mathcal{R}^2$ and $\mathcal{R}^3$ arose as a mathematical vehicle for modeling quantities that have magnitude and direction, but these two notions have not been mentioned so far in the present chapter. Here, a vector is nothing more than an element of an abstract vector space. The fact of the matter is that magnitude and direction are geometric concepts. They are not by-products of the algebraic structure of a vector space.

To have a sense of magnitude for elements of an abstract vector space $\mathcal{V}$ you need a way of assigning to each $\mathbf{v} \in \mathcal{V}$ a nonnegative real number $\|\mathbf{v}\|$. In other words, what is required is a function

$$\|\bullet\| : \mathcal{V} \to [0,\infty),$$

where $\|\bullet\|$ is the name of the function and $\|\mathbf{v}\|$ denotes its value at $\mathbf{v}$. This function should possess the basic properties of the usual length function in $\mathcal{R}^n$, and that requirement is satisfied by defining "magnitude" axiomatically, using features of Euclidean length as a guide. In the abstract context, it is customary to use the term norm instead of magnitude, so that will be the practice hereafter.

DEFINITION. If $\mathcal{V}$ is a vector space, then $\|\bullet\| : \mathcal{V} \to [0, \infty)$ is a *norm* on $\mathcal{V}$ if for all $\mathbf{u}, \mathbf{v} \in \mathcal{V}$ and for all $t \in \mathcal{R}$,
(1) $\|\mathbf{v}\| > 0$ whenever $\mathbf{v} \neq \mathbf{0}_{\mathcal{V}}$,
(2) $\|t\mathbf{v}\| = |t|\,\|\mathbf{v}\|$, and
(3) $\|\mathbf{u} + \mathbf{v}\| \leq \|\mathbf{u}\| + \|\mathbf{v}\|$.
A vector space with a norm is called a *normed vector space*.

The first Axiom states that all vectors except the additive identity have positive length. Since $\mathbf{0}_{\mathcal{V}} = 0\mathbf{v}$, $\mathbf{v} \in \mathcal{V}$, it follows from Axiom (2) that $\|\mathbf{0}_{\mathcal{V}}\| = |0|\,\|\mathbf{v}\| = 0$. Axioms (2) and (3) are motivated by the interaction between length and the algebraic operations in $\mathcal{R}^n$. Because of its geometric interpretation in $\mathcal{R}^2$ and $\mathcal{R}^3$, the third Axiom is called the triangle inequality.

Needless to say, $\|\bullet\| : \mathcal{R}^n \to [0,\infty)$ defined by $\|X\| = \sqrt{x_1^2 + \cdots + x_n^2}$ is a norm on $\mathcal{R}^n$. It is known as the *Euclidean norm.*

A vector space can have more than one norm, and to avoid confusion it may at times be necessary to distinguish notationally between two or more by attaching different subscripts to the norm symbol.

■ **EXAMPLE 7.39.** Define $\|\bullet\|_1 : \mathcal{R}^2 \to \mathcal{R}$ by

$$\|X\|_1 = |x_1| + |x_2|.$$

If $X \neq 0$, then at least one of x_1 and x_2 is not zero, so $\|X\|_1 > 0$. For $X,Y \in \mathcal{R}^2$ and $t \in \mathcal{R}$,

$$\|tX\|_1 = |tx_1| + |tx_2| = |t|(|x_1| + |x_2|) = |t|\,\|X\|_1,$$

and

$$\begin{aligned}\|X+Y\|_1 &= |x_1 + y_1| + |x_2 + y_2| \\ &\leq (|x_1| + |y_1|) + (|x_2| + |y_2|) \\ &= (|x_1| + |x_2|) + (|y_1| + |y_2|) \\ &= \|X\|_1 + \|Y\|_1,\end{aligned}$$

so $\|\bullet\|_1$ is a norm on $\mathcal{R}^2$. Of course, it is not the usual norm. If $\|\bullet\|$ denotes the Euclidean norm and $X = \begin{bmatrix} 1 \\ -2 \end{bmatrix}$, then $\|X\|_1 = |1| + |-2| = 3$, whereas $\|X\| = \sqrt{1^2 + (-2)^2} = \sqrt{5}$. Observe that

$$\|X\|_1^2 = (|x_1| + |x_2|)^2 = |x_1|^2 + 2|x_1|\,|x_2| + |x_2|^2 \geq x_1^2 + x_2^2 = \|X\|^2.$$

Thus, $\|X\|_1 \geq \|X\|$, with equality only when one of x_1 and x_2 is zero.

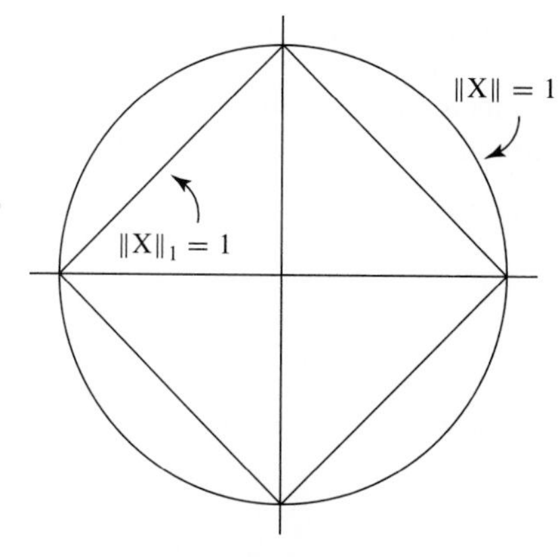

FIGURE 7.2

The unit vectors relative to the Euclidean norm are the points of the circle with center 0 and radius 1. By contrast, the unit vectors relative to $\|\bullet\|_1$ are the elements of $\mathcal{R}^2$ satisfying $|x_1| + |x_2| = 1$, that is, the points of the square with vertices $\pm E_1$ and $\pm E_2$ (Figure 7.2). □

■ **EXAMPLE 7.40.** Assume $a,b \in \mathcal{R}$, $a < b$, and define $\|\bullet\|_1 : \mathcal{C}[a,b] \to \mathcal{R}$ by

$$\|f\|_1 = \int_a^b |f(x)|\,dx.$$

Since $|f(x)| \geq 0$ for all $x \in [a,b]$, $\|f\|_1 \geq 0$. The only nonnegative continuous function on $[a,b]$ with integral 0 is the zero function, so if $\|f\|_1 = 0$, then $|f| = 0$, and consequently $f = 0$. For $t \in \mathcal{R}$ and $f \in \mathcal{C}[a,b]$,

$$\|tf\|_1 = \int_a^b |tf(x)|\,dx = |t| \int_a^b |f(x)|\,dx = |t|\|f\|_1.$$

Moreover, if $f,g \in \mathcal{C}[a,b]$, then $|f(x) + g(x)| \leq |f(x)| + |g(x)|$, $x \in [a,b]$, so

$$\begin{aligned}\|f+g\|_1 &= \int_a^b |f(x)+g(x)|\,dx \leq \int_a^b (|f(x)| + |g(x)|)\,dx \\ &= \int_a^b |f(x)|\,dx + \int_a^b |g(x)|\,dx = \|f\|_1 + \|g\|_1.\end{aligned}$$

Thus, $\|\bullet\|_1$ is a norm on $\mathcal{C}[a,b]$. In particular, if $a = 0$, $b = 2\pi$, and $f(x) = \sin(x)$, then

$$\|f\|_1 = \int_0^{2\pi} |\sin(x)|\,dx = 2\int_0^{\pi} \sin(x)\,dx = 4. \ \square$$

Direction for an abstract vector is a pretty nebulous idea. It is easy enough, however, to agree on what it should mean for two nonzero vectors to have the same direction, you simply insist that one be a positive scalar multiple of the other. Identifying different directions amounts to specifying a vector for each, which in a normed vector space can be done by describing all the unit vectors.

Consider now the problem of defining an angle between two nonzero abstract vectors. In $\mathcal{R}^n$ this concept is based on the scalar product, so it makes sense to generalize that notion first and then proceed as before.

Given an abstract vector space $\mathcal{V}$, the set of all ordered pairs of vectors in $\mathcal{V}$ is denoted by $\mathcal{V} \times \mathcal{V}$ (called the *Cartesian product* of $\mathcal{V}$ with $\mathcal{V}$), that is,

$$\mathcal{V} \times \mathcal{V} = \{(\mathbf{u},\mathbf{v}) \ : \ \mathbf{u},\mathbf{v} \in \mathcal{V}\}.$$

A scalar product on $\mathcal{V}$ is essentially a means of assigning a real number to each ordered pair of vectors, and as such it is a function from $\mathcal{V} \times \mathcal{V}$ to $\mathcal{R}$. Its customary symbol is

$$\langle \bullet,\bullet \rangle : \mathcal{V} \times \mathcal{V} \to \mathcal{R},$$

and $\langle \mathbf{u},\mathbf{v} \rangle$ denotes its value at $(\mathbf{u},\mathbf{v})$.

DEFINITION. An *inner product* or *scalar product* on a vector space $\mathcal{V}$ is a function $\langle \bullet,\bullet \rangle : \mathcal{V} \times \mathcal{V} \to \mathcal{R}$ such that for all $\mathbf{u},\mathbf{v},\mathbf{w} \in \mathcal{V}$ and $t \in \mathcal{R}$,

(1) $\langle \mathbf{u},\mathbf{v} \rangle = \langle \mathbf{v},\mathbf{u} \rangle$,
(2) $\langle \mathbf{u},\mathbf{v}+\mathbf{w} \rangle = \langle \mathbf{u},\mathbf{v} \rangle + \langle \mathbf{u},\mathbf{w} \rangle$,
(3) $\langle \mathbf{u},t\mathbf{v} \rangle = t\langle \mathbf{u},\mathbf{v} \rangle$, and
(4) $\langle \mathbf{u},\mathbf{u} \rangle > 0$ whenever $\mathbf{u} \neq \mathbf{0}_{\mathcal{V}}$.

A vector space with an inner product is called an *inner product space*.

Axioms (1) through (4) reflect the algebraic properties of the scalar product in $\mathcal{R}^n$ (compare with Theorem 3.1).

THEOREM 7.27. If $\mathcal{V}$ is an inner product space, $\mathbf{u},\mathbf{v},\mathbf{w} \in \mathcal{V}$, and $t \in \mathcal{R}$, then

(1) $\langle \mathbf{u},\mathbf{0}_{\mathcal{V}}\rangle = 0$,
(2) $\langle \mathbf{u}+\mathbf{v},\mathbf{w}\rangle = \langle \mathbf{u},\mathbf{w}\rangle + \langle \mathbf{v},\mathbf{w}\rangle$, and
(3) $\langle t\mathbf{u},\mathbf{v}\rangle = t\langle \mathbf{u},\mathbf{v}\rangle$.

Proof. By Axiom (3), $\langle \mathbf{u},\mathbf{0}_{\mathcal{V}}\rangle = \langle \mathbf{u},0\mathbf{v}\rangle = 0\langle \mathbf{u},\mathbf{v}\rangle = 0$. For $\mathbf{u},\mathbf{v},\mathbf{w} \in \mathcal{V}$,

$$\begin{aligned}\langle \mathbf{u}+\mathbf{v},\mathbf{w}\rangle &= \langle \mathbf{w},\mathbf{u}+\mathbf{v}\rangle && \text{(Axiom (1))}\\ &= \langle \mathbf{w},\mathbf{u}\rangle + \langle \mathbf{w},\mathbf{v}\rangle && \text{(Axiom (2))}\\ &= \langle \mathbf{u},\mathbf{w}\rangle + \langle \mathbf{v},\mathbf{w}\rangle. && \text{(Axiom (1))}\end{aligned}$$

Conclusion (3) is left as an exercise. ■

EXAMPLE 7.41. For $X,Y \in \mathcal{R}^2$ set

$$\langle X,Y\rangle = 4x_1y_1 + 3x_2y_2. \tag{7.16}$$

Certainly $\langle X,Y\rangle = \langle Y,X\rangle$. If $X,Y,Z \in \mathcal{R}^2$ and $t \in \mathcal{R}$, then

$$\begin{aligned}\langle X,Y+Z\rangle &= 4x_1(y_1+z_1) + 3x_2(y_2+z_2)\\ &= (4x_1y_1 + 3x_2y_2) + (4x_1z_1 + 3x_2z_2)\\ &= \langle X,Y\rangle + \langle X,Z\rangle,\end{aligned}$$

and

$$\langle X,tY\rangle = 4x_1(ty_1) + 3x_2(ty_2) = t(4x_1y_1 + 3x_2y_2) = t\langle X,Y\rangle.$$

Finally, $\langle X,X\rangle = 4x_1^2 + 3x_2^2 > 0$, provided $X \neq 0$. Thus, (7.16) is an inner product on $\mathcal{R}^2$. More generally, suppose $\langle X,Y\rangle = \lambda x_1y_1 + \mu x_2y_2$, $\lambda,\mu \in \mathcal{R}$. Which of the inner product axioms are satisfied by $\langle \bullet,\bullet\rangle$? What additional condition on λ and μ is needed for $\langle \bullet,\bullet\rangle$ to be an inner product on $\mathcal{R}^2$? □

EXAMPLE 7.42. Assume $a,b \in \mathcal{R}$, $a < b$. For $f,g \in \mathcal{C}[a,b]$, let

$$\langle f,g\rangle = \int_a^b f(x)g(x)\,dx. \tag{7.17}$$

Observe that

$$\langle f,g\rangle = \int_a^b f(x)g(x)\,dx = \int_a^b g(x)f(x)\,dx = \langle g,f\rangle.$$

For $f,g,h \in \mathcal{C}[a,b]$ and $t \in \mathcal{R}$,

$$\begin{aligned}\langle f,g+h\rangle &= \int_a^b f(x)[g(x)+h(x)]\,dx\\ &= \int_a^b [f(x)g(x) + f(x)h(x)]\,dx\\ &= \int_a^b f(x)g(x)\,dx + \int_a^b f(x)h(x)\,dx\\ &= \langle f,g\rangle + \langle f,h\rangle,\end{aligned}$$

and

$$\langle f, tg\rangle = \int_a^b f(x)[tg(x)]\,dx = t\int_a^b f(x)g(x)\,dx = t\langle f,g\rangle.$$

Finally, $\langle f,f\rangle = \int_a^b [f(x)]^2\,dx \geq 0$, and equality occurs only when $f^2 = 0$, that is, when $f = 0$. Thus, (7.17) is an inner product on $\mathcal{C}[a,b]$. □

If $\mathcal{V}$ is an inner product space, $\mathbf{u},\mathbf{v}_1,\ldots,\mathbf{v}_p \in \mathcal{V}$, and $t_1,\ldots,t_p \in \mathcal{R}$, then a standard induction argument produces

$$\langle \mathbf{u}, t_1\mathbf{v}_1 + \cdots + t_p\mathbf{v}_p\rangle = t_1\langle \mathbf{u},\mathbf{v}_1\rangle + \cdots + t_p\langle \mathbf{u},\mathbf{v}_p\rangle,$$

that is,

$$\left\langle \mathbf{u}, \sum_{k=1}^{p} t_k\mathbf{v}_k\right\rangle = \sum_{k=1}^{p} t_k\langle \mathbf{u},\mathbf{v}_k\rangle. \tag{7.18}$$

Similarly, it follows from Theorem 7.27(2), (3) that

$$\left\langle \sum_{j=1}^{q} s_j\mathbf{u}_j,\mathbf{v}\right\rangle = \sum_{j=1}^{q} s_j\langle \mathbf{u}_j,\mathbf{v}\rangle, \tag{7.19}$$

where $\mathbf{u}_1,\ldots,\mathbf{u}_q,\mathbf{v} \in \mathcal{V}$, $s_1,\ldots,s_q \in \mathcal{R}$. Combining (7.18) and (7.19) gives

$$\begin{aligned}\left\langle \sum_{j=1}^{q} s_j\mathbf{u}_j, \sum_{k=1}^{p} t_k\mathbf{v}_k\right\rangle &= \sum_{j=1}^{q} s_j\left\langle \mathbf{u}_j, \sum_{k=1}^{p} t_k\mathbf{v}_k\right\rangle = \sum_{j=1}^{q} s_j\sum_{k=1}^{p} t_k\langle \mathbf{u}_j,\mathbf{v}_k\rangle \\ &= \sum_{j=1}^{q}\sum_{k=1}^{p} s_jt_k\langle \mathbf{u}_j,\mathbf{v}_k\rangle.\end{aligned} \tag{7.20}$$

THEOREM 7.28. $\langle\bullet,\bullet\rangle : \mathcal{R}^n \times \mathcal{R}^n \to \mathcal{R}$ is an inner product if and only if there is a symmetric, positive definite $\mathrm{A} \in \mathcal{M}_{n\times n}$ such that for all $\mathrm{X},\mathrm{Y} \in \mathcal{R}^n$,

$$\langle \mathrm{X},\mathrm{Y}\rangle = \mathrm{X}^T\mathrm{AY}. \tag{7.21}$$

Proof. Assume $\langle\bullet,\bullet\rangle$ is an inner product on $\mathcal{R}^n$. If $\mathrm{X},\mathrm{Y} \in \mathcal{R}^n$, then $\mathrm{X} = \sum_{j=1}^{n} x_j\mathrm{E}_j$, $\mathrm{Y} = \sum_{k=1}^{n} y_k\mathrm{E}_k$, and by (7.20), $\langle \mathrm{X},\mathrm{Y}\rangle = \sum_{j=1}^{n}\sum_{k=1}^{n} x_jy_k\langle \mathrm{E}_j,\mathrm{E}_k\rangle$. Let

$$\mathrm{A} = \begin{bmatrix} \langle \mathrm{E}_1,\mathrm{E}_1\rangle & \cdots & \langle \mathrm{E}_1,\mathrm{E}_n\rangle \\ \vdots & & \vdots \\ \langle \mathrm{E}_n,\mathrm{E}_1\rangle & \cdots & \langle \mathrm{E}_n,\mathrm{E}_n\rangle \end{bmatrix}$$

and observe that

$$\begin{aligned}\mathrm{X}^T\mathrm{AY} &= [x_1 \quad \cdots \quad x_n]\begin{bmatrix} \langle \mathrm{E}_1,\mathrm{E}_1\rangle & \cdots & \langle \mathrm{E}_1,\mathrm{E}_n\rangle \\ \vdots & & \vdots \\ \langle \mathrm{E}_n,\mathrm{E}_1\rangle & \cdots & \langle \mathrm{E}_n,\mathrm{E}_n\rangle \end{bmatrix}\begin{bmatrix} y_1 \\ \vdots \\ y_n \end{bmatrix} \\ &= [x_1 \quad \cdots \quad x_n]\begin{bmatrix} \sum_{k=1}^{n} y_k\langle \mathrm{E}_1,\mathrm{E}_k\rangle \\ \vdots \\ \sum_{k=1}^{n} y_k\langle \mathrm{E}_n,\mathrm{E}_k\rangle \end{bmatrix}\end{aligned}$$

$$= \sum_{k=1}^{n} x_1 y_k \langle E_1, E_k \rangle + \cdots + \sum_{k=1}^{n} x_n y_k \langle E_n, E_k \rangle$$

$$= \sum_{j=1}^{n}\sum_{k=1}^{n} x_j y_k \langle E_j, E_k \rangle.$$

Thus, $\langle X,Y \rangle = X^T AY$. For $1 \le i,j \le n$,

$$[A]_{ij} = \langle E_i, E_j \rangle = \langle E_j, E_i \rangle = [A]_{ji},$$

so A is symmetric. By Axiom (4), $X \neq 0$ implies $X^T AX = \langle X,X \rangle > 0$, so $Q(X) = X^T AX$ is positive definite. Thus, A is positive definite. For the converse, suppose A is a symmetric, positive definite, $n \times n$ matrix and $\langle X,Y \rangle = X^T AY$. For $X,Y,Z \in \mathcal{R}^n$ and $t \in \mathcal{R}$,

$$\begin{aligned}\langle X, Y+Z \rangle &= X^T A(Y+Z) = X^T(AY + AZ) = X^T AY + X^T AZ \\ &= \langle X, Y \rangle + \langle X, Z \rangle,\end{aligned}$$

and

$$\langle X, tY \rangle = X^T A(tY) = t(X^T AY) = t\langle X, Y \rangle.$$

Moreover,

$$\langle X, Y \rangle = X^T AY = (X^T AY)^T = Y^T A^T (X^T)^T = Y^T AX = \langle Y, X \rangle.$$

Finally, since A is positive definite, $\langle X,X \rangle = X^T AX > 0$ whenever $X \neq 0$. Thus, $\langle X,Y \rangle = X^T AY$ defines an inner product on $\mathcal{R}^n$. ■

■ **EXAMPLE 7.43.** For $X,Y \in \mathcal{R}^3$, let

$$\langle X, Y \rangle = 3x_1y_1 + 2x_2y_2 - x_2y_3 - x_3y_2 + 2x_3y_3 = X^T \begin{bmatrix} 3 & 0 & 0 \\ 0 & 2 & -1 \\ 0 & -1 & 2 \end{bmatrix} Y.$$

Observe that $A = \begin{bmatrix} 3 & 0 & 0 \\ 0 & 2 & -1 \\ 0 & -1 & 2 \end{bmatrix}$ is symmetric. In the notation of Theorem 6.17, $\det(A_1) = \det[3] = 3$, $\det(A_2) = \det \begin{bmatrix} 3 & 0 \\ 0 & 2 \end{bmatrix} = 6$, and $\det(A_3) = \det(A) = 9$, so A is positive definite. Thus, $\langle \bullet, \bullet \rangle$ is an inner product on $\mathcal{R}^3$. □

■ **EXAMPLE 7.44.** The symmetric, positive definite, $n \times n$ matrix associated with the usual inner product on $\mathcal{R}^n$ is I_n. Indeed, $X \cdot Y = X^T Y = X^T I_n Y$. □

■ **EXAMPLE 7.45.** For $\lambda, \mu \in \mathcal{R}$, the function $\langle \bullet, \bullet \rangle : \mathcal{R}^2 \times \mathcal{R}^2 \to \mathcal{R}$ defined by

$$\langle X, Y \rangle = \lambda x_1 y_1 + \mu x_2 y_2 = X^T \begin{bmatrix} \lambda & 0 \\ 0 & \mu \end{bmatrix} Y$$

(recall Example 7.41) is an inner product on $\mathcal{R}^2$ precisely when $\begin{bmatrix} \lambda & 0 \\ 0 & \mu \end{bmatrix}$ is positive definite, that is, when $\lambda > 0$ and $\mu > 0$. □

The Euclidean norm on $\mathcal{R}^n$ and the usual scalar product on $\mathcal{R}^n$ are related by

$$\|X\| = \sqrt{X \cdot X},$$

which suggests the possibility of defining a norm on an abstract inner product space by

$$\|\mathbf{v}\| = \sqrt{\langle \mathbf{v},\mathbf{v}\rangle}.$$

The last equation is sensible, because $\langle \mathbf{v},\mathbf{v}\rangle \geq 0$ for all $\mathbf{v} \in \mathcal{V}$. Verification of the norm axioms involves an extension of the Cauchy-Schwarz inequality (Theorem 3.2).

THEOREM 7.29. **(Cauchy-Schwarz Inequality).** If $\mathcal{V}$ is an inner product space and $\mathbf{u},\mathbf{v} \in \mathcal{V}$, then

$$\langle \mathbf{u},\mathbf{v}\rangle^2 \leq \langle \mathbf{u},\mathbf{u}\rangle\langle \mathbf{v},\mathbf{v}\rangle.$$

Equality occurs if and only if $\mathbf{u}$ and $\mathbf{v}$ are linearly dependent.

Proof. The proof is identical to that of Theorem 3.2, with inner products replacing the dot products that occur in that argument. ■

THEOREM 7.30. If $\mathcal{V}$ is an inner product space, then $\|\bullet\| : \mathcal{V} \to \mathcal{R}$ defined by

$$\|\mathbf{v}\| = \sqrt{\langle \mathbf{v},\mathbf{v}\rangle} \tag{7.22}$$

is a norm.

Proof. If $\mathbf{v} \in \mathcal{V}$ and $\mathbf{v} \neq \mathbf{0}_{\mathcal{V}}$, then $\langle \mathbf{v},\mathbf{v}\rangle > 0$, so $\|\mathbf{v}\| > 0$. For $\mathbf{v} \in \mathcal{V}$ and $t \in \mathcal{R}$,

$$\|t\mathbf{v}\| = \sqrt{\langle t\mathbf{v},t\mathbf{v}\rangle} = \sqrt{t^2\langle \mathbf{v},\mathbf{v}\rangle} = |t|\sqrt{\langle \mathbf{v},\mathbf{v}\rangle} = |t|\|\mathbf{v}\|.$$

If $\mathbf{u},\mathbf{v} \in \mathcal{V}$, then

$$\begin{aligned}\|\mathbf{u}+\mathbf{v}\|^2 &= \langle \mathbf{u}+\mathbf{v},\mathbf{u}+\mathbf{v}\rangle = \langle \mathbf{u}+\mathbf{v},\mathbf{u}\rangle + \langle \mathbf{u}+\mathbf{v},\mathbf{v}\rangle \\ &= \langle \mathbf{u},\mathbf{u}\rangle + \langle \mathbf{v},\mathbf{u}\rangle + \langle \mathbf{u},\mathbf{v}\rangle + \langle \mathbf{v},\mathbf{v}\rangle \\ &= \|\mathbf{u}\|^2 + 2\langle \mathbf{u},\mathbf{v}\rangle + \|\mathbf{v}\|^2.\end{aligned}$$

By the Cauchy-Schwarz inequality,

$$\langle \mathbf{u},\mathbf{v}\rangle \leq |\langle \mathbf{u},\mathbf{v}\rangle| \leq \sqrt{\langle \mathbf{u},\mathbf{u}\rangle\langle \mathbf{v},\mathbf{v}\rangle} = \|\mathbf{u}\|\,\|\mathbf{v}\|,$$

so

$$\|\mathbf{u}+\mathbf{v}\|^2 \leq \|\mathbf{u}\|^2 + 2\|\mathbf{u}\|\,\|\mathbf{v}\| + \|\mathbf{v}\|^2 = (\|\mathbf{u}\| + \|\mathbf{v}\|)^2.$$

Thus, $\|\mathbf{u}+\mathbf{v}\| \leq \|\mathbf{u}\| + \|\mathbf{v}\|$. ■

Thanks to Theorem 7.30, every inner product space is a normed vector space. The induced norm (i.e. (7.22)) is considered the natural norm on the space, and unless mention is made to the contrary, $\|\bullet\|$ is reserved for this norm. The Cauchy-Schwarz inequality then reads $|\langle \mathbf{u},\mathbf{v}\rangle| \leq \|\mathbf{u}\|\;\|\mathbf{v}\|$.

■ **EXAMPLE 7.46.** The norm induced on $\mathcal{C}[a,b]$ by inner product (7.14) is

$$\|f\| = \sqrt{\langle f,f\rangle} = \sqrt{\int_a^b [f(x)]^2\,dx},$$

and in this setting the Cauchy-Schwarz inequality reads

$$\left|\int_a^b f(x)g(x)\,dx\right| \leq \sqrt{\int_a^b [f(x)]^2\,dx}\;\sqrt{\int_a^b [g(x)]^2\,dx}.$$

In particular, if $a = 0$, $b = 2\pi$, and $f(x) = \sin(x)$, then

$$\|f\|^2 = \langle f, f\rangle = \int_0^{2\pi} \sin^2(x)\,dx = \frac{1}{2}\int_0^{2\pi} [1 - \cos(2x)]\,dx = \pi,$$

so $\|f\| = \sqrt{\pi}$. □

An inner product induces a norm, but not all norms are produced in this manner. Consider $\mathcal{R}^2$. According to Theorem 7.28, an inner product on $\mathcal{R}^2$ has equation $\langle X,Y\rangle = X^T AY$, where $A \in \mathcal{M}_{2\times 2}$ is symmetric and positive definite. The induced norm is

$$\|X\| = \sqrt{X^T AX}.$$

Note that $Q(X) = X^T AX$ is a positive definite quadratic form. The set of unit vectors relative to $\|\bullet\|$ is the level curve $\Gamma = \{X : Q(X) = 1\}$, and its shape can be determined by diagonalizing A. If P is an orthogonal 2×2 matrix such that

$$P^T AP = \begin{bmatrix} \lambda & 0 \\ 0 & \mu \end{bmatrix},$$

and $X = PY$, then

$$X^T AX = (PY)^T A(PY) = Y^T (P^T AP)Y = \lambda y_1^2 + \mu y_2^2.$$

Thus, relative to a coordinate system generated by orthonormal eigenvectors of A, Γ has the equation

$$\lambda y_1^2 + \mu y_2^2 = 1. \tag{7.23}$$

Since A is positive definite, λ and μ are both positive, and (7.23) is then the equation of an ellipse. Thus, a norm on $\mathcal{R}^2$ that is induced by an inner product has an associated set of unit vectors that forms an ellipse. The norm discussed in Example 7.37 does not have this property and, therefore, is not induced by an inner product.

If **u** and **v** are nonzero elements of an inner product space $\mathcal{V}$, then by the Cauchy-Schwarz inequality,

$$-1 \le \frac{\langle \mathbf{u},\mathbf{v}\rangle}{\|\mathbf{u}\|\,\|\mathbf{v}\|} \le 1.$$

Thus, the inner product and its induced norm lend themselves to defining an angle between **u** and **v** in exactly the same manner as in Euclidean n-space.

DEFINITION. In an inner product space $\mathcal{V}$, the angle between nonzero vectors $\mathbf{u},\mathbf{v} \in \mathcal{V}$ is the unique $\theta \in [0,\pi]$ satisfying

$$\cos(\theta) = \frac{\langle \mathbf{u},\mathbf{v}\rangle}{\|\mathbf{u}\|\,\|\mathbf{v}\|}.$$

For $\mathbf{u},\mathbf{v} \in \mathcal{V}$, **u** is said to be orthogonal to **v** when $\langle \mathbf{u},\mathbf{v}\rangle = 0$.

■ **EXAMPLE 7.47.** According to Theorem 6.17, $A = \begin{bmatrix} 2 & 1 \\ 1 & 1 \end{bmatrix}$ is positive definite and therefore

$$\langle X,Y\rangle = X^T \begin{bmatrix} 2 & 1 \\ 1 & 1 \end{bmatrix} Y = 2x_1y_1 + x_1y_2 + x_2y_1 + x_2y_2$$

is an inner product on $\mathcal{R}^2$. The norm induced by $\langle \bullet,\bullet \rangle$ is

$$\|X\| = \sqrt{\langle X,X\rangle} = \sqrt{2x_1^2 + 2x_1x_2 + x_2^2}.$$

Observe that

$$\langle E_1,E_2\rangle = (2\cdot 1\cdot 0) + (1\cdot 1) + (0\cdot 0) + (0\cdot 1) = 1,$$
$$\|E_1\| = \sqrt{(2\cdot 1^2) + (2\cdot 1\cdot 0) + 0^2} = \sqrt{2},$$

and

$$\|E_2\| = \sqrt{(2\cdot 0^2) + (2\cdot 0\cdot 1) + 1^2} = 1.$$

The angle between E_1 and E_2, relative to this inner product, is the $\theta \in [0,\pi]$ satisfying

$$\cos(\theta) = \frac{\langle E_1, E_2\rangle}{\|E_1\|\|E_2\|} = \frac{1}{\sqrt{2}},$$

that is, $\theta = \pi/4$. □

As in $\mathcal{R}^n$, a set of mutually orthogonal vectors is called an *orthogonal set*, and orthogonal unit vectors are said to be *orthonormal*.

THEOREM 7.31. If $\mathcal{V}$ is an inner product space and $\mathcal{A}$ is an orthogonal subset consisting of nonzero vectors, then $\mathcal{A}$ is linearly independent.

Proof. Suppose $\mathcal{A}$ is an orthogonal set of nonzero vectors and $\mathbf{v}_1, \ldots, \mathbf{v}_p \in \mathcal{A}$. If $t_1, \ldots, t_p$ are scalars such that $\mathbf{0}_\mathcal{V} = t_1\mathbf{v}_1 + \cdots + t_p\mathbf{v}_p$, then

$$0 = \langle \mathbf{v}_k, \mathbf{0}_\mathcal{V}\rangle = \left\langle \mathbf{v}_k, \sum_{j=1}^{p} t_j\mathbf{v}_j \right\rangle = \sum_{j=1}^{p} t_j\langle \mathbf{v}_k,\mathbf{v}_j\rangle = t_k\langle \mathbf{v}_k,\mathbf{v}_k\rangle = t_k\|\mathbf{v}_k\|^2,$$

$k = 1, \ldots, p$. By assumption, $\mathbf{v}_k \neq \mathbf{0}_\mathcal{V}$, so $t_k = 0$. Thus, $\{\mathbf{v}_1, \ldots, \mathbf{v}_p\}$ is linearly independent. Since each finite subset of $\mathcal{A}$ is linearly independent, $\mathcal{A}$ itself is linearly independent. ■

The Gram-Schmidt process applies to any finite linearly independent subset of an inner product space to produce an orthonormal set that spans the same subspace.

■ **EXAMPLE 7.48.** Relative to the inner product on $\mathcal{R}^2$ defined in Example 7.47

$$\langle E_1,E_2\rangle = 1, \quad \|E_1\| = \sqrt{2}, \quad \text{and} \quad \|E_2\| = 1,$$

so $\mathcal{E} = \{E_1,E_2\}$ is not orthonormal. The Gram-Schmidt process gives $Y_1 = E_1$ and

$$Y_2 = E_2 - \frac{\langle E_2, Y_1\rangle}{\|Y_1\|^2}Y_1 = \begin{bmatrix} 0 \\ 1 \end{bmatrix} - \frac{1}{2}\begin{bmatrix} 1 \\ 0 \end{bmatrix} = \frac{1}{2}\begin{bmatrix} -1 \\ 2 \end{bmatrix}.$$

Now, $\|Y_1\| = \|E_1\| = \sqrt{2}$, and $\|Y_2\| = \sqrt{2(1/4) + 2(-1/2)1 + 1^2} = 1/\sqrt{2}$, so the resulting orthonormal basis for $\mathcal{R}^2$ is $\{U_1, U_2\}$, where

$$U_1 = \frac{Y_1}{\|Y_1\|} = \frac{1}{\sqrt{2}}\begin{bmatrix} 1 \\ 0 \end{bmatrix} \quad \text{and} \quad U_2 = \frac{Y_2}{\|Y_2\|} = \frac{1}{\sqrt{2}}\begin{bmatrix} -1 \\ 2 \end{bmatrix}. \ \square$$

■ **EXAMPLE 7.49.** Let $f(x) = \sin(x)$, $g(x) = \cos(x)$, and $\mathcal{V} = \mathcal{S}pan\{f,g\} \subseteq \mathcal{C}[0,2\pi]$. Relative to the inner product defined by (7.17),

$$\langle f,g \rangle = \int_0^{2\pi} \sin(x)\cos(x)\,dx = \frac{1}{2}\int_0^{2\pi} \sin(2x)\,dx = 0,$$

so f is orthogonal to g. According to Example 7.46, $\|f\| = \sqrt{\pi}$, and similar calculations show that $\|g\| = \sqrt{\pi}$, so $\left\{\frac{1}{\sqrt{\pi}}\sin(x), \frac{1}{\sqrt{\pi}}\cos(x)\right\}$ is an orthonormal basis for $\mathcal{V}$. □

Exercises 7.6

1. Define $\|\bullet\|_\infty : \mathcal{R}^2 \to \mathcal{R}$ by $\|X\|_\infty = \max\{|x_1|, |x_2|\}$.
   a. Verify that $\|\bullet\|_\infty$ is a norm on $\mathcal{R}^2$.
   b. Sketch $\{U \in \mathcal{R}^2 : \|U\|_\infty = 1\}$.
   c. Show that $\|X\| \geq \|X\|_\infty$, where $\|\bullet\|$ denotes the Euclidean norm.

2. Define $\|\bullet\|_\infty : \mathcal{C}[0,\pi] \to \mathcal{R}$ by $\|f\|_\infty = \max\{|f(x)| : 0 \leq x \leq \pi\}$.
   a. Find $\|f\|_\infty$ when $f(x) = \sin(x)$, $f(x) = 3\cos(2x)$, and $f(x) = 1 - x^2$.
   b. Describe the graph of a typical element of $\{f \in \mathcal{C}[0,\pi] : \|f\|_\infty = 1\}$.
   c. Verify that $\|\bullet\|_\infty$ is a norm on $\mathcal{C}[0,\pi]$.

3. Let $\mathcal{V} = \{\{x_k\}_0^\infty \in \mathcal{R}^\infty : \{x_k\}_0^\infty \text{ is convergent}\}$. For each $X = \{x_k\}_0^\infty \in \mathcal{V}$ there is an $x \in \mathcal{R}$ such that $x_k \to x$. Define $\|\bullet\| : \mathcal{V} \to \mathcal{R}$ by $\|X\| = |x|$. Which of the axioms for a norm are satisfied by $\|\bullet\|$?

4. Which norm axioms are satisfied by $\|\mathbf{v}\| = \begin{cases} 0, & \text{if } \mathbf{v} = \mathbf{0}_\mathcal{V} \\ 1, & \text{if } \mathbf{v} \neq \mathbf{0}_\mathcal{V} \end{cases}$, $\mathbf{v} \in \mathcal{V}$?

5. Define $\|\bullet\|_f : \mathcal{M}_{n\times n} \to \mathcal{R}$ by $\|A\|_f = \left(\sum_{i,j=1}^n a_{ij}^2\right)^{1/2}$.
   a. Find $\|A\|_f$ when $A = \begin{bmatrix} 1 & 2 \\ 3 & 4 \end{bmatrix}$. When $A = I_n$. When A is orthogonal.
   b. Show that $\|\bullet\|_f$ is a norm on $\mathcal{M}_{n\times n}$ (called the Frobenius norm).

   $\left(\text{Hint: observe that } \|A\|_f \text{ is the same as the Euclidean norm of } \begin{bmatrix} \text{Col}_1(A) \\ \hline \vdots \\ \hline \text{Col}_n(A) \end{bmatrix} \in \mathcal{R}^{n^2}.\right)$

   c. Show that $\|A\|_f^2 = \text{trace}(A^T A)$.
   d. Show that if P is orthogonal and $B = P^T AP$, then $\|B\|_f = \|A\|_f$.

e. Show that if A is symmetric, with eigenvalues $\lambda_1, \ldots, \lambda_n$, then

$$\|A\|_f = \sqrt{\lambda_1^2 + \cdots + \lambda_n^2}.$$

6. Suppose $\mathcal{V}$ is a vector space of dimension n and $T : \mathcal{V} \to \mathcal{R}^n$ is linear. Define $\|\bullet\|_T : \mathcal{V} \to \mathcal{R}$ by $\|\mathbf{v}\|_T = \|T(\mathbf{v})\|$, where $\|\bullet\|$ denotes the Euclidean norm.
   a. Which axioms for a norm are satisfied by $\|\bullet\|_T$?
   b. What additional hypothesis on T will ensure that $\|\bullet\|_T$ is a norm on $\mathcal{V}$?

7. Prove property (3) of Theorem 7.27.

8. For the given A, determine whether $\langle X,Y \rangle = X^T AY$ is an inner product. If not, which inner product axioms are not satisfied?

   a. $A = \begin{bmatrix} 3 & 2 \\ 2 & 1 \end{bmatrix}$ b. $A = \begin{bmatrix} 4 & 1 \\ 3 & 2 \end{bmatrix}$

   c. $A = \begin{bmatrix} 2 & 1 & 0 \\ 1 & 3 & 1 \\ 0 & 1 & 2 \end{bmatrix}$ d. $A = \begin{bmatrix} 2 & 0 & 1 & 1 \\ 0 & 1 & 1 & 0 \\ 1 & 1 & 2 & 0 \\ 1 & 0 & 0 & 1 \end{bmatrix}$

   e. $A = \begin{bmatrix} 1 & 1 & 1 & \cdot & \cdot & \cdot & 1 \\ 1 & 2 & 2 & \cdot & \cdot & \cdot & 2 \\ 1 & 2 & 3 & \cdot & \cdot & \cdot & 3 \\ \cdot & \cdot & \cdot & \cdot & & & \cdot \\ \cdot & \cdot & \cdot & & \cdot & & \cdot \\ \cdot & \cdot & \cdot & & & \cdot & \cdot \\ 1 & 2 & 3 & \cdot & \cdot & \cdot & n \end{bmatrix}$

9. Let $X_1 = \begin{bmatrix} 1 \\ 0 \\ 1 \end{bmatrix}$, $X_2 = \begin{bmatrix} 1 \\ 2 \\ 0 \end{bmatrix}$, and $X_3 = \begin{bmatrix} 2 \\ -2 \\ 3 \end{bmatrix}$. Relative to $\langle X,Y \rangle = X^T \begin{bmatrix} 2 & 1 & 0 \\ 1 & 3 & 1 \\ 0 & 1 & 2 \end{bmatrix} Y$,

   find

   a. the angle between X_1 and X_2.
   b. the angle between X_2 and X_3.
   c. an orthonormal basis for $\mathcal{S}pan\{X_1,X_2\}$.

10. Let $h(x) = x$ and $\mathcal{V} = \mathcal{S}pan\{h\} \subseteq \mathcal{P}_2$. Relative to $\langle f,g \rangle = \int_0^1 f(x)g(x)\,dx$, find a basis for $\mathcal{V}^\perp$.

11. Consider $\langle \bullet,\bullet \rangle : \mathcal{M}_{n\times n} \times \mathcal{M}_{n\times n} \to \mathcal{R}$ defined by $\langle A,B \rangle = \sum_{i=1}^n \sum_{j=1}^n a_{ij} b_{ij}$.
   a. Verify that $\langle \bullet,\bullet \rangle$ is an inner product on $\mathcal{M}_{n\times n}$.
   b. Show that the norm induced by $\langle \bullet,\bullet \rangle$ is the Frobenius norm (Problem 5).
   c. If $\mathcal{S}_{n\times n}$ and $\mathcal{SK}_{n\times n}$ are the subspaces of $\mathcal{M}_{n\times n}$ consisting of the symmetric and skew-symmetric matrices, respectively, show that $\mathcal{SK}_{n\times n} = (\mathcal{S}_{n\times n})^\perp$.

12. Show that $\left\{ \frac{1}{\sqrt{\pi}} \cos(nx) : n = 1, 2, 3, \ldots \right\}$ is an orthonormal set in $\mathcal{C}[0,2\pi]$ relative to the inner product in Example 7.42.

7.7
THE SPACE OF LINEAR FUNCTIONS

Sums and scalar multiples of functions from one vector space to another are defined in terms of addition and scalar multiplication of their values. That is, for $S,T : \mathcal{V} \to \mathcal{W}$ and $t \in \mathcal{R}$,

$$(S+T)(\mathbf{v}) = S(\mathbf{v}) + T(\mathbf{v}) \qquad \text{and} \qquad (tS)(\mathbf{v}) = tS(\mathbf{v}).$$

Also, $S - T = S + (-1)T$, so $(S - T)(\mathbf{v}) = S(\mathbf{v}) - T(\mathbf{v})$.

THEOREM 7.32. If $\mathcal{V}$ and $\mathcal{W}$ are vector spaces, $S,T : \mathcal{V} \to \mathcal{W}$ are linear, and $t \in \mathcal{R}$, then $S + T$ and tS are linear.

Proof. For $\mathbf{u},\mathbf{v} \in \mathcal{V}$ and $r \in \mathcal{R}$,

$$\begin{aligned}
(S+T)(\mathbf{u}+\mathbf{v}) &= S(\mathbf{u}+\mathbf{v}) + T(\mathbf{u}+\mathbf{v}) && \text{(defn. of } S+T) \\
&= [S(\mathbf{u}) + S(\mathbf{v})] + [T(\mathbf{u}) + T(\mathbf{v})] && \text{(linearity of } S \text{ and } T) \\
&= [S(\mathbf{u}) + T(\mathbf{u})] + [S(\mathbf{v}) + T(\mathbf{v})] && \text{(algebra in } \mathcal{W}) \\
&= (S+T)(\mathbf{u}) + (S+T)(\mathbf{v}), && \text{(defn. of } S+T)
\end{aligned}$$

and

$$\begin{aligned}
(S+T)(r\mathbf{v}) &= S(r\mathbf{v}) + T(r\mathbf{v}) && \text{(defn. of } S+T) \\
&= rS(\mathbf{v}) + rT(\mathbf{v}) && \text{(linearity of } S \text{ and } T) \\
&= r[S(\mathbf{v}) + T(\mathbf{v})] && \text{(algebra in } \mathcal{W}) \\
&= r[(S+T)(\mathbf{v})]. && \text{(defn. of } S+T)
\end{aligned}$$

Checking the linearity of tS is left as an exercise. ■

DEFINITION. Given vector spaces $\mathcal{V}$ and $\mathcal{W}$, the set of linear functions from $\mathcal{V}$ to $\mathcal{W}$ is denoted by $\mathcal{L}(\mathcal{V},\mathcal{W})$. Members of $\mathcal{L}(\mathcal{V},\mathcal{V})$ are called *linear operators* on $\mathcal{V}$.

Theorem 7.32 states that $\mathcal{L}(\mathcal{V},\mathcal{W})$ is closed under addition and scalar multiplication. With these operations, $\mathcal{L}(\mathcal{V},\mathcal{W})$ is in fact a vector space.

THEOREM 7.33. $\mathcal{L}(\mathcal{V},\mathcal{W})$ is a vector space.

Proof. Verification of Axioms (1)–(8) for $\mathcal{L}(\mathcal{V},\mathcal{W})$ is similar to that for $\mathcal{F}(\mathrm{J})$ in Example 7.2. The details are left for the reader. ■

If $\mathcal{V}$ and $\mathcal{W}$ are finite dimensional, with bases $\mathcal{A}$ and $\mathcal{B}$, respectively, then each $T \in \mathcal{L}(\mathcal{V},\mathcal{W})$ has an associated matrix $[T]_{\mathcal{BA}}$. The interaction between this association and the operations in $\mathcal{L}(\mathcal{V},\mathcal{W})$ is recorded in the next theorem. Its proof is the same as that of Theorem 4.17.

THEOREM 7.34. Suppose $\mathcal{V}$ and $\mathcal{W}$ are finite dimensional vector spaces with bases $\mathcal{A}$ and $\mathcal{B}$, respectively. If $S,T \in \mathcal{L}(\mathcal{V},\mathcal{W})$ and $t \in \mathcal{R}$, then

(1) $[S+T]_{\mathcal{BA}} = [S]_{\mathcal{BA}} + [T]_{\mathcal{BA}}$, and
(2) $[tT]_{\mathcal{BA}} = t[T]_{\mathcal{BA}}$.

THEOREM 7.35. If $\mathcal{V}$ and $\mathcal{W}$ are vector spaces, dim $\mathcal{V} = n$, and dim $\mathcal{W} = m$, then $\mathcal{L}(\mathcal{V},\mathcal{W})$ is isomorphic to $\mathcal{M}_{m\times n}$. In particular, dim $\mathcal{L}(\mathcal{V},\mathcal{W}) = mn$.

Proof. Let $\mathcal{A}$ and $\mathcal{B}$ be bases for $\mathcal{V}$ and $\mathcal{W}$, respectively, and define $L : \mathcal{L}(\mathcal{V},\mathcal{W}) \to \mathcal{M}_{m\times n}$ by $L(T) = [T]_{\mathcal{B}\mathcal{A}}$. For $S,T \in \mathcal{L}(\mathcal{V},\mathcal{W})$ and $t \in \mathcal{R}$,

$$L(S+T) = [S+T]_{\mathcal{B}\mathcal{A}} = [S]_{\mathcal{B}\mathcal{A}} + [T]_{\mathcal{B}\mathcal{A}} = L(S) + L(T),$$

and

$$L(tS) = [tS]_{\mathcal{B}\mathcal{A}} = t[S]_{\mathcal{B}\mathcal{A}} = tL(S),$$

so L is linear. If $L(S) = L(T)$, then for each $\mathbf{v} \in \mathcal{V}$, $[S(\mathbf{v})]_{\mathcal{B}} = [S]_{\mathcal{B}\mathcal{A}}[\mathbf{v}]_{\mathcal{A}} = [T]_{\mathcal{B}\mathcal{A}}[\mathbf{v}]_{\mathcal{A}} = [T(\mathbf{v})]_{\mathcal{B}}$, so $S(\mathbf{v}) = T(\mathbf{v})$. Thus, $S = T$. This establishes that L is one-to-one. To see that L is onto, suppose $\mathrm{A} \in \mathcal{M}_{m\times n}$. The object is to show that $\mathrm{A} = L(T)$ for some $T \in \mathcal{L}(\mathcal{V},\mathcal{W})$. To that end, consider the diagram below,

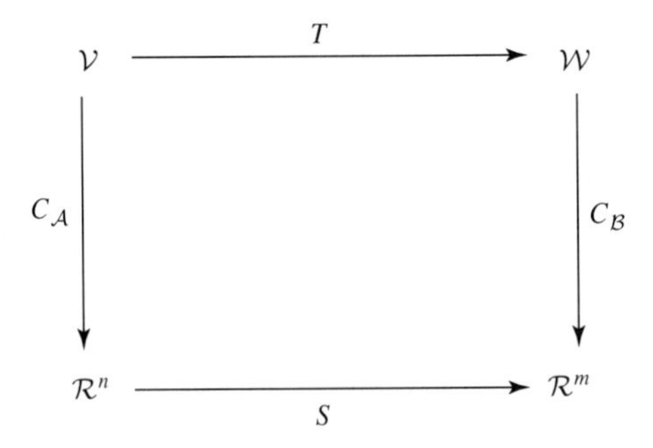

where $S : \mathcal{R}^n \to \mathcal{R}^m$ is the linear function defined by

$$S(\mathrm{X}) = \mathrm{AX}$$

and

$$T = C_{\mathcal{B}}^{-1} \circ S \circ C_{\mathcal{A}}.$$

Being a composition of linear functions, T is linear, and by Theorem 7.26,

$$[T]_{\mathcal{B}\mathcal{A}} = [C_{\mathcal{B}}^{-1}]_{\mathcal{B}\mathcal{E}_m}[S]_{\mathcal{E}_m\mathcal{E}_n}[C_{\mathcal{A}}]_{\mathcal{E}_n\mathcal{A}}. \tag{7.24}$$

Of course, $[S]_{\mathcal{E}_m\mathcal{E}_n} = \mathrm{A}$. If $\mathcal{A} = \{\mathbf{v}_1, \ldots, \mathbf{v}_n\}$, then the jth column of $[C_{\mathcal{A}}]_{\mathcal{E}_n\mathcal{A}}$ is

$$[C_{\mathcal{A}}(\mathbf{v}_j)]_{\mathcal{E}_n} = [\mathrm{E}_j]_{\mathcal{E}_n} = \mathrm{E}_j,$$

$1 \le j \le n$, that is, $[C_{\mathcal{A}}]_{\mathcal{E}_n\mathcal{A}} = \mathrm{I}_n$. Similarly, $[C_{\mathcal{B}}]_{\mathcal{E}_m\mathcal{B}} = \mathrm{I}_m$. Theorem 7.25 implies

$$[C_{\mathcal{B}}^{-1}]_{\mathcal{B}\mathcal{E}_m} = ([C_{\mathcal{B}}]_{\mathcal{E}_m\mathcal{B}})^{-1} = \mathrm{I}_m^{-1} = \mathrm{I}_m,$$

so by (7.24), $L(T) = [T]_{\mathcal{B}\mathcal{A}} = \mathrm{I}_m\mathrm{A}\mathrm{I}_n = \mathrm{A}$. Thus, L is an isomorphism. Isomorphic vector spaces have the same dimension, so dim $\mathcal{L}(\mathcal{V},\mathcal{W}) =$ dim $\mathcal{M}_{m\times n} = mn$. ■

The remainder of this section is concerned with the problem of defining a norm on $\mathcal{L}(\mathcal{R}^n,\mathcal{R}^m)$. In the discussion that follows, $\|\bullet\|$ refers to the Euclidean norm.

First consider the case when $n = 2$, that is, assume $T \in \mathcal{L}(\mathcal{R}^2,\mathcal{R}^m)$. Then

$$\Gamma = \{\mathrm{U} \in \mathcal{R}^2 \ : \ \|\mathrm{U}\| = 1\}$$

is the circle centered at 0 of radius 1, but for current purposes think of it as the collection of all directions in $\mathcal{R}^2$. Let $\mathrm{U} \in \Gamma$ and set $\mathcal{U} = \mathcal{S}pan\{\mathrm{U}\}$. For each $\mathrm{X} \in \mathcal{U}$ there is a $t \in \mathcal{R}$ such that $\mathrm{X} = t\mathrm{U}$, and since U is a unit vector, $\|\mathrm{X}\| = |t|\|\mathrm{U}\| = |t|$. Thus,

$$\|T(\mathrm{X})\| = \|tT(\mathrm{U})\| = |t|\|T(\mathrm{U})\| = \|T(\mathrm{U})\|\|\mathrm{X}\|, \tag{7.25}$$

that is, T changes the length of each vector in $\mathcal{U}$ by a factor of $\|T(\mathrm{U})\|$. This last quantity is called the *change of scale factor for T in the direction* U.

Now, Γ can be parameterized by

$$\mathrm{U}(t) = \cos(t)\mathrm{E}_1 + \sin(t)\mathrm{E}_2, \quad 0 \le t \le 2\pi,$$

so $T(\Gamma)$ is the curve in $\mathcal{R}^2$ with description

$$T(\mathrm{U}(t)) = \cos(t)T(\mathrm{E}_1) + \sin(t)T(\mathrm{E}_2), \quad 0 \le t \le 2\pi.$$

The change of scale factor, $\|T(\mathrm{U}(t))\|$, is therefore

$$\sqrt{\cos^2(t)\|T(\mathrm{E}_1)\|^2 + 2\,\cos(t)\sin(t)[T(\mathrm{E}_1)\cdot T(\mathrm{E}_2)] + \sin^2(t)\|T(\mathrm{E}_2)\|^2}. \tag{7.26}$$

Observe that (7.26) is a continuous function of t on $[0,2\pi]$ and, consequently, has a maximum and a minimum. Thus, T has a largest and smallest change of scale factor. Let

$$M(T) = \max\{\|T(\mathrm{U})\| \;:\; \mathrm{U} \in \Gamma\}$$

and

$$m(T) = \min\{\|T(\mathrm{U})\| \;:\; \mathrm{U} \in \Gamma\}.$$

Because of the linearity of T, $M(T)$ and $m(T)$ actually govern the size of $T(\mathrm{X})$ for every $\mathrm{X} \in \mathcal{R}^2$. Indeed, for each nonzero $\mathrm{X} \in \mathcal{R}^2$, $\mathrm{X}/\|\mathrm{X}\| \in \Gamma$, so

$$m(t) \le \left\| T\left(\frac{\mathrm{X}}{\|\mathrm{X}\|}\right) \right\| \le M(T),$$

and since

$$\left\| T\left(\frac{\mathrm{X}}{\|\mathrm{X}\|}\right) \right\| = \left\| \frac{1}{\|\mathrm{X}\|} T(\mathrm{X}) \right\| = \frac{\|T(\mathrm{X})\|}{\|\mathrm{X}\|},$$

$$m(T)\|\mathrm{X}\| \le \|T(\mathrm{X})\| \le M(T)\|\mathrm{X}\|. \tag{7.27}$$

If U_0 is a unit vector associated with $M(T)$, that is, if $\|T(\mathrm{U}_0)\| = M(T)$, then setting $\mathrm{U} = \mathrm{U}_0$ in (7.25) gives $\|T(\mathrm{X})\| = M(T)\|\mathrm{X}\|$, for all $\mathrm{X} \in \mathcal{S}pan\{\mathrm{U}_0\}$. Similarly, equality occurs on the left of (7.27) along any line through zero whose direction is associated with $m(T)$. Thus, $M(T)$ and $m(T)$ can be characterized as the smallest and largest numbers, respectively, in $[0,\infty)$ satisfying

$$m\|\mathrm{X}\| \le \|T(\mathrm{X})\| \le M\|\mathrm{X}\|$$

for all $\mathrm{X} \in \mathcal{R}^2$.

Most of the foregoing discussion applies equally well to functions $T \in \mathcal{L}(\mathcal{R}^n,\mathcal{R}^m)$. When $n > 2$ the existence of a largest and smallest change of scale factor depends on more sophisticated results in analysis, but given that fact, $M(T)$ is a reasonable measure of the "size" of T. It indicates the greatest extent to which T distorts the length of a vector.

THEOREM 7.36. $\|\bullet\|_* : \mathcal{L}(\mathcal{R}^n,\mathcal{R}^m) \to \mathcal{R}$ defined by $\|T\|_* = \max\{\|T(\mathrm{U})\| : \|\mathrm{U}\| = 1\}$ is a norm on $\mathcal{L}(\mathcal{R}^n,\mathcal{R}^m)$.

Proof. Let $T \in \mathcal{L}(\mathcal{R}^n,\mathcal{R}^m)$. If $T \neq 0$, then there is an $\mathrm{X}_0 \in \mathcal{R}^n$ such that $T(\mathrm{X}_0) \neq 0$, and since $\mathrm{X}_0 \neq 0$, $\mathrm{U}_0 = \mathrm{X}_0/\|\mathrm{X}_0\|$ is a unit vector. Thus,

$$\|T\|_* = \max\{\|T(\mathrm{U})\| : \|\mathrm{U}\| = 1\} \geq \|T(\mathrm{U}_0)\| = \frac{1}{\|\mathrm{X}_0\|}\|T(\mathrm{X}_0)\| > 0.$$

For $t \in \mathcal{R}$,

$$\begin{aligned}\|tT\|_* &= \max\{\|(tT)(\mathrm{U})\| : \|\mathrm{U}\| = 1\}\\ &= \max\{|t|\|T(\mathrm{U})\| : \|\mathrm{U}\| = 1\}\\ &= |t|\max\{\|T(\mathrm{U})\| : \|\mathrm{U}\| = 1\}\\ &= |t|\|T\|_*.\end{aligned}$$

Finally, for $S,T \in \mathcal{L}(\mathcal{R}^n,\mathcal{R}^m)$ and $\|\mathrm{U}\| = 1$,

$$\|(S+T)(\mathrm{U})\| = \|S(\mathrm{U}) + T(\mathrm{U})\| \leq \|S(\mathrm{U})\| + \|T(\mathrm{U})\| \leq \|S\|_* + \|T\|_*,$$

so $\|S+T\|_* = \max\{\|(S+T)(\mathrm{U})\| : \|\mathrm{U}\| = 1\} \leq \|S\|_* + \|T\|_*$. ■

■ **EXAMPLE 7.50.** If $T(\mathrm{X}) = \begin{bmatrix} 1 & -2 \\ 0 & 1 \\ -2 & 0 \end{bmatrix}\mathrm{X}$, $\mathrm{X} \in \mathcal{R}^2$, then $T(\mathrm{E}_1) = \begin{bmatrix} 1 \\ 0 \\ -2 \end{bmatrix}$, $T(\mathrm{E}_2) = \begin{bmatrix} -2 \\ 1 \\ 0 \end{bmatrix}$, $\|T(\mathrm{E}_1)\|^2 = 5$, $\|T(\mathrm{E}_2)\|^2 = 5$, and $T(\mathrm{E}_1) \cdot T(\mathrm{E}_2) = -2$. According to (7.26),

$$\|T(\mathrm{U}(t))\|^2 = 5\cos^2(t) - 4\cos(t)\sin(t) + 5\sin^2(t) = 5 - 2\sin(2t),$$

so $\|T\|_* = \max\{\sqrt{5 - 2\sin(2t)} : 0 \leq t \leq 2\pi\} = \sqrt{7}$, and $m(T) = \min\{\|T(\mathrm{U}(t))\| : 0 \leq t \leq 2\pi\} = \sqrt{3}$. □

■ **EXAMPLE 7.51.** The identity operator on $\mathcal{R}^n$ has norm 1, because $\max\{\|\mathrm{Id}(\mathrm{U})\| : \mathrm{U} \in \Gamma\} = \max\{\|\mathrm{U}\| : \|\mathrm{U}\| = 1\} = 1$. □

■ **EXAMPLE 7.52.** Let λ, μ, and ν be real numbers satisfying $|\lambda| \leq |\mu| \leq |\nu|$, and consider

$$T(\mathrm{X}) = \begin{bmatrix} \lambda & 0 & 0 \\ 0 & \mu & 0 \\ 0 & 0 & \nu \end{bmatrix}\mathrm{X}, \quad \mathrm{X} \in \mathcal{R}^3.$$

Observe that $\|T(\mathrm{X})\|^2 = \lambda^2x_1^2 + \mu^2x_2^2 + \nu^2x_3^2$. Since $|\lambda| \leq |\mu| \leq |\nu|$,

$$\|T(\mathrm{X})\|^2 \leq \nu^2x_1^2 + \nu^2x_2^2 + \nu^2x_3^2 = \nu^2\|\mathrm{X}\|^2,$$

and

$$\|T(\mathrm{X})\|^2 \geq \lambda^2x_1^2 + \lambda^2x_2^2 + \lambda^2x_3^2 = \lambda^2\|\mathrm{X}\|^2,$$

so

$$|\lambda|\|X\| \leq \|T(X)\| \leq |\nu|\|X\|.$$

In particular, if U is a unit vector, then $|\lambda| \leq \|T(U)\| \leq |\nu|$. Note also that $\|T(E_3)\| = |\nu|$ and $\|T(E_1)\| = |\lambda|$, so the largest and smallest change of scale factors for T are $|\nu|$ and $|\lambda|$, respectively. Thus, $\|T\|_* = M(T) = |\nu|$ and $m(T) = |\lambda|$. □

If T is a linear operator on $\mathcal{R}^n$, with $T(X) = AX$, then

$$\|T(X)\|^2 = (AX) \cdot (AX) = (AX)^T(AX) = X^T(A^TA)X.$$

Note that A^TA is symmetric and $Q(X) = X^T(A^TA)X$ is a quadratic form. Let $U_1, \ldots, U_n$ be orthonormal eigenvectors of A^TA, with corresponding eigenvalues $\mu_1, \ldots, \mu_n$. Since $Q(X) = \|T(X)\|^2 \geq 0$, Q is positive semidefinite, and therefore $\mu_k \geq 0$, $k = 1, \ldots, n$ (Theorem 6.16). By relabeling, if necessary, you may assume $0 \leq \mu_1 \leq \mu_2 \leq \cdots \leq \mu_n$. Set $P = [U_1 \ \cdots \ U_n]$ and $X = PY$. Then $Y = P^TX$,

$$D = P^T(A^TA)P = \begin{bmatrix} \mu_1 & & 0 \\ & \ddots & \\ 0 & & \mu_n \end{bmatrix},$$

and

$$\|T(X)\|^2 = X^T(A^TA)X = (PY)^T(A^TA)(PY) = Y^T[P^T(A^TA)P]Y = Y^TDY.$$

Thus,

$$\|T(X)\| = \sqrt{\mu_1 y_1^2 + \cdots + \mu_n y_n^2}. \tag{7.28}$$

Now, $0 \leq \mu_1 \leq \mu_2 \leq \cdots \leq \mu_n$ implies

$$\mu_1\|Y\|^2 \leq \mu_1 y_1^2 + \cdots + \mu_n y_n^2 \leq \mu_n\|Y\|^2,$$

so

$$\sqrt{\mu_1}\,\|P^TX\| = \sqrt{\mu_1}\,\|Y\| \leq \|T(X)\| \leq \sqrt{\mu_n}\,\|Y\| = \sqrt{\mu_n}\,\|P^TX\|. \tag{7.29}$$

Since P^T is orthogonal, $Y = P^TX$ is an isometry (Theorem 4.24). Thus, $\|P^TX\| = \|X\|$, and (7.29) becomes

$$\sqrt{\mu_1}\,\|X\| \leq \|T(X)\| \leq \sqrt{\mu_n}\,\|X\|.$$

In particular, if $\|U\| = 1$, then $\sqrt{\mu_1} \leq \|T(U)\| \leq \sqrt{\mu_n}$. When $Y = E_k$, $X = PE_k = U_k$, and (7.28) reads $\|T(U_k)\| = \sqrt{\mu_k}$, $k = 1, \ldots, n$. Thus, $\sqrt{\mu_n} = \|T(U_n)\|$ and $\sqrt{\mu_1} = \|T(U_1)\|$ are the largest and smallest change of scale factors, respectively, for T. This establishes the following result.

THEOREM 7.37. Suppose $A \in \mathcal{M}_{n\times n}$ and T is the linear operator on $\mathcal{R}^n$ defined by $T(X) = AX$. If $\mu_1, \ldots, \mu_n$ are the eigenvalues of A^TA, listed in increasing order, then $\mu_k \geq 0$, $k = 1, \ldots, n$, $\|T\|_* = M(T) = \sqrt{\mu_n}$, and $m(T) = \sqrt{\mu_1}$.

■ **EXAMPLE 7.53.** Let $T(X) = AX$, where $A = \begin{bmatrix} 3 & 0 \\ 4 & 5 \end{bmatrix}$. Then

$$A^TA = \begin{bmatrix} 3 & 4 \\ 0 & 5 \end{bmatrix}\begin{bmatrix} 3 & 0 \\ 4 & 5 \end{bmatrix} = \begin{bmatrix} 25 & 20 \\ 20 & 25 \end{bmatrix}$$

and

$$\det\begin{bmatrix} \lambda - 25 & -20 \\ -20 & \lambda - 25 \end{bmatrix} = \lambda^2 - 50\lambda + 225 = (\lambda - 5)(\lambda - 45),$$

so $m(T) = \sqrt{5}$ and $\|T\|_* = M(T) = \sqrt{45} = 3\sqrt{5}$. □

■ **EXAMPLE 7.54.** For the linear operator T, discussed in Example 7.49,

$$\mathrm{A} = \begin{bmatrix} \lambda & 0 & 0 \\ 0 & \mu & 0 \\ 0 & 0 & \nu \end{bmatrix} \quad \text{and} \quad \mathrm{A}^T\mathrm{A} = \mathrm{A}^2 = \begin{bmatrix} \lambda^2 & 0 & 0 \\ 0 & \mu^2 & 0 \\ 0 & 0 & \nu^2 \end{bmatrix}.$$

Since $|\lambda| \le |\mu| \le |\nu|$, the largest and smallest eigenvalues of $\mathrm{A}^T\mathrm{A}$ are ν^2 and λ^2, respectively. Thus, $m(T) = \sqrt{\lambda^2} = |\lambda|$ and $\|T\|_* = \sqrt{\nu^2} = |\nu|$. □

When $\mathrm{A} \in \mathcal{M}_{n\times n}$ is symmetric, with eigenvalues $\lambda_1, \ldots, \lambda_n$, there is an orthogonal $\mathrm{P} \in \mathcal{M}_{n\times n}$ such that

$$\mathrm{P}^T\mathrm{AP} = \mathrm{D} = \begin{bmatrix} \lambda_1 & & 0 \\ & \ddots & \\ 0 & & \lambda_n \end{bmatrix},$$

and therefore

$$\mathrm{P}^T(\mathrm{A}^T\mathrm{A})\mathrm{P} = (\mathrm{P}^T\mathrm{AP})(\mathrm{P}^T\mathrm{AP}) = \mathrm{D}^2 = \begin{bmatrix} \lambda_1^2 & & 0 \\ & \ddots & \\ 0 & & \lambda_n^2 \end{bmatrix}.$$

Thus, the eigenvalues of $\mathrm{A}^T\mathrm{A}$ are $\lambda_1^2, \ldots, \lambda_n^2$.

THEOREM 7.38. If $\mathrm{A} \in \mathcal{M}_{n\times n}$ is symmetric, with eigenvalues $\lambda_1, \ldots, \lambda_n$, and $T(\mathrm{X}) = \mathrm{AX}, \mathrm{X} \in \mathcal{R}^n$, then $\|T\|_* = \max\{|\lambda_1|, \ldots, |\lambda_n|\}$.

Proof. The discussion above shows that the eigenvalues of $\mathrm{A}^T\mathrm{A}$ are $\lambda_1^2, \ldots, \lambda_n^2$, and by Theorem 7.37, $\|T\|_* = \max\{\sqrt{\lambda_1^2}, \ldots, \sqrt{\lambda_n^2}\} = \max\{|\lambda_1|, \ldots, |\lambda_n|\}$. ■

Exercises 7.7

1. Assuming $S \in \mathcal{L}(\mathcal{V}, \mathcal{W})$ and $t \in \mathcal{R}$, show that tS is linear.

2. Assume $\mathcal{U}$, $\mathcal{V}$, and $\mathcal{W}$ are vector spaces and $t \in \mathcal{R}$. Verify that
   a. $R \circ (S + T) = R \circ S + R \circ T$, where $S,T \in \mathcal{L}(\mathcal{U},\mathcal{V})$ and $R \in \mathcal{L}(\mathcal{V}, \mathcal{W})$.
   b. $(R + S) \circ T = R \circ T + S \circ T$, where $T \in \mathcal{L}(\mathcal{U},\mathcal{V})$ and $R,S \in \mathcal{L}(\mathcal{V}, \mathcal{W})$.
   c. $S \circ (tT) = t(S \circ T) = (tS) \circ T$, where $T \in \mathcal{L}(\mathcal{U},\mathcal{V})$ and $S \in \mathcal{L}(\mathcal{V}, \mathcal{W})$.

3. Given $p,q \in \mathcal{R}$, $T = D^2 + pD + q\mathrm{Id}$ is referred to as a linear, second-order, differential operator with constant coefficients.
   a. Verify that T is a linear operator on $C^\infty(\mathcal{R})$ and that $\mathcal{K}er(T)$ consists of the functions $y = f(x)$ satisfying the differential equation

$$\frac{d^2y}{dx^2} + p\frac{dy}{dx} + qy = 0. \qquad (*)$$

b. Show that $e^{\lambda x} \in \mathcal{K}er(T)$ if and only if $\lambda^2 + p\lambda + q = 0$.

It can be shown that for each $\begin{bmatrix} a \\ b \end{bmatrix} \in \mathcal{R}^2$ there is a unique solution of $(*)$ satisfying the initial data $f(0) = a, f'(0) = b$. Suppose g and h are the unique solutions with initial data

$\begin{bmatrix} g(0) \\ g'(0) \end{bmatrix} = \begin{bmatrix} 1 \\ 0 \end{bmatrix}$ and $\begin{bmatrix} h(0) \\ h'(0) \end{bmatrix} = \begin{bmatrix} 0 \\ 1 \end{bmatrix}$.

c. Show that if f is a solution of $(*)$ with initial data $\begin{bmatrix} f(0) \\ f'(0) \end{bmatrix} = \begin{bmatrix} a \\ b \end{bmatrix}$, then $f = ag + bh$.

(Hint: Explain why $ag + bh$ is a solution of $(*)$ and check its initial data.)

d. Show that {g,h} is linearly independent.

e. What is the dimension of $\mathcal{K}er(T)$?

4. Assuming T(X) = AX, find $\|T\|_*$ when A is

a. $\begin{bmatrix} 1 & -2 \\ 0 & 3 \end{bmatrix}$. b. $\begin{bmatrix} 1 & 1 \\ \sqrt{2} & 0 \end{bmatrix}$. c. $\begin{bmatrix} 1 & -1 \\ -1 & 2 \end{bmatrix}$.

d. $\begin{bmatrix} 1 & 0 & 1 \\ 0 & 1 & 0 \\ 0 & 0 & 1 \end{bmatrix}$. e. $\begin{bmatrix} 1 & 0 & 0 \\ 1 & 1 & 1 \\ 0 & 0 & 1 \end{bmatrix}$. **f.** $\begin{bmatrix} -1 & 0 & 5 \\ 0 & 2 & 0 \\ 5 & 0 & -1 \end{bmatrix}$.

g. $\begin{bmatrix} 1 & 0 & -1 & -2 \\ 0 & -8 & 0 & 0 \\ -1 & 0 & 1 & 2 \\ -2 & 0 & 2 & 4 \end{bmatrix}$. **h.** $\begin{bmatrix} 1 & 0 & 0 & 0 \\ 2 & 1 & 0 & 0 \\ 0 & 0 & 1 & 2 \\ 0 & 0 & 0 & 1 \end{bmatrix}$. i. $\begin{bmatrix} 0 & 0 & -1 & 0 \\ 0 & 2 & 0 & 0 \\ 3 & 0 & 0 & 0 \\ 0 & 0 & 0 & -4 \end{bmatrix}$.

5. Find $\|T\|_*$ when $T \in \mathcal{L}(\mathcal{R}^n, \mathcal{R}^n)$ and $[T]_{\mathcal{E}}$ is
   a. $E_i(t)$. b. E_{ij}. c. $E_{ij}(t)$.

6. Suppose $\mathcal{B}$ is a basis for $\mathcal{R}^2$ and $[T]_{\mathcal{B}} = \begin{bmatrix} 1 & 0 \\ 0 & 2 \end{bmatrix}$. Find $\|T\|_*$ when

   a. $\mathcal{B} = \left\{ \begin{bmatrix} 1 \\ 0 \end{bmatrix}, \begin{bmatrix} 1 \\ 1 \end{bmatrix} \right\}$. b. $\mathcal{B} = \left\{ \begin{bmatrix} 1 \\ 1 \end{bmatrix}, \begin{bmatrix} -1 \\ 1 \end{bmatrix} \right\}$.

7. Suppose $\mathcal{A}$ is an orthonormal basis for $\mathcal{R}^n$ and $[T]_{\mathcal{A}} = D[\lambda_1, \ldots, \lambda_n]$. Show that $\|T\|_* = \max\{|\lambda_1|, \ldots, |\lambda_n|\}$.

8. What is $\|T\|_*$ when $T \in \mathcal{L}(\mathcal{R}^n, \mathcal{R}^n)$ is an isometry?

9. Suppose $\mathcal{V}$ is a subspace of $\mathcal{R}^n$. Find $\|T\|_*$ when
   a. $T(X) = \text{Refl}_{\mathcal{V}}(X)$. b. $T(X) = \text{Proj}_{\mathcal{V}}(X)$.

10. Assume T is a linear operator on $\mathcal{R}^n$.
    a. Show that T is one-to-one if and only if $m(T) > 0$.
    b. Show that if T is one-to-one, then $\|T^{-1}\|_* = 1/m(T)$.

11. Suppose S and T are linear operators on $\mathcal{R}^n$.
    a. Show that $\|S \circ T\|_* \leq \|S\|_* \|T\|_*$.
    b. Find an example in which strict inequality holds.

12. Assume $A \in \mathcal{M}_{n \times n}$, $T(X) = AX$, and $S(X) = A^T X$.
    a. Show that $A^T A$ and AA^T have the same eigenvalues.
    b. Show that $\|T\|_* = \|S\|_*$.

SELECTED ANSWERS

Section 1.1

1. a. linear **c.** linear **e.** not linear **g.** not linear **i.** not linear **k.** linear

2. a. $\left\{\left(\frac{-2s+t}{3}, s, t\right) : s,t \in \mathcal{R}\right\}$ **c.** $\left\{\left(\frac{1}{2}s + t + 2, s, t\right) : s,t \in \mathcal{R}\right\}$

e. $\{(-2t, s, t) : s,t \in \mathcal{R}\}$ **g.** $\{(s, 4/3, t) : s,t \in \mathcal{R}\}$

4. $x = a + by + cy^2$,
$$\begin{aligned} 1 &= a + b + c \\ 2 &= a + 3b + 9c \\ 3 &= a - b + c \end{aligned}$$

5. a. no

6. b. yes

7. a. $ma + dv + kx = 0$

Section 1.2

1. a. $(1, -1)$. The two lines intersect in a point.
c. Inconsistent. The equations represent parallel lines.
e. $(5 - 7t, 2 - 5t, t)$. The two planes intersect in a line.
g. Inconsistent. The first and second equations describe the same plane, which is parallel to the plane described by the third equation.
i. $(t, t, -t, t) = (x, y, z, w)$

4. Substituting the three points into the equation of the circle produces

$$\begin{aligned} a + c &= -1 \\ b + c &= -1 \\ -a - 3b + c &= -10, \end{aligned}$$

which has unique solution $a = b = 9/5$, $c = -14/5$.

6.
$$\begin{aligned} x_1 + x_2 + x_3 - 3x_4 &= 0 \\ x_1 + x_2 - 3x_3 + x_4 &= 0 \\ x_1 - 3x_2 + x_3 + x_4 &= 0 \\ -3x_1 + x_2 + x_3 + x_4 &= 0 \end{aligned}$$
Solution: (t,t,t,t), $t \in \mathcal{R}$

Section 1.3

1. a. $\left[\begin{array}{rr|r} 3 & -1 & 3 \\ 1 & 2 & 3 \end{array}\right] \longrightarrow \left[\begin{array}{rr|r} 1 & 2 & 3 \\ 0 & 1 & 6/7 \end{array}\right],\quad x = 9/7,\quad y = 6/7$

c. $\left[\begin{array}{rr|r} 1 & -2 & -6 \\ -2 & 4 & -5 \end{array}\right] \longrightarrow \left[\begin{array}{rr|r} 1 & -2 & -6 \\ 0 & 0 & -17 \end{array}\right],$ inconsistent

e. $\left[\begin{array}{rrr|r} 2 & 4 & -1 & 8 \\ -4 & 1 & 5 & -7 \\ 2 & 1 & -2 & 5 \end{array}\right] \longrightarrow \left[\begin{array}{rrr|r} 1 & 2 & -1/2 & 4 \\ 0 & 1 & 1/3 & 1 \\ 0 & 0 & 0 & 0 \end{array}\right],\quad x = 2 + 7t/6,\ y = 1 - t/3,\ z = t$

g. $\left[\begin{array}{rrrrr|r} 2 & -2 & 8 & 1 & 6 & -3 \\ -2 & -1 & -5 & 0 & -5 & 2 \\ 1 & -2 & 5 & 1 & 4 & 0 \\ 1 & 1 & 2 & 0 & 3 & 1 \end{array}\right] \longrightarrow \left[\begin{array}{rrrrr|r} 1 & 1 & 2 & 0 & 3 & 1 \\ 0 & 1 & -1 & 0 & 1 & 4 \\ 0 & 0 & 0 & 1 & 4 & 11 \\ 0 & 0 & 0 & 0 & 0 & 0 \end{array}\right],$

$x_1 = -3 - 3s - 2t,$
$x_2 = 4 + s - t,\quad x_3 = s,$
$x_4 = 11 - 4t,\quad x_5 = t$

2. a. $\begin{bmatrix} 1 & 0 \\ 0 & 1 \end{bmatrix}$ **c.** $\begin{bmatrix} 0 & 1 \\ 0 & 0 \end{bmatrix}$

e. $\begin{bmatrix} 1 & 0 & 0 \\ 0 & 1 & 0 \\ 0 & 0 & 1 \end{bmatrix}$ **g.** $\begin{bmatrix} 1 & 0 & 0 & 0 & 0 \\ 0 & 1 & 0 & 0 & 0 \\ 0 & 0 & 1 & 0 & -1 \\ 0 & 0 & 0 & 1 & 2 \end{bmatrix}$

i. $\begin{bmatrix} 1 & 0 & 1 \\ 0 & 1 & 1 \\ 0 & 0 & 0 \\ 0 & 0 & 0 \end{bmatrix}$ **k.** $\begin{bmatrix} 1 & 0 & -1 & -2 & -3 \\ 0 & 1 & 2 & 3 & 4 \\ 0 & 0 & 0 & 0 & 0 \\ 0 & 0 & 0 & 0 & 0 \\ 0 & 0 & 0 & 0 & 0 \end{bmatrix}$

4. $\begin{bmatrix} 2 & 1 & 0 & 3 \\ -1 & 0 & 1 & 2 \\ 0 & 2 & 1 & -1 \end{bmatrix} \longrightarrow \begin{bmatrix} 1 & 0 & -1 & -2 \\ 0 & 1 & 2 & 7 \\ 0 & 0 & -3 & -15 \end{bmatrix}\quad r_{12}, r_1(-1), r_{12}(-2), r_{23}(-2)$

$\longrightarrow \begin{bmatrix} 1 & 0 & 0 & 3 \\ 0 & 1 & 0 & -3 \\ 0 & 0 & 1 & 5 \end{bmatrix}\quad r_3(-1/3), r_{32}(-2), r_{31}(1)$

a. $r_{12}, r_1(-1), r_{12}(-2), r_{23}(-2), r_3(-1/3), r_{32}(-2), r_{31}(1)$
b. $r_{31}(-1), r_{32}(2), r_3(-3), r_{23}(2), r_{12}(2), r_1(-1), r_{12}$

9. a. $\left[\begin{array}{ccc|c} 1 & x_1 & x_1^2 & y_1 \\ 1 & x_2 & x_2^2 & y_2 \\ 1 & x_3 & x_3^2 & y_3 \end{array}\right]$

Section 1.4

1. a. $\begin{bmatrix} 3 & -2 & -1 \\ -7 & 4 & 5 \end{bmatrix}$ **c.** $\begin{bmatrix} 1 & 1 & -6 \\ -1 & 5 & 2 \end{bmatrix}$ **e.** $\begin{bmatrix} 4 & -2 & 6 \\ -4 & 2 & 2 \end{bmatrix}$ **g.** $\begin{bmatrix} 0 & 1 & -1 \\ 2 & 1 & -1 \end{bmatrix}$

2. a. $\begin{bmatrix} 5 & 2 & 1 \\ 3 & 2 & -1 \\ -1 & 1 & 3 \\ -2 & 2 & 1 \end{bmatrix}$ **c.** $\begin{bmatrix} 5 & -2 & -6 \\ 2 & 8 & 6 \\ -4 & -1 & 7 \\ 2 & 3 & -1 \end{bmatrix}$

4. $A+X=0 \Rightarrow 0=[0]_{ij}=[A+X]_{ij}=[A]_{ij}+[X]_{ij} \Rightarrow [X]_{ij}=-[A]_{ij}=[-A]_{ij}$, $1 \le i \le m$, $1 \le j \le n$. Thus, $X=-A$.

5. a. $\begin{bmatrix} -3 & 1 & -1 & 2 \\ 4 & -2 & 6 & 4 \\ -7 & 2 & 0 & 8 \end{bmatrix}$ **c.** $\begin{bmatrix} 1 & 1 \\ -4 & 8 \end{bmatrix}$ **e.** D

6. a. $\begin{bmatrix} 4 & -2 \\ -1 & -4 \end{bmatrix}$ **c.** $\begin{bmatrix} 2 & 3 \\ -2 & -3 \\ 4 & 6 \end{bmatrix}$ **e.** $\begin{bmatrix} 5 & -1 & 4 \\ -1 & -3 & -2 \\ -1 & 3 & 2 \end{bmatrix}$

8. Assume $B, C \in \mathcal{M}_{m\times n}$ and $A \in \mathcal{M}_{n\times p}$. For $1 \le i \le m$, $1 \le j \le p$,

$$\begin{aligned}
[(B+C)A]_{ij} &= \sum_{k=1}^{n} [B+C]_{ik}[A]_{kj} && \text{(defn. matrix mult.)}\\
&= \sum_{k=1}^{n} ([B]_{ik}+[C]_{ik})[A]_{kj} && \text{(defn. matrix addition)}\\
&= \sum_{k=1}^{n} ([B]_{ik}[A]_{kj}+[C]_{ik}[A]_{kj}) && \text{(algebra in } \mathcal{R})\\
&= \sum_{k=1}^{n} [B]_{ik}[A]_{kj} + \sum_{k=1}^{n} [C]_{ik}[A]_{kj} && \text{(algebra in } \mathcal{R})\\
&= [BA]_{ij}+[CA]_{ij} && \text{(defn. matrix mult.)}\\
&= [BA+CA]_{ij}. && \text{(defn. matrix addition)}
\end{aligned}$$

10. a. If $\text{Row}_i(A)=0$, then for $j=1,\ldots,p$, $[AB]_{ij}=[0 \;\cdots\; 0]\text{Col}_j(B)=0$.

11. b. $(A+B)(A-B)=A(A-B)+B(A-B)=AA-AB+BA-BB=A^2-AB+BA-B^2$, and the last expression reduces to A^2-B^2 if and only if $AB=BA$.

13. a. Since $AB=BA$, $(tB)A=t(BA)=t(AB)=A(tB)$.

15. $\left[\begin{array}{ccc|cc} 1 & 2 & 1 & 1 & 0 \\ 2 & 3 & 3 & 1 & 1 \\ 0 & 1 & -1 & 1 & -1 \end{array}\right] \longrightarrow \left[\begin{array}{ccc|cc} 1 & 0 & 3 & -1 & 2 \\ 0 & 1 & -1 & 1 & -1 \\ 0 & 0 & 0 & 0 & 0 \end{array}\right]$

a. $x_1=-1-3t, \quad x_2=1+t, \quad x_3=t$ **c.** $x_1=2-3t, \quad x_2=-1+t, \quad x_3=t$

17. a. $X_1 = AX_0$, $X_2 = AX_1 = A(AX_0) = A^2X_0$, $X_3 = AX_2 = A(A^2X_0) = A^3X_0$, and so on.

b. $A^5 \approx \begin{bmatrix} -0.166 & 1.382 \\ -0.276 & 2.045 \end{bmatrix}$, $A^{10} \approx \begin{bmatrix} -0.354 & 2.597 \\ -0.519 & 3.800 \end{bmatrix}$, $A^{20} \approx \begin{bmatrix} -1.222 & 8.945 \\ -1.789 & 13.089 \end{bmatrix}$,

$X_5 \approx \begin{bmatrix} 1365.414 \\ 2017.279 \end{bmatrix}$, $X_{10} \approx \begin{bmatrix} 2560.830 \\ 3747.760 \end{bmatrix}$, $X_{20} \approx \begin{bmatrix} 8822.778 \\ 12910.591 \end{bmatrix}$

19. a. $a_{ik}a_{kj} = 1$ when $a_{ik} = 1$ and $a_{kj} = 1$, that is, when there is a connection from city i to city k and from city k to city j. Otherwise, $a_{ik}a_{kj} = 0$. Thus, $[A^2]_{ij}$ is the number of ways to travel from city i to city j by passing through exactly one other city.

b. $A^2 = \begin{bmatrix} 2 & 1 & 1 & 0 & 1 & 1 & 2 \\ 1 & 4 & 1 & 2 & 1 & 2 & 2 \\ 1 & 1 & 3 & 1 & 1 & 2 & 2 \\ 0 & 2 & 1 & 3 & 0 & 2 & 1 \\ 1 & 1 & 1 & 0 & 2 & 0 & 2 \\ 1 & 2 & 2 & 2 & 0 & 4 & 1 \\ 2 & 2 & 2 & 1 & 2 & 1 & 4 \end{bmatrix}$

22. Assume $XA = A$ for all $A \in \mathcal{M}_{m\times n}$. Let $E_k = \text{Col}_k(I_m)$ and set $A = [E_k \,\vdots\, 0]$, where $0 \in \mathcal{M}_{m\times(n-1)}$. Then $[\text{Col}_k(X) \,\vdots\, 0] = X[E_k \,\vdots\, 0] = [E_k \,\vdots\, 0]$, so $\text{Col}_k(X) = E_k$, $k = 1, \ldots, m$. Thus, $X = I_m$.

Section 1.5

1. a. E_{12} **c.** not elementary **e.** $E_1(-2)$ **g.** $E_{31}(-3)$
i. $E_{12}(4)$ **k.** E_{23} **m.** not elementary

2. a. $\begin{bmatrix} 0 & 0 & 1 \\ 0 & 1 & 0 \\ 1 & 0 & 0 \end{bmatrix}$ **c.** $\begin{bmatrix} 1 & 0 & 0 \\ 0 & -2 & 0 \\ 0 & 0 & 1 \end{bmatrix}$

3. a. $E_{12}(-2)$, $E_{32}(-2)$, $E_{42}(-1)$ **c.** E_{12}

5. If $k \neq i$, then $\text{Row}_k(E_i(t)A) = \text{Row}_k(A)$. $\text{Row}_i(E_i(t)A) = \text{Row}_i(E_i(t))A = [t\text{Row}_i(I_m)]A = t[\text{Row}_i(I_m)A] = t\text{Row}_i(I_mA) = t\text{Row}_i(A)$.

6. a. I_n **c.** t^2 **e.** -1 **g.** E_{ij}

10. a. all three **c.** none **e.** all three **g.** all three

11. a. Assume $1 \leq i, j \leq n$, and $i \neq j$. Then $a_{ij} = 0 = b_{ij}$ and

$$[AB]_{ij} = \sum_{k=1}^{n} a_{ik}b_{kj}. \qquad (*)$$

For $k = 1, \ldots, n$, either $k \neq i$ or $k \neq j$, so at least one factor in $a_{ik}b_{kj}$ is 0. Thus, $[AB]_{ij} = 0$. When $i = j$ there is one nonzero term on the right of $(*)$ (corresponding to $k = i$), so $[AB]_{ii} = a_{ii}b_{ii}$.

13. a. $\text{trace}(tA) = \sum_{i=1}^{n} [tA]_{ii} = \sum_{i=1}^{n} t[A]_{ii} = t\left(\sum_{i=1}^{n} [A]_{ii}\right) = t[\text{trace}(A)]$.

Section 1.6

1. a. $\frac{1}{6}\begin{bmatrix} 1 & 3 \\ -2 & 0 \end{bmatrix}$ **c.** $\frac{1}{2}\begin{bmatrix} -2 & 4 \\ 3 & -5 \end{bmatrix}$ **d.** $\frac{1}{3}\begin{bmatrix} -5 & 0 & 2 \\ 1 & -3 & 2 \\ 3 & 3 & -3 \end{bmatrix}$ **f.** singular

h. $\frac{1}{3}\begin{bmatrix} 1 & 1 & 1 & -2 \\ -2 & 1 & 1 & 1 \\ 1 & -2 & 1 & 1 \\ 1 & 1 & -2 & 1 \end{bmatrix}$ **j.** $\begin{bmatrix} 1 & 0 & 0 \\ 0 & 1/2 & 0 \\ 0 & 0 & 1/3 \end{bmatrix}$ **l.** $\begin{bmatrix} 0 & -1 & 1 \\ -1 & 1 & 0 \\ 1 & 0 & 0 \end{bmatrix}$

2. $I_n I_n = I_n \Rightarrow I_n$ is nonsingular and $I_n^{-1} = I_n$.

6. a. $a,b,c \neq 0, \quad A^{-1} = \begin{bmatrix} 1/a & 0 & 0 \\ 0 & 1/b & 0 \\ 0 & 0 & 1/c \end{bmatrix}$

b. An $n \times n$ diagonal matrix A is nonsingular $\Leftrightarrow a_{ii} \neq 0,\ 1 \leq i \leq n,$ in which case

$$[A^{-1}]_{ij} = \begin{cases} 0\ , & i \neq j \\ 1/a_{ii}, & i = j \end{cases}.$$

7. a. If A is nonsingular, then so is $-A$, but $A + (-A) = 0$ is singular.

8. a. $X = \begin{bmatrix} 7 & -4 & 3 \\ 2 & 1 & -5 \\ 6 & -3 & 2 \end{bmatrix}\begin{bmatrix} 1 \\ 1 \\ 1 \end{bmatrix} = \begin{bmatrix} 6 \\ -2 \\ 5 \end{bmatrix}$ **c.** $X = \begin{bmatrix} -4 \\ -7 \\ -4 \end{bmatrix}$

9. a. A is nonsingular $\Leftrightarrow A \equiv_r I_n \Leftrightarrow a_{ii} \neq 0, 1 \leq i \leq n.$

12. $A^{-1} = E_j(-2)^{-1}E_{ij}^{-1}[E_{ji}(2)^{-1}]^{-1} = E_j(-1/2)E_{ij}E_{ji}(2)$

13. By Theorem 1.16, $B = PA$, where $P \in \mathcal{M}_{n\times n}$ is nonsingular. Being a product of nonsingular matrices, B is nonsingular.

17. If $ABAB = (AB)^2 = A^2B^2 = AABB$, then multiplying on the left and right by A^{-1} and B^{-1}, respectively, gives $BA = AB$.

18. a. $\frac{1}{3}\begin{bmatrix} -5 & 2 \\ 4 & -1 \end{bmatrix}$ **c.** $\begin{bmatrix} 0 & -1 & 1 \\ -1 & 1 & 0 \\ 1 & 0 & 0 \end{bmatrix}$

e. $\frac{1}{4}\begin{bmatrix} -4 & 12 & 18 & -4 & -32 \\ 12 & -8 & -18 & 0 & 32 \\ -12 & -4 & 1 & 6 & 2 \\ 0 & -4 & -5 & 2 & 10 \\ 4 & 0 & -1 & -2 & 2 \end{bmatrix}$

20. a. B singular $\Rightarrow$ there is an $X \in \mathcal{M}_{n\times 1}, X \neq 0$, such that $BX = 0$ (Theorem 1.14). Then $(AB)X = A(BX) = A0 = 0$, so AB is singular (Theorem 1.14).

Section 1.7

1. a. $\begin{bmatrix} 1 & -2 \\ -1 & 3 \\ 2 & -3 \end{bmatrix}$ **c.** $\begin{bmatrix} 3 & 2 & 1 \\ 2 & 2 & 1 \\ 1 & 1 & 1 \end{bmatrix}$ **e.** $\begin{bmatrix} 2 & -3 \\ 6 & -6 \end{bmatrix}$

2. b. $[(tA)^T]_{ij} = [tA]_{ji} = t[A]_{ji} = t[A^T]_{ij} = [tA^T]_{ij}$

7. a. $(A^2)^T = (AA)^T = A^TA^T = AA = A^2$. Similarly, A^p is symmetric.

9. a. $\begin{bmatrix} 1 & 0 & 0 \\ 0 & 1 & 0 \\ -1 & 2 & 0 \\ 0 & 0 & 1 \end{bmatrix}$ **c.** $\begin{bmatrix} 1 & 0 \\ 0 & 1 \\ -1 & 2 \end{bmatrix}$ **e.** $\begin{bmatrix} 1 & 0 & 0 \\ 0 & 1 & 0 \\ 5/3 & 4/3 & 0 \end{bmatrix}$

10. a. $\begin{bmatrix} 0 & 1 \\ 1 & 0 \end{bmatrix}$

11. a. $F_{34}(-3), F_{12}(-2), F_{14}(1), F_{23}$

12. a. $\frac{1}{6}\begin{bmatrix} 6 & 8 & 4 \\ 3 & 5 & 1 \\ 0 & -4 & -2 \end{bmatrix}$ **c.** $\frac{1}{4}\begin{bmatrix} -2 & 3 \\ 2 & -1 \end{bmatrix}$ **e.** $\frac{1}{3}\begin{bmatrix} 1 & -1 & -3 \\ 1 & 2 & 3 \\ 0 & 0 & 3 \end{bmatrix}$

13. $U_iU_j^T$ is the $n \times n$ matrix with 1 in the i,j position and zeros elsewhere. $I_n + tU_iU_j^T$ is the $n \times n$ matrix with ones on the diagonal, t in the i,j position, and zeros elsewhere.

Section 1.8

1. a. $C = \begin{bmatrix} 0 \\ 1 \end{bmatrix}[1 \quad 1 \quad 2 \quad 2] + \begin{bmatrix} 0 & 1 \\ 2 & 0 \end{bmatrix}\begin{bmatrix} 0 & 0 & -1 & -1 \\ 3 & 3 & 4 & 4 \end{bmatrix} = \begin{bmatrix} 0 & 0 & 0 & 0 \\ 1 & 1 & 2 & 2 \end{bmatrix}$

$+\begin{bmatrix} 3 & 3 & 4 & 4 \\ 0 & 0 & -2 & -2 \end{bmatrix} = \begin{bmatrix} 3 & 3 & 4 & 4 \\ 1 & 1 & 0 & 0 \end{bmatrix}$

c. $C_{11} = \begin{bmatrix} 1 & 0 \\ 0 & 1 \end{bmatrix}\begin{bmatrix} 1 & 2 \\ -2 & 1 \end{bmatrix} + \begin{bmatrix} -1 \\ 1 \end{bmatrix}[1 \quad 1] = \begin{bmatrix} 1 & 2 \\ -2 & 1 \end{bmatrix} + \begin{bmatrix} -1 & -1 \\ 1 & 1 \end{bmatrix} = \begin{bmatrix} 0 & 1 \\ -1 & 2 \end{bmatrix},$

$C_{12} = \begin{bmatrix} 1 & 0 \\ 0 & 1 \end{bmatrix}\begin{bmatrix} 0 & 1 \\ 3 & 2 \end{bmatrix} + \begin{bmatrix} -1 \\ 1 \end{bmatrix}[-1 \quad 0] = \begin{bmatrix} 0 & 1 \\ 3 & 2 \end{bmatrix} + \begin{bmatrix} 1 & 0 \\ -1 & 0 \end{bmatrix} = \begin{bmatrix} 1 & 1 \\ 2 & 2 \end{bmatrix},$

$C_{21} = [0 \quad 0]\begin{bmatrix} 1 & 2 \\ -2 & 1 \end{bmatrix} + [2][1 \quad 1] = [0 \quad 0] + [2 \quad 2] = [2 \quad 2],$

$C_{22} = [0 \quad 0]\begin{bmatrix} 0 & 1 \\ 3 & 2 \end{bmatrix} + [2][-1 \quad 0] = [0 \quad 0] + [-2 \quad 0] = [-2 \quad 0]$, so

$$C = \left[\begin{array}{cc|cc} 0 & 1 & 1 & 1 \\ -1 & 2 & 2 & 2 \\ \hline 2 & 2 & -2 & 0 \end{array}\right]$$

e. $\left[\begin{array}{cc|c|ccc} -1 & 1 & 0 & 0 & 0 & 0 \\ 2 & -1 & 0 & 0 & 0 & 0 \\ \hline 0 & 0 & 1 & 0 & 0 & 0 \\ \hline 0 & 0 & 0 & 1 & 0 & 1 \\ 0 & 0 & 0 & 0 & 0 & 1 \\ 0 & 0 & 0 & 2 & 1 & 0 \end{array}\right]$

2. a. $\left[\begin{array}{cc|c} -1 & 2 & 0 \\ 2 & -3 & 0 \\ \hline 0 & 0 & 1/4 \end{array}\right]$ **c.** $\left[\begin{array}{ccc|cc} 0 & -1 & 1 & 0 & 0 \\ -1 & 1 & 0 & 0 & 0 \\ 1 & 0 & 0 & 0 & 0 \\ \hline 0 & 0 & 0 & -1/2 & 3/2 \\ 0 & 0 & 0 & 1 & -2 \end{array}\right]$

4. $\begin{bmatrix} 0 & A \\ B & 0 \end{bmatrix}\begin{bmatrix} 0 & B^{-1} \\ A^{-1} & 0 \end{bmatrix} = I_{m+n} \Rightarrow \begin{bmatrix} 0 & A \\ B & 0 \end{bmatrix}^{-1} = \begin{bmatrix} 0 & B^{-1} \\ A^{-1} & 0 \end{bmatrix}$

5. b. $\left[\begin{array}{c|c} tI_m & 0 \\ \hline 0 & I_n \end{array}\right]$

Section 2.2

2. a. $\overrightarrow{PQ} \in \mathbf{x} \leftrightarrow [-3 \quad 3 \quad -3]^T$ and $\overrightarrow{PR} \in \mathbf{y} \leftrightarrow [1 \quad -1 \quad 1]^T$. Since $\mathbf{x} = -3\mathbf{y}$, P, Q, and R are collinear.

3. a. $\sqrt{3}$ **c.** $\sqrt{21}$

5. a. $X(t) = t\begin{bmatrix} -3 \\ -1 \\ -6 \end{bmatrix} + \begin{bmatrix} 1 \\ 2 \\ 3 \end{bmatrix}$ **c.** $X(t) = t\begin{bmatrix} -1 \\ 3 \\ 0 \end{bmatrix} + \begin{bmatrix} 2 \\ -1 \\ 1 \end{bmatrix}$ **e.** $X(t) = t\begin{bmatrix} 1 \\ -3 \\ 0 \end{bmatrix} + \begin{bmatrix} 0 \\ 1 \\ 1 \end{bmatrix}$

8. ($\Rightarrow$) If ℓ passes through 0, then there is a $t_0 \in \mathcal{R}$ such that $0 = X(t_0) = t_0M + B$, so $B = -t_0M$.
($\Leftarrow$) If $B = t_0M$ for some $t_0 \in \mathcal{R}$, then $X(t) = tM + B = (t + t_0)M$, so $0 = X(-t_0)$ lies on ℓ.

Section 2.3

1. a. $\begin{bmatrix} -11 \\ 10 \\ 3 \\ -9 \end{bmatrix}$ **c.** $\sqrt{22}$ **e.** $\dfrac{1}{\sqrt{23}}\begin{bmatrix} 3 \\ -1 \\ -3 \\ 2 \end{bmatrix}$

2. a. $\begin{bmatrix} -1 \\ 4 \\ 3 \\ -4 \\ -6 \end{bmatrix}$ **c.** $\begin{bmatrix} 0 \\ -t-1 \\ -2 \\ t-1 \\ t+2 \end{bmatrix}, \ t \in \mathcal{R}$

3. c. $r\mathrm{X} + s\mathrm{Y} = \begin{bmatrix} r+s \\ s-r \\ r+s \\ r-s \end{bmatrix}$, so $6 = \|r\mathrm{X} + s\mathrm{Y}\| = \sqrt{4r^2 + 4s^2} \Leftrightarrow r^2 + s^2 = 9$.

5. By Theorem 2.6(3), with $t = 1/\|\mathrm{X}\|$, $\|\mathrm{U}\| = (1/\|\mathrm{X}\|)\|\mathrm{X}\| = 1$.

6. $\mathrm{X}^T\mathrm{X} = [x_1 \quad \cdots \quad x_n] \begin{bmatrix} x_1 \\ \vdots \\ x_n \end{bmatrix} = [x_1^2 + \cdots + x_n^2] = \|\mathrm{X}\|^2$

Section 2.4

1. a. subspace
c. closed under scalar multiplication, not closed under addition
e. subspace
g. subspace
i. closed under scalar multiplication, not closed under addition
k. subspace

4. Suppose U,V $\in \mathcal{V}$ and $r \in \mathcal{R}$. Then $\mathrm{U} = sX + tY$ and $\mathrm{V} = \sigma\mathrm{X} + \tau\mathrm{Y}$ for some $s,t,\sigma,\tau \in \mathcal{R}$, so $\mathrm{U} + \mathrm{V} = (s\mathrm{X} + t\mathrm{Y}) + (\sigma\mathrm{X} + \tau\mathrm{Y}) = (s + \sigma)\mathrm{X} + (t + \tau)\mathrm{Y} \in \mathcal{V}$ and $r\mathrm{U} = r(s\mathrm{X} + t\mathrm{Y}) = (rs)\mathrm{X} + (rt)\mathrm{Y} \in \mathcal{V}$.

5. a. Suppose X,Y $\in \mathcal{U} \cap \mathcal{V}$, that is, X,Y $\in \mathcal{U}$ and X,Y $\in \mathcal{V}$. Since $\mathcal{U}$ and $\mathcal{V}$ are subspaces, $\mathrm{X} + \mathrm{Y} \in \mathcal{U}$ and $\mathrm{X} + \mathrm{Y} \in \mathcal{V}$, so $\mathrm{X} + \mathrm{Y} \in \mathcal{U} \cap \mathcal{V}$. If $t \in \mathcal{R}$, then $t\mathrm{X} \in \mathcal{U}$ and $t\mathrm{X} \in \mathcal{V}$, so $t\mathrm{X} \in \mathcal{U} \cap \mathcal{V}$.

6. b. $\mathcal{W}$ is the plane in $\mathcal{R}^3$ containing the two intersecting lines.

8. If A is nonsingular, then $\mathrm{AX} = 0$ if and only if $\mathrm{X} = 0$, that is, $\mathcal{N}(\mathrm{A}) = \{0\}$.

10. a. $\mathrm{X} \in \mathcal{N}(\mathrm{A}) \Rightarrow \mathrm{AX} = 0 \Rightarrow (\mathrm{PA})\mathrm{X} = \mathrm{P}(\mathrm{AX}) = \mathrm{P}0 = 0 \Rightarrow \mathrm{X} \in \mathcal{N}(\mathrm{PA})$

11. a. $\mathrm{X} = t\begin{bmatrix} -3 \\ -1 \\ 1 \end{bmatrix} + \begin{bmatrix} 5 \\ -2 \\ 0 \end{bmatrix}, \quad \mathrm{X}_0 = \begin{bmatrix} 5 \\ -2 \\ 0 \end{bmatrix},$

$$\mathcal{N}(\mathrm{A}) = \left\{ t\begin{bmatrix} -3 \\ -1 \\ 1 \end{bmatrix} : t \in \mathcal{R} \right\}$$

Section 2.5

1. b. $\mathcal{R}^2$

3. a. For each $X = \begin{bmatrix} a \\ b \\ c \end{bmatrix} \in \mathcal{R}^3$, $X = a\begin{bmatrix} 1 \\ 0 \\ 0 \end{bmatrix} + b\begin{bmatrix} 0 \\ 1 \\ 0 \end{bmatrix} + c\begin{bmatrix} 0 \\ 0 \\ 1 \end{bmatrix} = aE_1 + bE_2 + cE_3$.

4. a. $A \longrightarrow \begin{bmatrix} 1 & 0 & -1 \\ 0 & 1 & -2 \\ 0 & 0 & 0 \end{bmatrix} \Rightarrow \mathcal{N}(A) = \mathcal{S}pan\left\{\begin{bmatrix} 1 \\ 2 \\ 1 \end{bmatrix}\right\}$

c. $A \longrightarrow \begin{bmatrix} 1 & 0 & 3 \\ 0 & 1 & -1 \\ 0 & 0 & 0 \\ 0 & 0 & 0 \end{bmatrix} \Rightarrow \mathcal{N}(A) = \mathcal{S}pan\left\{\begin{bmatrix} -3 \\ 1 \\ 1 \end{bmatrix}\right\}$

e. $A \longrightarrow \begin{bmatrix} 1 & -3 & 0 & -1 & 0 & -1 \\ 0 & 0 & 1 & 2 & 0 & 2 \\ 0 & 0 & 0 & 0 & 1 & 1 \\ 0 & 0 & 0 & 0 & 0 & 0 \end{bmatrix} \Rightarrow \mathcal{N}(A) = \mathcal{S}pan\left\{\begin{bmatrix} 3 \\ 1 \\ 0 \\ 0 \\ 0 \\ 0 \end{bmatrix}, \begin{bmatrix} 1 \\ 0 \\ -2 \\ 1 \\ 0 \\ 0 \end{bmatrix}, \begin{bmatrix} 1 \\ 0 \\ -2 \\ 0 \\ -1 \\ 1 \end{bmatrix}\right\}$

5. a. $[X_1 \quad X_2 \quad X_3 \mid B_1 \quad B_2 \quad B_3] \longrightarrow \left[\begin{array}{ccc|ccc} 1 & 0 & 0 & 0 & 5/3 & -1 \\ 0 & 1 & 0 & 0 & 2/3 & 2 \\ 0 & 0 & 1 & 0 & -1/3 & -3 \\ 0 & 0 & 0 & 1 & 0 & 0 \end{array}\right]$,

$B_1 \notin \mathcal{S}pan\{X_1, X_2, X_3\}$, $B_2 = (5/3)X_1 + (2/3)X_2 - (1/3)X_3$,
$B_3 = -X_1 + 2X_2 - 3X_3$

b. $[X_1 \quad X_2 \quad X_3 \mid B_1 \quad B_2 \quad B_3] \longrightarrow \left[\begin{array}{ccc|ccc} 1 & 0 & 1 & 2 & 2 & 0 \\ 0 & 1 & 2 & -4 & 6 & 0 \\ 0 & 0 & 0 & 0 & 0 & 1 \end{array}\right]$,

$B_3 \notin \mathcal{S}pan\{X_1, X_2, X_3\}$, $B_1 = (2 - t)X_1 - (4 + 2t)X_2 + tX_3$,
$B_2 = (2 - t)X_1 + (6 - 2t)X_2 + tX_3$, $t \in \mathcal{R}$

8. a. Since $[A \mid X] \longrightarrow \left[\begin{array}{ccc|c} 1 & 2 & 5 & x \\ 0 & 1 & 3 & x - z \\ 0 & 0 & 0 & -2x + y + 3z \end{array}\right]$, $X \in \mathcal{C}(A) \Leftrightarrow -2x + y + 3z = 0$.

c. Since $[A \mid X] \longrightarrow \left[\begin{array}{cccc|c} 1 & 0 & -1 & -1 & x \\ 0 & 1 & -1 & -2 & x + y \\ 0 & 0 & 0 & -3 & y + z \end{array}\right]$, $\mathcal{C}(A) = \mathcal{R}^3$.

10. $B = [x \quad y \quad z]^T \in \mathcal{C}(A) \Leftrightarrow AT = B$ has a solution. $[A \mid B] \to [C \mid D]$, where $C = \text{rref}(A)$ and D is an element of $\mathcal{R}^3$ whose entries are linear combinations of x, y, and z. If $C = I$, then $AT = B$ is consistent for all $B \in \mathcal{R}^3$, so $\mathcal{C}(A) = \mathcal{R}^3$. Suppose C has two leading ones. Then $AT = B$ is consistent $\Leftrightarrow$ the third entry of D is 0. In this case, the coordinates of B satisfy a linear homogeneous equation, and $\mathcal{C}(A)$ is therefore a plane. Similarly, when C has one leading one, the second and third entries of D must be 0's. This time, $\mathcal{C}(A)$, being the intersection of two planes, is a line.

Section 2.6

1. Let A be the matrix whose columns are the vectors in $\mathcal{X}$.

a. $A \longrightarrow I_2$, $\mathcal{X}$ is linearly independent

c. $A \longrightarrow \begin{bmatrix} 1 & 0 & 0 & 0 \\ 0 & 1 & 0 & 4 \\ 0 & 0 & 1 & 2 \end{bmatrix}$, $\mathcal{Y} = \left\{ \begin{bmatrix} 1 \\ -1 \\ 0 \end{bmatrix}, \begin{bmatrix} 0 \\ 1 \\ -1 \end{bmatrix}, \begin{bmatrix} 2 \\ 1 \\ 3 \end{bmatrix} \right\}$,

$$\begin{bmatrix} 4 \\ 6 \\ 2 \end{bmatrix} = 0 \begin{bmatrix} 1 \\ -1 \\ 0 \end{bmatrix} + 4 \begin{bmatrix} 0 \\ 1 \\ -1 \end{bmatrix} + 2 \begin{bmatrix} 2 \\ 1 \\ 3 \end{bmatrix}$$

e. $A \longrightarrow \begin{bmatrix} 1 & -1 & 0 \\ 0 & 0 & 1 \end{bmatrix}$, $\mathcal{Y} = \left\{ \begin{bmatrix} 1 \\ -2 \end{bmatrix}, \begin{bmatrix} 2 \\ -2 \end{bmatrix} \right\}$, $\begin{bmatrix} -1 \\ 2 \end{bmatrix} = -\begin{bmatrix} 1 \\ -2 \end{bmatrix} + 0 \begin{bmatrix} 2 \\ -2 \end{bmatrix}$

g. $\begin{bmatrix} -1 & 1 & 2 \\ 2 & 0 & 1 \\ 0 & 2 & -1 \\ 1 & -1 & 0 \end{bmatrix} \longrightarrow \begin{bmatrix} 1 & 0 & 0 \\ 0 & 1 & 0 \\ 0 & 0 & 1 \\ 0 & 0 & 0 \end{bmatrix}$, $\mathcal{X}$ is linearly independent

i. $A \longrightarrow \begin{bmatrix} 1 & 0 & 2 & 0 \\ 0 & 1 & -1 & 0 \\ 0 & 0 & 0 & 1 \\ 0 & 0 & 0 & 0 \end{bmatrix}$, $\mathcal{Y} = \left\{ \begin{bmatrix} 1 \\ 0 \\ 2 \\ 2 \end{bmatrix}, \begin{bmatrix} 1 \\ 2 \\ -1 \\ 1 \end{bmatrix}, \begin{bmatrix} -1 \\ 2 \\ 4 \\ 5 \end{bmatrix} \right\}$,

$$\begin{bmatrix} 1 \\ -2 \\ 5 \\ 3 \end{bmatrix} = 2 \begin{bmatrix} 1 \\ 0 \\ 2 \\ 2 \end{bmatrix} - \begin{bmatrix} 1 \\ 2 \\ -1 \\ 1 \end{bmatrix} + 0 \begin{bmatrix} -1 \\ 2 \\ 4 \\ 5 \end{bmatrix}$$

k. $A \longrightarrow \begin{bmatrix} 1 & 0 & 0 & 0 \\ 0 & 1 & 0 & 0 \\ 0 & 0 & 1 & 0 \\ 0 & 0 & 0 & 1 \\ 0 & 0 & 0 & 0 \end{bmatrix}$, $\mathcal{X}$ is linearly independent

2. If {X,Y,Z} is linearly dependent, then one of the vectors is in the span of the other two. The other two span either a line through 0 or a plane through 0, and in either event, 0, X, Y, and Z lie in a plane. Conversely, suppose, 0, X, Y, and Z are coplanar. If they are collinear, then any one of X, Y, or Z is a scalar multiple of any of the others. Alternatively, two of them, say X and Y, don't lie on a line through 0. In that event, $\mathcal{Span}$\{X,Y\} is a plane and Z lies in that plane. In each case, one of X, Y, and Z is a linear combination of the others, so {X,Y,Z} is linearly dependent.

6. a. X not a scalar multiple of Y $\Rightarrow sX$ is not a scalar multiple of tY.

7. b. Suppose $0 = t\text{U} + s\text{V} = t(\text{X} + \text{Y}) + s(\text{X} - \text{Y}) = (t + s)\text{X} + (t - s)\text{Y}$. Linear independence of $\{\text{X,Y}\} \Rightarrow t + s = 0 = t - s$. The unique solution of this system is $t = 0 = s$, so U and V are linearly independent.

10. a. Some nontrivial linear combination of the elements of $\mathcal{Y}$ produces the zero vector, and that combination can be extended to a linear combination of the elements of $\mathcal{X}$ by inserting 0's as the coefficients of the vectors that are not in $\mathcal{Y}$. Thus, $\mathcal{X}$ is linearly dependent.

b. If $\mathcal{X}$ is linearly independent, then it is not linearly dependent, so by the previous argument, $\mathcal{Y}$ cannot be linearly dependent.

Section 2.7

1. a. $\begin{bmatrix} 1 & -3 \\ -2 & 4 \end{bmatrix} \longrightarrow \text{I}_2$. The two vectors are linearly independent and, therefore, form a basis for $\mathcal{R}^2$.

c. not a basis for $\mathcal{R}^2$; the two vectors are linearly dependent

e. not a basis for $\mathcal{R}^3$; two vectors cannot span $\mathcal{R}^3$

g. $\begin{bmatrix} 1 & 0 & -1 & 1 \\ 1 & -1 & 0 & 1 \\ -1 & 1 & 1 & 0 \\ 0 & 1 & 1 & -1 \end{bmatrix} \longrightarrow \text{I}_4$. The set is a basis.

i. not a basis; a basis for $\mathcal{R}^4$ must contain four vectors

2. a. $\begin{bmatrix} 1 & -1 & 2 & -3 \\ -2 & 1 & 0 & 4 \end{bmatrix}\begin{bmatrix} 2 \\ 4 \\ 1 \\ 0 \end{bmatrix} = \begin{bmatrix} 0 \\ 0 \end{bmatrix}$ and $\begin{bmatrix} 1 & -1 & 2 & -3 \\ -2 & 1 & 0 & 4 \end{bmatrix}\begin{bmatrix} 1 \\ -2 \\ 0 \\ 1 \end{bmatrix} = \begin{bmatrix} 0 \\ 0 \end{bmatrix}$

$\Rightarrow \mathcal{B} \subseteq \mathcal{V}$. Since $\begin{bmatrix} 1 & -1 & 2 & -3 \\ -2 & 1 & 0 & 4 \end{bmatrix} \longrightarrow \begin{bmatrix} 1 & 0 & -2 & -1 \\ 0 & 1 & -4 & 2 \end{bmatrix}$,

$\mathcal{V} = Span\left\{\begin{bmatrix} 2 \\ 4 \\ 1 \\ 0 \end{bmatrix}, \begin{bmatrix} 1 \\ -2 \\ 0 \\ 1 \end{bmatrix}\right\}$. Since $\mathcal{B}$ is also linearly independent, $\mathcal{B}$ is a basis for $\mathcal{V}$.

c. $\mathcal{V} = \left\{\begin{bmatrix} a \\ b \\ c \\ 0 \end{bmatrix} : a,b,c \in \mathcal{R}\right\} = Span\{\text{E}_1,\text{E}_2,\text{E}_3\}$. Since $\{\text{E}_1,\text{E}_2,\text{E}_3\}$ is linearly independent, dim $\mathcal{V} = 3$. Any set of three linearly independent vectors in $\mathcal{V}$ is a basis for $\mathcal{V}$, and $\mathcal{B}$ is such a set.

e. The coordinates of each vector in $\mathcal{B}$ satisfy $2x_1 - x_2 + 3x_3 - x_4 = 0$, so $\mathcal{B} \subseteq \mathcal{V}$. $\mathcal{V} = \mathcal{N}([2 \;\; -1 \;\; 3 \;\; -1])$. $[2 \;\; -1 \;\; 3 \;\; -1] \longrightarrow [1 \;\; -1/2 \;\; 3/2 \;\; -1/2] \Rightarrow$

$$\left\{\begin{bmatrix}1\\2\\0\\0\end{bmatrix}, \begin{bmatrix}-3\\0\\2\\0\end{bmatrix}, \begin{bmatrix}1\\0\\0\\2\end{bmatrix}\right\} \text{ is a basis for } \mathcal{V} \Rightarrow \dim \mathcal{V} = 3. \begin{bmatrix}-1 & 0 & 2\\0 & 1 & 1\\1 & 0 & -1\\1 & -1 & 0\end{bmatrix} \longrightarrow$$

$$\begin{bmatrix}1 & 0 & 0\\0 & 1 & 0\\0 & 0 & 1\\0 & 0 & 0\end{bmatrix} \Rightarrow \mathcal{B} \text{ is linearly independent. Thus, } \mathcal{B} \text{ is a basis for } \mathcal{V}.$$

g. $$\left[\begin{array}{rrr|rr}1 & -1 & 1 & -1 & 1\\1 & -2 & -1 & -3 & -3\\-1 & 1 & -1 & 1 & -1\end{array}\right] \longrightarrow \left[\begin{array}{rrr|rr}1 & 0 & 3 & 1 & 5\\0 & 1 & 2 & 2 & 4\\0 & 0 & 0 & 0 & 0\end{array}\right] \Rightarrow \begin{bmatrix}-1\\-3\\1\end{bmatrix} \in \mathcal{V},$$

$$\begin{bmatrix}1\\-3\\-1\end{bmatrix} \in \mathcal{V}, \text{ and } \left\{\begin{bmatrix}1\\1\\-1\end{bmatrix}, \begin{bmatrix}-1\\-2\\1\end{bmatrix}\right\} \text{ is a basis for } \mathcal{V}. \text{ Since } \dim \mathcal{V} = 2 \text{ and } \mathcal{B}$$

consists of two linearly independent vectors in $\mathcal{V}$, $\mathcal{B}$ is a basis.

7. a. If $\mathcal{V} = \mathcal{S}pan\{X_1, \ldots, X_p\}$, then by Theorem 2.17 some linearly independent subset of $\{X_1, \ldots, X_p\}$ also spans $\mathcal{V}$, so $\dim \mathcal{V} \leq p$.
b. $\{X_1, \ldots, X_p\}$ linearly independent implies $p \leq \dim \mathcal{V}$ (Theorem 2.20).

8. a. If $\mathcal{U} = \mathcal{S}pan\{X_1, \ldots, X_p\}$ is a proper subset of $\mathcal{V}$, then there is an $X_{p+1} \in \mathcal{V}$ such that $X_{p+1} \notin \mathcal{S}pan\{X_1, \ldots, X_p\}$. By Theorem 2.21, $\{X_1, \ldots, X_p, X_{p+1}\}$ is a linearly independent subset of $\mathcal{V}$. Continuing, as in the proof of Theorem 2.22, you obtain a linearly independent subset, $\mathcal{A}$, of $\mathcal{V}$ that also spans $\mathcal{V}$. Thus, $\mathcal{A}$ is a basis for $\mathcal{V}$, $\mathcal{B} \subseteq \mathcal{A}$, and $\dim \mathcal{U} \leq \dim \mathcal{V}$.
b. Certainly $\mathcal{U} = \mathcal{V} \Rightarrow \dim \mathcal{U} = \dim \mathcal{V}$. Suppose $\dim \mathcal{U} = \dim \mathcal{V}$. If $\mathcal{U} \neq \mathcal{V}$, then there is an $X \in \mathcal{V}$ such that $X \notin \mathcal{U} = \mathcal{S}pan\{X_1, \ldots, X_p\}$. By Theorem 2.21, $\{X_1, \ldots, X_p, X\}$ is a linearly independent subset of $\mathcal{V}$, contrary to Theorem 2.20. Thus, $\mathcal{U} = \mathcal{V}$.

10. a. No linearly independent subset of $\mathcal{V}$ contains $\mathcal{B}$ as a proper subset.

Section 2.8

1. a. rank(A) = 2, nul(A) = 0
b. rank(A) = 2, nul(A) = 2
e. rank(A) = 2, nul(A) = 1
g. rank(A) = 3, nul(A) = 0
i. rank(A) = 2, nul(A) = 1

2. $\dim \mathcal{N}(A) = 2 \Rightarrow \text{rank}(A) = 3 - \dim \mathcal{N}(A) = 1$

5. $AX = B$ consistent for all $B \in \mathcal{R}^{15} \Rightarrow \mathcal{C}(A) = \mathcal{R}^{15} \Rightarrow \text{rank}(A) = 15 \Rightarrow \text{nul}(A) = 23 - 15 = 8$.

8. a. $\left[\begin{array}{rr|rrr} 1 & 1 & 1 & 0 & 0 \\ -1 & 0 & 0 & 1 & 0 \\ 1 & 2 & 0 & 0 & 1 \end{array}\right] \underset{r}{\longrightarrow} \left[\begin{array}{rr|rrr} 1 & 0 & 0 & -1 & 0 \\ 0 & 1 & 1 & 1 & 0 \\ 0 & 0 & -2 & -1 & 1 \end{array}\right] \Rightarrow \mathrm{PAQ} = \left[\begin{array}{rr} 1 & 0 \\ 0 & 1 \\ \hdashline 0 & 0 \end{array}\right]$, where

$$\mathrm{P} = \begin{bmatrix} 0 & -1 & 0 \\ 1 & 1 & 0 \\ -2 & -1 & 1 \end{bmatrix}, \ \mathrm{Q} = \begin{bmatrix} 1 & 0 \\ 0 & 1 \end{bmatrix}$$

b. $\left[\begin{array}{rrrr|rrr} 1 & -1 & 0 & 2 & 1 & 0 & 0 \\ 1 & 0 & 1 & -1 & 0 & 1 & 0 \\ 2 & -1 & 1 & 1 & 0 & 0 & 1 \end{array}\right] \underset{r}{\longrightarrow} \left[\begin{array}{rrrr|rrr} 1 & 0 & 1 & -1 & 0 & 1 & 0 \\ 0 & 1 & 1 & -3 & -1 & 1 & 0 \\ 0 & 0 & 0 & 0 & -1 & -1 & 1 \end{array}\right]$, and

$$\left[\begin{array}{rrrr} 1 & 0 & 1 & -1 \\ 0 & 1 & 1 & -3 \\ 0 & 0 & 0 & 0 \\ \hdashline 1 & 0 & 0 & 0 \\ 0 & 1 & 0 & 0 \\ 0 & 0 & 1 & 0 \\ 0 & 0 & 0 & 1 \end{array}\right] \underset{c}{\longrightarrow} \left[\begin{array}{rrrr} 1 & 0 & 0 & 0 \\ 0 & 1 & 0 & 0 \\ 0 & 0 & 0 & 0 \\ \hdashline 1 & 0 & -1 & 1 \\ 0 & 1 & -1 & 3 \\ 0 & 0 & 1 & 0 \\ 0 & 0 & 0 & 1 \end{array}\right] \Rightarrow \mathrm{PAQ} = \left[\begin{array}{rr|rr} 1 & 0 & 0 & 0 \\ 0 & 1 & 0 & 0 \\ \hdashline 0 & 0 & 0 & 0 \end{array}\right], \text{ where}$$

$$\mathrm{P} = \begin{bmatrix} 0 & 1 & 0 \\ -1 & 1 & 0 \\ -1 & -1 & 1 \end{bmatrix}, \quad \mathrm{Q} = \begin{bmatrix} 1 & 0 & -1 & 1 \\ 0 & 1 & -1 & 3 \\ 0 & 0 & 1 & 0 \\ 0 & 0 & 0 & 1 \end{bmatrix}$$

e. $\mathrm{P} = \dfrac{1}{2}\begin{bmatrix} 0 & 0 & -2 & 1 \\ 0 & 0 & 4 & -1 \\ 2 & 0 & 6 & -2 \\ 0 & 2 & -8 & 1 \end{bmatrix}, \quad \mathrm{Q} = \begin{bmatrix} 1 & 0 & 1 \\ 0 & 1 & -2 \\ 0 & 0 & 1 \end{bmatrix}, \quad \text{and} \quad \mathrm{PAQ} = \left[\begin{array}{rr|r} 1 & 0 & 0 \\ 0 & 1 & 0 \\ \hdashline 0 & 0 & 0 \\ 0 & 0 & 0 \end{array}\right]$

g. $\mathrm{P} = \dfrac{1}{3}\begin{bmatrix} 0 & -6 & 0 & 3 \\ 0 & 1 & -3 & 2 \\ 0 & 4 & 3 & -4 \\ 3 & -3 & -3 & 3 \end{bmatrix}, \quad \mathrm{Q} = \begin{bmatrix} 1 & 0 & 0 \\ 0 & 1 & 0 \\ 0 & 0 & 1 \end{bmatrix}, \quad \text{and} \quad \mathrm{PAQ} = \left[\begin{array}{rrr} 1 & 0 & 0 \\ 0 & 1 & 0 \\ 0 & 0 & 1 \\ \hdashline 0 & 0 & 0 \end{array}\right]$

i. $\mathrm{P} = \begin{bmatrix} 0 & 0 & 3 & -1 \\ 0 & 0 & -2 & 1 \\ 1 & 0 & 0 & 0 \\ 0 & 1 & -2 & 1 \end{bmatrix}, \quad \mathrm{Q} = \begin{bmatrix} 0 & 0 & 1 \\ 1 & 0 & 0 \\ 0 & 1 & 0 \end{bmatrix}, \quad \text{and} \quad \mathrm{PAQ} = \left[\begin{array}{rr|r} 1 & 0 & 0 \\ 0 & 1 & 0 \\ \hdashline 0 & 0 & 0 \\ 0 & 0 & 0 \end{array}\right]$

10. a. Suppose $\mathrm{P} \in \mathcal{M}_{m\times m}$ is nonsingular and PA = rref(A). If $r = \mathrm{rank(A)}$, then PA has $m - r$ zero rows, so (PA)B has at least $m - r$ zero rows. Thus, rank ((PA)B) $\leq r$. By Theorem 2.29, rank(AB) = rank(P(AB)) = rank((PA)B) $\leq r =$ rank(A). [Alternatively: $\mathcal{C}$(AB) $\subseteq$ $\mathcal{C}$(A) (Problem 11, Exercises 2.5) $\Rightarrow$ rank(AB) $\leq$ rank(A)].

13. Since each column of A + B is the sum of two of the columns of [A ¦ B], each linear combination of columns of A + B is a linear combination of columns of [A ¦ B]. Thus, $\mathcal{C}$(A + B) $\subseteq$ $\mathcal{C}$([A ¦ B]), and rank(A + B) $\leq$ rank([A ¦ B]). The number of linearly independent columns of [A ¦ B] is no more than that of A plus that of B, so rank ([A ¦ B]) $\leq$ rank(A) + rank(B).

15. Assume rank(A) $= r$ and rank(B) $= s$. Let $P_1 \in \mathcal{M}_{m\times m}$, $Q_1 \in \mathcal{M}_{n\times n}$, $P_2 \in \mathcal{M}_{p\times p}$, and $Q_2 \in \mathcal{M}_{q\times q}$ be nonsingular matrices such that $P_1AQ_1 = \left[\begin{array}{c|c} I_r & 0 \\ \hline 0 & 0 \end{array}\right]$ and $P_2BQ_2 = \left[\begin{array}{c|c} I_s & 0 \\ \hline 0 & 0 \end{array}\right]$. Then $P = \left[\begin{array}{c|c} P_1 & 0 \\ \hline 0 & P_2 \end{array}\right]$ and $Q = \left[\begin{array}{c|c} Q_1 & 0 \\ \hline 0 & Q_2 \end{array}\right]$ are invertible and

$$P\left[\begin{array}{c|c} A & 0 \\ \hline 0 & B \end{array}\right]Q = \left[\begin{array}{c|c} P_1 & 0 \\ \hline 0 & P_2 \end{array}\right]\left[\begin{array}{c|c} A & 0 \\ \hline 0 & B \end{array}\right]\left[\begin{array}{c|c} Q_1 & 0 \\ \hline 0 & Q_2 \end{array}\right] = \left[\begin{array}{c|c} P_1AQ_1 & 0 \\ \hline 0 & P_2BQ_2 \end{array}\right]$$

$$= \left[\begin{array}{cc|cc} I_r & 0 & & \\ 0 & 0 & \multicolumn{2}{c}{0} \\ \hline & & I_s & 0 \\ \multicolumn{2}{c|}{0} & 0 & 0 \end{array}\right].$$

Thus,

$$\text{rank}\left(\left[\begin{array}{c|c} A & 0 \\ \hline 0 & B \end{array}\right]\right) = \text{rank}\left(P\left[\begin{array}{c|c} A & 0 \\ \hline 0 & B \end{array}\right]Q\right) = r + s.$$

Section 2.9

1. a. $\begin{bmatrix} -4 \\ 9 \\ 13 \end{bmatrix}$ **c.** $\begin{bmatrix} -4 \\ 5 \\ -7 \end{bmatrix}$ **d.** $\begin{bmatrix} 4 \\ 7 \\ -4 \end{bmatrix}$ **f.** $\begin{bmatrix} 3 \\ 11/2 \\ -7/2 \end{bmatrix}$

3. $X = (-3)\begin{bmatrix} -1 \\ 2 \\ -3 \end{bmatrix}$, so $[X]_{\mathcal{B}} = [-3]$

6. a. $\left[\begin{array}{cc|cc} 2 & -1 & -1 & 1 \\ 4 & 2 & -1 & -2 \end{array}\right] \longrightarrow \left[\begin{array}{cc|cc} 1 & 0 & -3/8 & 0 \\ 0 & 1 & 1/4 & -1 \end{array}\right] = [I_2 \mid P_{\mathcal{AB}}]$

c. $P_{\mathcal{A}\mathcal{E}} = P_{\mathcal{E}\mathcal{A}}^{-1} = \begin{bmatrix} 2 & -1 \\ 4 & 2 \end{bmatrix}^{-1} = (1/8)\begin{bmatrix} 2 & 1 \\ -4 & 2 \end{bmatrix}$

7. a. $\left[\begin{array}{ccc|ccc} 1 & 0 & 1 & 1 & -1 & -1 \\ 1 & 1 & 1 & 0 & 1 & -1 \\ 1 & 1 & 0 & 1 & 0 & 1 \end{array}\right] \longrightarrow \left[\begin{array}{ccc|ccc} 1 & 0 & 0 & 2 & -2 & 1 \\ 0 & 1 & 0 & -1 & 2 & 0 \\ 0 & 0 & 1 & -1 & 1 & -2 \end{array}\right] = [I_3 \mid P_{\mathcal{BA}}]$

c. $\left[\begin{array}{ccc|ccc} 1 & -1 & -1 & 1 & 1 & 2 \\ 0 & 1 & -1 & 1 & 2 & 1 \\ 1 & 0 & 1 & 2 & 1 & 1 \end{array}\right] \longrightarrow \left[\begin{array}{ccc|ccc} 1 & 0 & 0 & 2 & 5/3 & 5/3 \\ 0 & 1 & 0 & 1 & 4/3 & 1/3 \\ 0 & 0 & 1 & 0 & -2/3 & -2/3 \end{array}\right] = [I_3 \mid P_{\mathcal{AC}}]$

e. $P_{\mathcal{EA}} = \begin{bmatrix} 1 & -1 & -1 \\ 0 & 1 & -1 \\ 1 & 0 & 1 \end{bmatrix}$

9. a. $Y_1, Y_2 \in \mathcal{S}pan\{X_1, X_2\} = \mathcal{V}$. If s and t are scalars such that $0 = sY_1 + tY_2 = s(3X_1 - 2X_2) + t(-X_1 + 4X_2) = (3s - t)X_1 + (-2s + 4t)X_2$, then independence of $\{X_1, X_2\}$ implies $3s - t = 0 = -2s + 4t$, which in turn implies $s = t = 0$. Thus, $\{Y_1, Y_2\}$ is linearly independent and hence a basis for $\mathcal{V}$.

12. a. $\begin{bmatrix} 1 & 1 & 1 & 3 \\ 2 & 1 & 0 & 5 \end{bmatrix} \longrightarrow \begin{bmatrix} 1 & 0 & -1 & 2 \\ 0 & 1 & 2 & 1 \end{bmatrix} \Rightarrow \mathcal{A} = \left\{ \begin{bmatrix} 1 \\ 2 \end{bmatrix}, \begin{bmatrix} 1 \\ 1 \end{bmatrix} \right\}$ is a basis for $\mathcal{C}(A) \Rightarrow$ $\dim \mathcal{C}(A) = 2$. $\mathcal{B}$ consists of two linearly independent columns of A, so $\mathcal{B}$ is also a basis for $\mathcal{C}(A)$. Since $\left[\begin{array}{cc|cc} 1 & 1 & 1 & 3 \\ 2 & 1 & 0 & 5 \end{array}\right] \longrightarrow \left[\begin{array}{cc|cc} 1 & 0 & -1 & 2 \\ 0 & 1 & 2 & 1 \end{array}\right]$, $P_{\mathcal{AB}} = \begin{bmatrix} -1 & 2 \\ 2 & 1 \end{bmatrix}$ and $P_{\mathcal{BA}} = P_{\mathcal{AB}}^{-1} = \frac{1}{5}\begin{bmatrix} -1 & 2 \\ 2 & 1 \end{bmatrix}$.

14. For $X \in \mathcal{V}$, $(P_{\mathcal{CB}}P_{\mathcal{BA}})[X]_{\mathcal{A}} = P_{\mathcal{CB}}(P_{\mathcal{BA}}[X]_{\mathcal{A}}) = P_{\mathcal{CB}}[X]_{\mathcal{B}} = [X]_{\mathcal{C}}$. Since $P_{\mathcal{CA}}$ is the unique matrix satisfying $[X]_{\mathcal{C}} = P_{\mathcal{CA}}[X]_{\mathcal{A}}$ for all $X \in \mathcal{V}$, $P_{\mathcal{CA}} = P_{\mathcal{CB}}P_{\mathcal{BA}}$.

Section 2.10

1. $\mathcal{V}$. Certainly $\mathcal{V} \subseteq \mathcal{U} + \mathcal{V}$ (Theorem 2.35). On the other hand, if $X \in \mathcal{U} + \mathcal{V}$, then $X = U + V$ for some $U \in \mathcal{U}$ and $V \in \mathcal{V}$. Since $\mathcal{U} \subseteq \mathcal{V}$, $U \in \mathcal{V}$, so $X = U + V \in \mathcal{V}$ ($\mathcal{V}$ is closed under addition). Thus, $\mathcal{U} + \mathcal{V} \subseteq \mathcal{V}$.

3. a. $\mathcal{S}pan\{X, Y, Z\} = \{rX + sY + tZ : r, s, t \in \mathcal{R}\} = \{U + V + W : U \in \mathcal{S}pan\{X\}, V \in \mathcal{S}pan\{Y\}, W \in \mathcal{S}pan\{Z\}\}$.

b. $W \in \mathcal{S}pan\{X\} + \mathcal{S}pan\{Y\} + \mathcal{S}pan\{Z\} \Leftrightarrow W = rX + sY + tZ$. If $\{X, Y, Z\}$ is linearly independent, then the scalars r, s, and t are unique. On the other hand, if the decomposition is unique for each W, then $0 = 0X + 0Y + 0Z$ is the unique decomposition of 0, so $\{X, Y, Z\}$ is linearly independent.

6. $\mathcal{U}$, $\mathcal{V}$, $\mathcal{W}$ independent $\Rightarrow 0 = 0 + 0 + 0$ is the unique expression for 0 as a sum of elements of $\mathcal{U}$, $\mathcal{V}$, and $\mathcal{W}$. Conversely, if $U_1 + V_1 + W_1 = U_2 + V_2 + W_2$, where $U_1, U_2 \in \mathcal{U}$, $V_1, V_2 \in \mathcal{V}$, and $W_1, W_2 \in \mathcal{W}$, then $0 = (U_1 - U_2) + (V_1 - V_2) + (W_1 - W_2)$. Uniqueness of decomposition of 0 implies $U_1 = U_2$, $V_1 = V_2$, and $W_1 = W_2$.

7. a. For $a, b, c \in \mathcal{R}$, $\begin{bmatrix} 0 \\ 0 \\ c \end{bmatrix} \in \mathcal{U}$, $\begin{bmatrix} a \\ 0 \\ 0 \end{bmatrix} \in \mathcal{V}$, $\begin{bmatrix} 0 \\ b \\ 0 \end{bmatrix} \in \mathcal{W}$, and $\begin{bmatrix} a \\ b \\ c \end{bmatrix} = \begin{bmatrix} 0 \\ 0 \\ c \end{bmatrix} + \begin{bmatrix} a \\ 0 \\ 0 \end{bmatrix} + \begin{bmatrix} 0 \\ b \\ 0 \end{bmatrix}$, so every vector in $\mathcal{R}^3$ can be expressed as a sum of a vector in $\mathcal{U}$, a vector in $\mathcal{V}$, and a vector in $\mathcal{W}$.

8. a. Suppose $X \in \mathcal{U} \cap \mathcal{V}$. Then $X \in \mathcal{U}$, $-X \in \mathcal{V}$, $0 \in \mathcal{W}$, and $0 = X + (-X) + 0$ implies $X = 0$. Thus, $\mathcal{U} \cap \mathcal{V} = \{0\}$. Similarly, $\mathcal{U} \cap \mathcal{W} = \{0\} = \mathcal{V} \cap \mathcal{W}$.

Section 3.2

1. **a.** -2 **b.** 0 **d.** 0

2. **a.** $\theta = 3\pi/4$ **c.** $\theta = 2\pi/3$

5. If ϕ is the angle between X and $-$Y, then $\cos(\phi) = \dfrac{X \cdot (-Y)}{\|X\|\|-Y\|} = -\dfrac{X \cdot Y}{\|X\|\|Y\|} =$
$-\cos(\theta) = \cos(\pi - \theta)$, so $\phi = \pi - \theta$.

7. $[AB]_{ij} = \sum_{k=1}^{n} a_{ik} b_{kj} = \begin{bmatrix} a_{i1} \\ \vdots \\ a_{in} \end{bmatrix} \cdot \begin{bmatrix} b_{1j} \\ \vdots \\ b_{nj} \end{bmatrix} = [\mathrm{Row}_i(A)]^T \cdot \mathrm{Col}_j(B)$.

9. Suppose P, Q, R, and S are vertices of a parallelogram and let $\mathbf{x}$ and $\mathbf{y}$ be the geometric vectors represented by $\overrightarrow{PQ}$ and $\overrightarrow{PR}$, respectively. Then $\overrightarrow{PS} \in \mathbf{x}+\mathbf{y}$ and $\overrightarrow{RQ} \in \mathbf{x}-\mathbf{y}$. Let $X, Y \in \mathcal{R}^2$, where $X \leftrightarrow \mathbf{x}$, $Y \leftrightarrow \mathbf{y}$. Then $X + Y \leftrightarrow \mathbf{x}+\mathbf{y}$ and $X - Y \leftrightarrow \mathbf{x}-\mathbf{y}$. $\overrightarrow{PS}$ is perpendicular to $\overrightarrow{RQ} \Leftrightarrow (X+Y)\cdot(X-Y) = 0$, which by Problem 8 occurs exactly when $\|X\| = \|Y\|$, that is, when the parallelogram is a rhombus.

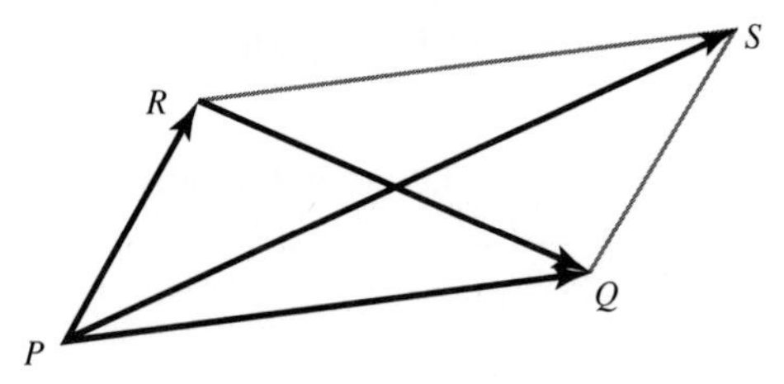

10. **a.** $\|U+V\|^2 = (U+V)\cdot(U+V) = \|U\|^2 + 2(U\cdot V) + \|V\|^2 = 1 + 2(0) + 1 = 2 \Rightarrow \|U+V\| = \sqrt{2}$.

13. $\|X+Y\| = \|tY+Y\| = \|(t+1)Y\| = |t+1|\|Y\| = (t+1)\|Y\| = t\|Y\| + \|Y\| = |t|\,\|Y\| + \|Y\| = \|tY\| + \|Y\| = \|X\| + \|Y\|$

14. **b.** Since every vector in $\mathcal{R}^n$ is orthogonal to 0, $\{0\}^\perp = \mathcal{R}^n$.

15. If $X \in \mathcal{V} \cap \mathcal{V}^\perp$, then $X \in \mathcal{V}$ and $X \in \mathcal{V}^\perp$, so $0 = X \cdot X = \|X\|^2$. Thus, $X = 0$.

Section 3.3

1. **a.** If $U_1 = \dfrac{1}{\sqrt{2}}\begin{bmatrix} 1 \\ 0 \\ -1 \end{bmatrix}$, $U_2 = \dfrac{1}{3}\begin{bmatrix} 2 \\ 1 \\ 2 \end{bmatrix}$, $U_3 = \dfrac{1}{3\sqrt{2}}\begin{bmatrix} 1 \\ -4 \\ 1 \end{bmatrix}$, then $U_1 \cdot U_2 = U_1 \cdot U_3 = U_2 \cdot U_3 = 0$ and $\|U_1\| = \|U_2\| = \|U_3\| = 1$. Being three linearly independent vectors in $\mathcal{R}^3$, U_1, U_2, and U_3 form a basis for $\mathcal{R}^3$.

b. $[X]_{\mathcal{A}} = \begin{bmatrix} X \cdot U_1 \\ X \cdot U_2 \\ X \cdot U_3 \end{bmatrix} = \begin{bmatrix} 1/\sqrt{2} \\ 5 \\ 1/\sqrt{2} \end{bmatrix}$ **c.** Let $X_1 = \begin{bmatrix} 1 \\ 0 \\ 0 \end{bmatrix}$, $X_2 = \begin{bmatrix} 1 \\ 1 \\ 0 \end{bmatrix}$, and $X_3 = \begin{bmatrix} 1 \\ 1 \\ 1 \end{bmatrix}$.

Then $P_{\mathcal{AB}} = \begin{bmatrix} [X_1]_{\mathcal{A}} & [X_2]_{\mathcal{A}} & [X_3]_{\mathcal{A}} \end{bmatrix} = \begin{bmatrix} 1/\sqrt{2} & 1/\sqrt{2} & 0 \\ 2/3 & 1 & 5/3 \\ 1/3\sqrt{2} & -1/\sqrt{2} & -\sqrt{2}/3 \end{bmatrix}$.

3. a. $\mathcal{V} = \mathcal{N}([0 \quad 1 \quad -1 \quad -1]) \Rightarrow \dim \mathcal{V} = 3$. $\mathcal{A}$ consists of three orthogonal unit vectors satisfying $x_2 - x_3 - x_4 = 0$, so $\mathcal{A}$ is a basis for $\mathcal{V}$.

b. $[X]_{\mathcal{A}} = \begin{bmatrix} \sqrt{3} \\ 3\sqrt{3} \\ 0 \end{bmatrix}$ **c.** $P_{\mathcal{AB}} = \dfrac{1}{\sqrt{3}} \begin{bmatrix} 1 & -1 & 1 \\ 1 & 2 & 1 \\ 1 & -1 & -2 \end{bmatrix}$

5. a. $\left\{ \dfrac{1}{3} \begin{bmatrix} 1 \\ 2 \\ -2 \end{bmatrix}, \dfrac{1}{\sqrt{2}} \begin{bmatrix} 0 \\ -1 \\ -1 \end{bmatrix} \right\}$ **d.** $\left\{ \dfrac{1}{\sqrt{3}} \begin{bmatrix} 1 \\ 1 \\ 0 \\ 1 \end{bmatrix}, \dfrac{1}{\sqrt{3}} \begin{bmatrix} -1 \\ 0 \\ 1 \\ 1 \end{bmatrix}, \dfrac{1}{\sqrt{51}} \begin{bmatrix} 1 \\ 3 \\ 5 \\ -4 \end{bmatrix} \right\}$

g. $\left\{ \dfrac{1}{\sqrt{3}} \begin{bmatrix} 1 \\ 0 \\ 1 \\ 0 \\ 1 \end{bmatrix}, \dfrac{1}{\sqrt{3}} \begin{bmatrix} 0 \\ 0 \\ 1 \\ 1 \\ -1 \end{bmatrix}, \dfrac{1}{2} \begin{bmatrix} 1 \\ 1 \\ 0 \\ -1 \\ -1 \end{bmatrix} \right\}$

i. $\left\{ \dfrac{1}{\sqrt{2}} \begin{bmatrix} 1 \\ 0 \\ 1 \\ 0 \\ 0 \\ 0 \end{bmatrix}, \dfrac{1}{\sqrt{2}} \begin{bmatrix} 0 \\ 0 \\ 0 \\ 0 \\ 1 \\ 1 \end{bmatrix}, \dfrac{1}{\sqrt{5}} \begin{bmatrix} 1 \\ 1 \\ -1 \\ -1 \\ 1 \\ -1 \end{bmatrix}, \dfrac{1}{\sqrt{6}} \begin{bmatrix} 0 \\ 2 \\ 0 \\ 0 \\ -1 \\ 1 \end{bmatrix} \right\}$

7. a. $\dfrac{2}{3} \begin{bmatrix} -1 \\ 1 \\ -1 \end{bmatrix}$ **c.** 0 **e.** $-\dfrac{1}{9} \begin{bmatrix} -1 \\ 1 \\ -1 \end{bmatrix}$

8. b. $\dfrac{1}{5} \begin{bmatrix} 2 \\ 1 \\ -1 \\ 2 \end{bmatrix}$

9. $\text{Proj}_{tY}(X) = \dfrac{X \cdot (tY)}{\|tY\|^2}(tY) = \dfrac{t(X \cdot Y)}{t^2\|Y\|^2}(tY) = \dfrac{X \cdot Y}{\|Y\|^2} Y = \text{Proj}_Y(X)$

12. $W = sX + tY$, and $W \cdot X = s(X \cdot X) + t(Y \cdot X) = s\|X\|^2$, so $s = (W \cdot X)/\|X\|^2$. Thus, $sX = \text{Proj}_X(W)$. Similarly, $tY = \text{Proj}_Y(W)$.

14. a. $(dU) \cdot U = d\|U\|^2 = d \Rightarrow dU \in \mathcal{H}$. $X \in \mathcal{H} \Rightarrow \text{Proj}_U(X) = \dfrac{X \cdot U}{\|U\|^2} U = dU$.

Section 3.4

1. a. $\mathcal{V}^\perp = \mathcal{N}\left(\begin{bmatrix} 1 & 2 & 3 \\ 3 & 2 & 1 \end{bmatrix}\right)$ and $\begin{bmatrix} 1 & 2 & 3 \\ 3 & 2 & 1 \end{bmatrix} \longrightarrow \begin{bmatrix} 1 & 0 & -1 \\ 0 & 1 & 2 \end{bmatrix}$, so

$$\mathcal{V}^\perp = \mathcal{S}pan\left\{\begin{bmatrix} 1 \\ -2 \\ 1 \end{bmatrix}\right\}$$

c. $\mathcal{V}^\perp = \mathcal{N}\left(\begin{bmatrix} 1 & 0 & -1 & 1 \\ 1 & -1 & -1 & 1 \\ 1 & 1 & -1 & 1 \end{bmatrix}\right) = \mathcal{S}pan\left\{\begin{bmatrix} 1 \\ 0 \\ 1 \\ 0 \end{bmatrix}, \begin{bmatrix} -1 \\ 0 \\ 0 \\ 1 \end{bmatrix}\right\}$

e. $\mathcal{V}^\perp = \mathcal{S}pan\left\{\begin{bmatrix} 1 \\ 0 \\ 1 \\ 0 \\ 0 \end{bmatrix}, \begin{bmatrix} 0 \\ 1 \\ 0 \\ 1 \\ 0 \end{bmatrix}, \begin{bmatrix} 0 \\ -1 \\ 0 \\ 0 \\ 1 \end{bmatrix}\right\}$

g. $\mathcal{V}^\perp = \mathcal{S}pan\left\{\begin{bmatrix} 1 \\ 0 \\ 0 \\ 0 \\ 0 \end{bmatrix}, \begin{bmatrix} 0 \\ 7 \\ 2 \\ -3 \\ 1 \end{bmatrix}\right\}$ **i.** $\mathcal{V}^\perp = \mathcal{S}pan\left\{\begin{bmatrix} 1 \\ 1 \end{bmatrix}\right\}$

2. Set $Y = \begin{bmatrix} 2 \\ 1 \\ -3 \end{bmatrix}$. $V = \text{Proj}_Y(X) = ((X \cdot Y)/\|Y\|^2)Y = -\frac{1}{7}\begin{bmatrix} 2 \\ 1 \\ -3 \end{bmatrix}$,

$$W = X - V = \frac{1}{7}\begin{bmatrix} 9 \\ -6 \\ 4 \end{bmatrix}$$

3. a. $\mathcal{V} = \mathcal{N}([1 \quad -1 \quad 1]) = \mathcal{S}pan\left\{\begin{bmatrix} 1 \\ 1 \\ 0 \end{bmatrix}, \begin{bmatrix} -1 \\ 0 \\ 1 \end{bmatrix}\right\}$. The Gram-Schmidt process yields an orthonormal basis $\{U_1, U_2\}$, where $U_1 = \frac{1}{\sqrt{2}}\begin{bmatrix} 1 \\ 1 \\ 0 \end{bmatrix}$, $U_2 = \frac{1}{\sqrt{6}}\begin{bmatrix} -1 \\ 1 \\ 2 \end{bmatrix}$. $\text{Proj}_{\mathcal{V}}(X) = (X \cdot U_1)U_1 + (X \cdot U_2)U_2 = \frac{1}{3}\begin{bmatrix} 1 \\ 8 \\ 7 \end{bmatrix}$, $\text{Proj}_{\mathcal{V}^\perp}(X) = X - \text{Proj}_{\mathcal{V}}(X) = \frac{2}{3}\begin{bmatrix} 1 \\ -1 \\ 1 \end{bmatrix}$

c. $\text{Proj}_{\mathcal{V}}(X) = X$, $\text{Proj}_{\mathcal{V}^\perp}(X) = 0$

4. a. $A \longrightarrow \begin{bmatrix} 1 & 0 & -2 & 1 \\ 0 & 1 & -1 & 1 \\ 0 & 0 & 0 & 0 \end{bmatrix} \Rightarrow \mathcal{N}(A) = \mathcal{S}pan\left\{\begin{bmatrix} -1 \\ -1 \\ 0 \\ 1 \end{bmatrix}, \begin{bmatrix} 2 \\ 1 \\ 1 \\ 0 \end{bmatrix}\right\}$. Gram-Schmidt produces the orthonormal basis $\left\{ \frac{1}{\sqrt{3}} \begin{bmatrix} -1 \\ -1 \\ 0 \\ 1 \end{bmatrix}, \frac{1}{\sqrt{3}} \begin{bmatrix} 1 \\ 0 \\ 1 \\ 1 \end{bmatrix} \right\}$. $\text{Proj}_{\mathcal{V}}(\mathrm{X}) = \frac{1}{3}\begin{bmatrix} 4 \\ 1 \\ 3 \\ 2 \end{bmatrix}$,

$\text{Proj}_{\mathcal{V}^{\perp}}(\mathrm{X}) = \frac{1}{3}\begin{bmatrix} -1 \\ 2 \\ 0 \\ 1 \end{bmatrix}$

6. $\mathrm{X} = \mathrm{X} + 0$. Since $\mathrm{X} \in \mathcal{V}$ and $0 \in \mathcal{V}^{\perp}$, $\text{Proj}_{\mathcal{V}}(\mathrm{X}) = \mathrm{X}$.

8. a. Let $\mathrm{U}_k = \mathrm{Y}_k/\|\mathrm{Y}_k\|$, $k = 1, \ldots, p$. Since $\{\mathrm{U}_1, \ldots, \mathrm{U}_p\}$ is an orthonormal basis for $\mathcal{V}$,

$$\text{Proj}_{\mathcal{V}}(\mathrm{X}) = \sum_{k=1}^{p} (\mathrm{X} \cdot \mathrm{U}_k)\mathrm{U}_k = \sum_{k=1}^{p} \left(\frac{\mathrm{X} \cdot \mathrm{Y}_k}{\|\mathrm{Y}_k\|^2} \right) \mathrm{Y}_k = \sum_{k=1}^{p} \text{Proj}_{\mathrm{Y}_k}(\mathrm{X}).$$

9. a. $\mathrm{Y} \in \mathcal{U} \cap \mathcal{V} \Rightarrow \mathrm{Y} \in \mathcal{U}$ and $\mathrm{Y} \in \mathcal{V} \Rightarrow 0 = \mathrm{Y} \cdot \mathrm{Y} = \|\mathrm{Y}\|^2 \Rightarrow \mathrm{Y} = 0$.

b. If $\mathcal{U} = \{0\}$, then $\mathcal{U} \oplus \mathcal{V} = \{0\} \oplus \mathcal{V} = \mathcal{V}$. In this case, $\text{Proj}_{\mathcal{U}\oplus\mathcal{V}}(\mathrm{X}) = \text{Proj}_{\mathcal{V}}(\mathrm{X})$ and $\text{Proj}_{\mathcal{U}}(\mathrm{X}) = 0$, so $\text{Proj}_{\mathcal{U}\oplus\mathcal{V}}(\mathrm{X}) = \text{Proj}_{\mathcal{U}}(\mathrm{X}) + \text{Proj}_{\mathcal{V}}(\mathrm{X})$. If $\mathcal{U} = \mathcal{R}^n$, then $\mathcal{V} = \{0\}$ (the only subspace orthogonal to $\mathcal{R}^n$ is $\{0\}$). In this case, $\mathcal{U} \oplus \mathcal{V} = \mathcal{R}^n \oplus \{0\} = \mathcal{R}^n = \mathcal{U}$, and $\text{Proj}_{\mathcal{V}}(\mathrm{X}) = 0$, so the conclusion holds. Assume now that $\mathcal{U}$ and $\mathcal{V}$ are nontrivial subspaces. If $\mathcal{A} = \{\mathrm{U}_1, \ldots, \mathrm{U}_p\}$ and $\mathcal{B} = \{\mathrm{V}_1, \ldots, \mathrm{V}_q\}$ are orthonormal bases for $\mathcal{U}$ and $\mathcal{V}$, respectively, then $\mathcal{A} \cup \mathcal{B}$ is an orthonormal basis for $\mathcal{U} \oplus \mathcal{V}$. By Theorem 3.13, $\text{Proj}_{\mathcal{U}\oplus\mathcal{V}}(\mathrm{X}) = \sum_1^p (\mathrm{X} \cdot \mathrm{U}_k)\mathrm{U}_k + \sum_1^q (\mathrm{X} \cdot \mathrm{V}_k)\mathrm{V}_k$, $\text{Proj}_{\mathcal{U}}(\mathrm{X}) = \sum_1^p (\mathrm{X} \cdot \mathrm{U}_k)\mathrm{U}_k$, and $\text{Proj}_{\mathcal{V}}(\mathrm{X}) = \sum_1^q (\mathrm{X} \cdot \mathrm{V}_k)\mathrm{V}_k$.

Section 3.5

1. $\mathrm{P}_{\mathcal{BA}} = \frac{1}{\sqrt{10}}\begin{bmatrix} 3 & -1 \\ 1 & 3 \end{bmatrix}$, $\mathrm{P}_{\mathcal{AB}} = \frac{1}{\sqrt{10}}\begin{bmatrix} 3 & 1 \\ -1 & 3 \end{bmatrix}$

2. $\mathrm{P}_{\mathcal{BA}} = \frac{1}{2}\begin{bmatrix} -\sqrt{2} & \sqrt{2} & 0 \\ 1 & 1 & \sqrt{2} \\ -1 & -1 & \sqrt{2} \end{bmatrix}$, $\mathrm{P}_{\mathcal{AB}} = \mathrm{P}_{\mathcal{BA}}^T$

3. b. I and $-$I are orthogonal, but $\mathrm{I} + (-\mathrm{I}) = 0$ is not orthogonal.

4. The first column of A, being a unit vector, is a point on the circle of radius 1 centered at 0, so there is a $\theta \in \mathcal{R}$ such that $\text{Col}_1(\mathrm{A}) = \begin{bmatrix} \cos(\theta) \\ \sin(\theta) \end{bmatrix}$. Then $\text{Col}_2(\mathrm{A})$, being perpendicular to $\text{Col}_1(\mathrm{A})$, is either

$$\begin{bmatrix} \cos(\theta + \pi/2) \\ \sin(\theta + \pi/2) \end{bmatrix} = \begin{bmatrix} -\sin(\theta) \\ \cos(\theta) \end{bmatrix} \text{ or } \begin{bmatrix} \cos(\theta - \pi/2) \\ \sin(\theta - \pi/2) \end{bmatrix} = \begin{bmatrix} \sin(\theta) \\ -\cos(\theta) \end{bmatrix}.$$

8. Suppose $\mathrm{A} = [\mathrm{X}_1 \ \cdots \ \mathrm{X}_n]$ is orthogonal and upper triangular. X_1, being a unit vector whose entries other than the first are all zeros, must be $\pm\mathrm{E}_1$. Then $0 = \mathrm{X}_2 \cdot \mathrm{X}_1 = \mathrm{X}_2 \cdot (\pm\mathrm{E}_1)$ implies the first entry of X_2 is 0. Entries 3 through n of X_2 are also zero, and since $\|\mathrm{X}_2\| = 1$, the second entry of X_2 is ±1. Thus, $\mathrm{X}_2 = \pm\mathrm{E}_2$. Continuing in this manner yields $\mathrm{X}_k = \pm\mathrm{E}_k$, $k = 1, \ldots, n$, that is, A is a diagonal matrix whose diagonal entries are ±1.

9. a. $X = (X \cdot U_1)U_1 + \cdots + (X \cdot U_p)U_p$ and $Y = (Y \cdot U_1)U_1 + \cdots + (Y \cdot U_p)U_p$, so

$$[X]_{\mathcal{A}} = \begin{bmatrix} X \cdot U_1 \\ \vdots \\ X \cdot U_p \end{bmatrix} \quad \text{and} \quad [Y]_{\mathcal{A}} = \begin{bmatrix} Y \cdot U_1 \\ \vdots \\ Y \cdot U_p \end{bmatrix}.$$

Since $U_i \cdot U_j = 0$ whenever $i \neq j$, and $U_i \cdot U_i = 1$, $1 \le i \le p$,

$$\begin{aligned} X \cdot Y &= [(X \cdot U_1)U_1 + \cdots + (X \cdot U_p)U_p] \cdot [(Y \cdot U_1)U_1 + \cdots + (Y \cdot U_p)U_p] \\ &= (X \cdot U_1)(Y \cdot U_1) + \cdots + (X \cdot U_p)(Y \cdot U_p) \\ &= [X]_{\mathcal{A}} \cdot [Y]_{\mathcal{A}}. \end{aligned}$$

11. a. $\begin{bmatrix} 1/\sqrt{2} & -1/\sqrt{2} \\ 1/\sqrt{2} & 1/\sqrt{2} \end{bmatrix}\begin{bmatrix} \sqrt{2} & \sqrt{2} \\ 0 & \sqrt{2} \end{bmatrix}$

d. $\begin{bmatrix} 1/\sqrt{3} & -1/\sqrt{3} & 1/\sqrt{15} \\ 1/\sqrt{3} & 0 & 1/\sqrt{15} \\ 0 & 1/\sqrt{3} & 3/\sqrt{15} \\ 1/\sqrt{3} & 1/\sqrt{3} & -2/\sqrt{15} \end{bmatrix}\begin{bmatrix} \sqrt{3} & 2\sqrt{3} & -\sqrt{3} \\ 0 & \sqrt{3} & \sqrt{3} \\ 0 & 0 & \sqrt{15} \end{bmatrix}$

12. a. $[Q_2^T Q_2]_{ij} = \mathrm{Row}_i(Q_2^T)\mathrm{Col}_j(Q_2) = \mathrm{Col}_i(Q_2) \cdot \mathrm{Col}_j(Q_2) = \begin{cases} 0, & \text{if } i \neq j \\ 1, & \text{if } i = j \end{cases}.$

Similarly, $Q_1^T Q_1 = I_p$.

b. $Q_2^T(Q_1R_1) = Q_2^T(Q_2R_2) = (Q_2^T Q_2)R_2 = I_pR_2 = R_2$. Multiplying by R_1^{-1} on the right gives $Q_2^T Q_1 = R_2R_1^{-1}$. Being a product of upper triangular matrices, $Q_2^T Q_1$ is upper triangular. Similarly, $Q_1^T Q_2$ is upper triangular.

Section 3.6

1. a. $\begin{bmatrix} 2 \\ -4 \\ 2 \end{bmatrix}$ **c.** $\begin{bmatrix} 4 \\ -8 \\ 4 \end{bmatrix}$ **e.** $\begin{bmatrix} 5 \\ -13 \\ 7 \end{bmatrix}$ **g.** $\begin{bmatrix} -5 \\ -2 \\ 3 \end{bmatrix}$

2. $E_1 \times E_2 = E_3$, $E_2 \times E_3 = E_1$, $E_3 \times E_1 = E_2$

4. a. $X \times Y = \begin{bmatrix} x_2y_3 - x_3y_2 \\ x_3y_1 - x_1y_3 \\ x_1y_2 - x_2y_1 \end{bmatrix} = -\begin{bmatrix} y_2x_3 - y_3x_2 \\ y_3x_1 - y_1x_3 \\ y_1x_2 - y_2x_1 \end{bmatrix} = -(Y \times X)$

b. $X \times (Y + Z) = \begin{bmatrix} x_2(y_3 + z_3) - x_3(y_2 + z_2) \\ x_3(y_1 + z_1) - x_1(y_3 + z_3) \\ x_1(y_2 + z_2) - x_2(y_1 + z_1) \end{bmatrix} = \begin{bmatrix} x_2y_3 - x_3y_2 \\ x_3y_1 - x_1y_3 \\ x_1y_2 - x_2y_1 \end{bmatrix} + \begin{bmatrix} x_2z_3 - x_3z_2 \\ x_3z_1 - x_1z_3 \\ x_1z_2 - x_2z_1 \end{bmatrix}$

$= X \times Y + X \times Z$

7. a. $Y \times Z$ is orthogonal to the plane spanned by Y and Z. $X \times (Y \times Z)$, being orthogonal to $Y \times Z$, must be in $\mathcal{Span}\{Y, Z\}$.

b. $X \times (Y \times Z) = \begin{bmatrix} x_1 \\ x_2 \\ x_3 \end{bmatrix} \times \begin{bmatrix} y_2z_3 - y_3z_2 \\ y_3z_1 - y_1z_3 \\ y_1z_2 - y_2z_1 \end{bmatrix} = \begin{bmatrix} x_2(y_1z_2 - y_2z_1) - x_3(y_3z_1 - y_1z_3) \\ x_3(y_2z_3 - y_3z_2) - x_1(y_1z_2 - y_2z_1) \\ x_1(y_3z_1 - y_1z_3) - x_2(y_2z_3 - y_3z_2) \end{bmatrix}$

$$= (x_1z_1 + x_2z_2 + x_3z_3)\begin{bmatrix} y_1 \\ y_2 \\ y_3 \end{bmatrix} - (x_1y_1 + x_2y_2 + x_3y_3)\begin{bmatrix} z_1 \\ z_2 \\ z_3 \end{bmatrix}$$

$$= (X \cdot Z)Y - (X \cdot Y)Z$$

Section 3.7

1. a. $A = \begin{bmatrix} 1 & 1 \\ 1 & 2 \\ 1 & 3 \end{bmatrix}$, $Y = \begin{bmatrix} 1 \\ 1 \\ 2 \end{bmatrix}$, and $(A^TA)Z = A^TY$ becomes $\begin{bmatrix} 3 & 6 \\ 6 & 14 \end{bmatrix}\begin{bmatrix} a \\ b \end{bmatrix} = \begin{bmatrix} 4 \\ 9 \end{bmatrix}$. Thus,

$\begin{bmatrix} a \\ b \end{bmatrix} = \begin{bmatrix} 3 & 6 \\ 6 & 14 \end{bmatrix}^{-1}\begin{bmatrix} 4 \\ 9 \end{bmatrix} = \begin{bmatrix} 1/3 \\ 1/2 \end{bmatrix}$, and $y = \frac{1}{3} + \frac{1}{2}x$.

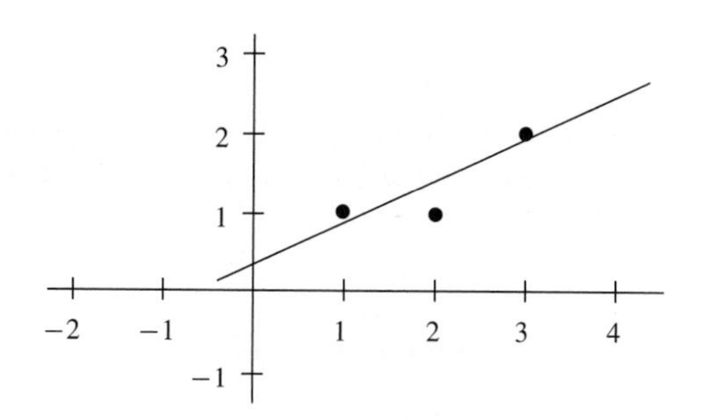

c. $A^TA = \begin{bmatrix} 5 & 5 \\ 5 & 15 \end{bmatrix}$, $A^TY = \begin{bmatrix} 9.18 \\ 2.44 \end{bmatrix}$, and $\begin{bmatrix} a \\ b \end{bmatrix} = \begin{bmatrix} 5 & 5 \\ 5 & 15 \end{bmatrix}^{-1}\begin{bmatrix} 9.18 \\ 2.44 \end{bmatrix} = \begin{bmatrix} 2.510 \\ -0.674 \end{bmatrix}$.

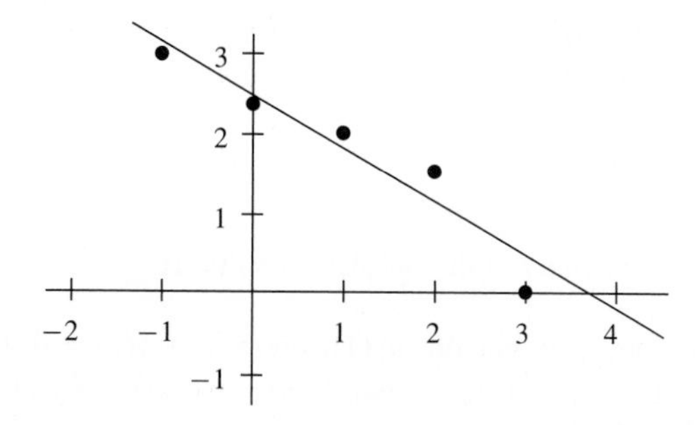

3. a. $A = \begin{bmatrix} 1 & -2 & 4 \\ 1 & -1 & 1 \\ 1 & 0 & 0 \\ 1 & 1 & 1 \end{bmatrix}$, $Y = \begin{bmatrix} 1 \\ 0 \\ -1 \\ 0 \end{bmatrix}$, $A^TA = \begin{bmatrix} 4 & -2 & 6 \\ -2 & 6 & -8 \\ 6 & -8 & 18 \end{bmatrix}$ and $A^TY = \begin{bmatrix} 0 \\ -2 \\ 4 \end{bmatrix}$.

The solution of the normal equations is $\begin{bmatrix} a \\ b \\ c \end{bmatrix} = \begin{bmatrix} 4 & -2 & 6 \\ -2 & 6 & -8 \\ 6 & -8 & 18 \end{bmatrix}^{-1} \begin{bmatrix} 0 \\ -2 \\ 4 \end{bmatrix}$

$= \begin{bmatrix} -0.7 \\ 0.1 \\ 0.5 \end{bmatrix}$, so the desired parabola has equation $y = -\frac{7}{10} + \frac{1}{10}x + \frac{1}{2}x^2$.

7. Let x be the number of years since 1987 and y the mean SAT score.

$$\mathrm{X} = \begin{bmatrix} 0 \\ 6 \\ 7 \\ 8 \\ 9 \end{bmatrix}, \mathrm{Y} = \begin{bmatrix} 521 \\ 527 \\ 531 \\ 535 \\ 535 \end{bmatrix}, \mathrm{A}^T\mathrm{A} = \begin{bmatrix} 5 & 30 \\ 30 & 230 \end{bmatrix}, \mathrm{A}^T\mathrm{Y} = \begin{bmatrix} 2649 \\ 15{,}974 \end{bmatrix}, \begin{bmatrix} a \\ b \end{bmatrix} = \begin{bmatrix} 520.2 \\ 1.6 \end{bmatrix},$$

$y = (1.6)x + 520.2$. In the year 2000, $x = 13$, and $y = 541$.

9. a. $0 = \sum_0^q t_k \mathrm{Col}_k(\mathrm{A}) = \begin{bmatrix} \sum_0^q t_k x_1^k \\ \vdots \\ \sum_0^q t_k x_n^k \end{bmatrix} = \begin{bmatrix} p(x_1) \\ \vdots \\ p(x_n) \end{bmatrix} \Leftrightarrow p(x_1) = \cdots = p(x_n) = 0$

b. $0 = \sum_0^q t_k \mathrm{Col}_k(\mathrm{A}) \Rightarrow p(x)$ vanishes at $x_1, \ldots, x_n$. Since p is a polynomial of degree $\leq q$, either $p(x) \equiv 0$ or p has at most q zeros. If $x_1, \ldots, x_n$ contains $q + 1$ distinct values, p must be identically zero, and therefore $t_0 = \cdots = t_n = 0$.

Section 4.2

1. a. linear, $\mathrm{A} = \begin{bmatrix} 2 & -3 & 1 \\ -1 & 0 & 4 \end{bmatrix}$

c. not linear. T satisfies neither of the linearity conditions.

f. linear, $\mathrm{A} = \begin{bmatrix} -1 & 0 & 0 \\ 0 & 1 & 0 \\ 0 & 0 & 1 \end{bmatrix}$

h. linear, $T(\mathrm{X}) = -\mathrm{X} = -\mathrm{I}_3\mathrm{X} \Rightarrow \mathrm{A} = -\mathrm{I}_3$.

k. linear, $\mathrm{A} = 5\mathrm{I}_n$

m. not linear. T satisfies neither of the linearity conditions.

3. $T(0) = b$. If $b \neq 0$, then T is not linear (Theorem 4.2). If $b = 0$, then for $x_1, x_2, x \in \mathcal{R}$, and $t \in \mathcal{R}$, $T(x_1 + x_2) = m(x_1 + x_2) = mx_1 + mx_2 = T(x_1) + T(x_2)$ and $T(tx) = m(tx) = t(mx) = tT(x)$, so T is linear.

4. a. $\begin{bmatrix} 1 \\ 1 \\ 1 \end{bmatrix} = \mathrm{E}_1 + \mathrm{E}_2 + \mathrm{E}_3$, $T\left(\begin{bmatrix} 1 \\ 1 \\ 1 \end{bmatrix}\right) = T(\mathrm{E}_1) + T(\mathrm{E}_2) + T(\mathrm{E}_3) = \begin{bmatrix} 4 \\ 2 \\ 0 \end{bmatrix}$ **c.** $\begin{bmatrix} -1 \\ -4 \\ 7 \end{bmatrix}$

5. a. $[X_1 \quad X_2 \quad X_3 \mid X] = \left[\begin{array}{rrr|r} 1 & -1 & 3 & 6 \\ 0 & 1 & 2 & 1 \\ 2 & 1 & -1 & 2 \end{array}\right] \longrightarrow \left[\begin{array}{rrr|r} 1 & 0 & 0 & 2 \\ 0 & 1 & 0 & -1 \\ 0 & 0 & 1 & 1 \end{array}\right]$

$$\Rightarrow X = 2X_1 - 1X_2 + 1X_3,$$

$$\Rightarrow T(X) = 2T(X_1) - T(X_2) + T(X_3)$$

$$= 2\begin{bmatrix} 1 \\ 2 \end{bmatrix} - \begin{bmatrix} -2 \\ 3 \end{bmatrix} + \begin{bmatrix} -1 \\ 0 \end{bmatrix} = \begin{bmatrix} 3 \\ 1 \end{bmatrix}.$$

c. $\begin{bmatrix} 4 \\ 1 \end{bmatrix}$

8. a. Let $U = \frac{1}{\sqrt{2}}\begin{bmatrix} 1 \\ -1 \end{bmatrix}$. $T(E_1) = (E_1 \cdot U)U = \frac{1}{2}\begin{bmatrix} 1 \\ -1 \end{bmatrix}$, $T(E_2) = (E_2 \cdot U)U = -\frac{1}{2}\begin{bmatrix} 1 \\ -1 \end{bmatrix}$,

and $A = [T(E_1) \quad T(E_2)] = \frac{1}{2}\begin{bmatrix} 1 & -1 \\ -1 & 1 \end{bmatrix}$.

c. $\mathcal{V} = \mathcal{N}([1 \quad 1 \quad 1]) = \mathcal{S}pan\left\{\begin{bmatrix} -1 \\ 1 \\ 0 \end{bmatrix}, \begin{bmatrix} -1 \\ 0 \\ 1 \end{bmatrix}\right\}$. Gram-Schmidt gives $Y_1 = \begin{bmatrix} -1 \\ 1 \\ 0 \end{bmatrix}$ and

$Y_2 = \begin{bmatrix} -1 \\ 0 \\ 1 \end{bmatrix} - \frac{1}{2}\begin{bmatrix} -1 \\ 1 \\ 0 \end{bmatrix} = \begin{bmatrix} -1/2 \\ -1/2 \\ 1 \end{bmatrix} = \frac{1}{2}\begin{bmatrix} -1 \\ -1 \\ 2 \end{bmatrix}$, so $U_1 = \frac{1}{\sqrt{2}}\begin{bmatrix} -1 \\ 1 \\ 0 \end{bmatrix}$ and $U_2 = \frac{1}{\sqrt{6}}\begin{bmatrix} -1 \\ -1 \\ 2 \end{bmatrix}$

are orthonormal basis vectors for $\mathcal{V}$. Thus,

$$T(X) = (X \cdot U_1)U_1 + (X \cdot U_2)U_2 = \frac{(-x_1 + x_2)}{2}\begin{bmatrix} -1 \\ 1 \\ 0 \end{bmatrix} + \frac{(-x_1 - x_2 + 2x_3)}{6}\begin{bmatrix} -1 \\ -1 \\ 2 \end{bmatrix}$$

$$= \frac{1}{6}\begin{bmatrix} 4x_1 - 2x_2 - 2x_3 \\ -2x_1 + 4x_2 - 2x_3 \\ -2x_1 - 2x_2 + 4x_3 \end{bmatrix} = \frac{1}{3}\begin{bmatrix} 2 & -1 & -1 \\ -1 & 2 & -1 \\ -1 & -1 & 2 \end{bmatrix}X.$$

Section 4.3

1. a. $A \longrightarrow \begin{bmatrix} 1 & 0 & -1 \\ 0 & 1 & 1 \\ 0 & 0 & 0 \\ 0 & 0 & 0 \end{bmatrix}$, $\mathcal{R}ng(T) = \mathcal{S}pan\left\{\begin{bmatrix} 1 \\ -1 \\ 0 \\ -1 \end{bmatrix}, \begin{bmatrix} 1 \\ 1 \\ 2 \\ 3 \end{bmatrix}\right\}$, $\mathcal{K}er(T) = \mathcal{S}pan\left\{\begin{bmatrix} 1 \\ -1 \\ 1 \end{bmatrix}\right\}$,

T is neither one-to-one nor onto.

c. $A \longrightarrow \begin{bmatrix} 1 & 0 \\ 0 & 1 \\ 0 & 0 \end{bmatrix}$, $\mathcal{R}ng(T) = \mathcal{S}pan\left\{\begin{bmatrix} 1 \\ 2 \\ 3 \end{bmatrix}, \begin{bmatrix} 4 \\ 5 \\ 6 \end{bmatrix}\right\}$, $\mathcal{K}er(T) = \{0\}$, T is one-to-one

but not onto.

f. $A \longrightarrow \begin{bmatrix} 1 & 0 & 0 & 1 \\ 0 & 1 & 0 & -1 \\ 0 & 0 & 1 & 1 \end{bmatrix}, \mathcal{R}ng(T) = \mathcal{S}pan\left\{\begin{bmatrix} 1 \\ 0 \\ 0 \end{bmatrix}, \begin{bmatrix} 1 \\ 1 \\ 0 \end{bmatrix}, \begin{bmatrix} 0 \\ 1 \\ 1 \end{bmatrix}\right\},$

$\mathcal{K}er(T) = \mathcal{S}pan\left\{\begin{bmatrix} -1 \\ 1 \\ -1 \\ 1 \end{bmatrix}\right\}$, T is onto but not one-to-one.

g. $A \longrightarrow \begin{bmatrix} 1 & 0 & 0 \\ 0 & 1 & 0 \\ 0 & 0 & 1 \end{bmatrix}, \mathcal{R}ng(T) = \mathcal{S}pan\left\{\begin{bmatrix} 1 \\ 1 \\ 0 \end{bmatrix}, \begin{bmatrix} 0 \\ 1 \\ 1 \end{bmatrix}, \begin{bmatrix} 1 \\ 0 \\ 1 \end{bmatrix}\right\}, \mathcal{K}er(T) = \{0\},$

T is one-to-one and onto.

3. $T(X) = [X]_{\mathcal{B}} = 0 \Rightarrow X = 0$, so $\mathcal{K}er(T) = \{0\}$ and T is one-to-one. By Theorem 4.9, T is also onto.

5. T is onto $\Leftrightarrow$ each column of rref(A) contains a leading one $\Leftrightarrow$ rref(A) $= I_n \Leftrightarrow$ A is nonsingular.

7. a. $\mathcal{R}ng(T) = \mathcal{V}$. For each $Y \in \mathcal{V}$, $Y = Y + 0$ is the decomposition of Y into the sum of a vector in $\mathcal{V}$ and a vector in $\mathcal{V}^{\perp}$, so $T(Y) = \text{Proj}_{\mathcal{V}}(Y) = Y$.

b. $\mathcal{K}er(T) = \mathcal{V}^{\perp}$. Suppose $X \in \mathcal{R}^n$ and $X = V + W$, where $V \in \mathcal{V}$ and $W \in \mathcal{V}^{\perp}$. Then $X \in \mathcal{K}er(T) \Leftrightarrow T(X) = 0 \Leftrightarrow V = 0 \Leftrightarrow X = W \in \mathcal{V}^{\perp}$.

9. a. $0 = t_1X_1 + \cdots + t_pX_p \Rightarrow 0 = T(0) = t_1T(X_1) + \cdots + t_pT(X_p) \Rightarrow t_1 = \cdots = t_p = 0$ (since $T(\mathcal{B})$ is linearly independent).

Section 4.4

1. a. $T(\mathcal{V}) = \mathcal{S}pan\left\{T\left(\begin{bmatrix} 1 \\ 0 \\ -1 \end{bmatrix}\right), T\left(\begin{bmatrix} 1 \\ -1 \\ -3 \end{bmatrix}\right)\right\} = \mathcal{S}pan\left\{\begin{bmatrix} -1 \\ 1 \\ 3 \end{bmatrix}, \begin{bmatrix} 1 \\ 2 \\ 5 \end{bmatrix}\right\},$

$\dim T(\mathcal{V}) = 2$

c. $T(\mathcal{V}) = \mathcal{S}pan\left\{\begin{bmatrix} -3 \\ 0 \\ 1 \end{bmatrix}, \begin{bmatrix} 6 \\ 0 \\ -2 \end{bmatrix}\right\}$, $\dim T(\mathcal{V}) = 1$

2. b. $T(\mathcal{V}) = \mathcal{S}pan\left\{\begin{bmatrix} -1 \\ 3 \\ 1 \end{bmatrix}, \begin{bmatrix} 2 \\ -6 \\ -2 \end{bmatrix}, \begin{bmatrix} 1 \\ -3 \\ -1 \end{bmatrix}\right\},$

$$\begin{bmatrix} -1 & 2 & 1 \\ 3 & -6 & -3 \\ 1 & -2 & -1 \end{bmatrix} \longrightarrow \begin{bmatrix} 1 & -2 & 1 \\ 0 & 0 & 0 \\ 0 & 0 & 0 \end{bmatrix}, \ \dim T(\mathcal{V}) = 1$$

d. $T(\mathcal{V}) = \mathcal{S}pan\left\{\begin{bmatrix} 1 \\ -1 \\ 1 \end{bmatrix}, \begin{bmatrix} 1 \\ 0 \\ 2 \end{bmatrix}, \begin{bmatrix} 1 \\ 1 \\ 3 \end{bmatrix}, \begin{bmatrix} 3 \\ -2 \\ 4 \end{bmatrix}\right\}$,

$$\begin{bmatrix} 1 & 1 & 1 & 3 \\ -1 & 0 & 1 & -2 \\ 1 & 2 & 3 & 4 \end{bmatrix} \longrightarrow \begin{bmatrix} 1 & 0 & -1 & 2 \\ 0 & 1 & 2 & 1 \\ 0 & 0 & 0 & 0 \end{bmatrix}, \quad \dim T(\mathcal{V}) = 2$$

3. a. $\mathcal{S} = \left\{\begin{bmatrix} (1/3)t - 2 \\ t \end{bmatrix} : t \in \mathcal{R}\right\} = \left\{\frac{t}{3}\begin{bmatrix} 1 \\ 3 \end{bmatrix} + \begin{bmatrix} -2 \\ 0 \end{bmatrix} : t \in \mathcal{R}\right\}$

$$= \begin{bmatrix} -2 \\ 0 \end{bmatrix} + \mathcal{S}pan\left\{\begin{bmatrix} 1 \\ 3 \end{bmatrix}\right\}$$

5. a.

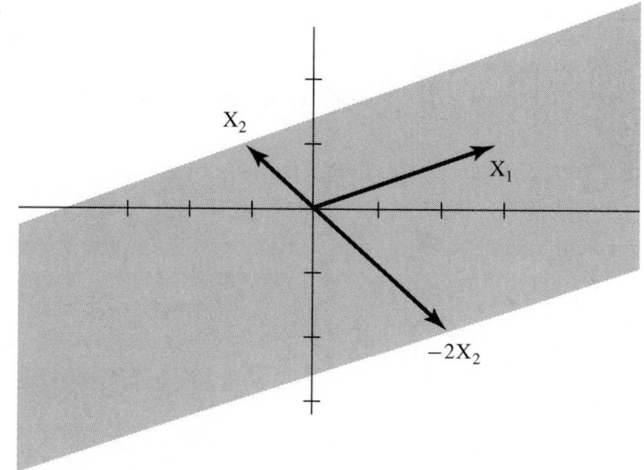

b.

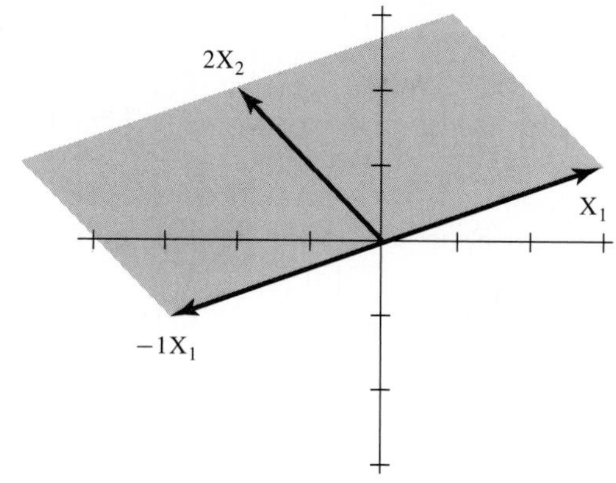

6. a. Let $Y_1 = T(E_1) = \begin{bmatrix} 1 \\ 1 \end{bmatrix}$ and $Y_2 = T(E_2) = \begin{bmatrix} -3 \\ 1 \end{bmatrix}$. Then $T(\mathcal{X}_1) = \mathcal{S}pan\{Y_1\}$, $T(\mathcal{X}_2) = \mathcal{S}pan\{Y_2\}$, $T(Q) = \{x_1Y_1 + x_2Y_2 : x_1, x_2 > 0\}$, and $T(H) = \{x_1Y_1 + x_2Y_2 : x_2 < 0\}$. $\mathcal{U} = \mathcal{S}pan\left\{\begin{bmatrix} 1 \\ 1 \end{bmatrix}\right\} \Rightarrow T(\mathcal{U}) = \mathcal{S}pan\left\{T\left(\begin{bmatrix} 1 \\ 1 \end{bmatrix}\right)\right\} = \mathcal{S}pan\left\{\begin{bmatrix} -2 \\ 2 \end{bmatrix}\right\}$, and

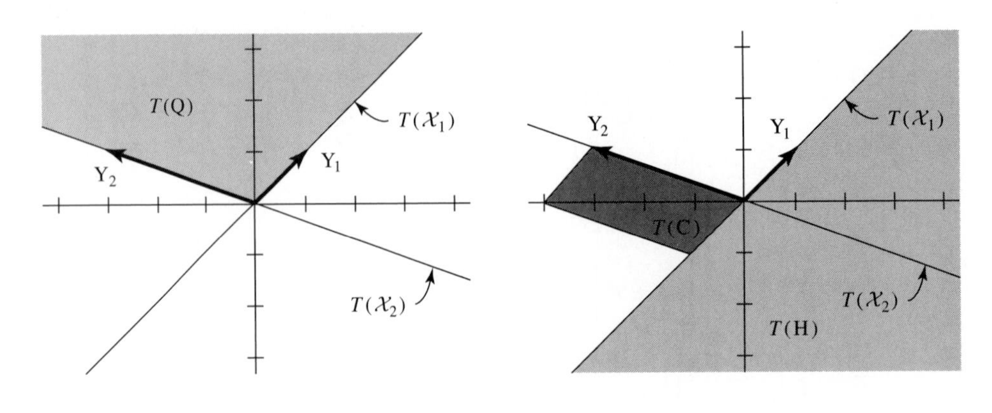

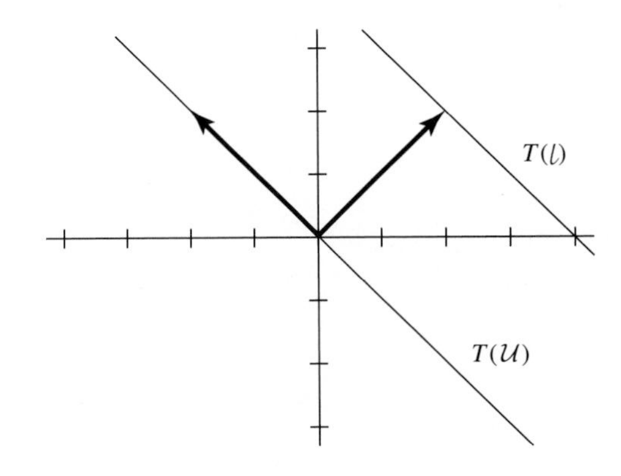

$\ell = \begin{bmatrix} 2 \\ 0 \end{bmatrix} + \mathcal{U} \Rightarrow T(\ell) = T\left(\begin{bmatrix} 2 \\ 0 \end{bmatrix}\right) + T(\mathcal{U}) = \begin{bmatrix} 2 \\ 2 \end{bmatrix} + T(\mathcal{U})$. $T(\text{C}) = \{x_1 Y_1 + x_2 Y_2 :$ $-1 \leq x_1 \leq 0, 0 \leq x_2 \leq 1\}$.

7. a. Let $Y_1 = T(E_1) = \begin{bmatrix} 1 \\ -2 \\ 0 \end{bmatrix}$, $Y_2 = T(E_2) = \begin{bmatrix} 0 \\ 2 \\ 0 \end{bmatrix}$, and $Y_3 = T(E_3) = \begin{bmatrix} -1 \\ 1 \\ 1 \end{bmatrix}$.
$T(\mathcal{X}_k) = \mathcal{S}pan\{Y_k\}$, $k = 1, 2, 3$. $\mathcal{U} = \mathcal{S}pan\{E_1, E_3\} \Rightarrow T(\mathcal{U}) = \mathcal{S}pan\{Y_1, Y_3\}$.
$p = \begin{bmatrix} 0 \\ 1 \\ 0 \end{bmatrix} + \mathcal{U} \Rightarrow T(p) = T\left(\begin{bmatrix} 0 \\ 1 \\ 0 \end{bmatrix}\right) + T(\mathcal{U}) = \begin{bmatrix} 0 \\ 2 \\ 0 \end{bmatrix} + T(\mathcal{U})$. $T(\text{H}) = \{x_1 Y_1 +$ $x_2 Y_2 + x_3 Y_3 : x_2 < 0\}$. $\mathcal{S}pan\{Y_1, Y_3\}$ divides $\mathcal{R}^3$ into two half-spaces, and $T(\text{H})$ is the one containing the negative y_2 axis.

$$T(\mathcal{V}) = \mathcal{S}pan\left\{T\left(\begin{bmatrix} 0 \\ -1 \\ 1 \end{bmatrix}\right)\right\} = \mathcal{S}pan\left\{\begin{bmatrix} -1 \\ -1 \\ 1 \end{bmatrix}\right\},$$

$$T(\ell) = T\left(\begin{bmatrix} 1 \\ 2 \\ 3 \end{bmatrix}\right) + T(\mathcal{V}) = \begin{bmatrix} -2 \\ 5 \\ 3 \end{bmatrix} + T(\mathcal{V}).$$

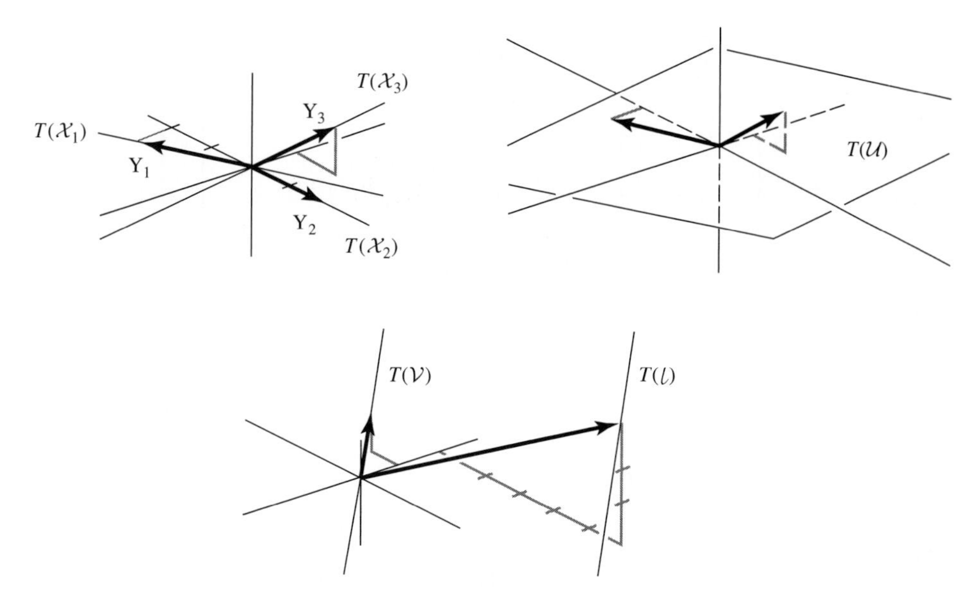

9. Assume $\mathcal{V} \cap \mathcal{K}er(T) \neq \{0\}$ (otherwise the conclusion follows from Problem 8). Suppose $\dim \mathcal{V} = p$, $\{X_1, \ldots, X_q\}$ is a basis for $\mathcal{V} \cap \mathcal{K}er(T)$, and $V_1, \ldots, V_{p-q} \in \mathcal{V}$ such that $\{X_1, \ldots, X_q, V_1, \ldots, V_{p-q}\}$ is a basis for $\mathcal{V}$. Then $T(\mathcal{V}) = \mathcal{S}pan\{T(X_1), \ldots, T(X_q), T(V_1), \ldots, T(V_{p-q})\} = \mathcal{S}pan\{T(V_1), \ldots, T(V_{p-q})\}$. If $0 = \sum_1^{p-q} t_k T(V_k) = T(\sum_1^{p-q} t_k V_k)$, then $\sum_1^{p-q} t_k V_k \in \mathcal{V} \cap \mathcal{K}er(T)$, so there are scalars $s_1, \ldots, s_q$ such that $\sum_1^{p-q} t_k V_k = \sum_1^q s_k X_k$. Then $\sum_1^{p-q} t_k V_k - \sum_1^q s_k X_k = 0$, and therefore the scalars are zeros. In particular, $t_1, \ldots, t_{p-q}$ are 0's, so $\{T(V_1), \ldots, T(V_{p-q})\}$ is linearly independent and, therefore, is a basis for $T(\mathcal{V})$. Thus, $\dim T(\mathcal{V}) = p - q = \dim \mathcal{V} - \dim \mathcal{V} \cap \mathcal{K}er(T)$.

10. **c.** $T(\mathcal{U}) = \mathcal{S}pan\{T(E_1), T(E_2)\} = \mathcal{S}pan\left\{\begin{bmatrix} -1 \\ 2 \\ 0 \end{bmatrix}, \begin{bmatrix} 1 \\ 0 \\ 2 \end{bmatrix}\right\} = \mathcal{R}ng(T)$. Since $T(B) \in \mathcal{R}ng(T)$, $T(B + \mathcal{U}) = T(B) + T(\mathcal{U}) = T(B) + \mathcal{R}ng(T) = \mathcal{R}ng(T)$.

11. **a.** scale change in the E_1 direction by a factor of a
b. scale change in E_1 and E_2 directions by factors of a and b, respectively
c. same as (b) together with a reflection in the x_2 axis
d. scale change in E_1 direction by a factor of a together with projection onto the x_1 axis

12. $\left[\begin{array}{rrr|r} 1 & 1 & 0 & -1 \\ -1 & 1 & -2 & 5 \end{array}\right] \longrightarrow \left[\begin{array}{rrr|r} 1 & 0 & 1 & -3 \\ 0 & 1 & -1 & 2 \end{array}\right]$.

a. $\mathcal{K}er(T) = \mathcal{S}pan\left\{\begin{bmatrix} -1 \\ 1 \\ 1 \end{bmatrix}\right\}$ **c.** $T^{-1}\left(\begin{bmatrix} -1 \\ 5 \end{bmatrix}\right) = \begin{bmatrix} -3 \\ 2 \\ 0 \end{bmatrix} + \mathcal{K}er(T)$

16. $B + \mathcal{V} = \mathcal{V} \Leftrightarrow$ if $B \in \mathcal{V}$.
($\Rightarrow$) Suppose $B + \mathcal{V} = \mathcal{V}$. Since $0 \in \mathcal{V}$, $B = B + 0 \in B + \mathcal{V} = \mathcal{V}$.
($\Leftarrow$) Suppose $B \in \mathcal{V}$. If $X \in B + \mathcal{V}$, then $X = B + V$, where $V \in \mathcal{V}$. Since $\mathcal{V}$ is closed under addition, $X \in \mathcal{V}$. Thus, $B + \mathcal{V} \subseteq \mathcal{V}$. If $X \in \mathcal{V}$, then $X - B \in \mathcal{V}$, so $X = B + (X - B) \in B + \mathcal{V}$. Thus, $\mathcal{V} \subseteq B + \mathcal{V}$.

Section 4.5

1. Set $X_1 = \begin{bmatrix} 2 \\ -1 \end{bmatrix}$, $X_2 = \begin{bmatrix} -1 \\ 2 \end{bmatrix}$, $Y_1 = \begin{bmatrix} 1 \\ 0 \\ -1 \end{bmatrix}$, $Y_2 = \begin{bmatrix} -1 \\ 1 \\ 0 \end{bmatrix}$, and $Y_3 = \begin{bmatrix} -1 \\ 1 \\ -1 \end{bmatrix}$.

a. $$[Y_1 \quad Y_2 \quad Y_3 \mid T(X_1) \quad T(X_2)] \longrightarrow \left[\begin{array}{ccc|cc} 1 & 0 & 0 & -4 & 2 \\ 0 & 1 & 0 & -3 & 3 \\ 0 & 0 & 1 & 4 & -5 \end{array}\right]$$

$$\Longrightarrow [T]_{\mathcal{BA}} = \begin{bmatrix} -4 & 2 \\ -3 & 3 \\ 4 & -5 \end{bmatrix}$$

c. $$[Y_1 \quad Y_2 \quad Y_3 \mid T(E_1) \quad T(E_2)] \longrightarrow \left[\begin{array}{ccc|cc} 1 & 0 & 0 & -2 & 0 \\ 0 & 1 & 0 & -1 & 1 \\ 0 & 0 & 1 & 1 & -2 \end{array}\right]$$

$$\Longrightarrow [T]_{\mathcal{BE}} = \begin{bmatrix} -2 & 0 \\ -1 & 1 \\ 1 & -2 \end{bmatrix}$$

2. b. $$\begin{bmatrix} 3 & 1 & 3 & -1 \\ 2 & -4 & -2 & -4 \\ -2 & -2 & 2 & -2 \end{bmatrix}$$

3. a. $$\begin{bmatrix} 1 & 0 & 1 \\ 0 & 1 & 0 \\ 1 & 0 & 1 \end{bmatrix}$$ **c.** $$\begin{bmatrix} 3 & 2 & 1 \\ -1 & -1 & -1 \\ 3 & 2 & 1 \end{bmatrix}$$

4. b. $$\begin{bmatrix} 2 & -1 & 0 & -1 \\ -1 & 2 & -1 & 0 \\ -1 & -2 & 1 & -1 \\ 1 & 1 & 0 & 1 \end{bmatrix}$$

5. b. $$\begin{bmatrix} 7 & 3 & 8 & 2 \\ 5 & 5 & 9 & 1 \\ -1 & 4 & 4 & -1 \end{bmatrix}$$

6. $$[T]_{\mathcal{E}} = P_{\mathcal{EB}}[T]_{\mathcal{B}}P_{\mathcal{BE}} = \begin{bmatrix} 1 & -1 \\ -3 & 1 \end{bmatrix}\begin{bmatrix} 1 & 0 \\ 0 & 2 \end{bmatrix}\left(-\frac{1}{2}\begin{bmatrix} 1 & 1 \\ 3 & 1 \end{bmatrix}\right) = \frac{1}{2}\begin{bmatrix} 5 & 1 \\ -3 & 1 \end{bmatrix}$$

7. b. $$\left[\begin{array}{cc|c} 1 & -1 & 2 \\ 1 & 1 & 0 \end{array}\right] \longrightarrow \left[\begin{array}{cc|c} 1 & 0 & 1 \\ 0 & 1 & -1 \end{array}\right] \Longrightarrow [X]_{\mathcal{A}}$$

$$= \begin{bmatrix} 1 \\ -1 \end{bmatrix} \Longrightarrow [T(X)]_{\mathcal{B}} = \begin{bmatrix} 1 & -6 \\ -2 & 5 \\ 3 & -4 \end{bmatrix}\begin{bmatrix} 1 \\ -1 \end{bmatrix} = \begin{bmatrix} 7 \\ -7 \\ 7 \end{bmatrix} \Longrightarrow T(X_2) = \begin{bmatrix} 21 \\ -7 \\ 21 \end{bmatrix}$$

9. a. Since $\mathcal{A} = \{X_1, X_2\}$ is linearly independent and T is one-to-one, $\{Y_1, Y_2\} = T(\mathcal{A})$ is linearly independent (Theorem 4.12).

10. a. $T(\mathrm{X}_1) = 0, T(\mathrm{X}_2) = \mathrm{X}_2, T(\mathrm{X}_3) = \mathrm{X}_3 \implies [T]_{\mathcal{B}} = \begin{bmatrix} 0 & 0 & 0 \\ 0 & 1 & 0 \\ 0 & 0 & 1 \end{bmatrix}$

c. $[T]_{\mathcal{BE}} = [T]_{\mathcal{B}}\mathrm{P}_{\mathcal{BE}} = \begin{bmatrix} 0 & 0 & 0 \\ 0 & 1 & 0 \\ 0 & 0 & 1 \end{bmatrix} \left(\frac{1}{3} \begin{bmatrix} 1 & 1 & 1 \\ -1 & 2 & -1 \\ -1 & -1 & 2 \end{bmatrix} \right) = \frac{1}{3} \begin{bmatrix} 0 & 0 & 0 \\ -1 & 2 & -1 \\ -1 & -1 & 2 \end{bmatrix}$

12. a. $[T]_{\mathcal{B}} = \mathrm{P}_{\mathcal{BA}}[T]_{\mathcal{A}}\mathrm{P}_{\mathcal{AB}} = \begin{bmatrix} 0 & 0 & 1 \\ 0 & 1 & 0 \\ 1 & 0 & 0 \end{bmatrix} \begin{bmatrix} 1 & 2 & 3 \\ 4 & 5 & 6 \\ 7 & 8 & 9 \end{bmatrix} \begin{bmatrix} 0 & 0 & 1 \\ 0 & 1 & 0 \\ 1 & 0 & 0 \end{bmatrix} = \begin{bmatrix} 9 & 8 & 7 \\ 6 & 5 & 4 \\ 3 & 2 & 1 \end{bmatrix}$

b. $[T]_{\mathcal{BA}} = \mathrm{P}_{\mathcal{BA}}[T]_{\mathcal{A}} = \begin{bmatrix} 0 & 0 & 1 \\ 0 & 1 & 0 \\ 1 & 0 & 0 \end{bmatrix} \begin{bmatrix} 1 & 2 & 3 \\ 4 & 5 & 6 \\ 7 & 8 & 9 \end{bmatrix} = \begin{bmatrix} 7 & 8 & 9 \\ 4 & 5 & 6 \\ 1 & 2 & 3 \end{bmatrix}$

c. $[T]_{\mathcal{AB}} = [T]_{\mathcal{A}}\mathrm{P}_{\mathcal{AB}} = \begin{bmatrix} 1 & 2 & 3 \\ 4 & 5 & 6 \\ 7 & 8 & 9 \end{bmatrix} \begin{bmatrix} 0 & 0 & 1 \\ 0 & 1 & 0 \\ 1 & 0 & 0 \end{bmatrix} = \begin{bmatrix} 3 & 2 & 1 \\ 6 & 5 & 4 \\ 9 & 8 & 7 \end{bmatrix}$

16. a. $\mathrm{X} \in \mathcal{K}er(T) \iff 0 = T(\mathrm{X}) \iff 0 = [T(\mathrm{X})]_{\mathcal{B}} = [T]_{\mathcal{BA}}[\mathrm{X}]_{\mathcal{A}} \iff [\mathrm{X}]_{\mathcal{A}} \in \mathcal{N}([T]_{\mathcal{BA}})$

Section 4.6

2. a. $\begin{bmatrix} 0 & 0 & -1 \\ 0 & 1 & 0 \\ 1 & 0 & 0 \end{bmatrix}$ **c.** $\begin{bmatrix} 0 & 1 & 0 \\ 1 & 0 & 0 \\ 0 & 0 & 1 \end{bmatrix}$ **e.** $\begin{bmatrix} 0 & 1 & 0 \\ 0 & 0 & -1 \\ 1 & 0 & 0 \end{bmatrix}$ **g.** $\frac{1}{2}\begin{bmatrix} 0 & 1 & 0 \\ -1 & 1 & 0 \\ 0 & 0 & 2 \end{bmatrix}$

3. Let $S(\mathrm{X}) = \mathrm{Proj}_{\mathcal{V}}(\mathrm{X})$. Then $T = 2S - \mathrm{Id} \Rightarrow [\mathrm{T}]_{\mathcal{E}} = 2[S]_{\mathcal{E}} - \mathrm{I}_3$.

a. $\mathrm{U} = \frac{1}{\sqrt{3}}\begin{bmatrix} 1 \\ 1 \\ 1 \end{bmatrix}$ is an orthonormal basis for $\mathcal{V}$. $S(\mathrm{X}) = (\mathrm{X} \cdot \mathrm{U})\mathrm{U} = \frac{x_1 + x_2 + x_3}{3}\begin{bmatrix} 1 \\ 1 \\ 1 \end{bmatrix}$

$$= \frac{1}{3}\begin{bmatrix} 1 & 1 & 1 \\ 1 & 1 & 1 \\ 1 & 1 & 1 \end{bmatrix} \mathrm{X} \Rightarrow [T]_{\mathcal{E}} = \frac{2}{3}\begin{bmatrix} 1 & 1 & 1 \\ 1 & 1 & 1 \\ 1 & 1 & 1 \end{bmatrix} - \begin{bmatrix} 1 & 0 & 0 \\ 0 & 1 & 0 \\ 0 & 0 & 1 \end{bmatrix} = \frac{1}{3}\begin{bmatrix} -1 & 2 & 2 \\ 2 & -1 & 2 \\ 2 & 2 & -1 \end{bmatrix}$$

c. $[T]_{\mathcal{E}} = \frac{1}{3}\begin{bmatrix} 1 & -2 & -2 \\ -2 & 1 & -2 \\ -2 & -2 & 1 \end{bmatrix}$.

4. a. $[T_1 \circ T_2 \circ T_3]_{\mathcal{E}} = [T_1]_{\mathcal{E}}[T_2]_{\mathcal{E}}[T_3]_{\mathcal{E}} = \begin{bmatrix} 1 & 0 & 0 \\ 0 & 1 & 0 \\ 0 & 0 & -1 \end{bmatrix} \begin{bmatrix} -1 & 0 & 0 \\ 0 & 1 & 0 \\ 0 & 0 & 1 \end{bmatrix} \begin{bmatrix} 1 & 0 & 0 \\ 0 & -1 & 0 \\ 0 & 0 & 1 \end{bmatrix} =$

$\begin{bmatrix} -1 & 0 & 0 \\ 0 & -1 & 0 \\ 0 & 0 & -1 \end{bmatrix}$ so $(T_1 \circ T_2 \circ T_3)(\mathrm{X}) = -\mathrm{I}_3\mathrm{X} = \mathrm{Refl}_{\{0\}}(\mathrm{X})$.

7. a. If $X \in \mathcal{V}$, then $\text{Proj}_{\mathcal{V}}(X) = X$, so $\text{Refl}_{\mathcal{V}}(X) = 2\text{Proj}_{\mathcal{V}}(X) - \text{Id}(X) = X$.
b. If $X \in \mathcal{V}^{\perp}$, then $\text{Proj}_{\mathcal{V}}(X) = 0$, so $\text{Refl}_{\mathcal{V}}(X) = 2\text{Proj}_{\mathcal{V}}(X) - \text{Id}(X) = -X$.

8. a. $[T]_{\mathcal{E}} = \begin{bmatrix} 2 & 1 & 3 \\ 1 & 0 & 2 \\ 0 & 1 & 1 \end{bmatrix}$ nonsingular $\Rightarrow T$ is one-to-one. By Theorem 4.18, $[T^{-1}]_{\mathcal{E}} =$

$$[T]_{\mathcal{E}}^{-1} = \frac{1}{2}\begin{bmatrix} 2 & -2 & -2 \\ 1 & -2 & 1 \\ -1 & 2 & 1 \end{bmatrix}.$$

10. a. Let $X \in \mathcal{R}^n$ and suppose $X = V + W$, where $V \in \mathcal{V}$ and $W \in \mathcal{V}^{\perp}$. Then $S^2(X) = \text{Proj}_{\mathcal{V}}(\text{Proj}_{\mathcal{V}}(X)) = \text{Proj}_{\mathcal{V}}(V) = V = \text{Proj}_{\mathcal{V}}(X) = S(X)$.

13. a. Assume $\mathcal{V} \neq \{0\}$ and $\mathcal{V} \neq \mathcal{R}^n$. Let $\mathcal{V}$ and $\mathcal{V}^{\perp}$ have bases $\{X_1, \ldots, X_p\}$ and $\{W_1, \ldots, W_{n-p}\}$, respectively. If $\mathcal{B} = \{X_1, \ldots, X_p, W_1, \ldots, W_{n-p}\}$, then $[S]_{\mathcal{B}} = \left[\begin{array}{c|c} I_p & 0 \\ \hline 0 & -I_{n-p} \end{array}\right]$ and $[T]_{\mathcal{B}} = \left[\begin{array}{c|c} -I_p & 0 \\ \hline 0 & I_{n-p} \end{array}\right]$, so $[S+T]_{\mathcal{B}} = 0$. Thus, $S + T = 0$. When $\mathcal{V} = \{0\}$, $\mathcal{V}^{\perp} = \mathcal{R}^n$, $S(X) = -X$ and $T(X) = X$, so $S + T = 0$. Similarly, $\mathcal{V} = \mathcal{R}^n \Rightarrow \mathcal{V}^{\perp} = \{0\} \Rightarrow S(X) = X$, $T(X) = -X$, and $S + T = 0$.

14. a. $([T]_{\mathcal{CB}}[S]_{\mathcal{BA}})[X]_{\mathcal{A}} = [T]_{\mathcal{CB}}([S]_{\mathcal{BA}}[X]_{\mathcal{A}}) = [T]_{\mathcal{CB}}[S(X)]_{\mathcal{B}} = [T(S(X))]_{\mathcal{C}} = [(T \circ S)(X)]_{\mathcal{C}}$
c. Suppose $Z \in \mathcal{R}^p$. Since T is onto, there is a $Y \in \mathcal{R}^m$ such that $T(Y) = Z$. Since S is onto, there is an $X \in \mathcal{R}^n$ such that $S(X) = Y$. Then $(T \circ S)(X) = T(S(X)) = T(Y) = Z$, so $Z \in \mathcal{Rng}(T \circ S)$.

Section 4.7

1. a. For $X, Y \in \mathcal{R}^2$, $T(X) = \begin{bmatrix} x_1 - x_2 \\ x_1 + x_2 \end{bmatrix}$, $T(Y) = \begin{bmatrix} y_1 - y_2 \\ y_1 + y_2 \end{bmatrix}$, $T(X) \cdot T(Y) = (x_1 - x_2)(y_1 - y_2) + (x_1 + x_2)(y_1 + y_2) = 2(x_1y_1 + x_2y_2) = 2(X \cdot Y)$, $\|T(X)\| = \sqrt{(x_1 - x_2)^2 + (x_1 + x_2)^2} = \sqrt{2(x_1^2 + x_2^2)} = \sqrt{2}\,\|X\|$, and $\|T(Y)\| = \sqrt{2}\,\|Y\|$. Thus

$$\frac{T(X) \cdot T(Y)}{\|T(X)\|\|T(Y)\|} = \frac{2(X \cdot Y)}{(\sqrt{2}\|X\|)(\sqrt{2}\|Y\|)} = \frac{X \cdot Y}{\|X\|\|Y\|},$$

that is, T is a similarity.

b. No, $\|T(X)\| \neq \|X\|$. Alternatively: $[T]_{\mathcal{E}} = \sqrt{2}\begin{bmatrix} 1/\sqrt{2} & -1/\sqrt{2} \\ 1/\sqrt{2} & 1/\sqrt{2} \end{bmatrix}$, and $\begin{bmatrix} 1/\sqrt{2} & -1/\sqrt{2} \\ 1/\sqrt{2} & 1/\sqrt{2} \end{bmatrix}$ is orthogonal.

2. a. $[S \circ T]_{\mathcal{E}} = [S]_{\mathcal{E}}[T]_{\mathcal{E}}$. The product of orthogonal matrices is orthogonal.

4. a. isometry **b.** similarity **c.** neither

7. a. $F(X) = G(T(G^{-1}(X))) = T(G^{-1}(X)) + C = T(X - C) + C = T(X) - T(C) + C$. Thus, $F(X) = T(X) + B$, where $B = C - T(C)$.

b. Same as part (a).

Section 5.2

1. a. (i) $-0 + 3(2-1) - 2(1+2) = -3$
(ii) $-1(0+3) - 2(1+2) + 2(3-0) = -3$
(iii) $1(6-2) - 0 + (-1)(4+3) = -3$

c. (i) $-2(3+4) + 1(-1-6) - 3(2-9) = 0$
(ii) $-2(4+3) - 3(2-9) + (-1)(1+6) = 0$
(iii) $1(-1-6) - 2(3+4) + (-3)(-9+2) = 0$

2. a. -14 **c.** 18 **e.** $-abc$ **g.** 10 **i.** 19 **k.** $-(ad - bc)$

6. a. Expansion by the first row produces $\det(A) = 1(-1)^{p+1}\det\begin{bmatrix} 0 & 1 & & 0 \\ & & \ddots & 1 \\ 1 & & & 0 \end{bmatrix}$, and expanding the last determinant by the first column yields

$$\det(A) = (-1)^{p+1}(-1)^{(p-1)+1}\det\begin{bmatrix} 1 & & 0 \\ & \ddots & \\ 0 & & 1 \end{bmatrix} = (-1)^{2p+1} = -1.$$

b. Assume $E_{ij} \in \mathcal{M}_{n\times n}$ and $i < j$. If $i = 1$ and $j = n$, then E_{ij} has the form of the matrix in part (a), so $\det(E_{ij}) = -1$. If $i > 1$ and/or $j < n$, then a few successive expansions by the first and/or last rows shows that $\det(E_{ij}) = \det(A)$, where A is a $p \times p$ matrix of the form discussed in part (a) $(p = j - i + 1)$. Thus, $\det(E_{ij}) = -1$.

7. a. If $A = \begin{bmatrix} a & b \\ c & d \end{bmatrix}$, then $tA = \begin{bmatrix} ta & tb \\ tc & td \end{bmatrix}$, so $\det(tA) = t^2ad - t^2bc = t^2\det(A)$.

b.
$$\det\begin{bmatrix} ta & tb & tc \\ td & te & tf \\ tg & th & ti \end{bmatrix} = (ta)\det\begin{bmatrix} te & tf \\ th & ti \end{bmatrix} - (tb)\det\begin{bmatrix} td & tf \\ tg & ti \end{bmatrix} + (tc)\det\begin{bmatrix} td & te \\ tg & th \end{bmatrix}$$
$$= (ta)t^2(ei - fh) - (tb)t^2(di - fg) + (tc)t^2(dh - eg)$$
$$= t^3[a(ei - fh) - b(di - fg) + c(dh - eg)]$$
$$= t^3\det\begin{bmatrix} a & b & c \\ d & e & f \\ g & h & i \end{bmatrix}.$$

8.
$$\det\left(\begin{bmatrix} a & b \\ c & d \end{bmatrix}^2\right) = \det\begin{bmatrix} a^2 + bc & ab + bd \\ ca + dc & cb + d^2 \end{bmatrix} = a^2cb + c^2b^2 + a^2d^2 + bcd^2 - a^2bc$$
$$-2abcd - bcd^2$$
$$= a^2d^2 - 2abcd + b^2c^2 = (ad - bc)^2 = \left(\det\begin{bmatrix} a & b \\ c & d \end{bmatrix}\right)^2.$$

Section 5.3

1. a. $\det\begin{bmatrix} 1 & 1 & 0 & -1 \\ 2 & 4 & 1 & -3 \\ -1 & 0 & 2 & 4 \\ 0 & 2 & 3 & 2 \end{bmatrix} = \det\begin{bmatrix} 1 & 1 & 0 & -1 \\ 0 & 2 & 1 & -1 \\ 0 & 1 & 2 & 3 \\ 0 & 2 & 3 & 2 \end{bmatrix} = \det\begin{bmatrix} 2 & 1 & -1 \\ 1 & 2 & 3 \\ 2 & 3 & 2 \end{bmatrix} =$

$-\det\begin{bmatrix} 1 & 2 & 3 \\ 2 & 1 & -1 \\ 2 & 3 & 2 \end{bmatrix} = -\det\begin{bmatrix} 1 & 2 & 3 \\ 0 & -3 & -7 \\ 0 & -1 & -4 \end{bmatrix} = -\det\begin{bmatrix} -3 & -7 \\ -1 & -4 \end{bmatrix} = -5$

c. $\det\begin{bmatrix} 1 & -1 & 1 & -1 \\ -1 & 1 & 1 & -1 \\ 1 & -1 & -1 & -1 \\ -1 & -1 & 1 & 1 \end{bmatrix} = \det\begin{bmatrix} 1 & -1 & 1 & -1 \\ 0 & 0 & 2 & -2 \\ 0 & 0 & -2 & 0 \\ 0 & -2 & 2 & 0 \end{bmatrix} =$

$\det\begin{bmatrix} 0 & 2 & -2 \\ 0 & -2 & 0 \\ -2 & 2 & 0 \end{bmatrix} = -2\det\begin{bmatrix} 2 & -2 \\ -2 & 0 \end{bmatrix} = 8$

3. $\det(tA) = \det\begin{bmatrix} t\text{Row}_1(A) \\ \vdots \\ t\text{Row}_n(A) \end{bmatrix} = t^n \det\begin{bmatrix} \text{Row}_1(A) \\ \vdots \\ \text{Row}_n(A) \end{bmatrix} = t^n\det(A)$ (n applications of Theorem 5.3(2))

5. b. $-\frac{3}{2}$

8. a. Not true. If $A = \begin{bmatrix} 1 & 1 \\ 0 & 1 \end{bmatrix}$ and $B = \begin{bmatrix} -1 & 1 \\ 0 & -1 \end{bmatrix}$, then $A + B = \begin{bmatrix} 0 & 2 \\ 0 & 0 \end{bmatrix}$, and $\det(A + B) = 0 \neq 1 + 1 = \det(A) + \det(B)$.

10. If A is a skew-symmetric $n \times n$ matrix, then $A^T = -A$, so $\det(A) = \det(A^T) = \det(-A) = (-1)^n\det(A)$. When n is odd, it follows that $\det(A) = 0$.

11. a. $F_{ij} = E_{ij} \Rightarrow \det(AF_{ij}) = \det(A)\det(F_{ij}) = \det(A)\det(E_{ij}) = -\det(A)$.
c. $F_{ij}(t) = E_{ji}(t) \Rightarrow \det(AF_{ij}(t)) = \det(A)\det(E_{ji}(t)) = \det(A)$.

15. If $X, Y \in \mathcal{R}^n$ and $s, t \in \mathcal{R}$, then expansion by the jth column produces

$$\begin{aligned} T(sX + tY) &= \det(A_j(sX + tY)) \\ &= \sum_{k=1}^{n} [A_j(sX + tY)]_{kj}(-1)^{k+j}\det(M_{kj}(A_j(sX + tY))) \\ &= \sum_{k=1}^{n} [sx_k + ty_k](-1)^{k+j}\det(M_{kj}(A)) \\ &= s\sum_{k=1}^{n} x_k(-1)^{k+j}\det(M_{kj}(A)) + t\sum_{k=1}^{n} y_k(-1)^{k+j}\det(M_{kj}(A)) \end{aligned}$$

$$= s\sum_{k=1}^{n} [\mathrm{A}_j(\mathrm{X})]_{kj}(-1)^{k+j}\det(\mathrm{M}_{kj}(\mathrm{A}_j(\mathrm{X})))$$

$$+\, t\sum_{k=1}^{n} [\mathrm{A}_j(\mathrm{Y})]_{kj}(-1)^{k+j}\det(\mathrm{M}_{kj}(\mathrm{A}_j(\mathrm{Y}))$$

$$= s\det(\mathrm{A}_j(\mathrm{X})) + t\det(\mathrm{A}_j(\mathrm{Y})) = sT(\mathrm{X}) + tT(\mathrm{Y}).$$

Section 5.4

1. a. $\mathrm{Adj}(\mathrm{A}) = \begin{bmatrix} 3 & 3 & -3 \\ 2 & -3 & 1 \\ -4 & 0 & 1 \end{bmatrix}$, $\det(\mathrm{A}) = -3$, $\mathrm{A}^{-1} = -\frac{1}{3}\mathrm{Adj}(\mathrm{A})$

c. $\mathrm{Adj}(\mathrm{A}) = \begin{bmatrix} -3 & 6 & -3 \\ 6 & -12 & 6 \\ -3 & 6 & -3 \end{bmatrix}$, $\det(\mathrm{A}) = 0$

e. $\mathrm{Adj}(\mathrm{A}) = \begin{bmatrix} 24 & 0 & -8 & 0 \\ 0 & 12 & 0 & -3 \\ 0 & 0 & 8 & 0 \\ 0 & 0 & 0 & 6 \end{bmatrix}$, $\det(\mathrm{A}) = 24$, $\mathrm{A}^{-1} = -\frac{1}{24}\mathrm{Adj}(\mathrm{A})$

g. $\mathrm{Adj}(\mathrm{A}) = \begin{bmatrix} -1 & -1 & -1 & 2 \\ -1 & -1 & 2 & -1 \\ -1 & 2 & -1 & -1 \\ 2 & -1 & -1 & -1 \end{bmatrix}$, $\det(\mathrm{A}) = -3$, $\mathrm{A}^{-1} = -\frac{1}{3}\mathrm{Adj}(\mathrm{A})$

3. Assume $\det(\mathrm{A}) = 0$. If $\mathrm{A} = 0$, then $\mathrm{Adj}(\mathrm{A}) = 0$, so assume $\mathrm{A} \neq 0$. If Adj(A) were nonsingular, then $\mathrm{A} = (\mathrm{A}\,\mathrm{Adj}(\mathrm{A}))[\mathrm{Adj}(\mathrm{A})]^{-1} = (\det(\mathrm{A})\mathrm{I}_n)[\mathrm{Adj}(\mathrm{A})]^{-1} = 0$, a contradiction. Thus, Adj(A) is singular.

6. $\mathrm{Adj}(\mathrm{A}^{-1}) = \det(\mathrm{A}^{-1})(\mathrm{A}^{-1})^{-1} = \dfrac{1}{\det(\mathrm{A})}\mathrm{A} = [\det(\mathrm{A})\mathrm{A}^{-1}]^{-1} = [\mathrm{Adj}(\mathrm{A})]^{-1}$

8. a. Claim: $\mathrm{Adj}(\mathrm{A}^T) = [\mathrm{Adj}(\mathrm{A})]^T$. First observe that $\mathrm{M}_{ji}(\mathrm{A}^T) = \mathrm{M}_{ij}(\mathrm{A})^T$. Then

$$[\mathrm{Adj}(\mathrm{A}^T)]_{ij} = \mathrm{cof}_{ji}(\mathrm{A}^T) = (-1)^{j+i}\det(\mathrm{M}_{ji}(\mathrm{A}^T)) = (-1)^{i+j}\det(\mathrm{M}_{ij}(\mathrm{A})^T)$$
$$= (-1)^{i+j}\det(\mathrm{M}_{ij}(\mathrm{A})) = \mathrm{cof}_{ij}(\mathrm{A}) = [\mathrm{Adj}(\mathrm{A})]_{ji} = [\mathrm{Adj}(\mathrm{A})^T]_{ij}.$$

10. a. $\det(\mathrm{A}) = -11$, $\det(\mathrm{A}_1) = \det\begin{bmatrix} 1 & 5 \\ 2 & -3 \end{bmatrix} = -13$, and $\det(\mathrm{A}_2) = \det\begin{bmatrix} 2 & 1 \\ 1 & 2 \end{bmatrix} = 3$.

Thus, $x_1 = 13/11$, and $x_2 = -3/11$.

Section 5.5

1. a. negative **c.** positive **e.** negative

3. a. $\alpha = \pi/2$, and $T(\mathrm{X}_1) = \begin{bmatrix} 1 & 0 & 0 \\ 0 & 0 & -1 \\ 0 & 1 & 0 \end{bmatrix} \begin{bmatrix} 1 \\ 1 \\ 0 \end{bmatrix} = \begin{bmatrix} 1 \\ 0 \\ 1 \end{bmatrix}$. $\beta = \pi/4$, and

$\mathrm{W}_1 = S(T(\mathrm{X}_1)) = \frac{1}{2}\begin{bmatrix} \sqrt{2} & 0 & \sqrt{2} \\ 0 & 2 & 0 \\ -\sqrt{2} & 0 & \sqrt{2} \end{bmatrix} \begin{bmatrix} 1 \\ 0 \\ 1 \end{bmatrix} = \begin{bmatrix} \sqrt{2} \\ 0 \\ 0 \end{bmatrix}$. $\mathrm{W}_2 = S(T(\mathrm{X}_2)) =$

$\frac{1}{2}\begin{bmatrix} \sqrt{2} & 0 & \sqrt{2} \\ 0 & 2 & 0 \\ -\sqrt{2} & 0 & \sqrt{2} \end{bmatrix} \begin{bmatrix} 1 & 0 & 0 \\ 0 & 0 & -1 \\ 0 & 1 & 0 \end{bmatrix} \begin{bmatrix} 1 \\ -1 \\ \sqrt{2} \end{bmatrix} = \begin{bmatrix} 0 \\ -\sqrt{2} \\ -\sqrt{2} \end{bmatrix}$. The angle of rotation for R

is therefore $3\pi/4$, and $R(\mathrm{W}_2) = \frac{1}{2}\begin{bmatrix} 2 & 0 & 0 \\ 0 & -\sqrt{2} & -\sqrt{2} \\ 0 & \sqrt{2} & -\sqrt{2} \end{bmatrix} \begin{bmatrix} 0 \\ -\sqrt{2} \\ -\sqrt{2} \end{bmatrix} = \begin{bmatrix} 0 \\ 2 \\ 0 \end{bmatrix}$. Finally,

$L = R \circ S \circ T$.

b. $L(\mathrm{X}_3) = \frac{1}{2}\begin{bmatrix} 2 & 0 & 0 \\ 0 & -\sqrt{2} & -\sqrt{2} \\ 0 & \sqrt{2} & -\sqrt{2} \end{bmatrix} \frac{1}{2}\begin{bmatrix} \sqrt{2} & 0 & \sqrt{2} \\ 0 & 2 & 0 \\ -\sqrt{2} & 0 & \sqrt{2} \end{bmatrix} \begin{bmatrix} 1 & 0 & 0 \\ 0 & 0 & -1 \\ 0 & 1 & 0 \end{bmatrix} \begin{bmatrix} 0 \\ \sqrt{2} \\ 1 \end{bmatrix} =$

$\begin{bmatrix} 1 \\ 0 \\ -\sqrt{2} \end{bmatrix}$. Since the third coordinate of $L(\mathrm{X}_3)$ is negative, $\mathcal{B}$ is negatively oriented.

4. b.

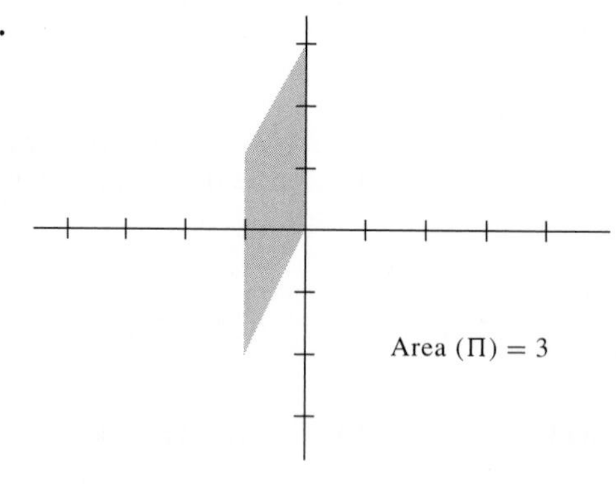

5. a. 4 **c.** 0

6. a. 33

8. Observe that $\mathrm{Vol}(\Pi) = \left| \det \begin{bmatrix} 1 & -1 & -1 \\ -1 & 1 & -1 \\ -1 & -1 & 1 \end{bmatrix} \right| = 4.$

a. Since $\det(A) = 3$, $\mathrm{Vol}(T(\Pi)) = 3 \cdot 4 = 12$
c. Since $\det(A) = 6$, $\mathrm{Vol}(T(\Pi)) = 6 \cdot 4 = 24$

10. $[X_{k_1} \ \cdots \ X_{k_n}]$ can be obtained from $[X_1 \ \cdots \ X_n]$ by a finite number of column interchanges. Thus, $\det([X_{k_1} \ \cdots \ X_{k_n}]) = \pm\det([X_1 \ \cdots \ X_n])$, and therefore $|\det([X_{k_1} \ \cdots \ X_{k_n}])| = |\det([X_1 \ \cdots \ X_n])|$.

12. a. $[T]_{\mathcal{E}} = \begin{bmatrix} 0 & 1 \\ 1 & 0 \end{bmatrix}$, $\det([T]_{\mathcal{E}}) = -1$, T is orientation reversing.

13. b. $[T]_{\mathcal{E}} = \begin{bmatrix} 1 & 0 & 0 \\ 0 & 1 & 0 \\ 0 & 0 & -1 \end{bmatrix}$, $\det([T]_{\mathcal{E}}) = -1$, T is orientation reversing.

15. b. $\det([tT]_{\mathcal{E}}) = \det(t[T]_{\mathcal{E}}) = t^n\det([T]_{\mathcal{E}})$. If $t > 0$ or n is even, then tT is orientation preserving.

17. a. Yes! $Y_2 = X_2 - \mathrm{Proj}_{X_1}(X_2)$

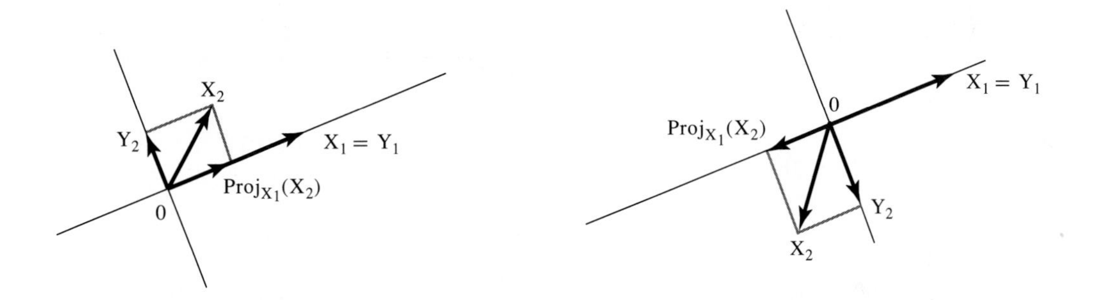

b. $Y_1 = X_1$ and $Y_2 = X_2 - \dfrac{X_2 \cdot X_1}{\|X_1\|^2}X_1 \Rightarrow [Y_1 \ \ Y_2] = [X_1 \ \ X_2]\begin{bmatrix} 1 & t \\ 0 & 1 \end{bmatrix}$, where $t = -\dfrac{X_2 \cdot X_1}{\|X_1\|^2}$.

Section 6.2

1. a. $\mathcal{V}(1) = \mathcal{V}$, $\mathcal{V}(-1) = \mathcal{V}^{\perp}$, $\mathcal{V}^{\perp} = \mathcal{N}([1 \ \ 2 \ \ 1]) = \mathcal{S}pan\left\{\begin{bmatrix} -2 \\ 1 \\ 0 \end{bmatrix}, \begin{bmatrix} -1 \\ 0 \\ 1 \end{bmatrix}\right\}$

2. b. $(tT)(X) = tT(X) = t(\lambda X) = (t\lambda)X$. The eigenvalue is $t\lambda$.

6. a. $AE_1 = E_1, AE_2 = 2E_2, AE_3 = 3E_3$; $\lambda_1 = 1, \lambda_2 = 2, \lambda_3 = 3$

7. $AX = \lambda X \Rightarrow A^2X = A(AX) = A(\lambda X) = \lambda(AX) = \lambda(\lambda X) = \lambda^2 X$. The eigenvalue is λ^2.

9. a. $0 = \det(A^k) = [\det(A)]^k \Rightarrow \det(A) = 0 \Rightarrow A$ is singular.
b. $AX = \lambda X, X \neq 0 \Rightarrow A^k X = \lambda^k X$. But $A^k X = 0X = 0$, so $\lambda^k = 0$, that is, $\lambda = 0$.

11. $T(\mathcal{K}er\,(T)) = \{0\} \subseteq \mathcal{K}er(T)$
$T(\mathcal{R}ng(T)) = \{T(X) : X \in \mathcal{R}ng(T)\} \subseteq \{T(X) : X \in \mathcal{R}^n\} = \mathcal{R}ng(T)$

13. a. $T = 2S - \text{Id}$. Thus, $S(X) = \lambda X \iff 2S(X) - \text{Id}(X) = 2\lambda X - X = (2\lambda - 1)X \iff T(X) = (2\lambda - 1)X$.
c. S has eigenvalues 1 and 0, so by part (a), T has eigenvalues $2 \cdot 1 - 1 = 1$ and $2 \cdot 0 - 1 = -1$. The eigenspaces of T associated with 1 and -1 are the same as the eigenspaces of S associated with 1 and 0, respectively.

14. a. $CX = \left[\begin{array}{c|c} A & 0 \\ \hline 0 & B \end{array}\right]\left[\begin{array}{c} U \\ \hline V \end{array}\right] = \left[\begin{array}{c} AU \\ \hline BV \end{array}\right]$ and $\lambda X = \left[\begin{array}{c} \lambda U \\ \hline \lambda V \end{array}\right]$, so $CX = \lambda X$ if and only if

$AU = \lambda U$ and $BV = \lambda V$.

Section 6.3

1. a. $c(\lambda) = (\lambda - 5)(\lambda + 1)$, $\mathcal{V}(5) = \mathcal{S}pan\left\{\begin{bmatrix} 1 \\ 2 \end{bmatrix}\right\}$, $\mathcal{V}(-1) = \mathcal{S}pan\left\{\begin{bmatrix} -1 \\ 1 \end{bmatrix}\right\}$

c. $c(\lambda) = (\lambda - 4)(\lambda - 3)$, $\mathcal{V}(4) = \mathcal{S}pan\left\{\begin{bmatrix} 3 \\ 2 \end{bmatrix}\right\}$, $\mathcal{V}(3) = \mathcal{S}pan\left\{\begin{bmatrix} 1 \\ 1 \end{bmatrix}\right\}$

e. $c(\lambda) = (\lambda - 1)(\lambda - 2)(\lambda + 1)$,

$$\mathcal{V}(1) = \mathcal{S}pan\left\{\begin{bmatrix} 1 \\ 2 \\ 2 \end{bmatrix}\right\}, \mathcal{V}(2) = \mathcal{S}pan\left\{\begin{bmatrix} 1 \\ 1 \\ 0 \end{bmatrix}\right\}, \mathcal{V}(-1) = \mathcal{S}pan\left\{\begin{bmatrix} -1 \\ 0 \\ 1 \end{bmatrix}\right\}$$

g. $c(\lambda) = (\lambda - 1)^2(\lambda - 3)$, $\mathcal{V}(1) = \mathcal{S}pan\left\{\begin{bmatrix} 2 \\ 1 \\ 0 \end{bmatrix}, \begin{bmatrix} 1 \\ 0 \\ 1 \end{bmatrix}\right\}$, $\mathcal{V}(3) = \mathcal{S}pan\left\{\begin{bmatrix} 1 \\ 0 \\ 2 \end{bmatrix}\right\}$

i. $c(\lambda) = (\lambda - 2)(\lambda - 3)^2$, $\mathcal{V}(2) = \mathcal{S}pan\left\{\begin{bmatrix} 1 \\ 0 \\ 0 \end{bmatrix}\right\}$, $\mathcal{V}(3) = \mathcal{S}pan\left\{\begin{bmatrix} 1 \\ 1 \\ 0 \end{bmatrix}, \begin{bmatrix} 0 \\ 0 \\ 1 \end{bmatrix}\right\}$

k. $c(\lambda) = (\lambda - 2)(\lambda - 1)(\lambda^2 + \lambda + 1)$, $\mathcal{V}(2) = \mathcal{S}pan\left\{\begin{bmatrix} 1 \\ 0 \\ 0 \\ 0 \end{bmatrix}\right\}$,

$$\mathcal{V}(1) = \mathcal{S}pan\left\{\begin{bmatrix} 0 \\ 1 \\ -1 \\ 1 \end{bmatrix}\right\}$$

4. a. $c(\lambda) = (\lambda - a)(\lambda - b)(\lambda - c)$

b. the diagonal entries of the matrix

7. $c_B(\lambda^2) = \det(\lambda^2 I - A^2) = \det[(\lambda I - A)(\lambda I + A)] = \det(\lambda I - A)\det(\lambda I + A)$
$= c_A(\lambda)(-1)^n \det(-\lambda I - A) = (-1)^n c_A(\lambda) c_A(-\lambda)$.

9. a. T and $[T]_{\mathcal{B}}$ have the same eigenvalues, namely, 1, 2, and -1. $\mathcal{V}(1) = \mathcal{S}pan\{X\}$, where

$$[X]_{\mathcal{B}} = \begin{bmatrix} 1 \\ 2 \\ 2 \end{bmatrix}, \text{ that is, } X = 1\begin{bmatrix} 1 \\ 1 \\ 0 \end{bmatrix} + 2\begin{bmatrix} 0 \\ 1 \\ 1 \end{bmatrix} + 2\begin{bmatrix} 1 \\ 0 \\ 1 \end{bmatrix} = \begin{bmatrix} 3 \\ 3 \\ 4 \end{bmatrix}. \text{ Similarly,}$$

$$\mathcal{V}(2) = \mathcal{S}pan\left\{\begin{bmatrix} 1 \\ 2 \\ 1 \end{bmatrix}\right\}, \text{ and } \mathcal{V}(-1) = \mathcal{S}pan\left\{\begin{bmatrix} 0 \\ -1 \\ 1 \end{bmatrix}\right\}.$$

11. $$\lambda I - A = \begin{bmatrix} \lambda & -1 & 0 & 0 & \cdots & 0 & 0 \\ 0 & \lambda & -1 & 0 & & & 0 \\ \vdots & & & & & & \vdots \\ 0 & 0 & 0 & & & \lambda & -1 \\ a_0 & a_1 & a_2 & a_3 & \cdots & a_{n-2} & \lambda + a_{n-1} \end{bmatrix}$$

Expanding $\det(\lambda I - A)$ by the last row yields

$$a_0(-1)^{n+1}\det(M_{n1}) + \sum_{2}^{n-1} a_{k-1}(-1)^{n+k}\det(M_{nk}) + (\lambda + a_{n-1})(-1)^{n+n}\det(M_{nn}). \quad (*)$$

M_{n1} is a lower triangular $(n-1) \times (n-1)$ matrix whose diagonal entries are all -1's, so $\det(M_{n1}) = (-1)^{n-1}$. The first term in $(*)$ is therefore $a_0(-1)^{n+1}(-1)^{n-1} = a_0$. M_{nn} is an upper triangular $(n-1) \times (n-1)$ matrix whose diagonal entries are all λ's, so $\det(M_{nn}) = \lambda^{n-1}$. Thus, the last term in $(*)$ is $(\lambda + a_{n-1})(-1)^{n+n}\lambda^{n-1} = \lambda^n + a_{n-1}\lambda^{n-1}$. For $k = 2, \ldots, n-1$,

$$M_{nk} = \left[\begin{array}{c|c} U_k & 0 \\ \hline 0 & L_k \end{array}\right],$$

where U_k and L_k are upper and lower triangular matrices of size $(k-1) \times (k-1)$ and $(n-k) \times (n-k)$, respectively. Moreover,

$$U_k = \begin{bmatrix} \lambda & -1 & 0 \\ & \ddots & -1 \\ 0 & & \lambda \end{bmatrix} \quad \text{and} \quad L_k = \begin{bmatrix} -1 & & 0 \\ \lambda & \ddots & \\ 0 & \lambda & -1 \end{bmatrix},$$

so $\det(M_{nk}) = \det(U_k)\det(L_k) = \lambda^{k-1}(-1)^{n-k}$. Thus,

$$\sum_{2}^{n-1} a_{k-1}(-1)^{n+k}\det(M_{nk}) = \sum_{2}^{n-1} a_{k-1}(-1)^{n+k}\lambda^{k-1}(-1)^{n-k} = \sum_{2}^{n-1} a_{k-1}\lambda^{k-1}.$$

Section 6.4

1. **a.** $c(\lambda) = (\lambda - 3)(\lambda - 4)$, $P = \begin{bmatrix} -2 & -1 \\ 3 & 1 \end{bmatrix}$, $D = \begin{bmatrix} 3 & 0 \\ 0 & 4 \end{bmatrix}$

c. $c(\lambda) = (\lambda + 1)^2$, $\mathcal{V}(-1) = \mathcal{S}pan\left\{\begin{bmatrix} 5 \\ 2 \end{bmatrix}\right\}$, A is not diagonalizable

e. $c(\lambda) = (\lambda + 1)(\lambda - 1)(\lambda - 2)$, $P = \begin{bmatrix} -1 & -5 & -1 \\ 0 & 2 & 3 \\ 1 & 3 & 1 \end{bmatrix}$, $D = \begin{bmatrix} -1 & 0 & 0 \\ 0 & 1 & 0 \\ 0 & 0 & 2 \end{bmatrix}$

g. $c(\lambda) = (\lambda - 5)^2(\lambda - 1)$, $\mathcal{V}(5) = \mathcal{S}pan\left\{\begin{bmatrix} -2 \\ 1 \\ 0 \end{bmatrix}\right\}$, A is not diagonalizable

i. $c(\lambda) = (\lambda - 2)^2(\lambda - 1)$, $P = \begin{bmatrix} -3 & -1 & -1 \\ 0 & 1 & -1 \\ 1 & 0 & 1 \end{bmatrix}$, $D = \begin{bmatrix} 2 & 0 & 0 \\ 0 & 2 & 0 \\ 0 & 0 & 1 \end{bmatrix}$

k. $c(\lambda) = \lambda^2(\lambda^2 + 4)$, A is not diagonalizable

m. $c(\lambda) = (\lambda - 2)^2(\lambda - 3)^3$, $\mathcal{V}(3) = \mathcal{S}pan\left\{\begin{bmatrix} 1 \\ 0 \\ 1 \\ 0 \\ 0 \end{bmatrix}, \begin{bmatrix} 0 \\ 1 \\ 0 \\ 1 \\ 0 \end{bmatrix}\right\}$, A is not diagonalizable

o. $c(\lambda) = (\lambda - 10)(\lambda - 3)^3(\lambda + 1)$, $P = \begin{bmatrix} -1 & 2 & 0 & 0 & 0 \\ 3 & 1 & 0 & 0 & 0 \\ 0 & 0 & 1 & 0 & 0 \\ 0 & 0 & 0 & 0 & -2 \\ 0 & 0 & 0 & 1 & 1 \end{bmatrix}$,

$$D = \begin{bmatrix} 10 & 0 & 0 & 0 & 0 \\ 0 & 3 & 0 & 0 & 0 \\ 0 & 0 & 3 & 0 & 0 \\ 0 & 0 & 0 & 3 & 0 \\ 0 & 0 & 0 & 0 & -1 \end{bmatrix}$$

2. **a.** $\dim \mathcal{N}(2I - A) = 3$, $\dim \mathcal{N}(2I - B) = 2$, $\dim \mathcal{N}(2I - C) = 1$

4. Assume P is nonsingular, D is diagonal, and $P^{-1}AP = D$.
a. $D = D^T = (P^{-1}AP)^T = P^TA^T(P^{-1})^T = P^TA^T(P^T)^{-1} = Q^{-1}A^TQ$, where $Q = (P^T)^{-1}$

7. Observe that for each $X \in \mathcal{R}^n$, $AX = (UV^T)X = U(V^TX) = U(V \cdot X) = (X \cdot V)U$.
a. Setting $X = U$ gives $AU = (U \cdot V)U$, that is, $AU = \lambda U$, where $\lambda = U \cdot V$. Since $[A]_{ii} = u_i v_i$, trace(A) $= \sum_1^n [A]_{ii} = \sum_1^n u_i v_i = U \cdot V$.
c. If $U \cdot V \neq 0$, then A has two distinct eigenvalues, namely, $U \cdot V$ and 0. Since $\dim \mathcal{V}(0) = n - 1$, $\dim \mathcal{V}(U \cdot V)$ must be 1, and by Theorem 6.10, A is diagonalizable.

Section 6.5

1. There are invertible matrices, P and Q, such that $B = P^{-1}AP$ and $C = Q^{-1}BQ$. Thus, $C = Q^{-1}(P^{-1}AP)Q = (Q^{-1}P^{-1})A(PQ) = (PQ)^{-1}A(PQ)$.

3. $\det(\lambda I - A) = \det\begin{bmatrix} \lambda & -1 \\ -1 & \lambda - 1 \end{bmatrix} = \lambda^2 - \lambda - 1 = 0 \Rightarrow \lambda = (1 \pm \sqrt{5})/2$. If $\mu = (1 - \sqrt{5})/2$ and $\nu = (1 + \sqrt{5})/2$, then $\dim \mathcal{V}(\mu) = 1 = \dim \mathcal{V}(\nu)$. Since $\mu + \nu = 1$ and $\mu\nu = -1$, $\begin{bmatrix} 0 & 1 \\ 1 & 1 \end{bmatrix}\begin{bmatrix} 1 \\ \mu \end{bmatrix} = \begin{bmatrix} \mu \\ 1 + \mu \end{bmatrix} = \begin{bmatrix} \mu \\ -\mu\nu + \mu \end{bmatrix} = \mu\begin{bmatrix} 1 \\ 1 - \nu \end{bmatrix} = \mu\begin{bmatrix} 1 \\ \mu \end{bmatrix}$. Thus, $\mathcal{V}(\mu) = \mathcal{S}pan\left\{\begin{bmatrix} 1 \\ \mu \end{bmatrix}\right\}$. Similarly, $\mathcal{V}(\nu) = \mathcal{S}pan\left\{\begin{bmatrix} 1 \\ \nu \end{bmatrix}\right\}$.

4. a. $P = \frac{1}{\sqrt{2}}\begin{bmatrix} 1 & -1 \\ 1 & 1 \end{bmatrix}, D = \begin{bmatrix} 1 & 0 \\ 0 & -1 \end{bmatrix}$ **c.** $P = \frac{1}{\sqrt{5}}\begin{bmatrix} 1 & -2 \\ 2 & 1 \end{bmatrix}, D = \begin{bmatrix} 3 & 0 \\ 0 & -2 \end{bmatrix}$

e. $c(\lambda) = (\lambda + 1)^2(\lambda - 8)$, $P = \begin{bmatrix} 2/3 & -1/\sqrt{2} & -1/\sqrt{18} \\ 1/3 & 0 & 4/\sqrt{18} \\ 2/3 & 1/\sqrt{2} & -1/\sqrt{18} \end{bmatrix}, D = \begin{bmatrix} 8 & 0 & 0 \\ 0 & -1 & 0 \\ 0 & 0 & -1 \end{bmatrix}$

g. $c(\lambda) = (\lambda + 2)(\lambda - 6)^2(\lambda - 4)$, $P = \frac{1}{\sqrt{2}}\begin{bmatrix} 0 & 0 & 1 & -1 \\ 1 & -1 & 0 & 0 \\ 0 & 0 & 1 & 1 \\ 1 & 1 & 0 & 0 \end{bmatrix}, D = \begin{bmatrix} -2 & 0 & 0 & 0 \\ 0 & 6 & 0 & 0 \\ 0 & 0 & 6 & 0 \\ 0 & 0 & 0 & 4 \end{bmatrix}$

i. $c(\lambda) = (\lambda - 3)^2(\lambda + 1)(\lambda - 1)$, $P = \frac{1}{\sqrt{2}}\begin{bmatrix} 1 & 0 & -1 & 0 \\ 0 & 1 & 0 & -1 \\ 1 & 0 & 1 & 0 \\ 0 & 1 & 0 & 1 \end{bmatrix}, D = \begin{bmatrix} 3 & 0 & 0 & 0 \\ 0 & 3 & 0 & 0 \\ 0 & 0 & -1 & 0 \\ 0 & 0 & 0 & 1 \end{bmatrix}$

5. A is similar to a diagonal matrix D, whose diagonal entries are the eigenvalues of A.
a. $\det(A) = \det(D) =$ the product of the eigenvalues
b. $\operatorname{trace}(A) = \operatorname{trace}(D) =$ the sum of the eigenvalues

7. a. Assume $B = P^{-1}AP$. By Example 6.20, $B^k = P^{-1}A^kP$, $k \geq 1$. Thus,

$$\begin{aligned} g(B) &= a_0I_n + a_1B + a_2B^2 + \cdots + a_qB^q \\ &= a_0(P^{-1}I_nP) + a_1(P^{-1}AP) + a_2(P^{-1}A^2P) + \cdots + a_q(P^{-1}A^qP) \\ &= P^{-1}(a_0I_n + a_1A + a_2A^2 + \cdots + a_qA^q)P \\ &= P^{-1}g(A)P. \end{aligned}$$

9. a. In the terminology of Example 6.21, $X_0 = \begin{bmatrix} 1 \\ 1 \end{bmatrix}$, and $X_k = A^kX_0$, where $A = \begin{bmatrix} 0 & 1 \\ 2 & 1 \end{bmatrix}$.

$\det(\lambda I - A) = \det\begin{bmatrix} \lambda & -1 \\ -2 & \lambda - 1 \end{bmatrix} = \lambda^2 - \lambda - 2 = (\lambda - 2)(\lambda + 1)$. $\mathcal{V}(2) = \mathcal{S}pan\left\{\begin{bmatrix} 1 \\ 2 \end{bmatrix}\right\}$,

$\mathcal{V}(-1) = \mathcal{S}pan\left\{\begin{bmatrix} -1 \\ 1 \end{bmatrix}\right\}$, and $\mathrm{P} = \begin{bmatrix} 1 & -1 \\ 2 & 1 \end{bmatrix}$, so $\mathrm{Y}_0 = \mathrm{P}^{-1}\mathrm{X}_0 = \dfrac{1}{3}\begin{bmatrix} 1 & 1 \\ -2 & 1 \end{bmatrix}\begin{bmatrix} 1 \\ 1 \end{bmatrix} = \begin{bmatrix} 2/3 \\ -1/3 \end{bmatrix}$. $\mathrm{X}_k = \dfrac{2}{3}2^k\begin{bmatrix} 1 \\ 2 \end{bmatrix} - \dfrac{1}{3}(-1)^k\begin{bmatrix} -1 \\ 1 \end{bmatrix}$, $a_k = \dfrac{2^{k+1} - (-1)^{k+1}}{3}$.

Section 6.6

1. a. $\begin{bmatrix} 2 & -1 \\ -1 & -3 \end{bmatrix}$ **c.** $\begin{bmatrix} 1 & 1/2 & -3/2 \\ 1/2 & -1 & -1 \\ -3/2 & -1 & 2 \end{bmatrix}$ **e.** $\begin{bmatrix} 2 & -1 & -1 \\ -1 & 2 & -1 \\ -1 & -1 & 2 \end{bmatrix}$ **g.** $\begin{bmatrix} a & 0 & 0 \\ 0 & b & 0 \\ 0 & 0 & c \end{bmatrix}$.

2. a. $\lambda = 10, 20$, $\mathcal{A} = \left\{\dfrac{1}{\sqrt{10}}\begin{bmatrix} -3 \\ 1 \end{bmatrix}, \dfrac{1}{\sqrt{10}}\begin{bmatrix} 1 \\ 3 \end{bmatrix}\right\}$, $Q(\mathrm{X}) = 10y_1^2 + 20y_2^2$, $\mathrm{Y} = [\mathrm{X}]_{\mathcal{A}}$

c. $\lambda = 4{,}2$, $\mathcal{A} = \left\{\dfrac{1}{\sqrt{2}}\begin{bmatrix} -1 \\ 0 \\ 1 \end{bmatrix}, \begin{bmatrix} 0 \\ 1 \\ 0 \end{bmatrix}, \dfrac{1}{\sqrt{2}}\begin{bmatrix} 1 \\ 0 \\ 1 \end{bmatrix}\right\}$, $Q(\mathrm{X}) = 4y_1^2 + 4y_2^2 + 2y_3^2$, $\mathrm{Y} = [\mathrm{X}]_{\mathcal{A}}$

e. $\lambda = 0, \pm\sqrt{2}$, $\mathcal{A} = \left\{\dfrac{1}{\sqrt{2}}\begin{bmatrix} 1 \\ 1 \\ 0 \end{bmatrix}, \dfrac{1}{2}\begin{bmatrix} -1 \\ 1 \\ \sqrt{2} \end{bmatrix}, \dfrac{1}{2}\begin{bmatrix} 1 \\ -1 \\ \sqrt{2} \end{bmatrix}\right\}$, $Q(\mathrm{X}) = \sqrt{2}y_2^2 - \sqrt{2}y_3^2$, $\mathrm{Y} = [\mathrm{X}]_{\mathcal{A}}$

3. $\lambda = 8, -2$, $\mathcal{A} = \left\{\dfrac{1}{\sqrt{5}}\begin{bmatrix} 1 \\ 2 \end{bmatrix}, \dfrac{1}{\sqrt{5}}\begin{bmatrix} -2 \\ 1 \end{bmatrix}\right\}$, $\mathrm{Y} = [\mathrm{X}]_{\mathcal{A}}$, $Q(\mathrm{X}) = 8y_1^2 - 2y_2^2$

a.

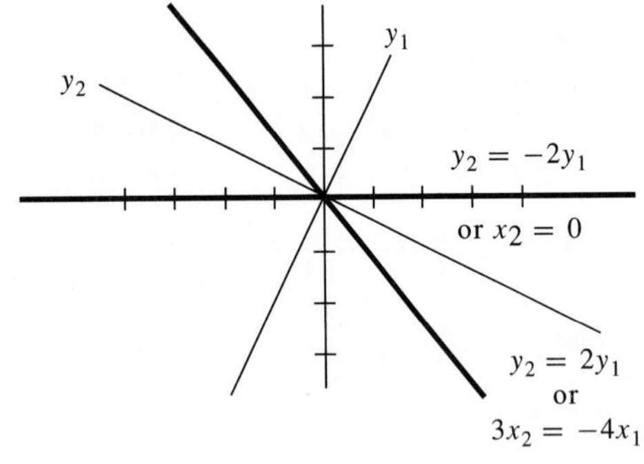

b.

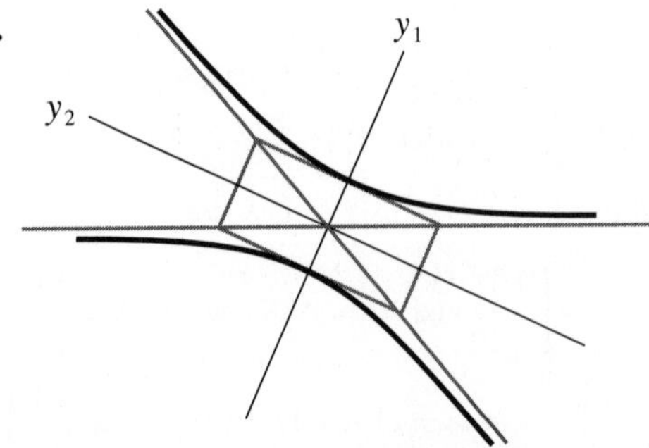

5. $\lambda = 10{,}20,\ \mathcal{A} = \left\{\frac{1}{\sqrt{10}}\begin{bmatrix}-3\\1\end{bmatrix}, \frac{1}{\sqrt{10}}\begin{bmatrix}1\\3\end{bmatrix}\right\}$, $\mathrm{Y} = [\mathrm{X}]_{\mathcal{A}}$, $Q(\mathrm{X}) = 10y_1^2 + 20y_2^2$. Sketches for both part (a) and part (b) are shown in the figure.

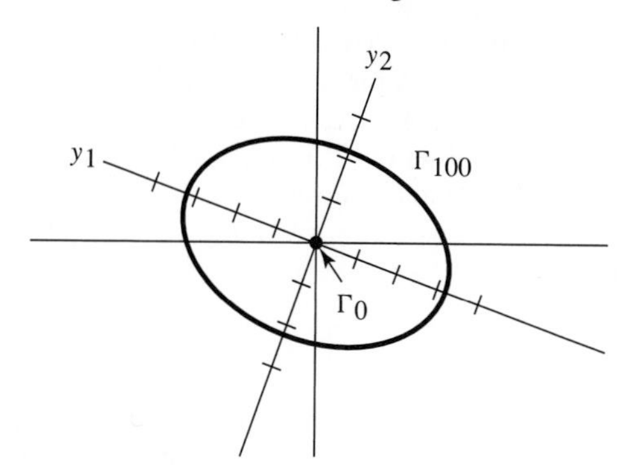

7. $\lambda = 4, 0, 2,\ \mathcal{A} = \left\{\begin{bmatrix}0\\0\\1\end{bmatrix}, \frac{1}{\sqrt{2}}\begin{bmatrix}1\\1\\0\end{bmatrix}, \frac{1}{\sqrt{2}}\begin{bmatrix}-1\\1\\0\end{bmatrix}\right\}$, $\mathrm{Y} = [\mathrm{X}]_{\mathcal{A}}$, $Q(\mathrm{X}) = 4y_1^2 + 2y_3^2$

a. Γ_0 is the y_2 axis.
b. Γ_4 is a cylinder. Its axis is the y_2 axis, and each cross section perpendicular to that axis is the ellipse with equation $y_1^2 + \frac{1}{2}y_3^2 = 1$.

Section 6.7

1. **a.** positive definite **b.** not semidefinite **c.** positive definite
d. negative semidefinite **e.** not semidefinite **f.** positive definite

4. **a.** $\det(\mathrm{A}_1) = 2$, $\det(\mathrm{A}_2) = 1$, $\det(\mathrm{A}_3) = 1 \Rightarrow$ A is positive definite.
c. $\det(\mathrm{A}_1) = -1$, $\det(\mathrm{A}_2) = 2$, $\det(\mathrm{A}_3) = -3 \Rightarrow$ A is negative definite.
e. $\det(\mathrm{A}_1) = 2$, $\det(\mathrm{A}_2) = 1$, $\det(\mathrm{A}_3) = 1$, $\det(\mathrm{A}_4) = 1 \Rightarrow$ A is positive definite.

5. **a.** positive definite. If $\mathrm{X} \neq 0$, then $\mathrm{X}^T(\mathrm{A}+\mathrm{B})\mathrm{X} = \mathrm{X}^T\mathrm{A}\mathrm{X} + \mathrm{X}^T\mathrm{B}\mathrm{X} > 0$ because $\mathrm{X}^T\mathrm{A}\mathrm{X} > 0$ and $\mathrm{X}^T\mathrm{B}\mathrm{X} > 0$.

8. **a.** For $\mathrm{X} \neq 0$, $\mathrm{X}^T(-\mathrm{A})\mathrm{X} = -\mathrm{X}^T\mathrm{A}\mathrm{X} > 0 \Leftrightarrow \mathrm{X}^T\mathrm{A}\mathrm{X} < 0$.
b. Set $\mathrm{B} = -\mathrm{A}$. Then $\mathrm{B}_k = -\mathrm{A}_k$ and $\det(\mathrm{B}_k) = (-1)^k\det(\mathrm{A}_k)$. Now,

$$\begin{aligned}\text{A is negative definite} &\Leftrightarrow \text{B is positive definite (part (a))}\\ &\Leftrightarrow \det(\mathrm{B}_k) > 0,\ 1 \le k \le n \text{ ((1) of Theorem 6.17)}\\ &\Leftrightarrow (-1)^k\det(\mathrm{A}_k) > 0,\ 1 \le k \le n.\end{aligned}$$

10. Det(A) is the product of the eigenvalues of A, which are all positive.

11. **a.** $Q(\mathrm{X}) = \mathrm{X}^T\mathrm{A}\mathrm{X} = [\mathrm{Z}^T \mid 0^T]\left[\begin{array}{c|c}\mathrm{A}_k & \mathrm{B}\\\hline \mathrm{B}^T & \mathrm{C}\end{array}\right]\left[\begin{array}{c}\mathrm{Z}\\\hline 0\end{array}\right] = [\mathrm{Z}^T\mathrm{A}_k \mid \mathrm{Z}^T\mathrm{B}]\left[\begin{array}{c}\mathrm{Z}\\\hline 0\end{array}\right] = \mathrm{Z}^T\mathrm{A}_k\mathrm{Z} =$ $Q_k(\mathrm{Z})$. Thus, $\mathrm{Z} \neq 0 \Rightarrow \mathrm{X} \neq 0 \Rightarrow Q(\mathrm{X}) > 0 \Rightarrow Q_k(\mathrm{Z}) > 0$.

Section 7.2

1. a. (1) $A + B = A \cup B = B \cup A = B + A$

(2) $A + (B + C) = A \cup (B \cup C) = (A \cup B) \cup C = (A + B) + C$

(3) $\mathbf{0}_{\mathcal{V}} = \varnothing$. $A + \varnothing = A \cup \varnothing = A$ for each $A \subseteq \mathcal{R}$. Moreover, if D is a subset of $\mathcal{R}$ such that $A \cup D = A$ for each $A \subseteq \mathcal{R}$, then setting $A = \varnothing$ yields $D = \varnothing \cup D = \varnothing$.

(4) If $A \neq \varnothing$ and B is any subset of $\mathcal{R}$, then $A \subseteq A \cup B = A + B$, so $A + B \neq \varnothing$. Thus, A does not have an "additive inverse."

(5) $(s + t)A = A = A \cup A = A + A = sA + tA$

(6) $s(A + B) = s(A \cup B) = A \cup B = A + B = sA + sB$

(7) $s(tA) = sA = A = (st)A$

(8) $1A = A$

2. a. By Theorem 1.11, the sum of two upper triangular matrices is upper triangular and a scalar multiple of an upper triangular matrix is upper triangular. $\mathcal{U}$ is a subspace.

c. From Problem 13, Exercises 1.5, $\text{trace}(A + B) = \text{trace}(A) + \text{trace}(B)$ and $\text{trace}\,(tA) = t[\text{trace}(A)]$. It follows that $\mathcal{T}$ is a subspace.

3. b. If $f, g \in \mathcal{W}$ and $t \in \mathcal{R}$, then $(f + g)(-x) = f(-x) + g(-x) = f(x) + g(x) = (f + g)(x)$ and $(tf)(-x) = tf(-x) = tf(x) = (tf)(x)$. $\mathcal{W}$ is a subspace.

4. a. The sum of two convergent series is convergent and a scalar multiple of a convergent series is convergent. $\mathcal{W}$ is a subspace.

c. If $x_k = 0$ for all but finitely many values of k and $y_k = 0$ for all but finitely many values of k, then $x_k + y_k = 0$ and $tx_k = 0$ for all but finitely many values of k. $\mathcal{W}$ is a subspace.

6. a. The sum of two continuous functions is continuous and a constant multiple of a continuous function is continuous.

c. If $f, g \in \mathcal{W}$ and $t \in \mathcal{R}$, then $\int_a^b f(x)\,dx = 0 = \int_a^b g(x)\,dx \Rightarrow \int_a^b (f + g)(x)\,dx = \int_a^b f(x)\,dx + \int_a^b g(x)\,dx = 0 + 0 = 0$ and $\int_a^b (tf)(x)\,dx = t\int_a^b f(x)\,dx = t0 = 0$.

e. If $f, g \in \mathcal{W}$ and $t \in \mathcal{R}$, then $(f + g)(c) = f(c) + g(c) = 0 + 0 = 0$, and $(tf)(c) = tf(c) = t0 = 0$.

7. a. $\left\{\begin{bmatrix} 1 & 0 & 0 \\ 0 & 0 & 0 \\ 0 & 0 & 0 \end{bmatrix}, \begin{bmatrix} 0 & 0 & 0 \\ 0 & 1 & 0 \\ 0 & 0 & 0 \end{bmatrix}, \begin{bmatrix} 0 & 0 & 0 \\ 0 & 0 & 0 \\ 0 & 0 & 1 \end{bmatrix}\right\}$

c. $\left\{\begin{bmatrix} 0 & 1 \\ 0 & 0 \end{bmatrix}, \begin{bmatrix} 1 & 0 \\ 0 & -1 \end{bmatrix}, \begin{bmatrix} 0 & 0 \\ 1 & 0 \end{bmatrix}\right\}$

e. $0 = \int_0^1 f(x)\,dx = \int_0^1 (a + bx + cx^2)\,dx = ax + \frac{b}{2}x^2 + \frac{c}{3}x^3\Big|_0^1 = a + \frac{b}{2} + \frac{c}{3} \Rightarrow$

$a = -\frac{b}{2} - \frac{c}{3} \Rightarrow f(x) = b\left(x - \frac{1}{2}\right) + c\left(x^2 - \frac{1}{3}\right) \Rightarrow \left\{x - \frac{1}{2}, x^2 - \frac{1}{3}\right\}$ spans the subspace

g. $\{X_1, X_2, \ldots, X_n\}$, where X_k has a 1 in the kth position and zeros elsewhere, $k = 1, \ldots, n$

Section 7.3

1. a. $0 = s(1-2x+x^2) + t(2-x+x^2) \Leftrightarrow \begin{bmatrix} 1 & 2 \\ -2 & -1 \\ 1 & 1 \end{bmatrix}\begin{bmatrix} s \\ t \end{bmatrix} = \begin{bmatrix} 0 \\ 0 \\ 0 \end{bmatrix}$. Since

$\begin{bmatrix} 1 & 2 \\ -2 & -1 \\ 1 & 1 \end{bmatrix} \longrightarrow \begin{bmatrix} 1 & 0 \\ 0 & 1 \\ 0 & 0 \end{bmatrix}$, $s = t = 0$, and therefore $\mathcal{A}$ is linearly independent.

c. Suppose $0 = a\sin(x) + b\cos(x)$ for all $x \in \mathcal{R}$. Setting $x = 0$ and $x = \pi/2$ yields $0 = b$ and $0 = a$, respectively, so $\mathcal{A}$ is linearly independent.

f. $\cos(2x) = \cos^2(x) - \sin^2(x) \Rightarrow \mathcal{A}$ is linearly dependent. If $0 = a\cos^2(x) + b\sin^2(x)$, $x \in \mathcal{R}$, then $x = 0$ and $x = \pi/2$ yield $a = 0$ and $b = 0$, respectively, so $\{\cos^2(x), \sin^2(x)\}$ is linearly independent.

i. $\begin{bmatrix} 0 & 0 \\ 0 & 0 \end{bmatrix} = a\begin{bmatrix} 1 & 1 \\ 0 & 1 \end{bmatrix} + b\begin{bmatrix} 1 & 0 \\ 1 & 1 \end{bmatrix} + c\begin{bmatrix} 0 & 1 \\ 1 & 0 \end{bmatrix} + d\begin{bmatrix} 1 & 0 \\ 0 & 1 \end{bmatrix} \Leftrightarrow$

$\begin{bmatrix} 0 \\ 0 \\ 0 \\ 0 \end{bmatrix} = \begin{bmatrix} 1 & 1 & 0 & 1 \\ 1 & 0 & 1 & 0 \\ 0 & 1 & 1 & 0 \\ 1 & 1 & 0 & 1 \end{bmatrix}\begin{bmatrix} a \\ b \\ c \\ d \end{bmatrix}$. $\begin{bmatrix} 1 & 1 & 0 & 1 \\ 1 & 0 & 1 & 0 \\ 0 & 1 & 1 & 0 \\ 1 & 1 & 0 & 1 \end{bmatrix} \longrightarrow \begin{bmatrix} 1 & 0 & 0 & 1/2 \\ 0 & 1 & 0 & 1/2 \\ 0 & 0 & 1 & -1/2 \\ 0 & 0 & 0 & 0 \end{bmatrix} \Rightarrow$

$\begin{bmatrix} 1 & 0 \\ 0 & 1 \end{bmatrix} = \frac{1}{2}\begin{bmatrix} 1 & 1 \\ 0 & 1 \end{bmatrix} + \frac{1}{2}\begin{bmatrix} 1 & 0 \\ 1 & 1 \end{bmatrix} - \frac{1}{2}\begin{bmatrix} 0 & 1 \\ 1 & 0 \end{bmatrix}$. $\left\{\begin{bmatrix} 1 & 1 \\ 0 & 1 \end{bmatrix}, \begin{bmatrix} 1 & 0 \\ 1 & 1 \end{bmatrix}, \begin{bmatrix} 0 & 1 \\ 1 & 0 \end{bmatrix}\right\}$ is a linearly independent subset of $\mathcal{A}$ with the same span as $\mathcal{A}$.

k. $\begin{bmatrix} 0 & 0 & 0 \\ 0 & 0 & 0 \end{bmatrix} = a\begin{bmatrix} 1 & 1 & 0 \\ 0 & 1 & 1 \end{bmatrix} + b\begin{bmatrix} 0 & 1 & 1 \\ 1 & 0 & 1 \end{bmatrix} + c\begin{bmatrix} 1 & 0 & 1 \\ 1 & 1 & 0 \end{bmatrix} + d\begin{bmatrix} 1 & 1 & 1 \\ 0 & 1 & 0 \end{bmatrix} \Leftrightarrow$

$\begin{bmatrix} 1 & 0 & 1 & 1 \\ 1 & 1 & 0 & 1 \\ 0 & 1 & 1 & 1 \\ 0 & 1 & 1 & 0 \\ 1 & 0 & 1 & 1 \\ 1 & 1 & 0 & 0 \end{bmatrix}\begin{bmatrix} a \\ b \\ c \\ d \end{bmatrix} = \begin{bmatrix} 0 \\ 0 \\ 0 \\ 0 \\ 0 \\ 0 \end{bmatrix}$. The coefficient matrix reduces to $\begin{bmatrix} \mathrm{I}_4 \\ \hline 0 \end{bmatrix}$, so $a = b = c = d = 0$ and $\mathcal{A}$ is linearly independent.

3. a. $0 = a_0 + \frac{a_1}{x} + \frac{a_2}{x^2} + \cdots + \frac{a_n}{x^n} = \frac{a_0x^n + a_1x^{n-1} + \cdots + a_n}{x^n} \Rightarrow$
$0 = a_0x^n + a_1x^{n-1} + \cdots + a_n \Rightarrow a_0 = a_1 = \cdots = a_n = 0$.

b. Apply the argument in part (a) to a finite subset $\{x^{-k_1}, \ldots, x^{-k_p}\}$, where $1 \le k_1 < \cdots < k_p$.

4. Consider a finite subset $\{\mathrm{Y}_{k_1}, \ldots, \mathrm{Y}_{k_p}\}$ of $\mathcal{A}$, $1 \le k_1 < \cdots < k_p$, $p \ge 2$, and suppose

$$t_1\mathrm{Y}_{k_1} + \cdots + t_p\mathrm{Y}_{k_p} = 0 = (0,0,0,\ldots). \qquad (*)$$

a. Comparing the entries in positions $k_1, \ldots, k_p$ on both sides $(*)$ produces

$$\begin{aligned} t_1 + t_2 + t_3 + \cdots + t_p &= 0 \\ t_2 + t_3 + \cdots + t_p &= 0 \\ t_3 + \cdots + t_p &= 0 \\ &\vdots \\ t_p &= 0, \end{aligned}$$

which implies $t_1 = t_2 = \cdots = t_p = 0$. $\mathcal{A}$ is linearly independent.

5. a. $\mathcal{B}$ is a linearly independent subset of $\mathcal{P}_2$ containing three elements, and dim $\mathcal{P}_2 = 3$, so $\mathcal{B}$ is a basis.

b. $\mathcal{B}$ is a linearly independent subset of $\mathcal{M}_{2\times 2}$ containing three elements, but dim $\mathcal{M}_{2\times 2} = 4$, so $\mathcal{B}$ is not a basis.

6. a. $$\left\{ \begin{bmatrix} 1 & 0 & 0 \\ 0 & 0 & 0 \\ 0 & 0 & 0 \end{bmatrix}, \begin{bmatrix} 0 & 0 & 0 \\ 0 & 1 & 0 \\ 0 & 0 & 0 \end{bmatrix}, \begin{bmatrix} 0 & 0 & 0 \\ 0 & 0 & 0 \\ 0 & 0 & 1 \end{bmatrix}, \begin{bmatrix} 0 & 1 & 0 \\ 1 & 0 & 0 \\ 0 & 0 & 0 \end{bmatrix}, \begin{bmatrix} 0 & 0 & 1 \\ 0 & 0 & 0 \\ 1 & 0 & 0 \end{bmatrix}, \begin{bmatrix} 0 & 0 & 0 \\ 0 & 0 & 1 \\ 0 & 1 & 0 \end{bmatrix} \right\}$$

is a basis for $\mathcal{W}$. dim $\mathcal{W} = 6$.

d. $\{2x - 1, 3x^2 - 1, 4x^3 - 1\}$ is a basis for $\mathcal{W}$. dim $\mathcal{W} = 3$.

Section 7.4

1. a. linear. $T(f+g)(x) = x(f+g)'(x) = x[f'(x) + g'(x)] = xf'(x) + xg'(x) = T(f)(x) + T(g)(x) = [T(f) + T(g)](x) \Rightarrow T(f+g) = T(f) + T(g)$.
$T(tf)(x) = x(tf)'(x) = xtf'(x) = t[xf'(x)] = tT(f)(x) \Rightarrow T(tf) = tT(f)$.

c. linear. $T(f+g)(x) = (f+g)(x-c) = f(x-c) + g(x-c) = T(f)(x) + T(g)(x)$ and $T(tf)(x) = (tf)(x-c) = tf(x-c) = tT(f)(x)$.

e. linear. $T(f+g) = (f+g)(c) = f(c) + g(c) = T(f) + T(g)$ and $T(tf) = (tf)(c) = tf(c) = tT(f)$.

2. a. $$T\left(\begin{bmatrix} a & b \\ c & d \end{bmatrix} + \begin{bmatrix} \alpha & \beta \\ \gamma & \delta \end{bmatrix}\right) = T\left(\begin{bmatrix} a+\alpha & b+\beta \\ c+\gamma & d+\delta \end{bmatrix}\right) = \begin{bmatrix} (a+\alpha)+(b+\beta) \\ (b+\beta)+(c+\gamma) \\ (c+\gamma)+(d+\delta) \end{bmatrix}$$

$$= \begin{bmatrix} a+b \\ b+c \\ c+d \end{bmatrix} + \begin{bmatrix} \alpha+\beta \\ \beta+\gamma \\ \gamma+\delta \end{bmatrix} = T\left(\begin{bmatrix} a & b \\ c & d \end{bmatrix}\right) + T\left(\begin{bmatrix} \alpha & \beta \\ \gamma & \delta \end{bmatrix}\right).$$

$$T\left(t\begin{bmatrix} a & b \\ c & d \end{bmatrix}\right) = T\left(\begin{bmatrix} ta & tb \\ tc & td \end{bmatrix}\right) = \begin{bmatrix} ta+tb \\ tb+tc \\ tc+td \end{bmatrix} = tT\left(\begin{bmatrix} a & b \\ c & d \end{bmatrix}\right).$$

$$\begin{bmatrix} a & b \\ c & d \end{bmatrix} \in \mathcal{K}er(T) \Leftrightarrow \begin{bmatrix} 0 \\ 0 \\ 0 \end{bmatrix} = T\left(\begin{bmatrix} a & b \\ c & d \end{bmatrix}\right) = \begin{bmatrix} a+b \\ b+c \\ c+d \end{bmatrix} = \begin{bmatrix} 1 & 1 & 0 & 0 \\ 0 & 1 & 1 & 0 \\ 0 & 0 & 1 & 1 \end{bmatrix}\begin{bmatrix} a \\ b \\ c \\ d \end{bmatrix}.$$

Since $\begin{bmatrix} 1 & 1 & 0 & 0 \\ 0 & 1 & 1 & 0 \\ 0 & 0 & 1 & 1 \end{bmatrix} \longrightarrow \begin{bmatrix} 1 & 0 & 0 & 1 \\ 0 & 1 & 0 & -1 \\ 0 & 0 & 1 & 1 \end{bmatrix}$, $\mathcal{K}er(T) = \mathcal{S}pan\left\{\begin{bmatrix} -1 & 1 \\ -1 & 1 \end{bmatrix}\right\}$.

T is not one-to-one. $\mathcal{Rng}(T) = \mathcal{Span}\left\{T\left(\begin{bmatrix}1 & 0\\ 0 & 0\end{bmatrix}\right), T\left(\begin{bmatrix}0 & 1\\ 0 & 0\end{bmatrix}\right), T\left(\begin{bmatrix}0 & 0\\ 1 & 0\end{bmatrix}\right),\right.$

$\left.T\left(\begin{bmatrix}0 & 0\\ 0 & 1\end{bmatrix}\right)\right\}, = \mathcal{Span}\left\{\begin{bmatrix}1\\0\\0\end{bmatrix}, \begin{bmatrix}1\\1\\0\end{bmatrix}, \begin{bmatrix}0\\1\\1\end{bmatrix}, \begin{bmatrix}0\\0\\1\end{bmatrix}\right\}.$

The first three vectors in this spanning set are linearly independent, so dim $\mathcal{Rng}(T) = 3$. Thus, T is onto.

c. T is linear by Problem 1(e). $f(x) = a + bx + cx^2 + dx^3 \Rightarrow T(f) = a + b + c + d$. Thus, $f \in \mathcal{Ker}(T)$ $a = -b - c - d \Leftrightarrow f(x) = b(x-1) + c(x^2-1) + d(x^3-1)$, that is, $\mathcal{Ker}(T) = \mathcal{Span}\{x-1, x^2-1, x^3-1\}$. The vectors in this spanning set are linearly independent and hence constitute a basis for $\mathcal{Ker}(T)$. $\mathcal{Rng}(T) = \mathcal{R}$. Indeed, if $t \in \mathcal{R}$ and $f(x) = tx$, then $f \in \mathcal{P}_3$ and $T(f) = t$. A basis for $\mathcal{Rng}(T)$ is $\{1\}$.

5. $T(\mathrm{X}+\mathrm{Y}) = (x_1+y_1, x_2+y_2, \ldots) = (x_1, x_2, \ldots) + (y_1, y_2, \ldots) = T(\mathrm{X}) + T(\mathrm{Y})$, and $T(t\mathrm{X}) = (tx_1, tx_2, \ldots) = t(x_1, x_2, \ldots) = tT(\mathrm{X})$, so T is linear. Observe that $T(\mathrm{X}) = 0 \Leftrightarrow x_1 = x_2 = x_3 = \cdots = 0$, so $\mathcal{Ker}(T) = \mathcal{Span}\{(1,0,0,\ldots)\} \neq \{0\}$ and T is not one-to-one. If $\mathrm{Y} = (y_0, y_1, y_2, \ldots)$ and $\mathrm{X} = (t, y_0, y_1, y_2, \ldots)$, $t \in \mathcal{R}$, then $T(\mathrm{X}) = \mathrm{Y}$, so T is onto.

9. a. If $f(x) = x^2$, then $D(f)(x) = 2x$ and $T(D(f))(x) = x(2x) = 2x^2$, but $T(f)(x) = x^3$ and $D(T(f))(x) = 3x^2$.

b. $(D \circ T)^2(f)(x) = a_0 + 4a_1x + 9a_2x^2 + \cdots + (n+1)^2a_nx^n$.

11. a. For $\mathbf{u}, \mathbf{v} \in \mathcal{V}$ and $t \in \mathcal{R}$, $\mathrm{Id}_{\mathcal{V}}(\mathbf{u}+\mathbf{v}) = \mathbf{u}+\mathbf{v} = \mathrm{Id}_{\mathcal{V}}(\mathbf{u}) + \mathrm{Id}_{\mathcal{V}}(\mathbf{v})$, and $\mathrm{Id}_{\mathcal{V}}(t\mathbf{v}) = t\mathbf{v} = t\mathrm{Id}_{\mathcal{V}}(\mathbf{v})$.

b. For $\mathbf{v} \in \mathcal{V}$, $(\mathrm{Id}_{\mathcal{V}} \circ S)(\mathbf{v}) = \mathrm{Id}_{\mathcal{V}}(S(\mathbf{v})) = S(\mathbf{v}) = S(\mathrm{Id}_{\mathcal{V}}(\mathbf{v})) = (S \circ \mathrm{Id}_{\mathcal{V}})(\mathbf{v})$.

c. If $S(\mathbf{u}) = S(\mathbf{v})$, then $\mathbf{u} = \mathrm{Id}_{\mathcal{V}}(\mathbf{u}) = T(S(\mathbf{u})) = T(S(\mathbf{v})) = \mathrm{Id}_{\mathcal{V}}(\mathbf{v}) = \mathbf{v}$, so S is one-to-one. If $\mathbf{v} \in \mathcal{V}$ and $\mathbf{u} = T(\mathbf{v})$, then $S(\mathbf{u}) = S(T(\mathbf{v})) = \mathrm{Id}_{\mathcal{V}}(\mathbf{v}) = \mathbf{v}$, so S is onto. Finally, if $\mathbf{w} = S(\mathbf{v})$, then $T(\mathbf{w}) = T(S(\mathbf{v})) = \mathrm{Id}_{\mathcal{V}}(\mathbf{v}) = \mathbf{v}$, so T is the inverse of S.

12. $\lambda = 0$ is an eigenvalue of $T \Leftrightarrow$ there is a nonzero $\mathbf{v} \in \mathcal{V}$ such that $T(\mathbf{v}) = 0\mathbf{v} = \mathbf{0}_{\mathcal{W}} \Leftrightarrow \mathcal{Ker}(T) \neq \{\mathbf{0}_{\mathcal{V}}\}$.

16. If $\lambda > 0$, and $f(x) = e^{\sqrt{\lambda}x}$, then $D^2(f)(x) = \lambda e^{\sqrt{\lambda}x} = \lambda f(x)$. If $\lambda = 0$ and $f(x) = x$, then $D^2(f)(x) = 0 = \lambda f(x)$. If $\lambda < 0$ and $f(x) = \sin(\sqrt{|\lambda|}x)$, then $D^2(f)(x) = -|\lambda|\sin(\sqrt{|\lambda|}x) = \lambda\sin(\sqrt{|\lambda|}x) = \lambda f(x)$.

Section 7.5

1. a. $f(x) = 3(1+x^2) - 2(2x+x^2) + (1+x+x^2) = 4 - 3x + 2x^2$

b. $-4 + x - 2x^2 = r(1+x^2) + s(2x+x^2) + t(1+x+x^2) \Leftrightarrow \begin{bmatrix}-4\\1\\-2\end{bmatrix} =$

$$\begin{bmatrix}1 & 0 & 1\\ 0 & 2 & 1\\ 1 & 1 & 1\end{bmatrix}\begin{bmatrix}r\\s\\t\end{bmatrix}. \left[\begin{array}{ccc|c}1 & 0 & 1 & -4\\ 0 & 2 & 1 & 1\\ 1 & 1 & 1 & -2\end{array}\right] \longrightarrow \left[\begin{array}{ccc|c}1 & 0 & 0 & -1\\ 0 & 1 & 0 & 2\\ 0 & 0 & 1 & -3\end{array}\right] \Rightarrow [f]_{\mathcal{A}} = \begin{bmatrix}-1\\2\\-3\end{bmatrix}.$$

3. Let $A = \begin{bmatrix} 1 & 1 \\ 1 & 0 \end{bmatrix}$, $B = \begin{bmatrix} 1 & 1 \\ 1 & 1 \end{bmatrix}$, and $C = \begin{bmatrix} 0 & 1 \\ 1 & 1 \end{bmatrix}$. If $\mathcal{A}$ is the usual basis for $\mathcal{M}_{2\times 2}$, then

$$\big[[A]_{\mathcal{A}} \quad [B]_{\mathcal{A}} \quad [C]_{\mathcal{A}}\big] = \begin{bmatrix} 1 & 1 & 0 \\ 1 & 1 & 1 \\ 1 & 1 & 1 \\ 0 & 1 & 1 \end{bmatrix} \longrightarrow \begin{bmatrix} 1 & 0 & 0 \\ 0 & 1 & 0 \\ 0 & 0 & 1 \\ 0 & 0 & 0 \end{bmatrix}$$, so $\mathcal{B}$ is linearly independent.

Since $\dim \mathcal{S}_{2\times 2} = 3$ and $\mathcal{B}$ contains three elements, $\mathcal{B}$ spans $\mathcal{S}_{2\times 2}$. Observe that $rA + sB + tC = I_2 \Leftrightarrow r[A]_{\mathcal{A}} + s[B]_{\mathcal{A}} + t[C]_{\mathcal{A}} = [I_2]_{\mathcal{A}} \Leftrightarrow \begin{bmatrix} 1 & 1 & 0 \\ 1 & 1 & 1 \\ 1 & 1 & 1 \\ 0 & 1 & 1 \end{bmatrix} \begin{bmatrix} r \\ s \\ t \end{bmatrix} = \begin{bmatrix} 1 \\ 0 \\ 0 \\ 1 \end{bmatrix}$. Since

$$\left[\begin{array}{ccc|c} 1 & 1 & 0 & 1 \\ 1 & 1 & 1 & 0 \\ 1 & 1 & 1 & 0 \\ 0 & 1 & 1 & 1 \end{array}\right] \longrightarrow \left[\begin{array}{ccc|c} 1 & 0 & 0 & -1 \\ 0 & 1 & 0 & 2 \\ 0 & 0 & 1 & -1 \\ 0 & 0 & 0 & 0 \end{array}\right], \quad [I_2]_{\mathcal{B}} = \begin{bmatrix} -1 \\ 2 \\ -1 \end{bmatrix}.$$

6. a. $T\left(\begin{bmatrix} 1 & 0 \\ 0 & 0 \end{bmatrix}\right) = 1 + x^2, \quad T\left(\begin{bmatrix} 0 & 1 \\ 0 & 0 \end{bmatrix}\right) = 1 + x, \quad T\left(\begin{bmatrix} 0 & 0 \\ 1 & 0 \end{bmatrix}\right) = -1 + x - 2x^2,$

and $T\left(\begin{bmatrix} 0 & 0 \\ 0 & 1 \end{bmatrix}\right) = x + x^2 \Rightarrow [T]_{\mathcal{BA}} = \begin{bmatrix} 1 & 1 & -1 & 0 \\ 0 & 1 & 1 & 1 \\ 1 & 0 & -2 & -1 \end{bmatrix}$.

c. $f_1(x) = 1, \quad f_2(x) = x, \quad f_3(x) = x^2, \quad f_4(x) = x^3 \Rightarrow T(f_1)(x) = 1, \quad T(f_2)(x) = 2x,$

$T(f_3)(x) = 5x^2, \quad T(f_4)(x) = 10x^3 \Rightarrow [T]_{\mathcal{A}} = \begin{bmatrix} 1 & 0 & 0 & 0 \\ 0 & 2 & 0 & 0 \\ 0 & 0 & 5 & 0 \\ 0 & 0 & 0 & 10 \end{bmatrix}$.

e. $[T]_{\mathcal{BA}} = [1 \quad 1/2 \quad 1/3]$.

g. $f_1(x) = 1, \quad f_2(x) = x, \quad f_3(x) = x^2 \Rightarrow T(f_1)(x) = (1/2)[1 - 1] = 0, \quad T(f_2)(x) = (1/2)[(x+1) - (x-1)] = 1, T(f_3)(x) = (1/2)[(x+1)^2 - (x-1)^2] = 2x \Rightarrow$

$[T]_{\mathcal{A}} = \begin{bmatrix} 0 & 1 & 0 \\ 0 & 0 & 2 \\ 0 & 0 & 0 \end{bmatrix}$.

7. a. $[T]_{\mathcal{BA}} = \begin{bmatrix} 1 & 1 & -1 & 0 \\ 0 & 1 & 1 & 1 \\ 1 & 0 & -2 & -1 \end{bmatrix} \longrightarrow \begin{bmatrix} 1 & 0 & -2 & -1 \\ 0 & 1 & 1 & 1 \\ 0 & 0 & 0 & 0 \end{bmatrix} \Rightarrow T$ is neither one-to-one nor onto. $\mathcal{K}er(T) = \mathcal{S}pan\left\{\begin{bmatrix} 2 & -1 \\ 1 & 0 \end{bmatrix}, \begin{bmatrix} 1 & -1 \\ 0 & 1 \end{bmatrix}\right\}$, $\mathcal{R}ng(T) = \mathcal{S}pan\{1 + x^2, 1 + x\}$.

c. $[T]_{\mathcal{A}} \to I_4 \Rightarrow T$ is one-to-one and onto. $\mathcal{K}er(T) = \{0\}$, $\mathcal{R}ng(T) = \mathcal{S}pan\{\mathcal{A}\}$

e. $[T]_{\mathcal{BA}} \to [1 \quad 1/2 \quad 1/3] \Rightarrow T$ is onto but not one-to-one. $\mathcal{K}er(T) = \mathcal{S}pan\{-1 + 2x, -1 + 3x^2\}$, $\mathcal{R}ng(T) = \mathcal{S}pan\{1\}$.

g. $[T]_{\mathcal{A}} = \begin{bmatrix} 0 & 1 & 0 \\ 0 & 0 & 2 \\ 0 & 0 & 0 \end{bmatrix} \longrightarrow \begin{bmatrix} 0 & 1 & 0 \\ 0 & 0 & 1 \\ 0 & 0 & 0 \end{bmatrix} \Rightarrow T$ is neither one-to-one nor onto.

$\mathcal{K}er\{T\} = \mathcal{S}pan\{1\}$, $\mathcal{R}ng\{T\} = \mathcal{S}pan\{1, x\}$.

9. By Theorem 7.25, $[T^{-1}]_{\mathcal{A}} = [T]_{\mathcal{A}}^{-1} = \begin{bmatrix} 1 & -1 & 1 & 0 \\ 0 & 1 & -1 & 1 \\ 1 & 0 & 1 & -1 \\ -1 & 1 & 0 & 1 \end{bmatrix}^{-1}$

$= \dfrac{1}{3}\begin{bmatrix} 1 & 2 & 1 & -1 \\ -1 & 1 & 2 & 1 \\ 1 & -1 & 1 & 2 \\ 2 & 1 & -1 & 1 \end{bmatrix}$. If $\mathrm{A} = \begin{bmatrix} a & b \\ c & d \end{bmatrix}$, then $[T^{-1}(\mathrm{A})]_{\mathcal{A}} = [T]_{\mathcal{A}}^{-1}[\mathrm{A}]_{\mathcal{A}}$

$= \dfrac{1}{3}\begin{bmatrix} 1 & 2 & 1 & -1 \\ -1 & 1 & 2 & 1 \\ 1 & -1 & 1 & 2 \\ 2 & 1 & -1 & 1 \end{bmatrix}\begin{bmatrix} a \\ b \\ c \\ d \end{bmatrix} = \dfrac{1}{3}\begin{bmatrix} a+2b+c-d \\ -a+b+2c+d \\ a-b+c+2d \\ 2a+b-c+d \end{bmatrix}$, so $T^{-1}(\mathrm{A})$

$= \dfrac{1}{3}\begin{bmatrix} a+2b+c-d & -a+b+2c+d \\ a-b+c+2d & 2a+b-c+d \end{bmatrix}$.

11. $f_1(x) = 1, f_2(x) = x, f_3(x) = x^2 \Rightarrow T(f_1)(x) = x, T(f_2)(x) = x^2/2, T(f_3)(x) = x^3/3 \Rightarrow$
$[T]_{\mathcal{AB}} = \begin{bmatrix} 0 & 0 & 0 \\ 1 & 0 & 0 \\ 0 & 1/2 & 0 \\ 0 & 0 & 1/3 \end{bmatrix}$. Moreover, $[D]_{\mathcal{BA}} = \begin{bmatrix} 0 & 1 & 0 & 0 \\ 0 & 0 & 2 & 0 \\ 0 & 0 & 0 & 3 \end{bmatrix}$ (Example 7.34), so

$$[D]_{\mathcal{BA}}[T]_{\mathcal{AB}} = \begin{bmatrix} 0 & 1 & 0 & 0 \\ 0 & 0 & 2 & 0 \\ 0 & 0 & 0 & 3 \end{bmatrix}\begin{bmatrix} 0 & 0 & 0 \\ 1 & 0 & 0 \\ 0 & 1/2 & 0 \\ 0 & 0 & 1/3 \end{bmatrix} = \begin{bmatrix} 1 & 0 & 0 \\ 0 & 1 & 0 \\ 0 & 0 & 1 \end{bmatrix}.$$

14. a. $[\mathrm{Id}_{\mathcal{V}}]_{\mathcal{BA}} = \big[[\mathrm{Id}_{\mathcal{V}}(\mathbf{v}_1)]_{\mathcal{B}} \ \cdots \ [\mathrm{Id}_{\mathcal{V}}(\mathbf{v}_n)]_{\mathcal{B}}\big] = \big[[\mathbf{v}_1]_{\mathcal{B}} \ \cdots \ [\mathbf{v}_n]_{\mathcal{B}}\big]$
c. $\mathrm{P}_{\mathcal{BA}}[\mathbf{v}]_{\mathcal{A}} = [\mathrm{Id}_{\mathcal{V}}]_{\mathcal{BA}}[\mathbf{v}]_{\mathcal{A}} = [\mathrm{Id}_{\mathcal{V}}(\mathbf{v})]_{\mathcal{B}} = [\mathbf{v}]_{\mathcal{B}}, \mathbf{v} \in \mathcal{V}$

15. $r(\mathbf{v}_1 + 2\mathbf{v}_2) + s(\mathbf{v}_1 - \mathbf{v}_2 + \mathbf{v}_3) + t(2\mathbf{v}_1 + \mathbf{v}_3) = \mathbf{0}_{\mathcal{V}} \Leftrightarrow (r + s + 2t)\mathbf{v}_1 + (2r - s)\mathbf{v}_2 +$
$(s + t)\mathbf{v}_3 = \mathbf{0}_{\mathcal{V}} \Leftrightarrow \begin{bmatrix} 0 \\ 0 \\ 0 \end{bmatrix} = \begin{bmatrix} r+s+2t \\ 2r-s \\ s+t \end{bmatrix} = \begin{bmatrix} 1 & 1 & 2 \\ 2 & -1 & 0 \\ 0 & 1 & 1 \end{bmatrix}\begin{bmatrix} r \\ s \\ t \end{bmatrix}$ (because $\mathbf{v}_1$, $\mathbf{v}_2$, and $\mathbf{v}_3$ are independent). $\begin{bmatrix} 1 & 1 & 2 \\ 2 & -1 & 0 \\ 0 & 1 & 1 \end{bmatrix} \longrightarrow \begin{bmatrix} 1 & 0 & 0 \\ 0 & 1 & 0 \\ 0 & 0 & 1 \end{bmatrix} \Rightarrow r = s = t = 0 \Rightarrow \mathcal{B}$ is independent. $\mathrm{P}_{\mathcal{AB}} = \big[[\mathbf{v}_1 + 2\mathbf{v}_2]_{\mathcal{A}} \ [\mathbf{v}_1 - \mathbf{v}_2 + \mathbf{v}_3]_{\mathcal{A}} \ [2\mathbf{v}_1 + \mathbf{v}_3]_{\mathcal{A}}\big] = \begin{bmatrix} 1 & 1 & 2 \\ 2 & -1 & 0 \\ 0 & 1 & 1 \end{bmatrix}$ and

$\mathrm{P}_{\mathcal{BA}} = \mathrm{P}_{\mathcal{AB}}^{-1} = \begin{bmatrix} -1 & 1 & 2 \\ -2 & 1 & 4 \\ 2 & -1 & -3 \end{bmatrix}$.

18. a. If $\mathcal{A}$ is the usual basis for $\mathcal{M}_{2\times 2}$, then $[T]_{\mathcal{A}} = \begin{bmatrix} 2 & 0 & 0 & 0 \\ 0 & 1 & 1 & 0 \\ 0 & 1 & 1 & 0 \\ 0 & 0 & 0 & 2 \end{bmatrix}$,

and

$$\det(\lambda I - [T]_{\mathcal{A}}) = \lambda(\lambda - 2)^3. \quad \mathcal{V}(0) = \mathcal{S}pan\left\{\begin{bmatrix} 0 & -1 \\ 1 & 0 \end{bmatrix}\right\},$$

and

$$\mathcal{V}(2) = \mathcal{S}pan\left\{\begin{bmatrix} 1 & 0 \\ 0 & 0 \end{bmatrix}, \begin{bmatrix} 1 & 0 \\ 0 & 1 \end{bmatrix}, \begin{bmatrix} 0 & 0 \\ 0 & 1 \end{bmatrix}\right\}.$$

c. $\mathcal{A} = \{1,x,x^2\} \Rightarrow [T]_{\mathcal{A}} = \begin{bmatrix} 1 & 0 & 1 \\ 0 & 1 & 0 \\ 0 & 0 & 1 \end{bmatrix} \Rightarrow \det(\lambda I - [T]_{\mathcal{A}}) = (\lambda - 1)^3.$

$\mathcal{V}(1) = \mathcal{S}pan\{1,x\}.$

e. $\mathcal{A}$ the usual basis for $\mathcal{M}_{2\times 2} \Rightarrow [T]_{\mathcal{A}} = \begin{bmatrix} 1 & 1 & 0 & 0 \\ 0 & 1 & 1 & 0 \\ 0 & 0 & 1 & 1 \\ 1 & 0 & 0 & 1 \end{bmatrix} \Rightarrow \det(\lambda I - [T]_{\mathcal{A}}) =$

$\lambda(\lambda - 2)(\lambda^2 - 2\lambda + 2)$, $\mathcal{V}(0) = \mathcal{S}pan\left\{\begin{bmatrix} -1 & 1 \\ -1 & 1 \end{bmatrix}\right\}$, and $\mathcal{V}(2) = \mathcal{S}pan\left\{\begin{bmatrix} 1 & 1 \\ 1 & 1 \end{bmatrix}\right\}.$

Section 7.6

1. a. Assume $X,Y \in \mathcal{R}^2$ and $t \in \mathcal{R}$.
(1) $X \neq 0 \Rightarrow$ one of x_1 and x_2 is not zero $\Rightarrow \|X\|_\infty = \max\{|x_1|, |x_2|\} > 0$.
(2) $\|tX\|_\infty = \max\{|tx_1|, |tx_2|\} = |t|\max\{|x_1|, |x_2|\} = |t|\|X\|_\infty$.
(3) $\|X + Y\|_\infty = \max\{|x_1 + y_1|, |x_2 + y_2|\} \leq \max\{|x_1| + |y_1|, |x_2| + |y_2|\}$. Since $|x_1| \leq \max\{|x_1|, |x_2|\} = \|X\|_\infty$ and $|y_1| \leq \max\{|y_1|, |y_2|\} = \|Y\|_\infty$, $|x_1| + |y_1| \leq \|X\|_\infty + \|Y\|_\infty$. Similarly, $|x_2| + |y_2| \leq \|X\|_\infty + \|Y\|_\infty$, so $\max\{|x_1| + |y_1|, |x_2| + |y_2|\} \leq \|X\|_\infty + \|Y\|_\infty$. Thus, $\|X + Y\|_\infty \leq \|X\|_\infty + \|Y\|_\infty$.

2. b. The graph of f lies between $y = 1$ and $y = -1$, and there is at least one $x \in [0,\pi]$ such that $f(x) = \pm 1$.

4. (1) If $\mathbf{v} \neq \mathbf{0}_{\mathcal{V}}$, then $\|\mathbf{v}\| = 1 > 0$.
(2) Suppose $\mathbf{v} \in \mathcal{V}$ and $t \in \mathcal{R}$. If either $\mathbf{v} = \mathbf{0}_{\mathcal{V}}$ or $t = 0$, then $t\mathbf{v} = \mathbf{0}_{\mathcal{V}}$, so $\|t\mathbf{v}\| = 0 = |t|\|\mathbf{v}\|$. If $\mathbf{v} \neq \mathbf{0}_{\mathcal{V}}$ and $t \neq 0$, then $t\mathbf{v} \neq \mathbf{0}_{\mathcal{V}}$, so $\|t\mathbf{v}\| = 1$ and $\|\mathbf{v}\| = 1$. Thus, $\|t\mathbf{v}\| = |t|\|\mathbf{v}\| \Leftrightarrow |t| = 1$.
(3) When $\mathbf{u} = \mathbf{0}_{\mathcal{V}}$, $\|\mathbf{u} + \mathbf{v}\| = \|\mathbf{v}\| = 0 + \|\mathbf{v}\| = \|\mathbf{u}\| + \|\mathbf{v}\|$. Similarly, $\mathbf{v} = \mathbf{0}_{\mathcal{V}} \Rightarrow \|\mathbf{u} + \mathbf{v}\| = \|\mathbf{u}\| + \|\mathbf{v}\|$. For $\mathbf{u} \neq \mathbf{0}_{\mathcal{V}}$ and $\mathbf{v} \neq \mathbf{0}_{\mathcal{V}}$, $\|\mathbf{u} + \mathbf{v}\| \leq 1 < 1 + 1 = \|\mathbf{u}\| + \|\mathbf{v}\|$. In all cases, $\|\mathbf{u} + \mathbf{v}\| \leq \|\mathbf{u}\| + \|\mathbf{v}\|$.

5. b. (1) $A \neq 0 \Rightarrow a_{ij} \neq 0$ for at least one pair $i,j \Rightarrow \|A\|_f \geq |a_{ij}| > 0$.

(2) $\|tA\|_f = \left(\sum_{i,j=1}^{n} (ta_{ij})^2\right)^{1/2} = |t|\left(\sum_{i,j=1}^{n} a_{ij}^2\right)^{1/2} = |t|\|A\|_f.$

$$(3)\ \|A+B\|_f = \left\|\begin{bmatrix} Col_1(A+B) \\ \hdashline \vdots \\ \hdashline Col_n(A+B) \end{bmatrix}\right\| = \left\|\begin{bmatrix} Col_1(A) \\ \hdashline \vdots \\ \hdashline Col_n(A) \end{bmatrix} + \begin{bmatrix} Col_1(B) \\ \hdashline \vdots \\ \hdashline Col_n(B) \end{bmatrix}\right\|$$

$$\leq \left\|\begin{bmatrix} Col_1(A) \\ \hdashline \vdots \\ \hdashline Col_n(A) \end{bmatrix}\right\| + \left\|\begin{bmatrix} Col_1(B) \\ \hdashline \vdots \\ \hdashline Col_n(B) \end{bmatrix}\right\| = \|A\|_f + \|B\|_f.$$

d. For C,D $\in \mathcal{M}_{n\times n}$, trace(CD) = trace(DC) (Problem 13, Exercises 1.5).

$$\begin{aligned} \|B\|_f^2 &= \text{trace}(B^TB) && \text{(part (c) above)} \\ &= \text{trace}[(P^TA^TP)(P^TAP)] && (B = P^TAP) \\ &= \text{trace}(P^TA^TAP) && (PP^T = I_n) \\ &= \text{trace}(PP^TA^TA) && (\text{trace}(CD) = \text{trace}(DC)) \\ &= \text{trace}(A^TA) && (PP^T = I_n) \\ &= \|A\|_f^2. \end{aligned}$$

8. Axioms (2) and (3) for an inner product are satisfied by $\langle X,Y\rangle = X^TAY$ for any $A \in \mathcal{M}_{n\times n}$. The argument is identical to the corresponding part of the proof of Theorem 7.28, which you will note uses neither symmetry nor positive definiteness of A. As in the proof of Theorem 7.28, A being symmetric implies that $\langle X,Y\rangle = X^TAY$ satisfies Axiom (1), and A being positive definite implies that $\langle X,Y\rangle = X^TAY$ satisfies Axiom (4).

a. Not an inner product. A is not positive definite. $X = \begin{bmatrix} 1 \\ -2 \end{bmatrix} \Rightarrow \langle X,X\rangle$

$$= [1 \quad -2]\begin{bmatrix} 3 & 2 \\ 2 & 1 \end{bmatrix}\begin{bmatrix} 1 \\ -2 \end{bmatrix} = -1 < 0, \text{ contrary to Axiom (4).}$$

c. $\det[2] = 2 > 0$, $\det\begin{bmatrix} 2 & 1 \\ 1 & 3 \end{bmatrix} = 5 > 0$, and $\det\begin{bmatrix} 2 & 1 & 0 \\ 1 & 3 & 1 \\ 0 & 1 & 2 \end{bmatrix} = 8 > 0 \Rightarrow$ A is a positive definite, symmetric matrix. Thus, $\langle \bullet,\bullet \rangle$ is an inner product.

9. $\|X_1\| = 2$, $\|X_2\| = 3\sqrt{2}$, $\|X_3\| = 3\sqrt{2}$, $X_1 \cdot X_2 = 6$, and $X_2 \cdot X_3 = 0$

a. $\cos(\theta) = 1/\sqrt{2} \Rightarrow \theta = \pi/4$

b. $\cos(\theta) = 0 \Rightarrow \theta = \pi/2$

11. a. $\langle A,B\rangle = \sum_{i=1}^{n}\sum_{j=1}^{n} a_{ij}b_{ij} = \sum_{i=1}^{n}\sum_{j=1}^{n} b_{ij}a_{ij} = \langle B,A\rangle,$

$$\langle A,B+C\rangle = \sum_{i=1}^{n}\sum_{j=1}^{n} a_{ij}(b_{ij}+c_{ij}) = \sum_{i=1}^{n}\sum_{j=1}^{n} a_{ij}b_{ij} + \sum_{i=1}^{n}\sum_{j=1}^{n} a_{ij}c_{ij}$$
$$= \langle A,B\rangle + \langle A,C\rangle,$$

$$\langle A,tB\rangle = \sum_{i=1}^{n}\sum_{j=1}^{n} a_{ij}(tb_{ij}) = t\left(\sum_{i=1}^{n}\sum_{j=1}^{n} a_{ij}b_{ij}\right) = t\langle A,B\rangle, \quad \text{and}$$

$$\langle A,A\rangle = \sum_{i,j=1}^{n} a_{ij}^2 > 0 \text{ whenever } A \neq 0.$$

b. $\|A\|_f = \left(\sum_{i,j=1}^{n} a_{ij}^2\right)^{1/2} = \sqrt{\langle A,A\rangle}$

Section 7.7

2. a. $[R \circ (S+T)](\mathbf{u}) = R[(S+T)](\mathbf{u})] = R[S(\mathbf{u}) + T(\mathbf{u})] = R(S(\mathbf{u})) + R(T(\mathbf{u})) = (R \circ S)(\mathbf{u}) + (R \circ T)(\mathbf{u}) = [R \circ S + R \circ T](\mathbf{u}), \mathbf{u} \in \mathcal{U} \Rightarrow R \circ (S+T) = R \circ S + R \circ T.$

c. $[S \circ (tT)](\mathbf{u}) = S[(tT)(\mathbf{u})] = S(tT(\mathbf{u})) = tS(T(\mathbf{u})) = t(S \circ T)(\mathbf{u}) = [t(S \circ T)](\mathbf{u})$, $\mathbf{u} \in \mathcal{U} \Rightarrow S \circ (tT) = t(S \circ T)$. Similarly, $(tS) \circ T = t(S \circ T)$.

3. a. Since D^2, D, and Id are linear operators on $\mathcal{C}^\infty(\mathcal{R})$, it follows from Theorem 7.32 that T is a linear operator on $\mathcal{C}^\infty(\mathcal{R})$. $f \in \mathcal{K}er(T) \Leftrightarrow 0 = T(f)$, and $T(f) = D^2(f) + pD(f) + q\mathrm{Id}(f) = f'' + pf' + qf$.

c. Since $\mathcal{K}er(T)$ is a subspace of $\mathcal{C}^\infty(\mathcal{R})$, $g,h \in \mathcal{K}er(T) \Rightarrow ag + bh \in \mathcal{K}er(T)$, $a,b \in \mathcal{R}$. $\begin{bmatrix} (ag+bh)(0) \\ (ag+bh)'(0) \end{bmatrix} = a\begin{bmatrix} g(0) \\ g'(0) \end{bmatrix} + b\begin{bmatrix} h(0) \\ h'(0) \end{bmatrix} = \begin{bmatrix} a \\ b \end{bmatrix}$. Since $ag + bh$ is a solution with the same initial data as f, $ag + bh = f$.

4. a. $\mathrm{A}^T\mathrm{A} = \begin{bmatrix} 1 & -2 \\ -2 & 13 \end{bmatrix}$. $\det(\lambda\mathrm{I} - \mathrm{A}^T\mathrm{A}) = \lambda^2 - 14\lambda + 9 = 0 \Leftrightarrow \lambda = 7 \pm 2\sqrt{10}$. By Theorem 7.37, $\|T\|_* = \sqrt{7 + 2\sqrt{10}}$.

d. $\mathrm{A}^T\mathrm{A} = \begin{bmatrix} 1 & 0 & 1 \\ 0 & 1 & 0 \\ 1 & 0 & 2 \end{bmatrix}$. $\det(\lambda\mathrm{I} - \mathrm{A}^T\mathrm{A}) = (\lambda - 1)(\lambda^2 - 3\lambda + 1) = 0 \Leftrightarrow \lambda = 1$, $(3 \pm \sqrt{5})/2$. By Theorem 7.37, $\|T\|_* = \sqrt{(3 + \sqrt{5})/2}$.

f. $\det(\lambda\mathrm{I} - \mathrm{A}) = (\lambda - 2)(\lambda - 4)(\lambda + 6)$. By Theorem 7.38, $\|T\|_* = 6$.

h. $\mathrm{A}^T\mathrm{A} = \begin{bmatrix} 5 & 2 & 0 & 0 \\ 2 & 1 & 0 & 0 \\ 0 & 0 & 1 & 2 \\ 0 & 0 & 2 & 5 \end{bmatrix}$. $\det(\lambda\mathrm{I} - \mathrm{A}^T\mathrm{A}) = 0 \Leftrightarrow \lambda = 3 \pm 2\sqrt{2}$, $\|T\|_* = \sqrt{3 + 2\sqrt{2}}$.

7. Let A be the matrix that represents T relative to $\mathcal{E}$. Then A, being orthogonally similar to $D[\lambda_1, \ldots, \lambda_n]$, is symmetric with eigenvalues $\lambda_1, \ldots, \lambda_n$. The conclusion then follows from Theorem 7.38.

10. b. Suppose U is a unit vector and set $\mathrm{V} = T^{-1}(\mathrm{U})$. Then $\mathrm{U} = T(\mathrm{V})$, and $1 = \|\mathrm{U}\| = \|T(\mathrm{V})\| = \left\| T\left(\frac{\mathrm{V}}{\|\mathrm{V}\|}\right) \right\| \|\mathrm{V}\| \geq m(T)\|\mathrm{V}\| = m(T)\|T^{-1}(\mathrm{U})\|$, so

$$\|T^{-1}(\mathrm{U})\| \leq \frac{1}{m(T)}. \qquad (*)$$

If V is a unit vector such that $\|T(\mathrm{V})\| = m(T)$ and $\mathrm{W} = T(\mathrm{V})$, then $\left\| T^{-1}\left(\frac{\mathrm{W}}{\|\mathrm{W}\|}\right) \right\| = \frac{\|T^{-1}(\mathrm{W})\|}{\|\mathrm{W}\|} = \frac{\|\mathrm{V}\|}{m(T)} = \frac{1}{m(T)}$. This, together with $(*)$ shows that $\|T^{-1}\|_* = \dfrac{1}{m(T)}$.

Index